Contemporary Accounts in Drug Discovery and Development

Contemporary Accounts in Drug Discovery and Development

Edited by

Xianhai Huang
InventisBio Co., Ltd., Florham Park, USA

Robert G. Aslanian
New Jersey City University, Jersey City, USA

Wayne H. Tang
Schrödinger Inc., New York, USA

WILEY

This edition first published 2022

Registered Office
John Wiley & Sons, Inc., 111 River Street, Hoboken, NJ 07030, USA

Editorial Office
111 River Street, Hoboken, NJ 07030, USA

For details of our global editorial offices, customer services, and more information about Wiley products visit us at www.wiley.com.

Wiley also publishes its books in a variety of electronic formats and by print-on-demand. Some content that appears in standard print versions of this book may not be available in other formats.

Library of Congress Cataloging-in-Publication Data

Names: Huang, Xianhai, editor. | Aslanian, Robert G., editor. | Tang, Wayne H. (Wayne Haifeng), editor.
Title: Contemporary accounts in drug discovery and development / edited by Xianhai Huang, Robert G. Aslanian, Wayne H. Tang.
Other titles: Case studies in modern drug discovery and development.
Description: Second edition. | Hoboken, NJ : Wiley, 2022. | Preceded by Case studies in modern drug discovery and development / edited by Xianhai Huang, Robert G. Aslanian. 2012. | Includes bibliographical references and index.
Identifiers: LCCN 2021055890 (print) | LCCN 2021055891 (ebook) | ISBN 9781119627715 (cloth) | ISBN 9781119627852 (adobe pdf) | ISBN 9781119627814 (epub) | ISBN 9781119627784 (oBook)
Subjects: MESH: Drug Discovery | Drug Development
Classification: LCC RM301.25 (print) | LCC RM301.25 (ebook) | NLM QV 745 | DDC 615.1/9–dc23/eng/20211228
LC record available at https://lccn.loc.gov/2021055890
LC ebook record available at https://lccn.loc.gov/2021055891

Cover Design: Wiley
Cover Image: © Sergey Nivens/Shutterstock

MIX
Paper from responsible sources
FSC® C013604

Contents

Preface

It has been almost nine years now since the publication of the first edition of *Case Studies in Modern Drug Discovery and Development* in 2012. The book has been well received, in part for its breadth of topics including accounts of the discovery of life-saving drugs. The purpose of the first edition of the book was to preserve knowledge of drug discovery and development at a time that small-molecule drug discovery was seemingly at a crossroad and looking for direction. In the years since the publication of the first edition, we have been fortunate to witness the approval of many new drug products. In fact, 2018 witnessed a historic number of novel drug approvals by the FDA (a total of 59). Technology has advanced at an amazing pace in the past few years, which has enabled many formally untenable modalities, such as peptides, drug conjugates, antibodies, RNA, and vaccines, to provide new options for drug discovery. In addition, advances in basic science have also helped to accelerate the drug discovery process. In particular, structure-based drug design has had a profound impact on modern drug discovery. For example, physics-based free-energy perturbation could finally be used to realistically predict relative ligand-binding potency after almost 30 years of effort. CryoEM technology, which was the subject of the 2017 Nobel Prize in Chemistry, has also shown an increasing impact on structure-based drug design and enabled access to previously unobtainable protein structures. Artificial intelligence has also started to play an important role in drug discovery, although more time is needed to realize its full impact. These are very encouraging results; however, it still takes a long time to transform the initial basic discoveries from academic labs into a drug product. For example, a number of studies have shown that on average, it still takes more than 10 years and several billion dollars to develop and bring a new drug to market from the hit-finding stage. Furthermore, the failure rate for clinical compounds remains unacceptably high at about 90% with some therapy areas suffering even higher rates. With these important advances and challenges in the drug discovery field, we feel that it is the right time to have a new edition of the book to showcase these new developments in driving the drug discovery process. In the first edition of the case study book, we mainly focused on marketed small molecule drugs and covered their discovery and development stories from initial biological rational to products on the market. In the new edition of the book, we not only cover the full stories of several marketed small molecule drugs but also include drug candidates with different modalities such as antibody drug conjugates, proteolysis targeting chimeras, and peptide drugs. The application of new technology such as DNA-encoded library and positron emission tomography in drug discovery will also be covered in the book. Improving the drug discovery success rate continues to be a key factor that will weigh heavily on the success, and perhaps even the survival, of the pharmaceutical industry. How to more efficiently apply drug discovery principles and technologies to improve the success rate becomes a critical question. Knowledge gained from the successful discovery and launch of marketed drugs, and lessons learned from failed drug discovery stories can provide a very useful and important template for future drug design and discovery. We hope that the new edition of *Case Studies in Modern Drug Discovery and Development* (newly named as *Contemporary Accounts in Drug Discovery and Development*) can serve this purpose in addition to bringing educational value to the readers. Like the first edition, each chapter will have a detailed biological rational and background information, drug design principles, SAR development, ADMET considerations, and clinical studies, with modern drug discovery principles and technologies incorporated.

Like the first edition, this edition of *Case Studies in Modern Drug Discovery and Development* covers almost all aspects of the drug discovery process with detailed examples to showcase the science and technology; thus, it essentially will be suitable for all levels of scientists who are involved in drug discovery. The main targeted readers are medicinal chemists, and undergraduate and graduate students who are interested in drug discovery. Since it not only traces the drug discovery process from idea to marketed drug with a detailed explanation of the application of modern drug design principles and discovery technologies, but also showcases the importance of cross-disciplinary applications of knowledge from different fields such as computer science and statistics, *Contemporary Accounts in Drug Discovery and Development* essentially will be beneficial to any scientist involved in the drug discovery process. It is also designed to be used as a textbook in a medicinal chemistry class. The book has 15 chapters and fits one full semester. With the comprehensive information included in each independent chapter, it is also suitable for professional seminars or courses that relate to drug design. Some of the drug discovery stories collected in this book are popular and life-saving medications, so it will be of interest to the public who want to learn more about how these drugs are discovered.

Editing a scientific book is not an easy process; it took more than three years from the conception of the book, author recruiting, and chapter editing to the publication of the book. During this long process, there are many friends and colleagues who helped to make it possible. We would like to thank our editor, Jonathan Rose, for initiating the process, giving us the opportunity, and trusting us in editing *Contemporary Accounts in Drug Discovery and Development.* We also want to thank Wiley Managing Editor Dr. Andreas Sendtko for his role as a consistent point of contact both during the writing phase and right through to publication. His support has made the book-editing process run very smoothly. We would also like to thank all the authors who dedicated their time to contribute the chapters and their respective companies for permission to publish their work. Without their unselfish contributions, especially during an unprecedentedly difficult period of the COVID-19 pandemic, it is almost impossible to accomplish such a daunting job. We salute them for their time, effort, persistence, and dedication. We continue to believe that all the chapters will have a very important impact on future drug discovery programs and benefit future scientists of this field for generations to come. We would like to thank the reviewers of our book proposal for their valuable suggestions and critiques. We would like to thank Dr. Wen-Lian Wu for his inputs/comments on the Alzheimer's disease chapter and Dr. Shuangping Shi for her expert comments on Appendix A. Xianhai is indebted to his wife Dr. Hongmei Li and his children Alexander and Angelina for their support and understanding. Bob would like to thank his wife Antoinette and his family for their encouragement and support over the years. Wayne is also grateful to the infinite support from his family.

The road to the discovery and development of life-saving drugs is not and will never be easy and straightforward. With scientific advancements, technological innovations, close collaborations and knowledge sharing, and the lifetime dedication of pharmaceutical scientists and academic medical researchers, it is possible to improve the efficiency of drug discovery and development to deliver the therapies that will alleviate suffering and extend lives. We hope that this book can serve as a small stepping stone in this complex process and help the readers to contribute to the improvement of the health of humanity.

New Jersey, February 2022

Xianhai Huang
Robert G. Aslanian
Wayne H. Tang

List of Contributors

Robert Abel
Drug Discovery Group
Schrödinger Inc.
New York
USA

Robert Aslanian
Department of Chemistry
New Jersey City University
Jersey City
USA

Corina Becker
Research & Development, Pharmaceuticals
Bayer AG
Wuppertal
Germany

Andrea Bortolato
Research and Development, Computer-Assisted Drug Design, Molecular Discovery Technologies
Bristol Myers Squibb
Princeton
USA

David Yu-Kai Chen
Department of Chemistry
Seoul National University
Seoul
South Korea

Yun Ding
Encoded Library Technologies/NCE Molecular Discovery
GSK
Cambridge
USA

Wu Du
Hinova Pharmaceuticals Inc.
Chengdu
People's Republic of China

Markus Follmann
Research & Development, Pharmaceuticals
Bayer AG
Wuppertal
Germany

Jeffrey J. Hale
Associate Vice President
Discovery Chemistry
Merck & Co., Inc.
West Point
USA

Amy Han
R&D Chemistry, Therapeutic Proteins
Regeneron Pharmaceuticals, Inc.
Tarrytown
USA

Junliang Hao
Discovery Chemistry Research
and Technologies
Lilly Research Laboratory
Eli Lilly and Company
Lilly Corporate Center
Indianapolis
USA

Timothy P. Heffron
Discovery Chemistry
Genentech, Inc.
South San Francisco
USA

Xianhai Huang
Discovery Chemistry
InventisBio Co., Ltd.
Florham Park
USA

Wenping Li
Translational Imaging
Merck Research Laboratories
West Point
USA

Jonathan S. Mason
Sosei Heptares
Steinmetz Building
Granta Park
Cambridge
UK

Lothar Roessig
Research & Development, Pharmaceuticals
Bayer AG
Wuppertal
Germany

Laurent Salphati
Drug Metabolism and Pharmacokinetics
Genentech, Inc.
South San Francisco
USA

Peter Sandner
Research & Development, Pharmaceuticals
Bayer AG
Wuppertal
Germany

Tomi Sawyer
Maestro Therapeutics
Southborough
USA

Sarah K. Scott
Encoded Library Technologies/
NCE Molecular Discovery
GSK
Cambridge
USA

Hong C. Shen
Department of Medicinal
Chemistry, Therapeutic Modalities
Roche Innovation Center Shanghai
Shanghai
People's Republic of China

Steven T. Staben
Discovery Chemistry
Genentech, Inc.
South San Francisco
USA

Johannes-Peter Stasch
Research & Development, Pharmaceuticals
Bayer AG
Wuppertal
Germany

Wayne Haifeng Tang
Drug Discovery Group
Schrödinger Inc.
New York
USA

Ling Tong
Discovery Chemistry
Merck Research Laboratories
Kenilworth
USA

Song Yang
Department of Medicinal Chemistry
Therapeutic Modalities
Roche Innovation Center Shanghai
Shanghai
People's Republic of China

Hongying Yun
Department of Medicinal Chemistry
Therapeutic Modalities
Roche Innovation Center Shanghai
Shanghai
People's Republic of China

Wei Zhu
Department of Medicinal Chemistry
Therapeutic Modalities
Roche Innovation Center Shanghai
Shanghai
People's Republic of China

Ning Zou
R&D Chemistry, Therapeutic Proteins
Regeneron Pharmaceuticals, Inc.
Tarrytown
USA

1

Current Drug Discovery: Great Challenges and Great Opportunity (an Introduction to *Contemporary Accounts in Drug Discovery and Development*)

Jeffrey J. Hale

Discovery Chemistry, Merck & Co., Inc., 770 Sumneytown Pike, PA 19486-004, West Point, PA, USA

As we approach the end of the second decade of the twenty-first century, global human health faces significant challenges. In May 2018, the World Health Organization (WHO) reported that more than half of recent deaths worldwide were attributable to 10 major causes [1]. Mortality following from cardiovascular disorders, such as ischemic heart disease and stroke, figured prominently on that list, while conditions such as chronic obstructive pulmonary disease, dementias, and diabetes were noted to have increased in frequency since 2000. While the prevalence of deaths from non-communicable diseases may be perceived to be of greater concern in high-income countries, the WHO noted that "diseases of aging" are significant globally. In September 2018, the International Agency for Research on Cancer noted the increasing prevalence of cancers worldwide, with an expectation at current rates that one in five men and one in six women will develop cancer during their lifetime, and one in eight men and one in 11 women die from it. Outbreaks of infectious diseases (e.g. dengue, Ebola, influenza, Zika) exacerbated by behavioral and social factors, as well as the associated issue of growing antimicrobial resistance, highlights another acute threat to human health [2]. The emergence of the severe acute respiratory syndrome coronavirus 2 (SARS-CoV-2) in late 2019/early 2020 and the subsequent world-wide pandemic that ensued was an extremely sobering reminder of this latter point [3]. A 2017 report from the United States Bureau of Vital Statistics indicated that life expectancy (all source mortality) may be leveling off or even decreasing, after close to a century of increase in this metric [4]. All told, a significant opportunity is before us to identify new methods to prevent, modify and treat disease and reverse the course of these human health trends.

In the first edition of *Case Studies in Modern Drug Discovery and Development*, my mentor and industry colleague Dr. Malcolm MacCoss posited that the pharmaceutical industry was facing a "perfect storm" of pressures that presented large challenges. These included patent expiries, the continued rising costs of research and development, the long discovery and development time cycles of the industry and the over-reliance on a "blockbuster business model". All of these are relevant at the present time with the addition of the current complexities of acute societal, political and economic pressures on healthcare allocation and delivery [5]. Since the middle of this decade, multiple analyses of industry-wide productivity have been published [6–9]. While the data sets and methodologies used by the respective authors differ somewhat, pipeline productivity (as measured

Contemporary Accounts in Drug Discovery and Development, First Edition. Edited by Xianhai Huang, Robert G. Aslanian, and Wayne H. Tang.

by transition rates through development phases or overall success rates to registration) has not manifestly improved. Questions about how best to model and organizationally scale basic biomedical research, the effectiveness of clinical trial design and execution practices, and how to navigate complex regulatory landscapes seem to regularly recur in an attempt to discern the root cause(s) of our inability to change productivity. A response in part to these challenges has been the continued activity within the biopharma industry in mergers, acquisitions and partnerships, driven by the desire to mitigate the effect of loss of exclusivity of key products, identify new portfolio synergies and hedge against the continued rise in operational costs [10]. “Mega-mergers” are still quite prevalent, as evidenced by 2019s numerous multi-billion-dollar deals (e.g. Takeda/Shire, Bristol-Myers Squibb/Celgene, AbbVie/Allergan, Pfizer’s merger of Upjohn/Mylan). For the biopharmaceutical industry today, this is on the background of additional questions regarding the nature of its interactions with partners and stakeholders (e.g. What should be the respective contributions and desired impact of biomedical research funded by governments vs. private industry? What is the value to patients and payors of “new” drugs that are or perceived to be “me too” products?) [11]. There is anticipation of enhanced pricing pressures on prescriptions drugs in the United States that will likely require creative solutions as that country moves into the new decade [12]. Isaac Stoner (COO, Octagon Therapeutics) recently commented on challenges specific to new antibiotics, but the call to have market reform drive research productivity and downstream patient access extends well beyond that arena:

> If our industry truly believes we have a responsibility to put patients ahead of profits, we need to work to fix this broken market rather than ignoring the problem in favor of more profitable disease areas. There are major externalities driven by access to new effective antibiotics. Without the ability to treat infections, simple procedures such as Caesarian-sections or hip replacements will present enormous risk, and cancer mortality rates will skyrocket. Saving lives should be good business but, in this case, it’s not. Without real market reform, antibiotic development will continue to be un-investable, and these medicines will not be available to patients who desperately need them. [13]

None of this is helped by the public’s low regard for the pharmaceutical industry, which in one recent analysis by Gallup lagged significantly behind perennial poor performers like the legal profession and the United States government [14]. While some of this may be attributed to decreasing science literacy [15], it is also undoubtedly driven by things like the poor handling by industry players of specific commercialization strategies (e.g. insulin pricing [16]) and the high-profile debacles involving extreme bad actors (e.g. the price gouging scandal involving the anti-toxoplasmosis drug Daraprim® by Martin Shkreli and Turing Pharmaceuticals [17]; the unethical and fraudulent promotion of its blood sample analysis technology by Elizabeth Holmes and Theranos [18]). A broad, holistic approach is likely needed to address the structural challenges that the pharmaceutical industry faces, and I would argue that broad reform of healthcare systems in the United States and abroad that encourages growth and protects its competitiveness is critical. Policymakers will need to surmount large political and economic challenges in order to foster continued robust biopharma research and development worldwide. Any solution that aspires to lower drug prices should also expand access to affordable insurance, streamline regulation, and spread cost and risks across all of the players in the biopharma ecosystem. While the reactive ramp up of world-wide efforts in reaction to the SARS-CoV-2 pandemic was not ideal, it may contain elements (e.g. aggressive repurposing of existing therapeutics; parallel, collaborative efforts interrogating new potential therapeutics and preventatives; rapid publishing of preclinical and clinical data on open-source platforms) that could inform on reforming drug discovery and development models [19].

Despite the somewhat dire picture painted by the challenges above, drug discoverers and developers worldwide have made some remarkable contributions to the pharmacopeia over the last decade. Advances in immuno-oncology that unleash a patient's own anti-tumor immunity to treat certain cancers have been revolutionary [20, 21]. The potential to couple new drugs with programmed cell death protein-1 (PD-1) inhibitors like Opdivo® (nivolumab) and Keytruda® (pembrolizumab) has led a veritable renaissance of research into novel immunomodulators and cellular metabolism regulators, all with the hope that they could be exploited as combination agents in cancer therapy, as well as provide insights into or be exploited for the treatment of cardiovascular and neurologic disorders. Dysfunctions in the humoral immune system that manifest as aggressive B-cell non-Hodgkin lymphomas are now treatable with inhibitors of Bruton's tyrosine kinase (Btk) inhibitors such as Imbruvica® (ibrutinib); Btk inhibitors with enhanced tolerability and resistance profiles are under active clinical investigation [22]. New anti-diabetic drugs like Jardiance® (empagliflozin), a selective inhibitor of sodium glucose co-transporter-2 (SGLT2), have been clinically demonstrated to provide significant cardiovascular benefit [23]. The discovery and development of the hepatitis C (HCV) NS5B protein RNA polymerase inhibitor Sovaldi® (sofosbuvir) and HCV NS5A inhibitor Harvoni® (ledipasvir) led the charge to identify cross-genotype HCV drugs that can cure infected patients in three months or less. Recent breakthroughs that have produced drugs effective against multi-drug resistant tuberculosis demonstrate that progress can be made against decades old problems [24]. These and other groundbreaking therapies have benefited from the innovations and insights of drug discovery and development chemists. I am confident that medical researchers and drug hunters have the potential to maintain this trajectory as we move into the 2020s.

While AstraZeneca's widely cited five-dimensional framework on research and development productivity [25] does not explicitly call out "the right chemistry" among its factors for long-term research success, many recent advances in chemical biology and medicinal chemistry have the potential to impact those factors as we move forward. Barriers to drugging "the right target" and establishing "the right safety" are on their way to being toppled by advanced tools and technologies that allow for deliberate exploitation of allostery, the modulation of RNA biology with small molecules [26], the therapeutically beneficial manipulation of transcription factors [27], the resetting of the autophagy "rheostat" [28] and the targeted degradation of disease-relevant proteins [29]. The last – which can be thought of conceptually as the post-translational knockdown of proteins – is particularly intriguing today, as the first "molecular glues" and proteolysis targeting chimeras (PROTACs) have reached the clinic [30] and the field of targeted degraders promises to produce more molecules that can be put in the context of "rule-of-five" chemical space [31]. Our rapidly developing understanding of the complexities of human microbiome and its connection to cardiovascular, immunologic, and neurologic "health" will likely expand target space and point to new approaches for therapeutic intervention. This is tied to the re-emergence of phenotypic drug discovery in general and all of this could be accelerated via the application of present-day genomics tools coupled with the state-of-the-art informatics capabilities.

A group of my MSD Chemistry colleagues recently rejected the position that the act of designing and synthesizing molecules is a commoditized aspect of pharmaceutical research, but rather "that excellence and innovation in synthetic chemistry continues to be critical to success in all phases of drug discovery and development [32]." This sentiment echoes the recent views of other chemists from both academic and industrial circles [33–36] and I can point to multiple times over my 35 year career where organic synthesis has indeed been a significant inspiration of my ideas (both good ones and bad ones). Looking forward, the capability exists for most industrial medicinal chemists to rapidly work through the more empirical aspects of their drug discovery projects using powerful automation tools [37] enabled by machine learning and artificial intelligence and this will likely become part of the standard academic training regimen for the next generation of drug discovery and development chemists [38]. The horizons of organic synthesis continue to expand via the continued innovation in areas of chemo- and biocatalysis [39], electrochemistry [40], flow chemistry [41], methods to productively disconnect carbon–carbon bonds [42], and photochemically induced oxidation/reduction chemistry [43]. These will not only afford access to new biologically interesting chemical space for medicinal chemists, but also pave future paths to design concepts. Synthesis is not (nor should it be) the only driver of molecular design and it is presently complemented by current capabilities of machine learning to understand and optimize molecular interactions as well as finely balance the molecular properties critical to bioperformance (absorption, distribution, metabolism, excretion, and toxicity) [44]. While there are currently growing pains with machine learning in drug research [45], it has the potential to positively impact future drug research efforts.

In the era when we see public figures staking claim to "alternative facts" and regularly hear claims that inconvenient or uncomfortable (but objectively verifiable) information is "fake news", broad societal pressures and the "human element" may pose the biggest threat to continued new discoveries in drug research. In his ground-breaking 2005 book *The World Is Flat: A Brief History of the Twenty-first Century*, Thomas Friedman lays out how the digitization of information and the broad reach of the internet has led to the rapid democratization of capabilities and knowledge [46]. Anyone connected to or embedded in drug research now regularly grapples with the resulting positive and negative impacts of Friedman's flat world. While we can marvel at the power and convenience of conducting broad literature searches in seconds by simply clicking a mouse, this comes at a cost to researchers of having to discern increasingly faint signals in an exponentially expanding sea of noise. As we enhance our ability to leverage artificial intelligence [47] and exploit big data we may be able to eventually corral large quantities of disparate sets of information that is relevant to drug discovery and development, but pitfalls that can be traced to the humans behind the computers have already come to the fore [48].

I also believe that there are significant risks to future innovations posed by the fracturing of the complex research environments into discrete units and continued efforts to optimize their individual parts. This has manifested itself in a pharmaceutical "gig economy" that can reinforce a short-term, transactional mindset and poses significant challenges to individuals who may be more accustomed to or who would thrive in more traditional research environments [49]. In order to ensure we are on the path to discovering future medicines, we must realize that the rigorous application of the scientific method coupled with savvy decision-making is critical. The disciplined fostering of individual and group behaviors that promote innovation should be a priority. Making sure that we tolerate failure, allow for experimentation, ensure psychological safety, promote collaboration and demand that we have leaders who support all of this is crucial to have if we seek to avoid having promising research undercut by non-productive human intervention [50].

While we in the field of medicinal chemistry and drug research and development find ourselves in an era of great challenges, I am optimistic about the future of the discipline. The chapters in this volume that detail efforts that resulted in new drugs to treat bacterial and viral infections, numerous metabolic disorders and various cancers serves as a testament to the creativity, persistence, technical skill and innovative insights of the many individuals and large teams required to make these therapeutics a reality. Beyond that, none of this would have been possible if not for the dedication and effort of those researchers who happen to have direct connections to the specific biologic targets and bioactive agents described herein. The contributions of many scientists to the disciplines of biochemistry, molecular biology, pharmacology, toxicology, and translational sciences are all necessary to advance biomedical research and is the foundation upon which all of the case studies stand. These previous words of Malcolm MacCoss still ring true to me – "... the noble endeavor of drug discovery must continue to move forward, even if the path is steep and the costs continue to rise. We must persevere because otherwise our children and their children will be restricted to using only the drugs of their parents to fight their battles with the same devastating diseases, despite all the wonderful discoveries in medicine and the biological sciences ..." I hope that the readers of this volume are inspired by the case studies described and conclude as I did that despite the great challenges in front of us in combatting human disease, the future provides us with the great opportunity to impact the course of human health.

References

1 World Health Organization (2018). The top ten causes of death. https://www.who.int/news-room/fact-sheets/detail/the-top-10-causes-of-death (accessed 29 September 2021).

2 Bloom, D.E. and Cadarette, D. (2019). Infectious disease threats in the twenty-first century: strengthening the global response. *Front. Immunol.* 10: 1–12.

3 Trypsteen, W., Van Cleemput, J., van Snippenberg, J. et al. (2020). On the whereabouts of SARS-CoV-2 in the human body: a systematic review. *PLoS Pathog.*: 1–26.

4 U.S. Department of Health and Human Services(2017). United States life tables, 2017. https://www.cdc.gov/nchs/data/nvsr/nvsr68/nvsr68_07-508.pdf (accessed 29 September 2021).

5 Vogenberg, F.R. (2019). U.S. Healthcare trends and contradictions in 2019. *Am. Health Drug Benefits* 12 (1): 40–47.

6 Hay, M., Thomas, D.W., Craighead, J.L. et al. (2014). Clinical development success rates for investigational drugs. *Nat. Biotechnol.* 32 (1): 40–51.

7 Smietana, K., Siatkowski, M., and Møller, M. (2016). Trends in clinical success rates. *Nat. Rev. Drug Discov.* 15: 379–380.

8 Thomas, D.W., Burns, J., Audette, J. et al. (2017). Clinical development success rates 2006–2015. https://www.bio.org/sites/default/files/legacy/bioorg/docs/Clinical%20Development%20Success%20Rates%202006-2015%20-%20BIO,%20Biomedtracker,%20Amplion%202016.pdf (accessed 29 September 2021).

9 Wong, C.H., Siah, W.W., and Lo, A.W. (2019). Estimation of clinical trial success rates and related parameters. *Biostatistics* 20 (2): 273–286.

10 Bansal, R., De Backer, R., and Ranade, V. (2018). What's behind the pharmaceutical sector's M&A push. https://www.mckinsey.com/business-functions/strategy-and-corporate-finance/our-insights/whats-behind-the-pharmaceutical-sectors-m-and-a-push# (accessed 29 September 2021).

11 Patterson, J.A. and Carroll, N.V. (2019). Should the United States regulate prescription drug prices? A critical review. *Res. Social Adm. Pharm* https://doi.org/10.1016/j.sapharm.2019.06.010.

12 Life Sci VC (2019). Venturing a perspective on the drug pricing debate. https://lifescivc.com/2019/12/venturing-a-perspective-on-the-drug-pricing-debate (accessed 29 September 2021).

13 Stoner, I. (2019). Biopharma has abandoned antibiotic development. Here's why we did, too. https://endpts.com/biopharma-has-abandoned-antibiotic-development-heres-why-we-did-too (accessed 29 September 2021).

14 McCarthy, J. (2019). Big pharma sinks to the bottom of U.S. industry rankings. https://news.gallup.com/poll/266060/big-pharma-sinks-bottom-industry-rankings.aspx (accessed 29 September 2021).

15 Lederman, L. (2019). The ever more pressing problem of science literacy. In: *Learning to Live Together: Promoting Social Harmony* (ed. J. Kelso), 139–147. Cham: Springer.

16 Alltucker, K. (2019). Struggling to stay alive: rising insulin prices cause diabetics to go to extremes https://www.usatoday.com/in-depth/news/50-states/2019/03/21/diabetes-insulin-costs-diabetics-drug-prices-increase/3196757002 (accessed 29 September 2021).

17 Smythe, C. and Geiger, K.(2015). Shkreli, drug price gouger, denies fraud and posts bail. https://www.bloomberg.com/features/2015-martin-shkreli-securities-fraud (accessed 29 September 2021).

18 Ramsey, L. (2018). The rise and fall of Theranos, the blood-testing startup that went from Silicon Valley darling to facing fraud charges. https://www.businessinsider.com/the-history-of-silicon-valley-unicorn-theranos-and-ceo-elizabeth-holmes-2018-5 (accessed 29 September 2021).

19 Ekins, S., Mottin, M., Ramos, P. et al. (2020). Déjà vu: stimulating open drug discovery for SARS-CoV-2. *Drug Discov. Today* 25 (5): 928–941.

20 Dubois, M., Ardin, C., André, F. et al. (2019). The revolution of immuno-oncology therapy: review of immune checkpoint inhibitors efficacy. *Med. Sci. (Paris)* 35 (12): 937–945.

21 Yu, J.X., Hubbard-Lucey, V.M., and Tang, J. (2019). Immuno-oncology drug development goes global. *Nat. Rev. Drug Discov.* 18: 899–900.

22 Owen, C., Berinstein, N.L., Christofides, A., and Sehn, L.H. (2019). Review of Bruton tyrosine kinase inhibitors for the treatment of relapsed or refractory mantle cell lymphoma. *Curr. Oncol.* 26 (2): e233–e240.

23 Zinman, B., Wanner, C., Lachin, J.M. et al. (2015). Empagliflozin, cardiovascular outcomes, and mortality in type 2 diabetes. *N. Engl. J. Med.* 373 (22): 2117–2128.

24 Pontali, E., Raviglione, M.C., Migliori, G.B., and the writing group members of the Global TB Network Clinical Trials Committee (2019). Regimens to treat multidrug-resistant tuberculosis: past, present and future perspectives. *Eur. Respir. Rev.* 28: 1–7.

25 Morgan, P., Brown, D.G., Lennard, S. et al. (2018). Impact of a five-dimensional framework on R&D productivity at AstraZeneca. *Nat. Rev. Drug Discov.* 17: 167–181.

26 Disney, M.D., Dwyer, B.G., and Childs-Disney, J.L. (2018). Drugging the RNA world. *Cold Spring Harb. Perspect. Biol.* 10: 1–15.

27 Dang, C.V., Reddy, E.P., Shokat, K.M., and Soucek, L. (2017). Drugging the 'Undruggable' cancer targets. *Nat. Rev. Cancer* 17: 502–508.

28 Netea-Maiera, R.T., Plantingaa, T.S., van de Veerdonka, F.L. et al. (2016). Modulation of Inflammation by Autophagy: Consequences for Human Disease. *Autophagy* 12 (2): 245–260.

29 Bondeson, D.P. and Crews, C.M. (2017). Targeted protein degradation by small molecules. *Annu. Rev. Pharmacol. Toxicol.* 57: 107–123.

30 Chamberlain, P.P. and Haman, L.G. (2019). Development of targeted protein degradation therapeutics. *Nat. Chem. Biol.* 15: 937–944.

31 Edmondson, S.D., Yanga, B., and Fallan, C. (2019). Proteolysis targeting chimeras (PROTACs) in 'Beyond Rule-of-Five' chemical space: recent progress and future challenges. *Bioorg. Med. Chem. Lett.* 29: 1555–1564.

32 Campos, K.R., Coleman, P.J., Alvarez, J.C. et al. (2019). The importance of synthetic chemistry in the pharmaceutical industry. *Science* 363 (6424).

33 Blakemore, D.C., Castro, L., Churcher, I. et al. (2018). Organic synthesis provides opportunities to transform drug discovery. *Nat. Chem.* 10: 383–394.

34 Taylor, A.P., Robinson, R.P., Fobian, Y.M. et al. (2016). Modern advances in heterocyclic chemistry in drug discovery. *Org. Biomol. Chem.* 14: 6611–6637.

35 Boström, J., Brown, D.G., Young, R.G., and Keserü, G.M. (2018). Expanding the medicinal chemistry synthetic toolbox. *Nat. Rev. Drug Discov.* 17: 709–727.

36 Denmark, S.E. (2018). Organic synthesis: wherefrom and whither? (some very personal reflections). *Isr. J. Chem.* 58 (1–2): 61–72.

37 Schneider, G. (2018). Automating drug discovery. *Nat. Rev. Drug Discov.* 17: 97–113.

38 Peplow, M. (2019). Automation for the people: training a new generation of chemists in data-driven synthesis. https://cen.acs.org/content/cen/articles/97/i42/Automation-people-Training-new-generation.html (accessed 29 September 2021).

39 Rudroff, F., Mihovilovic, M.D., Gröger, H. et al. (2018). Opportunities and challenges for combining chemo- and biocatalysis. *Nat. Biocatal.* 1: 12–22.

40 Halford, B. (2019). Amping up the pharma lab: drug companies explore the potential of electrochemistry. https://cen.acs.org/synthesis/medicinal-chemistry/Amping-pharma-lab-Drug-companies/97/i43 (accessed 29 September 2021).

41 Plutschack, M.B., Pieber, B., Gilmore, K., and Seeberger, P.H. (2017). The Hitchhiker's guide to flow chemistry. *Chem. Rev.* 117: 11796–11893.

42 Wang, B., Perea, M.A., and Sarpong, R. (2020). Transition metal-mediated C–C single bond cleavage: making the cut in total synthesis. *Angew. Chem., Int. Ed. Engl.* https://doi.org/10.1002/anie.201915657.

43 Prier, C.K. and MacMillan, D.W.C. (2018). Dual photoredox catalysis: the merger of photoredox catalysis with other catalytic activation modes. In: *Visible Light Photocatalysis in Organic Chemistry* (eds. R.J. Stephenson, T.P. Yoon and D.W.C. MacMillan), 299–334. Wiley & Sons.

44 Panteleeva, J., Gao, H., and Ji, L. (2018). Recent applications of machine learning in medicinal chemistry. *Bioorg. Med. Chem. Lett.* 28: 2807–2815.

45 Riley, P. (2019). Three pitfalls to avoid in machine learning. *Nature* 527: 27–28.

46 Friedman, T.L. (2005). *The World is Flat: A Brief History of the Twenty-first Century*. New York: Farrar, Straus and Giroux.

47 Jordan, A.M. (2018). Artificial intelligence in drug design: the storm before the calm? *ACS Med. Chem. Lett.* 18: 1150–1152.

48 Daniel, B.K. (2019). Big data and data science: a critical review of issues for educational research. *Br. J. Educ. Technol.* 50 (1): 101–113.

49 Petriglieri, G., Ashford, S.J., and Wrzesniewski, A. (2019). Agony and ecstasy in the gig economy: cultivating holding environments for precarious and personalized work identities. *Adm. Sci. Q.* 64: 124–170.

50 Pisano, G.P. (2019).The hard truth about innovative cultures. https://hbr.org/2019/01/the-hard-truth-about-innovative-cultures (accessed 29 September 2021).

2

Advanced Computational Modeling Accelerating Small-Molecule Drug Discovery: A Growing Track Record of Success

Robert Abel

Drug Discovery Group, Schrödinger Inc., 120 West 45th Street, NY 10036-4041, New York, NY, USA

2.1 Introduction

The last 10 years have marked a turning point in advanced computational modeling clearly demonstrating its value accelerating the discovery of newer and more efficacious small-molecule drug therapies suitable to be advanced into clinical trials. Much of this important history has not yet entered the peer-reviewed literature, and hopefully this chapter will serve as a reference to those wishing to better understand how the idea that computational analysis could drive pharmaceutical drug discovery forward from a position of hopeful optimism to a statement of objective fact.

It is important to acknowledge the hopeful optimism that computational modeling could accelerate drug discovery has been long standing. It has been almost 40 years since *Fortune* magazine in October of 1981 famously published the cover article "The next industrial revolution: designing drugs by computer at Merck" [1]. From the publication of that article to at least 2010, despite the efforts of many research groups and companies, disconcertingly little progress was made toward achieving these goals [2].

Evidence that computational modeling was starting to live up to its promise began to emerge in 2013 when Bruce Booth announced an innovative company he co-founded, Nimbus Discovery, having benefitted from advanced computational modeling technologies made available through a strategic partnership with Schrödinger, a novel virtual and globally distributed operating model and an innovative LLC-based asset centric business model, had succeeded in discovering an acetyl-CoA carboxylase (ACC) inhibitor suitable for advancement into clinical studies for the treatment of non-alcoholic steatohepatitis (NASH) in only 16 months [2, 3]. That announcement was followed by Nimbus initiating phase I clinical trials of this ACC inhibitor, and Gilead acquiring the asset to further advance the matter into phase II studies [4]. Post this watershed event, Nimbus has continued to extensively use advanced computational modeling to advance its discovery programs and has repeated this early success by developing a highly selective clinical stage tyrosine kinase 2 (TYK2) inhibitor, which it expects to advance into a phase IIB clinical trial for psoriasis pathogenesis in 2021 [5, 6].

Following in the footsteps of Nimbus, Morphic Therapeutics also utilized Schrödinger's advanced computational modeling techniques and their own considerable biology, structural biology, and medicinal chemistry expertise to advance their discovery pipeline for a broad range of highly

Contemporary Accounts in Drug Discovery and Development, First Edition. Edited by Xianhai Huang, Robert G. Aslanian, and Wayne H. Tang.

challenging integrin targets [7]. Those efforts have resulted in a clinical stage α4β7 integrin inhibitor that Morphic Therapeutics is advancing into clinical studies for inflammatory bowel disease [8, 9]. Similar success was also achieved by Relay Therapeutics utilizing computational methods to develop a clinical stage protein tyrosine phosphatase SHP2 inhibitor for patients with advanced or metastatic solid tumors [10–14]. The ability of these smaller biotech companies to successfully and rapidly identify chemical matter suitable for clinical studies using advanced computational modeling has motivated large pharmaceutical companies, such as Bayer and AstraZeneca, to bring these advanced computational methods in-house to more broadly accelerate their own discovery efforts [15–17]. These successes utilizing advanced computational modeling to accelerate discovery are also now being realized by other parties using different computational strategies, such as Exscientia and Sumitomo Dainippon Pharma, which will be jointly advancing a serotonin 5-HT1A receptor agonist into clinical trials for obsessive compulsive disorder [18].

The success of these smaller biotech companies using advanced computational modeling to accelerate their discovery efforts, and the growing investments of large pharmaceutical companies to build out similar capabilities, raises a couple important questions: what specific modeling technologies are being most used, and how are they delivering such value? In this chapter, we will review those technologies that have made the strongest contributions to these recent successes, as well as provide opinions regarding how these technologies and their deployment may be improved in the future.

2.2 Essential Techniques

Excitingly, over the last decade, advanced computational methods have broadly matured to the point where they now can be used to accelerate all stages of preclinical drug discovery, including (i) Target Validation and Feasibility Assessment, (ii) Hit Discovery, (iii) Hit-to-Lead and Lead Optimization, and (iv) Preclinical Development. In this section we will review which computational techniques, in our view, have become indispensable components of computationally driven drug discovery and their primary use cases.

2.2.1 Target Validation and Feasibility Assessment

The selection of a target for a drug discovery campaign is an important and highly consequential decision. Selection of an inappropriate target may doom an otherwise highly successful discovery project to failure in clinical trials, if connection of the target to the underlying disease biology is misassessed. Likewise, there are many targets where identification of drug-like small molecule ligands remains an unsolved challenge, and if the protein target structure presents the project team with no binding pockets appropriate to bind drug-like small molecules, alternative therapeutic modalities, such as biologics might be required. Advanced computational methods can assist this process by refining intermediate resolution structures to a higher apparent resolution. This allows a project team considering the target to better understand the fine details of the challenges likely to be posed by the protein earlier in the discovery project than would typically be possible [19–21]. Once these improved structures are in hand, they can be used to pursue computational solvent analysis of the protein surface to identify regions of unfavorable solvation that might be exploited to develop high-affinity small molecule ligands [22–26]. Likewise, computational fragment soak methodologies can be used both to infer the druggability of various sites at the surface of the protein, as well as investigate if cryptic pocket binding opportunities might be possible to explore [27–31]. Such analyses may provide the project team a firmer foundation from which to judge if a drug-like small-molecule therapeutic modality is an appropriate choice for a given target, and should further allow the project team to advance from target tractability assessment to hit discovery with confidence the identification of hit-like small-molecule ligands should be tractable.

2.2.2 Hit Discovery

If a project team intends to pursue a drug-like small-molecule therapeutic modality for a particular protein target, then an important first goal of the project will be to identify small molecules with appreciable affinity and characterizable structure–activity relationships. A wide variety of experimental techniques have historically been used to identify such developable ligand matter, including high-throughput screening, fragment screening, and DNA encoded library screening [32–35]. In each of these methods, a sizeable pool of small molecules, perhaps a few thousand molecules in a fragment-based screen to potentially billions of molecules in a DNA-encoded library screen, are tested for affinity for the target protein. The most potent of these initial hits will then typically be resynthesized and re-assayed to confirm their activity using lower-throughput and more reliable experimental techniques. Given the long timelines and significant expense associated with pursuing these experimental hit finding strategies, it is perhaps unsurprising that virtual screening, i.e. the use of computational modeling to similarly evaluate such a candidate pool of molecules for affinity, is gaining ever broader acceptance [36, 37]. These virtual screening techniques include:

1) Molecular docking technologies where each ligand is oriented into the binding site of a structurally characterized protein and its likelihood of binding to the protein with appreciable affinity is estimated by an analysis of the interactions it may form with the protein [38–41].
2) Shape-based screening where the 3D shape of the ligand is compared with the shape of other known active ligands for the protein [42–45].
3) Ligand-based methods where a machine learning algorithm using the 2D structures of known active molecules learns in the training of the model to identify other molecules likely to exhibit affinity based on this earlier provided information [46, 47].

It should be further mentioned that hybrid methods blurring the lines between these different approaches to virtual screening are under active development, and the above categories are meant more to orient the reader to this body of work, rather than to provide a rigid taxonomy of methods [48]. Once a virtual screen has been performed, much like a traditional experimental screen, putative active molecules will be purchased or synthesized and then assayed to confirm their activity using lower-throughput and more reliable experimental techniques. A successful virtual screening campaign will provide a diverse selection of hit molecules exhibiting affinity that will provide good starting points for the discovery project to initiate hit-to-lead efforts.

Rather than solely present experimental screening and virtual screening to find hit molecules as competing techniques, we also wish to highlight that hybrid methods using both experimental techniques and the broad toolkit provided by modern computational methods are being investigated to unlock challenging targets. For example, multiple groups have been exploring if the success rates of experimental DNA-encoded library screens can be enhanced through application of machine learning technologies [49]. Likewise, free energy calculations of fragment linking might be able increase the number of action hits identified in experimental fragment-based screens [50]. We anticipate further development of such mixed computational/experimental approaches to hit finding to be a very productive future research direction.

It is also appropriate to mention here that if small molecule ligands for a given protein have already been reported, it is common practice within the pharmaceutical industry to discover novel hits for the protein by way of scaffold hopping, where the central core of the earlier reported chemical matter is replaced by some other appropriate core either to resolve some deficiency in the earlier chemical matter or simply to avoid earlier intellectual property. Although this approach to hit finding is conceptually quite appealing, failure rates are quite high since the vast majority of alternative central core choices usually greatly reduce or eliminate the potency of the small molecule [51]. Highly predictive alchemical free energy calculations have emerged as a reliable solution to this challenge, and also appear well suited to guide macrocyclization modifications, and other more exotic approaches to derivative hit discovery [52–55].

2.2.3 Hit-to-lead and Lead Optimization

Once developable hits have been identified, a central goal for the project team will be to synthesize and characterize sets of congeneric compounds with developable structure–activity relationships and demonstrated *in vivo* efficacy, i.e. a lead series. And, once a lead series is identified, the project team will work to further identify a development candidate molecule within that lead series manifesting the property profile required for that molecule to be advanced with confidence into preclinical development and ultimately clinical trials. In practice, this can be slow, expensive, and pain-staking work where many small chemical modifications of the initial hit molecules are synthetically explored and profiled in a variety of *in vitro* and *in vivo* experimental assays to facilitate the project team learning what combination of modifications to the initial hits might lead to a molecule manifesting the desired property profile.

A major step forward has been the development of advanced computational methods to explicitly enumerate many of the chemical modifications a project team might consider to improve the chemical matter and to further accurately score the properties of these ideated molecules. The enumeration of project relevant design ideas can be performed by way of a variety of methods including classical rules based approaches, as well as more modern machine-learning and deep-learning based techniques, and hybrid approaches [56–63]. Of equal importance to the ability to enumerate such synthesizable and project-relevant design ideas is the ability to accurately evaluate their properties to facilitate the prioritization of this pool of molecules for synthesis. There are three major classes of scoring methods in common use:

1) Ligand-based methods where a machine learning algorithm is provided the 2D structures of molecules with experimentally measured properties to train a model that can be used to predict the properties of other molecules based on this earlier provided information [64–67].
2) Approximate physics-based methods, such as MM-GB/SA interaction energy analysis and molecular docking, in which the likelihood of the molecule manifesting a particular property of interest will be estimated on the basis of the interactions it may form with its environment [39–41, 68–75].
3) Rigorous physics-based methods, such as free energy calculations, where all atomistic contributions to the property of interest will be explicitly simulated [76–82].

Each of these scoring methods have different strengths and weaknesses and attempt to balance prediction accuracy, computational efficiency, and breadth of applicability.

A key advantage to ligand-based methods is the ease with which they can be employed to predict any property for which a representative training set can be assembled. This makes such methods especially attractive for ADMET property prediction, where many pharmaceutical companies have compiled extensive data sets for model parameterization [66, 83]. However, a key disadvantage of these methods is the extent to which model accuracy is expected to degrade as the evaluated molecules begin to differ substantially from the training set or manifest features not observed within the training set. Further, in practice it is quite difficult to determine whether or not a molecule under consideration for synthesis should be considered within or outside the domain of applicability of the parametrized ligand-based model [84, 85]. Lastly, we note such methods are highly computationally efficient, and may require only milliseconds of CPU time per molecule.

In contrast to ligand-based methods, approximate physics-based methods, such as molecular docking and MM-GB/SA scoring, do not require the construction of training sets and explicit parametrization, and instead proceed by way of an energetic analysis or three-dimensional contact analysis to estimate whether the molecule under consideration for synthesis might achieve the discovery project design goals. This class of methods has been most extensively developed for protein-ligand binding affinity analysis, and can be utilized to exclude molecules that are grossly sterically or electrostatically incompatible with the ligand binding site of the protein of interest, thereby enriching the quality of molecules under consideration by the discovery project team [68, 69]. This class of methods has also been adapted to address other ADMET properties such as membrane permeability, human serum albumin (HSA) binding, and hERG blockade, among others [86–90]. The computational efficiency of these methods is typically much lower and therefore more computationally expensive than ligand-based approaches, and will often require seconds to minutes of CPU time. A key disadvantage of this approach is that quantitative prediction accuracy is typically not possible, and instead the methods will typically be used to triage large sets of molecules under consideration for further analysis, either by the project team members, or by way of more sophisticated computational techniques.

Rigorous physics-based methods are distinguished from the earlier approximate methods by the explicit theoretically robust connection that can be made between the simulations performed and the experimentally measured quantities of interest. The benefit of this clear theoretical connection is that such methods can be quite accurate in practice, often with root-mean-square errors of ~1 kcal/mol, and do not require any system-specific parametrization [76]. Such calculation accuracy enables these methods to more robustly support discovery project decision making than would be possible with less accurate or less reliable techniques. This improved accuracy and reliability though comes at the price of a greatly increased computational cost, typically ~1 GPU day of simulation time per scored molecule. As with the earlier mentioned approximate physics-based methods, this class of methods has been most extensively developed for protein-ligand binding affinity analysis [77–82]. Likewise, important work is ongoing to extend this class of methods to more comprehensively support ADMET property optimization, including solubility and binding selectivity [91, 92].

With a better understanding of these essential techniques in-hand, we will now turn our attention to how these techniques have been used effectively by multiple parties to accelerate the discovery of novel and more efficacious drug therapies. We will illustrate these explanations through a variety of published case studies, which we hope will highlight the wide variety of ways such computational analyses can positively contribute to discovery project execution.

2.3 Illustrative Applications

2.3.1 Modeling Support of Target Validation, Feasibility Assessment, and Hit Discovery for Acetyl-CoA Carboxylase

De novo fatty acid synthesis is rate limited by the enzyme ACC [93, 94]. A wealth of animal study data has suggested that inhibition of ACC might be palliative for NASH, which is believed to be an increasing cause of liver failure [95–102]. The development of a drug therapy to inhibit this target has been a highly nontrivial challenge, and conventional drug discovery efforts targeting the carboxyltransferase (CT) domain of ACC were found to yield drug candidates with poor physical chemical properties [93, 103–110]. Nimbus therapeutics decided to take a different approach to drugging the target, and utilized a computational solvent analysis method, WaterMap, to investigate if it might be possible to develop improved drug-like inhibitors that bind to the biotin carboxylase (BC) domain of the enzyme [111, 112]. These WaterMap calculations indicated the presence of multiple high energy hydration sites adjacent to an earlier known binding site of a natural product inhibitor of the BC domain, which might be exploited to design tight-binding and ligand-efficient chemical matter [112] (Figure 2.1).

This information was viewed as crucial by the project team to support initiation of the discovery project, since the presence of these high energy hydration sites was expected to greatly increase the likelihood the project team would be able to find satisfactory chemical matter [111]. The displacement of the water bound to these hydration sites was used as a constraint in virtual screening efforts, where only compounds expected to displace the water molecules were purchased and assayed. In agreement with the expectations of the computational modeling, the virtual screening campaign identified compound ND-022 (Figure 2.2), which displaced the solvating water occupying these high-energy hydration sites and was observed to bind to the target with 3.9 μM affinity [112]. Although in this case study, we have focused on the contributions that computational modeling

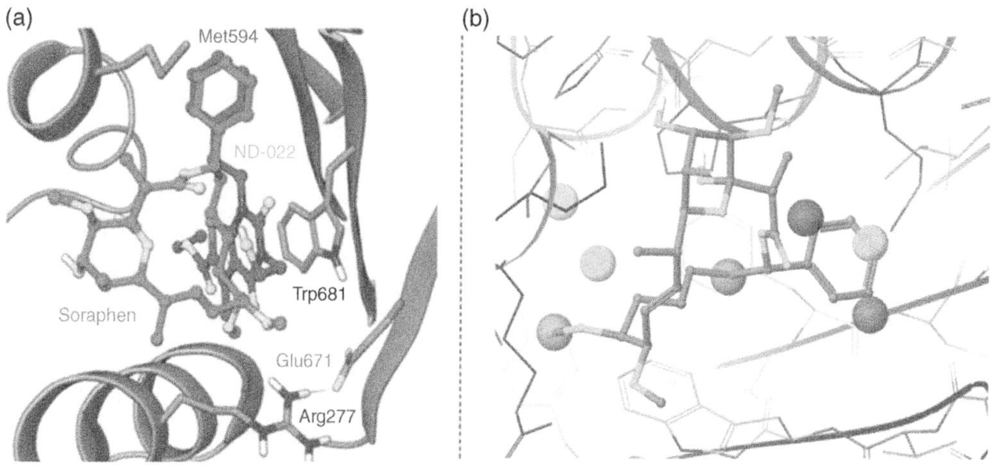

Figure 2.1 (a) An overlay of ND-022 with the natural product soraphen and (b) the high energy hydration sites that were crucial to the decision to initiate the ACC drug discovery project and the discovery of ND-022. *Source:* Reproduced with permission. Copyright© 2017, Elsevier [112].

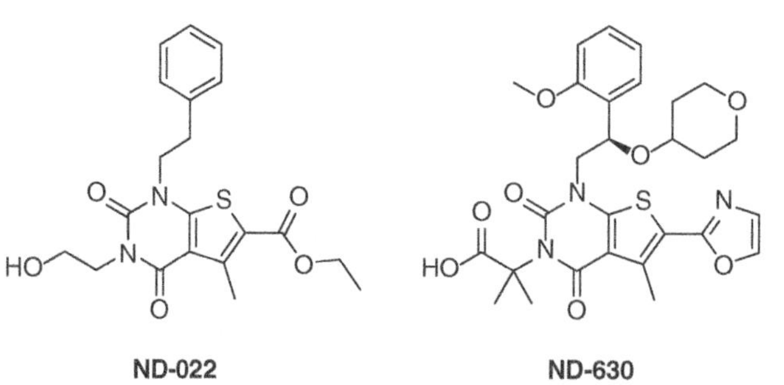

Figure 2.2 Structures of ND-022 and ND-630.

made to the target validation, feasibility assessment, and hit discovery phases of this discovery project, ND-022 was advanced into lead optimization, which resulted in the discovery of an exceptional development candidate molecule, ND-630, which has been advanced into phase II clinical trials [4, 112, 113].

2.3.2 Optimizing Selectivity in Lead Optimization for Tyrosine Kinase 2

The Janus kinase (JAK) family, comprised of Janus kinase 1 (JAK1), Janus kinase 2 (JAK2), Janus kinase 3 (JAK3), and tyrosine kinase 2 (TYK2), controls a variety of inflammation pathways, which may be relevant to the treatment of autoimmune disorders such as psoriasis, inflammatory bowel disease, and rheumatoid arthritis [114]. A variety of drug therapies inhibit these kinases, including Ruxolitinib, Tofacitinib, Baricitinib, Fedratinib, and Upadacitinib [114–118]. However, significant side effects are sometimes observed with these therapies. For example, inhibition of JAK2 can lead to anemia, and inhibition of JAK1 and JAK3 can lead to increased risk of infection [112]. Interestingly, genome-wide and phenome-wide association studies suggest the safety profile of a TYK2 selective drug therapy might have an improved safety profile while remaining efficacious for the treatment of autoimmune disorders [119–121].

The preceding opportunity notwithstanding, designing a highly selective TYK2 inhibitor is a substantial challenge. The four Janus family kinases have a high degree of structural and sequence similarity, especially so in the active site regions of the receptors [112, 122] (Figure 2.3). This has led to very few selective TYK2 drug therapies being advanced into clinical trials [123]. Nimbus therapeutics' approach to solving this difficult challenge in late-stage lead optimization was to employ rigorous physics-based methods at an unprecedented scale, including free energy calculation-based scoring of 4000 design ideas over a six month period to identify molecules predicted by the free energy calculations to be potent and selective inhibitors of TYK2 [112]. Of these 4000 evaluated compounds, 46 were advanced to synthesis and assay with 9 of the compounds meeting the targeted property profile. This computational modeling hit rate of 20% identifying satisfactory chemical matter should be viewed as outstanding given the difficulty of the challenge. Preclinical studies of several of these compounds further demonstrated outstanding efficacy in mouse model of psoriasis, and the discovery project has succeeded to advance a TYK2 inhibitor into clinical studies [112, 124].

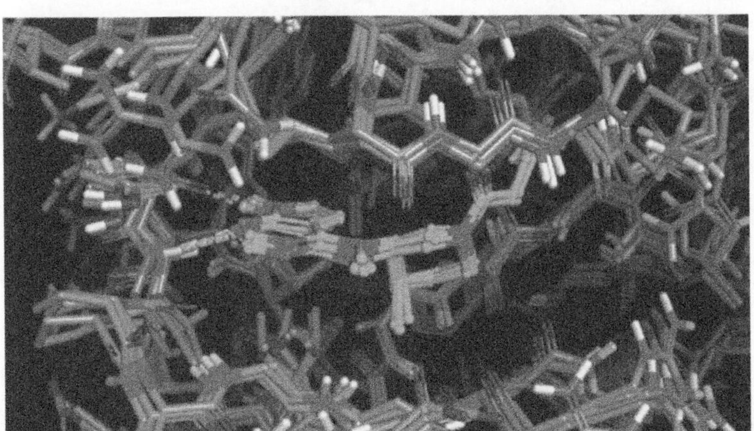

Figure 2.3 Superimposed crystal structures of TYK2, JAK1, JAK2, and JAK3 cocrystalized with tofacitinib. *Source:* Reproduced with permission. Copyright© 2017, Elsevier [112].

2.3.3 Discovery of Novel Allosteric Covalent Inhibitors of KRAS^{G12C}

KRAS mutations are widely observed in a variety of cancers [125]. The G12C mutation has received a great deal of recent attention, since it creates an allosteric pocket that can be exploited by covalent ligands to inactivate the protein [125]. Seeking to build on this growing body of work, researchers at Bayer Pharmaceuticals employed a modern computational approach to identify novel allosteric inhibitors, as depicted in Figure 2.4 [125]. In this workflow, seven million putative design ideas were first computationally enumerated and then advanced to a computational screening funnel where more approximate methods, i.e. pharmacophore analysis and MM-GB/SA scoring, were used to reject unpromising design ideas, and then more accurate and computationally intensive methods, i.e. free energy calculations, were used to finalize those molecules which were advanced to synthesis. The results of this effort led to only four of the original seven million design ideas being designated for synthesis and assay. This effort led directly to the development of a novel congeneric series with KRAS^{G12C} activity, and may constitute an important step along the path of the discovery of KRAS-targeted drug therapies.

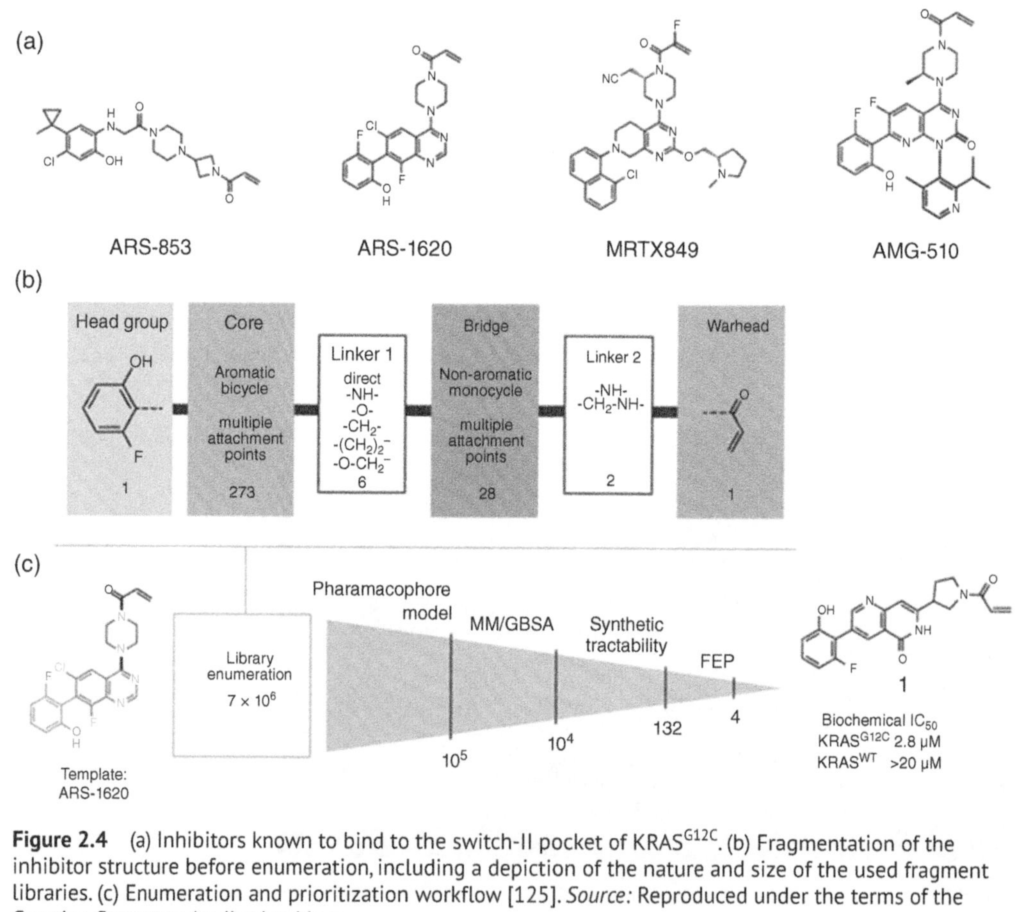

Figure 2.4 (a) Inhibitors known to bind to the switch-II pocket of KRAS^{G12C}. (b) Fragmentation of the inhibitor structure before enumeration, including a depiction of the nature and size of the used fragment libraries. (c) Enumeration and prioritization workflow [125]. *Source:* Reproduced under the terms of the Creative Commons Attribution License.

2.3.4 Supporting Hit to Lead Exploration for a Series of Phosphodiesterase 2A Inhibitors

Phosphodiesterase enzymes are important regulators of intracellular signal transduction and can be divided into 12 families [125–127]. Phosphodiesterase 2A (PDE2A) is expressed within the brain, and it is believed that inhibition of PDE2A may improve cognitive function [128, 129]. In order to exploit this emerging biology, researchers at Janssen Pharmaceutical utilized free energy calculations to optimize a hit compound identified in a high-throughput experimental screen to a lead series suitable for advancement into lead optimization [130]. The initial hit compound was found to have an IC_{50} value of 66 nM, but only an 8× selectivity versus antitarget PDE10A. As such, this hit compound was crystalized to facilitate a computational modeling drive design campaign where free energy calculations would be utilized to improve the properties of derivative molecules to establish a lead series suitable for initiation of lead optimization. During this campaign, 250 putative design ideas were scored with free energy calculations, and 100 of the top performing compounds were advanced to synthesis and assay (Figure 2.5). The calculations were found to be highly accurate with an R^2 value of 0.57 and mean unsigned error of 0.79 kcal/mol. In some ways more important than the accuracy of the calculations, these modeling efforts lead directly to the identification of a lead compound with single digit nanomolar potency and 100× selectivity versus antitarget PDE10A (Figure 2.6). This lead compound was further shown to have *in vivo* target occupancy and was of sufficient quality to justify initiation of lead optimization.

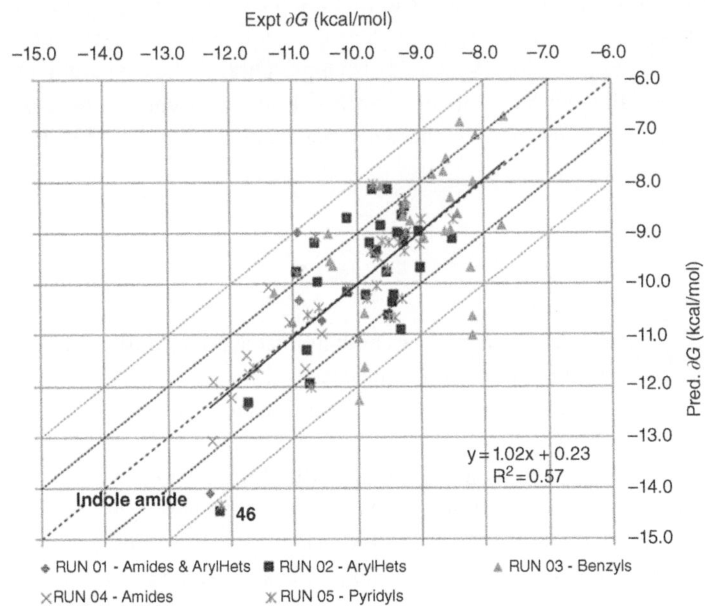

Figure 2.5 Correlation of experimental and calculated binding activity (∂G) for the compounds synthesized in each round of the FEP-guided design. The compound data points are shaded and shaped according to the rounds of FEP and synthesis. The diagonal lines of unity and ±1 or 2 kcal/mol errors are shown. The trendline (black, partly obscured by the line of unity), equation, and R^2 were calculated based on all the data combined, showing almost linear slope and very little offset. Source: Reproduced with permission. Copyright© 2020, American Chemical Society [130].

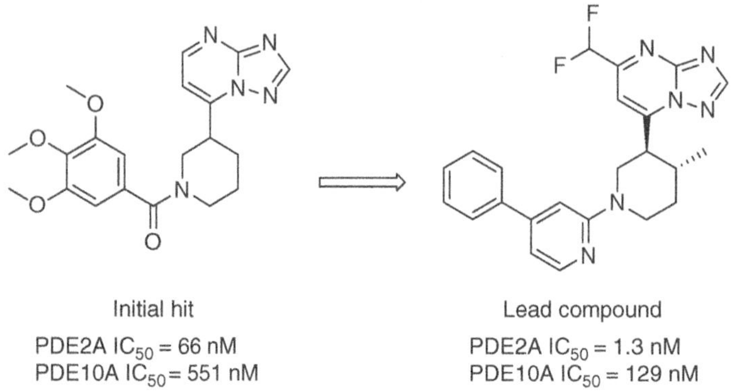

Figure 2.6 Structures of initial hit and lead compound [130].

2.4 Conclusion and Future Outlook

For the many researchers who have dedicated decades of their professional lives to the invention, development, and improvement of computational methods to aid drug discovery, it is deeply gratifying to now routinely see reports of these methods accelerating the discovery of novel and more effective drug therapies. These computational methods enable discovery teams to more comprehensively vet protein targets under consideration for project tractability and rapidly discover novel hits as was demonstrated for ACC and KRAS, and also accelerate hit-to-lead and lead optimization, as was shown for PDE2A and TYK2. Yet, despite these successes, there is clearly a great deal of work yet to do. We anticipate a major focus of future work to include the development of mixed experimental/computational approaches such as machine learning enhanced DNA-encoded library screening and FEP-guided fragment linking. We would also highlight that the development of predictive ADMET methods, especially related to rate of clearance, is an area clearly in need of improvement. We are excited though by the progress that has been made and look forward to what will be accomplished in the future.

References

1 *Fortune* (1981). The next industrial revolution: designing drugs by computer at Merck. https://www.backissues.com/issue/Fortune-October-05-1981 (accessed 29 August 2020).

2 Seddon, G., Lounnas, V., McGuire, R. et al. (2012). Drug design for ever, from hype to hope. *J. Comput. Aided Mol. Des.* 26: 137.

3 Atlas Venture (2013). The Nimbus experiment: structure-based drug deals. https://lifescivc.com/2013/06/the-nimbus-experiment-structure-based-drug-deals (accessed 29 August 2020).

4 Atlas Venture (2016). Nimbus delivers its Apollo Mission: a $1.2B Gilead partnership. https://lifescivc.com/2016/04/nimbus-delivers-apollo-mission-1-2b-gilead-partnership (accessed 29 August 2020)

5 Al Idrus, A. (2020). BIO: Nimbus CEO Keiper: "We're not just a one-hit wonder". https://www.fiercebiotech.com/biotech/bio-nimbus-ceo-we-re-not-just-a-one-hit-wonder (accessed 29 August 2020).

6 Nimbus Therapeutics (2020). Pipeline & Targets. https://www.nimbustx.com/pipeline-targets (accessed 30 August 2020).

7 BioSpace (2019). Schrödinger expands drug discovery partnership with Morphic Therapeutic. https://www.biospace.com/article/schrodinger-expands-drug-discovery-partnership-with-morphic-therapeutic (accessed 29 August 2020).
8 Morphic Therapeutic (2020). Morphic announces corporate highlights and second quarter 2020 financial results. https://investor.morphictx.com/news-releases/news-release-details/morphic-announces-corporate-highlights-and-second-quarter-2020 (accessed 29 August 2020).
9 Morphic Therapeutic (2020). Integrins in human disease. https://morphictx.com/pipeline (accessed 30 August 2020).
10 Bhat, S., Dahlgren, M.K., Giordanetto, F. et al. (2020). Pyrazolo[3,4-*b*]pyrazine derivatives as shp2 phosphatase inhibitors. https://uspto.report/patent/app/20200253969 (accessed 29 August 2020).
11 Bhat, S., Dahlgren, M.K., DiPietro, L.V. et al. (2020). Pyrazolo[3,4-*b*]pyrazine derivatives as shp2 phosphatase inhibitors. https://uspto.report/patent/app/20200172546 (accessed 29 August 2020).
12 Bhat, S., Dahlgren, M.K., DiPietro, L.V. et al. (2020). shp2 phosphatase inhibitors and methods of use thereof https://uspto.report/patent/app/20200062760 (accessed 29 August 2020).
13 Murcko, M. (2018).Workshop on free energy, kinetics, and Markov state models. https://www.youtube.com/watch?v=T4zEx-l10BQ (accessed 29 August 2020).
14 Relay Therapeutics (2020).Our focus on protein motion is creating new possibilities in drug discovery. https://relaytx.com/pipeline (accessed 30 August 2020).
15 Business WIre (2020).Schrödinger and Bayer collaborate to co-develop de novo design technology to accelerate drug discovery. https://www.businesswire.com/news/home/20200108005059/en/Schr%C3%B6dinger-Bayer-Collaborate-Co-Develop-de-novo-Design (accessed 29 August 2020).
16 Business WIre (2019).Schrödinger announces collaboration with AstraZeneca to deploy advanced computing technology for drug discovery. https://www.businesswire.com/news/home/20190904005166/en/Schr%C3%B6dinger-Announces-Collaboration-AstraZeneca-Deploy-Advanced-Computing (accessed 29 August 2020).
17 Business WIre (2020).Schrödinger announces expanded collaboration with AstraZeneca to extend computational modeling solutions to biologics. https://www.businesswire.com/news/home/20200323005050/en/Schr%C3%B6dinger-Announces-Expanded-Collaboration-AstraZeneca-Extend-Computational (accessed 29 August 2020).
18 Rees, V. (2020).Discovering and designing drugs with artificial intelligence. https://www.drugtargetreview.com/article/56366/discovering-and-designing-drugs-with-artificial-intelligence (accessed 30 August 2020).
19 Bell, J.A., Cao, Y., Gunn, J.R. et al. (2012). PrimeX and the Schrödinger computational chemistry suite of programs. In: *International. Tables for Crystallography*, 534–538. Wiley https://doi.org/10.1107/97809553602060000864.
20 Robertson, M.J., van Zundert, G.C.P., Borrelli, K., and Skiniotis, G. (2020). GemSpot: a pipeline for robust modeling of ligands into Cryo-EM maps. *Structure* 28: 707–716.e3.
21 van Zundert, G.C.P., Moriarty, N.W., Sobolev, O.V. et al. (2020). Macromolecular refinement of X-ray and cryo-electron microscopy structures with Phenix / OPLS3e for improved structure and ligand quality. bioRχiv. https://doi.org/10.1101/2020.07.10.198093.
22 Beuming, T., Che, Y., Abel, R. et al. (2012). Thermodynamic analysis of water molecules at the surface of proteins and applications to binding site prediction and characterization. *Proteins* 80: 871–883.
23 Young, T., Abel, R., Kim, B. et al. (2007). Motifs for molecular recognition exploiting hydrophobic enclosure in protein–ligand binding. *Proc. Natl. Acad. Sci.*: 808–813. https://doi.org/10.1073/pnas.0610202104.
24 Abel, R., Young, T., Farid, R. et al. (2008). Role of the active-site solvent in the thermodynamics of factor Xa ligand binding. *J. Am. Chem. Soc.* 130: 2817–2831.

25 Bayden, A.S., Moustakas, D.T., Joseph-McCarthy, D., and Lamb, M.L. (2015). Evaluating free energies of binding and conservation of crystallographic waters using SZMAP. *J. Chem. Inf. Model.* 55: 1552–1565.

26 Nittinger, E., Gibbons, P., Eigenbrot, C. et al. (2019). Water molecules in protein-ligand interfaces. Evaluation of software tools and SAR comparison. *J. Comput. Aided Mol. Des.* 33: 307–330.

27 Ghanakota, P., DasGupta, D., and Carlson, H.A. (2019). Free energies and entropies of binding sites identified by MixMD Cosolvent simulations. *J. Chem. Inf. Model.* 59: 2035–2045.

28 Ghanakota, P., van Vlijmen, H., Sherman, W., and Beuming, T. (2018). Large-scale validation of mixed-solvent simulations to assess hotspots at protein–protein interaction interfaces. *J. Chem. Inf. Model.*: 784–793. https://doi.org/10.1021/acs.jcim.7b00487.

29 Ghanakota, P. and Carlson, H.A. (2016). Driving structure-based drug discovery through Cosolvent molecular dynamics. *J. Med. Chem.* 59: 10383–10399.

30 Ghanakota, P. and Carlson, H.A. (2016). Moving beyond active-site detection: MixMD applied to allosteric systems. *J. Phys. Chem. B* 120: 8685–8695.

31 Bian, Y. and Xie, X.-Q.S. (2018). Computational fragment-based drug design: current trends, strategies, and applications. *AAPS J.* 20: 59.

32 Blay, V., Tolani, B., Ho, S.P., and Arkin, M.R. (2020). High-throughput screening: today's biochemical and cell-based approaches. *Drug Discov. Today* https://doi.org/10.1016/j.drudis.2020.07.024.

33 Price, A.J., Howard, S., and Cons, B.D. (2017). Fragment-based drug discovery and its application to challenging drug targets. *Essays Biochem.* 61: 475–484.

34 Favalli, N., Bassi, G., Scheuermann, J., and Neri, D. (2018). DNA-encoded chemical libraries – achievements and remaining challenges. *FEBS Lett.* 592: 2168–2180.

35 Yuen, L.H. and Franzini, R.M. (2017). Achievements, challenges, and opportunities in DNA-encoded library research: an academic point of view. *Chembiochem.* 18: 829–836.

36 Gimeno, A., Ojeda-Montes, M.J., Tomás-Hernández, S. et al. (2019). The light and dark sides of virtual screening: what is there to know. *Int. J. Mol. Sci.*: 20. https://doi.org/10.3390/ijms20061375.

37 Fradera, X. and Babaoglu, K. (2017). Overview of methods and strategies for conducting virtual small molecule screening. *Curr. Protoc. Chem. Biol.* 9: 196–212.

38 Pagadala, N.S., Syed, K., and Tuszynski, J. (2017). Software for molecular docking: a review. *Biophys. Rev.*: 91–102. https://doi.org/10.1007/s12551-016-0247-1.

39 Murphy, R.B., Repasky, M.P., Greenwood, J.R. et al. (2016). WScore: a flexible and accurate treatment of explicit water molecules in ligand-receptor docking. *J. Med. Chem.* 59: 4364–4384.

40 Friesner, R.A., Murphy, R.B., Repasky, M.P. et al. (2006). Extra precision glide: docking and scoring incorporating a model of hydrophobic enclosure for protein-ligand complexes. *J. Med. Chem.* 49: 6177–6196.

41 Friesner, R.A., Banks, J.L., Murphy, R.B. et al. (2004). Glide: a new approach for rapid, accurate docking and scoring. 1. Method and assessment of docking accuracy. *J. Med. Chem.*: 1739–1749. https://doi.org/10.1021/jm0306430.

42 McGaughey, G.B., Sheridan, R.P., Bayly, C.I. et al. (2007). Comparison of topological, shape, and docking methods in virtual screening. *J. Chem. Inf. Model.* 47: 1504–1519.

43 Sastry, G.M., Inakollu, V.S.S., and Sherman, W. (2013). Boosting virtual screening enrichments with data fusion: coalescing hits from two-dimensional fingerprints, shape, and docking. *J. Chem. Inf. Model.* 53: 1531–1542.

44 Hawkins, P.C.D., Skillman, A.G., and Nicholls, A. (2007). Comparison of shape-matching and docking as virtual screening tools. *J. Med. Chem.* 50: 74–82.

45 Grant, J.A., Gallardo, M.A., and Pickup, B.T. (1996). A fast method of molecular shape comparison: a simple application of a Gaussian description of molecular shape. *J. Comput. Chem.*: 1653–1666. https://doi.org/10.1002/(SICI)1096-987X(19961115)17:14<1653::AID-JCC7>3.0.CO;2-K.

46 Bahi, M. and Batouche, M. (2018). Deep learning for ligand-based virtual screening in drug discovery. In: *2018 3rd International Conference on Pattern Analysis and Intelligent Systems (PAIS)*. Tebessa, Algeria (24–25 October 2018). IEEE https://doi.org/10.1109/pais.2018.8598488.

47 Pérez-Sianes, J., Pérez-Sánchez, H., and Díaz, F. (2019). Virtual screening meets deep learning. *Curr. Comput. Aided Drug Des.* 15: 6–28.

48 Gentile, F., Agrawal, V., Hsing, M. et al. (2019). Deep docking – a deep learning approach for virtual screening of big chemical datasets. doi:10.1101/2019.12.15.877316.

49 McCloskey, K., Sigel, E.A., Kearnes, S. et al. (2020). Machine learning on DNA-encoded libraries: a new paradigm for hit finding. *J. Med. Chem.* 63: 8857–8866.

50 Yu, H.S., Modugula, K., Ichihara, O. et al. (2020). General theory of fragment linking in molecular design: why fragment linking rarely succeeds and how to improve outcomes. *J. Chem. Theory Comput.* https://doi.org/10.1021/acs.jctc.0c01004.

51 Zhao, H. (2007). Scaffold selection and scaffold hopping in lead generation: a medicinal chemistry perspective. *Drug Discov. Today* 12: 149–155.

52 Wang, L., Deng, Y., Wu, Y. et al. (2017). Accurate modeling of scaffold hopping transformations in drug discovery. *J. Chem. Theory Comput.* 13: 42–54.

53 Paulsen, J.L., Yu, H.S., Sindhikara, D. et al. (2020). Evaluation of free energy calculations for the prioritization of macrocycle synthesis. *J. Chem. Inf. Model.* 60: 3489–3498.

54 Wagner, V., Jantz, L., Briem, H. et al. (2017). Computational macrocyclization: from de novo macrocycle generation to binding affinity estimation. *ChemMedChem* 12: 1866–1872.

55 Yu, H.S., Deng, Y., Wu, Y. et al. (2017). Accurate and reliable prediction of the binding affinities of macrocycles to their protein targets. *J. Chem. Theory Comput.* 13: 6290–6300.

56 Ghanakota, P., Bos, P.H., Konze, K.D. et al. (2020). Combining cloud-based free-energy calculations, synthetically aware enumerations, and goal-directed generative machine learning for rapid large-scale chemical exploration and optimization. *J. Chem. Inf. Model.* https://doi.org/10.1021/acs.jcim.0c00120.

57 Konze, K.D., Bos, P.H., Dahlgren, M.K. et al. (2019). Reaction-based enumeration, active learning, and free energy calculations to rapidly explore synthetically tractable chemical space and optimize potency of cyclin-dependent kinase 2 inhibitors. *J. Chem. Inf. Model.* 59: 3782–3793.

58 Gómez-Bombarelli, R., Wei, J.N., Duvenaud, D. et al. (2018). Automatic chemical design using a data-driven continuous representation of molecules. *ACS Cent. Sci.* 4: 268–276.

59 Kwon, Y., Yoo, J., Choi, Y.-S. et al. (2019). Efficient learning of non-autoregressive graph variational autoencoders for molecular graph generation. *J. Cheminformatics* https://doi.org/10.1186/s13321-019-0396-x.

60 Popova, M., Isayev, O., and Tropsha, A. (2018). Deep reinforcement learning for de novo drug design. *Sci. Adv.* 4: eaap7885.

61 Zhou, Z., Kearnes, S., Li, L. et al. (2019). Optimization of molecules via deep reinforcement learning. *Sci. Rep.* 9: 10752.

62 Zhavoronkov, A., Ivanenkov, Y.A., Aliper, A. et al. (2019). Deep learning enables rapid identification of potent DDR1 kinase inhibitors. *Nat. Biotechnol.* 37: 1038–1040.

63 Abel, R. and Bhat, S. (2017). Free energy calculation guided virtual screening of synthetically feasible ligand R-group and scaffold modifications: an emerging paradigm for lead optimization. *Annu. Rep. Med. Chem.*: 237–262. https://doi.org/10.1016/bs.armc.2017.08.007.

64 Zhang, L., Tan, J., Han, D., and Zhu, H. (2017). From machine learning to deep learning: progress in machine intelligence for rational drug discovery. *Drug Discov. Today* 22: 1680–1685.

65 Wu, Z., Ramsundar, B., Feinberg, E.N. et al. (2018). MoleculeNet: a benchmark for molecular machine learning. *Chem. Sci.* 9: 513–530.

66 Feinberg, E.N., Joshi, E., Pande, V.S., and Cheng, A.C. (2020). Improvement in ADMET prediction with multitask deep featurization. *J. Med. Chem.* 63: 8835–8848.

67 Wieder, O., Kohlbacher, S., Kuenemann, M. et al. (2020). A compact review of molecular property prediction with graph neural networks. *Drug Discov. Today Technol.* https://doi.org/10.1016/j.ddtec.2020.11.009.

68 Knight, J.L., Krilov, G., Borrelli, K.W. et al. (2014). Leveraging data fusion strategies in multireceptor lead optimization MM/GBSA end-point methods. *J. Chem. Theory Comput.* 10: 3207–3220.

69 Wang, E., Sun, H., Wang, J. et al. (2019). End-point binding free energy calculation with MM/PBSA and MM/GBSA: strategies and applications in drug design. *Chem. Rev.* 119: 9478–9508.

70 Halgren, T.A., Murphy, R.B., Friesner, R.A. et al. (2004). Glide: a new approach for rapid, accurate docking and scoring. 2. Enrichment factors in database screening. *J. Med. Chem.* 47: 1750–1759.

71 Nissink, J.W.M., Murray, C., Hartshorn, M. et al. (2002). A new test set for validating predictions of protein-ligand interaction. *Proteins: Struct. Funct. Bioinformatics*: 457–471. https://doi.org/10.1002/prot.10232.

72 Verdonk, M.L., Cole, J.C., Hartshorn, M.J. et al. (2003). Improved protein-ligand docking using GOLD. *Proteins* 52: 609–623.

73 McGann, M. (2011). FRED pose prediction and virtual screening accuracy. *J. Chem. Inf. Model.*: 578–596. https://doi.org/10.1021/ci100436p.

74 Rarey, M., Kramer, B., Lengauer, T., and Klebe, G. (1996). A fast flexible docking method using an incremental construction algorithm. *J. Mol. Biol.* 261: 470–489.

75 Jain, A.N. (2003). Surflex: fully automatic flexible molecular docking using a molecular similarity-based search engine. *J. Med. Chem.* 46: 499–511.

76 Abel, R., Wang, L., Harder, E.D. et al. (2017). Advancing drug discovery through enhanced free energy calculations. *Acc. Chem. Res.* 50: 1625–1632.

77 Wang, L., Wu, Y., Deng, Y. et al. (2015). Accurate and reliable prediction of relative ligand binding potency in prospective drug discovery by way of a modern free-energy calculation protocol and force field. *J. Am. Chem. Soc.* 137: 2695–2703.

78 Kuhn, M., Firth-Clark, S., Tosco, P. et al. (2020). Assessment of binding affinity via alchemical free-energy calculations. *J. Chem. Inf. Model.* 60: 3120–3130.

79 Lee, T.-S., Allen, B.K., Giese, T.J. et al. (2020). Alchemical binding free energy calculations in AMBER20: advances and best practices for drug discovery. *J. Chem. Inf. Model.* 60: 5595–5623.

80 Lin, Z., Zou, J., Peng, C. et al. (2020).A Cloud Computing Platform for Scalable Relative and Absolute Binding Free Energy Prediction: New Opportunities and Challenges for Drug Discovery. ChemRxiv. doi:10.26434/chemrxiv.13096157.v1

81 Gapsys, V., Pérez-Benito, L., Aldeghi, M. et al. (2020). Large scale relative protein ligand binding affinities using non-equilibrium alchemy. *Chem. Sci.*: 1140–1152. https://doi.org/10.1039/c9sc03754c.

82 Song, L.F., Lee, T.-S., Zhu, C. et al. (2019). Using AMBER18 for relative free energy calculations. *J. Chem. Inf. Model.* 59: 3128–3135.

83 Göller, A.H., Kuhnke, L., Montanari, F. et al. (2020). Bayer's in silico ADMET platform: a journey of machine learning over the past two decades. *Drug Discov. Today* 25: 1702–1709.

84 Kar, S., Roy, K., and Leszczynski, J. (2019). On error measures for validation and uncertainty estimation of predictive QSAR models. *Comput. Nanotoxicol.*: 437–493. https://doi.org/10.1201/9780429341373-10.

85 Hanser, T., Barber, C., Marchaland, J.F., and Werner, S. (2016). Applicability domain: towards a more formal definition. *SAR QSAR Environ. Res.* 27: 893–909.

86 Leung, S.S.F., Sindhikara, D., and Jacobson, M.P. (2016). Simple predictive models of passive membrane permeability incorporating size-dependent membrane-water partition. *J. Chem. Inf. Model.* 56: 924–929.

87 Leung, S.S.F., Mijalkovic, J., Borrelli, K., and Jacobson, M.P. (2012). Testing physical models of passive membrane permeation. *J. Chem. Inf. Model.* 52: 1621–1636.

88 Hall, M.L., Jorgensen, W.L., and Whitehead, L. (2013). Automated ligand- and structure-based protocol for in silico prediction of human serum albumin binding. *J. Chem. Inf. Model.* 53: 907–922.

89 Lexa, K.W., Dolghih, E., and Jacobson, M.P. (2014). A structure-based model for predicting serum albumin binding. *PLoS One* 9: e93323.

90 Dickson, C.J., Velez-Vega, C., and Duca, J.S. (2020). Revealing molecular determinants of hERG blocker and activator binding. *J. Chem. Inf. Model.* 60: 192–203.

91 Mondal, S., Tresadern, G., Greenwood, J. et al. (2019). A Free Energy Perturbation Approach to Estimate the Intrinsic Solubilities of Drug-like Small Molecules. ChemRxiv. doi:10.26434/chemrxiv.10263077

92 Albanese, S.K., Chodera, J.D., Volkamer, A. et al. (2020). Is structure-based drug design ready for selectivity optimization? *J. Chem. Inf. Model.* https://doi.org/10.1021/acs.jcim.0c00815.

93 Harwood, H.J. Jr. (2005). Treating the metabolic syndrome: acetyl-CoA carboxylase inhibition. *Expert Opin. Ther. Targets* 9: 267–281.

94 Harwood, H.J. and James, H.H. (2012). The adipocyte as an endocrine organ in the regulation of metabolic homeostasis. *Neuropharmacology*: 57–75. https://doi.org/10.1016/j.neuropharm.2011.12.010.

95 Lucas, C., Lucas, G., Lucas, N. et al. (2018). A systematic review of the present and future of non-alcoholic fatty liver disease. *Clin. Exp. Hepatol.* 4: 165–174.

96 Abu-Elheiga, L., Matzuk, M.M., Abo-Hashema, K.A., and Wakil, S.J. (2001). Continuous fatty acid oxidation and reduced fat storage in mice lacking acetyl-CoA carboxylase 2. *Science* 291: 2613–2616.

97 Abu-Elheiga, L., Oh, W., Kordari, P., and Wakil, S.J. (2003). Acetyl-CoA carboxylase 2 mutant mice are protected against obesity and diabetes induced by high-fat/high-carbohydrate diets. *Proc. Natl. Acad. Sci. USA* 100: 10207–10212.

98 Savage, D.B., Choi, C.S., Samuel, V.T. et al. (2006). Reversal of diet-induced hepatic steatosis and hepatic insulin resistance by antisense oligonucleotide inhibitors of acetyl-CoA carboxylases 1 and 2. *J. Clin. Invest.* 116: 817–824.

99 Qi, L., Heredia, J.E., Altarejos, J.Y. et al. (2006). TRB3 links the E3 ubiquitin ligase COP1 to lipid metabolism. *Science* 312: 1763–1766.

100 Fullerton, M.D., Galic, S., Marcinko, K. et al. (2013). Single phosphorylation sites in Acc1 and Acc2 regulate lipid homeostasis and the insulin-sensitizing effects of metformin. *Nat. Med.* 19: 1649–1654.

101 Hoehn, K.L., Turner, N., Swarbrick, M.M. et al. (2010). Acute or chronic upregulation of mitochondrial fatty acid oxidation has no net effect on whole-body energy expenditure or adiposity. *Cell Metab.*: 70–76. https://doi.org/10.1016/j.cmet.2009.11.008.

102 Olson, D.P., Pulinilkunnil, T., Cline, G.W. et al. (2010). Gene knockout of Acc2 has little effect on body weight, fat mass, or food intake. *Proc. Natl. Acad. Sci. USA*: 7598–7603. https://doi.org/10.1073/pnas.0913492107.

103 Harwood, H.J. Jr., Petras, S.F., Shelly, L.D. et al. (2003). Isozyme-nonselective N-substituted bipiperidylcarboxamide acetyl-CoA carboxylase inhibitors reduce tissue malonyl-CoA concentrations, inhibit fatty acid synthesis, and increase fatty acid oxidation in cultured cells and in experimental animals. *J. Biol. Chem.* 278: 37099–37111.

104 Tong, L. and Harwood, H.J. Jr. (2006). Acetyl-coenzyme a carboxylases: versatile targets for drug discovery. *J. Cell. Biochem.* 99: 1476–1488.

105 Corbett, J.W., Freeman-Cook, K.D., Elliott, R. et al. (2010). Discovery of small molecule isozyme non-specific inhibitors of mammalian acetyl-CoA carboxylase 1 and 2. *Bioorg. Med. Chem. Lett.* 20: 2383–2388.

106 Bourbeau, M.P. and Bartberger, M.D. (2015). Recent advances in the development of acetyl-CoA carboxylase (ACC) inhibitors for the treatment of metabolic disease. *J. Med. Chem.* 58: 525–536.

107 Griffith, D.A., Kung, D.W., Esler, W.P. et al. (2014). Decreasing the rate of metabolic ketone reduction in the discovery of a clinical acetyl-CoA carboxylase inhibitor for the treatment of diabetes. *J. Med. Chem.* 57: 10512–10526.

108 Glund, S., Schoelch, C., Thomas, L. et al. (2012). Inhibition of acetyl-CoA carboxylase 2 enhances skeletal muscle fatty acid oxidation and improves whole-body glucose homeostasis in db/db mice. *Diabetologia*: 2044–2053. https://doi.org/10.1007/s00125-012-2554-9.

109 Tong, L. (2005). Acetyl-coenzyme a carboxylase: crucial metabolic enzyme and attractive target for drug discovery. *Cell. Mol. Life Sci.* 62: 1784–1803.

110 Zhang, H., Tweel, B., Li, J., and Tong, L. (2004). Crystal structure of the carboxyltransferase domain of acetyl-coenzyme a carboxylase in complex with CP-640186. *Structure* 12: 1683–1691.

111 Harriman, G., Greenwood, J., Bhat, S. et al. (2016). Acetyl-CoA carboxylase inhibition by ND-630 reduces hepatic steatosis, improves insulin sensitivity, and modulates dyslipidemia in rats. *Proc. Natl. Acad. Sci. USA* 113: E1796–E1805.

112 Abel, R., Mondal, S., Masse, C. et al. (2017). Accelerating drug discovery through tight integration of expert molecular design and predictive scoring. *Curr. Opin. Struct. Biol.* 43: 38–44.

113 Tong, A. (2019). Gilead shores up hope for NASH cocktail with a glimpse at positive proof-of-concept data. https://endpts.com/gilead-shores-up-hope-for-nash-cocktail-with-a-glimpse-at-proof-of-concept-data (accessed 29 December 2020).

114 Golan, D.E. (2008). *Principles of Pharmacology: The Pathophysiologic Basis of Drug Therapy*. Lippincott Williams & Wilkins.

115 Mesa, R.A. (2010). Ruxolitinib, a selective JAK1 and JAK2 inhibitor for the treatment of myeloproliferative neoplasms and psoriasis. *IDrugs* 13: 394–403.

116 Adis Editorial (2010). Tofacitinib. *Drugs R. D.* 10: 271–284.

117 Tuttle, K.R., Brosius, F.C. 3rd, Adler, S.G. et al. (2018). JAK1/JAK2 inhibition by baricitinib in diabetic kidney disease: results from a phase 2 randomized controlled clinical trial. *Nephrol. Dial. Transplant.* 33: 1950–1959.

118 Pardanani, A., Hood, J., Lasho, T. et al. (2007). TG101209, a small molecule JAK2-selective kinase inhibitor potently inhibits myeloproliferative disorder-associated JAK2V617F and MPLW515L/K mutations. *Leukemia* 21: 1658–1668.

119 Ishizaki, M., Muromoto, R., Akimoto, T. et al. (2014). Tyk2 is a therapeutic target for psoriasis-like skin inflammation. *Int. Immunol.* 26: 257–267.

120 Diogo, D., Bastarache, L., Liao, K.P. et al. (2015). TYK2 protein-coding variants protect against rheumatoid arthritis and autoimmunity, with no evidence of major pleiotropic effects on non-autoimmune complex traits. *PLoS One* 10: e0122271.

121 Gracey, E., Hromadová, D., Lim, M. et al. (2020). TYK2 inhibition reduces type 3 immunity and modifies disease progression in murine spondyloarthritis. *J. Clin. Invest.* 130: 1863–1878.

122 BioSpace (2020). Nimbus Therapeutics announces $60 million private financing from new investors to advance expanded pipeline. https://www.biospace.com/article/

nimbus-therapeutics-announces-60-million-private-financing-from-new-investors-to-advance-expanded-pipeline (accessed 29 December 2020).

123 Hammitzsch, A., Lorenz, G., and Moog, P. (2020). Impact of Janus kinase inhibition on the treatment of axial spondyloarthropathies. *Front. Immunol.* 11: 591176.

124 Associated Press (2020).Nimbus Therapeutics announces $60 million private financing from new investors to advance expanded pipeline. https://apnews.com/press-release/business-wire/business-corporate-news-massachusetts-drug-trials-products-and-services-e3c360003ea74eec9f626a3e82790b86 (accessed 29 December 2020).

125 Mortier, J., Friberg, A., Badock, V. et al. (2020). Computationally empowered workflow identifies novel covalent allosteric binders for KRAS. *ChemMedChem* 15: 827–832.

126 Conti, M. (2000). Phosphodiesterases and cyclic nucleotide signaling in endocrine cells. *Mol. Endocrinol.*: 1317–1327. https://doi.org/10.1210/mend.14.9.0534.

127 Maurice, D.H., Ke, H., Ahmad, F. et al. (2014). Advances in targeting cyclic nucleotide phosphodiesterases. *Nat. Rev. Drug Discov.* 13: 290–314.

128 Gomez, L. and Guy, B.J. (2013). PDE2 inhibition: potential for the treatment of cognitive disorders. *Bioorg. Med. Chem. Lett.*: 6522–6527. https://doi.org/10.1016/j.bmcl.2013.10.014.

129 Wu, Y., Li, Z., Huang, Y.-Y. et al. (2018). Novel phosphodiesterase inhibitors for cognitive improvement in Alzheimer's disease. *J. Med. Chem.* 61: 5467–5483.

130 Tresadern, G., Velter, I., Trabanco, A.A. et al. (2020). [1,2,4]Triazolo[1,5-]pyrimidine phosphodiesterase 2A inhibitors: structure and free-energy perturbation-guided exploration. *J. Med. Chem.* 63: 12887–12910.

3

Discovery and Development of the Soluble Guanylate Cyclase Stimulator Vericiguat for the Treatment of Chronic Heart Failure

Markus Follmann, Corina Becker, Lothar Roessig, Peter Sandner, and Johannes-Peter Stasch

Research & Development, Pharmaceuticals, Bayer AG, Aprather Weg 18a, 42096, Wuppertal, Germany

3.1 Introduction

Despite better therapies, cardiovascular (CV) diseases remain the leading cause of death worldwide. Among them, heart failure (HF) is a leading cause of hospitalization and CV death [1]. More than 63 million people worldwide suffered from HF in 2016 [2], and around 50% of patients with HF die within five years [3]. HF, by definition, is a condition where the heart is unable to pump sufficiently in order to provide adequate blood flow to other organs such as the kidneys and the brain. Overall, it is a complex clinical syndrome caused by cardiac structural abnormalities and/or functional impairments. In general, HF is a chronic, progressive condition associated with significant morbidity and mortality. Main treatments include healthy lifestyle changes, in some cases surgery and implantation of devices to control heart rhythm, and pharmacotherapy. Prescribed drugs for heart failure with reduced ejection fraction (HFrEF) include angiotensin-converting enzyme inhibitors, angiotensin-2 receptor blockers (ARBs) or an angiotension receptor neprilysin inhibitor (ARNI), beta blockers, mineralocorticoid receptor antagonists, diuretics, and more recently SGLT-2 (sodium glucose co-transporter-2) inhibitors. However, even with these new entrants and optimal guideline-directed medical therapy, event rates in HFrEF remain as high as 15% per year of the composite outcome CV death or HF hospitalization in the empagliflozin arm of EMPEROR-REDUCED [4] and even 38% in patients with HFrEF after a previous event in the placebo group of VICTORIA [5]. Therefore, the unmet need in HF is still high and drugs targeting other pathways involved in disease progression are sought. No less than 1 out of 6 patients suffering from HFrEF develop worsening HF on average 1.5 years following initial HF diagnosis. The high risk for two-year mortality and recurrent HF hospitalizations in these patients indicates particularly high unmet medical need for novel treatment strategies [1]. Here we present the discovery and development of the soluble guanylate cyclase (sGC) stimulator vericiguat (BAY 1021189; MK-1242) for the treatment of chronic HF. Data from the phase 3 VICTORIA trial in patients with worsening chronic HF and ejection fraction <45% were published in the *New England Journal of Medicine* [5, 6]. Vericiguat 10 mg once daily significantly reduced the combined risk of first HF hospitalization or CV death (Figure 3.1).

Contemporary Accounts in Drug Discovery and Development, First Edition. Edited by Xianhai Huang, Robert G. Aslanian, and Wayne H. Tang.

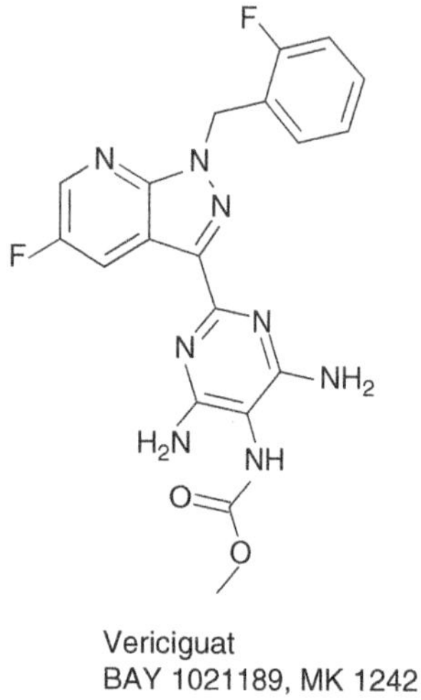

Figure 3.1 sGC stimulator vericiguat.

3.2 Soluble Guanylate Cyclase Stimulators as Treatment Option for Heart Failure

The nitric oxide (NO)/sGC/cGMP pathway is one of the major signaling cascades for regulation of cardiopulmonary, vascular and renal function (Figure 3.2) [7, 8]. This signaling pathway is triggered by the production of NO via NO synthases. NO, which is a gas, can pass cellular membranes and binds to the cytosolic sGC which carries an NO-binding heme structure in the beta subunit of the enzyme. NO binding to sGC triggers a conformational change of the enzyme and an activation of the catalytic site which transfers GTP in the second messenger molecule cGMP which binds and activates several downstream targets, e.g. the cGMP-activated protein kinases (PKG1 and PKG2) [9]. Impairment and disruption of the NO/cGMP pathway which is characterized by a decline in cGMP production causing the abrogation of cGMP downstream signaling, contributes to the development and progression of CV diseases including HF [10]. The pathophysiology of CV diseases includes endothelial cell dysfunction which impairs NO production leading to decreased NO availability and reduced cGMP tissue levels. This reduced NO availability and insufficient stimulation of sGC results in systemic, vascular dysregulation, affecting the coronary, pulmonary, and renal circulation. Moreover, impaired sGC signaling is leading to organ damage and dysfunction of the heart. Therefore, it is an intriguing therapeutic concept to restore cGMP signaling via direct stimulation

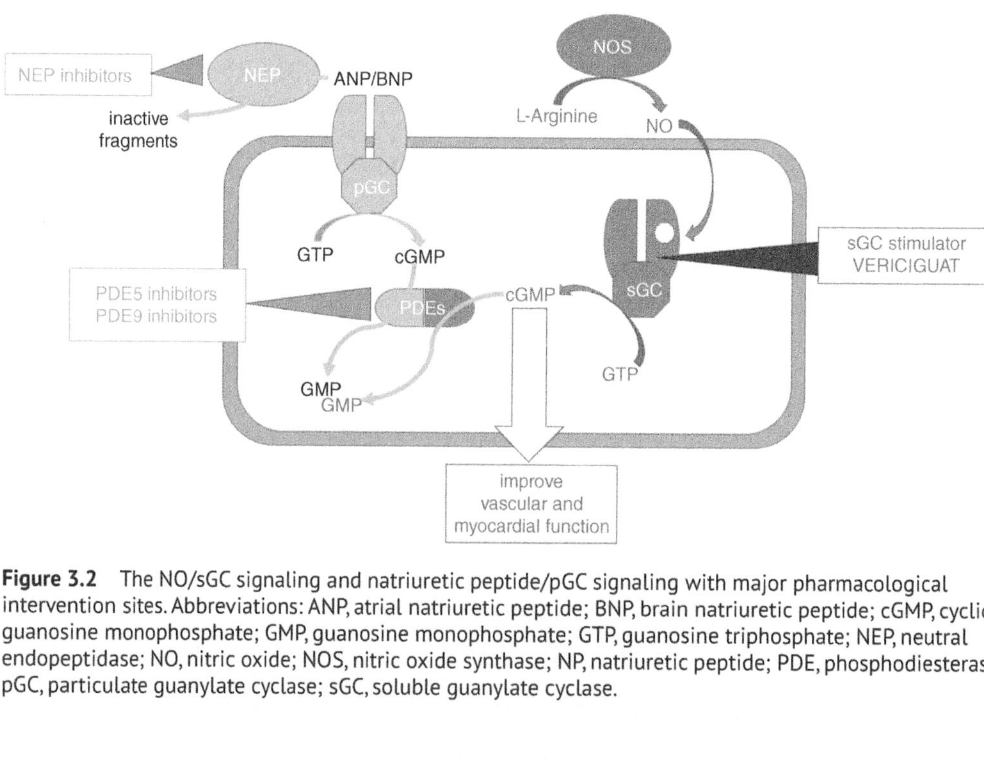

Figure 3.2 The NO/sGC signaling and natriuretic peptide/pGC signaling with major pharmacological intervention sites. Abbreviations: ANP, atrial natriuretic peptide; BNP, brain natriuretic peptide; cGMP, cyclic guanosine monophosphate; GMP, guanosine monophosphate; GTP, guanosine triphosphate; NEP, neutral endopeptidase; NO, nitric oxide; NOS, nitric oxide synthase; NP, natriuretic peptide; PDE, phosphodiesterase; pGC, particulate guanylate cyclase; sGC, soluble guanylate cyclase.

of sGC with the novel, NO-independent sGC stimulator vericiguat [11, 12]. Vericiguat has been shown to directly bind to sGC and stimulates the enzyme also in the absence of NO. In addition, vericiguat could enhance the efficacy of endogenous NO. Thus, vericiguat is able to restore impaired cGMP signaling in chronic HF. This unique mode of action makes vericiguat a promising molecule for the treatment of CV disease including HF.

Despite this strong rationale for cGMP elevation as an effective treatment option for HF, previous attempts by NO donors which also stimulate sGC and cGMP production, and also by PDE5 inhibitors which inhibit degradation of cGMP, mostly failed [13].

sGC activators which generate cGMP via interaction with the dysfunctional form of sGC, apo-sGC, have not been explored in chronic HF; however, cinaciguat (BAY 58-2667) has been characterized in acute HF [13]. Based on promising preclinical results, a phase 2 study in patients with acute decompensated heart failure (ADHF) was initiated. Continuous intravenous infusion of cinaciguat was well tolerated and resulted in an improvement of cardiopulmonary hemodynamics. The subsequent clinical phase 2b program studied the effects of cinaciguat in three randomized, double-blind, placebo-controlled studies in ADHF patients; however, the clinical development of cinaciguat was terminated prematurely because of hypotensive events without clear benefit [12].

Very recently, another cGMP-elevating principle, sacubitril combined with the angiotensin receptor blocker valsartan, was studied successfully as an oral treatment in HF [14]. The neprilysin inhibitor sacubitril is hypothesized to work via natriuretic peptide elevation and associated stimulation of the particulate guanylate cyclase (pGC) [15–17]. The endopeptidase inhibitor sacubitril may affect other potential downstream effectors such as substance P, and the actual contribution to the improved outcomes in HFrEF of the observed elevation of natriuretic peptide in response to neprilysin inhibition with subsequent cGMP generation via pGC remains open [16]. However, as discussed elsewhere in more detail, since the cellular signaling and biologic effects of pGC-derived and sGC-derived cGMP differ [12], stimulation of sGC with vericiguat is likely to act complementarily and may add further efficacy to the remaining high residual event rates still seen in HFrEF even with sacubitril/valsartan treatment [14].

3.2.1 Persistent High Medical Need in High-Risk Patients with Chronic HF

A unique population of patients with worsening chronic HF was studied in VICTORIA. All patients were enrolled within a time window of six months after a qualifying previous HF decompensation (defined in the protocol as HF hospitalization or use of IV diuretics for HF [without hospitalization]) [18]. Worsening chronic HF was identified in registries of patients admitted for HF about two decades ago. In order to overcome the misleading distinction between acute and stable chronic HF, the term worsening chronic HF was coined when registry data indicated that the majority of events previously classified as acute decompensated HF events occurred due to exacerbations of chronic HF in more than 70% of patients admitted with acute HF [19, 20]. In these patients, worsening of hemodynamic function and further activation of neurohormones during acute HF syndromes contribute to the progression of HF (Figure 3.3) [21].

To reflect the recurrent nature of worsening HF episodes within the course of the disease, worsening chronic HF denominates the specific population of those patients with HF who had at least one worsening HF event as opposed to those who never required hospitalization. This has important prognostic implications. Even after just one worsening HF event, subsequent clinical event rates of recurrent worsening HF or CV death are substantially higher than in patients without a previous decompensation [22–24]. The independent prognostic value of the previous worsening

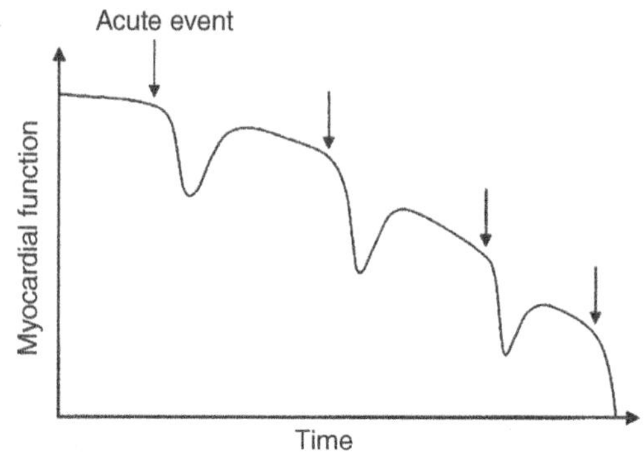

Figure 3.3 Recurrent acute events contribute to heart failure progression. With each admission for acute heart failure syndrome, there is a short-term improvement; however, the patient leaves the hospital with a further decrease in cardiac. *Source:* Reproduced with permission. Copyright© 2005, Elsevier [21].

HF event becomes evident when comparing the clinical characteristics of patients with a previous hospitalization as opposed to those without who were enrolled in the CHARM trial [24]. Subtle differences were seen between these patient groups for several traditionally established risk factors. Patients with a previous HF hospitalization had suffered more frequently with diabetes and atrial fibrillation, were on average in a higher New York Heart Association (NYHA) class, though less often had a previous myocardial infarction. However, the presence or absence of these established factors was less clearly differentiating between populations at different levels of risk than the single attribute of the presence or absence of a previous HF event [24].

As worsening HF events occur despite established treatment for HF, a large residual medical need persists in patients with worsening chronic HF. New therapies are urgently needed to reduce the unacceptably high mortality and morbidity rates in this segment of the overall population of patients with HF at particular risk of future events. The pivotal event-driven phase 3 trial studied the efficacy of vericiguat in the population of patients with worsening chronic heart failure and EF <45% in order to reduce the residual burden of CV death and HF hospitalization in these patients [18].

Upon the successful completion of phase 2 studies with the sGC stimulator riociguat in patients with pulmonary artery hypertension (PAH) and chronic thromboembolic pulmonary hypertension (CTEPH), we hypothesized that direct sGC stimulation may also benefit patients in whom pulmonary hypertension was associated with left ventricular dysfunction (Figure 3.4) [25]. Therefore, the LEft ventricular systolic dysfunction associated with Pulmonary Hypertension riociguat Trial (LEPHT) was conducted in parallel with the phase 3 studies PATENT and CHEST in PAH and in CTEPH, respectively. We tested the hypothesis that direct sGC stimulation with riociguat would reduce elevated mean pulmonary artery pressures (PAPs; inclusion criterion: above 25 mmHg) in patients suffering from heart failure with reduced ejection fraction (HFrEF) who were symptomatic despite optimized medical therapy at a stable dose regimen for >30 days before randomization. Although the primary endpoint reduction in PAPmean upon 16 weeks of treatment with riociguat compared with placebo both added to standard of care was not met, consistent dose-dependent hemodynamic improvements were found in invasive hemodynamic measurements [26]. These included balanced reductions in both systemic and pulmonary resistance, which were associated with significant increases in cardiac index and stroke volume at the highest tested dose regimen, and a parallel decrease in N-terminal pro–B-type natriuretic peptide

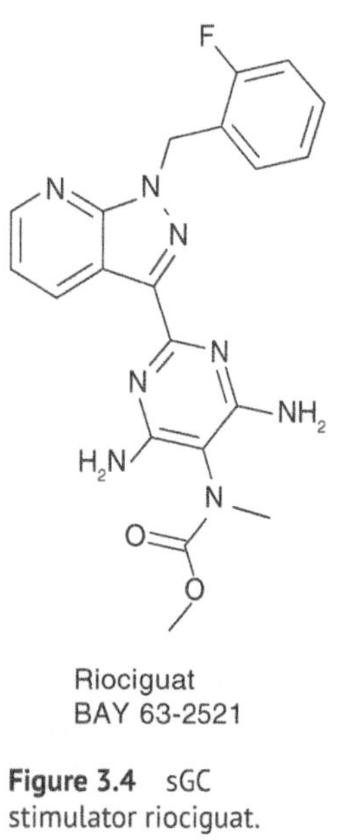

Riociguat
BAY 63-2521

Figure 3.4 sGC stimulator riociguat.

(NT-proBNP) as marker of cardiac wall stress. Riociguat titration up to 2 mg was well tolerated, and the observed dose-dependent hemodynamic improvements did not come at the expense of an increase in heart rate (HR) or decreases in blood pressure (BP) at 16 weeks of treatment.

Mechanistically, we postulated that balanced reductions in pulmonary and systemic resistances may facilitate unloading of the dysfunctional left ventricle, improving CV performance and flow without adverse compromise in BP. However, as patients with HFrEF in the LEPHT study were only eligible with elevated PAPmean including both pre- and post-capillary pulmonary hypertension (PH), we considered whether the observed improvements were dependent on reductions in the precapillary component of PH in patients with high baseline pulmonary vascular resistance (PVR), or more generalizable resulting from decongestion addressing the post-capillary passive PH component that occurs secondary to backward failure of the left ventricle. Subgroup analyses presented at the 18th Annual Scientific Meeting of the HFSA 2014 indicated that riociguat improved pulmonary capillary wedge pressure (PCWP), pulmonary artery compliance (PAC), stroke volume (SV), and quality of life (QoL) to a greater extent in patients with HFrEF, secondary PH, and low baseline PVR than in those with high baseline PVR (Table 3.1 [27]).

Since the LEPHT study was confined to patients with PH associated with left ventricular dysfunction and did not meet its primary endpoint, it was unclear at that time whether a benefit of direct sGC stimulation may be observed in patients with HFrEF and if this would be observed irrespective of the documented presence of secondary PH. Application of riociguat in other CV indications is limited by its short half-life, necessitating a three times daily dosing regimen. In our efforts to further optimize the compound class, we have uncovered interesting structure–activity relationships and were able to decrease oxidative metabolism significantly. These studies resulted in the discovery of once daily sGC stimulator vericiguat (BAY 1021189; MK-1242) that we decided to develop for chronic HF (see Section 3.6).

Table 3.1 Hemodynamic changes and QoL at 16 weeks in patients with high vs low PVR at baseline.

	PVR ≤240 dyn·s/cm^5		PVR >240 dyn·s/cm^5		
Parameter[a]	n^b	**LS mean diff. (95% CI) Riociguat 2 mg vs placebo**	n^b	**LS mean diff. (95% CI) Riociguat 2 mg vs placebo**	**Inter- action *p*-value**
SVR (dyn·s/cm^5)	21/25	−276 (−459 to −93)	32/31	−210 (−386 to −34)	0.6
SBP (mmHg)	21/25	−0.96 (−9.5 to 7.6)	32/31	+1.4 (−5.2 to 7.9)	0.7
Cardiac index (l/min/m^2)	22/25	+0.6 (0.3 to 0.9)	32/31	+0.2 (−0.02 to 0.4)	0.03
Stroke volume (ml)	22/25	+17 (7 to 28)	32/31	+7 (−1 to 14)	0.07
PCWP (mmHg)	22/25	−6.55 (−10.65 to −2.44)	31/31	+1.18 (−2.47 to 4.83)	0.005
mPAP (mmHg)	22/25	−8.8 (−13.3 to −4.3)	32/31	+1.1 (−3.4 to 5.6)	0.002
PAC (ml/mmHg)	22/25	+1.7 (0.9 to 2.6)	32/30	+0.4 (−0.2 to 1.0)	<0.005
MLHF QoL total score	27/30	−12 (−21 to −3)	40/39	−6 (−12 to 0.2)	0.3

[a] Per-protocol population used for hemodynamics; ITT used for QoL. Estimates and CIs are based on an ANCOVA model with baseline value, treatment effect and region as fixed effects. *p*-value is based on an ANCOVA model with baseline value, treatment effect, region and subgroup as fixed effects and interaction of subgroup and treatment (placebo and riociguat 2 mg).
[b] Riociguat 2 mg/placebo [27].

3.3 Medicinal Chemistry Program

Optimization strategy. The goal of this medicinal chemistry program resulting in vericiguat **2** was to identify an orally bioavailable sGC stimulator with a longer duration of action than riociguat **1**. In order to support a profile allowing for once daily oral dosing, the strategy is to reduce oxidative metabolism which would also lower drug–drug interaction potentials. Riociguat is a very potent sGC stimulator *in vitro* and *in vivo*; however, it has a moderate half-life in different animal species [28] and this pharmacokinetic profile translated into a three times daily dosing regimen in patients. *N*-demethylation to compound **10** appears to be the main biotransformation pathway of riociguat as described by Gnoth et al. [29] in 2015. This step is mainly catalyzed by CYP1A1, but also by CYP3A4, CYP3A5, and CYP2J2 [29–35]. Our optimization strategy was aiming for further optimizing the metabolic stability of riociguat and hence reducing blood clearance in order to achieve a longer half-life. We began by varying the substituents on the 5-carbamoyl residue on the pyrimidine ring, aiming to achieve metabolically more stable derivatives while maintaining good potency. In a second optimization step, we later focused on variations of the central pyrazolopyridine scaffold.

SAR and DMPK optimization. First, the SAR of a series of novel N-substituted methyl carbamates **1, 3–9** was explored using a cGMP formation assay with sGC-overexpressing Chinese hamster ovary (CHO) cells [36]. In addition, the metabolic stability of the analogs was assessed *in vitro* by incubation with rat hepatocytes. Initially, different carbamate N-substituents were synthesized with the aim of achieving higher metabolic stability than **1** while maintaining high potency (Table 3.2). Increasing steric bulk (e.g. ethyl **3**) or the introduction of polarity (e.g. a hydroxyethyl functionality, as in **4**) led to a significant loss of potency relative to **1**, along with decreased metabolic stability. Also, the introduction of fluorine atoms at the terminal position of the N-substituent, a common strategy to block metabolism, was not met by success. Interestingly, both the 2,2-difluoroethyl and 2,2,2-trifluoroethyl derivatives, **5** and **6**, exhibited high potency in the cGMP assay but unfortunately had a 2- to 4.5-fold higher clearance in rat hepatocytes. *N*-fluorobenzyl substitution revealed that the *ortho*-fluoro derivative **7** is significantly more potent (MEC = 0.1 μM) than the meta and para isomers, **8** and **9** (MEC = 0.2 and 0.7 μM, respectively). Nevertheless, all the benzylic derivatives were characterized by high clearance in rat hepatocytes. These results and the testing of many further derivatives (data not shown) suggested that improving the metabolic stability by altering the N-substitution might be very difficult to achieve. As a consequence, optimization efforts were focused on the main metabolite of riociguat, the *N*-desmethyl derivative **10**. This compound is less potent than **1** but displayed a somewhat higher metabolic stability in the rat hepatocyte assay, with a clearance of 0.1 l/h/kg. When tested in rats (i.v. dosing), compound **10** had insufficient pharmacokinetic properties (moderate clearance of 1.2 l/h/kg and a short half-life of only 1.2 hours).

Therefore, variations in the carbamate alkoxy group were examined next while leaving the N-position unsubstituted, with the goal of improving DMPK properties relative to compound **10** (Table 3.3).

The introduction of steric bulk in the form of an isopropyl residue (compound **11**) led to a strong decrease in potency as well as decreased metabolic stability. Interestingly, potency was slightly increased in cyclobutyl derivative **12**; however, the analog was also unfavorable with respect to *in vitro* clearance. In some cases, the oxetanyl group has improved metabolic stability [38, 39] compared with related alkyl derivatives. Remarkably, the exchange of cyclobutyl for oxetanyl (compound **13**) led to a 14-fold higher stability in rat hepatocytes, but sGC stimulation was weaker than with derivative **10** or **12**. Since the permeability across Caco-2 cell monolayers was also very low (P_{app} A–B = 2 nm/s), combined with a high efflux ratio of 74, compound **13** was not pursued further. Additional efforts to improve the overall profile, by the introduction of fluorine or steric bulk

Table 3.2 Properties of the N-substituted methyl carbamates **1, 3–9**.

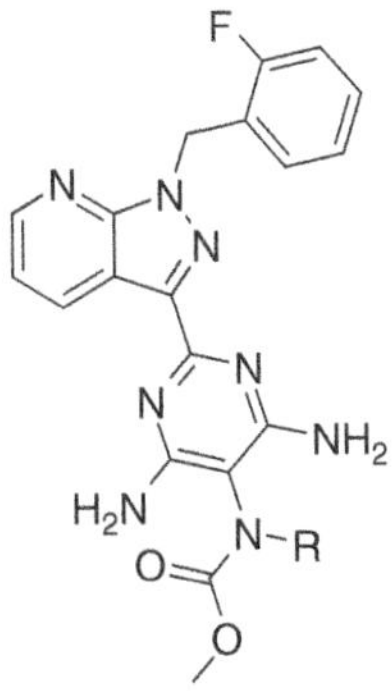

Compound	R	cGMP formation MEC[a] (μM)	ClogD [37] pH 7.5	*In vitro* clearance (rat hepatocytes) CL_b (l/h/kg)
1		0.03	1.99	0.2
3		0.2	2.29	0.7
4	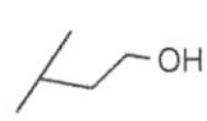	0.3	1.52	3.7
5	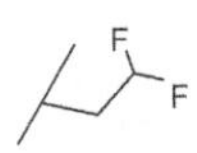	0.1	2.21	0.4
6	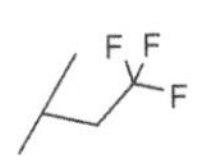	0.1	2.48	0.9
7	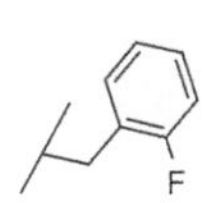	0.1	3.15	3.2
8	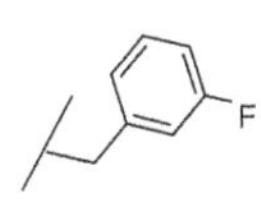	0.2	3.20	3.2
9	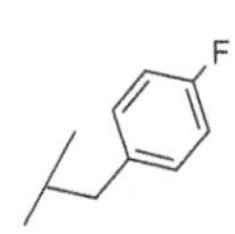	0.7	3.29	2.4

[a] MEC, minimal effective concentration to achieve stimulation of cGMP formation (≥3-fold increase in basal luminescence) in a recombinant sGC-overexpressing cell line [36].

(compounds **14** and **15**), led to a slight increase in potency (MEC = 0.2 μM vs 0.3 μM for **10**); however, metabolic stability was in all cases dramatically reduced.

Summing up, the carbamate portion of the analogs tolerates various substituents with respect to sGC stimulation but seems to be a crucial region for influencing metabolic stability. All our different strategies, introduction of steric bulk, polarity and substitution by fluorine seemed to offer no promising path forward for optimizing *in vitro* clearance. Still, the N-unsubstituted carbamate **10**

Table 3.3 Properties of the N-H alkyl carbamates **10–15**.

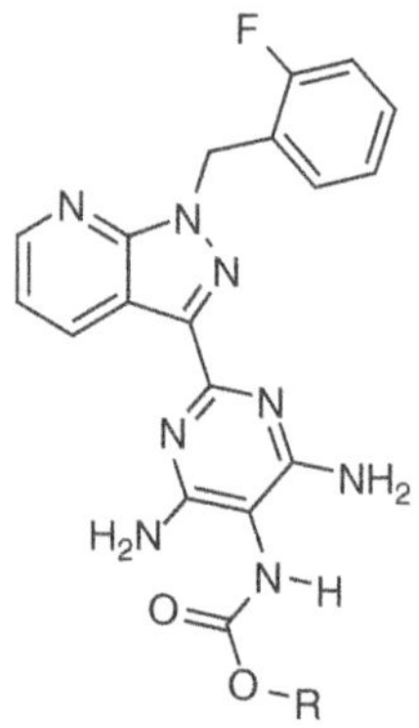

Compound	R	cGMP formation MEC[a] (μM)	ClogD [37] pH 7.5	*In vitro* clearance (rat hepatocytes) CL_b (l/h/kg)	Caco-2 P_{app} A–B (nm/s) (Efflux ratio)
10		0.3	1.49	0.1	79 (5)
11		0.8	2.10	0.9	n.d.[b]
12	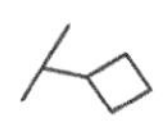	0.2	2.05	1.4	23 (25)
13	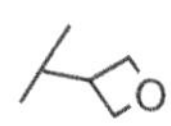	2.3	1.21	0.1	2 (74)
14	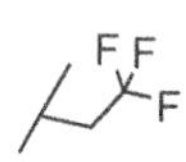	0.2	2.05	2.1	n.d.[b]
15		0.2	2.09	1.4	n.d.[b]

[a] MEC, minimal effective concentration to achieve stimulation of cGMP formation (≥3-fold increase in basal luminescence) in a recombinant sGC-overexpressing cell line [36].
[b] n.d., not determined.

appeared to be optimally constructed in this respect. Thus, the focus of the optimization strategy was shifted away from the carbamate moiety and our next efforts were directed to the central scaffold and the identification of alternative cores that could lead to a potentially superior overall pharmacokinetic profile (modifications summarized in Table 3.4).

At first, changes were made to the pyrazolo portion of the molecule, leading to the 1*H*-pyrazolo[4,3-*b*]pyridine derivative **16**. In addition to the benefit of a shorter synthetic route than that for compound **10**, **16** proved to be a moderately potent sGC stimulator with an MEC of 1.2 μM and had high metabolic stability when tested in rat hepatocytes. The introduction of substituents at the 6-position of the 1*H*-pyrazolo[4,3-*b*]pyridine core was, in general, not well tolerated leading to a dramatic loss of potency (not shown), with the exception of the 6-fluoro derivative **17** which exhibited an MEC of 0.5 μM and good metabolic stability. To our surprise, compounds **16** and **17** had very different *in vivo* clearances when compared in a rat pharmacokinetic experiment (i.v.

Table 3.4 Properties of the core variation compounds **16–20, 2**.

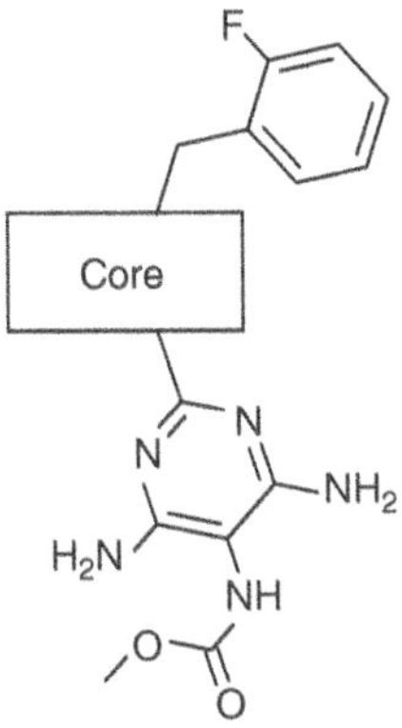

Compound	Core	cGMP formation MEC[a] (μM)	*In vitro* clearance (rat hepatocytes) CL_b (l/h/kg)	*In vivo* clearance (rat) CL_b (l/h/kg)
16	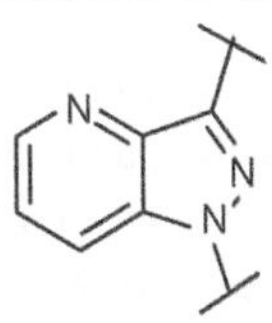	1.2	<0.1	1.0
17	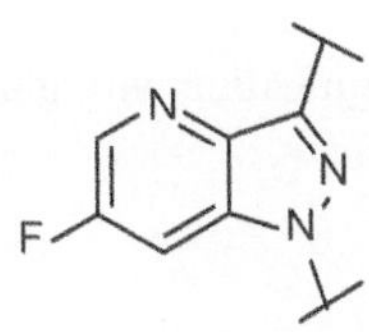	0.5	0.1	0.3
18	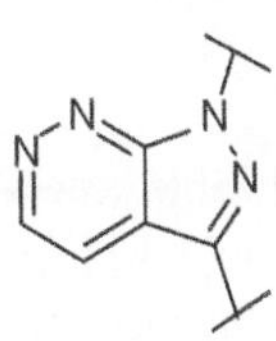	2.0	<0.1	3.8
19	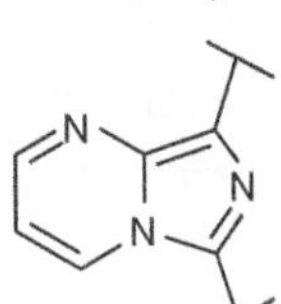	1.7	n.d.[b]	1.8
20	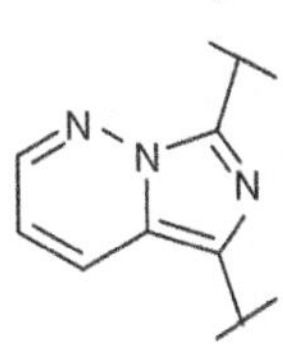	0.7	<0.1	0.9
2	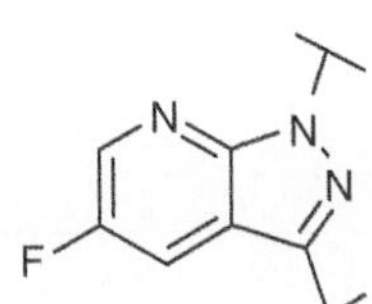	0.3	0.1	0.3

[a] MEC, minimal effective concentration to achieve stimulation of cGMP formation (≥3-fold increase in basal luminescence) in a recombinant sGC-overexpressing cell line [36].
[b] n.d., not determined.

dosing): fluoro derivative **17** had a low clearance of 0.3 l/h/kg versus 1.0 l/h/kg for compound **16**. As metabolite identification did not point to metabolism occurring at the pyridine core, the rationale for this threefold reduction in blood clearance remains unclear.

Novel 1*H*-pyrazolo[3,4-*c*]pyridazine **18** proved to be a reasonably potent sGC stimulator and seemed to be stable in rat hepatocytes (although solubility was limited and that might have impacted the assay) but exhibited an unacceptable high clearance of 3.8 l/h/kg when tested *in vivo* in rats and thus was not further profiled, along with imidazo[1,5-*a*]pyrimidine **19** for similar reasons. In contrast, imidazo[1,5-*b*]pyridazine **20** exhibited potent sGC stimulator properties (MEC = 0.7 μM), good metabolic stability in rat hepatocytes and a low to moderate clearance of 0.9 l/h/kg after i.v. dosing to rats. Finally, revisiting 1*H*-pyrazolo[3,4-*b*]pyridine **10** with an additional fluorine at the 5-position resulted in the potent sGC stimulator **2** (vericiguat) (MEC = 0.3 μM) with good metabolic stability in rat hepatocytes and a surprisingly low clearance of 0.3 l/h/kg after i.v. dosing to rats (Table 3.4). Thus, the blood clearance of derivative **10** in rats (1.2 l/h/kg) was reduced fourfold by fluorination at position 5.

Finally, compounds **17**, **20** and **2** were selected for further pharmacokinetic profiling across different species (Table 3.5). Fluoropyrazolo[3,4-*b*]pyridine derivative **2** exhibited the best overall pharmacokinetic profile by far [40], with a low clearance and long half-life in rats and dogs after i.v. dosing, as well as high oral bioavailability. In addition, **2** had no inhibitory effects on major CYP isoforms (1A2, 2A6, 2B6, 2C8, 2C9, 2C19, 2D6, and 3A4), as indicated by IC_{50} values of >50 μM. Metabolite identification in human hepatocytes was characterized by a low turnover, with the glucuronide (major) and the debenzylated compound (minor) being the only metabolites found. Thus, the main biotransformation pathway shifted from predominantly phase 1 oxidative CYP-mediated metabolism (**1**) to primarily phase 2 UGT-mediated conjugation with glucuronic acid.

3.4 Synthesis Routes toward Vericiguat

3.4.1 Medicinal Chemistry Route to Vericiguat

The first vericiguat synthesis stemed from a time when rapid and flexible access to many derivatives was key in the project. For this purpose we employed the pyrazolo[3,4-*b*]pyridine-3-iodide **21** and coupled with 2-chloro-5-nitropyrimidine-4,6-diamine, **34**, by means of a modified-Stille

Table 3.5 *In vivo* pharmacokinetic properties of **17**, **20**, vericiguat **2** in comparison with riociguat **1**.

Compound	Species	V_{ss} (l/kg)	CL_b (l/h/kg)	$t_{1/2}$ (h)	Bioavailability (%)
17	Rat	0.5	0.3	1.5	26
	Dog	1.0	0.2	4.1	56
20	Rat	0.3	0.9	0.5	37
	Dog	2.0	0.9 (Cl_p)[a]	1.8	n.d.[b]
2	Rat	1.0	0.3	3.4	65
	Dog	1.4	0.2	6.2	75
1	Rat	1.2	1.3	1.4	46
	Dog	0.7	0.3 (Cl_p)[a]	2.4	79

[a] Cl_p, plasma clearance.
[b] n.d., not determined.

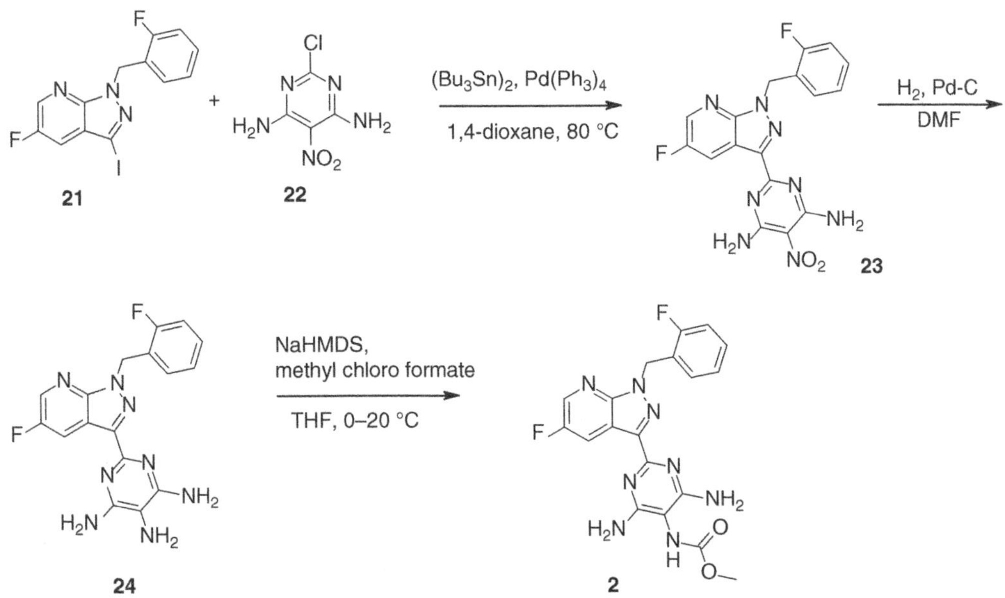

Scheme 3.1 Early days synthesis of vericiguat **2**.

coupling using excess amounts of hexabutylditin and 32 mol% of palladium catalyst. Although yielding the desired product **23** in low yields and moderate purity, it was very clear already at that time that this route will not be suitable for scale-up campaigns for obvious reasons. Nevertheless, after reduction of the respective nitro group to the amine **24**, a chemoselective acylation could be performed leading to **2** in overall good yield for the last two steps (Scheme 3.1).

In the further course of the project, we decided to abandon the Stille coupling, which appeared rather unpredictable in terms of yields depending on the scale and thought of a more sustainable and scalable synthetic route with more reliable yields. Thus, the iodide **21** was treated with copper(I) cyanide at 150 °C in DMF to yield the cyano derivative **25**. In order to facilitate the formation of the pyrimidine, the cyano functional group was converted into an amidine using standard methodology. Based on experience from the former riociguat project it was planned to transform the amidine **26** to the trisamino pyrimidine derivative **24** using a reaction sequence highlighting a rarely used malonodinitrile derivative **27** for condensation with the amidine **26** and subsequent reduction to the desired trisamino derivative **24**. We were pleased to find that this sequence worked nicely also for the novel fluoro core and it proved to be a more efficient route from iodide **21** in our hands (Scheme 3.2).

The synthesis leading to iodide **21** started with a chemoselective dehalogenation of 2,6-dichloro -5-fluoro-3-cyanopyridine **29** which can be achieved by the transformations shown in Scheme 3.3. The 5-fluoro-1*H*-pyrazolo[3,4-*b*]pyridine **33** was constructed by reaction of **32** with hydrazine. The amino group was then modified to the corresponding iodide **34** by standard diazotization and subsequent treatment with sodium iodide. Alkylation with 2-fluorobenzylbromide mediated by cesium carbonate in DMF finally yielded iodide intermediate **21**.

3.4.2 Development Chemistry Route to Vericiguat

While the synthesis route to vericiguat described in Section 3.4.1 was straightforward and high yielding, it nevertheless did not fulfill all criteria for an efficient process development synthesis. After significant experimentation and route scouting, our chemical development team finally

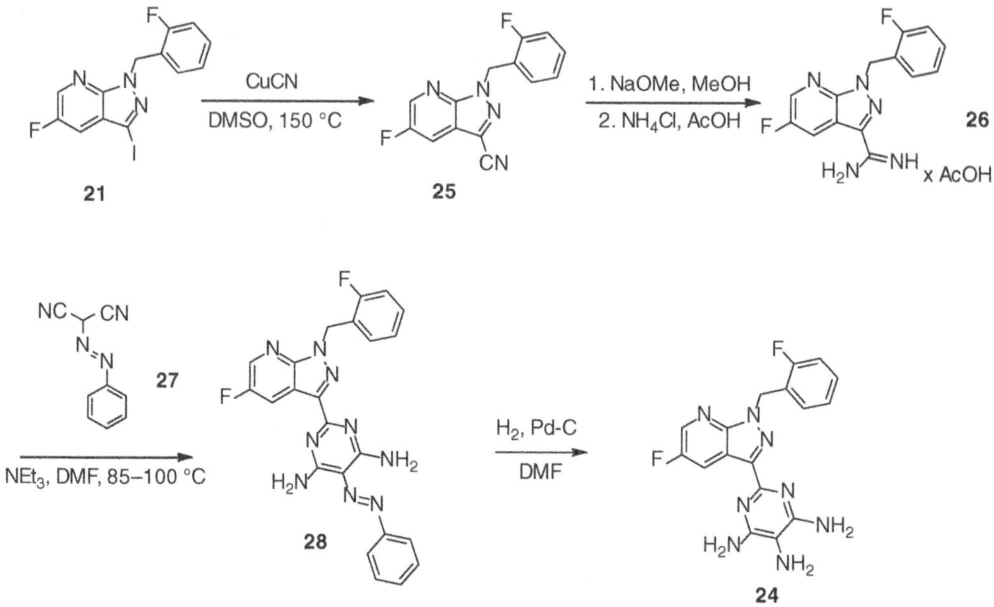

Scheme 3.2 High yielding route to trisamine **24**.

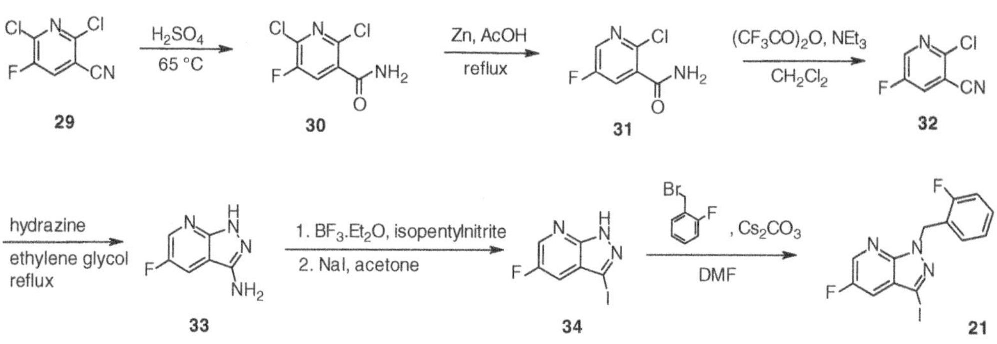

Scheme 3.3 Scale-up synthesis of iodide **21**.

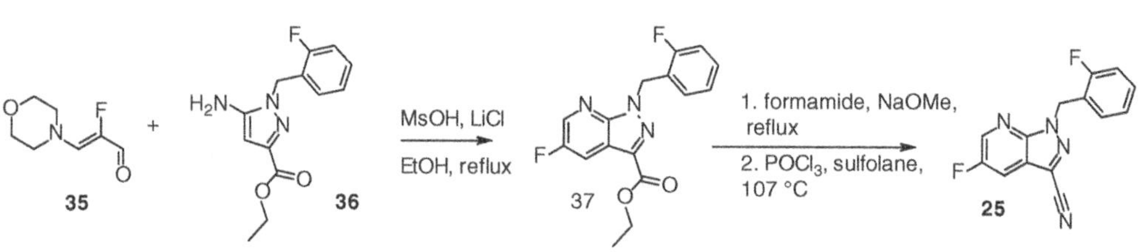

Scheme 3.4 Development route to key intermediate **25**.

established a very short, high yielding and more economic synthesis toward the cyano derivative **25** (Scheme 3.4). The key step in this route is the condensation reaction of acrylaldehyde derivative **35** with ethyl 5-amino-1-(2-fluorobenzyl)-1*H*-pyrazole-3-carboxylate [28] **36** in ethanol yielding ethyl ester intermediate **37**. This ester was then converted in two steps to the nitrile **25**.

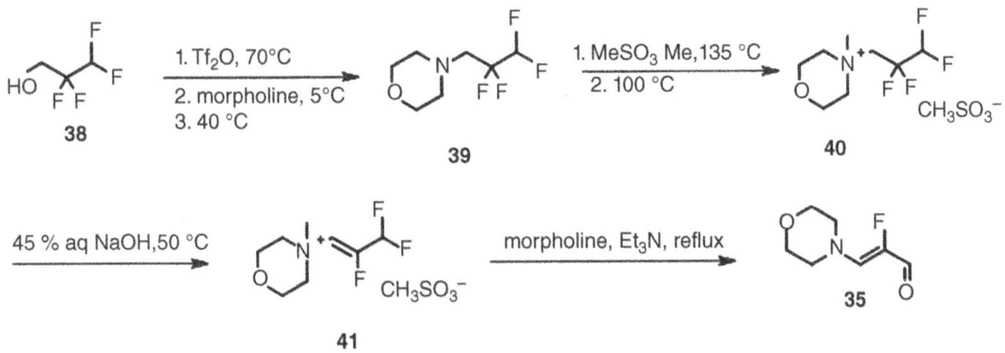

Scheme 3.5 Optimized scheme for the synthesis of acrylaldehyde derivative **35**.

The synthesis of the acrylaldehyde derivative **35** started with activation of 2,2,3,3-tetrafluoro-1-propanol **38** with trifluoromethanesulfonic anhydride and subsequent treatment with morpholine to yield **39**. At 135 °C, **39** was alkylated with methyl methanesulfonate leading to quaternized derivative **40** which was converted into **41** using sodium hydroxide. Further reaction with morpholine and triethylamine yielded **35** after crystallization from toluene (Scheme 3.5).

3.5 Preclinical Studies

The pathophysiology of CV diseases including HF includes endothelial cell dysfunction which impairs NO production leading to decreased NO availability and reduced cGMP tissue levels. This reduced NO availability and insufficient stimulation of sGC results in systemic, vascular dysregulation, affecting the coronary, pulmonary, and renal circulation. Moreover, impaired sGC signaling leads to organ damage and dysfunction of the heart. Therefore, sGC stimulation by vericiguat could provide a novel approach for the treatment of CV diseases and different forms of HF. Vericiguat has a unique mode of action and was extensively profiled preclinically *in vitro*, on the isolated sGC enzyme and on isolated vessels, *ex vivo* in isolated hearts and *in vivo* in a rat model of CV disease associated with cardiorenal syndrome.

3.5.1 *In vitro* Effects on Recombinant sGC and sGC Overexpressing Cells

In vitro, vericiguat led to a concentration-dependent stimulation of the sGC enzyme. Even in the absence of NO, vericiguat stimulated the sGC by 1.7-fold to 57.6-fold (Figure 3.5). When vericiguat was combined with the NO donor diethylamine/nitric oxide complex (DEA/NO), a synergistic effect on sGC activity and cGMP stimulation was observed over a wide range of concentrations (Figure 3.5). In addition, the stimulation of sGC by vericiguat was examined *in vitro* in recombinant CHO cells overexpressing sGC. Vericiguat stimulated these sGC reporter cell lines concentration dependently, with an EC_{50} of 1005 ± 145 nM. In the presence of the NO donor *S*-nitroso-*N*-acetyl-D,L-penicillamine (SNAP) (30 and 100 nM), the EC_{50} value shifted to 39.0 ± 5.1 nM and 10.6 ± 1.7 nM, respectively (Table 3.6). Thus, vericiguat exhibits the characteristics of a potent and selective sGC stimulator, stimulating sGC NO independently and in synergy with NO.

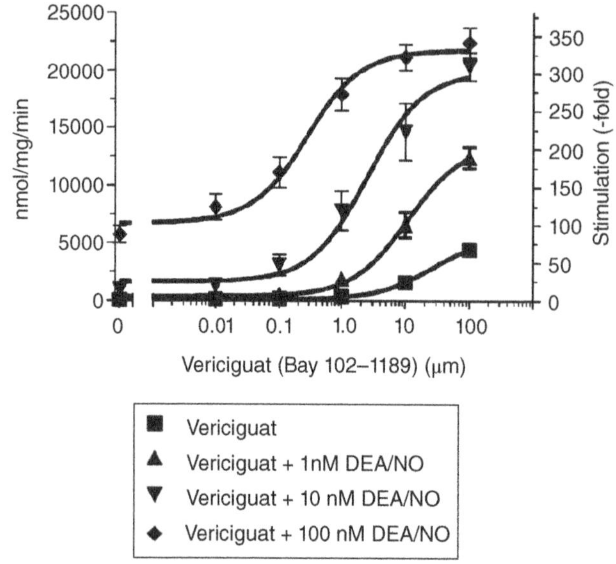

Figure 3.5 Concentration-dependent effects of vericiguat and NO (DEA/NO) on the stimulation of highly purified sGC.

3.5.2 *Ex vivo* Effects on Isolated Blood Vessels and Hearts

As shown in Table 3.6, vericiguat significantly inhibited phenylephrine-induced contractions of rabbit saphenous artery rings, rabbit aortic rings and canine femoral vein rings concentration dependently, with IC_{50} values of 798, 692 and 3072 nM, respectively. Accordingly, vericiguat inhibited the contractions induced by the thromboxane A2 agonist U46619 of porcine coronary artery rings concentration dependently, with an IC_{50} of 956 nM. Since it is known that NO is leading to nitrate tolerance and tachyphylaxia, vericiguat was also investigated on isolated saphenous artery rings taken from nitrate-tolerant rabbits. In contrast to the NO-donor glycerol trinitrate (GTN) vericiguat also relaxed the nitrate-tolerant saphenous artery rings, with IC_{50} values of 5.6 and 5.8 nM, respectively (not shown).

Ex vivo in an isolated rat heart preparation (Langendorff experiment), vericiguat reduced the coronary perfusion pressure in a concentration-dependent manner with a maximal reduction of 30% at 10 μM (Table 3.6). In these rat heart experiments, no effects on HR and contractility (dp/dt, left ventricular diastolic pressure), were observed (Table 3.6).

3.5.3 *In vivo* Effects in a Disease Model with CV Disease and HF and Kidney Failure

Although predictive animal models for HF including HFrEF and HFpEF (heart failure with preserved ejection fraction) are scarce, there are models reflecting presenting with CV morbidity and some features of HF. The renin transgenic (RenTG) rats carrying the additional mouse renin gene (mRenR2)27 are characterized by hypertension. If these RenTG(mRenR2)27 rats were additionally treated with the NO synthase inhibitor N_{ω}-nitro-L-arginine methyl ester (L-NAME) in the drinking water, they suffer endothelial dysfunction, NO decoupling and rapidly progressing hypertension-induced end-organ damage resulting in high morbidity and mortality.

Table 3.6 Nonclinical pharmacodynamics of vericiguat *in vitro*, *ex vivo* and *in vivo*.

Experiments	Assay/Test system	Effects of vericiguat
In vitro effects on sGC activity in transfected cells	Recombinant CHO cell line overexpressing sGC(sGC reporter cell line)	sGC activity: vericiguat alone: $EC_{50} = 1005 \pm 145$ nM vericiguat +30 nM SNAP: $EC_{50} = 39.0 \pm 5.1$ nM vericiguat +100 nM SNAP: $EC_{50} = 10.6 \pm 1.7$ nM
Ex vivo effects on relaxation of precontracted blood vessels of different vascular beds	Rabbit saphenous artery Rabbit aortic artery Porcine coronary artery Canine femoral vein	Inhibition of contraction with vericiguat: rabbit saphenous artery: $IC_{50} = 798$ nM rabbit aorta: $IC_{50} = 692$ nM porcine coronary artery: $IC_{50} = 956$ nM canine femoral vein: $IC_{50} = 3072$ nM
Ex vivo effects in the Langendorff heart preparations	Langendorff-perfused rat heart	Reduction of coronary perfusion pressure by up to 30% (10 μM); no influence on heart rate and contractility
In vivo effects on the cardiovascular system in a disease model	L-NAME treated renin transgenic rats (RenTG)	Significant decrease of mortality rates: Placebo: 75% mortality, 3 mg/kg: 33%, 10 mg/kg: 7% Significant decrease of heart hypertrophy (% from placebo) 3 mg/kg: 8%, 10 mg/kg: 11% Significant decrease of plasma ANP (% from placebo) 3 mg/kg: 45%, 10 mg/kg: 64% Significant decrease of proteinuria (% from placebo) 3 mg/kg: 65%, 10 mg/kg: 88%

Chronic once daily oral treatment with 3 or 10 mg/kg vericiguat resulted in a significant attenuation of mortality. The mortality rate in the placebo arm was 75%, whereas mortality in the vericiguat treatment arms were 33 and 7% for the 3 and 10 mg/kg vericiguat dose arm, respectively (Table 3.6). In addition, vericiguat could significantly prevent the development of heart hypertrophy. In comparison with the placebo group, the heart/body weight ratio decreased by 8 and 11% under vericiguat treatment in the 3 and 10 mg/kg treatment arm, respectively (Table 3.6). In parallel, the plasma atrial natriuretic peptide (ANP) levels decreased significantly in both vericiguat treatment arms by 45 and 65%, respectively when compared with placebo treatment. (Table 3.6). The L-NAME-treated RenTG rats also develop severe kidney damage characterized by a substantial proteinuria. The proteinuria was also significantly and dose-dependently decreased by 65 and 88% in the 3 and 10 mg/kg vericiguat treatment group, respectively (Table 3.6).

In summary, these results suggested that vericiguat is a stimulator of sGC and induces a cGMP increase by stimulating sGC NO independently but also in synergy with endogenous NO. Due to this mode of action and pharmacological properties, vericiguat showed pronounced effects on the CV system *ex vivo* on blood vessels and heart. In addition, vericiguat was effective *in vivo* in CV disease model. Thus, vericiguat has potential for the treatment of CV diseases and HF [41].

3.6 Clinical Studies

The clinical development of vericiguat started on the 29th of July 2011 with the first-in-human study. Overall, 28 clinical phase 1 trials to evaluate vericiguat, which enrolled more than 650 participants, have been designed, conducted and completed to characterize the safety and pharmacodynamics (PD) in healthy volunteers and HFrEF patients, pharmacokinetics (PK), biopharmaceutical properties, metabolism, and PK and PD drug–drug interaction (DDI) potential, influence of intrinsic factors such as renal and hepatic impairment on vericiguat pharmacokinetics and dynamics in support of the development of vericiguat in HFrEF in combination with other HF therapies. In addition, population-pharmacokinetic (POP-PK)/PD analyses and physiologically based pharmacokinetic (PBPK) evaluations have been carried out. All studies were performed according to GCP (Good Clinical Practice) guidelines.

3.6.1 Safety, PD, PK and PK/PD in Healthy Volunteers

The safety, tolerability, PD, and PK of vericiguat were investigated in six phase 1, randomized, single/double-blind studies conducted between July 2011 and May 2017 [42]. The studies included: two single dose escalation studies in Germany and Singapore; three multiple dose escalation studies conducted in Germany, Japan and China and a relative bioavailability and food effect study to characterize the vericiguat immediate release (IR) tablet. All studies comprised a screening, treatment, and follow-up period.

Across the vericiguat clinical pharmacology program, vericiguat (IR) tablets administered once daily at doses of 1.25, 5, and 10 mg or 5 mg twice daily up to a treatment period of seven days were well tolerated in healthy volunteers. A single dose of 15 mg administered as an oral solution (fasted) was not well tolerated due to the occurrence of hemodynamic effects (sinus bradycardia, orthostatic hypotension, orthostatic dizziness, and syncope during the standing BP procedure) reported by 3 of the 4 subjects tested at this dose range and prevented further dose escalation in the single rising dose study. However, administration of a single dose of 15 mg vericiguat as an IR tablet (fasted) was safe and well tolerated most likely because the slope of the plasma concentration time curves was less steep as compared with the oral solution. All treatment emergent adverse events (AEs) were of mild or moderate intensity. Within the phase 1 program, no severe or serious AEs occurred. The most common AEs after vericiguat treatment were headache with an incidence of 15.0% (placebo: 5.1%), followed by dyspepsia with 3.8% (placebo: 0%), nasopharyngitis with 3.5% (placebo: 1.7%) and ocular hyperemia with 2.8% (placebo: 1.7%). Overall, the AE profile of vericiguat was predominantly associated with its pharmacodynamic mode of action, (i.e. relaxation of smooth muscle leading to hemodynamic changes and gastrointestinal side effects). Neither AE analysis, nor the laboratory results, vital signs or ECG evaluations identified a safety signal.

The PD effects of vericiguat were evaluated after single and multiple dose administrations in healthy subjects and are consistent with the mode of action of a sGC stimulator, i.e. relaxation of the smooth muscles in the vasculature leading to changes in hemodynamics. In accordance with the established pharmacological profile of vericiguat as a direct sGC stimulator in preclinical experiments, the expected hemodynamic effects were observed in healthy subjects. Specifically, in healthy subjects, an increase in HR was observed as a compensatory reaction to the BP lowering activity of vericiguat through the baroreflex.

Generally, immediate and direct correlations between vericiguat plasma concentration and PD measures (hemodynamics and vasoactive hormones) were observed (Figure 3.6). Consistent changes in pharmacodynamic parameters were observed at doses of 5 mg vericiguat and higher

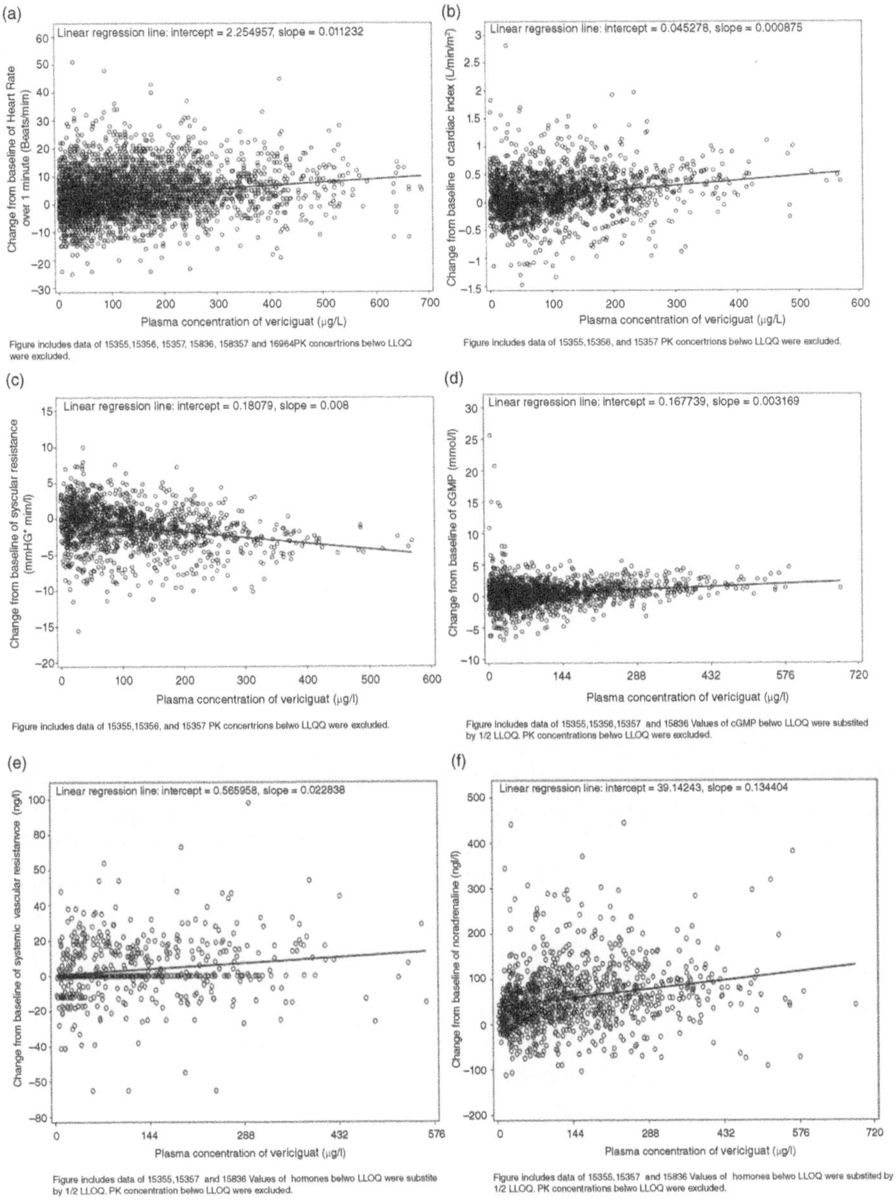

Figure 3.6 Relationship between vericiguat PK and (a) HR over one minute, (b) cardiac index, (c) SVR, (d) cGMP, (e) adrenaline and (f) noradrenaline.

with an increase in HR, in vasoactive hormones, plasma cGMP and changes in impedance cardiographic parameters (increased stroke volume, cardiac output and cardiac index, decreased systemic vascular resistance). Effects on BP were less consistent and not seen in a dose-dependent manner. Mostly, slight decreases were observed in systolic BP and in diastolic BP (decreases in the range of 2–3 mmHg), associated with the increase in HR.

In a dedicated QT study in patients with stable coronary artery disease, administration of vericiguat 10 mg at steady-state did not prolong the QT interval to a clinically relevant extent, i.e.

the maximum mean prolongation of the QTcF interval did not exceed 6 ms (upper bound of the 90%CI <10 ms).

3.6.2 Clinical Pharmacokinetics

Vericiguat shows slightly less than dose proportional, time-independent PK, with low to moderate variability when administered with food. Vericiguat accumulates in plasma up to 155–171% and reaches pharmacokinetic steady-state after approximately six days. The mean steady-state population PK parameters of vericiguat in HF patients are summarized (Table 3.7).

3.6.2.1 Absorption

The absolute bioavailability of vericiguat is high (93%) when taken with food. Bioavailability (*F*) (area under the plasma concentration–time curve, AUC) and peak plasma levels (C_{max}) of vericiguat administered orally as a crushed tablet in water is comparable with that of a whole tablet.

3.6.2.2 Effect of Food

Administration of vericiguat with a high-fat, high-calorie meal increases T_{max} from about one hour (fasted) to about four hours (fed), reduces PK variability, and increases vericiguat exposure by 19% (AUC) and 9% (C_{max}) for the 5 mg tablet and by 44% (AUC) and 41% (C_{max}) for the 10 mg tablet as compared with the fasted state. Similar results were obtained when vericiguat was administered with a low-fat, high-carbohydrate meal. Based on the results of these studies, vericiguat was recommended to be taken with food in the phase 2 and 3 studies.

3.6.2.3 Distribution

The mean steady-state volume of distribution of vericiguat in healthy subjects is approximately 44 l. Plasma protein binding of vericiguat is about 98%, with serum albumin being the main binding component. Plasma protein binding of vericiguat is not altered by renal or hepatic impairment [43].

3.6.2.4 Metabolism

Glucuronidation is the major biotransformation pathway of vericiguat to form an N-glucuronide, which is pharmacologically inactive and the major drug related component in plasma. N-glucuronidation is catalyzed predominantly by UGT1A9, as well as UGT1A1. CYP-mediated metabolism is a minor clearance pathway (<5%).

3.6.2.5 Elimination

Vericiguat is a low-clearance drug (1.6 l/h in healthy subjects). The half-life is about 20 hours in healthy subjects and 30 hours in HF patients. Following oral administration of [^{14}C]vericiguat to

Table 3.7 Population pharmacokinetic model based steady-state geometric mean (CV%) plasma pharmacokinetic parameters of vericiguat 2.5, 5, or 10 mg in heart failure patients (*N* = 2321).

PK parameters	2.5 mg	5 mg	10 mg
C_{max} (μg/l)	120 (29.0)	201 (29.0)	350 (29.0)
AUC (μg•h/l)	2300 (33.9)	3850 (33.9)	6680 (33.9)

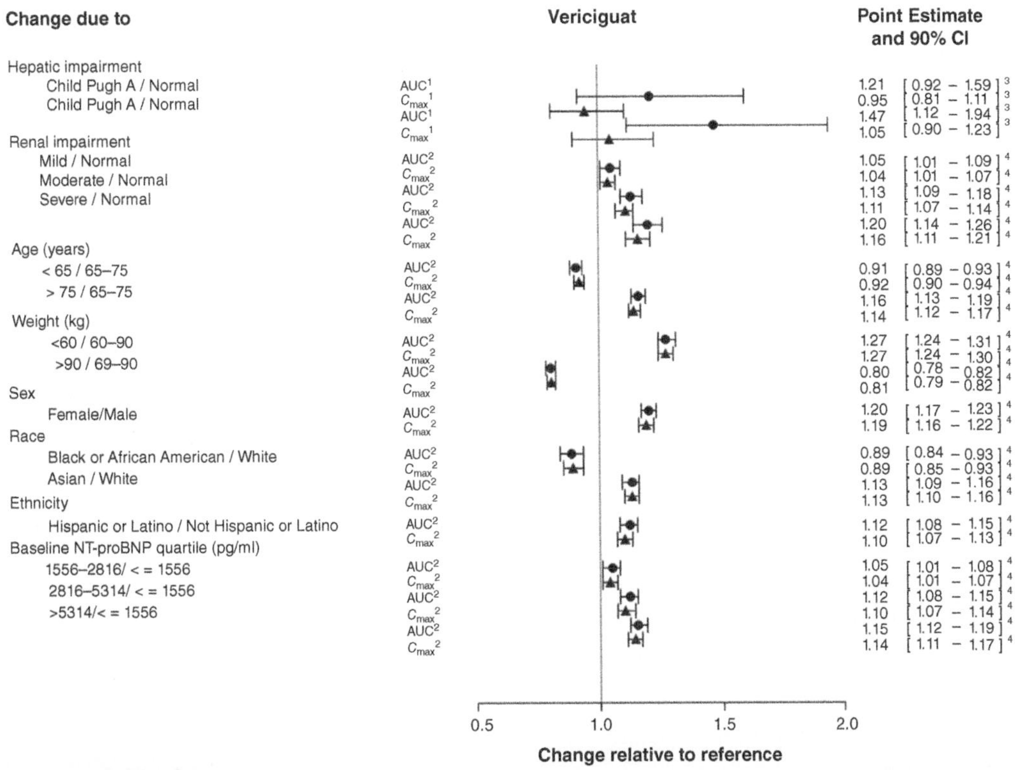

CI: Confidence Interval.

1. Dose and body weight normalized AUC and dose and body weight normalized C_{max} of unbound concentrations after single dose administration.
2. AUC over the dosing interval after multiple dose administration. C_{max} after multiple dose administration.
3. Based on data from healthy subjects (phase I trial).
4. Based on population-pharmacokinetic (POP-PK) modeling of VICTORIA and SOCRATES-REDUCED.

Figure 3.7 Pharmacokinetics of vericiguat in special populations.

healthy subjects, approximately 53% of the dose was excreted in urine (primarily as the N-glucuronide) and 45% of the dose was excreted in feces (primarily as vericiguat).

3.6.2.6 Special Populations

Based on a population pharmacokinetic analysis, age, gender, ethnicity, race, and baseline NT-proBNP have not been identified as significant co-variates and thus do not have a clinically meaningful effect on the PK of vericiguat in HFrEF patients. The effect of body weight on vericiguat exposure was not clinically meaningful. No clinically relevant increase in exposure was observed for subjects with mild hepatic impairment (Child-Pugh A and B). The PK of vericiguat have not been studied in patients with severe hepatic impairment (Child-Pugh C) (Figure 3.7).

3.6.2.7 Drug Interactions

In vitro studies indicate that vericiguat and its N-glucuronide are neither inhibitors of major CYP isoforms (CYP1A2, 2B6, 2C8, 2C9, 2C19, 2D6, and 3A4) or UGT isoforms (UGT1A1, 1A4, 1A6, 1A9, 2B4, and 2B7), nor inducers of CYP1A2, 2B6, and 3A4, at clinically relevant concentrations.

Vericiguat is a substrate of P-glycoprotein (P-gp) and breast cancer resistance protein (BCRP) transporters and is not a substrate of organic cation transporter (OCT1), or organic anion transporting polypeptides (OATP1B1 and OATP1B3). Vericiguat and its N-glucuronide are not inhibitors of drug transporters, including P-gp, BCRP, BSEP, OATP1B1/1B3, OAT1, OAT3, OCT1, OCT2, MATE1, and MATE2K, at clinically relevant concentrations. Overall, these data indicate that the administration of vericiguat is unlikely to affect the PK of concurrently administered medications that are substrates of these enzymes or transporters.

3.6.2.8 *In vivo* Assessment of Drug Interactions

Effects of Other Drugs on the Pharmacokinetics of Vericiguat

The effects of coadministered drugs on the PK of vericiguat have been assessed in 10 clinical drug–drug interaction studies. Overall, there was no clinically relevant effect on vericiguat PK with coadministration of drugs increasing gastric pH (e.g. proton pump inhibitors, H2-receptor antagonists, antacids) in HF patients; or with coadministration of mefenamic acid, ketoconazole, rifampicin, digoxin, warfarin, aspirin, sildenafil, or the combination of sacubitril/valsartan in healthy subjects [43]. There was no clinically relevant effect on vericiguat PK with coadministration of atazanavir based on physiologically based pharmacokinetic (PBPK) modeling. The results are summarized in Figure 3.8.

These data indicate that no dose adjustment of vericiguat is recommended when coadministered with commonly prescribed medicinal products [44].

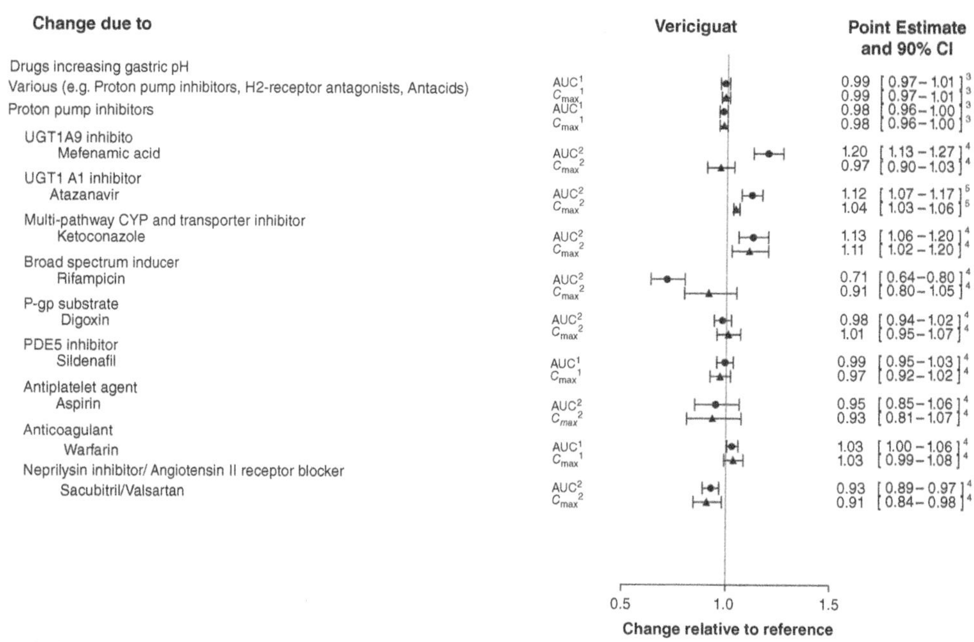

CI: Confidence interval.
1. AUC over the dosing interval after multiple dose administration. C_{max} after multiple dose administration.
2. AUC and C_{max} after single dose administration.
3. Based on populationpharmacokinetic (POP-pharmacokinetic) modeling of VICTORIA and SOCRATES-REDUCED.
4. Based on data from healthy subjects (phase I trial).
5. Based on physiologically-based PK (PBPK) modeling, interval represents 90% population interval.

Figure 3.8 Effects of other drugs on the pharmacokinetics of vericiguat.

Effects of Vericiguat on the Pharmacokinetics of Other Drugs

The effects of vericiguat on the PK of coadministered drugs have also been assessed in clinical drug–drug interaction studies [43, 45]: vericiguat had no clinically relevant effect on the PK of midazolam, digoxin, warfarin, sildenafil, and the combination of sacubitril/valsartan when coadministered in healthy subjects (Figure 3.9).

3.6.3 Pharmacodynamic Interactions

The effects of vericiguat on the PD of coadministered drugs have also been assessed in 6 clinical drug–drug interaction studies: vericiguat had no clinically relevant effect on the PD of aspirin, warfarin, sildenafil, and the combination of sacubitril/valsartan when coadministered in healthy subjects; or on the pharmacodynamics of organic nitrates when coadministered in patients with coronary artery disease [43, 45].

3.6.4 Vericiguat Phase 2 and Phase 3 studies in HFrEF patients

The transition from healthy volunteers to HFrEF patients investigated in SOCRATES-REDUCED was supported by a modeling-based bridging approach using clinical data from the sGC stimulators riociguat and nelociguat (BAY 60-4552) replacing a standard proof-of-concept study and enabling a direct transition into Phase 2b [46].

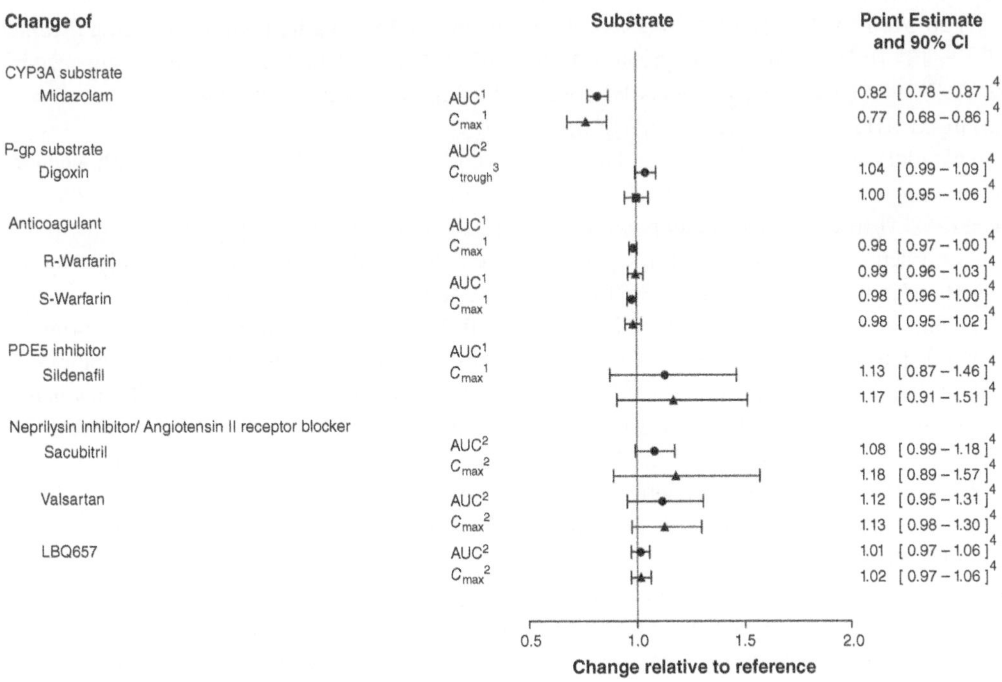

CI: Confidence Interval.
1. AUC and C_{max} after single dose administration.
2. AUC over the dosing interval after multiple dose administration. C_{max} after multiple dose administration.
3. C_{trough} of digoxin on day 10.
4. Based on data from healthy subjects (phase I trial).

Figure 3.9 Effects of vericiguat on the pharmacokinetics of other drugs [44].

The SOCRATES program was designed to study in two parallel dose-ranging trials vericiguat at titrated dosing schemes up to 10 mg once per day in the highest target dose arm over 12 weeks in patients with stabilized worsening chronic HFrEF (SOCRATES-REDUCED) and in patients with stabilized worsening chronic HFpEF (SOCRATES-PRESERVED). Based on insights from the riociguat LEPHT study [25, 26, 46], patients with symptomatic worsening chronic HF were enrolled irrespective of documented elevations in pulmonary artery pressure. Since very high mortality and morbidity rates persist in patients who experience HF events despite receiving optimal medical therapy, these studies specifically enrolled patients within a one month window after a preceding HF hospitalization, or alternatively after a HF event that had to be treated with i.v. diuretics in an outpatient setting [47]. Rather than transient management with additional i.v. therapy, long-term treatment with an oral sGC stimulator was deemed to carry promise [48] to meet this medical need in patients suffering from HF events despite available therapies [49].

Primary endpoint in SOCRATES-REDUCED was reduction in NT-proBNP at 12 weeks, and the study was powered to detect a decrease in the 10 mg arm as previously observed in patients with HFrEF and concomitant PH in the 2 mg riociguat target dose arm of the LEPHT study [47]. Although the primary analysis of the pooled three higher dose arms did not reach statistical significance, the pre-specified secondary analysis of the primary endpoint NT-proBNP comparing the individual dose arms versus placebo showed a statistically and clinically significant reduction in NT-proBNP in the 10 mg target dose arm at 12 weeks [50]. Based upon the established association of short-term NT-proBNP lowering with long-term decreases in clinical events in patients with HFrEF, the Vericiguat Global Study in Subjects with Heart Failure with Reduced Ejection Fraction (VICTORIA) phase 3 event-driven outcome trial was designed to study whether vericiguat at a target dose of 10 mg once daily reduces mortality and morbidity in patients with worsening chronic HFrEF [18]. Different from other previous trials in patients with stable HFrEF, patients enrolled in the VICTORIA trial represent a broadly generalizable high-risk population suffering from worsening chronic HFrEF despite very good HF therapy [51].

In parallel with the development in HFrEF, vericiguat was also studied in patients with worsening chronic HFpEF. Conducted back-to-back with the phase 2 HFrEF study, the SOCRATES-PRESERVED trial enrolled patients within one month after an HF event who had a preserved ejection fraction (LVEF >45%) [47]. Opposed to the clinically meaningful reduction observed in patients with HFrEF in SOCRATES-REDUCED, the same dose and duration of treatment with vericiguat did not reduce NT-proBNP in patients with HFpEF, and changes in the other primary endpoint, left atrial volume, did not differ between treatment arms [52]. However, exploratory analyses showed an improved health status assessed by the Kansas City Cardiomyopathy Questionnaire (KCCQ) and QoL measured with the EQ-5D in patients treated with vericiguat [53]. Therefore, the phase 2b study VITALITY-HFpEF has been designed to determine the efficacy and safety of vericiguat on QoL and exercise tolerance in patients with HFpEF [54], and results of this study showed that vericiguat did not improve the physical limitation score (KCCQ PLS) at 24 weeks [55].

The phase III VICTORIA trial successfully met its primary endpoint. Vericiguat treatment significantly reduced the composite endpoint CV death and HF hospitalization upon 10.8 months median treatment duration in patients with HF and reduced ejection fraction. The placebo event rate in these patients who were enrolled within six months of a previous HF event (that could have been an HF hospitalization or i.v. diuretic treatment for HF) was 37.8 per 100 patient years. In the vericiguat arm, this was significantly reduced to 33.6, reflecting an absolute risk reduction of 4.2% per year for the dual composite primary endpoint of CV death or HF hospitalization [5]. Thus, in this population of patients at high risk of events, veriguat treatment is the first new drug candidate directly targeting the sGC pathway that showed a significant improvement in outcomes when combined with other HF therapies. Importantly, background HF therapy in 15% of studied patients included the combination with sacubitril/valsartan, in whom the efficacy of vericiguat to reduce the risk for the primary composite was no less than in the overall population.

3.7 Summary

Vericiguat belongs to a novel class of drugs termed sGC stimulators. Vericiguat selectively and specifically binds to sGC leading to concentration-dependent production of the second messenger cGMP. Since NO/sGC signaling is impaired in CV diseases and HF, vericiguat can restore this dysfunctional NO/sGC signaling. Preclinical and clinical data strongly suggest that vericiguat as confirmed in the recent VICTORIA phase 3 trial does improve clinical outcomes in patients with HF and maybe other CV diseases.

References

1 Butler, J., Yang, M., Manzi, M.A. et al. (2019). *J. Am. Coll. Cardiol.* 73: 935–944.
2 Vos, T., Abajobir, A.A., Abate, K.H. et al. (2017). *The Lancet* 390: 1211–1259.
3 Roger, V.L. (2013). *Circ. Res.* 113: 646–659.
4 Packer, M., Anker, S.D., Butler, J. et al. (2020). *N. Engl. J. Med.* 383: 1413–1424.
5 Armstrong, P.W., Pieske, B., Anstrom, K.J. et al. (2020). *N. Engl. J. Med.* 382: 1883–1893.
6 Burnett, J.C. Jr. (2020). *N. Engl. J. Med.* 382: 1952–1953.
7 Evgenov, O.V., Pacher, P., Schmidt, P.M. et al. (2006). *Nat. Rev. Drug Discov.* 5: 755–768.
8 Follmann, M., Griebenow, N., Hahn, M.G. et al. (2013). *Angew. Chem. Int. Ed. Engl.* 52: 9442–9462.
9 Murad, F. (2006). *N. Engl. J. Med.* 355: 2003–2011.
10 Greene, S.J., Gheorghiade, M., Borlaug, B.A. et al. (2013). *J. Am. Heart Assoc.* 2: e000536.
11 Gheorghiade, M., Marti, C.N., Sabbah, H.N. et al. (2013). *Heart Fail. Rev.* 18: 123–134.
12 Breitenstein, S., Roessig, L., Sandner, P., and Lewis, K.S. (2017). *Handb. Exp. Pharmacol.* 243: 225–247.
13 Sandner, P., Zimmer, D.P., Milne, G.T. et al. (2018). Soluble guanylate cyclase stimulators and activators. In: *Reactive Oxygen Species*, Handbook of Experimental Pharmacology, vol. 264 (eds. H.H.H.W. Schmidt, P. Ghezzi and A. Cuadrado). Cham: Springer https://doi.org/10.1007/164_2018_197.
14 McMurray, J.J., Packer, M., Desai, A.S. et al. (2014). *N. Engl. J. Med.* 371: 993–1004.
15 Packer, M., McMurray, J.J., Desai, A.S. et al. (2015). *Circulation* 131: 54–61.
16 Nougué, H., Pezel, T., Picard, F. et al. (2019). *Eur. J. Heart Fail.* 21: 598–605.
17 Goetze, J.P., Bruneau, B.G., Ramos, H.R. et al. (2020). *Nat. Rev. Cardiol.* 17 (11): 698–717.
18 Armstrong, P.W., Roessig, L., Patel, M.J. et al. (2018). *JACC Heart Fail.* 6: 96–104.
19 Felker, G.M., Adams, K.F. Jr., Konstam, M.A. et al. (2003). *Am. Heart J.* 145: S18–S25.
20 Gheorghiade, M. and Pang, P.S. (2009). *J. Am. Coll. Cardiol.* 53: 557–573.
21 Gheorghiade, M., De Luca, L., Fonarow, G.C. et al. (2005). *Am. J. Cardiol.* 96: 11g–17g.
22 Solomon, S.D., Dobson, J., Pocock, S. et al. (2007). *Circulation* 116: 1482–1487.
23 Abrahamsson, P., Swedberg, K., Borer, J.S. et al. (2013). *Eur. J. Heart Fail.* 15: 885–891.
24 Bello, N.A., Claggett, B., Desai, A.S. et al. (2014). *Circ. Heart Fail.* 7: 590–595.
25 Ghio, S., Bonderman, D., Felix, S.B. et al. (2012). *Eur. J. Heart Fail.* 14: 946–953.

26 Bonderman, D., Ghio, S., Felix, S.B. et al. (2013). *Circulation* 128: 502–511.

27 Bonderman, D., Ghio, S., Felix, S.B. et al. (2014). *J. Card. Fail.* 20: S9.

28 Mittendorf, J., Weigand, S., Alonso-Alija, C. et al. (2009). *ChemMedChem* 4: 853–865.

29 Gnoth, M.J., Hopfe, P.M., and Czembor, W. (2015). *Bioanalysis* 7: 193–205.

30 Becker, C., Frey, R., Thomas, D. et al. (2016). *Pulm. Circ.* 6: S49–S57.

31 Frey, R., Becker, C., Unger, S. et al. (2016). *Pulm. Circ.* 6: S15–S26.

32 Frey, R., Mueck, W., Unger, S. et al. (2008). *J. Clin. Pharmacol.* 48: 926–934.

33 Saleh, S., Frey, R., Becker, C. et al. (2016). *Pulm. Circ.* 6: S66–S74.

34 Zhao, X., Wang, Z., Wang, Y. et al. (2016). *Clin. Pharmacokinet.* 55: 615–624.

35 FDA (2013). NDA 204819. http://www.accessdata.fda.gov/drugsatfda_docs/label/2014/204819s002lbl.pdf. (accessed 20 September 2021).

36 Wunder, F., Stasch, J.P., Huetter, J. et al. (2005). *Anal. Biochem.* 339: 104–112.

37 Hillisch, A., Heinrich, N., and Wild, H. (2015). *ChemMedChem* 10: 1958–1962.

38 Burkhard, J.A., Wuitschik, G., Rogers-Evans, M. et al. (2010). *Angew. Chem. Int. Ed. Engl.* 49: 9052–9067.

39 Wuitschik, G., Carreira, E.M., Wagner, B. et al. (2010). *J. Med. Chem.* 53: 3227–3246.

40 Follmann, M., Ackerstaff, J., Redlich, G. et al. (2017). *J. Med. Chem.* 60: 5146–5161.

41 Dubin, R.F. and Shah, S.J. (2016). *Curr. Heart Fail. Rep.* 13: 132–139.

42 Boettcher, M., Thomas, D., Mueck, W. et al. (2021). *Eur. J. Clin. Pharmacol.* 77 (4): 527–537. https://doi.org/10.1007/s00228-020-03023-7.

43 Boettcher, M., Loewen, S., Gerrits, M., and Becker, C. (2021). *Clin. Pharmacokinet.* 60 (3): 337–351. https://doi.org/10.1007/s40262-020-00935-6.

44 FDA (2021). VERQUVO. https://www.accessdata.fda.gov/drugsatfda_docs/label/2021/214377s000lbl.pdf (accessed 26 September 2021).

45 Boettcher, M., Gerisch, M., Lobmeyer, M. et al. (2020). *Clin. Pharmacokinet.* 59 (11): 1407–1418. https://doi.org/10.1007/s40262-020-00895-x.

46 Meyer, M., Schneckener, S., Loosen, R. et al. (2021). Leveraging translational approaches for clinical development of vericiguat. Poster presented at the European Society of Cardiology Congress 2021, The Digital Experience (27–30 August 2021).

47 Pieske, B., Butler, J., Filippatos, G. et al. (2014). *Eur. J. Heart Fail.* 16: 1026–1038.

48 Voors, A.A. and van Veldhuisen, D.J. (2012). *Eur. J. Heart Fail.* 14: 955–956.

49 Butler, J., Braunwald, E., and Gheorghiade, M. (2014). *JAMA* 312: 789–790.

50 Gheorghiade, M., Greene, S.J., Butler, J. et al. (2015). *JAMA* 314: 2251–2262.

51 Pieske, B., Patel, M.J., Westerhout, C.M. et al. (2019). *Eur. J. Heart Fail.* 21: 1596–1604.

52 Pieske, B., Maggioni, A.P., Lam, C.S.P. et al. (2017). *Eur. Heart J.* 38: 1119–1127.

53 Filippatos, G., Maggioni, A.P., Lam, C.S.P. et al. (2017). *Eur. J. Heart Fail.* 19: 782–791.

54 Butler, J., Lam, C.S.P., Anstrom, K.J. et al. (2019). *Circ. Heart Fail.*: 12, e005998.

55 Armstrong, P.W., Lam, C.S.P., Anstrom, K.J. et al. (2020). *JAMA* 324: 1512–1521.

4

Finding Cures for Alzheimer's Disease: From γ-Secretase Inhibitors to γ-Secretase Modulators and β-Secretase Inhibitors

Xianhai Huang[1] *and Robert Aslanian*[2]

[1] *Discovery Chemistry, InventisBio Co., Ltd., 25A Hanover Road #340, NJ 07932, Florham Park, NJ, USA*
[2] *Department of Chemistry, New Jersey City University, S550J, 2039 John F. Kennedy Boulevard, NJ 07305, Jersey City, NJ, USA*

4.1 Introduction

4.1.1 Alzheimer's Disease

Dementia is caused by abnormal alterations to the brain with associated damage to brain cells and is a general term for loss of memory, language, problem-solving and other thinking abilities that are severe enough to interfere with daily life. It is not a single disease and includes Alzheimer's disease (AD), Lewy body dementia (LBD), vascular dementia (VD), frontotemporal dementia (FD), mixed dementia, and reversible causes. According to the US Centers for Disease Control and Prevention (CDC), there was an estimated five million adults 65 years old or older with dementia in 2014 in the United States, and it is projected to be nearly 14 million by 2060. In 2015, there were estimated to be 50 million people worldwide to have AD and other dementias with an estimated global cost of dementia care of US$818 billion, equivalent to 1.1% of global gross domestic product (GDP) [1]. The cost of caring for people with dementia could rise to some US$2 trillion by 2030 when an estimated 74.7 million people will have the disease [2, 3]. The estimated number of people with dementia will reach 115–131.5 million by 2050 without effective therapies [4]. Although lifestyle changes such as diet and exercise can have a positive impact on the risk of developing dementia, therapeutic modalities will be necessary to slow the progression or cure the disease in those who are stricken. Dementia represents one of most critical unmet medical needs, and novel therapies and new targets are urgently needed.

AD is a degenerative brain disease characterized by progressive memory and neuronal loss combined with cognitive impairment with symptoms gradually worsening over a number of years, and it is the most common cause of dementia in late life, accounting for 60–80% of dementia cases. Although memory loss is mild in its early stages, individuals lose the ability to carry on a conversation and respond to their environment with late-stage AD. AD is the sixth leading cause of death in the United States and the most common neurodegenerative disease worldwide affecting one-fifth of people over 85 years old. It is a complex and multifactorial disorder where premature neuronal death is induced by pathogenically converged genomic defects, epigenetic aberrations, cerebrovascular damage, and environmental risk factors [5–7]. There are no cures currently available for AD and only palliative treatments for symptoms are prescribed, and these are of

Contemporary Accounts in Drug Discovery and Development, First Edition. Edited by Xianhai Huang, Robert G. Aslanian, and Wayne H. Tang.

limited utility. Current Alzheimer's treatments cannot stop Alzheimer's from progressing but can temporarily slow the worsening of dementia symptoms and improve the quality of life. There are five neuropathological hallmarks of AD including acetylcholine deficiency, glutamate excitotoxicity, extracellular deposition of amyloid-β, formation of intraneuronal neurofibrillary tangles (NFTs), and neuroinflammation. The neuropathological findings of AD related to these hallmarks are targeted by many current putative therapies in clinical trials. However, the development of new and effective drugs targeting central nervous system diseases is a very challenging and time-consuming process with a very high failure rate. Over two hundred drug candidates have reached phase 2 clinical trials, but no new drugs have been approved for the treatment of AD as a result of these efforts since 2003 [2, 8–10]. Many pharmaceutical companies worldwide continue to aggressively look for curative medications for AD, however, there is still a long way to go to discover effective drugs for the treatment of this disease from delaying its onset to preventing it from getting worse.

4.1.2 Alzheimer's Disease and Amyloid Beta Theory

The formation of two types of pathological lesions in the brain, neuritic plaques and NFTs, is postulated to be the main cause of AD. Neuritic plaques are thought to be the primary lesion in AD. They are spherical extracellular aggregates of amyloid-β protein (Aβ) lesions with an array of abnormal dendrites and axons surrounding these plaques [11, 12]. NFTs, on the other hand, consist mainly of the protein tau and the tangle formation is suggested to be due to neuronal responses to the formation of plaques [13, 14]. However, it should be noted that some recent research and clinical data suggests that AD may in fact result from soluble Aβ oligomers which can induce inflammation and oxidative stress.

Since the identification of AD by Alois Alzheimer in 1906, the amyloid-tau hypotheses dominated its history [15]. Genetic mutations in chromosome 21 can lead to forms of familial AD which often occurs as early-onset AD although many AD patients seem to have no genetic predisposition to the disease and these forms of AD are very similar or identical phenotypically [14]. Based on the observations that the familial cases of AD can be the result of missense mutations in both amyloid-beta precursor protein (APP) and the presenilins, and Aβ aggregates are formed in sporadic AD, the amyloid hypothesis of the progression of AD was developed suggesting that Aβ plaque formation initiates a cascade of events eventually leading to dementia in AD patients [16]. In this regard, Aβ accumulation may affect the efficiency and function of synapses at the beginning, and ultimately leads to neuronal loss and the development of dementia [17–21]. A simplified process for the involvement of Aβ leading to the development of dementia is exemplified in Figure 4.1 [17, 22–24]. Aβ proteins are generated from the processing of APP and overall levels of Aβ in the brain have been shown to correlate to the degree of dementia in AD patients [25–30].

APP is a 695–770 residue transmembrane protein consisting of an N-terminal signaling peptide, a large ectodomain, a hydrophobic transmembrane domain (TMD), and cytoplasmic domain that is expressed throughout the body in many tissues with broad functions. APP695 is the main form

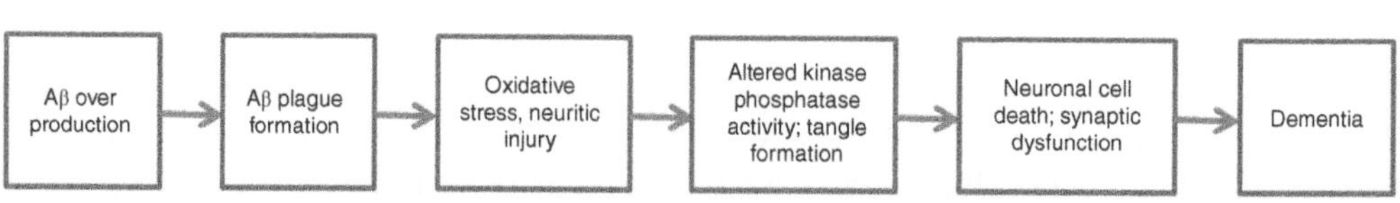

Figure 4.1 The amyloid hypothesis of Aβ aggregation leading to the development of AD.

expressed in neuronal cells. APP goes through different proteolytic events to release a variety of soluble and membrane bound fragments when it is moved through the secretory pathway [21, 31]. Enzymes including α-secretase, β-secretase (BACE), and γ-secretase (GS) are responsible for these proteolytic cleavages as illustrated in Figure 4.2.

Full length APP is cleaved by α-secretase in the luminal Aβ region between residues 16 and 17 which results in the release of a soluble ectodomain fragment (sAPPα), and a membrane bound C-terminal fragment (C83). Since this cleavage precludes the formation of Aβ thus eliminating the possibility of the formation of aggregates and plaques, α-secretase activity is considered nonamyloidogenic. BACE1 cleaves full length APP at the N-terminus where it is 16 residues down from the α-secretase cleavage site leading to the formation of a smaller soluble ectodomain fragment (sAPPβ) and a C-terminal fragment (CTF) C99. Sequentially, γ-secretase cleaves at the C-terminus of the Aβ region in both C83 (to give the p3 fragment) and C99 [33]. For CTF C99, it is consecutively cleaved at ε-, ζ-, and γ sites and releases the APP intracellular domain (AICD) into the cytosol and 37–43 Aβ peptide species into the extracellular space or lumen of secretory pathway organelles. Among these Aβ peptides, the longer ones such as $Aβ_{42}$ and $Aβ_{43}$ are thought to be toxic and more prone to aggregation and the shorter ones such as $Aβ_{37}$ and $Aβ_{38}$ are thought to be harmless or even protective against plaque formation (Figure 4.2) [32, 34]. The increased production of $Aβ_{42}$ over $Aβ_{40}$ due to mutations in the catalytic components of γ-secretase presenilin (PSEN or PS) 1 and 2 or in APP itself is the main cause of rare familial forms of Alzheimer's disease (fAD) [35, 36].

Reducing the amount of $Aβ_{42}$ peptides produced and passive immunization with therapeutic antibodies are the main drug development efforts on reducing the plaque load in the brain of AD patients, and several cellular and animal models were developed to answer questions about the pathological process of Aβ peptides in the development of AD [37–41]. Due to the important roles of β-secretase and γ-secretase in APP processing that generates Aβ peptides, β-secretase (BACE) inhibitors, γ-secretase inhibitors (GSIs) or γ-secretase modulators (GSMs) may be employed to

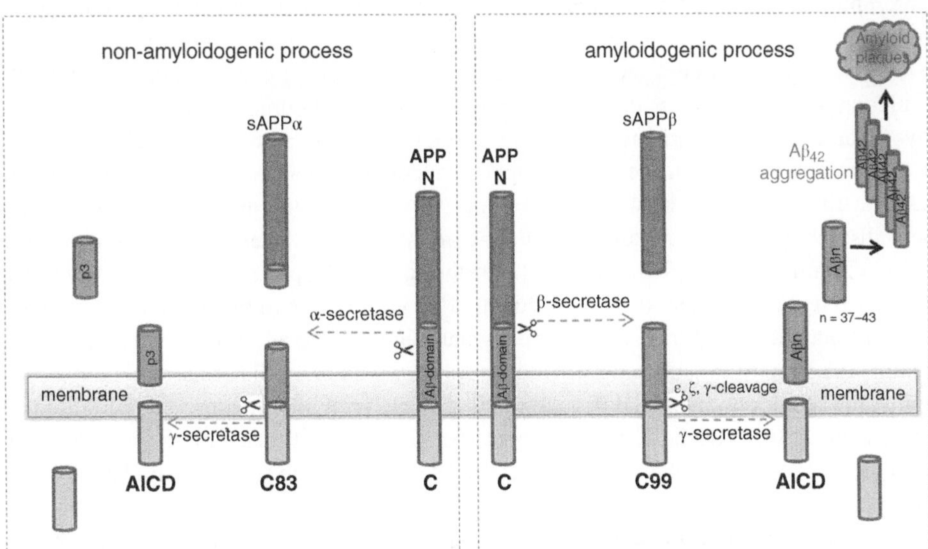

Figure 4.2 Schematic illustration of APP process and generation of Aβ peptides by α-secretase, β-secretase, and γ-secretase. AICD, APP intracellular domain. *Source:* Adapted with permission [32], Copyright© 2020 Zhao, Liu, Xia, Zhang and Wang. This is an open-access article distributed under the terms of the Creative Commons Attribution License (CC BY).

reduce the formation of Aβ peptides. BACE inhibitors block the entire amyloidogenic pathway, leading to APP being processed by α-secretase avoiding the formation of β amyloid. GSIs inhibit the production of all Aβ species which may impact other GS substrates. GSMs modulate the processivity of the enzymatic complex and shift the production of Aβ peptides to shorter non-toxic pieces and do not affect other GS substrates [42–44].

In this chapter, we will employ case studies to showcase the drug discovery efforts in the identification of potent GSIs, GSMs and BACE inhibitors for the potential treatment of AD. While numerous pharmaceutical companies and institutions have had such efforts on relevant targets over the past almost 30 years, the case studies discussed here will be limited to what has been done at Merck/Schering-Plough Research Institute with emphasis on different design strategies that were employed to drive each program to a conclusive stage. Since the authors were heavily involved in the GSM project, the discovery story is more deeply discussed. On the other hand, the GSI and BACE projects involved two big separate teams with the BACE inhibitor discovery story heavily reported in the literature, thus the discovery of GSI is presented as a review and BACE inhibitor is summarized very concisely to showcase the tremendous efforts involved in these programs.

4.2 γ-Secretase Inhibitors Drug Discovery and Development

4.2.1 GSIs Rationale

γ-Secretase is an intramembranous multiprotein complex and an unusual aspartyl protease that cleaves a wide range of type I membrane proteins with critical roles in neuronal function in addition to APP within the lipid bilayer. Since it regulates the final step of Aβ formation, it is a rational target for the treatment of AD. The γ-secretase complex is composed of four components that are required for enzymatic activity: presenilin (PS which includes PS-1 and PS-2), Aph-1 (aph-1a and aph-1b), presenilin enhancer-2 (Pen-2), and nicastrin (Figure 4.3). Although all four components are essential for enzymatic activity, extensive studies suggest that PS contains both the catalytic site and the allosteric inhibitor binding sites [46–49]. PS-1 and PS-2, showing a high degree of homology and functional redundancy, are nine-TMD proteins that must undergo endoproteolysis to generate active N- and C-terminal fragments, each of which contains an active site closely associated aspartate. The cleavage occurs in the large cytoplasmic loop within a short hydrophobic domain that dives into the membrane between TMD6 and TMD7. In aggressive early onset familial Alzheimer's disease (FAD), more than 175 PS1 gene mutations have been identified to date with most of them elevating the $A\beta_{42}/A\beta_{40}$ ratio and interfering with the processing of APP and other γ-secretase substrates. A logical strategy for reducing Aβ accumulation in the brain of AD patients is to inhibit the catalytic unit of the γ-secretase enzymatic complex. In addition to APP, γ-secretase cleaves many other important substrates including Notch which is a cell surface receptor that is involved in many cell developments, differentiations, and cell–cell interactions. The inhibition of Notch is the main cause of mechanism based toxicity of GSIs in preclinical testing and is a major toxicological concern which could result in undesired side effects and toxicities in clinical studies [45, 50, 51].

Most early GSIs are peptidomimetics which are aspartyl protease transition-state analogs based on the APP substrate cleavage site. These inhibitors are usually not orally available let alone brain penetrant for central nervous system (CNS) targets, so further development of these inhibitors for the treatment of AD is difficult. During the past few years, a number of potent more drug-like small molecules have been disclosed in the literature including clinical stage compounds such as

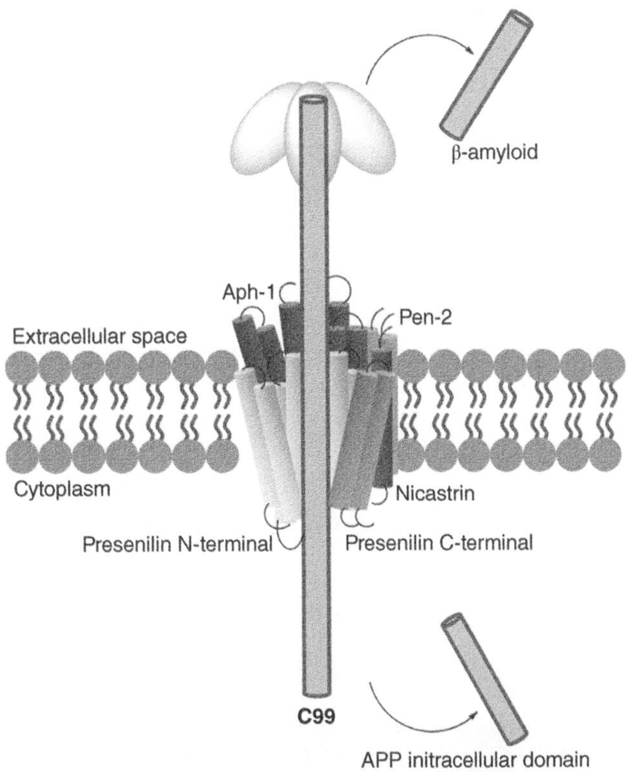

Figure 4.3 γ-secretase enzymatic complex and its processing of the C99 substrate. Aph-1, anterior pharynx-defective-1; APP, amyloid precursor protein; Pen-2, presenilin enhancer-2. *Source:* Adapted with permission from [45], Copyright© 2009, American Chemical Society.

LY-450139, MK-0752, E2012, BMS-708163, PF-3084014, and GSI-953. In contrast to the early strategy of inhibiting proteases by blocking their active sites, these small molecules are non-transition-state inhibitors and selectively interfere with APP binding by targeting the substrate-docking site of γ-secretase which is distinct from the transition-state analog binding site [52, 53].

4.2.2 The Discovery of GSI SCH 900229

4.2.2.1 The Discovery of 2,6-Disubstituted Piperidine Sulfonamide GSIs

The GSI program at Schering-Plough Research Institute started with screening of an in-house library which provided several hits with modest γ-secretase inhibitory activity, one of which was the bicyclic sulfonamide **1** (IC_{50} = 2.5 μM). Compared with a recently reported GSI **2** (IC_{50} = 25 nM) [54], compound **1** could represent a ring constrained version of sulfonamide **2**. Thus, the combination of the key features from these two structures resulted in the design of scaffold **3** which represented a novel class of tetrahydroquinoline GSI. This design coupled with rapid structure–activity relationship (SAR) studies at the R1 and R2 positions of the side chain and tetrahydroquinoline core substitutions resulted in the identification of a potent inhibitor (**4**). Unfortunately, selected tetrahydroquinoline sulfonamide analogs from this series displayed poor pharmacokinetic (PK) properties and further work was needed to discover more potent GSIs with improved PK profiles [55].

The team decided to modify the tetrahydroquinoline core and designed the 2,6-disubstituted *N*-arylsulfonylpiperidine core **5**. Through modification of the R3 and R1/R2 groups, a potent compound (**6**) was identified with a 3 nM $A\beta_{40}$ IC_{50}. This SAR study formed the basis of this GSI series which led to an advanced tool compound (SCH 697466) that helped to support necessary in vitro and in vivo studies of the program. cis-Relative stereochemistry of the substituents with absolute stereochemistry shown at the piperidine ring was established as a requirement for the potency [56] (Figures 4.4 and 4.5).

Compound **6** was assessed in vivo in a transgenic animal model of AD based on TgCRND8 mice [57], and 100 mg/kg intravenous dosing was required to achieve a significant reduction of $A\beta_{40}$ in the plasma. The low efficacy of **6** was believed to be due to the poor exposure which might have resulted from the rapid in vivo hydrolysis of the carbamate linker. In order to slow down this metabolic process, compound **7** featuring a gem-dimethyl group was designed to sterically hinder the carbamate moiety. This modification turned out to be fruitful as it improved the area under the curve (AUC) by twofold as designed. However, the chemical stability of this compound in an acidic environment like the stomach was not adequate, which led to the introduction of the cyclopropyl moiety whose stability was proposed to be better. This modification proved to be productive and a few very potent GSIs were identified with good exposure in rat when dosed orally at 10 mg/kg. One of the compounds (**8**) had a membrane $A\beta_{40}$ IC_{50} of 2.4 nM and cell $A\beta_{40}$ IC_{50} of 5 nM [58].

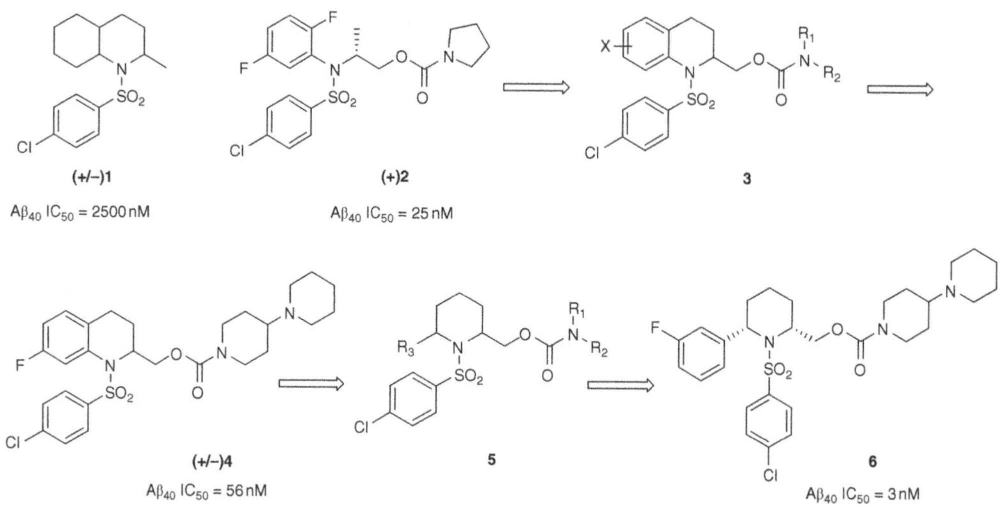

Figure 4.4 Initial SAR development of the tetrahydroquinoline and 2,6-disubstituted *N*-arylsulfonylpiperidine series.

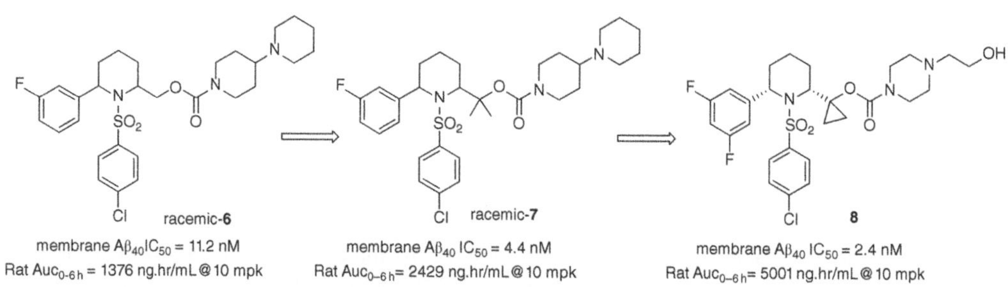

Figure 4.5 The SAR leading to the discovery of SCH 697466.

Based on these encouraging results, compound **8** (SCH 697466) was selected as a lead tool compound for further in vivo characterization in rodents to evaluate on target potency and Notch-related side effects [59, 60]. SCH 697466 dose-dependently reduced Aβ in plasma, cortex, and/or cerebrospinal fluid (CSF) in rats and TgCRND8 and nontransgenic mice following acute administration. In rats, $A\beta_{40}$ levels in plasma, cortex, and CSF, were significantly reduced (Figure 4.6a) three hours after administration of SCH 697466. In TgCRND8 mice, dose-dependent reduction of $A\beta_{40}$ levels in the plasma, cortex, and CSF was observed three and six hours after SCH 697466 administration (Figure 4.6b). In nontransgenic mice, a significant increase in plasma Aβ was observed unexpectedly at 3 and 10 mg/kg and significant lowering of Aβ was observed at 100 mg/kg (Figure 4.6c) three hours after drug administration; In the cortex, significant lowering of Aβ in a dose-dependent manner was observed for the 10, 30, and 100 mg/kg groups. The concentrations of SCH 697466 in plasma and brain has no major species differences, and the concentrations in plasma are usually greater than in brain with both concentrations increasing dose proportionally.

SCH 697466 was tested in a TgCRND8 rapid six-day therapeutic index assay in an effort to determine the potential mechanism-based Notch-related side effects including a reduction in thymus weight and an increase in goblet cell hyperplasia in the intestine, and the therapeutic window in rodents [59, 60]. Notch-related side effects in the intestine and thymus were observed when SCH 697466 was dosed subchronically at 100 mg/kg *b.i.d.* that provided sustained and complete lowering of Aβ. However, no evidence of Notch-related side effects was observed when a *b.i.d.* dose (30 mg/kg) of SCH 697466 was used that provided ~50% reduction of Aβ which was sustained over 24 hours. In addition, although the reasons are not entirely clear, *q.d.* dosing with SCH 697466 provided another way to maintain high level and continuous inhibition of Aβ (>80%), but without causing Notch-related side effects. Further, changes in several Notch-related biomarkers such as HES-1 expression in blood paralleled the side effect observations [60].

In summary, SCH 697466 is a novel GSI with good potency at reducing Aβ in plasma, CSF, and brain in rodents. A small but clear therapeutic index between Aβ lowering and Notch-related side effects was observed in the ileum and thymus in mice in in vivo studies. These results demonstrated that robust and sustained lowering of Aβ without Notch-related side effects using GSIs may be achievable by carefully varying the extent and duration of Aβ lowering by GSIs. This discovery may provide potential strategies to manage Notch-related side effects in preclinical lead compound selection and in clinical studies of GSI [60].

4.2.2.2 The Discovery of Tricyclic Sulfone GSIs and a Preclinical Candidate SCH 900229

As discussed previously, the tetrahydroquinoline analogs were identified as potent GSIs, however those early compounds suffered from an undesirable PK profile. To further improve the profile of this series of GSIs, attention was turned to the carbon analog, sulfone **9**, and its conformationally restricted analog (**10**) which were reported to have excellent γ-secretase activity [61, 62]. Based on a detailed conformational analysis of these compounds, together with in-house data, a novel series of tricyclic sulfone GSIs were designed to further constrain the conformation (Figure 4.7, **11**). The goal of this design was to improve potency through conformational constraint and achieve an improved PK profile by blocking metabolic sites through steric hindrance and simultaneously introduce multiple positions for SAR modification. In order to quickly validate this new series, synthetically simpler bicyclic analogs (**12**) were synthesized. A quick SAR investigation of this bicyclic series showed that this conformational constraint indeed was tolerated and a few potent analogs such as **12** were identified with similar potency to earlier in-house GSIs.

With the confirmed activity of the bicyclic series, a SAR study of the tricyclic series was carried out. Different C-ring sizes (3- to 6-membered rings) were explored first, and it turned out that the

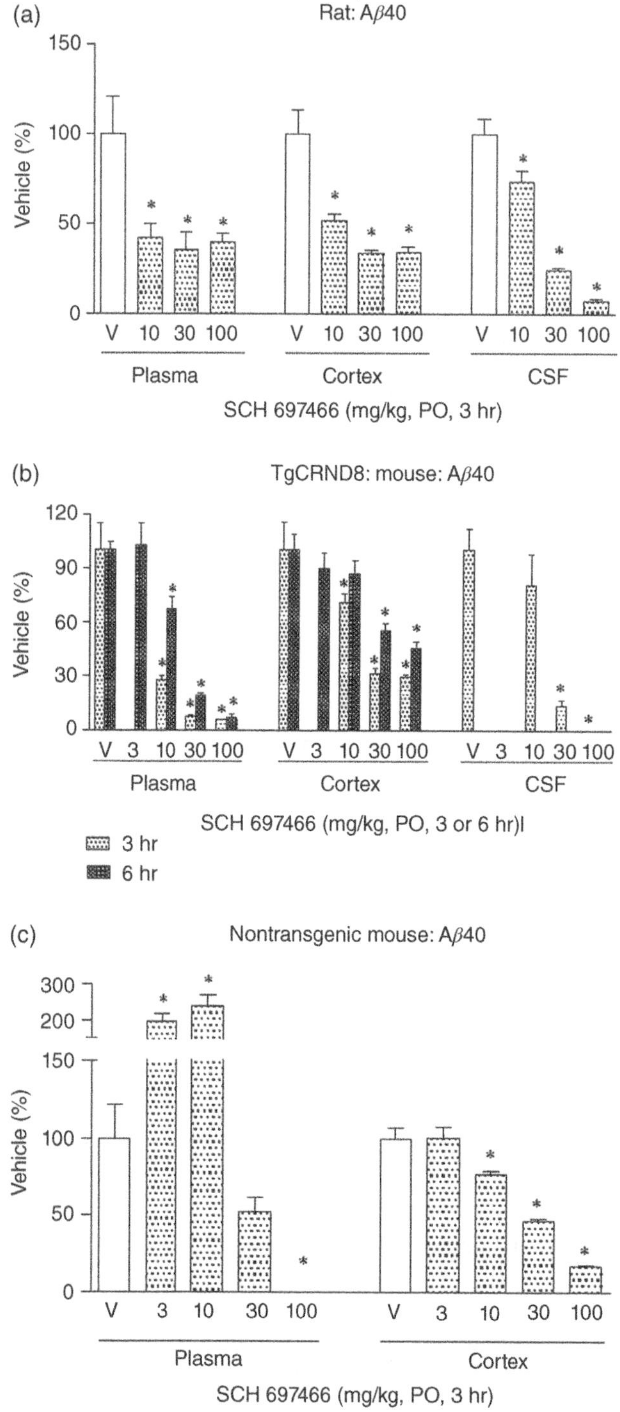

Figure 4.6 Aβ40 levels expressed as a percent of vehicle in plasma, cortex, and CSF following acute administration of SCH 697466 to (a) rats, (b) TgCRND8 mice, and (c) nontransgenic B6C3F1 mice. *$P < 0.02$ versus vehicle, Dunnett post hoc test. *Source:* Reproduced with permission from [60], Copyright© 2013 Lynn A. Hyde *et al.* This is an open access article distributed under the Creative Commons Attribution License, which permits unrestricted use, distribution, and reproduction in any medium, provided the original work is properly cited.

6-membered ring was identified as the optimal size with pyran and cyclohexane as the best while piperidine was also tolerated. Additional systematic SAR showed that the 7- and 8-positions were sweet spots on the tricyclic core to introduce polar groups for fine tuning the physicochemical properties of the analogs [63]. Many different substituents at the 8-positon of the C ring were

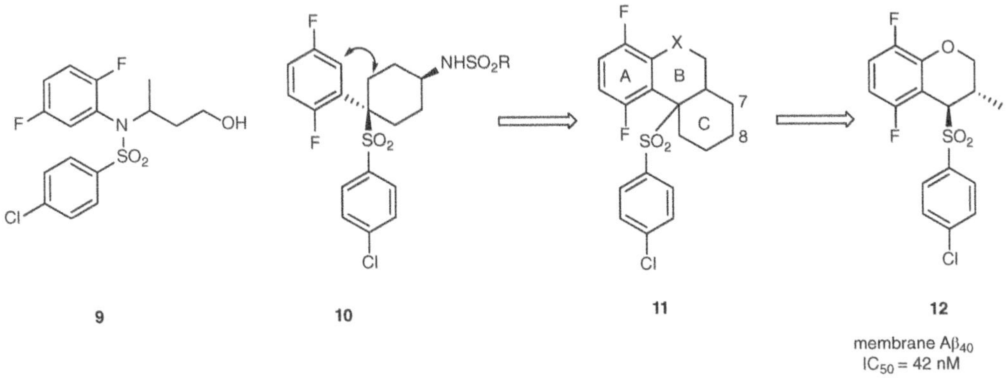

Figure 4.7 The design of tricyclic sulfone γ-secretase inhibitors.

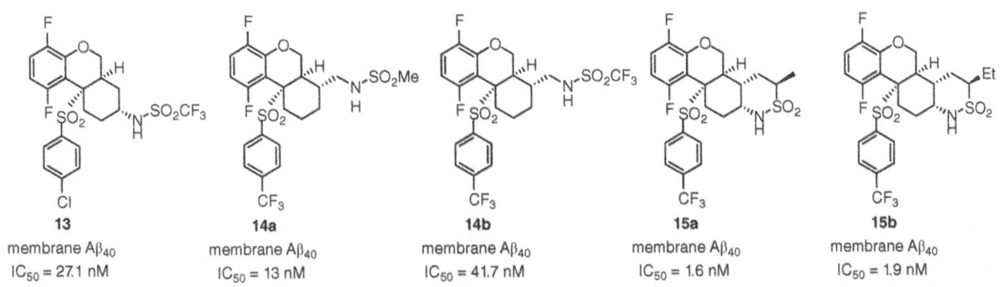

Figure 4.8 SAR evolution of the tricyclic sulfone GSI inhibitors.

tolerated, resulting in compounds with good to excellent in vitro activities. This provided a very good handle to adjust the overall profile of the final compounds. Further testing showed that these compounds had no CYP inhibition or liver toxicity issues as seen in the earlier series (data not shown). Through this traditional SAR effort, compound **13** (Figure 4.8) was identified as having good potency (membrane $A\beta_{40}$ IC_{50} of 27.1 nM and cell $A\beta_{40}$ IC_{50} of 108 nM and $A\beta_{42}$ IC_{50} of 46 nM) and PK profile in rat (5052 ng.h/ml *AUC* at 10 mg/kg oral dosing). This compound was tested in an acute efficacy assay in a CRND8 mouse model and inhibited plasma $A\beta_{40}$ by 90% and cortex $A\beta_{40}$ by 42% three hours after a single oral dose of 30 mg/kg. This compound also significantly lowered plasma and cortical $A\beta_{40}$ after chronic dosing in the CRND8 mouse and did not produce Notch-related gastrointestinal toxicity under conditions where the earlier piperidine sulfonamide series produced significant gastrointestinal toxicity [64]. Thus, 8-substituted tricyclic compounds have comparable or better in vitro activities and in vivo efficacies compared with earlier in-house compounds. Additionally, they have eliminated P450 enzyme inhibition and the liver toxicity problem that was seen in the sulfonamide series discussed previously [65].

The SAR at the 7-position was also systematically explored including modifications at the C-7 side chain and substitutions on the sulfone phenyl group [66]. This led to the identification of a number of potent GDIs. Mouse efficacy and PK studies were used as key studies to discriminate between similar analogs. Two of the best compounds, **14a** and **14b**, are identified and the data are summarized in Table 4.1. These compounds are orally bioavailable and showed excellent in vivo efficacy in a nontransgenic mouse model. Compound **14a** showed 96% inhibition of $A\beta_{40}$ in the plasma and 69% inhibition in the brain at 10 mg/kg PO. Compound **14b** also showed excellent inhibition both in plasma (101%) and brain (71%). Compounds **14a** and **14b** had no significant CYP450 inhibition nor hERG activity when tested at 30 μM. Both compounds are well absorbed

Table 4.1 Profiles of compounds **14a**, **14b**, **15a**, and **15b**.

Parameters	14a	14b	15a	15b
Membrane $A\beta_{40}$ IC_{50} (nM)	13	41.7	1.6	1.9
Whole cell $A\beta_{40}$ and $A\beta_{42}$ IC_{50} (nM)	12, 3.6	40.6, 18.7	1.0, 0.5	5, 2.0
Caco-2 permeability ($x10^{-6}$ cm/s)	42.1	23.0	NT	NT
P450 3A4 inhibition IC_{50} (uM)	>30	>30	>30	>30
Rat oral PK *AUC*@10 mg/kg (ng.h/ml)	2006	702	139	40
Brain concentration@6 h (ng/g)	80	91	<10	<10
Brain/plasma ratio	0.73	1.01	NT	1.0
Plasma protein binding (human, %bound)	98.1	99.9	97.3	98.1
Nontransgenic mouse %reduction of $A\beta_{40}$ @3 h with 10 mg/kg oral dosing	96 (pl.), 69 (br.)	101 (pl.), 71 (br.)	72 (pl.), 57 (br.)	84 (pl.), 42 (br.)

br., brain; NT, not tested; pl., plasma.

and are not P-glycoprotein (PGP) efflux substrates (Table 4.1). In an attempt to combine the SAR of both the 7- and 8-position modifications, tetracyclic sulfones analogs such as **15a** and **15b** were also prepared [67]. These modifications resulted in analogs with excellent cell potency and good ancillary profiles (Table 4.1), however, the overall profiles including PK profiles needed further improvement.

To further improve upon the target potency of the 7-subsituted tricyclic analogs and the physicochemical properties of the 8-substituted tricyclic series, such as the extremely low unbound plasma protein fraction (0.1%) of compound **14b**, the tricyclic bis-pyran sulfone was designed and the SAR was explored [68–70]. Small substituents on the sulfone aryl ring such as 4-Cl, Br, and CF_3 were well tolerated. Other analogs containing electron-rich and electron-poor groups such as 4-OMe and 4-CN were much less active. Eventually, the 4-chlorophenyl group was selected as the fixed piece of the arylsulfone side chain to support exploration of the right-hand side chain SAR. On the right-hand side, SAR exploration identified an alkyl sulfone as a well-tolerated piece. An ethylene linker between the tricyclic core and the alkylsulfone group proved to be optimal for in vitro and in vivo activity. Many substituents on the right-hand alkylsulfone conferred good in vitro potency, and a selected few demonstrated excellent in vivo efficacy and PK properties. Out of these analogs, compound **16** showed one of the best in vitro potencies (membrane $A\beta_{40}$ $IC_{50} = 1.3$ nM; cell $A\beta_{40}$ $IC_{50} = 4$ nM; and cell $A\beta_{42}$ $IC_{50} = 1$ nM) and in vivo efficacy which was fully profiled and eventually selected as a preclinical candidate (SCH 900229). The PK properties of SCH 900229 were determined in rat, monkey, and dog. Overall, it displayed good to excellent exposure with sufficient oral bioavailability in all species tested (Table 4.2). SCH 900229 also showed excellent efficacy at lowering of $A\beta_{40}$ in plasma, CSF, and cortex in rat in vivo model studies (ED_{50} values of 1.1, 3.7, and 0.3 mg/kg, respectively).

The ability of SCH 900229 to reduce Aβ was studied in different preclinical species. $A\beta_{40}$ in plasma and cortex was reduced significantly at the three hour time point, with ED_{50} values of 0.5 and 0.4 mg/kg, respectively when the compound was dosed orally in an acute in vivo model with CRND8 mice. In a subchronic study, when SCH 900229 was dosed *b.i.d.* orally for six days in CRND8 mice, a similar lowering of Aβ was observed in both plasma (ED_{50} 0.4 mg/kg) and cortex (ED_{50} 0.3 mg/kg). Notch-related effects, such as a reduction in thymus size or intestinal goblet cell hyperplasia, were not observed. In an in vitro cell based Notch cleavage assay, SCH 900229 showed

Table 4.2 PK profile of SCH 900229.

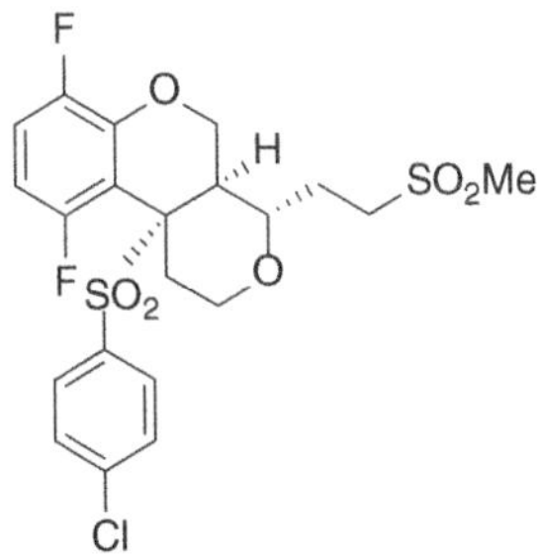

16 (SCH900229)

Species (vehicle)	Dose (mg/kg)	Cl (ml/min/kg)	V_{dss} (l/kg)	$t_{1/2}$ (h)	*AUC* (μM.h)	C_{max} (μM)	t_{max} (h)	*F* (%)
Rat (20% HPβCD)	10	69	5.4	1.1	1.4	0.65	0.7	29
Monkey (0.4% HPMC)	5	8.2	1.9	4.6	8.9	1.39	1.7	45
Dog (0.4% HPMC)	5	1.7	3.1	23.9	85.6	2.74	2.0	86

Cl and V_{dss} obtained with intravenous administration (5 mg/kg for rat, 2.5 mg/kg each for monkey and dog).
AUC, area under the curve; C_{max}, maximal drug concentration; Cl, in vivo clearance; *F*, bioavailability; $t_{1/2}$: half-life; t_{max}, time to reach C_{max}; V_{dss}, volume of distribution at steady state.

an IC_{50} of 46 nM and this level of selectivity between Notch and $A\beta_{40}$ processing was comparable with that of the clinical compound semagacestat. SCH 900229 is a PS1 selective inhibitor of γ-secretase as it displayed a high PS2/PS1 ratio of 25 when tested with reconstituted γ-secretase complexes [71], which may have contributed to its higher therapeutic window observed in rodents as compared with many earlier GSIs, such as semagacestat (PS2/PS1 = 1) and begacestat (PS2/PS1 = 1).

To this end, starting with early, non-optimal leads from an in-house screen and literature reports, a series of tricyclic benzopyrans bearing two sulfone groups were identified as potent and selective GSIs after a few rounds of rational design and through systematic SAR studies. Among them, sulfone **16** (SCH 900229) has excellent in vitro potency and demonstrated robust lowering of Aβ after oral administration in rodents and large animals (data not shown) and exhibited desirable selectivity against Notch processing, as well as a generally favorable PK profile across species. On the basis of these promising results, SCH 900229 was selected as a preclinical candidate and advanced to human clinical trials for the treatment of AD.

4.2.3 Summary of GSIs

Multiple GSIs such as semagacestat, begacestat, and avagacestat have gone through clinical trials (Figure 4.9). At the time this research was performed, the most advanced GSI, LY-450139 (semagacestat), was being studied in a two-year phase 3 study in patients with mild-to-moderate AD. It was reported that patients treated with this drug developed skin and gastrointestinal side effects that are characteristic of the inhibition of γ-secretase processing of Notch. This led to the eventual discontinuation of the clinical trial in 2010 [72, 73]. Subsequently, most GSIs in clinical development were halted including avagacestat and begacestat [74]. Starting from a weak GSI hit compound coupled with the SAR of known GSI structures, the Schering-Plough team identified

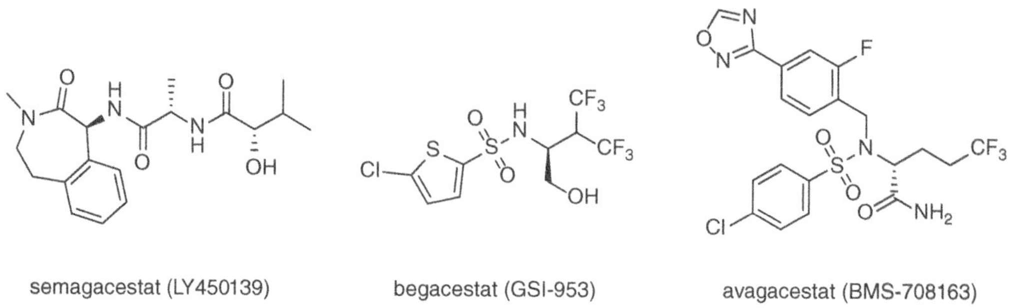

Figure 4.9 Structures of representative GSIs that were halted in clinical trials.

SCH 900229 as a potent GSI with a highly desirable PK profile and in vivo activity in animal studies through exhaustive SAR studies. The compound was selected as a preclinical development candidate and started a first in human clinical trial. However, further development of this compound was halted due to the reported safety concerns with GSIs in the clinical trials mentioned above.

Thus, although GSIs represented the most clinically advanced effort at the time to slow the rate of decline of AD and to modify the natural progression of this devastating disease, no GSI moved forward due to mechanism based toxicity and lack of clinical efficacy. GSIs with improved selectivity against other γ-secretase substrates such as Notch are needed to identify safe and effective drugs to treat AD. In this regard, GSMs may fit this purpose well and are discussed in the next section.

4.3 γ-Secretase Modulator Drug Discovery and Development

4.3.1 GSM Rationale

As discussed in the previous section, γ-secretase cleaves Notch and other type I membrane proteins. This non-discriminatory inhibition of Notch signaling, overall APP processing and the cleavage of other substrates may have resulted in the adverse effects of the GSIs that entered clinical trials. Therefore, these findings suggest the development of disease modifying agents that target γ-secretase activity to lower $A\beta_{42}$ production without blocking the overall processing of γ-secretase substrates. GSMs are believed to act at an allosteric site to shift the predominant site of γ-secretase cleavage toward shorter, nonamyloidogenic peptides such as $A\beta_{38}$ by selectively inhibiting $A\beta_{42}$ formation without affecting ε-site cleavage of γ-secretase substrates including Notch1 and without blocking overall γ-secretase function. This potentially offers a better selectivity window over GSIs versus, for example, Notch processing [75–81]. Recently, animal models have shown that short peptides such as $A\beta_{38}$ are not only nontoxic but could even be protective by attenuating the toxicity of $A\beta_{42}$ in vivo [82]. Mechanistically, GSMs enhance the processivity of γ-secretase by prolonging the residence time of $A\beta_{42}$ at the enzyme [83], which is probably due to stabilization of the enzyme–substrate transition state complex through a conformational change in PS upon GSM binding such that the longer Aβ can be processed to the shorter Aβ [84–87]. As discussed in Section 4.1, after APP is cleaved by BACE, GS catalyzes a series of cleavages within the TMD of the remaining fragment (C99), termed epsilon, zeta and gamma cleavages. The epsilon cleavage releases the intracellular domain of APP (AICD) and produces $A\beta_{49}$ or $A\beta_{48}$ followed by the carboxypeptidase cleavages at the zeta and gamma positions to generate both $A\beta_{40}$ and $A\beta_{42}$ in a stepwise manner [88–93]. As illustrated in Figure 4.10, γ-secretase performs four cleavage cycles to generate $A\beta_{40}$ (49-46-43-40)

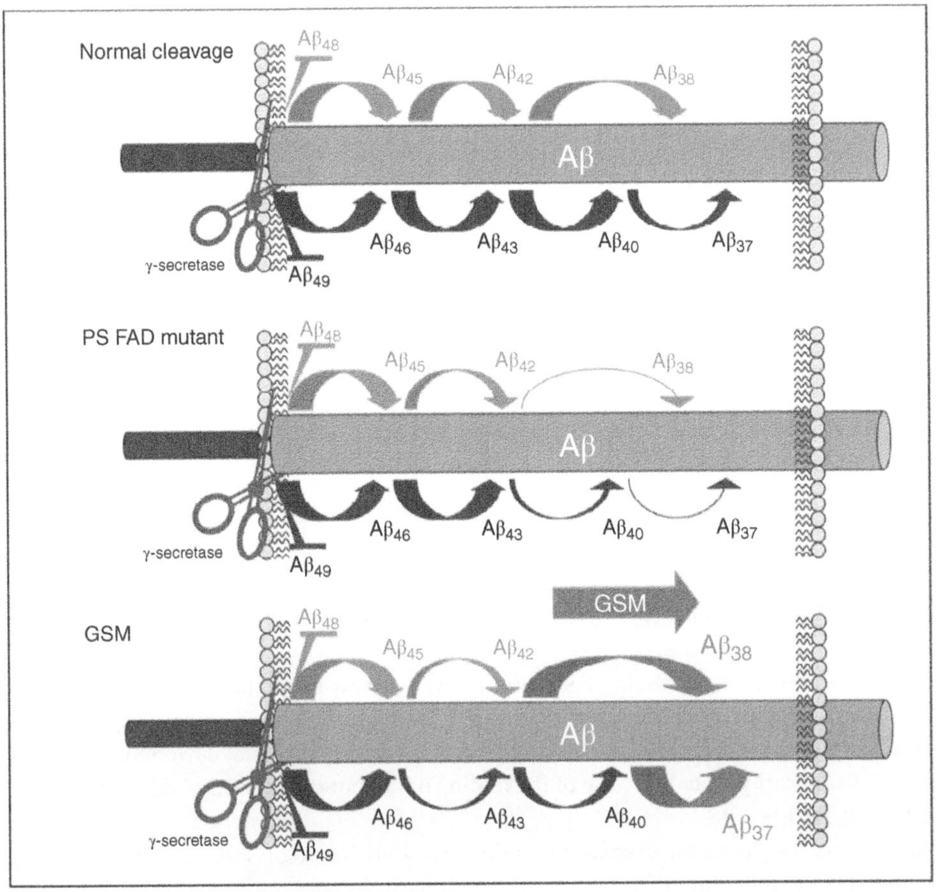

Figure 4.10 Stepwise cleavage model for Aβ generation and potential effect of PS FAD mutations and GSMs. Thickness of the arrows depicts the cleavage efficiency. *Source:* Reproduced with permission [42], Copyright© 2016, American Chemical Society.

and three cycles to generate $A\beta_{42}$ (48-45-42) where many PS mutations affect the fourth cleavage cycle of the complex leading to the increased $A\beta_{42}$:$A\beta_{40}$ ratio observed in FAD. GSMs increase the processivity of the complex leading to an increased production of the shorter, nonamyloidogenic peptides $A\beta_{37}$ and $A\beta_{38}$.

A class of small molecules was reported in the literature in 2001 that alters the known Aβ species product distribution without inhibition of the overall function of the γ-secretase enzyme including APP and Notch processing. A subclass of these GSMs was exemplified by some nonsteroidal anti-inflammatory drugs (NSAIDs) such as ibuprofen, indomethacin, and sulindac sulfide representing the first-generation GSMs (Figure 4.11) [94]. The NSAID derivative Flurizan was tested in a phase 3 trial in mild to moderate AD patients. However, this compound has very poor distribution into the brain and is a very weak modulator of γ-secretase (IC_{50} of around 250 μM) [43] which resulted in no exposure in the brain and led to a failed phase 3 trial [95, 96]. This highlighted the need for a more potent and brain penetrant GSM. A new second generation of non-NSAID class of GSMs (**17**) was reported in 2004 with good potency reducing $A\beta_{42}$ in cells and in animal models [97]. Following this lead, a biarylcinnamide class of GSM was soon reported and the lead compound **18** (E-2012) entered phase 1 clinical studies in 2006 (Figure 4.11) [98, 99]. E-2012 showed good in vitro potency

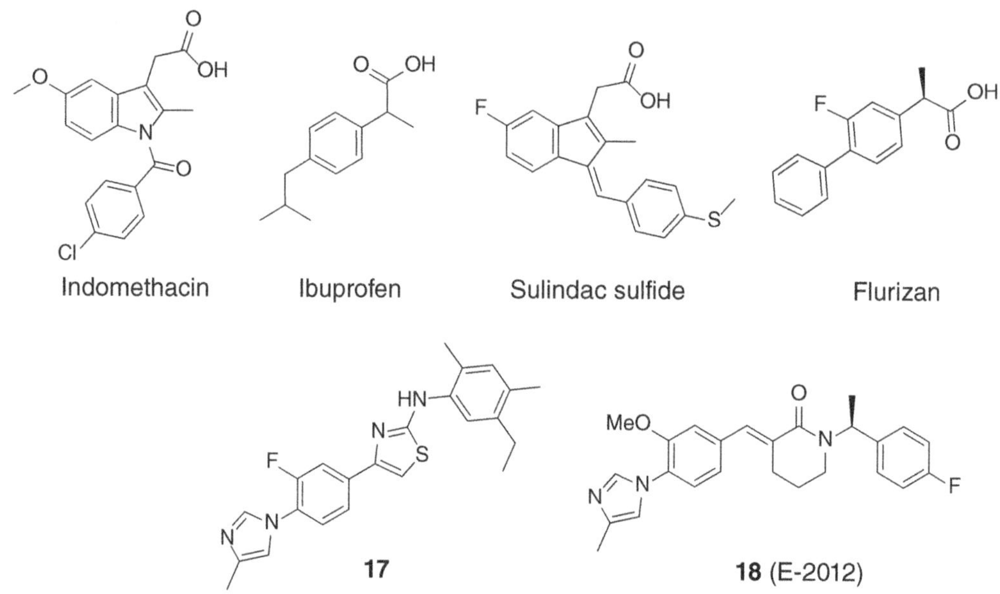

Figure 4.11 Structure of representative NSAID GSMs and non-NSAID GSMs.

(cell $A\beta_{42}$ IC_{50} = 64 nM) and a desirable GSM profile ($A\beta_{total}/A\beta_{42}$ IC_{50} ratio of 226) based on in-house data. It also exhibited good efficacy in reducing CSF $A\beta_{42}$ by 55% and cortical $A\beta_{42}$ by 44% in in vivo rat studies after a single oral dose of 30 mg/kg. However, the clinical development of this compound was temporarily halted because of the finding of rat lenticular toxicity after administration of high doses for 13 weeks.

In this section, we will focus the discussion on the SAR studies using E-2012 as a starting point for the design and identification of GSMs with overall desirable properties suitable for clinical development that avoid the safety liabilities from traditional GSIs.

4.3.2 The Discovery of Oxadiazoline and Oxadiazine GSMs

4.3.2.1 The Pyrazolopyridine Series of GSMs

Based on the reported SAR of the E-2012 series, the left-hand biaryl group is essential for retaining GSM activity, so core modifications of the α,β-unsaturated lactam were focused on at the beginning. The initial designs were to replace the double bond in E-2012 by mimicking the double bond conformation with an intramolecular hydrogen bond between the hydroxyl and the pyridazone carbonyl groups in compound **19** to lock the conformation of the molecule (Figure 4.12) [100]. This design change offered the additional benefit of removing a potential toxicophore due to the α,β-unsaturated lactam, a potential site for nucleophilic addition. Although this compound showed only weak $A\beta_{42}$ IC_{50} of 7 uM, it provided additional design venues. When a 5-membered ring intramolecular H-bond was employed instead of a 6-membered ring H-bond, a significant potency improvement was obtained in pyridazone **20** (IC_{50} = 503 nM) with a good Aβ total/$A\beta_{42}$ ratio of 40 was obtained. Additional SAR focusing on core and side chain modifications led to the identification of pyridine compound **21** which further improved $A\beta_{42}$ potency (IC_{50} = 44 nM) and Aβ total selectivity (a ratio of 450).

Compound **21** showed good in vivo efficacy in a CRND8 mouse model, giving over 85% reduction of $A\beta_{42}$ in plasma at 30 mg/kg with little effect on the $A\beta_{total}$. In the nontransgenic in vivo rat model,

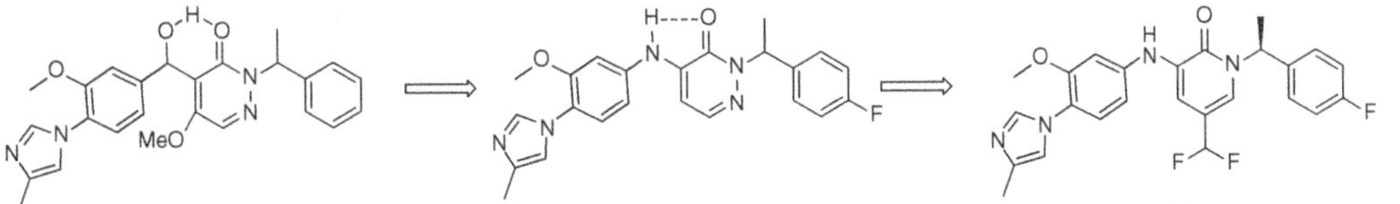

19	20	21
$A\beta_{42}$ IC_{50} (nM) =7099	$A\beta_{42}$ IC_{50} (nM) = 503	$A\beta_{42}$ IC_{50} (nM) = 44
Aβtotal/42 ratio = 2.8	Aβtotal/42 ratio = 40	Aβtotal/42 ratio = 450

Figure 4.12 Conformational constraint design leading to the identification of selective GSM pyridine **21**.

this compound displayed a 40% reduction of $A\beta_{42}$ in the CSF at 100 mg/kg and a 26% reduction of $A\beta_{42}$ in brain, while the $A\beta_{total}$ reduction was only 7% in the CSF. This compound had good rat PK with an $AUC_{1\text{-}6h}$ of 7.5 μM.h at 10 mg/kg and favorable brain concentration (347.3 ng/g) at the six hour time point. No abnormal behavior or side effects were observed in those studies. This compound had a good overall profile in terms of rat PK and ancillary profile such as clean hERG, clean hPXR, acceptable P450 inhibition profile, and good human hepatocyte clearance data (2.9 μl/m/M cell).

To further improve the in vivo efficacy at lower doses, additional core modifications were carried out based on this compound. Unfortunately, most of these changes only resulted in moderate in vitro potency and in vivo efficacy. Further improvement was therefore needed. The team decided to further constrain the conformation of the core by forming a second ring (tetrahydropyridine and tetrahydropyrimidine derivatives) as illustrated in Figure 4.13 [101, 102]. However, these modifications did not result in a significant improvement in potency. In addition, these compounds exhibited a relatively high efflux ratio which limited brain penetration and in vivo efficacy.

The pyrazolopyridine core was explored next in the hope that this modification might offer better physicochemical properties and efficacy than the tetrahydropyrazolopyridine series (Figure 4.13) [103]. This turned out to be a good strategy and compound **22** was identified with good in vitro binding potency and selectivity. However, when dosed orally in rats at 30 mg/kg, this compound did not demonstrate statistically significant efficacy in reducing CSF $A\beta_{42}$. A PK study of **22** revealed that it had limited brain exposure three hours after dosing (brain C3h = 0.17 μM). Not surprisingly, this compound was found to have poor permeability and was a strong PGP efflux substrate (P_{app}(A-B) = 0.9 x 10^{-6} cm/s, P_{app}(B-A) = 31.4 x 10^{-6} cm/s, and efflux ratio = 33.8).

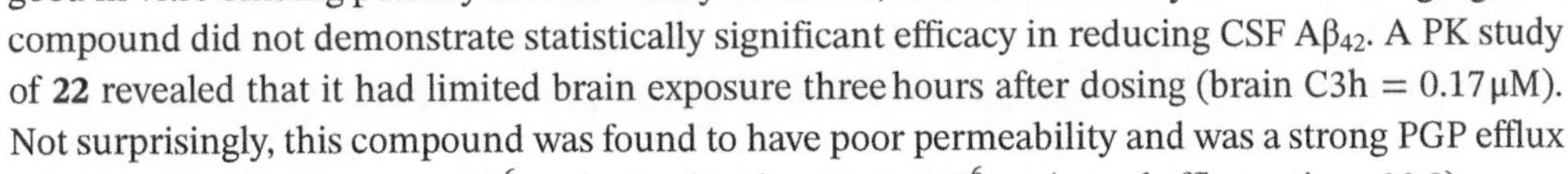

tetrahydrotriazolopyrimidine	tetrahydroisoxazolo[3,4-b]pyridine	tetrahydroisoxazolo[5,4-b]pyridine	tetrahydropyrazolopyridine

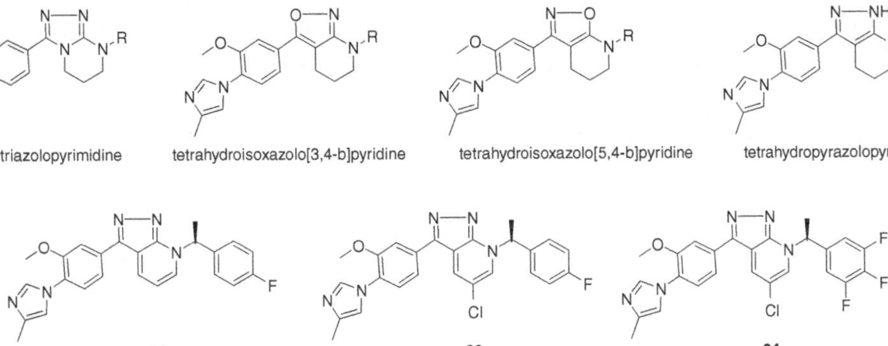

22	23	24
$A\beta_{42}$ IC_{50} (nM) = 97	$A\beta_{42}$ IC_{50} (nM) = 70	$A\beta_{42}$ IC_{50} (nM) = 107
Aβtotal/42 ratio = 205	Aβtotal/42 ratio = 288	Aβtotal/42 ratio = 186

Figure 4.13 SAR development of the pyrazolopyridine series.

To improve the permeability and PGP profile, additional lipophilic groups were introduced to modulate the lipophilicity of the compounds. As a result of this effort, compound **23** was identified with improved brain exposure (brain C3h = 0.92 μM at 10 mg/kg oral dosing), and it was not a PGP efflux substrate. Eventually, compound **24** was identified with a desirable in vitro profile, and it was tested in the rat in vivo model. This compound demonstrated good plasma (C3h = 5.02 μM) and brain exposure (brain C3h = 1.22 μM) at 10 mg/kg oral dosing and it was efficacious at reducing CSF $A\beta_{42}$ (45% $A\beta_{42}$ reduction three hours after 30 mg/kg oral dosing). It did not affect Notch cleavage at concentrations up to 20 μM. This result was consistent with the finding in the $A\beta_{total}$ assay. Notch signaling, assessed by HES-1 expression, was also not affected in studies conducted in mouse and dog. While compound **24** showed statistically significant in vivo efficacy comparable with compound E-2012, with the greatest reduction of CSF $A\beta_{42}$ in rats in this series, more efficacious compounds with improved in vivo activity were desirable to improve the chance of success in clinical trials. These SAR data also demonstrated that it is important to balance the modifications designed to improve physicochemical properties and those which improve in vivo brain exposure for CNS drug SAR development.

4.3.2.2 The Discovery of Oxadiazoline, Oxadiazine, and Oxadiazepine GSMs

To further improve the overall profile of the GSM compounds, the second design strategy employed bioisosteres to replace the key lactam functional group. E-2012 is a GSM with a highly optimized PK and pharmacodynamic (PD) profile. According to the literature, one analog prepared during the SAR exploration leading to E-2012 was compound **25** ($A\beta_{42}$ IC_{50} = 56 nM). The good potency of this compound indicates a high tolerance in the enzyme binding pocket for large structural changes around the right-hand benzylic position of the molecule. However, it was speculated that the steric bulkiness at this position may preclude the lone pair of electrons on the right side of the lactam carbonyl oxygen from interacting with any potential H-bond donor of the enzyme. These observations prompted the design of a molecule in which a heterocyclic ring was introduced from the position of the lactam carbonyl oxygen to the benzylic position while maintaining the overall conformation of the molecule. To test this idea, an oxadiazoline ring was selected as a reasonable bioisosteric replacement of the lactam carbonyl in E-2012 (Figure 4.14) [104].

With this design rationale in mind, the SAR was quickly explored focusing on the introduction of different right-hand side chains. Compound **26** was quickly identified which demonstrated a pharmacological profile indistinguishable from E-2012 in vitro, with an IC_{50} of 58 nM for reduction of $A\beta_{42}$ and an $A\beta_{total}/A\beta_{42}$ IC_{50} ratio of 237 in a HEK293 cell assay. Compound **26** possesses an excellent PK profile in all animal species examined, no inhibition of human cytochrome P450s, and no induction of human PXR. However, in a voltage clamp cellular assay measuring hERG liability, **26** had an IC_{50} of 740 nM which may carry some risk of QTc prolongation in humans. Compound **26** dose dependently reduced CSF levels of $A\beta_{42}$ by 51% and 30% three hours after

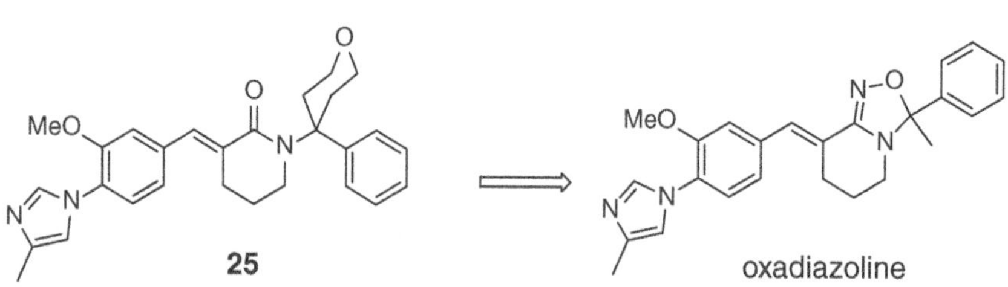

Figure 4.14 The rational design of oxadiazoline GSMs.

single oral doses of 10 and 3 mg/kg, respectively, with corresponding brain exposure of 5 and 0.8 μM when administered in rats. Due to concerns about the chemical stability of the oxadiazoline core, compound **26** was tested in aqueous 0.1 N HCl at room temperature overnight, and indeed deterioration of the enantiomeric excess from 99.6 to 55.7% was observed. In an attempt to improve the chemical stability and hERG profile, the more polar hydroxylmethyl group was introduced to adjust the pKa of the cyclic hydroxyamidine moiety (compound **26** has a pKa of 6.7) and lipophilicity of the compounds.

With the introduction of the hydroxyl methyl group coupled with modifications of the phenyl substituents, compound **27** (pKa of 5.8) was found to be more stable compared with **26**. It was stable for seven days at room temperature in 0.1 N HCl/MeOH (1:1) with no sign of decomposition or deterioration of enantiomeric excess. This compound has highly desirable in vitro potency ($A\beta_{42}$ IC_{50} of 29 nM) and excellent $A\beta_{total}$ and $A\beta_{42}$ selectivity (a ratio of 488). Additional SAR in this series led to the identification of one of the most potent GSM analogs **28** (Figure 4.15, $A\beta_{42}$ IC_{50} of 15 nM; $A\beta_{total}$ and $A\beta_{42}$ selectivity of 1023 fold).

With the full validation of 5-membered cyclic hydroxyamidines as lactam carbonyl isosteric replacement, the SAR was further explored using 6-member ring oxadiazines and 7-member ring oxadiazepine to replace the lactam [105–107]. These expanded ring systems should be more stable chemically and offer additional positions for SAR modifications. Surprisingly, it turned out that the SAR clearly did not translate from the oxadiazoline to the oxadiazine series although the ring expanded with only one carbon atom (Figure 4.16). Most dramatically, hydroxylmethyl oxadiazine **30** was much less active than oxadiazoline compound **30′**. This SAR suggests that the two series have different conformational interactions with the enzyme such that substitution at C3 of the oxadiazine core was preferred. This also reminds us that medicinal chemists should not take things for granted in terms of drug design and SAR development, as a single atom change can sometimes result in dramatic changes in potency and PK profile. Through extensive SAR development varying the substituents at the indicated positions in Figure 4.15 and focusing on in vitro potency, ancillary profile, brain exposure, and in vivo efficacy, four compounds stood out with the best potency and selectivity profiles, and these were selected for additional profiling (Figure 4.17). Eventually compounds **27** and **32** were profiled extensively as final preclinical development candidates as discussed in the next section.

4.3.2.3 Profiles of GSM Preclinical Candidates

Compounds **27** and **32** were fully profiled in the preclinical candidates (PCCs) workup process based on their initial promising activity and physicochemical property profiles. Compound **32**

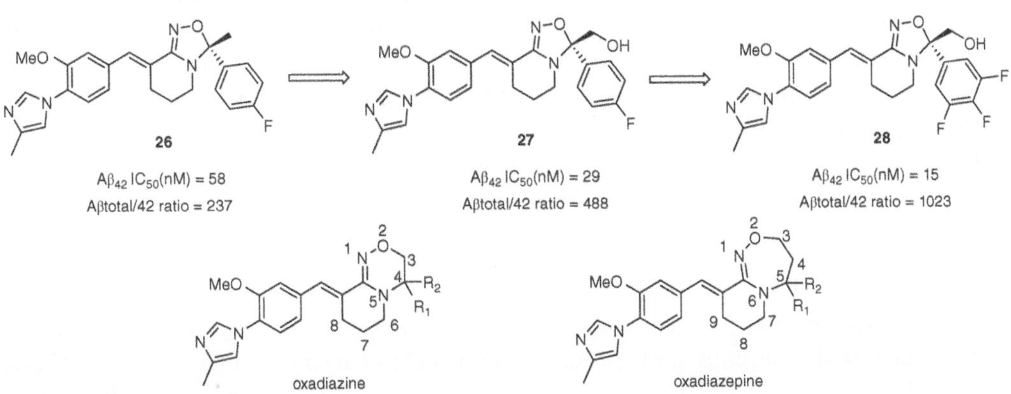

Figure 4.15 SAR development of oxadiazoline series and the design of oxadiazine and oxadiazepine series.

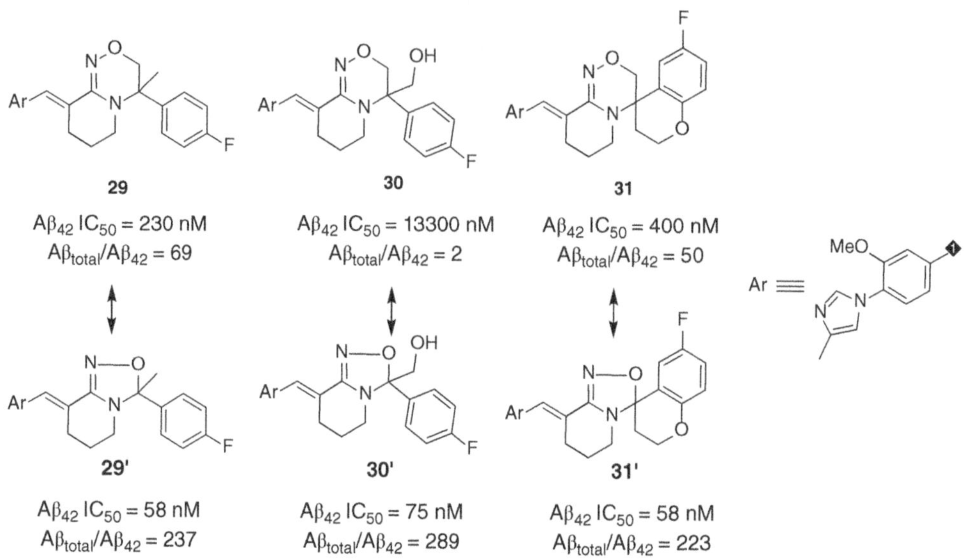

Figure 4.16 Initial SAR comparison between oxadiazoline and oxadiazine series.

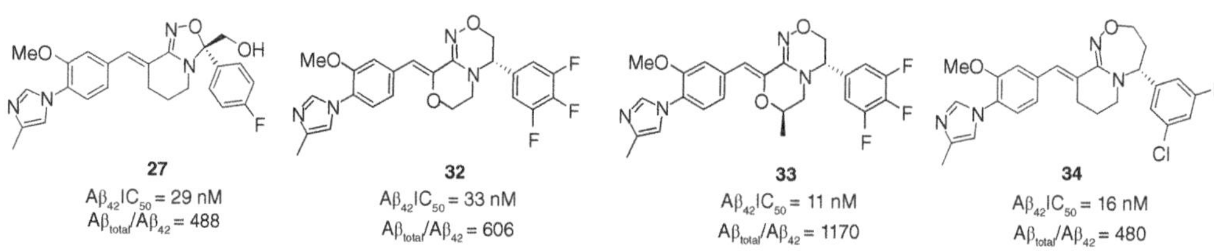

Figure 4.17 Lead compounds from oxadiazoline, oxadiazine, and oxadiazepine series.

displayed two pKa values (7.4 and 2.7) suggesting that the oxadiazine ring is slightly more basic than the oxadiazoline ring which gave **32** a highly desirable aqueous solubility of greater than 100 mM at pH 3.5 as the HCl salt. The in vitro characterization and in vivo PK profiles of these two compounds are summarized in Tables 4.3 and 4.4.

Compounds **27** and **32** are not substrates for PGP efflux with high permeability in a Caco-2 membrane permeability assay (Table 4.3). A better overall hERG profile was observed with these two compounds than compound **26**. No prolongation of the QTc interval was observed with these two compounds in dogs at sufficient multiples over the exposures required for a pharmacological effect (data not shown). Compounds **27** and **32** did not inhibit APP processing in vitro as total Aβ inhibition was not observed. They did not affect Notch processing at concentrations up to 50 μM in HEK293 cells expressing the human Notch1 protein. As assessed by mass spectrometry, the GSM characteristics of **27** and **32** were further confirmed by the changes they induced in the Aβ species produced by HEK293$^{Swe\text{-}Lon}$ cells in vitro. Incubation of HEK293$^{Swe\text{-}Lon}$ with 10 μM **32** profoundly increased production of $A\beta_{37}$ and $A\beta_{38}$ while selectively reducing $A\beta_{42}$, $A\beta_{40}$, and $A\beta_{39}$ (Figure 4.18). Compound **27** showed a similar modulator profile in this assay.

The shift in specificity of APP cleavage rather than a reduction in overall γ-secretase activity and the resulting lack of changes in substrate accumulation and Notch processing confirmed that both compounds are potent GSMs. Following repeated administration of the highest tolerable doses of **27** and **32** to rats in in vivo studies, there was no evidence of Notch-related side effects or biomarker

Table 4.3 In vitro profiles of **27** and **32** in comparison with **18** (E-2012).

Compound	27	32	18 (E-2012)
$A\beta_{42}$ IC_{50} (nM)	29	33	64
$A\beta_{total}$ IC_{50} (nM)	14130	>20000	14000
Notch IC_{50} (nM)	>50000	>50000	ND
hERG IC_{50} (nM)	3300	1000	1200
$P_{appA\text{-}B}$ ($\times 10^{-6}$ cm/s)	30.2	24.6	57.9
$P_{B\text{-}A}/P_{A\text{-}B}$	1.4	2.0	1.2

ND, not determined.

Table 4.4 PK parameters of compounds **27** and **32** in rat, dog, and monkey[a].

Compound	Species	Cl (ml/min/kg)	V_{dss} (l/kg)	$t_{1/2}$ (h)	*AUC* (μM.h)	C_{max} (μM)	*F* (%)
27	Rat	6.1	1.6	3.4	61	7.5	98
	Monkey	6	1.4	3	30	8.8	48
	Dog	5	0.4	1.4	26	4.0	38
32	Rat	1.4	0.4	3.5	265	19.8	100
	Monkey	11	0.8	0.9	11	4.9	34
	Dog	9	0.6	0.8	27	7.9	69

a 10 mg/kg for p.o.; 3 mg/kg for intravenous in rat, dog, and monkey.
AUC, area under the curve; C_{max}, maximal drug concentration; Cl, in vivo clearance; *F*, bioavailability; $t_{1/2}$: half-life; V_{dss}, volume of distribution at steady state.

changes. For common side effects of Notch-related γ-secretase cleavage, these two compounds did not produce goblet cell meta/hyperplasia in the ileum, and did not decrease thymus size or dramatically affect body weight either. In addition, Notch activity related biomarkers including HES-1 and adipsin expression were unaffected in the jejunum of treated rats.

These two compounds possess excellent PK profiles in dog, rat, and monkey with excellent oral exposure and bioavailability (Table 4.4). Both compounds were tested in in vivo PD model studies. Oral dosing of compound **32** at 3 mg/kg reduced cortical $A\beta_{42}$ to 63% of the level seen after vehicle administration, and **27** dosed at 10 mg/kg orally reduced cortical $A\beta_{42}$ levels to 37% (Table 4.5). The better in vivo activity of **32** may be attributed to the better brain/plasma exposure of this compound. Compound **32** dose-dependently reduced $A\beta_{42}$ preferentially over $A\beta_{total}$ and $A\beta_{40}$ upon acute administration to rat (Figure 4.19) and similar results were also

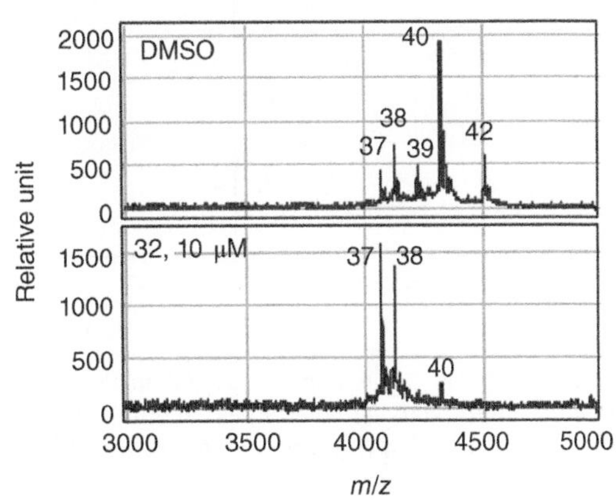

Figure 4.18 Mass spectrometric analysis of Aβ peptides generated from HEK293[Swe-Lon] cells after treatment with vehicle (DMSO) or 10 μM **32**. *Source:* Adapted with permission [104]. Copyright© 2012, American Chemical Society.

Table 4.5 In vivo PK/PD profile of **27** and **32** in rat.

				CSF Aβ (%)[a]			Cortex Aβ (%)[a]		Plasma Aβ (%)[a]		
Compound	Dose (mg/kg)	Plasma (μM)	Brain/plasma ratio	$A\beta_{42}$	$A\beta_{40}$	$A\beta_{total}$	$A\beta_{42}$	$A\beta_{40}$	$A\beta_{42}$	$A\beta_{40}$	$A\beta_{total}$
27	10	3.6	0.15	66	104	120	37	77	41	44	77
32	3	3	0.49	65	74	88	63	84	32	50	86

[a] Expressed as % vehicle in CSF, cortex, and plasma at 3 h after dosing.

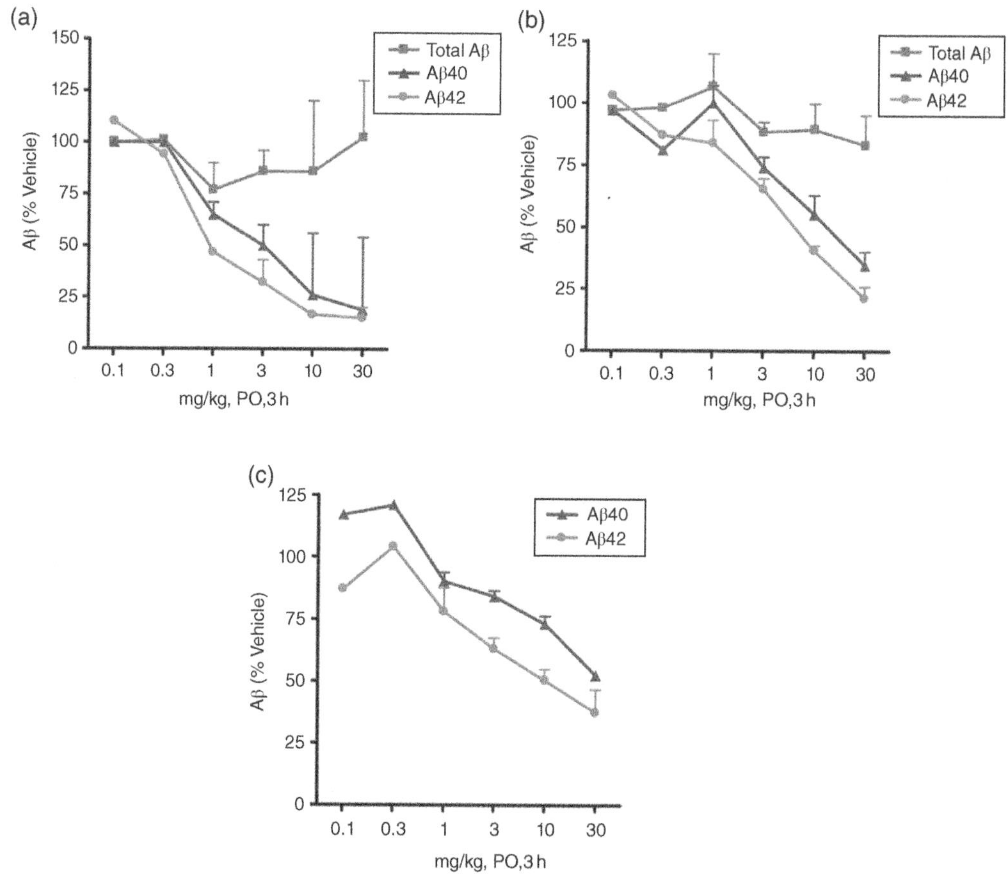

Figure 4.19 $A\beta_{total}$, $A\beta_{40}$, and $A\beta_{42}$ in (a) plasma and (b) CSF, and $A\beta_{40}$ and $A\beta_{42}$ in (c) cortex from rats treated acutely with various single oral doses of **32** (0.1–30 mg/kg). *Source:* Reproduced with permission [104]. Copyright© 2012, American Chemical Society.

observed with subchronic (six-day) dosing in rat (data not shown). In an in vivo study in cynomolgus monkeys, **32** reduced $A\beta_{42}$ in the cortex and CSF to 44% and 70%, respectively, of the values seen after vehicle administration at four hours after a single oral dose of 30 mg/kg with a plasma exposure of ~6 μM and an average brain/plasma ratio of 1.4.

In conclusion, using the known compound GSM E-2012 as a lead, two different design strategies were taken based on conformational constraints and bioisostere replacement. Through a well-executed SAR effort, oxadiazoline and oxadiazine lead compounds were selected as PCCs which displayed highly desirable profiles suitable for clinical development. Although further development of these two compounds was halted due to a strategic development consideration to avoid overlapping efforts with the in-house BACE inhibitor program (discussed in the next section), these novel GSMs did show that it was possible to modulate γ-secretase activity through GSMs to achieve the desired Aβ peptide cleavage profile with an improved safety window without impacting Notch processing. In addition, the knowledge and experience gained from the SAR studies undertaken during this process will be very valuable for the discovery of future CNS-related drugs. For example, 5-membered oxadiazolines, 6-membered oxadiazines and 7-membered oxadiazepines were found to be excellent bioisostere replacements for a lactam group and possess highly desirable PK and toxicological profiles which may be employed in other drug discovery programs.

4.3.2.4 On-going GSM Discovery

GSMs are promising disease modifying agents towards AD due to their improved safety profile while maintaining on-target potency. However, significant questions remain to be answered and more work is needed to determine their precise binding sites, the molecular basis for their mechanism of action, and the impact of GSMs on long term safety. Development of GSMs for the treatment of AD not only relies on the discovery of effective drug candidates, but also the progress of AD research in molecular pathogenesis, biomarkers, diagnosis, and other therapeutic developments all play important roles during this process. Without the availability of effective biomarkers for the early diagnosis of AD, it is not clear whether successful GSM clinical trials could be conducted efficiently. Despite all these considerations, a number of newer generation GSMs were/are in clinical development [108] which include PF-06648671 [109], RO7185876 [110], BPN-15606 [111], and BMS-932481 [112] (Figure 4.20). The results of these trials will shine a light on the future of GSMs in the treatment of AD [43].

4.4 Overview of β-Secretase Inhibitors

4.4.1 Rationale of β-Secretase Inhibitors

The search for GSM and GSI inhibitors has been extensive, but the major approach to the treatment of AD has been the effort to identify potent, selective brain-penetrant BACE1 inhibitors. BACE belongs to the family of aspartic proteases among which only BACE1 and BACE2 are localized on the cell membrane via the type I TMD [113, 114]. BACE1 and BACE2 exhibit different enzymatic specificity although they share 59% homology. As discussed earlier, GS and BACE1 play a critical role in APP processing. BACE1 mainly cleaves APP at both the N-terminus of the Aβ

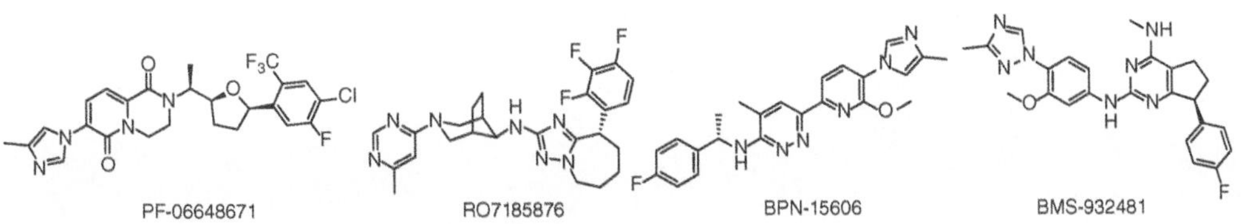

Figure 4.20 Structures of representative clinical stage GSMs.

domain at the β-secretase site and the β′-site (between E10 and V11), on the other hand, BACE2 preferentially cleaves APP between F19 and F20 or between F20 and A21 within the Aβ region [115–118]. In mouse models, BACE1 deficiency almost abolishes the production of Aβ, but no significant difference in Aβ generation was observed in BACE2-deficient mice. These findings suggest BACE1 is the sole enzyme for initiating Aβ generation [119]. Studies on BACE1 knockout mice suggested that BACE1 deficiency in mice posed no major concern, as these BACE1-null mice appeared to be healthy but display a moderate peripheral nerve hypomyelination caused by loss of BACE1-mediated processing of Neuregulin-1 [75, 76, 120, 121]. Thus, BACE1 inhibition to prevent Aβ peptide formation is considered to be a highly attractive amyloid-lowering approach to a potentially disease-modifying AD treatment.

The BACE1 protein structure complexed with inhibitors was resolved in the early 2000s which provided detailed information on the extensive interactions in each of the subsites of the protease (Figure 4.21) [122, 123]. The BACE1 active site is located at the groove between two lobes covered by the residues 69–74 flap and contains two conserved aspartic acid residues at D32 and D228. There are typically eight subsites at the active sites of aspartic peptidases which interact with different portions of the substrates [124]. For BACE1, there are 11 subsites (S7 to S4′) which could form more interactions with a peptidic substrate, but 3 subsites (S7 to S5) are rarely occupied by small molecular inhibitors [125]. This has made the BACE1 inhibitor discovery amenable to structure-based drug design and promoted intense efforts to identify potent and selective BACE1 inhibitors that effectively inhibit Aβ peptide production in the CNS. Initial research carried out on peptidomimetic inhibitors met with limited success since these inhibitors are usually PGP substrates and suffer insufficient brain exposure due to restricted brain penetration [123, 126, 127]. Conceptually, nonpeptidic small molecule BACE1 inhibitors that penetrate the CNS and lower CNS-derived Aβ offered the most promising avenue to identify potential drugs for treating AD. Tremendous efforts from both academic and industrial laboratories have been devoted toward the development of drug-like β-secretase inhibitors. However, the identification of such orally bioavailable, brain penetrant BACE1 inhibitors has proved to be exceptionally challenging [21, 32, 128–130].

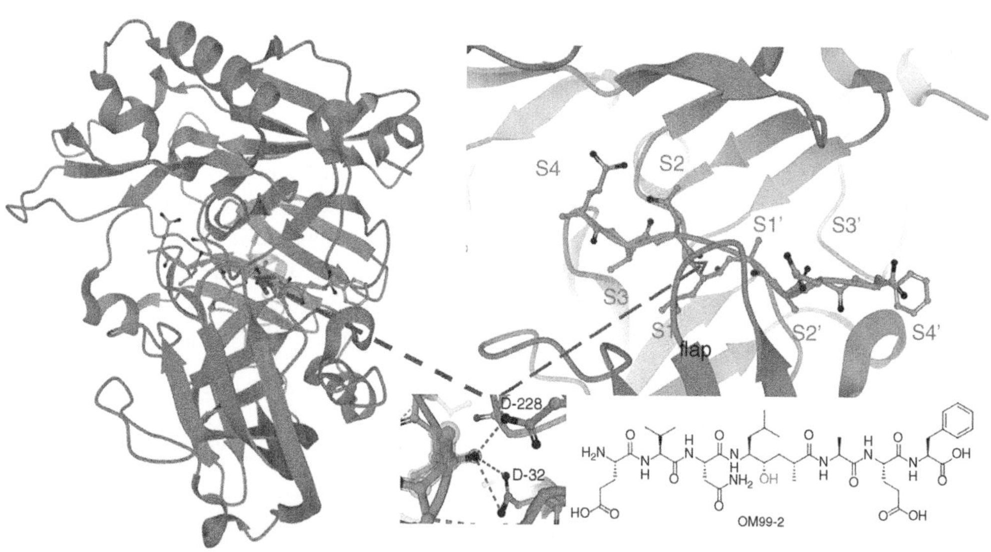

Figure 4.21 Structure of BACE1 in complex with inhibitor OM99-2 (PDB code: 1FKN [122]). The hydroxyl in OM99-2 is making key interactions with aspartic acids D-32 and D-228.

4.4.2 Brief Summary of Verubecestat (MK-8931) Discovery and Clinical Development

Perhaps the best compound to test the BACE hypothesis, and the one that proceeded farthest into the clinic, is MK-8931 (also known as SCH 900931, Figure 4.22) whose discovery was reported extensively in the literature and will be only briefly summarized here. The discovery story of MK-8931 is a classic example of how the drug discovery scientists can employ vastly different techniques to design the best molecules suitable for development. Starting from very weak NMR fragment-based screening hits [131], through strong insights into the properties for CNS drugs and structure-based drug design employing extensive protein crystallography, and generations of rational design processes [132–142], MK-8931 was eventually selected as the best candidate with overall desirable properties that entered clinical trials. Unfortunately, the clinical results with MK-8931were not promising. After initial positive results in early clinical trials evaluating safety and PK [143], long term efficacy trials failed. The EPOCH study was a randomized 78-week trial to evaluate verubecestat (MK-8931) at doses of 12 and 40 mg per day, as compared with placebo, in patients who had a clinical diagnosis of mild-to-moderate AD [144]. The coprimary outcomes were the change from baseline to week 78 in the score on the cognitive subscale of the Alzheimer's Disease Assessment Scale (ADAS-cog; higher scores indicating worse dementia with scores ranging from 0 to 70) and in the score on the Alzheimer's Disease Cooperative Study Activities of Daily Living inventory scale (ADCS-ADL; lower scores indicating worse function with scores ranging from 0 to 78). In the ADAS-cog score, the estimated mean change from baseline to week 78 was 7.9 in the 12 mg group, 8.0 in the 40 mg group, and 7.7 in the placebo group. In the ADCS-ADL score, the estimated mean change from baseline to week 78 was −8.4 in the 12 mg group, −8.2 in the 40 mg group, and − 8.9 in the placebo group. Adverse events were more common in the verubecestat groups than in the placebo group. Thus, verubecestat did not reduce cognitive or functional decline in patients with mild-to moderate AD and was associated with treatment-related adverse events. In February 2017, Merck halted this late-stage trial of verubecestat for mild to moderate AD due to lack of efficacy.

Prodromal AD offers an opportunity to test the effect of drugs that modify the deposition of amyloid in the brain before the onset of dementia. The phase 3 APECS trial was a randomized 104-week trial to evaluate verubecestat at doses of 12 and 40 mg per day, as compared with

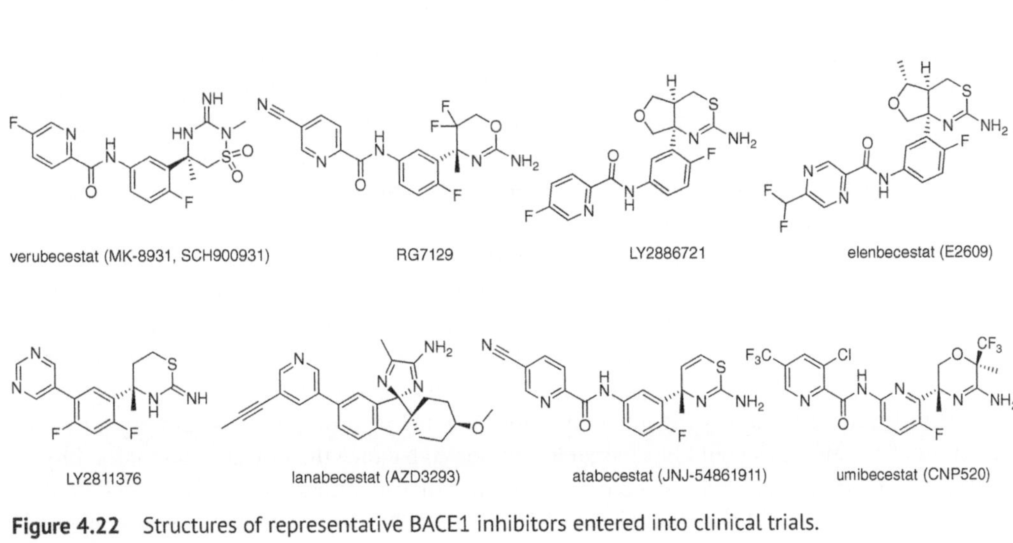

Figure 4.22 Structures of representative BACE1 inhibitors entered into clinical trials.

placebo, in patients who had memory impairment and elevated brain amyloid levels but whose condition did not meet the case definition of dementia [145]. The primary outcome was the change from baseline to week 104 in the score on the Clinical Dementia Rating Scale–Sum of Boxes (CDR-SB; higher scores indicating worse cognition and daily function with scores ranging from 0 to 18) and the secondary outcomes included other assessments of cognition and daily function. In the CDR-SB score, the estimated mean change from baseline to week 104 was 1.65 in the 12 mg group, 2.02 in the 40 mg group, and 1.58 in the placebo group which suggest a worse outcome in the higher dose group than in the placebo group. The estimated rate of progression to dementia due to AD was 24.5, 25.5, and 19.3 events per 100 patient-years in the 12 mg group, the 40 mg group, and the placebo group favoring placebo. Adverse events were less common in the placebo group than in the verubecestat groups. Thus, verubecestat did not improve clinical ratings of dementia among patients with prodromal AD, and some measures suggested that cognition and daily function were worse among patients who received verubecestat than among those who received placebo. With this clinical result, in February 2018, Merck announced the discontinuation of the phase 3 APECS study evaluating verubecestat for the treatment of patients with prodromal AD.

4.4.3 Summary of BACE1 Inhibitors

Besides MK-8931 (verubecestat), many BACE1 inhibitors entered clinical trials including BI-1181181, RG7129, LY2811376, LY2886721, elenbecestat (E2609), lanabecestat (AZD3293), umibecestat (CNP520) [146], and atabecestat (JNJ-54861911) (Figure 4.22). Unfortunately, clinical trials of the majority of these drug candidates were halted due to either safety concerns or lack of efficacy [147]. It was disappointing to find out that these treatments designed to stop amyloid peptide formation seem to cause cognitive impairment rather than prevent it, which seems to be related to a class effect when high level of BACE inhibition is achieved. More studies and a greater understanding of the amyloid hypothesis are necessary to determine if Aβ lowering is a viable strategy to treat Alzheimer's patients.

4.5 Summary

As discussed in this chapter, extensive drug discovery campaigns were carried out around lowering Aβ for treating AD using GSI, GSM, and BACE1 inhibitors within Merck/Schering Plough in addition to other pharmaceutical companies. These intensive efforts have resulted in the identification of potent drug candidates that entered clinical trials. However, there were no fruitful outcomes generated from these studies. This has highlighted the challenges facing the industry to identify efficacious drugs for treating AD and also demonstrates that discovering a successful drug is a very long and difficult process. In terms of the amyloid hypothesis, although it seems a broadly accepted general scheme of the pathophysiological events in AD [148], it does not adequately explain all observations which may warrant more vigorous scientific research [16, 21]. Based on the evidence from the clinical trials to date that the higher the dose, the worse the cognitive impairment suggests that a too aggressive inhibition of γ-secretase or BACE might interfere with cognitive function and is too much and not tolerated. A dosing strategy aiming for lower inhibition of the enzymes could be a possible solution to balance the efficacy and tolerability of the drugs. In this regard, GSMs may hold promise since they modulate the enzyme activity rather than totally inhibiting it [149]. Clinical data from future GSM clinical studies may shed light on this theory. It may also be important to treat the patients at an early stage of the disease for many years before the symptoms appear and before it is too late and too difficult to reverse the neurodegenerative process.

Based on a report in 2017, there have been over 100 drug candidates documented for AD in clinicaltrials.gov for phases 1–3, including disease-modifying therapies, symptomatic cognitive enhancers, and symptomatic agents for neuropsychiatric disorders [150]. Over 40 unsuccessful drugs were reported in registered clinical trials, including serotonergic 5-HT6 receptor antagonists, histaminergic H3 receptor antagonists, glutamatergic modulators, BACE inhibitors, GSIs, anti-Aβ agents, anti-tau agents, and microglial activation inhibitors [151]. Given the magnitude of the problem, and despite all these failures, the pharmaceutical industry and academic laboratories continue the search for effective AD drugs. In addition to Aβ lowering and the above-mentioned targets, a few companies are working on neuro inflammation for the treatment of AD with therapies targeting the genes CD33, TREM2 (triggering receptor expressed on myeloid cells), and the protein RIPK1 (receptor-interacting serine/threonine-protein kinase). New technology developments including positron emission tomography (PET) imaging which have made it possible to monitor amyloid and tau levels in the brain [152–154] and the ability to measure amyloid peptides in the blood for tracking AD patients at any stage of their disease [155], along with well-designed clinical trials with efficient execution and more effective data sharing should all help and give hope to find cures for AD despite all the frustrating, unsuccessful attempts [149]. Aβ amyloid is probably still a good target as extensive scientific research and genetics support it. Recent positive news from phase 2 clinical trial results of Eli Lilly's N3pG beta amyloid targeting antibody drug donanemab (LY3002813) has boosted optimism in this direction [156]. However, there are still many unknowns, and the search for effective AD drugs continues.

Acknowledgement

The authors would like to thank Dr. Wen-Lian Wu for proofreading the manuscript, especially the GSI discovery story. His expert suggestions and comments are invaluable.

References

1 Prashad, S. and Filbey, F.M. (2017). Cognitive motor deficits in cannabis users. *Curr. Opin. Behav. Sci.* 13: 1–7.

2 Cummings, J., Aisen, P.S., DuBois, B. et al. (2014). Drug development in Alzheimer's disease: the path to 2025. *Alzheimers Res. Ther.* 6 (4): 37. https://doi.org/10.1186/alzrt269.

3 Cacabelos, R., Teijido, O., and Carril, J.C. (2016). Can cloud-based tools accelerate Alzheimer's disease drug discovery? *Expert Opin. Drug Discovery* 11 (3): 215–223.

4 Dubois, B., Feldman, H.H., Jacova, C. et al. (2014). Advancing research diagnostic criteria for Alzheimer's disease: the IWG-2 criteria. *Lancet Neurol.* 13: 614–629.

5 GBD (2015). Neurological Disorders Collaborator Group. Global, regional, and national burden of neurological disorders during 1990–2015: a systematic analysis for the Global Burden of Disease Study 2015. *Lancet Neurol.* 16 (11): 877–897.

6 Chan, K.Y., Wang, W., Wu, J.J. et al. (2013). Global Health Epidemiology Reference Group (GHERG). Epidemiology of Alzheimer's disease and other forms of dementia in China, 1990–2010: a systematic review and analysis. *Lancet.* 381 (9882): 2016–2023.

7 Fiest, K.M., Roberts, J.I., Maxwell, C.J. et al. (2016). The prevalence and incidence of dementia due to Alzheimer's disease: a systematic review and meta-analysis. *Can. J. Neurol. Sci.* 43 (Suppl 1): S51–S82.

8 Godyn, J., Jonczyk, J., Panek, D., and Malawska, B. (2016). Therapeutic strategies for Alzheimer's disease in clinical trials. *Pharmacol. Rep.* 68: 127–138.

9 Hung, S.-Y. and Fu, W.-M. (2017). Drug candidates in clinical trials for Alzheimer's disease. *J. Biomed. Sci.* 24: 47. https://doi.org/10.1186/s12929-017-0355-7.

10 Alzheimer's Association (2019). Alzheimer's disease treatment horizons. https://www.alz.org/media/HomeOffice/Teaser%20Image/alzheimers-dementia-disease-treatment-horizons-ts.pdf (accessed 5 December 2020).

11 Kang, J., Lemaire, H.G., Unterbeck, A. et al. (1987). The precursor of Alzheimer's disease amyloid A4 protein resembles a cell-surface receptor. *Nature* 325: 733–736.

12 Selkoe, D.J. (1994). Cell biology of the amyloid beta-protein precursor and the mechanism of Alzheimer's disease. *Annu. Rev. Cell Biol.* 10: 373–403.

13 Selkoe, D.J. (1991). In the beginning. *Nature* 354: 432–433.

14 Selkoe, D.J. (2001). Alzheimer's disease: genes, proteins, and therapy. *Physiol. Rev.* 81: 741–766.

15 Schneider, L.S., Mangialasche, F., Andreasen, N. et al. (2014). Clinical trials and late-stage drug development for Alzheimer's disease: an appraisal from 1984 to 2014. *J. Intern. Med.* 275 (3): 251–283.

16 Hardy, J. and Selkoe, D.J. (2002). The amyloid hypothesis of Alzheimer's disease: progress and problems on the road to therapeutics. *Science* 297: 353–356.

17 Selkoe, D.J. (2002). Alzheimer's disease is a synaptic failure. *Science* 298: 789–791.

18 Karran, E., Mercken, M., and De Strooper, B. (2011). The amyloid cascade hypothesis for Alzheimer's disease: an appraisal for the development of therapeutics. *Nat. Rev. Drug Discovery* 10 (9): 698–712.

19 Selkoe, D.J. and Hardy, J. (2016). The amyloid hypothesis of Alzheimer's disease at 25 years. *EMBO Mol. Med.* 8 (6): 595–608.

20 Tanzi, R.E. and Bertram, L. (2005). Twenty years of the Alzheimer's disease amyloid hypothesis: a genetic perspective. *Cell* 120 (4): 545–555.

21 Ghosh, A.K. and Osswald, H.L. (2014). BACE1 (β-Secretase) inhibitors for the treatment of Alzheimer's disease. *Chem. Soc. Rev.* 43 (19): 6765–6813. https://doi.org/10.1039/c3cs60460h.

22 Citron, M. (2004). Beta-secretase inhibition for the treatment of Alzheimer's disease - promise and challenge. *Trends Pharmacol. Sci.* 25: 92–97.

23 Ghosh, A.K., Cárdenas, E.L., and Osswald, H.L. (2017). The design, development, and evaluation of BACE1 inhibitors for the treatment of Alzheimer's disease. *Springer Int. Publ. Ag.* 24: 27–86.

24 Menting, K.W. and Claassen, J.A.H.R. (2014). β-secretase inhibitor; a promising novel therapeutic drug in Alzheimer's disease. *Front. Aging Neurosci.* 6: 165.

25 Ohno, M., Cole, S.L., Yasvoina, M. et al. (2007). BACE1 gene deletion prevents neuron loss and memory deficits in 5XFAD APP/PS1 transgenic mice. *Neurobiol. Dis.* 26: 134–145.

26 Christie, R.H., Bacskai, B.J., Zipfel, W.R. et al. (2001). Growth arrest of individual senile plaques in a model of Alzheimer's disease observed by in vivo multiphoton microscopy. *J. Neurosci.* 21: 858–864.

27 Bussiere, T., Friend, P.D., Sadeghi, N. et al. (2002). Stereologic assessment of the total cortical volume occupied by amyloid deposits and its relationship with cognitive status in aging and Alzheimer's disease. *Neuroscience.* 112: 75–91.

28 Querfurth, H.W. and LaFerla, F.M. (2010). Alzheimer's disease. *N. Engl. J. Med.* 362 (4): 329–344.

29 Scheltens, P., Blennow, K., Breteler, M.M. et al. (2016). Alzheimer's disease. *Lancet.* 388 (10043): 505–517.

30 Naslund, J., Haroutunian, V., Mohs, R. et al. (2000). Correlation between elevated levels of amyloid β-peptide in the brain and cognitive decline. *J. Am. Med. Assoc.* 283: 1571–1577.

31 MacLeod, R., Hillert, E.-K., Cameron, R.T., and Baillie, G.S. (2015). The role and therapeutic targeting of α-, β- and γ-secretase in Alzheimer's disease. *Future Sci. OA.* 1 (3): FSO11.

32 Zhao, J., Liu, X., Xia, W. et al. (2020). Targeting amyloidogenic processing of APP in Alzheimer's disease. *Front. Mol. Neurosci.* 13 article 137. https://doi.org/10.3389/fnmol.2020.00137.

33 Nhan, H.S., Chiang, K., and Koo, E.H. (2015). The multifaceted nature of amyloid precursor protein and its proteolytic fragments: friends and foes. *Acta Neuropathol.* 129 (1): 1–19. https://doi.org/10.1007/s00401-014-1347-2.

34 Saito, T., Suemoto, T., Brouwers, N. et al. (2011). Potent amyloidogenicity and pathogenicity of Abeta43. *Nat. Neurosci.* 14 (8): 1023–1032.

35 Szaruga, M., Veugelen, S., Benurwar, M. et al. (2015). Qualitative changes in human gamma-secretase underlie familial Alzheimer's disease. *J. Exp. Med.* 212 (12): 2003–2013.

36 Veugelen, S., Saito, T., Saido, T.C. et al. (2016). Familial Alzheimer's disease mutations in presenilin generate amyloidogenic abeta peptide seeds. *Neuron.* 90 (2): 410–416.

37 Van Dam, D. and De Deyn, P.P. (2006). Drug discovery in dementia: the role of rodent models. *Nat. Rev. Drug Discovery* 5 (11): 956–970.

38 Israel, M.A., Yuan, S.H., Bardy, C. et al. (2012). Probing sporadic and familial Alzheimer's disease using induced pluripotent stem cells. *Nature* 482 (7384): 216–220.

39 Kondo, T., Asai, M., Tsukita, K. et al. (2013). Modeling Alzheimer's disease with iPSCs reveals stress phenotypes associated with intracellular Abeta and differential drug responsiveness. *Cell Stem Cell* 12 (4): 487–496.

40 Moore, S., Evans, L.D., Andersson, T. et al. (2015). APP metabolism regulates tau proteostasis in human cerebral cortex neurons. *Cell Rep.* 11 (5): 689–696.

41 Yagi, T., Ito, D., Okada, Y. et al. (2011). Modeling familial Alzheimer's disease with induced pluripotent stem cells. *Hum. Mol. Genet.* 20 (23): 4530–4539.

42 Bursavich, M.G., Harrison, B.A., and Blain, J.F. (2016). Gamma secretase modulators: new Alzheimer's drugs on the horizon? *J. Med. Chem.* 59 (16): 7389–7409.

43 Crump, C.J., Johnson, D.S., and Li, Y.M. (2013). Development and mechanism of gamma-secretase modulators for Alzheimer's disease. *Biochemistry* 52 (19): 3197–3216.

44 Cusulin, C., Wells, I., Badillo, S. et al. (2019). Gamma secretase modulators and BACE inhibitors reduce Aβ production without altering gene expression in Alzheimer's disease iPSC-derived neurons and mice. *Mol. Cell. Neurosci.* 100: 103392. https://doi.org/10.1016/j.mcn.2019.103392.

45 Kreft, A.F., Martone, R., and Porte, A. (2009). Recent advances in the identification of g-secretase inhibitors to clinically test the Ab oligomer hypothesis of Alzheimer's disease. *J. Med. Chem.* 52 (20): 6169–6188.

46 Li, Y.M., Lai, M.T., Xu, M. et al. (2000). Presenilin 1 is linked with gamma-secretase activity in the detergent solubilized state. *Proc. Natl. Acad. Sci. USA* 97: 6138–6143.

47 Li, Y.M., Xu, M., Lai, M.T. et al. (2000). Photoactivated gamma-secretase inhibitors directed to the active site covalently label presenilin 1. *Nature* 405: 689–694.

48 Schroeter, E.H., Ilagan, M.X., Brunkan, A.L. et al. (2003). A presenilin dimer at the core of the gamma-secretase enzyme: insights from parallel analysis of Notch 1 and APP proteolysis. *Proc. Natl. Acad. Sci. USA* 100: 13075–13080.

49 Cervantes, S., Saura, C.A., Pomares, E. et al. (2004). Functional implications of the presenilin dimerization. Reconstitution of gamma -secretase activity by assembly of a catalytic site at the dimer interface of two catalytically inactive presenilins. *J. Biol. Chem.* 279: 36519–36529.

50 Wolfe, M. and Structure, S. (2010). Mechanism and inhibition of γ-secretase and presenilin-like proteases. *Biol. Chem.* 391 (8): 839–847. https://doi.org/10.1515/BC.2010.086.

51 Panza, F., Frisardi, V., Imbimbo, B.P. et al. (2011). γ-secretase inhibitors for treating Alzheimer's disease: rationale and clinical data. *Clin. Invest.* 1 (8): 1175–1194.

52 Wu, W.L. and Zhang, L. (2009). γ-Secretase inhibitors for the treatment of Alzheimer's disease. *Drug Dev. Res.* 70: 94–100.

53 Imbimbo, B.P. (2008). Alzheimer's disease: γ-secretase inhibitors. *Drug Discov. Today Ther. Strateg.* 5 (3): 169–175. https://doi.org/10.1016/j.ddstr.2008.10.002.

54 Smith, D.W., Munoz, B., and Srinivasan, K. et al. (2000). Novel sulfonamide compounds and uses thereof. WO200050391, CAN133:207678.

55 Asberom, T., Bara, T.A., Clader, J.W. et al. (2007). Tetrahydroquinoline sulfonamides as γ-secretase inhibitors. *Bioorg. Med. Chem. Lett.* 17: 205–207.

56 Pissarnitski, D.A., Asberom, T., Bara, T.A. et al. (2007). 2,6-disubstituted N-arylsulfonyl piperidines as γ-secretase inhibitors. *Bioorg. Med. Chem. Lett.* 17: 57–62.

57 Chishti, M.A., Yang, D.S., Janus, C. et al. (2001). Early-onset amyloid deposition and cognitive deficits in transgenic mice expressing a double mutant form of amyloid precursor protein 695. *J. Biol. Chem.* 276: 21562–21570.

58 Asberom, T., Zhao, Z., Bara, T.A. et al. (2007). Discovery of γ-secretase inhibitors efficacious in a transgenic animal model of Alzheimer's disease. *Bioorg. Med. Chem. Lett.* 17: 511–516.

59 Hyde, L.A., McHugh, N.A., Chen, J. et al. (2006). Studies to investigate the in vivo therapeutic window of the γ-secretase inhibitor N2-[(2S)-2-(3,5-difluorophenyl)-2-hydroxyethanoyl]-N1-[(7S)-5-methyl-6-oxo-6,7-dihydro-5H-dibenzo[b,d] azepin-7-yl]-Lalaninamide (LY411,575) in the CRND8 mouse. *J. Pharmacol. Exp. Ther.* 319 (3): 1133–1143.

60 Hyde, L.A., Zhang, Q., Del Vecchio, R.A. et al. (2013). in vivo characterization of a novel γ-secretase inhibitor SCH 697466 in rodents and investigation of strategies for managing notch-related side effects. *Int. J. Alzheimers Dis.*: 823528. http://dx.doi.org/10.1155/2013/823528.

61 Teall, M., Oakley, P., Harrison, T. et al. (2005). Aryl sulfones: a new class of γ-secretase inhibitors. *Bioorg. Med. Chem. Lett.* 15: 2685–2688.

62 Churcher, I., Beher, D., Best, J.D. et al. (2006). 4-Substituted cyclohexyl sulfones as potent, orally active gamma-secretase inhibitors. *Bioorg. Med. Chem. Lett.* 16: 280–284.

63 Xu, R., Cole, D., Asberom, T. et al. (2011). SAR of tricyclic sulfones as γ-secretase inhibitors. *Sci. China Chem.* 54 (11): 1688–1701.

64 Wong, G.T., Manfra, D., Poulet, F.M. et al. (2004). Chronic treatment with the gamma-secretase inhibitor LY-411,575 inhibits beta-amyloid peptide production and alters lymphopoiesis and intestinal cell differentiation. *J. Biol. Chem.* 279: 12876–12882.

65 Xu, R., Cole, D., Asberom, T. et al. (2010). Design and synthesis of tricyclic sulfones as γ-secretase inhibitors with greatly reduced Notch toxicity. *Bioorg. Med. Chem. Lett.* 20: 2591–2596.

66 Sasikumar, T.K., Qiang, L., Burnett, D.A. et al. (2010). Tricyclic sulfones as orally active c-secretase inhibitors: synthesis and structure–activity relationship studies. *Bioorg. Med. Chem. Lett.* 20: 3632–3635.

67 Sasikumar, T.K., Burnett, D.A., Asberom, A. et al. (2010). Tetracyclic sulfones as potent c-secretase inhibitors: Synthesis and structure–activity relationship studies. *Bioorg. Med. Chem. Lett.* 20: 3645–3648.

68 Wu, W.-L., Domalski, M., Burnett, D.A. et al. (2012). Discovery of SCH 900229, a potent presenilin 1 selective γ-secretase inhibitor for the treatment of Alzheimer's disease. *ACS Med. Chem. Lett.* 3: 892–896.

69 Wu, W.-L., Asberom, T., Bara, T. et al. (2013). Structure activity relationship studies of tricyclic bispyran sulfone γ-secretase inhibitors. *Bioorg. Med. Chem. Lett.* 23: 844–849.

70 Wu, W.-L., Burnett, D.A., Clader, J. et al. (2016). Design and synthesis of water soluble β-aminosulfone analogues of SCH 900229 as γ-secretase inhibitors. *Bioorg. Med. Chem. Lett.* 23: 5836–5841.

71 Lee, J., Song, L., Terracina, G. et al. (2011). Identification of presenilin 1-selective γ-secretase inhibitors with reconstituted γ-secretase complexes. *Biochemi.* 50: 4973–4980.

72 Kieburtz, K., He, F., Sun, X. et al. (2013). A Phase 3 trial of semagacestat for treatment of Alzheimer's disease. *N. Engl. J. Med.* 369: 341–350. https://doi.org/10.1056/NEJMoa1210951.

73 Tagami, S., Yanagida, K., Kodama, T.S. et al. (2017). Semagacestat is a pseudo-inhibitor of γ-secretase. *Cell Rep.* 21: 259–273. https://doi.org/10.1016/j.celrep.2017.09.032.

74 Tate, B., McKee, T.D., Loureiro, R.M.B. et al. (2012). Modulation of gamma-secretase for the treatment of Alzheimer's disease. *Int. J. Alzheimers Dis.*: 210756. https://doi.org/10.1155/2012/210756.

75 Cai, H., Wang, Y., McCarthy, D. et al. (2001). BACE1 is the major β-secretase for generation of Aβ peptides by neurons. *Nat. Neurosci.* 4: 233–234.

76 Luo, Y., Bolon, B., Kahn, S. et al. (2001). Mice deficient in BACE1, the Alzheimer's β-secretase, have normal phenotype and abolished β-amyloid generation. *Nat. Neurosci.* 4: 231–232.

77 Weggen, S., Eriksen, J.L., Das, P. et al. (2001). A subset of NSAIDs lower amyloidogenic Aβ42 independently of cyclooxygenase activity. *Nature* 414 (6860): 212–216. https://doi.org/10.1038/35102591.

78 Weggen, S., Eriksen, J.L., Sagi, S.A. et al. (2003). Aβ42-lowering nonsteroidal anti-inflammatory drugs preserve intramembrane cleavage of the amyloid precursor protein (APP) and ErbB-4 receptor and signaling through the APP intracellular domain. *J. Biol. Chem.* 278 (33): 30748–30754. https://doi.org/10.1074/jbc.M304824200.

79 Weggen, S., Eriksen, J.L., Sagi, S.A. et al. (2003). Evidence that nonsteroidal anti-inflammatory drugs decrease amyloid β 42 production by direct modulation of γ-secretase activity. *J. Biol. Chem.* 278 (34): 31831–31837. https://doi.org/10.1074/jbc.M303592200.

80 Lesné, S., Koh, M.T., Kotilinek, L. et al. (2006). A specific amyloid-β protein assembly in the brain impairs memory. *Nature* 440: 352–357.

81 Shankar, G.M., Li, S., Mehta, T.H. et al. (2008). Amyloid-β protein dimers isolated directly from Alzheimer's brains impair synaptic plasticity and memory. *Nat. Med.* 14: 837–842.

82 Moore, B.D., Martin, J., de Mena, L. et al. (2018). Short Aβ peptides attenuate Aβ42 toxicity in vivo. *J. Exp. Med.* 215 (1): 283–301. https://doi.org/10.1084/jem.20170600.

83 Okochi, M., Tagami, S., Yanagida, K. et al. (2013). γ-Secretase modulators and presenilin 1 mutants act differently on presenilin/γ-secretase function to cleave Aβ42 and Aβ43. *Cell Rep.* 3 (1): 42–51. https://doi.org/10.1016/j.celrep.2012.11.028.

84 Szaruga, M., Munteanu, B., Lismont, S. et al. (2017). Alzheimer's-causing mutations shift Aβ length by destabilizing γ-secretase-Aβn Interactions. *Cell* 170 (3): 443–456. e414. https://doi.org/10.1016/j.cell.2017.07.004.

85 Lleo, A., Berezovska, O., Herl, L. et al. (2004). Nonsteroidal anti-inflammatory drugs lower Aβ42 and change presenilin 1 conformation. *Nat. Med.* 10 (10): 1065–1066. https://doi.org/10.1038/nm1112 161.

86 Takeo, K., Tanimura, S., Shinoda, T. et al. (2014). Allosteric regulation of γ-secretase activity by a phenylimidazole-type γ-secretase modulator. *Proc. Natl. Acad. Sci. USA* 111 (29): 10544–10549. https://doi.org/10.1073/pnas.1402171111.

87 Steiner, H., Fukumori, A., Tagami, S., and Okochi, M. (2018). Making the final cut: pathogenic amyloid-β peptide generation by γ-secretase. *Cell Stress* 2 (11): 292–310.

88 Wolfe, M.S. (2012). Processive proteolysis by gamma-secretase and the mechanism of Alzheimer's disease. *Biol. Chem.* 393: 899–905.

89 Zhao, G., Cui, M.Z., Mao, G. et al. (2005). Gamma-cleavage is dependent on zeta-cleavage during the proteolytic processing of amyloid precursor protein within its transmembrane domain. *J. Biol. Chem.* 280: 37689–37697.

90 Funamoto, S., Morishima-Kawashima, M., Tanimura, Y. et al. (2004). Truncated carboxyl-terminal fragments of beta-amyloid precursor protein are processed to amyloid beta-proteins 40 and 42. *Biochem.* 43: 13532–13540.

91 Yagishita, S., Morishima-Kawashima, M., Ishiura, S., and Ihara, Y. (2008). Abeta46 is processed to abeta40 and abeta43, but not to abeta42, in the low density membrane domains. *J. Biol. Chem.* 283: 733–738.

92 Takami, M., Nagashima, Y., Sano, Y. et al. (2009). Gamma-secretase: Successive tripeptide and tetrapeptide release from the transmembrane domain of beta-carboxyl terminal fragment. *J. Neurosci.* 29: 13042–13052.

93 Qi-Takahara, Y., Morishima-Kawashima, M., Tanimura, Y. et al. (2005). Longer forms of amyloid beta protein: implications for the mechanism of intramembrane cleavage by gamma-secretase. *J. Neurosci.* 25: 436–445.

94 Golde, T.E., Koo, E.H., Felsenstein, K.M. et al. (2013). γ-Secretase inhibitors and modulators. *Biochim. Biophys. Acta* 1828: 2898–2907.

95 Imbimbo, B.P. (2009). Why did tarenflurbil fail in Alzheimer's disease? *J. Alzheimers Dis.* 17: 757–760.

96 Eriksen, J.L., Sagi, S.A., Smith, T.E. et al. (2003). NSAIDs and enantiomers of flurbiprofen target γ-secretase and lower Aβ42 in vivo. *J. Clin. Invest.* 112: 440–449.

97 Kounnas, M.Z., Danks, A.M., Cheng, S. et al. (2010). Modulation of γ-secretase reduces β-amyloid deposition in a transgenic mouse model of Alzheimer's disease. *Neuron* 67: 769–780.

98 Kimura, T., Kawano, K., Doi, E. et al. (2006). Cinnamide compound. US 20060004013.

99 Kimura, T., Kawano, K., Doi, E. et al. (2007). Morpholine type cinnamide compound. US 20070117798.

100 Huang, X., Aslanian, R., Zhou, W. et al. (2010). The discovery of pyridone and pyridazone heterocycles as γ-secretase modulators. *ACS Med. Chem. Lett.* 1: 184–187.

101 Qin, J., Dhondi, P., Huang, X. et al. (2011). Discovery of fused 5,6-bicyclic heterocycles as γ-secretase modulators. *Bioorg. Med. Chem. Lett.* 21: 664–669.

102 Zhao, Z., Pissarnitski, D.A., Huang, X. et al. (2017). Discovery of a tetrahydrobenzisoxazole series of γ-secretase modulators. *ACS Med. Chem. Lett.* 8 (10): 1002–1006.

103 Qin, J., Zhou, W., Huang, X. et al. (2011). Discovery of a potent pyrazolopyridine series of γ-secretase modulators. *ACS Med. Chem. Lett.* 2: 471–476.

104 Sun, Z.-Y., Asberom, T., Bara, T. et al. (2012). Cyclic hydroxyamidines as amide isosteres: discovery of oxadiazolines and oxadiazines as potent and highly efficacious γ-secretase modulators in vivo. *J. Med. Chem.* 55: 489–502.

105 Huang, X., Zhou, W., Liu, X. et al. (2012). Synthesis and SAR studies of fused oxadiazines as γ-secretase modulators for treatment of Alzheimer's disease. *ACS Med. Chem. Lett.* 3: 931–935.

106 Mandal, M.B., Buevich, A.V., Caldwell, J.P. et al. (2020). Generation of leads for γ-secretase modulation. *J. Med. Chem.* 63 (15): 8216–8230. https://doi.org/10.1021/acs.jmedchem.0c00446.

107 Li, H., Qin, J., Dhondi, P. et al. (2013). The discovery of fused oxadiazepines as gamma secretase modulators for treatment of Alzheimer's disease. *Bioorg. Med. Chem. Lett.* 23: 466–471.

108 Hall, A. and Velter, A.I. (2017). Targeting γ-secretase for the treatment of Alzheimer's disease. In: *Comprehensive Medicinal Chemistry III* (eds. S. Chackalamannil, D. Rotella and S.E. Ward), 280–325. Elsevier Ltd.

109 Ahn, J.E., Carrieri, C., Dela Cruz, F. et al. (2020). Pharmacokinetic and pharmacodynamic effects of a γ-secretase modulator, PF-06648671, on CSF amyloid-β peptides in randomized phase I studies. *Clin. Pharmacol. Ther.* 107 (1): 211–220.

110 Ratni, H., Alker, A., Bartels, B. et al. (2020). Discovery of RO7185876, a highly potent γ-secretase modulator (GSM) as a potential treatment for Alzheimer's disease. *ACS Med. Chem. Lett.* 11 (6): 1257–1268.

111 Wagner, S.L., Rynearson, K.D., Duddy, S.K. et al. (2017). Pharmacological and toxicological properties of the potent oral g-Secretase modulator BPN-15606s. *J. Pharmacol. Exp. Ther.* 362: 31–44.

112 Soares, H.D., Gasior, M., Toyn, J.H. et al. (2016). The γ-secretase modulator, BMS-932481, modulates Aβ peptides in the plasma and cerebrospinal fluid of healthy volunteers. *J. Pharmacol. Exp. Ther.* 358: 138–150.

113 Szecsi, P.B. (1992). The aspartic proteases. *Scand. J. Clin. Lab. Invest. Suppl.* 210: 5–22.

114 Tang, J. and Lin, X. (1994). Engineering aspartic proteases to probe structure and function relationships. *Curr. Opin. Biotechnol.* 5: 422–427.

115 Farzan, M., Schnitzler, C.E., Vasilieva, N. et al. (2000). BACE2, a beta -secretase homolog, cleaves at the beta site and within the amyloid-beta region of the amyloid-beta precursor protein. *Proc. Natl. Acad. Sci. USA* 97: 9712–9717.

116 Southan, C. and Hancock, J.M. (2013). A tale of two drug targets: the evolutionary history of BACE1 and BACE2. *Front. Genet.* 4: 293.

117 Sun, X., He, G., and Song, W. (2006). BACE2, as a novel APP theta-secretase, is not responsible for the pathogenesis of Alzheimer's disease in Down syndrome. *FASEB J.* 20: 1369–1376.

118 Yan, R., Munzner, J.B., Shuck, M.E., and Bienkowski, M.J. (2001). BACE2 functions as an alternative alpha-secretase in cells. *J. Biol. Chem.* 276: 34019–34027.

119 Dominguez, D., Tournoy, J., Hartmann, D. et al. (2005). Phenotypical and biochemical analysis of BACE1 and BACE2 deficient mice. *J. Biol. Chem.* 280: 30797–30806.

120 Roberds, S.L., Anderson, J., Basi, G. et al. (2001). BACE knockout mice are healthy despite lacking the primary beta-secretase activity in brain: implications for Alzheimer's disease therapeutics. *Hum. Mol .Genet.* 10: 1317–1324.

121 Yan, R. (2016). Stepping closer to treating Alzheimer's disease patients with BACE1 inhibitor drugs. *Transl. Neurodegener.* 5: 13. https://doi.org/10.1186/s40035-016-0061-5.

122 Hong, L., Koelsch, G., Lin, X. et al. (2000). Structure of the protease domain of memapsin 2 (beta-secretase) complexed with inhibitor. *Science* 290: 150–153.

123 Ghosh, A.K., Shin, D., Koelsch, G. et al. (2000). Design of potent inhibitors for human brain memapsin 2 (β-secretase). *J. Am. Chem. Soc.* 122 (14): 3522–3523.

124 Schechter, I. and Berger, A. (1968). On the active site of proteases. 3. Mapping the active site of papain; specific peptide inhibitors of papain. *Biochem. Biophys. Res. Commun.* 32: 898–902.

125 Hu, H., Chen, Z., Xu, X., and Xu, Y. (2019). Structure-based survey of the binding modes of BACE1 inhibitors. *ACS Chem. Neurosci.* 10 (2): 880–889.

126 Ghosh, A.K., Bilcer, G., Harwood, C. et al. (2001). Structure-based design: potent inhibitors of human brain memapsin 2 (β-secretase). *J. Med. Chem.* 44: 2865–2868.

127 Stachel, S. (2009). Progress toward the development of a viable BACE-1 inhibitor. *Drug Dev. Res.* 70: 101–110.

128 Ghosh, A.K., Brindisi, M., and Tang, J. (2012). Developing β-secretase inhibitors for treatment of Alzheimer's disease. *J. Neurochem.* 120 (Suppl. 1): 71–83.

129 Prati, F., Bottegoni, G., Bolognesi, M.L., and Cavalli, A. (2018). BACE-1 inhibitors: from recent single-target molecules to multitarget compounds for Alzheimer's disease. *J. Med. Chem.* 61 (3): 619–637.

130 Moussa-Pacha, N.M., Abdin, S.M., Omar, H.A. et al. (2019). BACE1 inhibitors: current status and future directions in treating Alzheimer's disease. *Med. Res. Rev.*: 1–46.

131 Wyss, D.F., Wang, Y.-S., Eaton, H.L. et al. (2012). Combining NMR and X-ray crystallography in fragment-based drug discovery: discovery of highly potent and selective BACE-1 inhibitors. *Top. Curr. Chem.* 317: 83–114.

132 Wang, Y.-S., Strickland, C., Voigt, J.H. et al. (2010). Application of fragment-based NMR screening, X-ray crystallography, structure-based design, and focused chemical library design to identify novel μM leads for the development of nM BACE-1 (β-site APP Cleaving Enzyme 1) inhibitors. *J. Med. Chem.* 53: 942–950.

133 Rajamani, R. and Reynolds, C.H. (2004). Modeling the protonation states of the catalytic aspartates in beta-secretase. *J. Med. Chem.* 47: 5159–5166.

134 Zhu, Z., Sun, Z.-Y., Ye, Y. et al. (2010). Discovery of cyclic acylguanidines as highly potent and selective β-site amyloid cleaving enzyme (BACE) inhibitors: Part I. Inhibitor design and validation. *J. Med. Chem.* 53: 951–965.

135 Cumming, J.N., Smith, E.M., Wang, L. et al. (2012). Structure based design of iminohydantoin BACE1 inhibitors: Identification of an orally available, centrally active BACE1 inhibitor. *Bioorg. Med. Chem. Lett.* 22: 2444–2449.

136 Stamford, A.W., Scott, J.D., Li, S.W. et al. (2012). Discovery of an orally available, brain penetrant BACE1 inhibitor that affords robust CNS Aβ reduction. *ACS Med. Chem. Lett.* 3 (11): 897–902.

137 Mandal, M., Zhu, Z., Cumming, J.N. et al. (2012). Design and validation of bicyclic iminopyrimidinones as beta amyloid cleaving enzyme-1 (BACE1) inhibitors: conformational constraint to favor a bioactive conformation. *J. Med. Chem.* 55: 9331–9345.

138 Mandal, M., Wu, Y., Misiaszek, J. et al. (2016). Structure-based design of an iminoheterocyclic β-site amyloid precursor protein cleaving enzyme (BACE) inhibitor that lowers central Aβ in nonhuman primates. *J. Med. Chem.* 59: 3231–3248.

139 Koike, M., Nakanishi, H., Saftig, P. et al. (2000). Cathepsin D deficiency induces lysosomal storage with ceroid lipofuscin in mouse CNS neurons. *J. Neurosci.* 20: 6898–6906.

140 Siintola, E., Partanen, S., Strömme, P. et al. (2006). Cathepsin D deficiency underlies congenital human neuronal ceroid-lipofuscinosis. *Brain* 129: 1438–1445.

141 Scott, J.D., Li, S.W., Brunskill, A.P.J. et al. (2016). Discovery of the 3-imino-1,2,4-thiadiazinane 1,1-dioxide derivative verubecestat (MK-8931)–A β-site amyloid precursor protein cleaving enzyme 1 inhibitor for the treatment of Alzheimer's disease. *J. Med. Chem.* 59: 10435–10450.

142 Thaisrivongs, D.A., Morris, W.J., Tan, L. et al. (2018). A next generation synthesis of BACE1 inhibitor verubecestat (MK-8931). *Org. Lett.* 20 (6): 1568–1571.

143 Kennedy, M.E., Stamford, A.W., Chen, X. et al. (2016). The BACE1 inhibitor verubecestat (MK-8931) reduces CNS β-amyloid in animal models and in Alzheimer's disease patients. *Sci. Transl. Med.* 8: 363ra150.

144 Egan, M.F., Kost, J., Tariot, P.N. et al. (2018). Randomized trial of verubecestat for mild-to-moderate Alzheimer's disease. *N. Engl. J. Med.* 378: 1691–1703.

145 Egan, M.F., Kost, J., Voss, T. et al. (2019). Randomized trial of verubecestat for prodromal Alzheimer's disease. *N. Engl. J. Med.* 380: 1408–1420.

146 Neumann, U., Ufer, M., Jacobson, L.H. et al. (2018). The BACE-1 inhibitor CNP520 for prevention trials in Alzheimer's disease. *EMBO Mol. Med.* 10: e9316.

147 Graham, W.V., Bonito-Oliva, A., and Sakmar, T.P. (2017). Update on Alzheimer's disease therapy and prevention strategies. *Annu. Rev. Med.* 68: 413–430.

148 Selkoe, D.J. (2000). The origins of Alzheimer disease: a is for amyloid. *JAMA* 283: 1615–1617.

149 Houlton, S. (2019). Sixth in a line of Alzheimer's drugs fails in trials. https://www.chemistryworld.com/news/sixth-in-a-line-of-alzheimers-drugs-fails-in-trials/4010459.article (accessed 5 December 2020).

150 Cummings, J., Lee, G., Mortsdorf, T. et al. (2017). Alzheimer's disease drug development pipeline: 2017. *Alzheimers Dement. (NY)* 3 (3): 367–384.

151 Cacabelos, R. (2018). Have there been improvements in Alzheimer's disease drug discovery over the past 5 years? *Expert Opin. Drug Discovery* 13 (6): 523–538. https://doi.org/10.1080/17460441.2018.1457645.

152 Lee, J.C., Kim, S.J., Hong, S., and Kim, Y.S. (2019). Diagnosis of Alzheimer's disease utilizing amyloid and tau as fluid biomarkers. *Exp. Mol. Med.* 51 (5): 53.

153 Kim, Y., Rosenberg, P., and Oh, E. (2018). A review of diagnostic impact of amyloid positron emission tomography imaging in clinical practice. *Dement. Geriatr. Cogn. Disord.* 46: 154–167.

154 Uzuegbunam, B.C., Librizzi, D., and Yousefi, B.H. (2020). PET radiopharmaceuticals for Alzheimer's disease and Parkinson's disease diagnosis, the current and future landscape. *Molecules* 25 (4): 977.

155 Ridler, C. (2018). Blood amyloid-β successfully signals AD. *Nat. Rev. Neurol.* 14: 195.

156 Eli Lilly and Company (2021). Lilly's donanemab slows clinical decline of Alzheimer's disease in positive phase 2 trial. https://investor.lilly.com/node/44226/pdf (accessed 12 January 2021).

5

Discovery of Novel Antiviral Agents Enabled by Structural Biology, Compact Modules and Phenotypic Screening

Wei Zhu, Song Yang, Hongying Yun, and Hong C. Shen

Department of Medicinal Chemistry, Therapeutic Modalities, Roche Innovation Center Shanghai, 720 Cai Lun Road, 201203, Shanghai, People's Republic of China

5.1 Introduction

Medicinal chemistry has played a key role in modern antiviral drug discovery. Numerous antiviral agents have been discovered and developed to ultimately curb global pandemics and save the lives of millions. At the Roche Innovation Center in Shanghai, we have utilized cutting-edge technologies in the past decade to generate drug design insights and enabled the discovery of several compounds that are in various stages of clinical development. In this chapter, we wish to provide an account of the biological context, and lead identification and optimization progress that showcase the impact of structural biology, phenotypic screening, and compact modules. We will highlight the medicinal chemistry solutions identified to address various key challenges during the lead optimization process in three programs where active molecules against hepatitis B virus (HBV) and respiratory syncytial virus (RSV) were discovered.

5.2 Discovery and Early Development of Novel Core Protein Assembly Modulators for the Treatment of Chronic Hepatitis B Virus Infection

5.2.1 Introduction

HBV infection remains a major public health problem despite the global vaccination program. It is estimated that 240 million people worldwide are chronic HBV carriers [1]. If left untreated, approximately one-third of these individuals will die from the end stage liver diseases caused by HBV infection, such as cirrhosis, hepatocellular carcinoma, and liver failure [2]. Although the current standard of care, such as nucleos(t)ide analogues (lamivudine, adefovir, tenofovir, telbivudine, and entecavir) (Figure 5.1) and interferon alpha (IFN-α, including non-pegylated and pegylated), can reduce HBV DNA and normalize liver enzymes, neither treatment leads to a functional cure as defined by the complete loss of HBV surface antigen (HBsAg) (with or without seroconversion) and HBV DNA after a finite treatment. Furthermore, these treatments are also

Contemporary Accounts in Drug Discovery and Development, First Edition. Edited by Xianhai Huang, Robert G. Aslanian, and Wayne H. Tang.

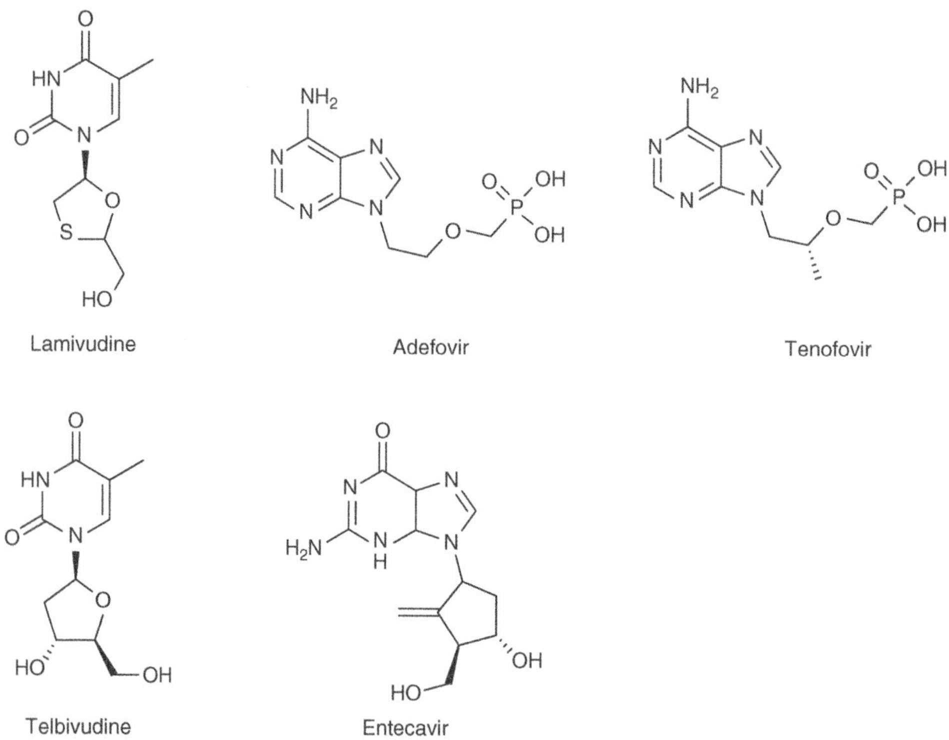

Figure 5.1 Chemical structures of lamivudine, adefovir, tenofovir, telbivudine, and entecavir.

associated with either side effects or risk of drug resistance [3]. As such, there remains a highly unmet medical need to identify novel anti-HBV agents to improve disease cure rates as a standalone or combination treatment.

As shown in Figure 5.2, encapsidation of HBV core proteins into a nucleocapsid is an essential step in the HBV life cycle [4, 5]. Interruption of capsid assembly will block HBV replication and thus represents a promising approach for HBV treatment. There are two types of core protein assembly modulators (CpAMs) [6]. The class I CpAMs, e.g. Bay 41-4109 and GLS-4, can induce the aggregation of core proteins into aberrant capsids, subject to subsequent degradation by host protease [5]. In contrast, the class II CpAMs represented by **AT-130** [7] and **DVR-23** [8] can only inhibit HBV replication by forming empty capsids without HBV genomes (pgRNA and polymerase) (Figure 5.3). The additional effect on core protein degradation by class I CpAMs is postulated to restore human immune response against HBV [9, 10].

5.2.2 Lead Generation and Optimization

Currently, there is only one chemical class, namely the heteroaryldihydropyrimidines (HAPs), that are known to be class I CpAM. Bay 41-4109 (compound **1**) is the first HAP compound reported by Bayer in 2003 [5]. Bay 41-4109 demonstrated moderate activity of inhibiting HBV replication in HepG2.2.15 cells with EC_{50} values in the range of 0.05–0.08 μM (internal and literature data) and a relatively high cytotoxicity (CC_{50} is 13 μM in HepDE19 cells). Despite the improvement in both EC_{50} and CC_{50}, the second-generation HAPs represented by GLS-4 [11] and an internal compound **2** suffered from the issues of poor metabolic stability and significant hERG inhibition. We then

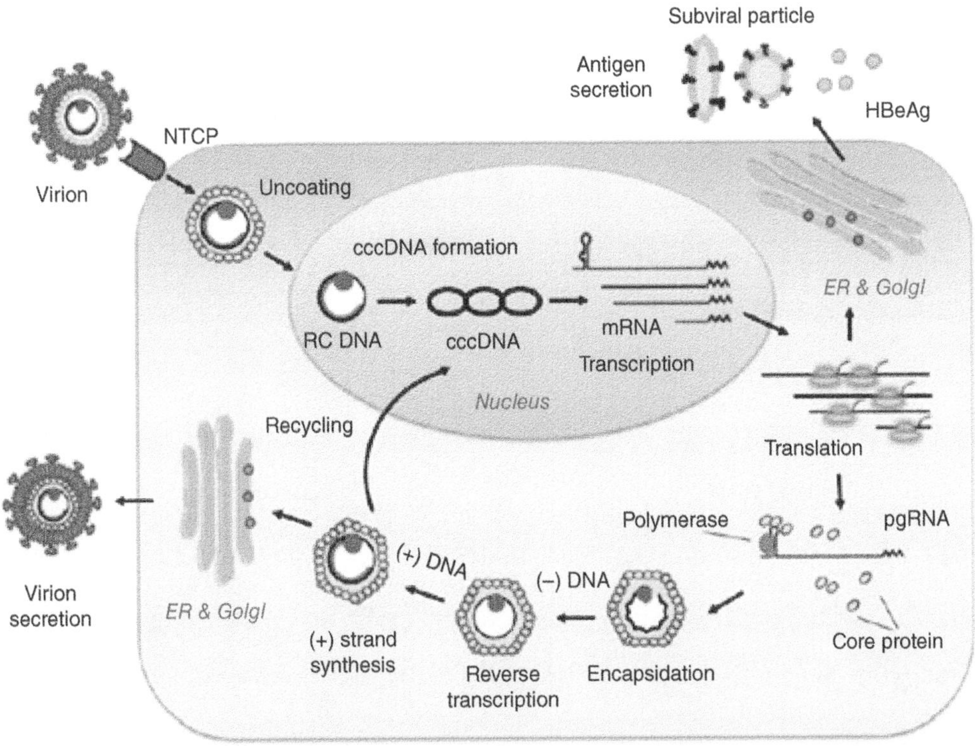

Figure 5.2 Schematic representation of the HBV life cycle. *Source:* Reproduced from [4] (figure 5.17.1); with permission from the publisher (Elsevier, https://www.sciencedirect.com).

aimed to discover a new HAP compound with improved anti-HBV activity, pharmacokinetic and safety profiles.

The second-generation HAP, e.g. compound **2**, has relatively high lipophilicity (clogP is 3.2), which is presumed to account for both the low metabolic stability and hERG issues like that seen in the first-generation HAPs. The basicity of the morpholine nitrogen (pKa = 6.3) can also contribute to potent hERG inhibition (IC_{20} of compound **2** is 0.48 μM). In the design of the third-generation HAPs, we wanted to improve the metabolic stability and reduce hERG inhibition through the introduction of a polar substituent. However, it was unclear whether the desired anti-HBV activity could be maintained with the incorporation of such a polar substituent, as the binding pocket located at the interface of the core protein dimer is largely hydrophobic judging from a low resolution crystal structure (2G34, 5 Å) at that time [12].

Indeed, introducing various polar substituents on the C2 thiazolyl, C4 phenyl or C5 positions of the second-generation HAPs was unsuccessful, as none of these analogues demonstrated acceptable anti-HBV activity. However, the C6 position tolerated polar substituents off the morpholine ring, such as ester, hydroxyl, sulfonamide, carboxamide, and carboxyl. All these analogues showed higher anti-HBV activity than the prototype compound **2**, but most of them (examples **4–7**), except the carboxyl analogue **3**, suffered from the same issue of low metabolic stability due to still higher lipophilicity. In contrast, the carboxyl analogue **3** showed significantly reduced lipophilicity with a clogP of 0.3. This low lipophilicity brought the favorable properties of high metabolic stability in human and mouse liver microsomes, high aqueous solubility (>666 μg/ml) and low hERG inhibition (IC_{20} >10 μM). The hERG improvement was likely due to a dual effect

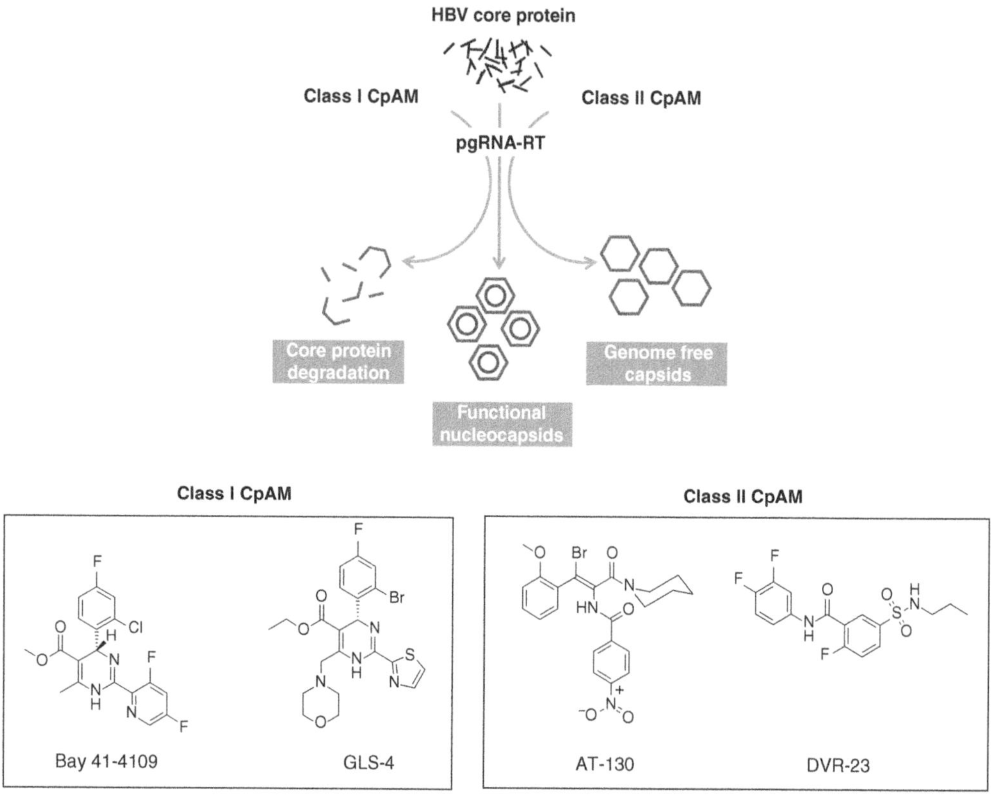

Figure 5.3 Two types of CpAMs (hexagon = capsid; circle = viral replication machinery including pgRNA and polymerase).

of both reduced lipophilicity and reduced basicity of compound **3** (later known as RG7689, pKa = 5.8). Together with a high CC_{50} (>100 μM), compound **3** demonstrated a higher selectivity index (SI, defined by CC_{50}/EC_{50}) than that of analogue **2** (Table 5.1).

Further structure–activity relationship (SAR) exploration revealed that the introduction of the carboxyl group was regio-, stereo- and steric-sensitive with compound **3** exhibiting the best potency, compared with examples **8–13**. The *gem*-difluoroproline was identified as a good bioisostere of the morpholine acid with the corresponding analogue **14** showing comparable anti-HBV activity with that of compound **3**. However, compound **14** had a less favorable CC_{50} and metabolic stability. We also did extensive SAR explorations at the C2, C4, and C5 positions based on compound **3** (data not shown). Although we were able to increase potency to sub-nM range by fine-tuning the substituents on the C4 phenyl group, compound **3** still appeared to be the best candidate with the most balanced profiles of potency, cytotoxicity and *in vitro* ADMET (absorption, distribution, metabolism, excretion, and toxicity) profiles.

In order to understand how these new HAP compounds bind to the core proteins and why the introduction of a carboxyl group improved potency, we solved the co-crystal structure of a hexamer of Cp149 Y132A in complex with compound **14** (PDB code: 5WRE) [6]. As shown in Figure 5.4, the carboxyl group forms strong bidentate hydrogen-bonding interactions with the backbone NH and the sidechain hydroxyl of S141 [13]. In addition, two additional hydrogen-bonding interactions are formed with a conserved water as a bridge to connect the C6 carboxyl and the N atom of the C2 thiazolyl. Clearly, these cooperative hydrogen bonds increase the binding

Table 5.1 Discovery of C6 morpholine carboxylic acid 4-H HAP analogues.

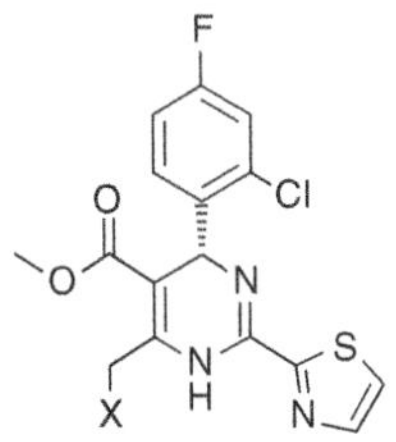

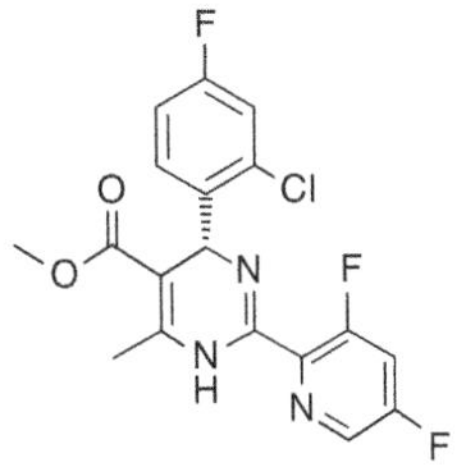

1

ID	X	EC_{50} (μM)[a]	CC_{50} (μM)[b]	SI[c]	HLM/MLM[d] (ml/min/kg)	LYSA[e] (μg/ml)	clogP[f]
1	H	0.05-0.08	13	260	6/30	38	4.5
2	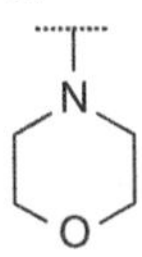	0.016	>100	>6250	21/88	195	3.2
3	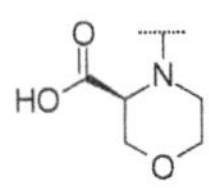	0.003	>100	>33333	5/10	>666	0.3
4	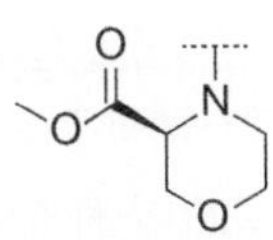	0.004	—	—	22/88	<1.0	2.6
5	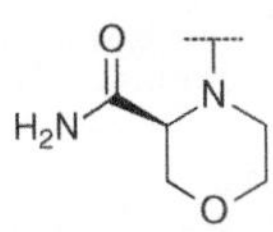	0.003	>100	>33333	18/82	205	1.5
6	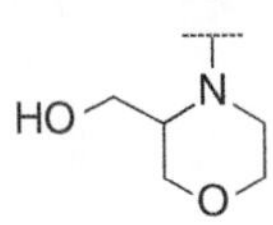	0.005	74	14800	21/88	475	2.1
7	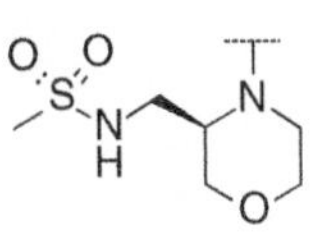	0.004	88	22000	22/89	48	2.1
8	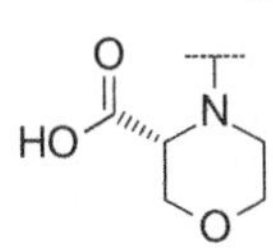	0.079	>100	>1266	—	—	—
9	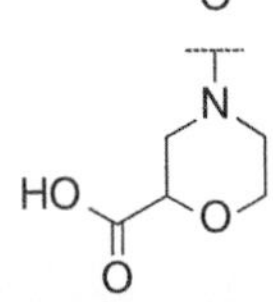	0.333	>100	> 300	2/3	>592	—

(Continued)

Table 5.1 (Continued)

ID	X	EC_{50} (μM)[a]	CC_{50} (μM)[b]	SI[c]	HLM/MLM[d] (ml/min/kg)	LYSA[e] (μg/ml)	clogP[f]
10	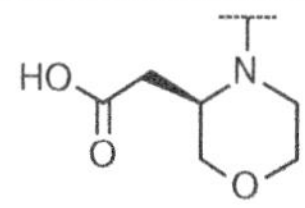	0.60	>100	>167	0.1/25	>638	—
11	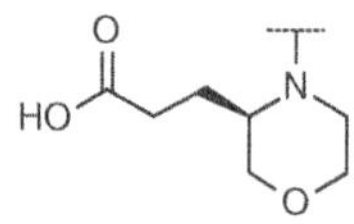	0.15	>100	>667	0/12	>697	—
12	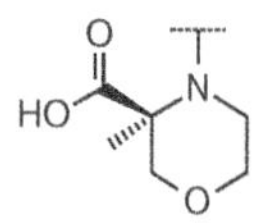	0.02	> 100	> 5000	1.4/50	600	0.8
13	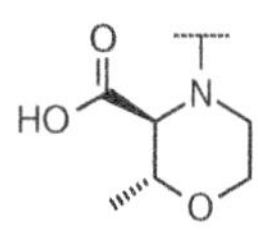	0.007	>100	>14286	5.4/32	>667	0.7
14	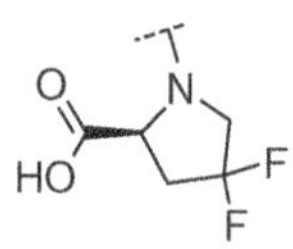	0.003	65	28333	4.4/43	>644	1.4

[a] EC_{50} is the mean value for reduction of HBV DNA by 50% in HepG.2.2.15 cells. Experiments were run in duplicate, with variation <15%.
[b] CC_{50} is the mean value for cytotoxicity to HepDE19 cells. Experiments were run in duplicate with variation <15%.
[c] Selectivity index (SI) is the ratio of CC_{50}/EC_{50}.
[d] Predicted hepatic clearance obtained with human liver microsome (HLM) and mouse liver microsome (MLM).
[e] LYSA, lyophilization solubility assay.
[f] clogP, predicted partition coefficient between the octanol phase and the water phase (KowlogP).

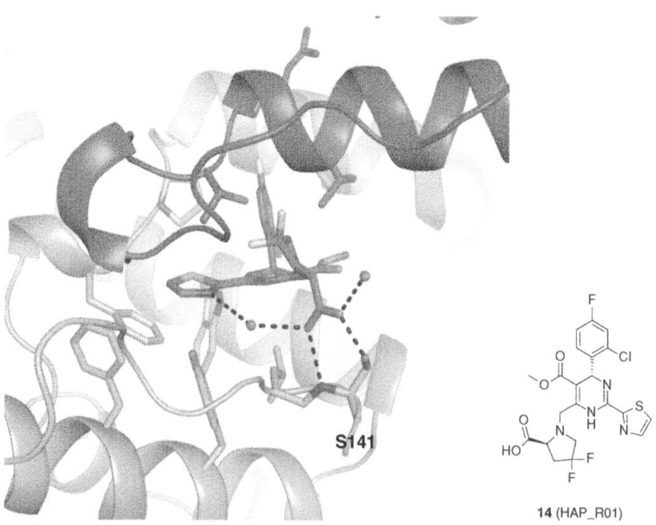

Figure 5.4 Crystal structure of Cp149 Y132A in complex with compound **14** (highlighted in stick). Black dashed lines represent the key hydrogen bonds. PDB code 5WRE. *Source:* Reproduced from [13] (Figure 5.2); used with the permission of the publisher (ACS publications).

affinity and thus the anti-HBV activity of the carboxyl analogues compared with the non-carboxyl analogues. We also concluded from the co-crystal structure that the C2 thiazolyl, C4 phenyl and C5 methyl ester bind to a largely hydrophobic pocket, which explains why the introduction of a polar substituent at these positions led to a drop in potency. Furthermore, the structural biology information suggested that at the solvent-exposed region of the molecular scaffold there is room to potentially optimize properties via significant structural modifications while minimizing the negative effect on potency, which will be described in the following.

5.2.3 Profile of Compound 3

Compound **3** (named **HAP_R10**) was further characterized in physicochemical and ADME assays in order to support preclinical PK and PD studies. It showed a logD of 0.06 and two pKa values of 1.8 (carboxylic acid) and 5.8 (nitrogen of morpholine), respectively. Interestingly, despite the zwitterionic structure, compound **3** still exhibited acceptable permeability (1.4×10^{-6} cm/s) in PAMPA (parallel artificial membrane permeability assay) test. In cryopreserved human and mouse hepatocytes incubation studies, the unchanged parent was the major form. The detectable metabolites were mainly from oxidation of the C6-morpholine ring and phase II glucuronidation of the carboxyl group. Hydrolysis of the C5-ester was almost negligible, especially in human hepatocytes. Moreover, compound **3** showed acceptable free fractions in human/mouse plasma as reflected by its PPB (plasma protein binding) values. The good solubility, permeability, metabolic stability, and unbound fraction characteristics indicate a potentially favorable PK and PD profile for compound **3** (Table 5.2).

In mouse *in vivo* PK studies (Figure 5.5 and Table 5.3), total plasma clearance (CL) was low following an IV bolus injection of compound **3**. After a single oral dose of 3 mg/kg and 10 mg/kg, respectively, compound **3** was well absorbed with a moderate plasma exposure (AUC, area under the curve) and bioavailability (*F*%). Moreover, compound **3** showed a much higher exposure in liver than that in plasma. Liver enrichment is considered a desirable attribute for an anti-HBV agent in order to maximize the therapeutic benefit in the liver while minimizing undesired side effects in plasma. The high liver exposure might point to a potential involvement of uptake transporters on the surface of hepatocytes. In subsequent *in vitro* studies, it was determined that compound **3** was a substrate for the liver transporter proteins OATP1B1 and OATP1B3. It is widely accepted that *in vivo* anti-HBV activity is driven by the free drug concentration in liver (drug concentration at trough needs to cover the *in vitro* antiviral EC_{50}). At 12 hours post dosing, although the plasma concentration was lower than *in vitro* anti-HBV EC_{50}, the liver concentration was much higher than the EC_{50} at both 3 mg/kg and 10 mg/kg dose levels. As such, 3 mg/kg, BID of compound **3** was proposed as the starting dose in the *in vivo* PD study in the hydrodynamic injection (HDI) mouse model [14].

Table 5.2 Physicochemical and ADME properties of compound 3.

ID	LogD[a]	pKa[b] (acid/base)	LYSA[c] (μg/ml)	HLM/MLM[d] (ml/min/kg)	PAMPA[e] (10^{-6} cm/s)	PPB[f] (human/mouse)
3	0.06	1.8/5.8	> 666	5/10	1.4	5.7/7.4

[a] Apparent partition coefficient at pH 7.4.
[b] Ionization constant.
[c] LYSA, lyophilization solubility assay.
[d] HLM, human liver microsome; MLM, mouse liver microsome.
[e] PAMPA, parallel artificial membrane permeability assay.
[f] PPB, plasma protein binding (% unbound fraction).

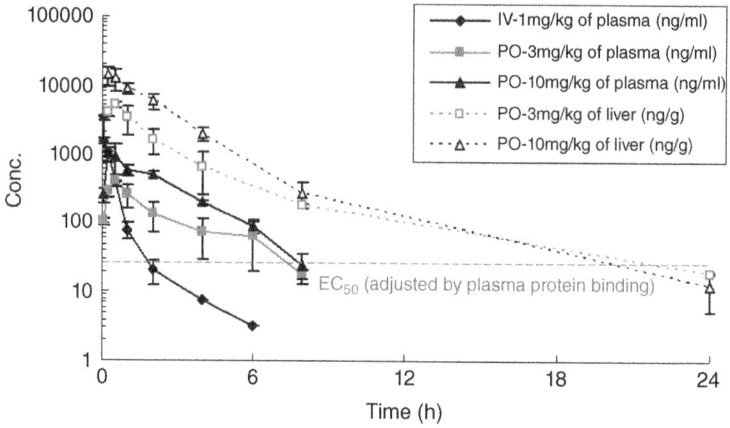

Figure 5.5 PK of compound **3** in mice at various doses.

Table 5.3 Selected pharmacokinetics parameters of compound **3** in male ICR mice following intravenous (IV) and oral (PO) administration.

	$AUC_{(0\text{-}\infty)}$	$t_{1/2}$	CL	C_{max}	F
	(μg/l*h)	(h)	(ml/min/kg)	(μg/l)	(%)
IV (1 mg/kg)	878	1.5	19	3310	—
PO (3 mg/kg, plasma)	962	—	—	413	37
PO (3 mg/kg, liver)	11800	—	—	5160	—
PO (10 mg/kg, plasma)	2430	—	—	993	28
PO (10 mg/kg, liver)	32400	—	—	13700	—

The single dose pharmacokinetics (SDPK) study of **3** was carried out in mice as indicated according to the standard procedures (IV: 1 mg/kg; PO: 3 and 10 mg/kg). Major parameters, including plasma clearance (CL), $t_{1/2}$ (IV), maximal concentration (C_{max}), area under the curve (*AUC*), and oral bioavailability (*F*) are reported.

In the PD studies (Figure 5.6), female BALB/c mice were hydrodynamically injected with replication-competent HBV DNA plasmid through the tail vein. Twenty hours post HDI, compound **3** was administered orally at 3 and 12.5 mg/kg per dose twice daily for five days. Entecavir (ETV, see Figure 5.1) was used as a positive control (0.1 mg/kg, once daily). Viral inhibition was measured using quantitative polymerase chain reaction (qPCR) by determining viral DNA copy numbers in both plasma and liver tissue. As shown in Figure 5.6, viral DNA copy numbers were rapidly and markedly reduced (by about 100-fold) in plasma and were no longer detectable in liver at doses of 3 and 12.5 mg/kg BID. The maximum efficacy of compound **3** was comparable with the maximum efficacy of ETV in this model.

In further *in vitro* ADMET studies, compound **3** demonstrated good drug-like properties. It showed no glutathione (GSH) adduct formation in liver microsomes, no mutagenicity (Ames assay) or clastogenicity (MNT assay). Although migration of acyl-glucuronide conjugates was observed in human hepatocyte incubation studies, the reactivity of the conjugates was determined to be low. The systemic toxicity of compound **3** was studied in rats and dogs with oral daily treatment up to four weeks. In all studies, sufficient safety margins were observed as defined by the

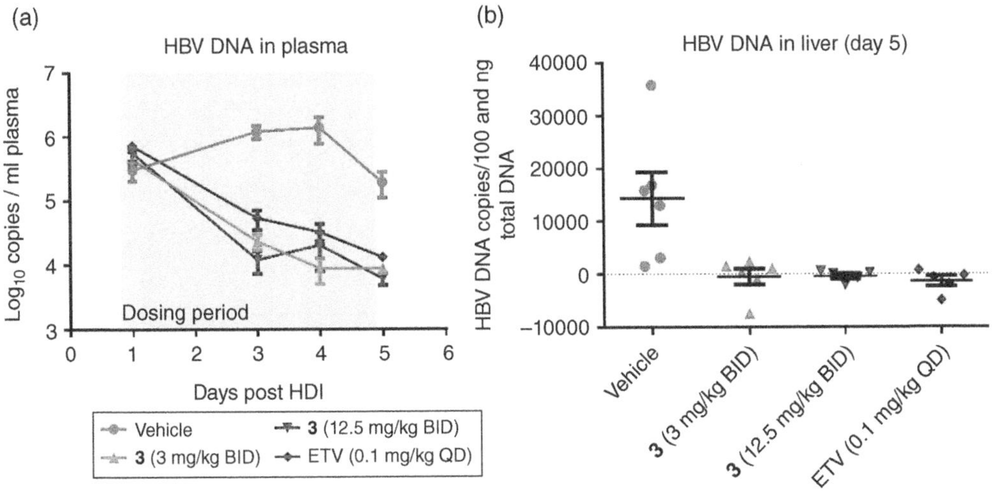

Figure 5.6 Levels of HBV DNA in plasma (a) and liver (b) of HBV HDI mice upon treatment with vehicle control, ETV control, and compound **3**. The HBV DNA copy numbers were quantified by real-time qPCR. Error bars indicate standard error of the mean (6 mice per group). *Source:* Modified from [13] (Figure 5.3); used with permission of the publisher (ACS publications).

ratio of plasma exposure (C_{max} and AUC) in animals at NOAEL (no observed adverse effect level) versus plasma exposure at a projected human efficacious dose.

Inhibition of CYPs by compound **3** was minimal. The concentration required for 50% inhibition of CYPs 1A2, 2B6, 2C8, 2C9, 2C19, 2D6, and 3A4 activity by compound **3** was greater than 50 μM in HLMs. It did not cause time and metabolism-dependent inhibition of CYP3A4 activity. However, compound **3** was found to be a strong inducer of CYP3A4 in primary human hepatocytes (PHHs) as reflected by the increase of CYP3A4 mRNA levels compared with a positive control, rifampicin (111, 83, and 22% induction at 10, 1, and 0.1 μM, relative to the induction of rifampicin at 10 μM) [15]. Therefore, there is a risk for compound **3** to be involved in drug–drug interactions with substrates of CYP3A4. The CYP3A4 auto-induction potential of compound **3** at predicted human efficacious doses was estimated to be acceptable as liver metabolism of compound **3** by CYP3A4 was low.

The issue of CYP3A4 induction is associated with the chemical class. More than 250 HAP analogues were screened and most of them demonstrated some level of CYP3A4 induction liability. In view of the overall good drug-like properties of compound **3** and the uncertainty of discovering a new candidate with equally promising properties but without the issue of CYP3A4 induction in a short period of time, the team decided to move compound **3** (named RG7689) into clinical trial to evaluate its CYP3A4 induction in humans. Concurrently with the clinical program, the team started a backup program to identify a new candidate with low/no CYP3A4 induction.

In a clinical study to evaluate the impact of RG7689 on CYP3A4 induction, a small amount of midazolam (100 μg) was used as the CYP3A4 substrate. Healthy volunteers were administered either midazolam alone or a combination of midazolam with different doses of RG7689 (60, 300, and 900 mg BID covering the predicted human efficacious dose from optimistic to conservative scenarios). Figure 5.7 shows the concentration of midazolam after two weeks of treatment with test article. The upper line represents the cohorts of midazolam and the lower line represents the cohorts of the coadministered midazolam and RG7689 at different dose levels. At 60 mg BID of RG7689, the exposure of midazolam in the midazolam cohort was much higher than that in the

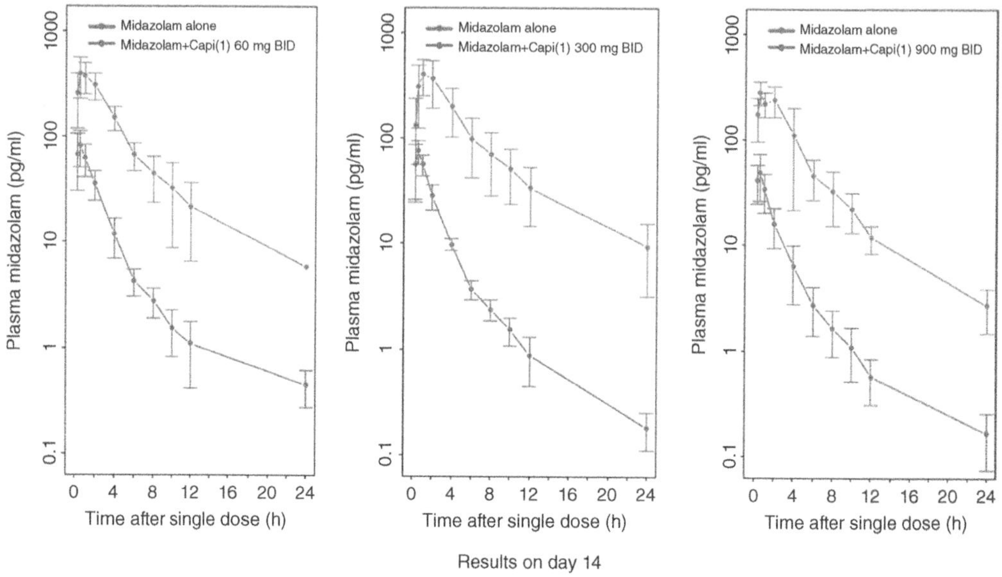

Figure 5.7 Evaluation of CYP3A4 induction of compound **3** (RG7689) in healthy volunteers.

cohort coadministered midazolam and RG7689. The higher metabolism of midazolam in the cohort coadministered midazolam and RG7689 indicated a significantly higher metabolic capacity of CYP3A4 because of induction by RG7689. Considering its strong CYP3A4 induction, plus the finding of additional induction pathways it triggered, the development of RG7689 was halted.

5.2.4 Approaches to Address CYP Induction Liability

In order to mitigate CYP3A4 induction and maintain the desired anti-HBV activity in an aggressive timeline, two chemistry approaches, supported by co-crystallization and modeling, were explored in parallel. One approach focused on fine-tuning the current chemical series to identify a candidate with a balanced profile of anti-HBV potency and CYP3A4 induction. The other aimed to improve anti-HBV activity of a new chemical series with an intrinsically lower risk of CYP3A4 induction.

The first approach is depicted in Figure 5.8. By adjusting the dihedral angle between the C4 phenyl and the dihydropyrimidine core, we were able to identify compound **15**, a close analogue of compound **3**, with good anti-HBV activity and significantly reduced CYP3A4 induction [16].

This seemingly minor adjustment of the substitution pattern on the phenyl ring from compound **3** to compound **15** led to a drastic conformational change as shown in Figure 5.9. In the overlay of the central cores of compounds **3** and **15**, derived from the single crystal X-ray structures, the dihedral angle between the C4 phenyl and the dihydropyrimidine core in compound **3** was determined to be 64.5°; the corresponding dihedral angle in compound **15** was 1.3°. Apparently, CYP3A4 induction, but not the anti-HBV activity, is sensitive to the position of the phenyl ring to the dihydropyridine. The optimized compound **15** demonstrated a much-reduced risk of CYP3A4 induction in the *in vitro* test in human primary hepatocytes (25% at 10 μM, 7.5% at 1 μM, and 2% at 0.1 μM, relative to the induction of rifampicin at 10 μM).

As revealed by the crystal structure of Cp149 Y132A in complex with compound **14**, the C6 position is largely exposed to a solvent environment and is thus considered tolerable to structural change. Indeed, during the initial SAR studies at the C6 position, it was found that some analogues

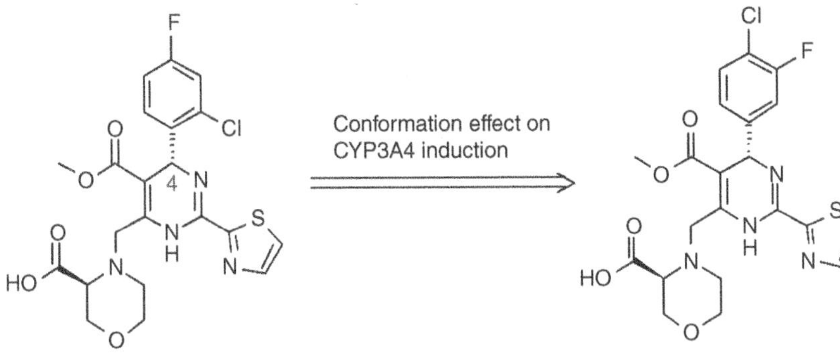

3 (HAP_R10)

EC_{50}/CC_{50}: 0.003 / >100 μM
CYP3A4 induction: 111% of rifampicin

15

EC_{50}/CC_{50}: 0.010 / >100 μM
CYP3A4 induction: 25% of rifampicin

Figure 5.8 Utilization of conformational adjustment to influence CYP3A4 induction.

with large and rigid substituents at C6 position could also provide an acceptable anti-HBV activity in the sub-μM range. Moreover, these analogues showed much lower or no CYP3A4 induction, which was in good agreement with the general strategy of mitigating CYP3A4 induction by enlarging and rigidifying a relevant substituent [16, 17]. Chemistry efforts were then focused on improving the anti-HBV activity guided by crystallography and modeling. Two chemical subseries, namely the bridged and fused, were successfully discovered with the representative compounds shown in Figure 5.10. They all maintained potent anti-HBV activity and showed significantly lower or no CYP3A4 induction [18, 19]. It should be noted that these interesting bridged and fused ring systems, named compact modules, provided a highly useful class of building blocks which possess two useful functionalities: acid and amine [18–21]. These compact modules have not been widely utilized in medicinal chemistry in part due to the synthetic challenges for their preparation. Our approach of preparing them

Figure 5.9 Overlay of the cores of compounds **3** and **15** derived from the single crystal X-ray structures.

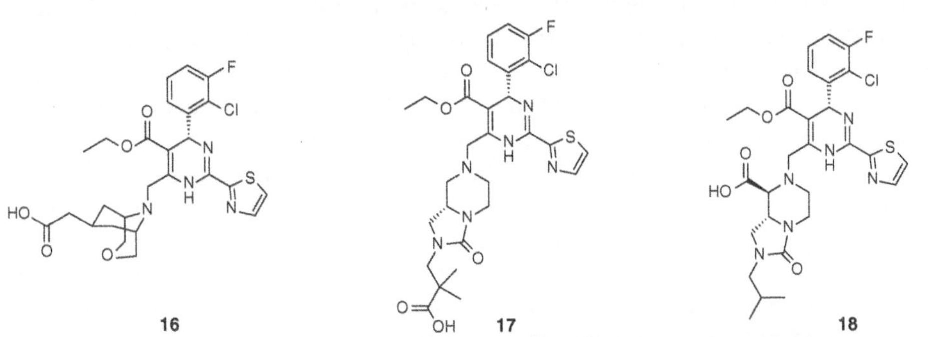

16

EC_{50}/CC_{50}: 0.004 / >100 μM
CYP3A4 induction: 6% of rifampicin

17

EC_{50}/CC_{50}: 0.003 / >100 μM
CYP3A4 induction: 1% of rifampicin

18

EC_{50}/CC_{50}: 0.001 / >100 μM
CYP3A4 induction: 1% of rifampicin

Figure 5.10 Designing new analogues to mitigate CYP3A4 induction potential.

systematically has paved the path to solve the CYP induction issue, and they could be applied to other drug discovery programs in the future.

A new candidate with the most balanced profile of potency, drug metabolism and pharmacokinetics (DMPK) and safety was then selected for clinical testing. Indeed, in line with the *in vitro* CYP3A4 induction result, this new candidate did not show any signs of drug–drug interaction (DDI) and auto-induction related to CYP3A4 induction in a phase I clinical trial. The results will be reported in due course.

5.2.5 Conclusion

In summary, the discovery of the new CpAMs described herein is highlighted by several tactics. First, the introduction of a carboxyl group improved not only the anti-HBV activity, but also the DMPK and safety profiles. Second, structure-based drug design was leveraged to provide clues to mitigate CYP3A4 induction while maintaining the desired anti-HBV activity. Lastly, rationally designed compact modules are successfully utilized to solve the issue of CYP3A4 induction, which was the major liability of the first clinical compound RG7689. As such, a new CpAM has been discovered and advanced into clinical testing with great potential of transforming the lives of HBV patients.

5.3 RG7834: The First-in-Class Selective and Orally Bioavailable Small Molecule HBV Expression Inhibitor with a Novel Mode of Action

5.3.1 Introduction

In the HBV life cycle, HBsAg plays a very important role, and a high level of HBsAg is considered as one of the hallmarks of chronic HBV infection in clinics. HBsAg may negatively regulate HBV-specific immune responses [22–24] and increase the risk of developing chronic liver diseases and hepatocellular carcinoma [1]. HBsAg loss is defined as an HBV "functional cure" and HBsAg seroconversion is considered an ideal end point for HBV therapy [25, 26]. As such, agents that reduce HBsAg levels have the potential to achieve a "functional cure" and thereby reduce the risk of developing severe liver diseases.

HBsAg loss is rarely achieved in the clinic with current therapies, including both nucleos(t)ide analogues (such as ETV) and interferons (IFNs), although they can efficiently reduce viral load (HBV DNA) [27]. Moreover, tolerability of IFN-based therapy is poor and the nucleos(t)ides therapies require lifelong treatment. Thus, agents with a differentiated mechanism of action (MOA) are required to achieve HBsAg eradication and improve the HBV cure rate. However, there is no molecular target reported to achieve HBsAg reduction.

In general, there are two different approaches to discover a lead molecule for chemical optimization – target based screening and phenotypic screening. It has been reported that the contribution of phenotypic screening to the discovery of first-in-class small-molecule drugs exceeds that of target-based approaches [28]. To identify agents that can reduce HBsAg levels, a Roche team of scientists used a phenotypic screen to discover RG7834, a first-in-class selective small molecule HBV expression inhibitor from dihydroquinolizinone (DHQ) chemical series. RG7834 has a novel MOA, distinct from current therapies and significantly reduces HBsAg, HBV DNA and HBeAg levels *in vitro* and in various animal models. The antiviral effect is highly specific for HBV when tested in a panel of 15 DNA and RNA viruses [29, 30]. The target identification effort revealed that

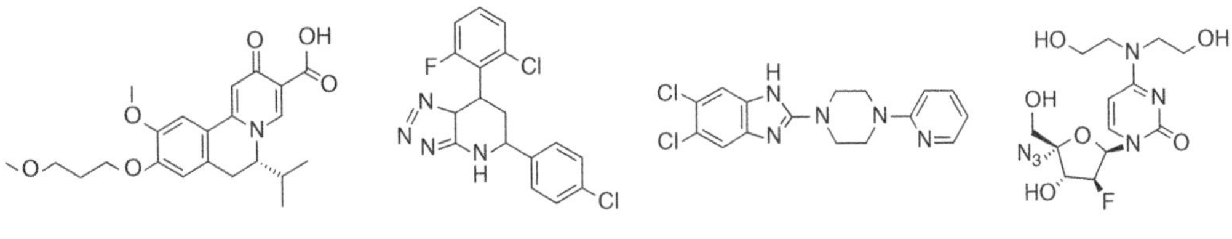

Figure 5.11 Structures of RG7834 and reported HBsAg inhibitors.

the anti-HBV activity of RG7834 is through the inhibition of non-canonical poly(A) RNA polymerases PAP-associated domain-containing protein 5 and 7 (PAPD5 and 7), which were never reported previously for anti-HBV therapy [31].

The MOA and phenotype of RG7834 is clearly different from several other molecules which were reported as HBsAg inhibitors (Figure 5.11). For example, HB-0259 inhibits HBsAg secretion but has no effect on HBV DNA or HBeAg secretion [32, 33]. Another reported HBsAg inhibitor, BM601, inhibits HBsAg secretion but not production, and the compound had no effect on HBV DNA replication and HBeAg secretion [34]. A series of fluorine substituted nucleoside analogues [35–37] were reported as inhibitors of HBsAg, HBV DNA and HBeAg, but they were also active against HIV and HCV, and not specific for HBV as is RG7834.

Herein, we summarize the discovery of RG7834 [29], its ADMET and antiviral profiles and target identification for the DHQ series.

5.3.2 The Discovery of RG7834

5.3.2.1 Lead Generation

To identify feasible chemical starting points, we performed a cell-based phenotypic high-throughput screening (HTS) of a library containing approximately one million compounds. Extracellular HBsAg levels were detected in this assay. Various chemotypes reduced HBsAg levels. After the hit validation process, compound **19** (Figure 5.12) incorporating a dihydroquinolizinone (DHQ) scaffold was selected for further investigation due to its promising biological profile. In HBV-expressing HepG2.2.15 cells, compound **19** reduced both HBsAg and HBV DNA levels with IC_{50} values of 0.5 μM and 0.2 μM, respectively. The antiviral activity of compound **19** was also confirmed in differentiated HepaRG (dHepaRG) cells and showed inhibition of HBsAg expression with a similar IC_{50} value (0.88 μM).

The primary SAR was explored with analogues containing variations at different positions. The data in Table 5.4 indicated that substituents at C6, C9, and C10 had a significant impact on potency. Removal of the C6 methyl or the C10 methoxy resulted in a significant potency reduction (compounds **20, 21** vs **19**), and further removal of the C9 methoxy (**22**) resulted in a complete loss of activity.

At the C10 position, both methyl and chloro groups are well tolerated (**23** and **24**), but larger groups result in reduced activity (**25, 26** vs **19**). Although removal of the C9 methoxy resulted in a dramatic potency drop (**27** vs **19**), diversified and larger substituents were well tolerated at this position and gave significantly improved potency (**28–30** vs **19**).

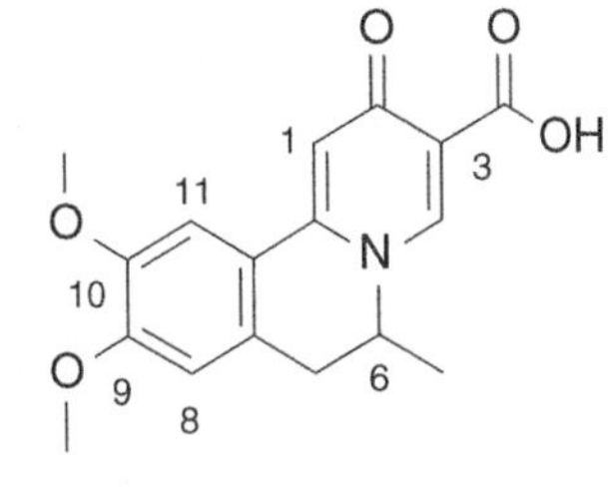

19

Figure 5.12 Hit generated from phenotypic screening against HBsAg.

Table 5.4 Effects on HBsAg reduction of analogues **19–35**.

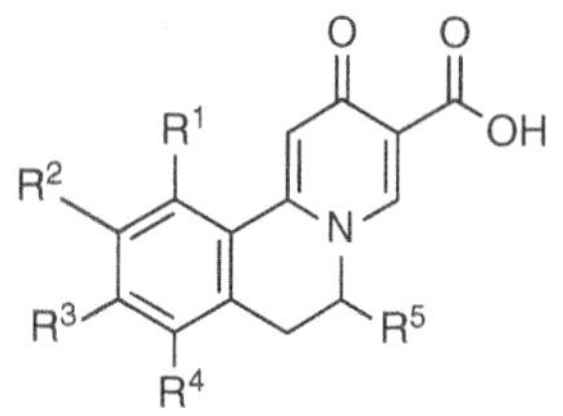

ID	R^1	R^2	R^3	R^4	R^5	IC_{50} (μM)[a]	CC_{50} (μM)[b]
19	-H	-OMe	-OMe	-H	-Me	0.5~0.87	>50
20	-H	-OMe	-OMe	-H	-H	2.64	>50
21	-H	-H	-OMe	-H	-Me	4.97	>50
22	-H	-H	-H	-H	-Me	52	>50
23	-H	-Me	-OMe	-H	-Me	0.24	>50
24	-H	-Cl	-OMe	-H	-Me	0.29	>50
25	-H	-OEt	-OMe	-H	-Me	1.2	>50
26	-H	-OBn	-OMe	-H	-Me	29	>50
27	-H	-OMe	-H	-H	-Me	7.67	>50
28	-H	-OMe	-OEt	-H	-Me	0.097	>50
29	-H	-OMe	-OBn	-H	-Me	0.10	>50
30	-H	-H	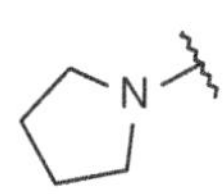	-H	-Me	0.081	>50
31	-H	-H	-OMe	-Cl	-Me	5.03	>50
32	-OMe	-H	-OMe	-H	-Me	4.56	>50
33	-H	-OMe	-OMe	-H	-Et	0.11	>50
(*R*)-33	-H	-OMe	-OMe	-H	(*R*)-Et	0.049	>50
(*S*)-33	-H	-OMe	-OMe	-H	(*S*)-Et	10.9	>50
34	-H	-OMe	-OMe	-H	-Pr	0.068	>50
35	-H	-OMe	-OMe	-H	-*iso*Bu	0.052	>50

[a] IC_{50} is the geometric mean value for reduction of HBsAg by 50% in HepG.2.2.15 cells.
[b] CC_{50} is the geometric mean value for cytotoxicity measured by CCK-8 in HepG.2.2.15 cells.

At the other two positions on the left ring, the C8 chloro analogue **31** and C11 methoxy analogue **32** showed similar potency to that of compound **21**. Potency might be less sensitive to modifications at these two positions compared with those at C9 and C10 positions.

Antiviral activity was extremely sensitive to modifications on the C6 position. While removing the C6 methyl group resulted in a reduced activity, larger alkyl groups at C6 gave significantly improved activity. The C6 ethyl compound **33** showed significantly increased potency vs **19**, while analogues incorporating an *n*-propyl or isobutyl group at the C6 position (**34, 35**) showed even better potency than **33**.

Table 5.5 Further SAR at the C9 and C10 positions: effects on HBsAg reduction and physicochemical properties of **36–46**.

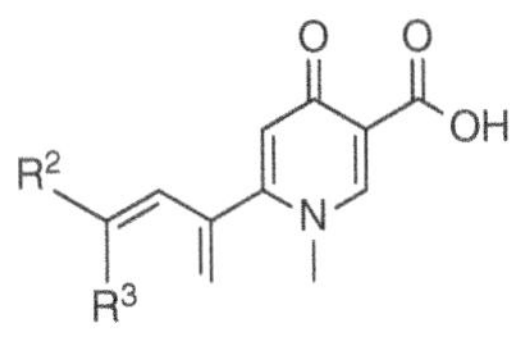

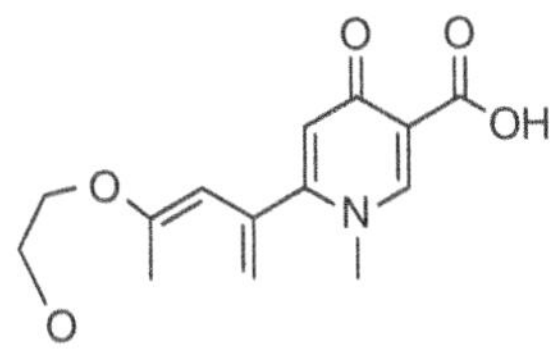

40

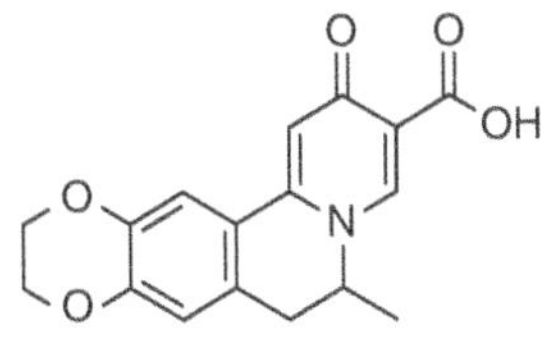

41

ID	R^2	R^3	IC_{50} (μM)[a]	CC_{50} (μM)[a]	LYSA (μg/ml)[b]	MLM (ml/min/mg)[c]	HLM (ml/min/mg)[d]
36	-OMe	-OBn	0.039	>50	<1.0	3.6	1.4
37	-OMe	-OEt	0.043	>50	25.0	5.3	1.3
38	-OMe	-OPr	0.020	>50	36.0	2.3	0.33
39	-OMe	-O-*iso*Bu	0.013	>50	3.0	2.5	1.3
40	—	—	1.94	>50			
41	—	—	0.67	>50			
42	-Et	-OMe	0.024	>50	10.0	22.2	2.5
43	-OMe	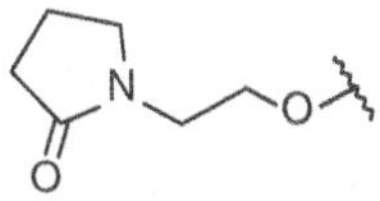	0.18	>50	>470	2.3	3.3
44	-OMe	$-OCH_2CH_2CH_2OH$	0.2	>50	101	9.5	1.8
45	-OMe	$-OCH_2CH_2OMe$	0.064	>50	225	0	2.4
46	-OMe	$-OCH_2CH_2CH_2OMe$	0.019	>50	63.5	4.0	2.5

[a] The definitions are the same as those in Table 5.4.
[b] LYSA, lyophilization solubility assay.
[c] MLM, mouse liver microsome.
[d] HLM, human liver microsome.

To understand the impact of the chiral center at the C6 position on antiviral activity, compound **33** was resolved. The two enantiomers showed marked difference in anti-HBV activities. (*R*)-**33** showed a more than 20-fold higher potency than (*S*)-**33** whose configuration was confirmed by single crystal X-ray studies.

5.3.2.2 Lead Optimization

The initial SAR exploration of **19** demonstrated the that the DHQ series provided ample chemical space for optimization (Table 5.5). Further SAR study confirmed that the C9 position could tolerate various substituents of different sizes and polarity. As shown in Table 5.5, many compounds with modifications at the C9 position showed antiviral activity with an IC_{50} <50 nM. At the C10 position, replacing methoxy by ethyl gave a fourfold increase in activity (**42** vs **33**), but the mouse liver microsome stability of **42** was not favorable. Cyclization of the C9 and C10 substituents (**40, 41**) resulted in a significant potency drop.

A hydrophobic group at the C9 position could be beneficial for potency (**36, 39**), but in general, this kind of group also led to unfavorable physiochemical properties such as reduced solubility. With a methoxypropoxy at the C9 position, analogue **46** demonstrated a balanced profile of potency and ADME.

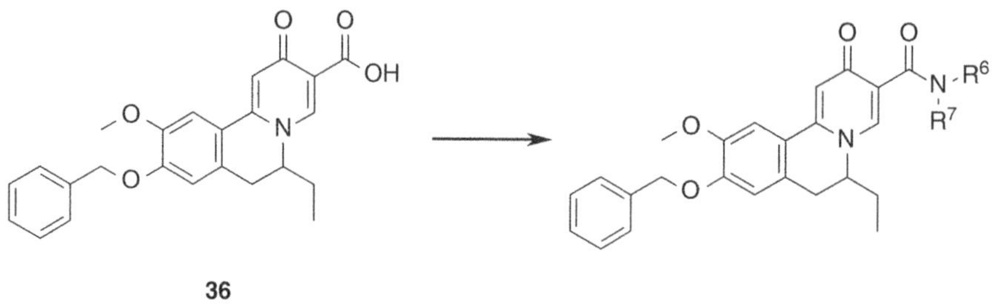

Figure 5.13 SAR exploration of the acid moiety.

The SAR at the C3 position was also investigated and the results demonstrated that the 3-carboxylic acid group played a critical role for antiviral activity (Figure 5.13). With compound **36** as the parent compound, a series of amide derivatives were prepared and several carboxylic acid bioisosteres were also tried. Unfortunately, all compounds showed dramatically diminished activity compared with compound **36,** and some amide modifications led to a complete loss of activity.

With favorable substituents fixed at the C3, C9, and C10 positions, the SAR at the C6 position was further explored (Table 5.6). Bulky and hydrophobic groups were well tolerated at this position. The isopropyl analogue **47** showed a twofold increase in potency compared with the corresponding ethyl analogue **46**, while the *t*-butyl group **(48)** gave the most potent activity with an IC_{50} value of 0.001 μM even as a mixture of two enantiomers. Aromatic groups were also tolerated at this position. Phenyl is acceptable (**52**) and the 2-thienyl analogue (**53**) showed very potent

Table 5.6 Further SAR at the C6 position: effects on HBsAg reduction and physicochemical properties of compounds **47–53**.

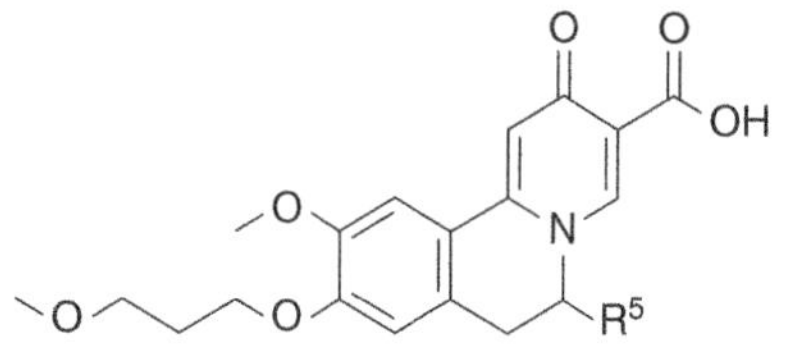

ID	R5	IC_{50} (μM)[a]	CC_{50} (μM)[a]	LYSA (μg/ml)[a]	MLM (ml/min/kg)[a]	HLM (ml/min/kg)[a]
47	- *iso*Pr	0.009	>50	71	5.0	1.1
(*S*)-**47 (RG7834)**	-(*S*)-isoPr	0.003	>50	82	13	3.7
(*R*)-**47**	- (*R*)-*iso*Pr	3.15	>50	85	9.3	0.77
48	-*t*Bu	0.001	>50	5.0	0	0.63
49	$-CH_2CF_3$	0.039	>50	119	0.92	0.76
50	$-CH_2OMe$	0.074	>50	192	8.4	4.6
51	$-C(Me)_2CH_2OMe$	0.005	>50	137	6.2	0
52	phenyl	0.064	>50	7.0	5.5	0
53	2-thienyl	0.003	>50	5.0	10	1.1

[a] The definitions are the same as those in Table 5.5.

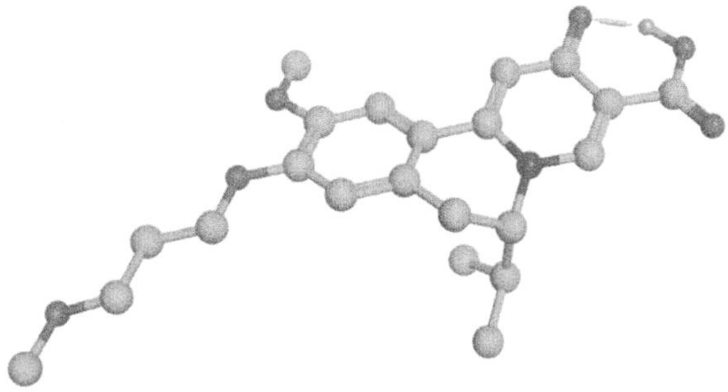

Figure 5.14 X-ray structure of RG7834 (*S*)-**47**.

antiviral activity with an IC_{50} value of 0.003 μM. Although the potency of **48** and **53** was very good, the increased hydrophobicity unfortunately also led to a reduction in solubility.

The two enantiomers of compound **47** were separated and the absolute configuration was confirmed by a single-crystal X-ray study. The crystal structure indicated that an intramolecular hydrogen bond was formed between the carboxyl group and the carbonyl at the C2 position (Figure 5.14). The enantioselective anti-HBV activity was confirmed as observed on the two enantiomers of **33**. The enantiomer (*S*)-**47** (RG7834) had an IC_{50} value of 0.003 μM, which was >1000-fold more potent than the enantiomer (*R*)-**47**. Considering the balanced profile, RG7834 was advanced into more *in vitro* and *in vivo* studies for further biology, ADME and toxicity profiling.

5.3.2.3 Profile of RG7834

It was encouraging to find that the DHQ compounds could not only reduce the HBsAg expression levels, but also the HBeAg and viral DNA levels in different cell lines. In HepaG2.2.15 cells, RG7834 displayed very potent activity in the inhibition of HBV DNA production with an IC_{50} <0.13 nM [30].

The potent activity of RG7834 against HBsAg, HBeAg and HBV DNA was also observed in HBV-infected dHepaRG cells, in which RG7834 showed an IC_{50} around 3 nM against all the three viral markers, while its enantiomer had no significant inhibition at concentrations up to 1 μM. In the same assay, the nucleoside drug ETV showed no activity against HBsAg or HBeAg although it was very potent in the inhibition of HBV DNA production [30].

Another report also confirmed the single digital nM activity of RG7834 against HBsAg and HBeAg in HepaG2.2.15 cells while ETV had no activity against the two antigens at concentrations of up to 50 μM [38].

In another test evaluating anti-HBV genotype coverage in PHHs, RG7834 inhibited all four prevalent HBV genotypes A to D with similar activity, reducing HBV DNA levels with single digit nM IC_{50} values [30].

The *in vitro* antiviral profile of RG7834 indicated that DHQ compounds had a differentiated MOA compared with nucleoside drugs. To understand the antiviral selectivity, RG7834 was tested in a panel of 15 different DNA and RNA viruses. The IC_{50} of RG7834 against all these viruses was greater than 10 μM which was >1000-fold higher than its anti-HBV activity. The high specificity for HBV was also confirmed in another study in which RG7834 did not exhibit detectable antiviral activity against 9 viruses up to the highest tested concentration of 30 μM [38]. Taken together, these data demonstrate the high specificity of RG7834 toward HBV

Besides favorable antiviral activity, solubility, and liver microsome stability, RG7834 also showed a good cytotoxicity profile with a CC_{50} value >50 μM in HepG2.2.15 cells. The permeability of RG7834 was high and no efflux issues were observed in the Caco-2 assay. The unbound fractions of RG7834 in human and mouse plasma were determined to be 32.8 and 35.2%, respectively [29].

In a mouse SDPK study, RG7834 showed moderate plasma clearance and good oral bioavailability ($F = 62\%$) providing satisfactory oral exposure for efficacy studies. The liver exposure was fourfold higher than that in plasma, which is considered a desirable attribute for anti-HBV drugs because liver is the target organ of HBV infection [29].

In a cynomolgus monkey SDPK study, RG7834 showed low plasma clearance following intravenous administration. The oral bioavailability in monkey was also good (57%) with a half-life of 4.9 hours following oral administration which was significantly longer than that in mouse (1.48 hour) [29].

RG7834 also showed a favorable profile in early safety assessments. In a broad panel of different human and rodent primary cells and cell lines representing various organs and tissues, RG7834 did not influence cell viability up to the highest tested concentration of 25 μM [30]. It did not show liability in the *in vitro* DDI assessment including both CYP inhibition and induction. Moreover, RG7834 showed no addition of glutathione (GSH), no inhibition of the hERG channel, and was clean in Ames assay (mutagenicity) and micronucleus test.

In preclinical *in vivo* toxicology studies, RG7834 demonstrated good tolerability in a four-day study in rats, as well as a four-week Good Laboratory Practice (GLP) study in rat and monkey.

In the HBV-infected human liver chimeric uPA-SCID mice model, RG7834 reduced both HBsAg (0.91 log units) and HBeAg (0.43 log units) significantly after 21 days of treatment at 4 mg/kg BID dosing. The result with the 10 mg/kg dose was similar which suggested that the maximal HBsAg reduction compared with baseline in this model is about 1.0 log unit. RG7834 also reduced serum HBV DNA by 0.6 log units with the 4 mg/kg dose. As a control in this study, ETV showed good efficacy in reducing HBV DNA levels but did not show clear efficacy reducing either HBsAg or HBeAg (Figure 5.15) [30]. To the best of our knowledge, RG7834 is the first small molecule known to significantly reduce HBsAg in the human liver chimeric uPA-SCID mouse model [30].

In another efficacy study in AAV-HBV1.2 mouse model, similar efficacy in HBsAg reduction was observed on day 4 with 3 mg/kg BID daily oral dosing [38].

5.3.2.4 Target Identification

The DHQ chemical class represented by RG7834 was discovered by phenotypic screening and displayed a distinct MOA compared with the approved anti-HBV drugs and other reported anti-HBV agents. To get a better understanding of the unique MOA, different approaches were tried to identify the molecular target(s) for the DHQ chemical series. Finally, with the DHQ compounds e.g. Y3H-DHQ (Figure 5.16), RG7834 and its enantiomer (*S*)-**47** as tools, we identified the non-canonical poly(A) RNA polymerases, PAP-associated domain-containing protein 5 and 7 (PAPD5 and PAPD7) as the protein targets for DHQs using a yeast three-hybrid (Y3H) platform. DHQ compounds could inhibit both PAPD5 and 7 simultaneously, which led to significant anti-HBV activity. Knockdown of both PAPD5 and PAPD7 led to destabilization and degradation of HBV mRNA, similar to that seen with RG7834 treatment. Prior to our research, these two proteins were never reported as cellular host factors for HBV therapy. The detailed information of target identification can be found in our publication in 2019 [31].

5.3.3 Conclusion

Starting from a phenotypic screen, we discovered the DHQ chemical series which showed a unique anti-HBV MOA distinct from current therapies. Target identification efforts by the Y3H approach identified PAPD5 and 7 as molecular targets for this chemical series.

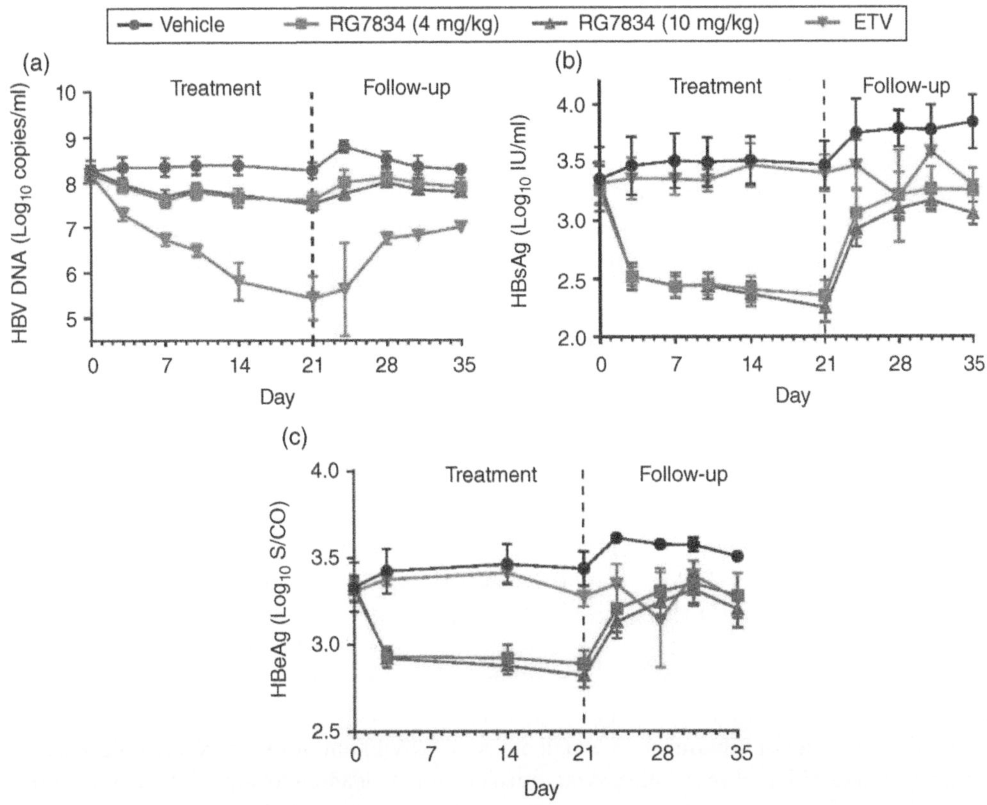

Figure 5.15 Effect of RG7834 and ETV on HBV viral markers in HBV-infected human liver chimeric uPA-SCID mice. (a) HBV DNA, (b) HBsAg and (c) HBeAg were determined in serum from HBV-infected uPA/SCID mice during the 21-day treatment and follow-up period up to 35 days. *Source:* Figure reprinted from *Journal of Hepatology* [30].

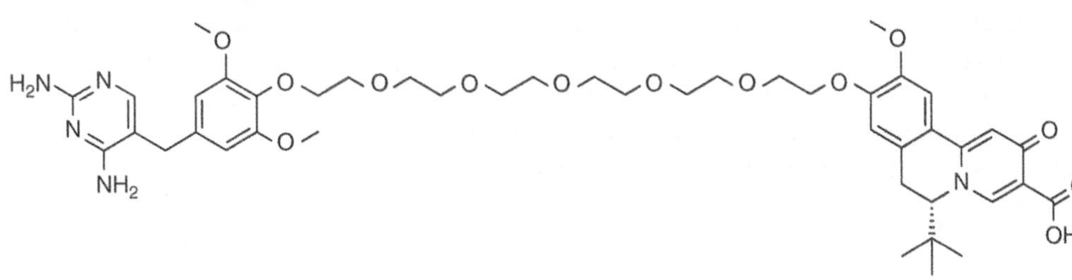

Figure 5.16 Structure of Y3H-DHQ.

RG7834, the representative lead compound from the DHQ series, reduces HBsAg, HBeAg and HBV DNA levels at single digit nM concentrations *in vitro*, and its *in vivo* efficacy was demonstrated in different animal models. The compound was highly specific for HBV. RG7834 also showed favorable *in vitro/in vivo* DMPK properties and was well tolerated in preclinical toxicity studies including a four-week GLP study in rat and monkey. Taken together, these data supported moving RG7834 forward into clinical investigation.

The discovery of the DHQ chemical series provided a new direction for HBV drug discovery. Compounds from this series or their derivatives have the potential to increase the HBV cure rate via monotherapy or most likely in combination with other therapies. The promising antiviral profile of the DHQ compounds has attracted the interest of the community working on anti-HBV treatments. An increasing number of patents related to DHQ series or series derived from the DHQ series have been published in recent years.

5.4 Ziresovir: The Discovery of a Highly Potent, Selective and Orally Bioavailable RSV Fusion Protein Inhibitor

5.4.1 Introduction

As a negative-sense single-stranded RNA virus, human RSV belongs to the family of Pneumoviridae [39]. Children under the age of two may be infected and re-infected by RSV during the RSV season [40]. In high-risk populations, such as premature infants and immunocompromised adults, RSV infection could develop into a lower respiratory tract infection, which is associated with mortality and morbidity in some cases. Severe RSV infection causes bronchiolitis in infants and pneumonia in adult patients, leading to increased morbidity or mortality [40, 41]. Severe RSV infection causes nearly 3.4 million hospitalizations and almost 160 000 deaths in children under five years old worldwide every year [42, 43].

RSV fusion (RSV F) protein is expressed on the surface of the viral envelope. It mediates both viral entry to healthy cells and syncytia formation among infected and uninfected cells, which is the hallmark of RSV infection [44]. Inhibition of the RSV fusion process provides an important target for the treatment of RSV infection. Small molecule RSV F inhibitors are expected to decrease the severity and duration of respiratory symptoms, eventually leading to a quick cure of the viral infection.

Several small molecule RSV F protein inhibitors have been discovered and moved into clinical trials in the last two decades (Figure 5.17) [45–49]. JNJ-53718678, a picomolar RSV F inhibitor, reduced viral load and clinical severity in viral challenge studies in phase IIa [50]. GS-5806, another potent RSV F inhibitor, significantly reduced viral load and disease symptom scores in viral challenge studies [51, 52]. RV521, a single digit nanomolar RSV F protein inhibitor, reduced viral load and mucous production in a phase IIa human viral challenge study [53]. Different from all of the other reported orally bioavailable RSV F inhibitors, VP-14637, a nanomolar RSV F protein inhibitor, was formulated as an inhaled dry powder and named MDT-637 in a phase I clinical trial [54].

In 2008, we started our own RSV F inhibitor program. Originating from a promising hit, we identified a novel chemical series [55]. After lead generation and optimization, Ziresovir (RO-0529, AK0529) was discovered. Ziresovir is a nanomolar RSV F inhibitor with good orally bioavailability and is currently in phase III clinical studies [56].

5.4.2 The Discovery of Ziresovir (RO-0529 OR ARK0529)

5.4.2.1 Lead Generation

In our endeavors to develop small molecule RSV F inhibitors, we conducted a similarity-based virtual screening approach and generated several hits as chemistry starting points. A cytopathic effect (CPE) assay in long strain was used to develop the SAR and rank the compounds. Compound **54** was identified as one of the validated hits (Figure 5.18) [55]. SAR exploration on compound **54** revealed a very narrow SAR on the head portion defined as substitutions at the C2 position of quinoline. Expansion from the tetrahydroisoquinoline moiety in compound **54** to the benzoazepine in

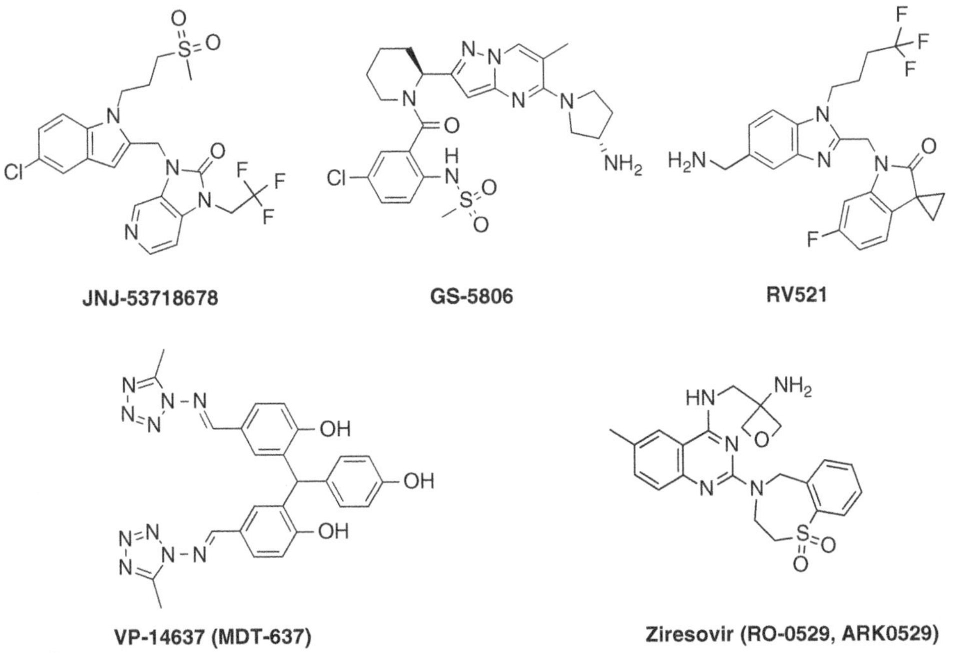

Figure 5.17 RSV fusion inhibitors in clinical development.

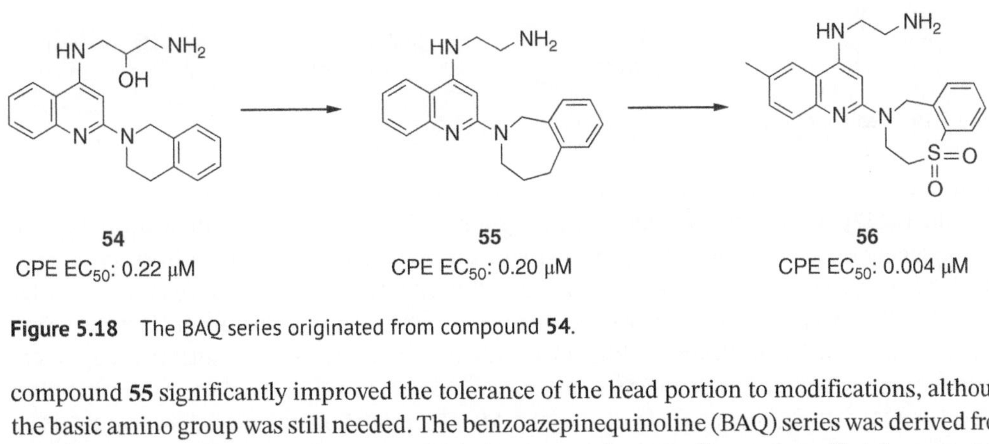

54
CPE EC_{50}: 0.22 μM

55
CPE EC_{50}: 0.20 μM

56
CPE EC_{50}: 0.004 μM

Figure 5.18 The BAQ series originated from compound **54**.

compound **55** significantly improved the tolerance of the head portion to modifications, although the basic amino group was still needed. The benzoazepinequinoline (BAQ) series was derived from compound **55**. Further introducing a 6-methyl group on the quinoline and a sulfonyl group at the benzylic position of the benzoazepine led to the discovery of compound **56**, a very potent RSV F inhibitor. The sulfonyl group also dramatically reduced metabolism on the benzoazepine.

5.4.2.2 Lead Optimization

During lead optimization, the DMPK properties of the lead compound were optimized (Table 5.7). When compound **56** was dosed to male Wistar rats, negligible oral exposure was observed. In order to improve the oral bioavailability, various modifications were carried out based on the structure of compound **56** (Figure 5.19). The most successful modification was to increase lipophilicity on the head portion as in compound **57**. However, one issue with compound **57** was decreased anti-RSV activity, and one potential issue was a significantly high V_{ss} (Table 5.7). It is known that the oxetane ring, a "compact module," can be used as bioisostere of a gem-di-methyl group. Replacement of the

Table 5.7 Rat SDPK profiles of selected BAQ compounds[a].

Compound		56	57	58	RO-0529 (Ziresovir)
CPE (Long)	IC_{50}/CC_{50} (μM)	0.004/66	0.012/24	0.002/>100	0.003/>100
Plasma	$t_{1/2}$ (IV, h)	4.5	14.0	1.7	1.2
	V_{ss} (IV, l/kg)	17	60	4.1	3.9
	CL (IV) (ml/min/kg)	97	40	51	58
	$AUC_{(0-\infty)}$ (PO, μg/l*h)	0	1938	442	906
	F (%)	0	54	13	32

[a] The SDPK study in male Wistar rats was carried out according to the standard procedures (IV: 2 mg/kg; PO: 10 mg/kg). Major parameters, including plasma clearance (CL), volume of distribution at steady state (V_{ss}), $t_{1/2}$ (IV), area under the cure (*AUC*), and oral bioavailability (*F*) are reported.

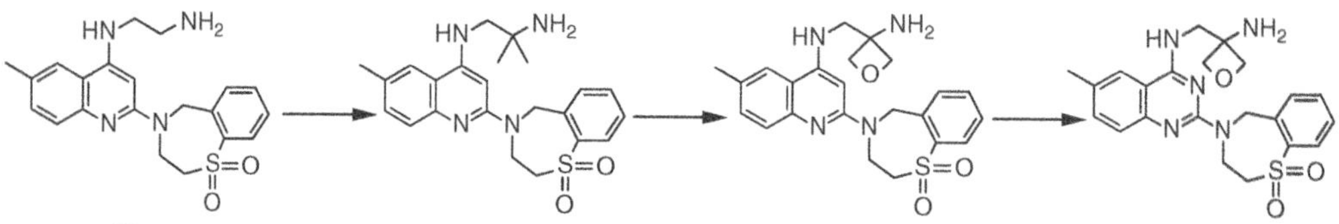

Figure 5.19 Lead optimization resulting in the discovery of Ziresovir.

gem-di-methyl with an oxetane ring reduces the basicity of a nearby amine group and modulates lipophilicity [57]. As an early example of the adoption of the oxetanyl moiety in medicinal chemistry, replacing the gem-di-methyl group of **57** with an oxetane ring led to the discovery of compound **58**. The oxetanyl analogue regained the anti-RSV activity (2 nM) and reduced the V_{SS} to 4.1 l/kg. The much reduced V_{SS} can be explained by the reduced basicity of the terminal amine group (ΔpKa = 2) by the adjacent electron-withdrawing oxetanyl group. The last optimization step was to convert the quinoline of compound **58** to a quinazoline, which led to the discovery of RO-0529 (Ziresovir) [58]. After thorough evaluation of the ADMET profiles of compound **58** and RO-0529, RO-0529 was selected as the clinical candidate.

5.4.2.3 Profile of Ziresovir

Ziresovir was tested against RSV strains of Long, A2, B18537 and clinical isolates from Chongqing Children's Hospital during the RSV season from 2011 to 2012. In the CPE assays, Ziresovir showed activity in the clinical strains similar to that of the laboratory strains of RSV (Long and B18537). RO-0529 did not demonstrate activity against influenza H1N1, human parainfluenza virus and rhinovirus in the respective CPE assays. Therefore, its antiviral activity is likely RSV specific [58].

An *in vivo* efficacy study was executed in RSV infected female BALB/c mice [58]. RO-0529 was dosed orally at 12.5 mg/kg and 50 mg/kg, respectively, to 12 mice twice a day for four days. One log unit of viral titer reduction was detected in the lung of infected mice at the oral dose of 12.5 mg/kg. An oral dose of 50 mg/kg reduced viral titer by 1.9 log units (Figure 5.20).

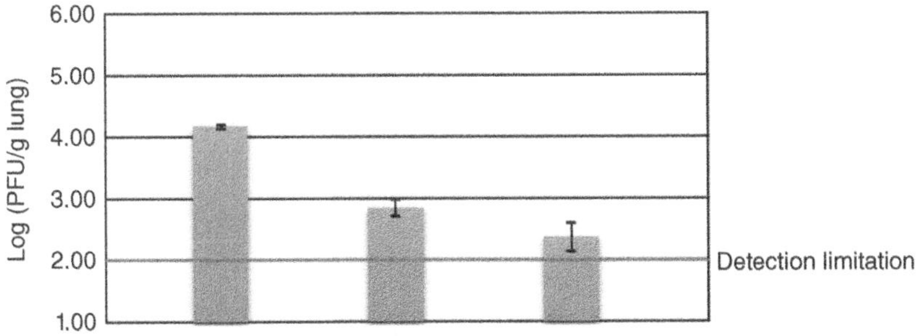

Figure 5.20 *in vivo* reduction of RSV titer in mouse lung with RO-0529. Female BALB/c mice were used for the *in vivo* efficacy assay. Animals were anesthetized and intranasally administrated RSV long strain (5×10^5 plaque-forming units [PFUs]). After oral administration of RO-0529 BID for four days, the mice were killed with CO_2, and lungs were harvested and analyzed for viral titer. The viral titer detection limit was 2 log units (PFU/g lung) in the plaque assay.

5.4.2.4 Mode of Action of Ziresovir

Drug resistant mutation studies were performed for RO-0529. The isolated resistant viruses were obtained after multiple passages in the presence of increasing concentrations of Ziresovir. The RSV genomes from the resistant isolates were amplified and sequenced. The F gene mutations were harbored in all resistant isolates, and consisted of D486N, D489A, D489V, and D489Y, respectively. All the mutations were located in the HRC region of the C-terminal of the RSV F protein. The results indicated that the target of Ziresovir is the RSV F protein. The antiviral activity of RO-0529 was tested in the CPE assays using plaque-purified virus with D486N or D489A mutants, and the results are shown in Table 5.8 [58].

The mode of action of Ziresovir was also confirmed by the inhibition of the RSV F protein-induced cell–cell fusion process. In the presence of 0.1 μM of Ziresovir, the syncytia formation induced by the RSV F protein was completely inhibited (Figure 5.21b) as compared with that in the DMSO control (Figure 5.21a). These results further confirmed that Ziresovir was a RSV F inhibitor [58].

5.4.3 Clinical Studies of Ziresovir

In 2014, Ark Biosciences licensed RO-0529 from Roche and renamed it AK0529, which later obtained the WHO INN Ziresovir. Multiple clinical studies with Ziresovir have been conducted. Three phase I clinical studies were conducted in healthy adults in Australia [59], China [60], and the UK, respectively [61]. In the middle of 2019, Ark Biosciences completed a phase II proof-of-concept study of Ziresovir (AK0529) in 72 infant patients with RSV infection [56]. Based on the results, Ark Biosciences plans to advance the antiviral into phase III trials. The clinical outcome will be published in due course.

Table 5.8 Effect of Ziresovir on the wild type and mutant strains[a].

CPE assay	EC_{50}/EC_{90} (μM) (WT)	EC_{50}/EC_{90} (μM) (D486N)	EC_{50}/EC_{90} (μM) (D489A)
RO-0529	0.003/0.005	>10/>10	2.1/10.0

[a] For the selected mutant strains of RSV, the CPE assay was conducted as follows: HEp-2 cells were seeded at 6000 cells/well in 96-well plates. The viral inoculation volume was 50 μl and MOI was 0.04. The compound volume and the plates were read at the absorbance of reagent CCK-8, 450 nm/650 nm.

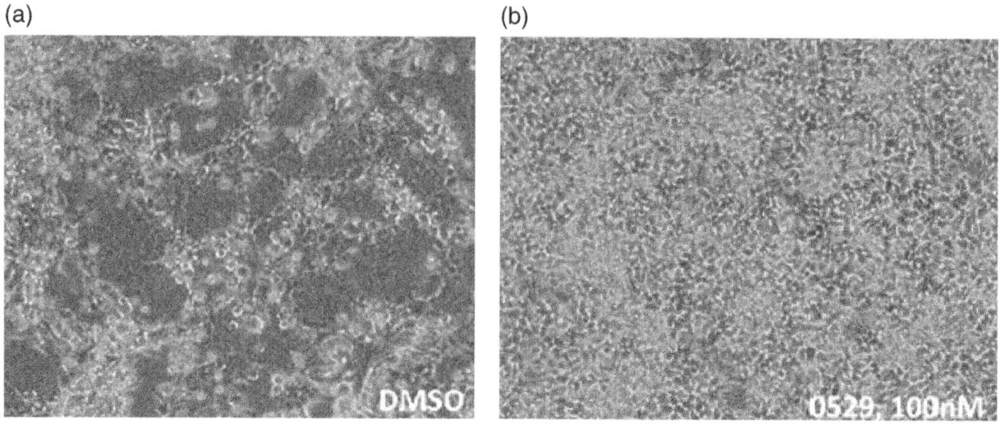

Figure 5.21 RO-0529 inhibits the RSV F protein-induced cell–cell fusion process. RSV-F expression was induced by 1 μg/ml tetracycline in a stable RSV F protein cell line. Compound was serially diluted and added into the RSV-F expressing cells. Cell–cell fusion was observed by microscopy at three days after compound addition.

5.5 Conclusion

In the three antiviral drug discovery projects introduced in this chapter, it is evident that issue-driven problem solving plays a critical role in advancing projects. With the aid of technologies such as structure-based drug design, application of novel compact modules, and careful analysis of SAR, not only was the HBV capsid modulator team able to improve PK and reduce hERG liability, but also very importantly, successfully dial out the CYP3A4 induction issue. In the HBV expression inhibitor program, the team leveraged an ingenious phenotypic HTS platform to identify novel chemical series, which eventually led to the discovery of a clinical compound and also a novel MOA to counter HBV. Lastly, in the RSV fusion inhibitor program, a novel compact module, an oxetane, was utilized to modulate the basicity of an amine, resulting in a less lipophilic molecule, and favorably influencing the tissue distribution leading to the discovery of a compound now in phase III clinical trials. In summary, the proficient use of technologies and hypothesis-driven problem solving have enabled the three antiviral programs to successfully deliver clinical compounds, which are advancing in various stages of development.

References

1 World Health Organization (2021). Hepatitis B. https://www.who.int/en/news-room/fact-sheets/detail/hepatitis-b (accessed 24 September 2021).
2 Papatheodoridis, G., Buti, M., Cornberg, M. et al. (2012). EASL clinical practice guidelines: management of chronic hepatitis B virus infection. *J. Hepatol.* 57: 167–185.
3 Feng, S., Gao, L., Han, X. et al. (2018). Discovery of small molecule therapeutics for treatment of chronic HBV infection. *ACS Infect. Dis.* 4: 257–277.
4 Hu, Y., Zhu, W., Tang, G. et al. (2013). Novel therapeutics in discovery and development for treatment of chronic HBV infection. *Annu. Rep. Med. Chem.* 48: 265–281.
5 Deres, K., Schröder, C.H., Paessens, A. et al. (2003). Inhibition of hepatitis B virus replication by drug-induced depletion of nucleocapsids. *Science* 299: 893–896.

6 Zhou, Z., Hu, T., Zhou, X. et al. (2017). Heteroaryldihydropyrimidine (HAP) and sulfamoylbenzamide (SBA) inhibit hepatitis B virus replication by different molecular mechanisms. *Sci. Rep.* 7: 42374.

7 Perni, R.B., Conway, S.C., Ladner, S.K. et al. (2000). Phenylpropenamide derivatives as inhibitors of hepatitis B virus replication. *Bioorg. Med. Chem. Lett.* 10: 2687.

8 Campagna, M.R., Liu, F., Mao, R. et al. (2013). Sulfamoylbenzamide derivatives inhibit the assembly of hepatitis B virus nucleocapsids. *J. Virol.* 87: 6931–6942.

9 Twu, J.S., Lee, C.H., Lin, P.M., and Schloemer, R.H. (1988). Hepatitis B virus suppresses expression of human β-interferon. *Proc. Natl. Acad. Sci. USA* 85: 252–256.

10 Fernandez, M., Quiroga, J.A., and Carreno, V. (2003). Hepatitis B virus downregulates the human interferon-inducible MxA promoter through direct interaction of precore/core proteins. *J. Gen. Virol.* 84: 2073–2082.

11 Wang, X.Y., Wei, Z.M., Wu, G.Y. et al. (2012). in vitro inhibition of HBV replication by a novel compound, GLS4, and its efficacy against adefovir-dipivoxil-resistant HBV mutations. *Antivir. Ther.* 17: 793–803.

12 Bourne, C.R., Finn, M.G., and Zlotnick, A. (2006). Global structural changes in hepatitis B virus capsids induced by the assembly effector HAP1. *J. Virol.* 80: 11055–11061.

13 Qiu, Z., Lin, X., Zhang, W. et al. (2017). Discovery and pre-clinical characterization of third-generation 4-H heteroaryldihydropyrimidine (HAP) analogues as hepatitis B virus (HBV) capsid inhibitors. *J. Med. Chem.* 60: 3352–3371.

14 Yang, P.L., Althage, A., Chung, J., and Chisari, F.V. (2002). Hydrodynamic injection of viral DNA: a mouse model of acute hepatitis B virus infection. *Proc. Natl. Acad. Sci. USA* 99: 13825–13830.

15 Lin, J.H. (2006). CYP induction-mediated drug interactions: in vitro assessment and clinical implications. *Pharm. Res.* 23: 1089–1116.

16 Lin, X., Shi, H., Zhang, W. et al. (2019). A new approach of mitigating CYP3A4 induction led to the discovery of potent hepatitis B virus (HBV) capsid inhibitor with optimal ADMET profiles. *J. Med. Chem.* 62: 10352–10361.

17 Gao, Y.-D., Olson, S.H., Balkovec, J.M. et al. (2007). Attenuating pregnane X receptor (PXR) activation: a molecular modelling approach. *Xenobiotica* 37: 124–138.

18 Guo, L., Hu, T.-S., Hu, Y.-M. et al. (2014). 6-Bridged heteroaryldihydropyrimidines for the treatment and prophylaxis of hepatitis B virus infection. WO2014184328 A1.

19 Guo, L., Hu, T.-S., Kou, B.-Y. et al. (2016). Novel 6-fused heteroaryldihydropyrimidines for the treatment and prophylaxis of hepatitis B virus infection. US20160083383.

20 Kou, B., Zhu, W., Liu, H. et al. (2017). Synthesis of bridged bicyclic morpholine amino acids as compact modules for medicinal chemistry. *Chem. Lett.* 46: 566–568.

21 Wu, G., Kou, B., Tang, G. et al. (2016). Synthesis of novel and conformationally constrained bridged amino acids as compact modules for drug discovery. *Tetrahedron Lett.* 57: 599–602.

22 Kondo, Y., Ninomiya, M., Kakazu, E. et al. (2013). Hepatitis B surface antigen could contribute to the immunopathogenesis of hepatitis B virus infection. *ISRN Gastroenterol.* 2013: 935295.

23 den Brouw, M.L.O., Binda, R.S., van Roosmalen, M.H. et al. (2009). Hepatitis B virus surface antigen impairs myeloid dendritic cell function: a possible immune escape mechanism of hepatitis B virus. *Immunology* 126: 280–289.

24 Woltman, A.M., den Brouw, M.L.O., Biesta, P.J. et al. (2011). Hepatitis B virus lacks immune activating capacity, but actively inhibits plasmacytoid dendritic cell function. *PLos One* 6: e15324.

25 Zoulim, F. and Durantel, D. (2015). Antiviral therapies and prospects for a cure of chronic hepatitis B. *CSH Perspect. Med.* 5: a021501.

26 Scaglione, S.J. and Lok, A.S.F. (2012). Effectiveness of hepatitis B treatment in clinical practice. *Gastroenterology* 142: 1360–1368.

27 Wu, D. and Ning, Q. (2017). Toward a cure for hepatitis B virus infection: combination therapy involving viral suppression and immune modulation and long-term outcome. *J. Infect. Dis.* 216: S771–S777.

28 Swinney, D.C. and Anthony, J. (2011). How were new medicines discovered? *Nat. Rev.-Drug Discov.* 10: 507–519.

29 Han, X., Zhou, C., Jiang, M. et al. (2018). Discovery of RG7834: the first-in-class selective and orally available small molecule hepatitis B virus expression inhibitor with novel mechanism of action. *J. Med. Chem.* 61: 10619–10634.

30 Mueller, H., Wildum, S., Luangsay, S. et al. (2018). A novel orally available small molecule that inhibits hepatitis B virus expression. *J. Hepatol.* 68: 412–420.

31 Mueller, H., Lopez, A., Tropberger, P. et al. (2019). PAPD5/7 are novel dependency host factors for high-level HBV gene expression. *Hepatology* 69: 1398–1411.

32 Dougherty, A.M., Guo, H., Westby, G. et al. (2007). A substituted tetrahydro-tetrazolo-pyrimidine is a specific and novel inhibitor of hepatitis B virus surface antigen secretion. *Antimicrob. Agents Chemother.* 51: 4427–4437.

33 Yu, W.Q., Goddard, G., Clearfield, E. et al. (2011). Design, synthesis, and biological evaluation of triazolo-pyrimidine derivatives as novel inhibitors of hepatitis B virus surface antigen (HBsAg) secretion. *J. Med. Chem.* 54: 5660–5670.

34 Xu, Y.B., Yang, L., Wang, G.F. et al. (2014). Benzimidazole derivative, BM601, a novel inhibitor of hepatitis B virus and HBsAg secretion. *Antivir. Res.* 107: 6–15.

35 Lv, Z.G., He, W., Tian, X.H. et al. (2015). Design, synthesis, and biological evaluation of new N-4-substituted 2 '-deoxy-2 '-fluoro-4 '-azido cytidine derivatives as potent anti-HBV agents. *Eur. J. Med. Chem.* 101: 103–110.

36 Zheng, L.Y., Wang, Q., Yang, X.R. et al. (2012). Antiviral activity of FNC, 2'-deoxy-2'-beta-fluoro-4'-azidocytidine, against human and duck HBV replication. *Antivir. Ther.* 17: 679–687.

37 Zhou, Y.B., Zhang, Y., Yang, X. et al. (2012). Novel nucleoside analogue FNC is effective against both wild-type and lamivudine-resistant HBV clinical isolates. *Antivir. Ther.* 17: 1593–1599.

38 Zhou, T., Block, T., Liu, F. et al. (2018). HBsAg mRNA degradation induced by a dihydroquinolizinone compound depends on the HBV posttranscriptional regulatory element. *Antivir. Res.* 149: 191–201.

39 Borchers, A.T., Chang, C., Gershwin, M.E., and Gershwin, L.J. (2013). Respiratory syncytial virus-a comprehensive review. *Clin. Rev. Allergy Immunol.* 45: 331–379.

40 Piedimonte, G. and Perez, M.K. (2014). Alternative mechanisms for respiratory syncytial virus (RSV) infection and persistence: could RSV be transmitted through the placenta and persist into developing fetal lungs? *Curr. Opin. Pharmacol.* 16: 82–88.

41 Wright, M. and Piedimonte, G. (2011). Respiratory syncytial virus prevention and therapy: past, present, and future. *Pediatr. Pulmonol.* 46: 324–347.

42 Falsey, A.R., Hennessey, P.A., Formica, M.A. et al. (2005). Respiratory syncytial virus infection in elderly and high-risk adults. *N. Engl. J. Med.* 352 (17): 1749–1759.

43 Boeckh, M., Englund, J., Li, Y. et al. (2007). Randomized controlled multicenter trial of aerosolized ribavirin for respiratory syncytial virus upper respiratory tract infection in hematopoietic cell transplant recipients. *Clin. Infect. Dis.* 44: 245–249.

44 Zhao, X., Singh, M., Malashkevich, V.N., and Kim, P.S. (2000). Structural characterization of the human respiratory syncytial virus fusion protein core. *PNAS* 97: 14172–14177.

45 Ding, W.D., Mitsner, B., Krishnamurthy, G. et al. (1998). Novel and specific respiratory syncytial virus inhibitors that target virus fusion. *J. Med. Chem.* 41: 2671–2675.

46 Andries, K., Moeremans, M., Gevers, T. et al. (2003). Substituted benzimidazoles with nanomolar activity against respiratory syncytial virus. *Antivir. Res.* 60: 209–219.

47 Bonfanti, J.F., Meyer, C., Doublet, F. et al. (2008). Selection of a respiratory syncytial virus fusion inhibitor clinical candidate. 2. Discovery of a morpholinopropylaminobenzimidazole derivative (TMC353121). *J. Med. Chem.* 51: 875–896.

48 Yu, K.L., Sin, N., Civiello, R.L. et al. (2007). Respiratory syncytial virus fusion inhibitors. Part 4: optimization for oral bioavailability. *Bioorg. Med. Chem. Lett.* 17: 895–901.

49 Sun, Z., Pan, Y., Jiang, S., and Lu, L. (2013). Respiratory syncytial virus entry inhibitors targeting the F protein. *Viruses* 5: 211–225.

50 Stevens, M., Rusch, S., Huntjens, D. et al. (2018). Antiviral activity of oral JNJ-53718678 in healthy adult volunteers challenged with respiratory syncytial virus: a placebo-controlled study. *J. Infect. Dis.* 218 (5): 748–756.

51 Mackman, R.L., Sangi, M., Sperandio, D. et al. (2015). Discovery of an oral respiratory syncytial virus (RSV) fusion inhibitor (GS-5806) and clinical proof of concept in a human RSV challenge study. *J. Med. Chem.* 58: 1630–1643.

52 Perron, M., Stray, K., Kinkade, A. et al. (2016). GS-5806 inhibits a broad range of respiratory syncytial virus clinical isolates by blocking the virus-cell fusion process. *Antimicrob. Agents Chemother.* 60: 1264–1273.

53 DeVincenzo, J., Tait, D., Oluwayi, O. et al. (2018). Safety and efficacy of oral RV521 in a human respiratory syncytial virus (RSV) phase 2a challenge study. B14 late breaking clinical trials and first reports in asthma and COPD. *Am. J. Respir. Crit. Care Med.* 197: A7715.

54 Douglas, J.L., Panis, M.L., Ho, E. et al. (2003). Inhibition of respiratory syncytial virus fusion by the small molecule VP-14637 via specific interactions with F protein. *J. Virol.* 77: 5054–5064.

55 Zheng, X., Liang, C., Wang, L. et al. (2018). Discovery of benzoazepinequinoline (BAQ) derivatives as novel, potent, orally bioavailable respiratory syncytial virus fusion inhibitors. *J. Med. Chem.* 61 (22): 10228–10241.

56 Ark Biosciences Inc. (2016). Viral inhibition in children for treatment of RSV (VICTOR). https://clinicaltrials.gov/cts/show/NCT02654171 (accessed 24 September 2021).

57 Wuitschik, G., Rogers-Evans, M., Müller, K. et al. (2006). Oxetanes as promising modules in drug discovery. *Angew. Chem. Int. Ed.* 45: 7736–7739.

58 Zheng, X., Gao, L., Wang, L. et al. (2019). Discovery of Ziresovir as a potent, selective, and orally bioavailable respiratory syncytial virus fusion protein inhibitor. *J. Med. Chem.* 62 (13): 6003–6014.

59 Ark Biosciences Inc. (2014). Safety, tolerability and PK study of AK-0529 in healthy human. http://clinicaltrials.gov/ct2/show/ NCT02297594 (accessed 24 September 2021).

60 Ark Biosciences Inc. (2017).A study of AK0529 to evaluate pharmacokinetics and safety in Chinese healthy volunteers. http://clinicaltrials.gov/ct2/show/ NCT03322800 (accessed 24 September 2021).

61 Ark Biosciences Inc. (2018). A study to evaluate the pharmacokinetics, metabolism, and excretion of AK0529. http://clinicaltrials.gov/ct2/show/NCT03400995 (accessed 24 September 2021).

6

Discovery of Subtype Selective Agonists of the Group II Metabotropic Glutamate Receptors

Junliang Hao

Discovery Chemistry Research and Technologies, Lilly Research Laboratory, Eli Lilly and Company, Lilly Corporate Center, IN 46285, Indianapolis, IN, USA

6.1 Background

6.1.1 The Dopamine and Glutamate Hypotheses of Schizophrenia

Schizophrenia, coined from Greek roots *schizein* (to split) and *phren* (mind) [1], is a chronic and severely disabling mental disorder that requires lifelong treatment and affects about 0.30–0.66% of the population during their lifetime [2]. People with schizophrenia may display positive and negative symptoms and cognitive deficits. Positive symptoms include delusions, hallucinations (e.g. auditory, visual, and olfactory), and disorganized thinking and behaviors, while negative symptoms consist of social withdrawal and apathy. Cognitive deficits may encompass difficulties in attention, learning and memory. These symptoms can vary in severity, duration and frequency over time, with periods of exacerbation and remission. Causes of schizophrenia are not well understood, but are thought to be a combination of genetics, environment, and developmental factors. The financial cost of schizophrenia to society in the United States is staggeringly high (e.g. estimated to be $62.7 billion in 2002) [3]. Antipsychotic medications are the cornerstone of treatments of this complex disorder. Other treatments include psychosocial interventions and electroconvulsive therapy [4, 5].

The serendipitous discovery and subsequent introduction of typical antipsychotics (i.e. chlorpromazine in the 1950s [6]) revolutionized schizophrenia treatment, allowing large numbers of schizophrenia patients to be deinstitutionalized and return to relatively normal lives. However, serious side effects of these typical treatments, such as tardive dyskinesia and akathisia, led to the introduction of newer, second-generation atypical antipsychotics such as clozapine [7], which were claimed to be less likely to cause extrapyramidal side effects [8–11]. Current antipsychotic treatments often reduce positive symptoms but usually fail to significantly improve negative symptoms and cognitive deficits. The observation that the efficacy of most of these treatments versus positive symptoms was strongly correlated with their binding affinity for the dopamine D2 receptor [12, 13], together with other observations (e.g. amphetamine-induced psychosis [14]), led to the introduction of the dopamine hypothesis of schizophrenia [15–17]. This hypothesis suggested that specific dopamine pathways may be hyperactive in schizophrenic brains, and blockade of dopamine D2 receptors in the mesolimbic pathway is responsible for the relief of the positive

Contemporary Accounts in Drug Discovery and Development, First Edition. Edited by Xianhai Huang, Robert G. Aslanian, and Wayne H. Tang.

symptoms. More recently, the revised dopamine hypothesis [16, 18] takes into account the complexity of dopamine transmission in the brain and adds hypoactive dopamine transmission in the prefrontal cortex (mesocortical pathway), which may be related to negative symptoms and cognitive deficits. In support of this revised hypothesis, pharmacological magnetic resonance imaging (phMRI) studies have confirmed enhanced dopaminergic activity in the prefrontal cortex in schizophrenia following atypical antipsychotic treatment [18].

The inability of dopamine receptor-blocking drugs to efficiently improve the negative symptoms and cognitive deficits of schizophrenia suggested that dopamine dysregulation cannot fully account for the underlying pathology and led to the exploration of alternative hypotheses. One such hypothesis that first emerged in the 1990s is related to the primary excitatory neurotransmitter glutamate [19]. This glutamatergic hypothesis can be traced back to the observation that administration of *N*-methyl-D-aspartate (NMDA) receptor non-competitive antagonists, such as phencyclidine (PCP) and ketamine, induced not only positive symptoms but also negative symptoms and cognitive deficits in healthy individuals, and exacerbated pre-existing symptoms in schizophrenic patients [20–22]. Furthermore, the psychosis-inducing potency of a range of compounds has been shown to be positively correlated to their rank order potency at the NMDA receptors [23]. NMDA receptor hypofunction forms the cornerstone for the glutamatergic hypothesis of schizophrenia. Studies have shown that NMDA receptor antagonists increase glutamate release in the prefrontal cortex, which then leads to excessive activation of α-amino-3-hydroxy-5-methyl-4-isoxazolepropionic acid (AMPA) receptors triggering subsequent excitotoxic processes, likely the result of disinhibition of cortical inhibitory GABAergic interneurons [20, 24].

6.1.2 The Ionotropic and Metabotropic Glutamate Receptors

The NMDA receptor belongs to the family of ionotropic glutamate receptors (iGluRs) through which glutamate mediates fast excitatory synaptic transmission. Other iGluRs include the AMPA receptor and kainate receptor [25]. NMDA receptors are ligand-gated ion channels that allow Ca^{2+} influx into cells upon activation, and are therefore critical for synaptic plasticity and memory function. They are heteromeric assemblies of four subunits selected from GluN1 and GluN2A-D [26]. Each NMDA receptor has two glycine-binding GluN1 and two glutamate-binding GluN2 subunits. Each subunit contains four semiautonomous domains (Figure 6.1a): (i) an extracellular amino-terminal domain (ATD); (ii) an agonist binding domain (ABD) where the orthosteric agonists glycine or glutamate bind; (iii) a pore-forming transmembrane domain (TMD); and (iv) an intramolecular carboxyl-terminal domain (CTD). Their activation is unique in that it requires simultaneous binding of two agonists, glutamate and glycine (or D-serine). Since glycine (or D-serine)

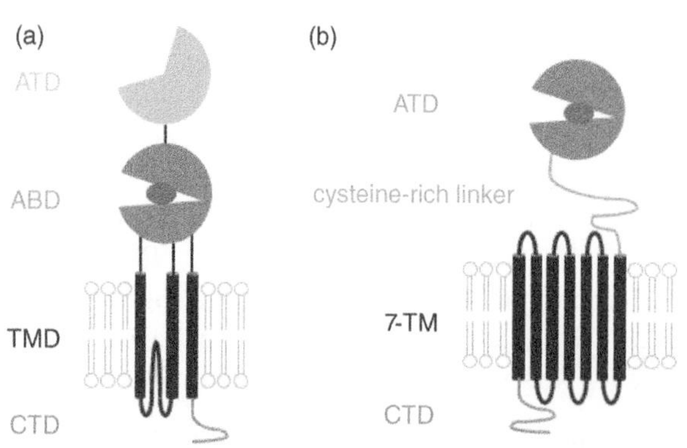

Figure 6.1 Cartoon representations of a subunit of NMDA receptors (a) and mGlu receptors (b). Spheres represent orthosteric agonists. *Source:* Adapted from figure 2 in ref. [27]. Reproduced with permission of Elsevier.

is thought to be continuously present at a relatively high concentration, the NMDA receptors are temporally activated by the synaptic release of glutamate. NMDA antagonists such as ketamine and PCP enter and bind inside the channel in the TMD after the receptor has been activated, and block the influx of Ca^{2+} [28].

Another family of glutamate receptors are the metabotropic glutamate receptors (mGluRs), which are class C G protein-coupled receptors (GPCRs) acting to regulate synaptic glutamatergic transmission [25]. They are classified into three subgroups based on sequence homology, pharmacology, expression location and signal transduction mechanism: group I ($mGlu_1$, $mGlu_5$); group II ($mGlu_2$ and $mGlu_3$); and group III ($mGlu_4$, $mGlu_6$, $mGlu_7$, and $mGlu_8$). The group I mGlu receptors are predominantly postsynaptic and are coupled to $G\alpha_q$. Stimulating these receptors leads to activation of phospholipase C (PLC). The groups II and III mGlu receptors are mainly presynaptic and transduce signals via $G\alpha_{i/o}$. They function as autoreceptors, limiting excessive transmitter release. Activation of these receptors leads to the inhibition of adenylate cyclase and a concomitant reduction of cyclic adenosine monophosphate (cAMP) formation. The mGlu receptors are homodimers with each protomer (Figure 6.1b) possessing a very large extracellular ATD, a 7-transmembrane (7-TM) domain, and an intramolecular CTD. The ATD of the mGluRs is in a shape similar to the ABD of the NMDA receptor, containing two globular lobes with a hinge region. The orthosteric ligand binding pocket is situated between the two lobes next to the hinge region. Activation of mGlu receptors by glutamate results in the closure of the bi-lobed ATD and subsequent conformational rearrangement of the 7-TM domain. These mGlu receptors have the ability to directly or indirectly alter NMDA receptor function or other aspects of glutamatergic signaling [29]. For example, postsynaptic $mGlu_5$ receptor activation can increase NMDA receptor function in forebrain regions [30].

6.1.3 Orthosteric Agonists of the Group II mGlu Receptors

According to the glutamatergic hypothesis of schizophrenia, pharmacological agents that enhance NMDA receptor function directly or indirectly could be therapeutically beneficial in treating all symptomatic aspects of schizophrenia. Direct activation of the NMDA receptor by either increasing synaptic glycine (induced by blockade of glycine transporter GlyT-1) [31, 32] or exogenous introduction of the more potent agonist D-serine has been explored in the clinic and summarized elsewhere [33]. At Lilly, the group II mGlu receptors were considered as promising new targets for treating schizophrenia [34] since these presynaptic autoreceptors limit glutamate release into the synaptic cleft when activated and can reduce the excessive glutamatergic tone induced by NMDA receptor antagonists in the prefrontal cortex [35].

The journey to synthesize a series of highly selective agonists for the group II mGlu receptors started with (2*S*,1'*S*,2'*S*)-2-(carboxycyclopropyl)glycine (L-CCG-1, **1**, Figure 6.2) [36, 37] and its analog (2*S*,2'*R*,3'*R*)-2-(2′,3′-dicarboxycyclopropyl)glycine (DCG-IV, **2**) [38, 39]. Both compounds

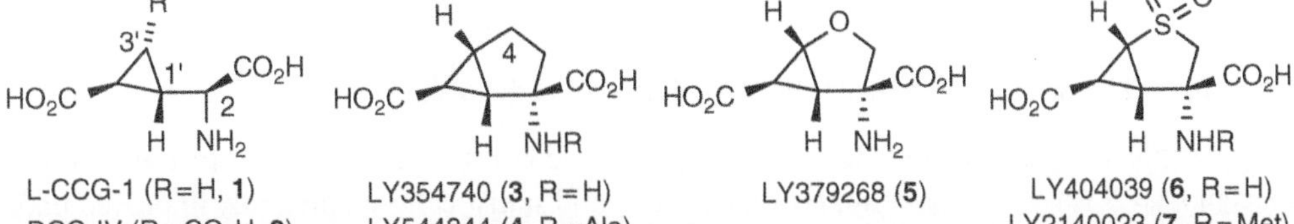

L-CCG-1 (R = H, **1**)
DCG-IV (R = CO_2H, **2**)

LY354740 (**3**, R = H)
LY544344 (**4**, R = Ala)

LY379268 (**5**)

LY404039 (**6**, R = H)
LY2140023 (**7**, R = Met)

Figure 6.2 A series of highly selective agonists for the group II mGlu receptors from further conformation restriction of L-CCG-1/DCG-IV.

possessed high potency at group II mGlu receptors but were also active at other glutamate receptors at higher concentrations. Structurally, they were conformationally constrained L-glutamic acid (L-Glu) analogs. It was thought then that activation of the group II mGlu receptors would require the L-Glu backbone to adopt a fully extended conformation (the conformation hypothesis) [27, 40] and this active conformation could be mimicked by further constraining the conformation of these two molecules. This was achieved by tying the C2 and C3′ positions of L-CCG-1 with a two-carbon linker, resulting in LY354740 (**3**, also known as eglumetad, Figure 6.2) [40]. This bicyclo[3.1.0]-hexane scaffold oriented the α-amino acid and the distal acid of the L-Glu in a fully extended arrangement confirmed by the X-ray single crystal structure of **3**. At that time, incorporation of this linker to retain the agonist activity was a risky proposition because the α-amino acid portion was known to be sensitive to steric modifications and addition of a small methyl group to the α-carbon could result in conversion of an agonist to an antagonist [27, 41]. Nevertheless, **3** was demonstrated to be a highly potent and selective group II mGlu receptor agonist with minimum agonist or antagonist activities at other mGlu or iGlu receptors (Table 6.1) [42, 43]. This could be explained by the recently proposed steric hypothesis that suggested these two methylene units could be accommodated in the ligand-binding pocket in the group II mGlu receptors but not in the group I or III mGlu receptors [27]. Further structure–activity relationship (SAR) studies found the C4-methylene unit of the all carbon **3** could be replaced with either an oxygen atom (LY379268) [44] or a sulfone group (LY404039, also known as pomaglutemad) (Figure 6.2) [45, 49] while retaining the potency and group II selectivity (Table 6.1).

6.1.4 Prodrug Approach to Improve Oral Bioavailability

The highly polar nature of these amino acid ligands resulted in relatively low bioavailability in different animal species (e.g 10% in rats and 40% in dogs for **3**), necessitating a search for an effective prodrug form that could yield higher bioavailability of the active (or parent) drug following oral dosing [50]. It was found that targeting the human peptide transporter hPepT1 that actively transports di- or tripeptides in the small intestine could result in improved absorption, and a dipeptide *N*-alanine prodrug (**4**, LY544344, also known as talaglumetad, Figure 6.2) was also effectively converted to the parent drug by intestinal peptidases before its appearance in the hepatic portal vein [50–52]. Compared with **3**, the sulfone analog **6** demonstrated much improved bioavailability in preclinical animal studies (ca. 63% in rats) following oral administration without the need of a prodrug approach [49]. However, its oral absorption was later found to be low (3% in humans) [53, 54]. A methionine prodrug LY2140023 (**7**, also known as pomaglutemad methionil, Figure 6.2) of **6** was then synthesized [53, 55]. This prodrug was found to be a substrate for both hPepT1 (for active transport to improve absorption in the intestine) and hPepT2 (for peptide efflux from the brain and reuptake in the kidney). Following oral administration in humans, it was rapidly absorbed and converted (both prodrug and active were detectable at the first sampling time point at 0.5 h, ~70% conversion) to active **6**, resulting in an improved oral bioavailability of 49% [53, 55].

6.1.5 Clinical Studies of 6 in Schizophrenia (via its Prodrug 7)

In an initial phase 2, randomized, double-blind, placebo-controlled, inpatient, multicenter, proof-of-concept trial, **7** was found to be superior to placebo and statistically significantly improved both positive and negative symptoms of schizophrenia as measured by the positive and negative syndrome scale (PANSS) total scores [54]. Most treatment-emergent adverse events (TEAEs) were mild to moderate in severity. In contrast to the statistically significant weight gain (0.74 kg) observed

Table 6.1 Binding affinities and functional agonist or *antagonist* responses in cells expressing recombinant human $mGlu_2$ and $mGlu_3$ receptors.

	^{3}H-LY459477 binding K_i ± SEM (nM)		**Whole cell functional agonist or *antagonist* responses EC_{50} or *IC_{50}* ± SEM (nM), (E_{max})**			
			$hmGlu_2$		**$hmGlu_3$**	
	$hmGlu_2$	**$hmGlu_3$**	**cAMP[b]**	**Ca^{2+} FLIPR[c]**	**cAMP[b]**	**Ca^{2+} FLIPR[c]**
3 [42, 43]	71.7 ± 5.9	107 ± 12.5	7.0 ± 1.2 (100%)	34.4 ± 4.0 (94%)	27.9 ± 3.0 (100%)	140 ± 4.6 (103%)
5 [44]	24.3 ± 3.9[a]	9.7 ± 0.5[a]	2.7 ± 0.3 (−)	-	4.6 ± 0.1 (−)	-
6 [45]	149 ± 11[a]	92 ± 14[a]	23 ± 1 (−)	-	48 ± 10 (−)	-
8 [46]	1040 ± 371	631 ± 129	62.5 ± 13.4 (98%)	-	*4460 ± 2620* *(73%)*	-
9 [47]	-	-	0.56 ± 0.12 (98%)	-	0.24 ± 0.04 (100%)	-
10 [43]	420 ± 30.3	138 ± 15.3	*7000 ± 64.5* *(80%)*	*>12 500*	*1070 ± 258* *(92%)*	*7690 ± 7260* *(62%)*
11 [43]	144 ± 20	156 ± 23.5	5.6 ± 0.5 (99%)	21.1 ± 2.2 (95%)	129 ± 34.9 (61%) *2730 ± 1430* *(40%)*	>25 000 *25 200 ± 15 600* *(58%)*
13 [47]	966 ± 40.5	637 ± 49	28.6 ± 7.0 (98%)	-	166 ± 66 (88%)	-
14 [43]	1230 ± 139	272 ± 25.1	1170 ± 232 (49%) *15 300 ± 5400* *(53%)*	>25 000 *>12 500*	*3370 ± 1210* *(77%)*	*6710* *(56%)*
15 [43]	60.4 ± 6.2	126 ± 14.5	11.2 ± 1.3 (99%)	69.1 ± 21.9 (89%)	227 ± 39.1 (52%) *4260 ± 4090* *(54%)*	>25 000 *3100 ± 170* *(72%)*
16 [48]	4410 ± 598	284 ± 15.4	21.2 ± 2.2 (99%)	-	24.9 ± 3.7 (98%)	-
17 [48]	11 200 ± 2420	1070 ± 172	508 ± 92 (100%)	-	356 ± 88 (94%)	-
18 [48]	294 ± 13	6.8 ± 1.8	19.8 ± 3.3 (100%)	-	5.0 ± 1.1 (92%)	-
19 [48]	412 ± 111	0.93 ± 0.23	47.5 ± 3.6 (98%)	277 ± 41 (86%)	0.47 ± 0.09 (88%)	11.6 ± 2.0 (54%)

[a] ^{3}H-LY341495 binding.

[b] The cAMP assays measures G_i-mediated reduction in cyclic nucleotide formation after activation of adenylate cyclase.

[c] The Ca^{2+} FLIPR assay measures downstream intracellular calcium mobilization coupled through an artificially expressed calcium-coupling G protein, $G\alpha_{15}$.

in the olanzapine group in this study, a mean weight loss (0.51 kg) from baseline was observed in the **7** treatment group. Also, treatment with this compound did not appear to be associated with the common D2-related adverse events, such as extrapyramidal symptoms and hyperprolactinemia.

These initial results were impressive and exciting since **7**, as the first investigational agent that acted on the mGlu receptors for schizophrenia treatment [54], could potentially lead to a new class of antipsychotics that is mechanistically different from all current dopaminergic treatments [56]. However, subsequent phase 2 and phase 3 clinical studies failed to replicate these impressive preliminary results as either monotherapy or add-on therapy with standard-of-care treatments [57–61]. These results led to the discontinuation of the development program of **7** at Lilly. The failure raised an immediate question whether the group II mGlu receptors were still promising therapeutic targets for the treatment of schizophrenia [62]. A post hoc meta-analysis of five of the double-blind, placebo-controlled efficacy studies showed that **7** exhibited therapeutic efficacy in subgroups of patients early-in-disease or those previously treated with dopamine D2 drugs (but not in patients previously treated with 5HT2A agents) [63]. These patients may be in a hyper-glutamatergic state or have sufficient $mGlu_2$ receptors in the frontal cortex that is required for **7** to produce an antipsychotic response. In contrast, patients late-in-disease may already be in a hypo-glutamatergic state and patients previously treated with 5HT2A agents may have reduced $mGlu_2$ receptor expression likely due to a newly discovered interplay between the 5HT2A and $mGlu_2$ receptors [64–69].

6.1.6 Rationale for Subtype Selective Agonists of the Group II mGlu Receptors

While compounds **3**, **5**, and **6** were potent agonists capable of selectively activating the group II mGlu receptors (vs other glutamate receptors), they activated both the $mGlu_2$ and $mGlu_3$ receptors with similar potency. It is known in the literature that these two receptors are localized differentially in the central nervous system and are involved in distinct biological processes [48, 70]. For example, genetic studies have implicated an association of GRM3 (gene encoding $mGlu_3$ receptor) polymorphisms with psychosis [71], but the antipsychotic responses to $mGlu_{2/3}$ agonists in preclinical animal models of psychosis appeared to be mediated by the $mGlu_2$ receptor through studies in either $mGlu_2$ or $mGlu_3$ knock-out animals [72–74]. However, compensatory changes could occur over the course of development in these genetically modified animals, potentially confounding the interpretation of study results. Therefore, small molecules that are capable of pharmacologically differentiating the $mGlu_2$ and $mGlu_3$ subtypes in both *in vitro* and *in vivo* systems would better serve the purpose of delineating the diverse roles for the $mGlu_2$ and $mGlu_3$ receptor subtypes.

It is well known that the mGlu receptors have two ligand-binding sites [75]. Compounds **3**, **5**, and **6** are glutamate analogs and bind at the orthosteric site in the ATD of the mGlu receptors. In addition to this orthosteric ligand-binding site, an allosteric site that has more diverse sequence homology – with the potential to provide modulators with potentially higher subtype selectivity – exists for the mGlu receptors in the 7-TM domain [76–79]. Although the first selective $mGlu_2$ positive allosteric modulators (PAMs) appeared in the literature in 2003 [80, 81], a selective $mGlu_3$ negative allosteric modulators (NAMs) with only modest potency was first disclosed in 2011 [82, 83] and no selective $mGlu_3$ PAMs have been disclosed [84]. Therefore, to further define the diverse roles for the mGlu2 and mGlu3 receptor subtypes *in vivo*, it was of interest to identify potent and subtype selective tools for these two structurally related receptors. The following sections summarize efforts to discover and design molecules capable of selectively activating either the $mGlu_2$ or the $mGlu_3$ receptors at the orthosteric site.

6.2 Discovery of Subtype Selective Agonist LY2812223 of the MGLU$_2$ Receptor

6.2.1 Barriers to Achieve High Subtype Selectivity at the Orthosteric Site

At the outset of the project to identify new mGlu$_2$ subtype selective agonists, several crystal structures of the mGlu ATD proteins had been reported in the literature [85–87]. These included four agonist-bound rat mGlu$_3$ ATD structures [87]. As shown in the L-Glu-rmGlu$_3$ ATD structure (PDB ID: 2E4U) in Figure 6.3a, seven residues interact with L-Glu either directly through a H-bond or salt-bridge, or indirectly through a H-bonded water molecule (Table 6.2). Given the high sequence homology between these two receptor subtypes [43], the residues in the mGlu$_2$ receptor that are involved in interactions with L-Glu were expected to be identical to those observed in the mGlu$_3$ crystal structures. This was confirmed by the L-Glu-hmGlu$_2$ ATD crystal structure (PDB ID: 5CNI) [43]. Furthermore, in the mGlu$_3$ crystal structures, it appeared that there was a ceiling in the binding pocket formed by Y150 from the top lobe (LB1) and R277 from the bottom lobe (LB2) which interact with each other through a π-cation interaction. Larger substitutions pointing in this direction will be limited by the Y150-R277 pair in the mGlu$_3$ ATD structure. Given the almost identical orthosteric binding pocket between the mGlu$_2$ and mGlu$_3$ receptors, it would be expected that the corresponding Y144-R271 amino acids of mGlu$_2$ would limit larger substitutions in the mGlu$_2$ pocket as well [43].

These structural features were consistent with our previous SAR observations that only small substitutions at the C4 position of **3** were tolerated for agonist activity [47]. For example, addition of a C4β-F (**9**, Figure 6.4) resulted in improved potency at both the mGlu$_2$ and the mGlu$_3$ receptors (Table 6.1). Addition of either an α- or β-substituted methyl group (**8** and **13**) at the C4 position resulted

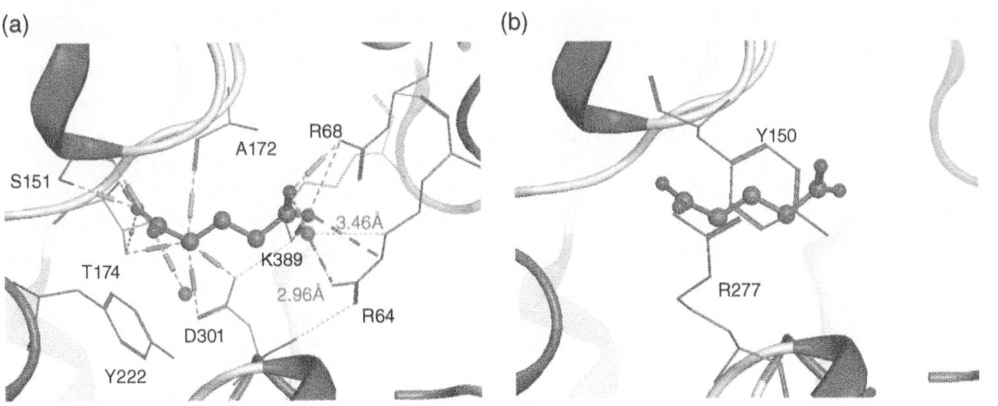

Figure 6.3 (a) Key interactions in the L-Glu-mGlu$_3$ ATD structure (PDB ID: 2E4U). (b) The ceiling of the ligand binding pocket formed by the Y150-R277 pair. The dotted lines (with cylinders) indicate favorable interactions between the ligand and residues using MOE software [88]. The dashed line shows the distance between the atoms at the ends of the line. *Source:* (a) Adapted from figure 7a in ref. [75]. Licensed under CC BY 4.0.

Table 6.2 Residues interacting with L-Glu in the orthosteric site.

Receptor	Residues						
mGlu$_2$	R57	R61	S145	A166	T168	D295	K377
mGlu$_3$	R64	R68	S151	A172	T174	D301	K389

in significantly decreased binding affinity at both receptors (Table 6.1). Further increasing the size of the substituents (e.g. **10** and **14**) resulted in loss of functional agonist activity. Both C4α- and C4β-substituted SPh analogs were weak antagonists at both the $mGlu_2$ and the $mGlu_3$ receptors (Table 6.1).

6.2.2 Discovery of Subtype Selective Agonists for the $mGlu_2$ Receptor

While the SAR exercise with small substitutions at the C4 positions was very limited in scope, it did provide an interesting hit. In the cAMP functional assay, compound **8** with the C4α-Me displayed weak antagonist activity (IC_{50} = 4460 nM) at the $mGlu_3$ receptor with reasonable agonist activity (EC_{50} = 62.5 nM) at the $mGlu_2$ receptor. This was in direct contrast to the C4β-Me analog **13**, which displayed agonist activity at both the $mGlu_2$ and the $mGlu_3$ receptors (EC_{50} of 28.5 and 166 nM, respectively) in the cAMP assay. Intrigued by the unique and remarkable $mGlu_2$ agonist/$mGlu_3$ antagonist activity of **8**, additional and expansive SAR at the C4 position was conducted. To this end, a series of C4α- and C4β-thiotriazolyl substituted compounds was identified with surprising potency at the $mGlu_2$ receptors in the cAMP agonist assay (Figure 6.4) [43]. This was in sharp contrast to the pharmacology exhibited by thiophenylethers **10** and **14** when tested under identical assay conditions (Table 6.1). For example, both C4α (**11**, LY2812223) and C4β unsubstituted triazoles (**15**) showed potent full agonist activity at the $mGlu_2$ receptor in the cAMP assay (EC_{50} of 5.6 and 11.2 nM, respectively). Furthermore, these compounds displayed partial agonist activity at the $mGlu_3$ receptor in the cAMP assay (EC_{50} of 129 and 227 nM, E_{max} of 61 and 52%, respectively), representing over 20× selectivity at the $mGlu_2$ receptor in the cAMP assay. In the less-amplified calcium mobilization (FLIPR) assay, both compounds also displayed very high subtype selectivity with minimum agonist activity at the $mGlu_3$ receptor. While compounds **11** and **15** showed potent full agonist activity at the $mGlu_2$ receptors (EC_{50} of 21.1 and 69.1 nM, respectively), they had no measurable agonist activity, but very weak partial antagonist activity, for the $mGlu_3$ receptors in the FLIPR assay (IC_{50} of 25 200 and 3100 nM, E_{max} of 58 and 71%, respectively).

6.2.3 Additional *in vitro* Characterization of 11

LY2812223 (**11**) was further characterized in various *in vitro* assays [89]. In addition to the cAMP and FLIPR assays, the CellKey assay, which provides an indication of general cellular activation, was employed to assess the agonist selectivity in human $mGlu_2$- and $mGlu_3$-expressing AV12 cells.

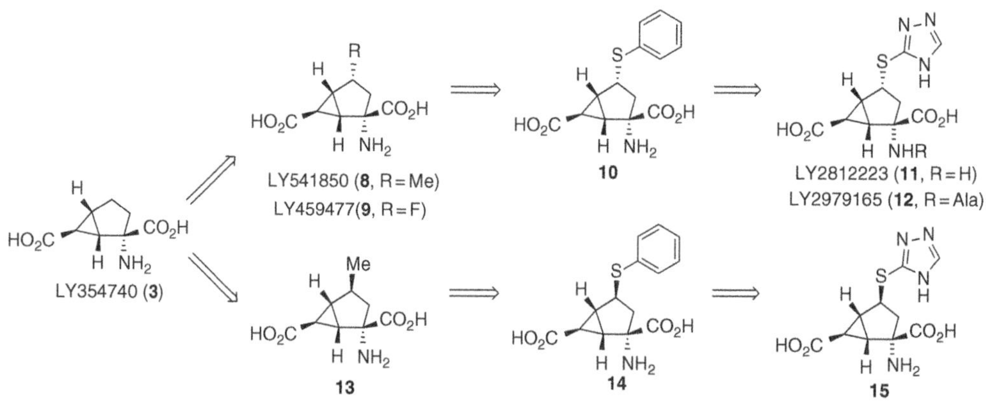

Figure 6.4 SAR evolution leading to discovery of $mGlu_2$ subtype selective agonists.

As shown in Figure 6.5, **11** was active in the mGlu$_2$-expressing cell line (EC$_{50}$ = 29.4 nM), but not in the mGlu$_3$-expressing cell line (EC$_{50}$ >25 000 nM). For comparison, the concentration–response curves in the cAMP and FLIPR assays are also shown in the same graphs. Moreover, **11** was found to be active (EC$_{50}$ = 13.6 nM) in the GTPγS binding assay in the mGlu$_2$-expressing cell line, which measures direct G$_i$ protein activation (Figure 6.5a).

To further validate the selectivity of **11**, it was tested in native membranes derived from brain cortex from wild-type mGlu$_2$-KO and mGlu$_3$-KO mice with GTPγS binding as the functional readout (Figure 6.6a) [89]. Agonist activity was almost abolished in the mGlu$_2$-KO preparations, but not in the mGlu$_3$-KO preparations (EC$_{50}$ = 46.4 nM, E$_{max}$ = 60.6%), where it showed similar activity as in the wild-type preparations (EC$_{50}$ = 23.3 nM, E$_{max}$ = 36.8%). Similarly, in wild-type rat cortical membranes (Figure 6.6b) [89], **11** was a partial agonist with similar efficacy (EC$_{50}$ = 40.1 nM, E$_{max}$ = 37.3%). In contrast, compound **5** was a full agonist in this assay (EC$_{50}$ = 7.6 nM, E$_{max}$ = 92.8%). The partial agonism was further demonstrated in the rat cortical synaptosome preparation measuring changes in cAMP production (Figure 6.6c) [89], where **11** behaved as a partial agonist (EC$_{50}$ = 16.1 nM, E$_{max}$ = 62.9%) compared with the full agonist response of LY379268 (EC$_{50}$ = 6.8 nM, E$_{max}$ = 96.4%). Finally, **11** was assessed in the GTPγS binding assays in cortical membranes derived from both nonhuman primates and human postmortem brain samples (Figure 6.6d) [89]. Similar to the partial agonism observed in rodent brain tissue preparations, **11** elicited partial agonist responses in both nonhuman primate (EC$_{50}$ = 13.5 nM, E$_{max}$ = 39.8%) and human (EC$_{50}$ = 7.57 nM, E$_{max}$ = 49.2%) brain membranes. Taken together, these data suggested that **11** is an mGlu$_2$-selective agonist with likely low partial mGlu$_3$ agonist activity. The observation of a partial agonist response of **11** but full agonist response of **5** in tissues derived from wild-type mice, rats, monkey and humans might suggest a functional interplay between the mGlu$_2$ and mGlu$_3$ receptors in the cortex [89].

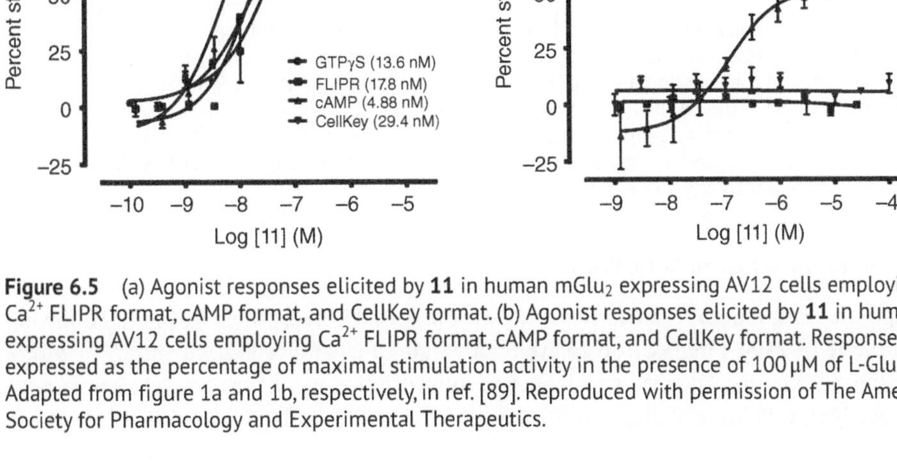

Figure 6.5 (a) Agonist responses elicited by **11** in human mGlu$_2$ expressing AV12 cells employing GTPγ, Ca^{2+} FLIPR format, cAMP format, and CellKey format. (b) Agonist responses elicited by **11** in human mGlu$_3$ expressing AV12 cells employing Ca^{2+} FLIPR format, cAMP format, and CellKey format. Response is expressed as the percentage of maximal stimulation activity in the presence of 100 μM of L-Glu. *Source:* Adapted from figure 1a and 1b, respectively, in ref. [89]. Reproduced with permission of The American Society for Pharmacology and Experimental Therapeutics.

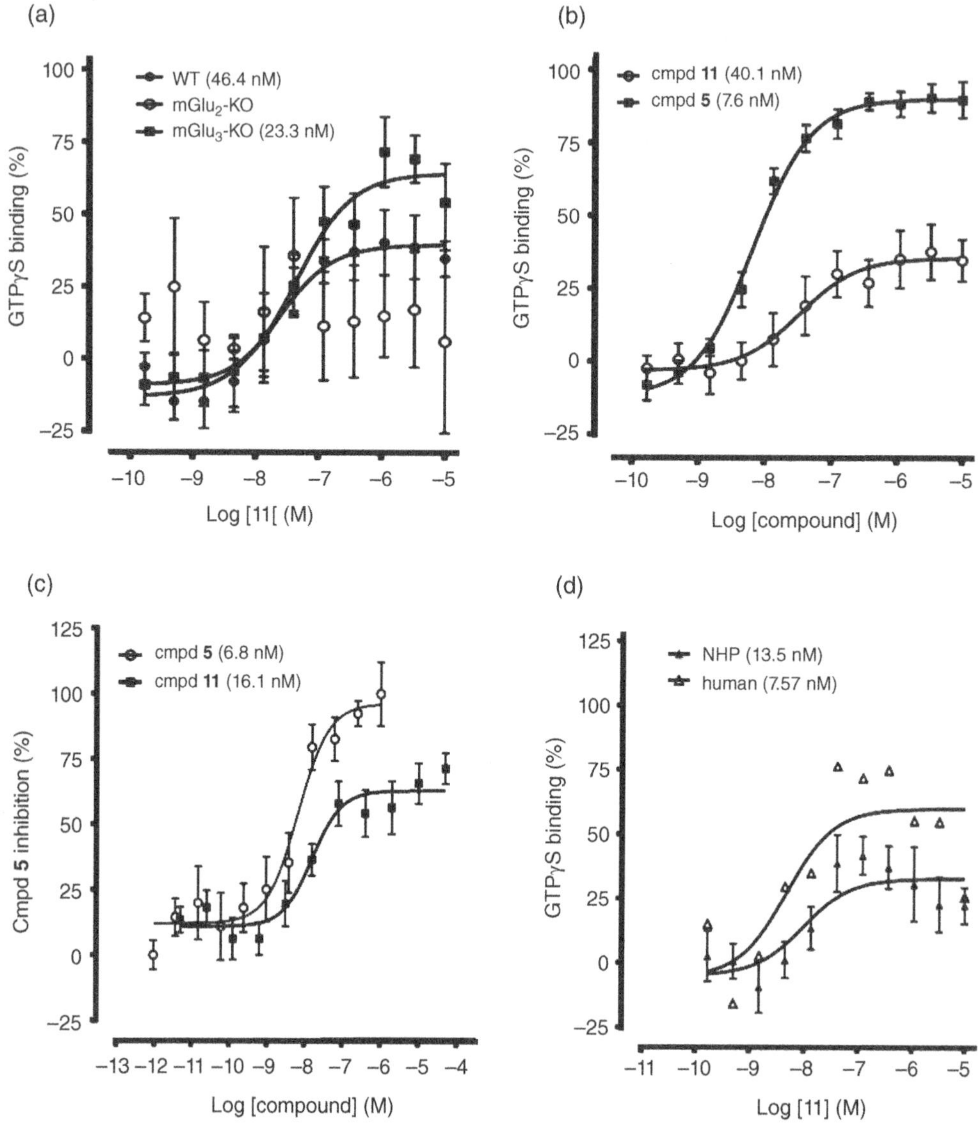

Figure 6.6 (a) Agonist response of **11** in cortical membranes derived from wild-type, mGlu$_2$-KO, and mGlu$_3$-KO mice in the functional GTPγS binding assay. (b) Agonist response of **11** and **5** in rat cortical membrane preparations in the functional GTPγS binding assay. (c) Effects of **11** and **5** in rat cortical synaptosome preparations in the cAMP assay. (d) Agonist response of **11** in native tissue cortical membranes derived from nonhuman primates and humans in the functional GTPγS binding assay. *Source:* Adapted from figures 2a, 3b, 3c, and 4a, respectively, in ref. [89]. Reproduced with permission of The American Society for Pharmacology and Experimental Therapeutics.

6.2.4 Preclinical Pharmacokinetic Profile of 11

The *in vivo* pharmacokinetic (PK) attributes of **11** were assessed in rats (Table 6.3) [43]. Following i.v. administration of **11** at 1 mg/kg, it displayed a low rate of clearance (12.3 ml/min/kg) with a plasma half-life of 3.3 hours. The mean peak plasma concentration was approximately 9.5 μM with a moderate steady state volume of distribution (1 l/kg). Similar to **3**, oral (p.o.) bioavailability in rats

Table 6.3 Mean plasma pharmacokinetic parameters for **11** following single i.v., oral (p.o.), or i.p. dosing in rats.

	Dose (mg/kg)/route		
PK parameter	1 / i.v.	5 / p.o.	1 / i.p.
AUC 0–24 h (nM·h)	4910	1010	6010
C_0 or C_{max} (nM)	9540	220	4920
T_{max} (h)	–	0.5	0.4
Cl (ml/min/kg)	12.3	–	–
V_{dss} (l/kg)	1.01	–	–
$T_{1/2}$ (h)	3.3	3.7	1.2
F (%)	–	4	122

PK, pharmacokinetic.

Table 6.4 Mean plasma and cerebrospinal fluid concentrations of **11** following single i.p. dosing in rats.

	Mean plasma (nM)			Mean CSF (nM)			CSF concentration fold over mGlu$_2$ FLIPR EC$_{50}$		
Dose (mg/kg)	0.5 h	1 h	4 h	0.5 h	1 h	4 h	0.5 h	1 h	4 h
1	6000	3010	314	48.8	39.9	9.7	2	2	0.5
3	10 100	4980	700	97.3	86.4	27.5	4	4	1
10	40 100	23 600	3020	235.0	272	86.7	11	13	4

CSF, cerebrospinal fluid.

following the 5 mg/kg dose was very low (4%). In contrast, i.p. administration significantly increased the apparent bioavailability at 1 mg/kg (122%). This suggested that a prodrug approach would be required to improve oral bioavailability.

In a separate experiment, cerebrospinal fluid (CSF) and plasma concentrations of **11** were determined at 0.5, 1, and 4 hours in rats following an ip dose of 1, 3, and 10 mg/kg (Table 6.4) [43]. Concentrations of **11** in rat CSF increased proportionately with dose, resulting in mean peak CSF concentrations of approximately 48.8 nM (1 mg/kg, 0.5 hours), 97.3 nM (3 mg/kg, 0.5 hours), and 272 nM (10 mg/kg, 1 hours). At all doses and time intervals, CSF concentrations of **11** were more than onefold over the more conservative estimate of hmGlu$_2$ potency as measured in the FLIPR assay with the exception of 1 mg/kg at the four hour time point (only 0.5×). Since **11** has very low protein binding (free fraction $f_u > 0.98$ in rat and human plasma) and does not penetrate effectively into cells, it was posited that CSF drug levels would be a reasonable estimate for drug concentrations at central extracellular mGlu$_2$ receptor binding sites.

6.2.5 Preclinical Animal Model of Psychosis

It has been shown previously that the antipsychotic response in the phencyclidine (PCP)-evoked locomotion assay in rats (an animal model of psychosis) was mediated by the mGlu$_2$ receptors instead of the $mGlu_3$ receptors [72, 73]. When **11** was evaluated in this assay, it elicited a dose-dependent decrease in horizontal ambulations (measured over 60 minutes immediately following PCP dosing) after i.p. administration (1, 3, 10, 30 mg/kg, 30 minutes prior to 5 mg/kg s.c. dosing of PCP), resulting in a calculated ED_{50} of 1.2 mg/kg (Figure 6.7a) [43]. This result was consistent with the results from previous studies in $mGlu_2$ or $mGlu_3$ knock-out animals. A rat rotarod test was employed to rule out the potentially confounding effects of nonspecific motor impairment in the PCP locomotor assay. As shown in Figure 6.7b, a statistically significant impairment was observed following a 10 mg/kg i.p. dose of **11** in rats, but not at the lower 3 mg/kg i.p. dose. Furthermore, this was only observed at the 1 hour time point but not at later time points for this dose. Therefore, the efficacious doses of 1 and 3 mg/kg in the PCP locomotor assay are not confounded by nonspecific motor effects.

6.3 Discovery of Subtype Selective Agonist LY2794193 OF THE MGLU$_3$ Receptor

6.3.1 Discovery of Subtype Selective Agonists for the mGlu$_3$ Receptor

As discussed previously in Section 6.2.1, the same obstacles existed for designing subtype selective agonists for the $mGlu_3$ receptor with minimal agonist activity for the $mGlu_2$ receptor. The amino acid residues binding to the L-Glu at the orthosteric binding sites of the $mGlu_2$ and the $mGlu_3$ receptors are 100% identical (Table 6.2), and the ligand binding pocket was perceived to be limited by the Y150-R277 pair (Figure 6.3b). The agonist-bound rat $mGlu_3$ ATD crystal structures [87] did not provide any design clues for the possibility of achieving subtype selectivity. As part of the C4 SAR studies with larger substituents, the synthesis and characterization of C4-nitrogen linked analogs was pursued (Figure 6.8) [48]. It was observed that the C4β-benzamide analog **18** possessed surprisingly high selectivity for the $mGlu_3$ receptors in both binding (43-fold) and cAMP functional (4-fold) assays when compared with unsubstituted amine analog **16** (15- and 0.85-fold for binding and functional, respectively) and benzylamine analog **17** (10- and 1.4-fold for binding and

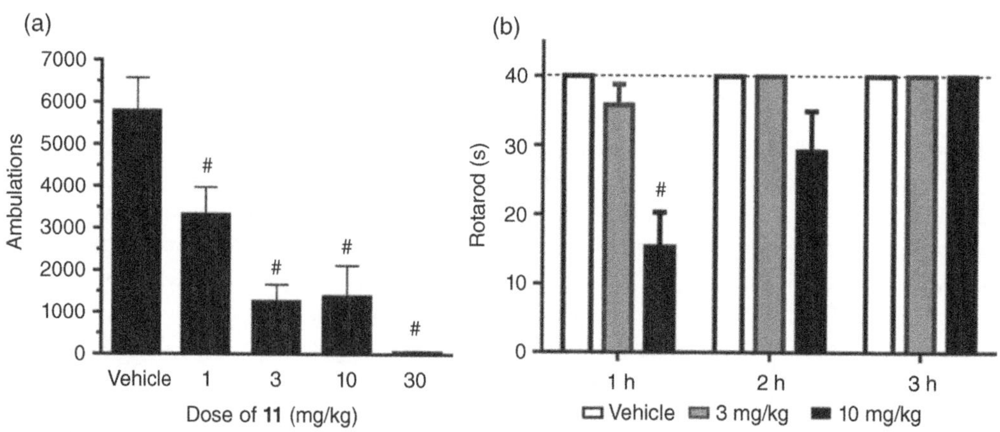

Figure 6.7 (a) Dose response of **11** in the rat PCP locomotor activity assay. # Statistically significant ($p < 0.05$, ANOVA followed by Dunnett's test) compared with rats treated with PCP alone. (b) Effect of **11** on rotarod performance in rats. # Statistically significant ($p < 0.05$) compared with vehicle group at 1 hour. *Source:* Adapted from figure 13 in ref. [43]. Reproduced with permission of The American Chemical Society.

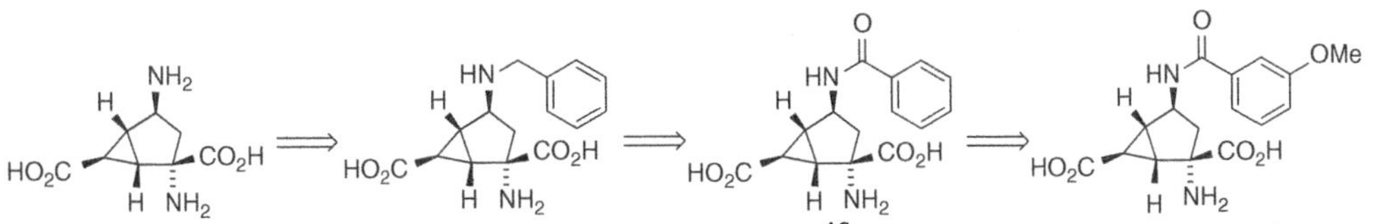

Figure 6.8 SAR evolution leading to discovery of the mGlu$_3$ subtype selective agonist LY2794193.

functional, respectively, Table 6.1) [48]. Further investigation of substituents on the phenyl ring resulted in the identification of LY2794193 (**19**) as a highly potent mGlu$_3$ agonist with unprecedented subtype selectivity (mGlu$_3$ EC_{50} = 0.47 nM in the cAMP assay, 444- and 100-fold for binding and functional, respectively) [48]. In the FLIPR assay format, **19** retained high mGlu$_3$ selectivity (24-fold), which was, however, lower than that measured in the cAMP format. Furthermore, the maximal efficacy observed for **19** was also reduced (hmGlu$_3$ E_{max} = 54% in FLIPR compared with 88% in cAMP). When assessed for agonist and antagonist activity against other mGlu receptors, it displayed low potency only against the hmGlu$_6$ agonist activity (EC_{50} of 1950 nM).

6.3.2 Additional *in vitro* Characterization of 19

To establish the efficacy of **19** in native brain tissue, it was tested in the Ca^{2+} oscillation assay in rat primary cortical neurons [48]. Interestingly, **19** displayed a biphasic inhibition of spontaneous Ca^{2+} transients (high affinity EC_{50} = 0.44 nM; 48% of the total agonist response; low affinity EC_{50} = 43.6 nM; 52% of the total agonist response) with combined maximal agonist efficacy (E_{max}) of 66% (Figure 6.9a). Both high and low potency sites for **19** were sensitive to 1 μM mGlu$_{2/3}$ antagonist LY341495 (Figure 6.9b). It has been demonstrated that mGlu$_2$ and mGlu$_3$ contributed about equally to the inhibition of Ca^{2+} oscillations by pharmacologically balanced mGlu$_{2/3}$ receptor agonists [90]. Therefore, the close agreement of these rodent native tissue derived potencies with those observed for **19** in the *in vitro* cAMP assays (hmGlu$_3$ EC_{50} = 0.47 nM, hmGlu$_2$ EC_{50} = 47.5 nM) was particularly intriguing. We therefore postulated that the high affinity component of the Ca^{2+}

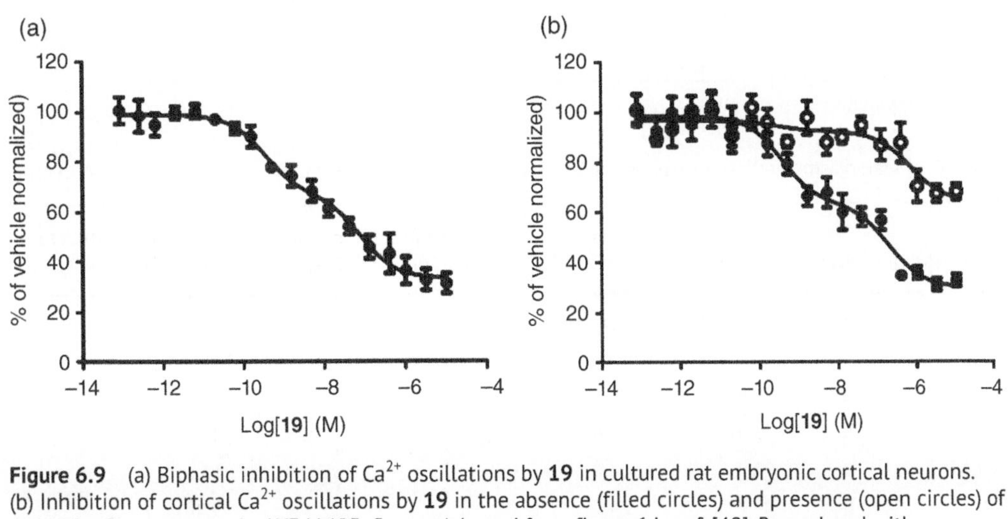

Figure 6.9 (a) Biphasic inhibition of Ca^{2+} oscillations by **19** in cultured rat embryonic cortical neurons. (b) Inhibition of cortical Ca^{2+} oscillations by **19** in the absence (filled circles) and presence (open circles) of 1 μM of mGlu$_{2/3}$ antagonist LY341495. *Source:* Adapted from figure 6 in ref. [48]. Reproduced with permission of The American Chemical Society.

oscillation inhibition by **19** in rat cortical neurons was mediated by $mGlu_3$, while the low affinity component was mediated by $mGlu_2$.

6.3.3 Preclinical Pharmacokinetic Profile of 19

The *in vivo* PK attributes of **19** in rats are shown in Table 6.5 [48]. Following i.v. dosing, it exhibited a moderate rate of clearance (15.3 ml/min/kg) with a plasma half-life of 3.1 hours. The mean peak plasma concentration was approximately 6.4 μM with a moderate steady state volume of distribution (1.21/kg). Subcutaneous dosing lead to the rapid appearance of **19** in rat plasma (T_{max} = 0.44 hours) with a high peak concentration (6.8 μM) and a calculated bioavailability of 121%. Furthermore, no significant metabolites were observed.

To identify the doses that would allow selective activation of the central $mGlu_3$ receptor, the CSF disposition of **19** in rats was assessed in two separate studies (dose levels of 0.1, 0.3 and 1 mg/kg for study A, and of 1, 3, and 10 mg/kg for study B). Good concordance was observed between the two studies with the overlapping 1 mg/kg dose group. In both studies, plasma and CSF concentrations of **19** were determined 1 hour after s.c. dosing (Table 6.6) [48]. At all dose levels, CSF levels of **19**

Table 6.5 Mean plasma pharmacokinetic parameters of **19** following single i.v., or s.c. dosing in rats.

	Dose (mg/kg)/route	
PK parameter	**1/i.v.**	**3/s.c.**
AUC 0–24 h (nM·h)	2750	9880
C_0 or C_{max} (nM)	6390	6780
T_{max} (h)	–	0.44
Cl (ml/min/kg)	18.3	–
V_{dss} (l/kg)	1.17	–
$T_{1/2}$ (h)	3.1	2.9
F (%)	–	121

PK, pharmacokinetic.

Table 6.6 Mean plasma and cerebrospinal fluid concentrations of **19** at 1 h following single s.c. dosing in rats.

Dose (mg/kg)	Study	Mean plasma (nM)	Mean CSF (nM)	CSF concentration fold under (over) $mGlu_2$ cAMP EC_{50}	CSF concentration fold over $mGlu_3$ cAMP EC_{50}
0.1	A	286	2.5	19	5
0.3	A	771	6.9	7	15
1	A	1580	22.5	2	48
1	B	1510	12.5	4	27
3	B	7180	73.6	0.7 (1.5)	157
10	B	25 900	249	0.2 (5)	530

CSF, cerebrospinal fluid.

were well above the EC_{50} value of the $mGlu_3$ agonist activity (the cAMP assay). It was noted that CSF levels of **19** at 1 hour post s.c. dosing was above the EC_{50} value of the $mGlu_2$ agonist activity (cAMP assay) when **19** was dosed at or over 3 mg/kg. Assuming the linear relationship between dose and CSF levels of **19**, it was predicted that an acute s.c. dose of 0.02 mg /kg would be sufficient to activate the high affinity component of the neuronal Ca^{2+} oscillation assay ($mGlu_3$) and acute s.c. doses of 2 mg/kg or greater would activate the low affinity component ($mGlu_2$). [48] Additionally, it was predicted that a s.c. infusion regiment of 0.1 mg/kg/day of **19** in rat would only engage the $mGlu_3$ receptors while a s.c. infusion regiment of 10 mg/kg/day or higher of **19** would be required to engage the $mGlu_2$ receptors [48].

6.3.4 Preclinical Animal Model

In the $mGlu_2$-sensitive PCP-evoked locomotion assay in rats, s.c. administration of **19** (10 or 30 mg/kg) 30 minutes prior to PCP dosing resulted in a dose-dependent reduction in ambulations (Figure 6.10) [48]. This was consistent with the PK studies which suggested the $mGlu_2$ receptor engagement could be achieved when dosed over 3 mg/kg s.c. At doses lower than 3 mg/kg s.c., only $mGlu_3$ receptors were engaged, and no statistically significant reduction in ambulations was observed in this assay. Taken together with the results from the rat PK studies, these data serve to guide future *in vivo* studies aiming to further understand $mGlu_3$ receptor activation using **19** as a tool compound.

6.4 Structural Basis for Subtype Selectivity

6.4.1 Crystal Structures of $hmGlu_2$ and $hmGlu_3$ ATDs in Complex with 3 and L-Glu

The high subtype selectivity observed for **11** and **19** was difficult to rationalize based on the existing rat $mGlu_3$ ATD crystal structures [87], given the residues binding to the L-Glu at the orthosteric binding pocket are identical for both receptors. At the beginning of the project, we first sought to obtain both $mGlu_2$ and $mGlu_3$ ATD crystal structures in complex with a compound based on the bicyclo[3.1.0]hexane scaffold to better understand how these glutamate analogs bind to the receptors. To this end, we were able to obtain both the human $mGlu_2$ and $mGlu_3$ ATD in complex with

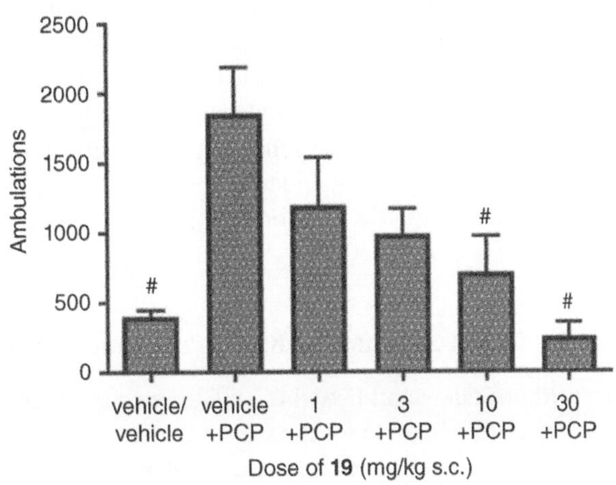

Figure 6.10 Dose response of **19** in the rat PCP locomotor activation assay. #Statistically significant ($p < 0.05$, ANOVA followed by Dunnett's multiple comparison test) compared with rats treated with PCP alone. *Source:* Adapted from figure 13 in ref. [48]. Reproduced with permission of The American Chemical Society.

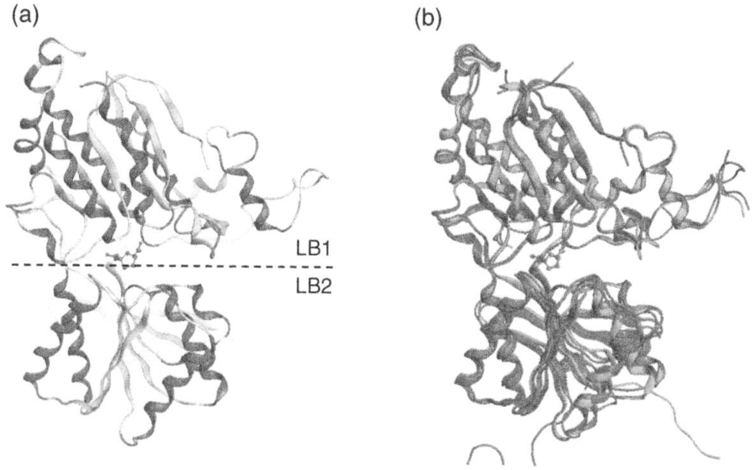

Figure 6.11 (a) A protomer from the **3**-hmGlu_2 ATD structure (PDB ID: 4XAQ) [46]. (b) Overlay of protomers from the **3**-hmGlu_2 (PDB ID: 4XAQ), **3**-hmGlu_3 (PDB ID: 4XAR) [46], and L-Glu-hmGlu_3 (PDB ID: 5CNK) [43] ATD structures.

3 and L-Glu [43, 46]. As shown in Figure 6.11a, the mGlu_2 ATD protein contains two lobes (top LB1 and bottom LB2) with **3** situated between the two lobes next to the hinge. Figure 6.11b shows the overlay of the **3**-hmGlu_2, **3**-hmGlu_3 and L-Glu-hmGlu_3 ATD structures. These three structures are virtually identical. Furthermore, the residues interacting with hmGlu_2 and hmGlu_3 were identical to those observed in the rmGlu3 ATD structures [27]. Other than confirming the mGlu_2 and mGlu_3 receptors have an identical orthosteric binding pocket, these new crystal structures in complex with **3** did not provide additional information to that already available from the L-Glu-rmGlu_3 ATD structure.

As observed previously in the L-Glu-rmGlu_3 ATD structure [87], as well as in both the **3**-hmGlu_2 and **3**-hmGlu_3 ATD structures, the Y144-R271 (of mGlu_2) and Y150-R277 (of mGlu_3) pairs formed the ceiling of the orthosteric binding pocket, which could restrict further substitutions at the C4 position (Figure 6.12a and b). Additionally, the Y144-R271 pair was further stabilized directly by D146 from LB1 and indirectly by S272 and E273 from LB2 in the **3**-hmGlu_2 structure. In the **3**-hmGlu_3 structure, the Y150-R277 pair was also stabilized indirectly by S152 from LB1 and directly by D279 from LB2, as observed in the L-Glu-rmGlu_3 structure. As expected, when **11** and **19** were docked into the holo-forms of the **3**-bound hmGlu_2 and hmGlu_3 structures, respectively, steric clashes were predicted between the C4α-thiotriazolyl group and the Y144-R271 pair in the hmGlu_2 structure (Figure 6.12c) [43] and between the C4β-benzamide group and the Y150-R277 pair in the hmGlu_3 structure (Figure 6.12d) [48]. As discussed before, these possible negative steric interactions were ascribed as the underlying reason why relatively large (e.g. SPh in **10** and **14**) substituents attached to the C4 position lack agonist activity while smaller ones (e.g. F in **9**) retain this pharmacology.

6.4.2 Crystal Structures of hmGlu_2 and hmGlu_3 ATDs in Complex with 11 and 19

To further understand how these mGlu receptors were able to accommodate bulkier ligands and the structural rationale for the subtype selectivity observed for **11** and **19**, co-crystal structures of these ligands with both the mGlu_2 and mGlu_3 ATD proteins were sought. To this end, three

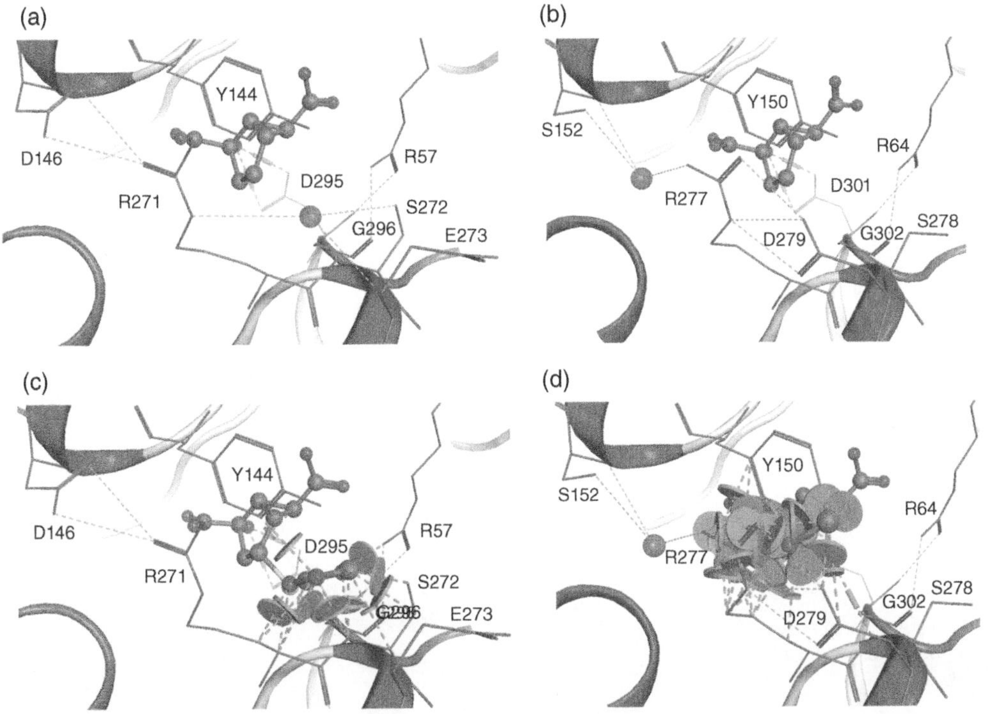

Figure 6.12 (a) The ceiling formed by the Y144-R271 pair in the **3**-hmGlu_2 ATD structure (PDB ID: 4XAQ). (b) The ceiling formed by the Y150-R277 pair in the **3**-hmGlu_3 ATD structure (PDB ID: 4XAR). (c) Steric clashed predicted when docking **11** to the **3**-hmGlu_2 structure holo form [43]. (d) Steric clashed predicted when docking **19** to the **3**-hmGlu_3 structure holo form.[48] The dotted lines (with cylinders) indicate favorable interactions between the ligand and residues using MOE software [88]. Disks indicate steric clashes between the ligand and residues in the binding pocket.

structures of ligand-bound proteins were obtained: **11**-hmGlu_2 ATD (PDB ID: 5CNJ) [43], **11**-hmGlu_3 ATD (PDB ID: 5CNM) [43], and **19**-hmGlu_3 ATD (PDB ID: 6B7H) [48]. As shown in Figure 6.13, the **11**-hmGlu_2 and **19**-hmGlu_3 ATD structures were in the closed form, while the **11**-hmGlu_3 ATD structure was in the open form, consistent with their observed pharmacology [75].

In the **11**-hmGlu_2 ATD structure, there was no gross difference between the two protomers. Furthermore, the residues interacting with the α-amino acid and distal acid portions of the ligand were the same. However, close inspection of the binding pocket residues revealed that significant difference existed for the conformation of the Y144-R271 pair (Figure 6.14). It appeared that in one protomer Y144 rotated away from the ligand to yield additional space to accommodate the bulky C4-substitution (Figure 6.14a). In the other protomer, the Y144-R271 pair appeared to be disturbed but there was no significant movement of the Y144 side chain when compared with that in the **3**-hmGlu_2 structure (Figure 6.14b).

Since the **11**-hmGlu_3 ATD structure was in the open form, the C4α-thiotriazolyl group did not appear to have any appreciable interactions with any residues in the binding pocket (not shown). In the **19**-hmGlu_3 ATD structure (Figure 6.15), the Y150-R277 pair was completely interrupted, and a highly unexpected triad was formed with the benzamide of the ligand intercalated between the side chains of Y150 and R277 (Figure 6.15a). These structures exposed a very interesting dynamic relationship between the tyrosine residue in LB1 (Y144 in $mGlu_2$ and Y150 in $mGlu_3$) and the arginine residue in LB2 (R271 in $mGlu_2$ and 277 in $mGlu_3$). The relative movements of the

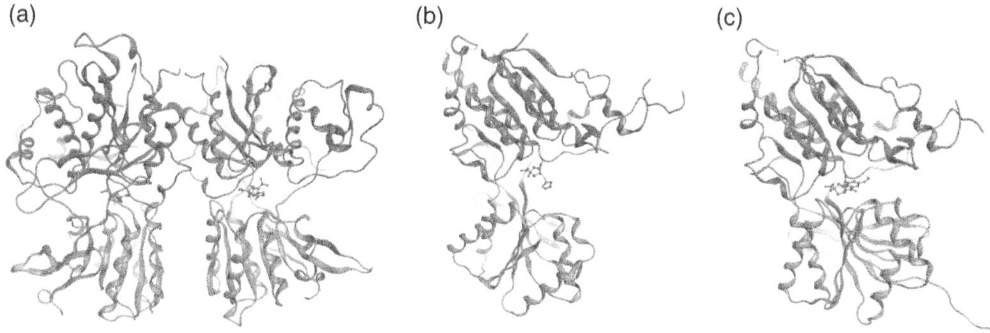

Figure 6.13 (a) Crystal structure of the **11**-hmGlu$_2$ ATD complex (PDB ID: 5CNJ) [43]. (b) Crystal structure of the **11**-hmGlu$_3$ ATD complex (PDB ID: 5CNM) [43]. (c) Crystal structure of the **19**-hmGlu$_3$ ATD complex (PDB ID: 6B7H) [48].

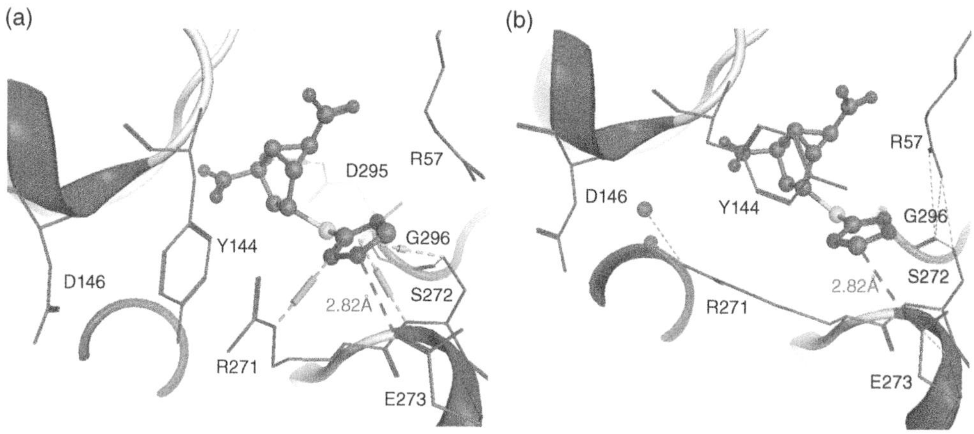

Figure 6.14 (a) Key interactions between the C4α-thiotriazolyl group and residues in one protomer of the **11**-hmGlu$_2$ ATD structure (PDB ID: 5CNJ) [43]. (b) Key interactions between the C4α-thiotriazolyl group and residues in the other protomer of the **11**-hmGlu$_2$ ATD structure (PDB ID: 5CNJ). The dotted lines (with cylinders) indicate favorable interactions between the ligand and residues using MOE software [88]. The dashed line shows the distance between the atoms at the ends of the line. *Source:* Adapted from figure 16c and 16d in ref. [75]. Licensed under CC BY 4.0.

Y144-R271 of mGlu$_2$ and Y150-R277 of mGlu$_3$ pairs are summarized in Figure 6.15b. In one protomer of the **11**-hmGlu$_2$ ATD structure, Y144 rotates around 100° clockwise along the Cα-Cβ bond (relative to the Y144 position in the **3**-hmGlu$_2$ structure) that breaks the D146-R271 interactions. However, R271 also rotates to keep the π-cation interactions with Y144 intact. In the other protomer of the **11**-hmGlu$_2$ ATD structure, Y144 rotates around 25° clockwise along the Cβ-C_{Ar} (relative to the Y144 position in the **3**-hmGlu$_2$ structure) that also breaks the D146-R271 interactions and perhaps weakens the π-cation interactions between Y144 and R271 as well. In the **19**-hmGlu$_3$ structure, Y150 rotates around 46° counterclockwise along the Cβ-C_{Ar} bond (relative to the Y150 position in the **3**-hmGlu$_3$ structure). R277 also moves downwards to provide an expanded pocket for the benzamide substitution.

In addition to the movements of the Y144-R271 of mGlu$_2$ and Y150-R277 of mGlu$_3$ pairs, the cleft opening between the two lobes also increases to accommodate agonists with bulkier groups [75]. In the **11**-mGlu$_2$ ATD structure, both protomers have much larger opening angles

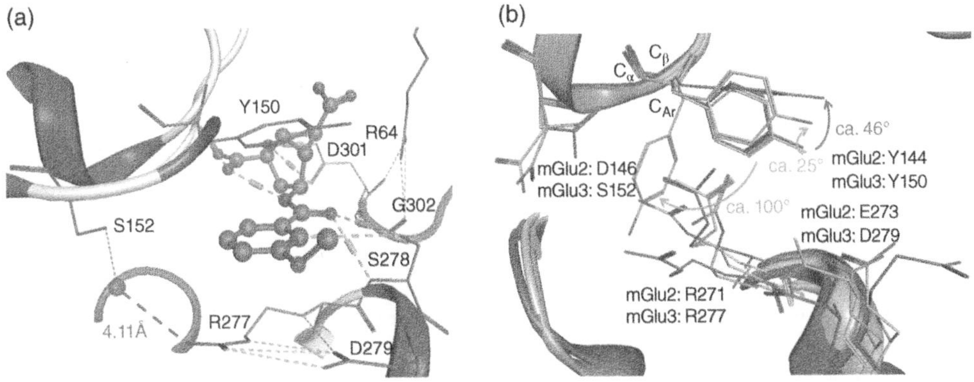

Figure 6.15 (a) Key interactions between the C4β-benzamide group and residues in the **19**-hmGlu$_3$ ATD structure (PDB ID: 6B7H) [48]. (b) Relative positions of the Y144-R271 of mGlu$_2$ and Y150-R277 of mGlu$_3$ pairs in the **3**-hmGlu$_2$ (PDB ID: 4XAR), **11**-hmGlu$_2$ (PDB ID: 5CNJ), **3**-mGlu$_3$ (PDB ID: 4XAQ), and **19**-mGlu$_3$ (PDB ID: 6B7H) structures. The dotted lines (with cylinders) indicate favorable interactions between the ligand and residues using MOE software [88]. The dashed line shows the distance between the atoms at the ends of the line. *Source:* Adapted from figure 16e and 16f in ref. [75]. Licensed under CC BY 4.0.

(27.5 and 28.2°) than the **3**-bound (25.0 and 25.3°) hmGlu$_2$ ATD structures. In the protomer with the smaller opening angle of 27.5°, the Y144 side chain rotates significantly to accommodate the bulky **11** (ca. 100° clockwise along the Cα-Cβ bond, Figure 6.14a). In the protomer with the larger opening angle of 28.2°, the Y144 side chain rotates only about 25° clockwise along Cβ-C$_{Ar}$ to stay roughly in the same location as in the **3**-hmGlu$_2$ crystal structure (Figure 6.14b). In the **19**-mGlu$_3$ structure, the protomer has an opening angle of 27.0°, which is larger than those of the **3**-bound (25.1°) hmGlu$_3$ ATD structures.

6.4.3 Structural Basis for the mGlu$_2$ Subtype Selectivity of 11 and the mGlu$_3$ Subtype Selectivity of 19

The subtype selectivity observed with **9** and **11** has been partially attributed to the E273 (mGlu$_2$)/D279 (mGlu$_3$) mutation based on the crystallographic data [43, 48, 75]. This subtle one-carbon difference makes the interaction between this acidic residue and its adjacent arginine (R271 of mGlu$_2$, and R277 of mGlu$_3$) significantly different. The D279–R277 interaction appeared much more frequently in the mGlu$_3$ ATD structures than the E273–R271 interaction in the mGlu$_2$ ATD structures [48, 75]. As a result, the longer E273 in mGlu$_2$ appeared to be able to directly interact with **9** while the shorter D279 in mGlu$_3$ did not appear to have any direct interaction but to assist R277 to interact with **11**. Thus, for distinct reasons, the E273 (mGlu$_2$)/D279 (mGlu$_3$) residues likely are major contributors to the observed agonist selectivity profiles of these two compounds.

This was further supported by mutagenesis studies [43]. In addition to the E273 (mGlu$_2$)/D279 (mGlu$_3$), the mutation of L300 (mGlu$_2$)/Q306 (mGlu$_3$) were also probed because Q306 could engage S278 in mGlu$_3$ (this serine residue engages with both **11** and **19** in the mGlu$_2$ and mGlu$_3$ structures, respectively). Therefore, they could potentially affect the potency and selectivity as well. Cells transiently transfected with hmGlu$_2$ wild-type, E273D, L300Q and E273D + L300Q receptors as well as cells expressing hmGlu$_3$ wild-type, D279E, Q306L, and D279E + Q306L receptors were constructed. Although the response to LY2812223 (**11**) appeared to be modestly affected by these point mutations for the mGlu$_2$ receptors (Figure 6.16a, Table 6.7), it was more pronounced for the mGlu$_3$ receptors (Figure 6.16b, Table 6.7) with regards to both agonist potency and efficacy.

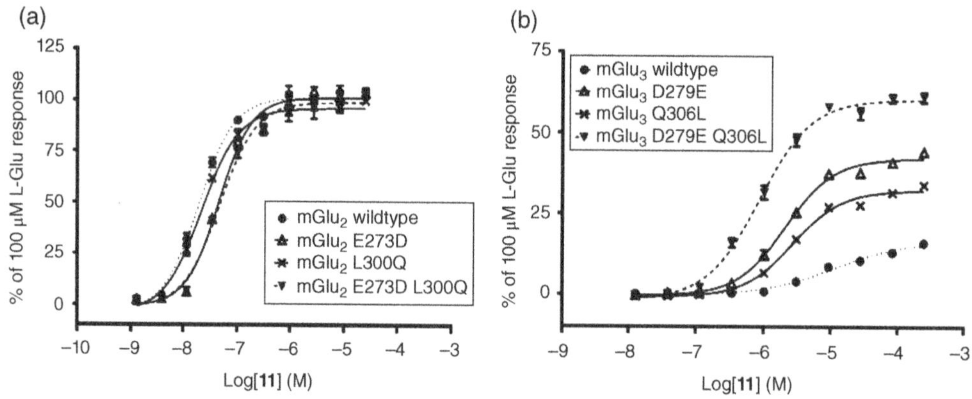

Figure 6.16 (a) FLIPR agonist responses produced by **11** in $hmGlu_2$ wild type and mutant receptors transiently expressed in AV12 cells. (b) FLIPR agonist responses produced by **11** in $hmGlu_3$ wild type and mutant receptors transiently expressed in AV12 cells. *Source:* Adapted from figure 11b and 11d, respectively, in ref. [43]. Reproduced with permission of The American Chemical Society.

Table 6.7 FLIPR agonist responses produced by **11** in AV12 cells transiently transfected with wild type and mutant $hmGlu_2$ and $hmGlu_3$ receptors.

	$mGlu_2$				$mGlu_3$			
Receptors	**Wild type**	**E273D**	**L300Q**	**E273D L300Q**	**Wild type**	**D277E**	**Q306L**	**D277E Q306L**
EC_{50} (nM)	20.1	46.3	22.1	48.2	5676	2181	2831	1175
E_{max} (%)	108	107	100	105	20	47	41	66

Both D279E and Q306L mutations in the $mGlu_3$ receptors lead to a gain in agonist potency and efficacy of **11**. A combination of these mutations (D279E + Q306L) resulted in a further gain of agonist potency and efficacy of **11**. These data suggested the E273 ($mGlu_2$)/D279 ($mGlu_3$) and L300 ($mGlu_2$)/Q306 ($mGlu_3$) mutations played some role in the observed pharmacological profiles of **11** and **19**. However, it was also clear additional and yet-to-be-identified residues in the binding pocket might also contribute to their observed subtype selectivity.

6.5 Divergent Synthesis of 11 and 19

The synthesis of both **11** and **19** started from *tert*-butyl methylmalonate (**20**, Scheme 6.1) [91]. Alkylation of this mixed malonate ester with *cis*-1,4-dichloro-2-butene using LiH resulted in a cyclic diester, which was hydrolyzed with aqueous NaOH to obtain mono-acid **21** in 79% yield over two steps. This acid was converted to the Boc-protected amine **22** in a three-step sequence (71% overall yield): acid chloride formation with thionyl chloride, conversion to acyl azide, Curtius rearrangement and trapping of the resulting isocyanate with *t*-BuOH. Epoxidation of alkene **22** was achieved facial-selectively with water-soluble magnesium monoperoxyphthalate (MMPP) in which the epoxide oxygen was delivered from the same face of the ring as the NHBoc group, presumably due to the steric bulk of the *tert*-butyl ester (71% yield).

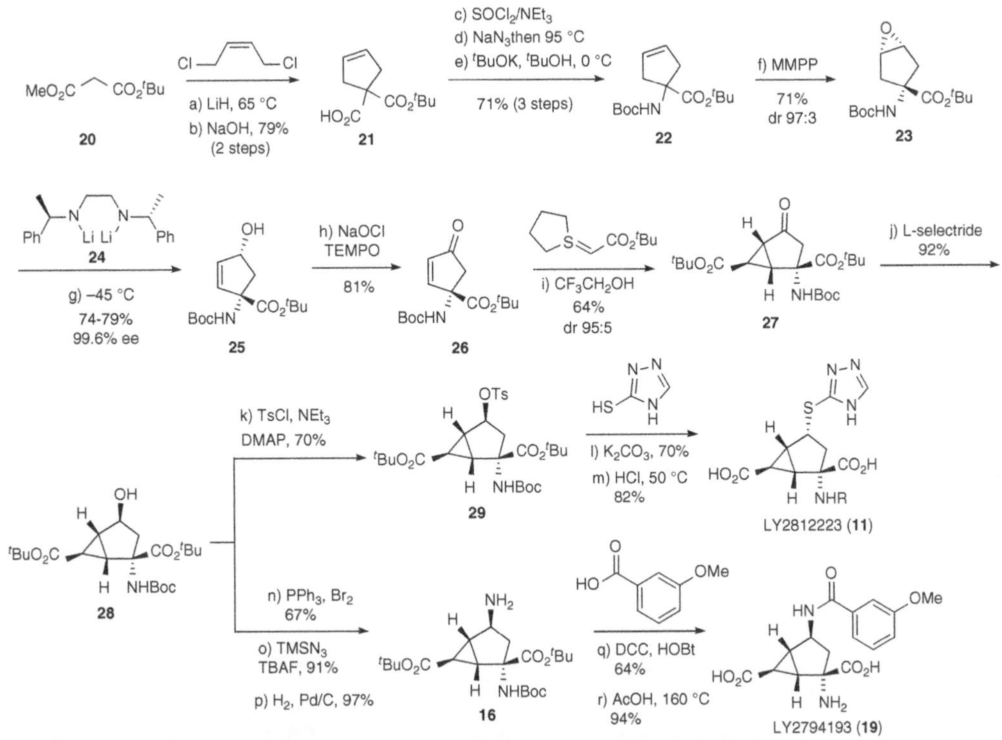

Scheme 6.1 Divergent synthesis of **11** and **19** through a common intermediate

A desymmetrization of the cyclic *meso*-epoxide **23** was achieved in high enantiomeric excess using a chiral lithium amide **24** (74–79% yield). Oxidation of the chiral alcohol **25** with aqueous NaOCl and TEMPO resulted in ketone **26** in 81% yield. Subsequent cyclopropanation of enone **26** was achieved using *tert*-butyl 2-(tetrahydro-1λ^4-thiophen-1-ylidene)acetate [92]. Addition of trifluoroethanol or other proton sources with a similar pKa was found to be essential to achieve high selectivity for the desired diastereomer. Again, the cyclopropanation proceeded preferentially from the face opposite the bulky *tert*-butyl ester to give ketone **27** (64% yield), which was reduced with L-selectride. This reduction proceeded with hydride attack exclusively from the concave face of the bicyclic ring system to give the common intermediate β-alcohol **28** (92% yield) [47].

To prepare LY2812223 (**11**) with the C4α stereochemistry, alcohol **28** was converted to β-tosylate **29** [43]. Subsequent nucleophilic displacement of the tosylate proceeded with complete inversion of stereochemistry to yield the fully protected **11**. Finally, simultaneous removal of *tert*-butyloxycarbonyl and *tert*-butyl ester protecting groups with HCl gave **11** in 82% overall yield. [43] To prepare LY2794193 (**19**) with the C4β stereochemistry [48], the stereochemistry of β-alcohol **28** was inverted twice to obtain β-amine **16** through a three-step sequence: inversion to α-bromide (67% yield) with PPh_3 and bromine, second inversion to β-azide with an organic azide $TMSN_3$ (91% yield), and hydrogenation of the azide (97% yield). Coupling of this amine with *m*-methoxybenzoic acid (64%) followed by full deprotection of the resulting amide with AcOH resulted in **19** (94% yield). This operationally simple and convenient deprotection method under mild HOAc conditions was developed to achieve the direct deprotection without the need for either isoelectric precipitation or ion exchange chromatography [93].

6.6 Clinical Experience with MGLU$_2$ Selective Agonist 11 (Via its Prodrug 12)

6.6.1 Human Plasma and CSF PK Profiles of 11

To assess the safety, tolerability, pharmacokinetics and pharmacodynamics of the mGlu$_2$ subtype selective agonist **11**, it was evaluated via its alanine prodrug **12** (LY2979165, Figure 6.4, designed to enhance the oral bioavailability of **11**) in phase 1 double-blind, randomized, placebo-controlled clinical studies [94]. In the single ascending dose (SAD, 20–150 mg, $N = 30$) and multiple ascending doses (MAD, 20–400 mg, $N = 84$) studies in healthy subjects, the plasma and urine concentrations of both prodrug **12** and active **11** were measured.

As shown in plasma PK curves in the SAD (part A) and MAD studies (Figure 6.17), systemic uptake of prodrug **12**, and conversion to **11** was rapid and extensive with appearance of measurable **11** in plasma within 30 minutes of dosing. The exposure of prodrug **12** was estimated to be less than 1% of active **11** exposure in plasma. The T_{max} was reached in about 3–4 hours after dosing, with the majority of drug eliminated within the first 24 hours. As shown in Table 6.8 and Figure 6.17, exposure of **11** increased proportionally as dose increased, and minimal

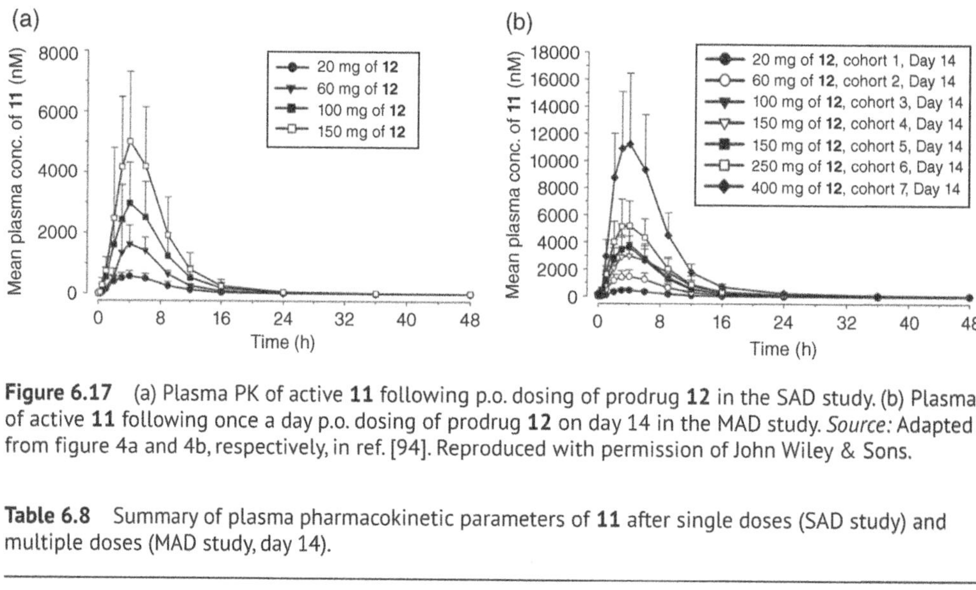

Figure 6.17 (a) Plasma PK of active **11** following p.o. dosing of prodrug **12** in the SAD study. (b) Plasma PK of active **11** following once a day p.o. dosing of prodrug **12** on day 14 in the MAD study. *Source:* Adapted from figure 4a and 4b, respectively, in ref. [94]. Reproduced with permission of John Wiley & Sons.

Table 6.8 Summary of plasma pharmacokinetic parameters of **11** after single doses (SAD study) and multiple doses (MAD study, day 14).

	SAD dose (mg)				MAD dose (mg)						
PK parameter	**20**	**60**	**100**	**150**	**20**	**60**	**100**	**150**[a]	**150**[b]	**250**	**400**
T_{max} (h)	4.00	4.00	4.00	4.00	4.00	4.00	4.00	4.00	4.00	3.00	4.00
C_{max} (nM)	588	1580	2790	4600	433	1500	3640	3260	3440	5340	10 900
AUC (nM·h)	4340	10 900	20 200	31 700	3240	11 300	25 400	25 000	24 100	38 600	80 100
$T_{1/2}$ (h)	9.91	12.5	17.8	14.1	11.0	10.6	9.0	10.0	10.9	11.4	10.9

[a] Cohort 4.
[b] Cohort 5.
PK, pharmacokinetic.

accumulation of **11** was observed with repeated once-daily oral dosing of **12**. Furthermore, the mean fraction of the dose excreted as **11** in the urine over 24 hours was estimated to be 44–62% across the dose groups. A preliminary assessment of the effect of food on the plasma PK of **11** and **12**, conducted at dose levels of 20 and 60 mg, found that food did not have a significant impact on **11** exposure or T_{max}.

Exposure of **11** in CSF was assessed in the SAD study (part B) and at a single time point on day 14 in the MAD study. As shown in Figure 6.18 and Table 6.9, C_{max} was delayed in CSF when compared with plasma. The CSF to plasma ratio of **11** was estimated to be 0.02 to 0.03, consistent across studies. Given the relatively flat CSF PK, good CSF exposure of **11** was achieved within 3–4 hours and the CSF levels of **11** at 24 hours after a single dose of 60 mg **12** were still significantly higher than the *in vitro* mGlu$_2$ agonist EC_{50} value (cAMP, 5.6 nM, or human cortical membrane, 7.57 nM), supporting q.d. oral dosing of prodrug **12** in future clinical studies. Furthermore, no **12** was detected in CSF. In the MAD study with the 150 mg cohort, the mean CSF level of **11** was over 10-fold higher than both *in vitro* mGlu$_2$ EC_{50} values 7–8 hours post-dose on day 14.

6.6.2 Biomarker

Several biogenic amines (dopamine, DOPAC, HVA, 5-HIAA, L-DOPA, MHPG, 5-HT, and VMA) were also measured in the 60 mg cohort in CSF to explore whether they were suitable as biomarkers of central pharmacodynamic (PD) effects exerted by **11**. There were no statistically significant changes in the levels of DOPAC and HVA when compared with placebo in the SAD study, in

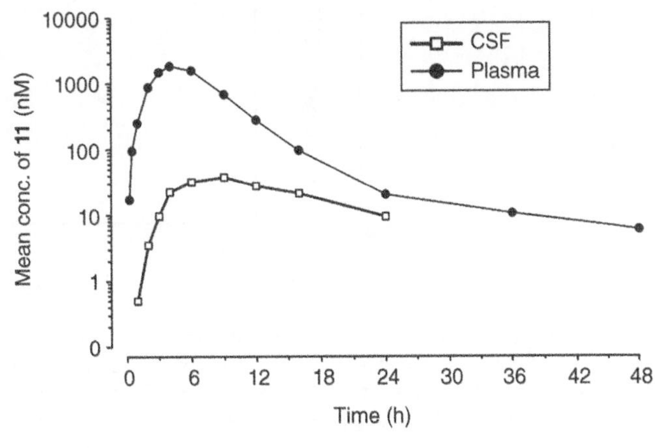

Figure 6.18 Plasma and CSF PK of active **11** after a single oral dose of 60 mg prodrug **12**. *Source:* Adapted from figure 5 in ref. [94]. Reproduced with permission of John Wiley & Sons.

Table 6.9 Summary of plasma and cerebrospinal fluid pharmacokinetic parameters of **11** after a single dose of 60 mg **12** (SAD) and after multiple doses of 150 mg **12** on day 14 (MAD study).

		Plasma	CSF	CSF/plasma ratio
SAD study	C_{max} (nM)	1740	29.8	0.02
	AUC 0–24 h (nM·h)	11 400	342	0.03
	T_{max} (h)	3.5	9.1	–
MAD study	Single concentration 7–8 h post-dose on day 14 (nM)	2808	87.1	0.03

CSF, cerebrospinal fluid.

contrast to previous results with the balanced $mGlu_{2/3}$ agonist prodrug **7** [95]. Although the reason for this finding was not clear, it could be due to the high variability in placebo effect and/or the small sample size.

To identify an alternative translational PD biomarker, prodrug **12** was tested in healthy volunteers in a ketamine-challenge phMRI model [96], which had been previously validated by attenuation of ketamine-evoked blood oxygenation level dependent (BOLD) signal increase by antipsychotic glutamatergic agents [97, 98]. In healthy subjects, single doses of **12** (20 and 60 mg) were administered 6.5 hours prior to ketamine infusion during MRI scanning. It was found that **12** was able to attenuate the ketamine-evoked BOLD phMRI signal increase in a dose- and exposure-dependent manner. The least-square mean results of the BOLD phMRI response across all regions of interest are shown in Figure 6.19a and Table 6.10. A significant mean decrease of 45% of the BOLD response was observed in the 60 mg cohort compared with placebo. While the 20 mg cohort did not separate statistically from placebo, a decrease was observed. As shown in Figure 6.19b, a significant relationship was observed between the BOLD response reduction and the plasma

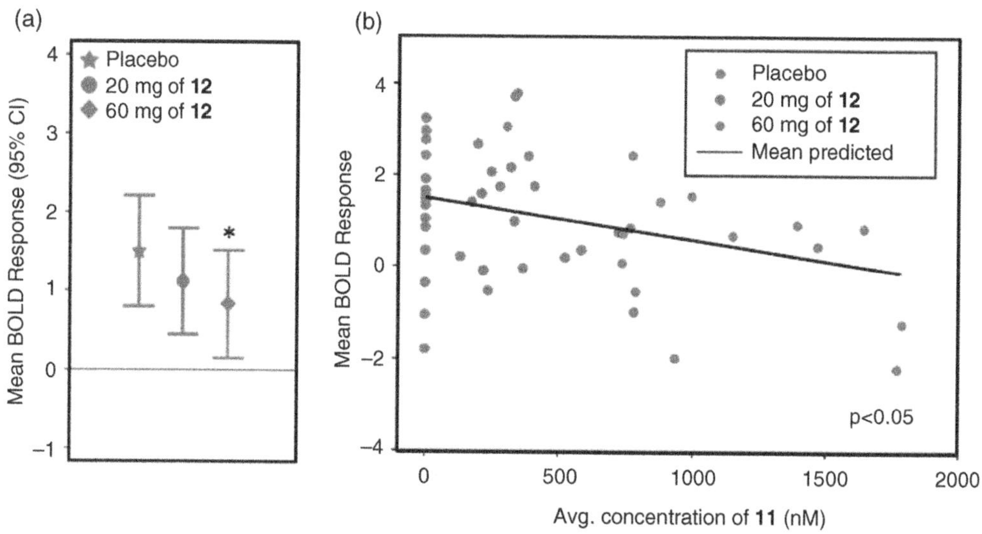

Figure 6.19 (a) Dose response for **12** in BOLD response to ketamine challenge across all regions of interests. (b) Relationship between BOLD response to ketamine challenge and measured plasma concentration of **11** (the average of the two samples immediately before and after the scanning session). The placebo response data have been assigned a value of 0 nM for average plasma concentration. The straight line indicates linear regression (slope = −0.000905 nM, 95% confidence interval = [−0.0015, −0.000299]). *Source:* Adapted from figure 4a and 4b, respectively, in ref. [96]. Reproduced with permission of Springer Nature.

Table 6.10 Primary statistical analysis of BOLD response to ketamine challenge across predefined atlas regions of interest.

Dose (mg)	Sample size	LS mean (95% CI)	Difference from placebo (95% CI)	Posterior probability
0	16	1.5 (0.80, 2.21)	–	–
20	18	1.12 (0.44, 1.79)	−0.38 (−1.10, 0.33)	0.862
60	17	0.83 (0.15, 1.52)	−0.67 (−1.39, 0.05)	0.967

concentration of **11** at the time of scanning. These results are consistent with a body of experimental data suggesting group II mGlu receptor agonists have the ability to reverse the effects of NMDA receptor antagonists in both rodents and humans [35, 99–101]. Furthermore, compared with the effects of other antipsychotic glutamatergic agents observed in similar studies at the clinically relevant doses, these results suggest clinically meaningful doses of **12** maybe be close to 60 mg or higher [96].

6.6.3 Safety

In the phase 1 SAD and MAD clinical studies, there were no serious adverse events (SAEs) [94]. Dizziness, vomiting, nausea, headache and somnolence were the most commonly reported treatment-emergent AEs, with the majority of them rated as mild in intensity. Single oral doses of up to 60 mg **12** were well tolerated with nausea and vomiting as dose limiting AEs at higher doses in the SAD study. In the MAD study, nausea and vomiting accommodated during titration, with the majority of events occurring on day 1. During the course of both studies, there were no clinically relevant alterations in hematology, clinical chemistry, vital signs, urinalysis or ECG measurements. In summary, oral doses of prodrug **12**, up to 60 mg as a single dose and 400 mg as multiple doses (using a titration regimen), were well tolerated.

6.7 Conclusion

Over the past two decades, multiple generations of group II mGlu receptor agonists have been designed to test the glutamatergic hypothesis of schizophrenia. First, the balanced group II mGlu receptor agonist LY2140023 (**7**, as the methionine prodrug of LY404039 [**6**]) was advanced to clinical studies. Although the initial phase 2 POC study was impressive and exciting, subsequent phase 3 studies failed to replicate these preliminary results. Ensuing meta-analysis of five double-blind, placebo-controlled efficacy studies suggested subpopulations of schizophrenia patients could significantly benefit from the LY2140023 treatment, and Denovo Biopharma is currently exploring options to identify and target these subpopulations [102]. We subsequently designed potent and subtype selective agonists toward either the $mGlu_2$ or the $mGlu_3$ receptors, which should provide opportunities to further delineate the individual physiological roles of these receptors. To this end, LY2812223 (**11**) was found to be a highly potent $mGlu_2$ subtype selective agonist with minimal $mGlu_3$ agonist activity in multiple *in vitro* systems. In phase 1 clinical studies, LY2812223 (via its prodrug LY2979165 [**12**]) exhibited excellent human PK properties, and achieved sufficient CSF exposure. With no SAEs, this compound is currently being evaluated in a phase 2 study in dementia-related psychosis and/or agitation and aggression (NCT03044249). Despite the best efforts, a potent $mGlu_3$ subtype selective agonist without discernable $mGlu_2$ activity remains elusive. However, LY2791493 (**19**), with high subtype selectivity, should allow *in vivo* studies to be designed to explore the role of the $mGlu_3$ receptor subtype.

A significant SAR design lesson was also learned from these projects, eventually leading to potent ligands with high subtype selectivity. Given the almost identical orthosteric binding pocket between the $mGlu_2$ and $mGlu_3$ receptors, and the ceiling formed by the cooperative Y144-R271 ($mGlu_2$) and Y152-R277 ($mGlu_3$) pairs in the pocket, it was considered close to impossible to identify ligands capable of selectively activating these two receptors. Yet, multiple potent compounds with high subtype selectivity toward each individual receptor were identified. This was not achieved by rational design but rather by strong desire and insight into the subtle SAR based on a singleton (**8**). Subsequent crystallographic studies revealed a dynamic relationship between the two residues in the tyrosine–arginine pairs that permitted a significantly enlarged binding pocket, enabling interactions with residues outside the canonical pocket where differences exist between these two closely related targets. Designing selective ligands for closely related targets is a highly challenging endeavor in medicinal chemistry, but insight and persistency with some luck can still be fruitful.

References

1 Kuhn, R. (2004). Eugen Bleuler's concepts of psychopathology. *Hist. Psychiatry* 15: 361–366.

2 van Os, J. and Kapur, S. (2009). Schizophrenia. *Lancet* 374: 635–645.

3 McEvoy, J.P. (2007). The costs of schizophrenia. *J. Clin. Psychiatry* 68 (Suppl 14): 4–7.

4 Faden, J. (2019). Treatment-resistant schizophrenia: a brief overview of treatment options. *J. Clin. Psychiatry* 80: 18ac12394.

5 Patel, K.R., Cherian, J., Gohil, K., and Atkinson, D. (2014). Schizophrenia: overview and treatment options. *P&T* 39: 638–645.

6 Ban, T.A. (2007). Fifty years of chlorpromazine: a historical perspective. *Neuropsychiatr. Dis. Treat.* 3: 495–500.

7 Crilly, J. (2007). The history of clozapine and its emergence in the US market: a review and analysis. *Hist. Psychiatry* 18: 39–60.

8 Leucht, S., Corves, C., Arbter, D. et al. (2009). Second-generation versus first-generation antipsychotic drugs for schizophrenia: a meta-analysis. *Lancet* 373: 31–41.

9 Lieberman, J.A. (2007). Effectiveness of antipsychotic drugs in patients with chronic schizophrenia: efficacy, safety and cost outcomes of CATIE and other trials. *J. Clin. Psychiatry* 68, e04.

10 Lieberman, J.A., Stroup, T.S., McEvoy, J.P. et al. (2005). Clinical antipsychotic trials of intervention effectiveness, I. Effectiveness of antipsychotic drugs in patients with chronic schizophrenia. *New Engl. J. Med.* 353: 1209–1223.

11 Stroup, T.S., Lieberman, J.A., McEvoy, J.P. et al. (2006). Effectiveness of olanzapine, quetiapine, risperidone, and ziprasidone in patients with chronic schizophrenia following discontinuation of a previous atypical antipsychotic. *Am. J. Psychiatry* 163: 611–622.

12 Miyamoto, S., Duncan, G.E., Marx, C.E., and Lieberman, J.A. (2005). Treatments for schizophrenia: a critical review of pharmacology and mechanisms of action of antipsychotic drugs. *Mol. Psychiatry* 10: 79–104.

13 Creese, I., Burt, D.R., and Snyder, S.H. (1976). Dopamine receptor binding predicts clinical and pharmacological potencies of antischizophrenic drugs. *Science* 192: 481–483.

14 Shoptaw, S.J., Kao, U., and Ling, W. (2009). Treatment for amphetamine psychosis. *Cochrane Database Syst. Rev.* (2) (Art. No.: CD003026) https://doi.org/10.1002/14651858.CD003026.pub2.

15 Seeman, P. (1987). Dopamine receptors and the dopamine hypothesis of schizophrenia. *Synapse* 1: 133–152.

16 Brisch, R., Saniotis, A., Wolf, R. et al. (2014). The role of dopamine in schizophrenia from a neurobiological and evolutionary perspective: old fashioned, but still in vogue. *Front. Psychiatry* 5: 47.

17 Stahl, S.M. (2007). Beyond the dopamine hypothesis to the NMDA glutamate receptor hypofunction hypothesis of schizophrenia. *CNS Spectr.* 12: 265–268.

18 da Silva Alves, F., Figee, M., van Amelsvoort, T. et al. (2008). The revised dopamine hypothesis of schizophrenia: evidence from pharmacological MRI studies with atypical antipsychotic medication. *Psychopharmacol. Bull.* 41: 121–132.

19 Olney, J.W. and Farber, N.B. (1995). Glutamate receptor dysfunction and schizophrenia. *Arch. Gen. Psychiatry* 52: 998–1007.

20 Moghaddam, B. and Javitt, D. (2012). From revolution to evolution: the glutamate hypothesis of schizophrenia and its implication for treatment. *Neuropsychopharmacol.* 37: 4–15.

21 Hu, W., MacDonald, M.L., Elswick, D.E., and Sweet, R.A. (2015). The glutamate hypothesis of schizophrenia: evidence from human brain tissue studies. *Ann. N.Y. Acad. Sci.* 1338: 38–57.

22 Stone, J.M., Morrison, P.D., and Pilowsky, L.S. (2007). Glutamate and dopamine dysregulation in schizophrenia – a synthesis and selective review. *J. Psychopharmacol.* 21: 440–452.

23 Javitt, D.C. and Zukin, S.R. (1991). Recent advances in the phencyclidine model of schizophrenia. *Am. J. Psychiatry* 148: 1301–1308.

24 Howes, O., McCutcheon, R., and Stone, J. (2015). Glutamate and dopamine in schizophrenia: an update for the 21st century. *J. Psychopharmacol.* 29: 97–115.

25 Kew, J.N. and Kemp, J.A. (2005). Ionotropic and metabotropic glutamate receptor structure and pharmacology. *Psychopharmacol.* 179: 4–29.

26 Hansen, K.B., Yi, F., Perszyk, R.E. et al. (2018). Structure, function, and allosteric modulation of NMDA receptors. *J. Gen. Physiol.* 150: 1081–1105.

27 Hao, J. and Chen, Q. (2019). On the origin of the 2-aminobicyclo[3.1.0]hexane-2,6-dicarboxylate scaffold's unique group II selectivity for the mGlu receptors. *Bioorg. Med. Chem. Lett.* 29: 297–301.

28 Chen, H.S. and Lipton, S.A. (2006). The chemical biology of clinically tolerated NMDA receptor antagonists. *J. Neurochem.* 97: 1611–1626.

29 Maksymetz, J., Moran, S.P., and Conn, P.J. (2017). Targeting metabotropic glutamate receptors for novel treatments of schizophrenia. *Mol. Brain* 10: 15.

30 Ayala, J.E., Chen, Y., Banko, J.L. et al. (2009). mGluR5 positive allosteric modulators facilitate both hippocampal LTP and LTD and enhance spatial learning. *Neuropsychopharmacol.* 34: 2057–2071.

31 Umbricht, D., Alberati, D., Martin-Facklam, M. et al. (2014). Effect of bitopertin, a glycine reuptake inhibitor, on negative symptoms of schizophrenia: a randomized, double-blind, proof-of-concept study. *JAMA Psychiatry* 71: 637–646.

32 Goff, D.C. (2014). Bitopertin: the good news and bad news. *JAMA Psychiatry* 71: 621–622.

33 Iwata, Y., Nakajima, S., Suzuki, T. et al. (2015). Effects of glutamate positive modulators on cognitive deficits in schizophrenia: a systematic review and meta-analysis of double-blind randomized controlled trials. *Mol. Psychiatry* 20: 1151–1160.

34 Noetzel, M.J., Jones, C.K., and Conn, P.J. (2012). Emerging approaches for treatment of schizophrenia: modulation of glutamatergic signaling. *Discov. Med.* 14: 335–343.

35 Lorrain, D.S., Baccei, C.S., Bristow, L.J. et al. (2003). Effects of ketamine and N-methyl-D-aspartate on glutamate and dopamine release in the rat prefrontal cortex: modulation by a group II selective metabotropic glutamate receptor agonist LY379268. *Neuroscience* 117: 697–706.

36 Hayashi, Y., Tanabe, Y., Aramori, I. et al. (1992). Agonist analysis of 2-(carboxycyclopropyl)glycine isomers for cloned metabotropic glutamate receptor subtypes expressed in Chinese hamster ovary cells. *Br. J. Pharmacol.* 107: 539–543.

37 Nakagawa, Y., Saitoh, K., Ishihara, T. et al. (1990). (2S,3S,4S) alpha-(carboxycyclopropyl)glycine is a novel agonist of metabotropic glutamate receptors. *Eur. J. Pharmacol.* 184: 205–206.

38 Ohfune, Y., Shimamoto, K., Ishida, M., and Shinozaki, H. (1993). Synthesis of L-2-(2,3-dicarboxyclopropyl)glycines. Novel conformationally restricted glutamate analogues. *Bioorg. Med. Chem. Lett.* 3: 15–18.

39 Hayashi, Y., Momiyama, A., Takahashi, T. et al. (1993). Role of a metabotropic glutamate receptor in synaptic modulation in the accessory olfactory bulb. *Nature* 366: 687–690.

40 Monn, J.A., Valli, M.J., Massey, S.M. et al. (1997). Design, synthesis, and pharmacological characterization of (+)-2-aminobicyclo[3.1.0]hexane-2,6-dicarboxylic acid (LY354740): a potent, selective, and orally active group 2 metabotropic glutamate receptor agonist possessing anticonvulsant and anxiolytic properties. *J. Med. Chem.* 40: 528–537.

41 Sekiyama, N., Hayashi, Y., Nakanishi, S. et al. (1996). Structure-activity relationships of new agonists and antagonists of different metabotropic glutamate receptor subtypes. *Br. J. Pharmacol.* 117: 1493–1503.

42 Schoepp, D.D., Johnson, B.G., Wright, R.A. et al. (1997). LY354740 is a potent and highly selective group II metabotropic glutamate receptor agonist in cells expressing human glutamate receptors. *Neuropharmacol.* 36: 1–11.

43 Monn, J.A., Prieto, L., Taboada, L. et al. (2015). Synthesis and pharmacological characterization of C4-(thiotriazolyl)-substituted-2-aminobicyclo[3.1.0]hexane-2,6-dicarboxylates: identification of (1R,2S,4R,5R,6R)-2-Amino-4-(1H-1,2,4-triazol-3-ylsulfanyl)bicyclo[3.1.0]hexane-2, 6-dicarboxylic acid (LY2812223), a highly potent, functionally selective mGlu2 receptor agonist. *J. Med. Chem.* 58: 7526–7548.

44 Monn, J.A., Valli, M.J., Massey, S.M. et al. (1999). Synthesis, pharmacological characterization, and molecular modeling of heterobicyclic amino acids related to (+)-2-aminobicyclo[3.1.0] hexane-2,6-dicarboxylic acid (LY354740): identification of two new potent, selective, and systemically active agonists for group II metabotropic glutamate receptors. *J. Med. Chem.* 42: 1027–1040.

45 Rorick-Kehn, L.M., Johnson, B.G., Burkey, J.L. et al. (2007). Pharmacological and pharmacokinetic properties of a structurally novel, potent, and selective metabotropic glutamate 2/3 receptor agonist: in vitro characterization of agonist (−)-(1R,4S,5S,6S)-4-amino-2-sulfonylbicyclo[3.1.0]-hexane-4,6-dicarboxylic acid (LY404039). *J. Pharmacol. Exp. Ther.* 321: 308–317.

46 Monn, J.A., Prieto, L., Taboada, L. et al. (2015). Synthesis and pharmacological characterization of C4-disubstituted analogs of 1S,2S,5R,6S-2-aminobicyclo[3.1.0]hexane-2,6-dicarboxylate: identification of a potent, selective metabotropic glutamate receptor agonist and determination of agonist-bound human mGlu2 and mGlu3 amino terminal domain structures. *J. Med. Chem.* 58: 1776–1794.

47 Monn, J.A., Valli, M.J., Massey, S.M. et al. (2013). Synthesis and pharmacological characterization of 4-substituted-2-aminobicyclo[3.1.0]hexane-2,6-dicarboxylates: identification of new potent and selective metabotropic glutamate 2/3 receptor agonists. *J. Med. Chem.* 56: 4442–4455.

48 Monn, J.A., Henry, S.S., Massey, S.M. et al. (2018). Synthesis and pharmacological characterization of C4beta-amide-substituted 2-aminobicyclo[3.1.0]hexane-2,6-dicarboxylates: identification of (1 S,2 S,4 S,5 R,6 S)-2-amino-4-[(3-methoxybenzoyl)amino]bicyclo[3.1.0]hexane-2,6-dicarboxylic acid (LY2794193), a highly potent and selective mGlu3 receptor agonist. *J. Med. Chem.* 61: 2303–2328.

49 Monn, J.A., Massey, S.M., Valli, M.J. et al. (2007). Synthesis and metabotropic glutamate receptor activity of S-oxidized variants of (−)-4-amino-2-thiabicyclo-[3.1.0]hexane-4,6-dicarboxylate: identification of potent, selective, and orally bioavailable agonists for mGlu2/3 receptors. *J. Med. Chem.* 50: 233–240.

50 Bueno, A.B., Collado, I., de Dios, A. et al. (2005). Dipeptides as effective prodrugs of the unnatural amino acid (+)-2-aminobicyclo[3.1.0]hexane-2,6-dicarboxylic acid (LY354740), a selective group II metabotropic glutamate receptor agonist. *J. Med. Chem.* 48: 5305–5320.

51 Perkins, E.J. and Abraham, T. (2007). Pharmacokinetics, metabolism, and excretion of the intestinal peptide transporter 1 (SLC15A1)-targeted prodrug (1S,2S,5R,6S)-2-[(2'S)-(2-amino) propionyl]aminobicyclo[3.1.0.]hexen-2,6-dicarboxy lic acid (LY544344) in rats and dogs: assessment of first-pass bioactivation and dose linearity. *Drug Metab. Dispos.* 35: 1903–1909.

52 Rorick-Kehn, L.M., Perkins, E.J., Knitowski, K.M. et al. (2006). Improved bioavailability of the mGlu2/3 receptor agonist LY354740 using a prodrug strategy: in vivo pharmacology of LY544344. *J. Pharmacol. Exp. Ther.* 316: 905–913.

53 Mezler, M., Geneste, H., Gault, L., and Marek, G.J. (2010). LY-2140023, a prodrug of the group II metabotropic glutamate receptor agonist LY-404039 for the potential treatment of schizophrenia. *Curr. Opin. Invest. Drugs* 11: 833–845.

54 Patil, S.T., Zhang, L., Martenyi, F. et al. (2007). Activation of mGlu2/3 receptors as a new approach to treat schizophrenia: a randomized phase 2 clinical trial. *Nat. Med.* 13: 1102–1107.

55 Kinon, B.J. and Gómez, J.-C. (2013). Clinical development of pomaglumetad methionil: a non-dopaminergic treatment for schizophrenia. *Neuropharmacol.* 66: 82–86.

56 *The New York Times* (2007). New schizophrenia drug shows promise in trials. https://www.nytimes.com/2007/09/03/business/03drug.html (accessed 18 December 2019).

57 Stauffer, V.L., Millen, B.A., Andersen, S. et al. (2013). Pomaglumetad methionil: no significant difference as an adjunctive treatment for patients with prominent negative symptoms of schizophrenia compared to placebo. *Schizophr. Res.* 150: 434–441.

58 Adams, D.H., Kinon, B.J., Baygani, S. et al. (2013). A long-term, phase 2, multicenter, randomized, open-label, comparative safety study of pomaglumetad methionil (LY2140023 monohydrate) versus atypical antipsychotic standard of care in patients with schizophrenia. *BMC Psychiatry* 13: 143.

59 Kinon, B.J., Zhang, L., Millen, B.A. et al.; Group, H. S. (2011). A multicenter, inpatient, phase 2, double-blind, placebo-controlled dose-ranging study of LY2140023 monohydrate in patients with DSM-IV schizophrenia. *J. Clin. Psychopharmacol.* 31: 349–355.

60 Adams, D.H., Zhang, L., Millen, B.A. et al. (2014). Pomaglumetad methionil (LY2140023 monohydrate) and aripiprazole in patients with schizophrenia: a phase 3, Multicenter, Double-Blind Comparison. *Schizophr. Res. Treatment* 2014: 758212.

61 Downing, A.M., Kinon, B.J., Millen, B.A. et al. (2014). A double-blind, placebo-controlled comparator study of LY2140023 monohydrate in patients with schizophrenia. *BMC Psychiatry* 14: 351.

62 Li, M.-L., Hu, X.-Q., Li, F., and Gao, W.-J. (2015). Perspectives on the mGluR2/3 agonists as a therapeutic target for schizophrenia: still promising or a dead end? *Prog. Neuropsychopharmacol. Biol. Psychiatry* 60: 66–76.

63 Kinon, B.J., Millen, B.A., Zhang, L., and McKinzie, D.L. (2015). Exploratory analysis for a targeted patient population responsive to the metabotropic glutamate 2/3 receptor agonist pomaglumetad methionil in schizophrenia. *Biol. Psychiat.* 78: 754–762.

64 Gonzalez-Maeso, J., Ang, R.L., Yuen, T. et al. (2008). Identification of a serotonin/glutamate receptor complex implicated in psychosis. *Nature* 452: 93–97.

65 Kurita, M., Holloway, T., Garcia-Bea, A. et al. (2012). HDAC2 regulates atypical antipsychotic responses through the modulation of mGlu2 promoter activity. *Nat. Neurosci.* 15: 1245–1254.

66 Fribourg, M., Moreno, J.L., Holloway, T. et al. (2011). Decoding the signaling of a GPCR heteromeric complex reveals a unifying mechanism of action of antipsychotic drugs. *Cell* 147: 1011–1023.

67 Benneyworth, M.A., Xiang, Z., Smith, R.L. et al. (2007). A selective positive allosteric modulator of metabotropic glutamate receptor subtype 2 blocks a hallucinogenic drug model of psychosis. *Mol. Pharmacol.* 72: 477–484.

68 Gewirtz, J.C. and Marek, G.J. (2000). Behavioral evidence for interactions between a hallucinogenic drug and group II metabotropic glutamate receptors. *Neuropsychopharmacol.* 23: 569–576.

69 Winter, J.C., Eckler, J.R., and Rabin, R.A. (2004). Serotonergic/glutamatergic interactions: the effects of mGlu2/3 receptor ligands in rats trained with LSD and PCP as discriminative stimuli. *Psychopharmacol.* 172: 233–240.

70 Gu, G., Lorrain, D.S., Wei, H. et al. (2008). Distribution of metabotropic glutamate 2 and 3 receptors in the rat forebrain: implication in emotional responses and central disinhibition. *Brain Res.* 1197: 47–62.

71 Harrison, P.J., Lyon, L., Sartorius, L.J. et al. (2008). The group II metabotropic glutamate receptor 3 (mGluR3, mGlu3, GRM3): expression, function and involvement in schizophrenia. *J. Psychopharmacol.* 22: 308–322.

72 Fell, M.J., Svensson, K.A., Johnson, B.G., and Schoepp, D.D. (2008). Evidence for the role of metabotropic glutamate (mGlu)2 not mGlu3 receptors in the preclinical antipsychotic pharmacology of the mGlu2/3 receptor agonist (−)-(1R,4S,5S,6S)-4-amino-2-sulfonylbicyclo[3.1.0]hexane-4,6-dicarboxylic acid (LY404039). *J. Pharmacol. Exp. Ther.* 326: 209–217.

73 Spooren, W.P., Gasparini, F., van der Putten, H. et al. (2000). Lack of effect of LY314582 (a group 2 metabotropic glutamate receptor agonist) on phencyclidine-induced locomotor activity in metabotropic glutamate receptor 2 knockout mice. *Eur. J. Pharmacol.* 397: R1–R2.

74 Woolley, M.L., Pemberton, D.J., Bate, S. et al. (2008). The mGlu2 but not the mGlu3 receptor mediates the actions of the mGluR2/3 agonist, LY379268, in mouse models predictive of antipsychotic activity. *Psychopharmacol.* 196: 431–440.

75 Hao, J. and Chen, Q. (2019). Insights into the structural aspects of the mGlu receptor orthosteric binding site. *Curr. Top. Med. Chem.* 19: 2421–2446.

76 Christopher, J.A., Orgován, Z., Congreve, M. et al. (2019). Structure-based optimization strategies for G protein-coupled receptor (GPCR) allosteric modulators: a case study from analyses of new metabotropic glutamate receptor 5 (mGlu5) X-ray structures. *J. Med. Chem.* 62: 207–222.

77 Dore, A.S., Okrasa, K., Patel, J.C. et al. (2014). Structure of class C GPCR metabotropic glutamate receptor 5 transmembrane domain. *Nature* 511: 557–562.

78 Wu, H., Wang, C., Gregory, K.J. et al. (2014). Structure of a class C GPCR metabotropic glutamate receptor 1 bound to an allosteric modulator. *Science* 344: 58–64.

79 Harpsøe, K., Isberg, V., Tehan, B.G. et al. (2015). Selective negative allosteric modulation of metabotropic glutamate receptors – a structural perspective of ligands and mutants. *Sci. Rep.* 5: 13869.

80 Johnson, M.P., Baez, M., Jagdmann, G.E. Jr. et al. (2003). Discovery of allosteric potentiators for the metabotropic glutamate 2 receptor: synthesis and subtype selectivity of N-(4-(2-methoxyphenoxy)phenyl)-N-(2,2,2- trifluoroethylsulfonyl)pyrid-3-ylmethylamine. *J. Med. Chem.* 46: 3189–3192.

81 Trabanco, A.A., Cid, J.M., Lavreysen, H. et al. (2011). Progress in the developement of positive allosteric modulators of the metabotropic glutamate receptor 2. *Curr. Med. Chem.* 18: 47–68.

82 Caraci, F., Molinaro, G., Battaglia, G. et al. (2011). Targeting group II metabotropic glutamate (mGlu) receptors for the treatment of psychosis associated with Alzheimer's disease: selective activation of mGlu2 receptors amplifies beta-amyloid toxicity in cultured neurons, whereas dual activation of mGlu2 and mGlu3 receptors is neuroprotective. *Mol. Pharmacol.* 79: 618–626.

83 Engers, J.L., Bollinger, K.A., Weiner, R.L. et al. (2017). Design and synthesis of N-aryl Phenoxyethoxy pyridinones as highly selective and CNS penetrant mGlu3 NAMs. *ACS Med. Chem. Lett.* 8: 925–930.

84 Schann, S. (2014). mGlu3 PAM as novel neuroprotective therapeutic strategy for Parkinson's disease. *Current Neuropharmacol.* 12: 54–54.

85 Tsuchiya, D., Kunishima, N., Kamiya, N. et al. (2002). Structural views of the ligand-binding cores of a metabotropic glutamate receptor complexed with an antagonist and both glutamate and Gd3+. *Proc. Natl. Acad. Sci. USA* 99: 2660–2665.

86 Kunishima, N., Shimada, Y., Tsuji, Y. et al. (2000). Structural basis of glutamate recognition by a dimeric metabotropic glutamate receptor. *Nature* 407: 971–977.

87 Muto, T., Tsuchiya, D., Morikawa, K., and Jingami, H. (2007). Structures of the extracellular regions of the group II/III metabotropic glutamate receptors. *Proc. Natl. Acad. Sci. USA* 104: 3759–3764.

88 Chemical Computing Group (2016).CCG Molecular Operating Environment (MOE), Version 2016. Montreal, QC, Canada.

89 Felder, C.C., Schober, D.A., Tu, Y. et al. (2017). Translational pharmacology of the metabotropic glutamate 2 receptor-preferring agonist LY2812223 in the animal and human brain. *J. Pharmacol. Exp. Ther.* 361: 190–197.

90 Sanger, H., Hanna, L., Colvin, E.M. et al. (2013). Pharmacological profiling of native group II metabotropic glutamate receptors in primary cortical neuronal cultures using a FLIPR. *Neuropharmacol.* 66: 264–273.

91 Varie, D.L., Beck, C., Borders, S.K. et al. (2007). Design, development, and scale-up of a selective meso-epoxide desymmetrization process. *Org. Process Res. Dev.* 11: 546–559.

92 Henry, S.S., Brady, M.D., Laird, D.L.T. et al. (2012). Improved synthesis of C4α- and C4β-methyl analogues of 2-Aminobicyclo[3.1.0]hexane-2,6-dicarboxylate. *Org. Lett.* 14: 2662–2665.

93 Hao, J., Reinhard, M., Henry, S.S. et al. (2012). Simple conversion of fully protected amino acids to zwitterions. *Tetrahedron Lett.* 53: 1433–1434.

94 McColm, J., Brittain, C., Suriyapperuma, S. et al. (2017). Evaluation of single and multiple doses of a novel mGlu2 agonist, a potential antipsychotic therapy, in healthy subjects. *Br. J. Clin. Pharmacol.* 83: 1654–1667.

95 Lowe, S., Dean, R., Ackermann, B. et al. (2012). Effects of a novel mGlu(2)/(3) receptor agonist prodrug, LY2140023 monohydrate, on central monoamine turnover as determined in human and rat cerebrospinal fluid. *Psychopharmacol.* 219: 959–970.

96 Mehta, M.A., Schmechtig, A., Kotoula, V. et al. (2018). Group II metabotropic glutamate receptor agonist prodrugs LY2979165 and LY2140023 attenuate the functional imaging response to ketamine in healthy subjects. *Psychopharmacol.* 235: 1875–1886.

97 De Simoni, S., Schwarz, A.J., O'Daly, O.G. et al. (2013). Test-retest reliability of the BOLD pharmacological MRI response to ketamine in healthy volunteers. *NeuroImage* 64: 75–90.

98 Doyle, O.M., De Simoni, S., Schwarz, A.J. et al. (2013). Quantifying the attenuation of the ketamine pharmacological magnetic resonance imaging response in humans: a validation using antipsychotic and glutamatergic agents. *J. Pharmacol. Exp. Ther.* 345: 151–160.

99 Gozzi, A., Large, C.H., Schwarz, A. et al. (2008). Differential effects of antipsychotic and glutamatergic agents on the phMRI response to phencyclidine. *Neuropsychopharmacol.* 33: 1690–1703.

100 Krystal, J.H., Abi-Saab, W., Perry, E. et al. (2005). Preliminary evidence of attenuation of the disruptive effects of the NMDA glutamate receptor antagonist, ketamine, on working memory by pretreatment with the group II metabotropic glutamate receptor agonist, LY354740, in healthy human subjects. *Psychopharmacol.* 179: 303–309.

101 Moghaddam, B. and Adams, B.W. (1998). Reversal of phencyclidine effects by a group II metabotropic glutamate receptor agonist in rats. *Science* 281: 1349–1352.

102 Denovo Biopharma (2019). Denovo's pipeline. https://www.denovobiopharma.com/index8.html (accessed 18 December 2019).

7

Discovery of Taselisib (GDC-0032): An Inhibitor of PI3Kα with Selectivity over PI3Kβ

Timothy P. Heffron[1], Laurent Salphati[2], and Steven T. Staben[1]

[1] *Discovery Chemistry, Genentech, Inc., 1 DNA Way, MS 252A, CA 94080, South San Francisco, CA, USA*
[2] *Drug Metabolism and Pharmacokinetics, Genentech, Inc., 1 DNA Way, MS 412A, CA 94080, South San Francisco, CA, USA*

7.1 Introduction

There are four Class I phosphatoinositide 3-kinase (PI3K) isoforms (PI3Kα, PI3Kβ, PI3Kδ, PI3Kγ) [1]. Each of the four isoforms are composed of both a catalytic (p110α, β, δ, γ) and a regulatory subunit (p85, p110). These Class I PI3 kinases are activated upstream by distinct upstream mechanisms and have distinct expression profiles leading to different potential applications of inhibitors. In each case the activated Class I PI3K functions to phosphorylate phosphatidylinositol-4,5-bisphosphate (PIP2) to the corresponding phosphatidylinositol-3,4,5-trisphosphate (PIP3). PIP3, in turn, facilitates downstream signaling in the PI3K/AKT/mTOR pathway. Working in the opposite direction is the phosphatase and tensin homolog (PTEN). This tumor suppressor, as a phosphatase, is a negative regulator of Class I PI3K signaling. Signaling through PI3K can be dysregulated through several mechanisms including through activating mutations of PI3Kα, loss of PTEN, and increased upstream activation. As the PI3K/mTOR/AKT pathway regulates cell growth, survival and proliferation, hallmarks of cancer, it is easy to appreciate how such aberrant signaling could contribute to malignancy [1]. While each of the individual kinase isoforms has merit as a drug target, the PI3Kα isoform has been the primary target of our own drug discovery efforts. Indeed, the Class I PI3K isoform PI3Kα has been identified as one of the most frequently mutated oncogenes [2]. Additionally, a significant percentage of specific breast cancers have an associated dysregulation of PI3Kα signaling [3]. Accordingly, inhibitors of PI3Kα have been intensely pursued for the treatment of cancer, and alpelisib, a selective inhibitor of the PI3Kα isoform, has received FDA approval [1, 4, 5] (Figure 7.1).

With four distinct Class I isoforms as well as the related lipid kinase mTOR, a range of desirable pharmacological selectivity profiles was identified (Figure 7.2). On one extreme, a molecule that inhibits each of the four Class I PI3K isoforms as well as mTOR, albeit with high selectivity over other kinases, was identified as being most capable of inhibiting cancer growth (left side in Figure 7.2). However, such a profile was also anticipated to have the greatest risk of lacking an adequate safety profile to realize efficacy. One step away, an inhibitor of each of the Class I PI3K isoforms (referred to as a pan-PI3K inhibitor) that spared inhibition of mTOR as well as other kinases might provide less comprehensive pathway blockade but we speculated that it might be

Contemporary Accounts in Drug Discovery and Development, First Edition. Edited by Xianhai Huang, Robert G. Aslanian, and Wayne H. Tang.

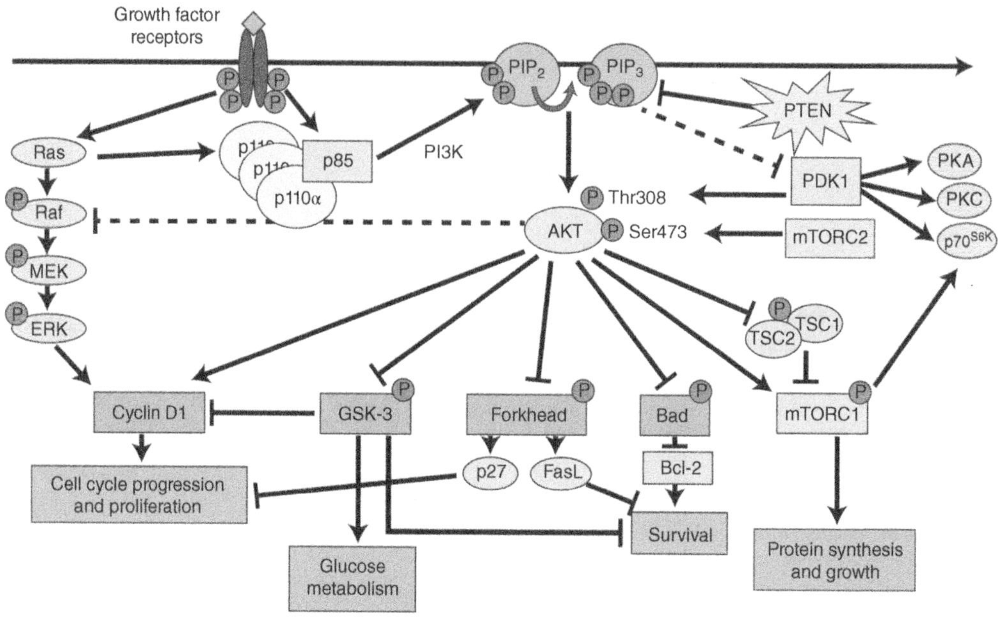

Figure 7.1 PI3K signaling pathway.

better tolerated in a clinical setting. At the far right end of the selectivity spectrum of PI3Kα inhibitors in Figure 7.2 would be a molecule that specifically inhibits only the PI3Kα isoform. As we started our efforts in the discovery of PI3Kα inhibitors we desired molecules across the selectivity spectrum to advance to clinical studies to determine what selectivity profile would allow for a desirable clinical profile.

The first Genentech efforts in the discovery and development of inhibitors of PI3Kα started via a collaboration with Piramed Pharma whose efforts had led to the discovery of the pan-Class I PI3K inhibitor GDC-0941 (Figure 7.2) [6]. Subsequently, we collaborated on the discovery of the dual mTOR/pan-class I PI3K inhibitor GDC-0980 (Figure 7.2) [7]. Having identified clinical molecules that inhibited each of the Class I PI3K isoforms, we became interested in identifying an inhibitor of PI3Kα that achieved selectivity over one or more of the other isoforms. Furthermore, in such efforts we also sought to mitigate any potential unknown risk that might be associated with the chemical class shared by our two clinical inhibitors (GDC-0941 and GDC-0980, Figure 7.2). Owing to these combined interests we initiated a high-throughput screening campaign to identify a new class of PI3Kα inhibitors that might be able to achieve selectivity over other isoforms. In particular, we were intrigued by the potential of molecules that would achieve selectivity over PI3Kβ due to its reported role in insulin signaling which led us to speculate that sparing inhibition of PI3Kβ might mitigate safety risk. These efforts ultimately led to the discovery of taselisib (GDC-0032), an inhibitor of PI3Kα that selectively spares inhibition of the PI3Kβ isoform [8].

7.2 Hit to Lead Efforts

In an effort to identify new chemical matter, in early 2007 we completed a high-throughput screen. This screen utilized the full length PI3Kα (p110α/p85α) enzyme and we evaluated a single concentration of each member of our internal screening library for inhibition via a fluorescence

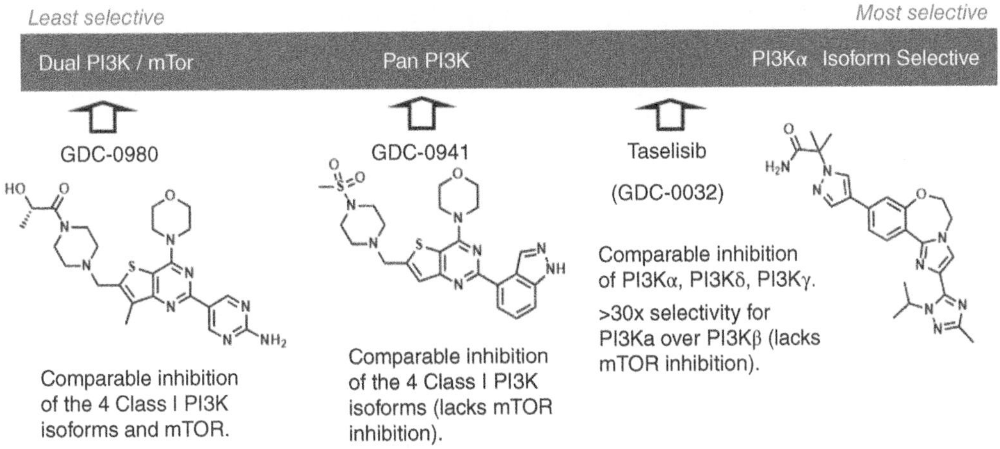

Figure 7.2 Spectrum of PI3Kα inhibitor selectivity profiles of Genentech developed PI3Kα inhibitors.

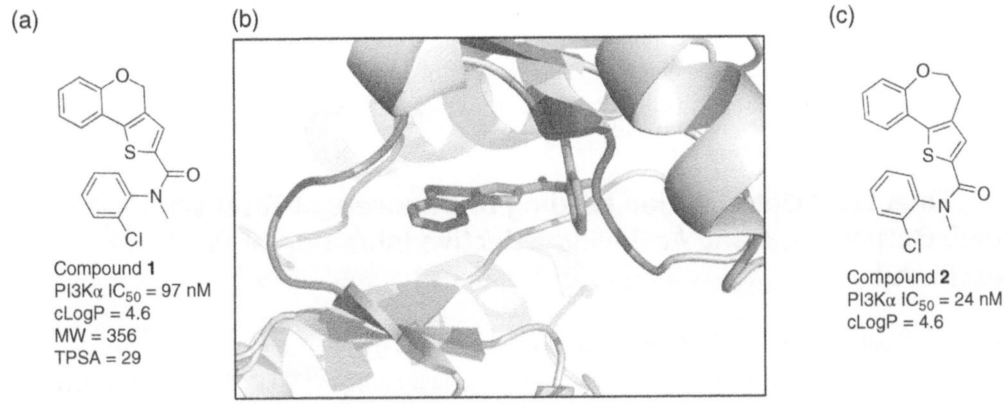

Figure 7.3 (a) Structure and properties of HTS hit **1**. (b) X-ray structure of **1** in p110γ (PDB ID: 3R7Q) highlighting select, proximal regions of differentiation in other p110 isoforms. (c) Structure and properties of benzoxepin **2**.

polarization assay. Subsequent to the initial screen of a single concentration, we selected hits identified as novel at that time for follow up and IC_{50}'s were generated using the same PI3Kα fluorescence polarization assay. While we were interested in identifying isoform selective inhibitors of PI3Kα we did not counterscreen other PI3K isoforms at this stage of our discovery efforts. A number of interesting hits were pursued in parallel to thienobenzopyran **1** (Figure 7.3a), a modestly active compound with a confirmed $IC_{50} = 97$ nM [9]. Although highly lipophilic (cLogP = 4.6), and possessing a number of structural alerts (thiophene, aniline), we were drawn to follow-up given this scaffold was not represented across the kinase literature. In addition, we were able to solve a crystal structure of compound **1** in p110γ which revealed vectors for substitution that could be directed toward some of the few areas near the ATP site where there were amino acid residue differences across the PI3K isoforms (Figure 7.3b, PDB 3R7Q).

The ADME profile of early compounds was consistent with their physicochemical properties and structural liabilities. Molecules in this series had high permeability, but low solubility and high clearance *in vitro* and *in vivo*. We also noted some challenges with chemical stability in solution that were influencing our assay data. After prolonged storage in DMSO, solvolysis at the pyran methylene was leading to ring-opened (inactive) products. Ring-expansion to thienobenzoxepin (i.e. **2**, Figure 7.3c) eliminated this liability and gave us a reliable three- to fivefold improvement in biochemical potency (IC_{50} = 24 nM).

Exploratory work at the 8-position of the benzoxepin, aimed at direct and water-mediated contacts with the lower hinge region, delivered additional potency improvements as exemplified in primary amide **3** (IC_{50} = 1.5 nM, Figure 7.4). A combination of *in vitro* metabolite ID studies (±cofactors) and computational metabolite prediction indicated the *N*-methyl aniline amide as a metabolic hot-spot contributing to the high clearance of this series of compounds (amide hydrolysis and/or *N*-demethylation) [10]. Compounds **2** (Figure 7.3c) and **3** (Figure 7.4) were both unstable in human liver microsomes (HLMs; clearance, Cl = 18 ml/min/kg) and highly cleared in rodents (rat plasma clearance, Cl_p = 60 and 88 ml/min/kg, respectively). Direct replacement of the aniline with alkyl or cycloalkyl substituted amines led to unsurprising losses in activity due to low energy conformations that were incompatible with the p110 binding site. We thus focused on replacement of this formally cis amide with aromatic heterocycles with the goals of improving metabolic stability and increasing potency by creating and/or improving interaction with polar residues (i.e. pseudo-gatekeeper Tyr, catalytic Lys, C-helix Asp). The 1,2,4-triazole (e.g. compound **4**, Figure 7.4) stood out among these giving us improved potency and *in vitro* and *in vivo* stability. Importantly, this conformational lock allowed replacement of the 2-haloarene with branched alkyl substituents to decrease the molecular weight and lipophilicity resulting in lead molecules such as **5** (Figure 7.4).

7.3 Final Lead Optimization Leading to Discovery of Taselisib: ADME Optimization and Achieving Selective Inhibition of PI3Kα over PI3Kβ

With a successful structure-based drug design effort having led to PI3Kα inhibitors of remarkable potency such as **5**, our efforts turned to further optimization of the ADME properties of these inhibitors as well as ensuring selective inhibition of PI3Kα over another isoform.

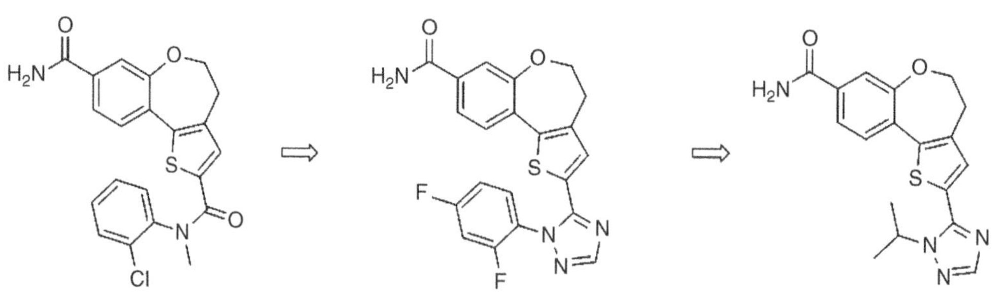

Compound **3**
PI3Kα IC_{50} = 1.5 nM
cLogP = 3.5
Rat Cl_p = 88 ml/min/kg
HLM = 18 ml/min/kg

Compound **4**
PI3Kα IC_{50} < 0.5 nM
cLogP = 4.4
Rat Cl_p = 38 ml/min/kg
Rat $Cl_{p,u}$ = 2300 ml/min/kg
HLM = 8 mL/min/kg

Compound **5**
PI3Kα IC_{50} = 5 nM
cLogP = 3.6
MW = 354
Rat Cl_p = 13 ml/min/kg
Rat $Cl_{p,u}$ = 230 ml/min/kg
HLM = 10 ml/min/kg

Figure 7.4 Replacement of the *N*-methyl aniline amide with a 1-isopropyl-1,2,4-triazole.

Initial hit-to-lead efforts had realized roughly 1000-fold improvements in potency in biochemical assays and eliminated initial challenges with chemical instability. Yet molecules at this stage of discovery efforts were not suitable for clinical advancement. As reasonable biochemical inhibition of PI3Kα was regularly achieved, we started to evaluate these inhibitors in cellular assays. The first cellular assay we utilized was a cellular proliferation assay in which we measured the affect of the molecule on inhibition of the proliferation of MCF7 cells. That MCF7 cell line harbors a PI3Kα mutation and had been demonstrated to be sensitive to the inhibition by the previous, clinical, PI3Kα inhibitors we had discovered. However, the benzoxepins we were studying at this time frequently lacked adequate cellular potency in the MCF7 anti-proliferation assay, which we attributed in part to exceptionally high plasma protein binding. Additionally, the benzoxepins at this stage frequently suffered from poor solubility and poor metabolic stability. We recognized that, in principle, each of these challenges could be overcome through reductions in lipophilicity. Therefore, our discovery team focused on retaining the key interactions made between the ligand and protein to maintain potency while reducing the lipophilicity of the ligands relative to what had been made previously. To this end, we prepared and evaluated thiazole and imidazole (and other) scaffolds to replace the original thiophene-containing core (Figure 7.5, i.e. compounds **6–8**). In contrast to compounds **3–5**, to initially evaluate different core heterocycles the phenyl rings in compounds **6–8** were substituted with a pyrazole. This change in phenyl ring substitution was made as the pyrazole was identified as a more versatile handle to attenuate physicochemical properties and PI3Kα over PI3Kβ selectivity (see below).

These efforts to evaluate alternate scaffolds led us to quickly settle on the benzoxazepin scaffold (e.g. compound **8**, Figure 7.5), as it led to improved metabolic stability and unbound Cl in preclinical species as well as solubility when compared with analogs with other scaffolds. Critically, we also found that expected cellular potency was also consistently realized as we transitioned away from the thiophene containing core to those with reduced lipolphilicity.

In parallel with the lead optimization efforts described above we had been monitoring isoform selectivity and had a dedicated medicinal chemistry effort directed toward achieving isoform selective inhibition of PI3Kα. These efforts began by identifying the several residues which differ across the Class I PI3K isoforms (Figure 7.3b). Initial efforts were directed toward achieving hydrogen bonding interactions with residues unique to PI3Kα. These efforts, however, were not successful in realizing PI3Kα inhibitors with selectivity over other isoforms. However,

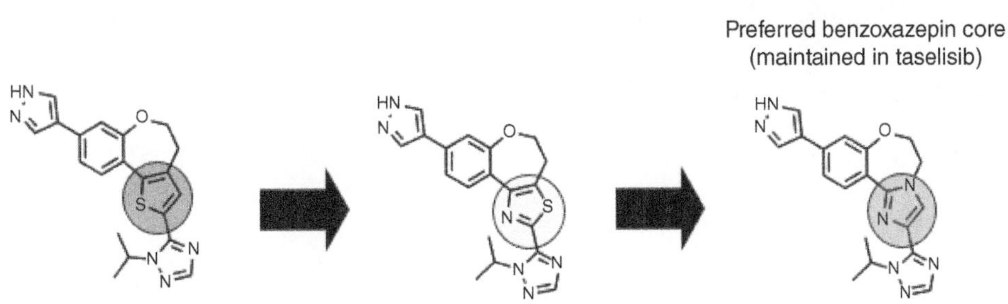

Compound **6**
PI3Kα IC_{50} = 1.6 nM
cLog P = 5.0

Compound **7**
PI3Kα IC_{50} = 0.5 nM
cLog P = 4.3

Compound **8**
PI3Kα IC_{50} = 0.5 nM
cLog P = 3.1

Figure 7.5 Representative molecules spanning various oxepin scaffolds evaluated during the discovery of taselisib. A comparison of PI3Kα inhibition and cLogP is provided.

simultaneously we assessed inhibition of all Class I isoforms for molecules produced in the hit-to-lead efforts described above. Some of these inhibitors demonstrated diminished inhibition of PI3Kβ relative to PI3Kα and the other two Class I isoforms. We postulated that molecules that attained this partial selectivity profile could be desirable as they would occupy an as yet unevaluated portion of the PI3K selectivity spectrum. In an attempt to understand the basis of this selectivity we developed models of the active sites of PI3Kα and PI3Kβ. These models were built from crystal structures of PI3Kγ as no crystal structures of PI3Kα or PI3Kβ were available at the time. The models built indicated an electrostatic potential difference that might be exploited to enhance selectivity for inhibition of PI3Kα over PI3Kβ. The electrostatic potential maps (Figure 7.6a,b) suggested that positioning non-basic functionality in the region of isoform specific difference should result in an improvement in selectivity for PI3Kα over PI3Kβ. Indeed structure–activity relationships suggested that this was the case [11]. Additionally, throughout this program we leveraged co-crystal structures of our ligands in p110γ. Complementing the electrostatic hypothesis for selectivity, we also recognized through empirical structure–activity relationships that bulky substitution off the 8-position of the benzoxepin core provided high selectivity over PI3Kβ. While we were unable to develop a strong rationale for this observation initially, subsequent X-ray crystal structures of all isoforms of the catalytic p110 subunits of PI3Ks were published and led us to formulate a hypothesis for selectivity over PI3Kβ. We believe this selectivity is driven by a unique conformations and mobility of a conserved tryptophan in PI3Kβ relative to the other isoforms (Figure 7.6c,d). Second shell residues (glutamic acid in PI3Kα and PI3Kγ, methionine in PI3Kδ) promote similar conformation of this conserved tryptophan in these isoforms. However, in PI3Kβ this residue is a valine (bulky, branched and hydrophobic) that appears to promote a distinct conformation of the conserved tryptophan within

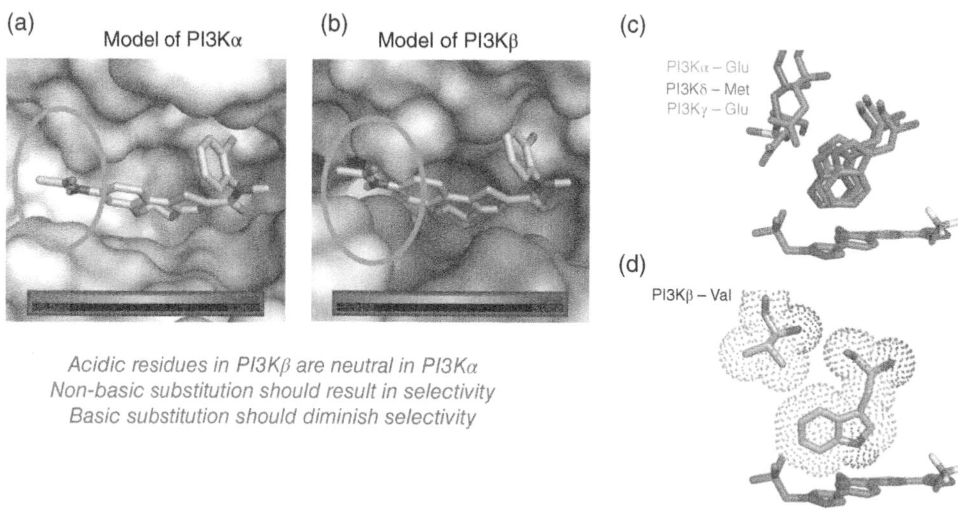

Figure 7.6 Structural differences contributing to selectivity for PI3Kα over PI3Kβ. Electrostatic potential differences between the active sites of (a) PI3Kα and (b) PI3Kβ. (c) Crystal structure of a representative thiazolobenzoxepin in p110γ (PDB ID 4JI6 [11]). Aligned structures of p110α (PDB ID 3HIZ [12]), p110β (PDB ID 2Y3A [13]) and p110δ (PDB ID 2XWM [14]). (d) Second shell residues may dictate the orientation and mobility of a conserved tryptophan residue. Source: (a, b) Adapted with permission from [11]. Copyright 2011, American Chemical Society.

PI3Kβ that would clash with our ligand [15]. As a result of these combined approaches, appropriate substitution of the benzene ring in the benzoxepin core could realize the desired isoform selectivity.

At this stage of the program, we had identified the benzoxazepin series of molecules as the preferred series, and had developed a structural hypothesis using multiple approaches to predictably achieve meaningful levels of selectivity for the inhibition of PI3Kα over PI3Kβ. Shortly after these advances, taselisib was identified as having the right balance of overall characteristics. In preclinical evaluations taselisib was a highly potent inhibitor of PI3Kα, with selectivity over PI3Kβ, as well as highly selective over related PIKK family kinases (Figure 7.7). Additionally, taselisib also had acceptable solubility and highly desirable ADME properties (Figure 7.7).

7.4 Preclinical *in vivo* Pharmacology of Taselisib

In preclinical *in vivo* studies taselisib demonstrated significant tumor growth inhibition in a xenograft tumor model in which the MCF7-neo-HER2 tumor cell line was driven by aberrant PI3Kα signaling (Figure 7.8). In addition to the antitumor efficacy achieved by taselisib, a consistent pharmacodynamic effect demonstrated on-target pathway inhibition that correlated with efficacy in that same model [7]. Subsequent preclinical safety profiling of taselisib supported its advancement to clinical studies.

	Ki_{app}(nM)
p110α	0.2
p110α–H1047R	0.1
p110α–E545K	0.1
p110β	7.8 (39 X)
p110δ	0.1
p110γ	0.7
mTOR	1200 (>4000X)

PIKK Family	IC50 (nM)
C2α	10,000
C2β	292
VPS34	374
DNA-PK	>4800

>230 Kinases : <50% inhibition @ 1um

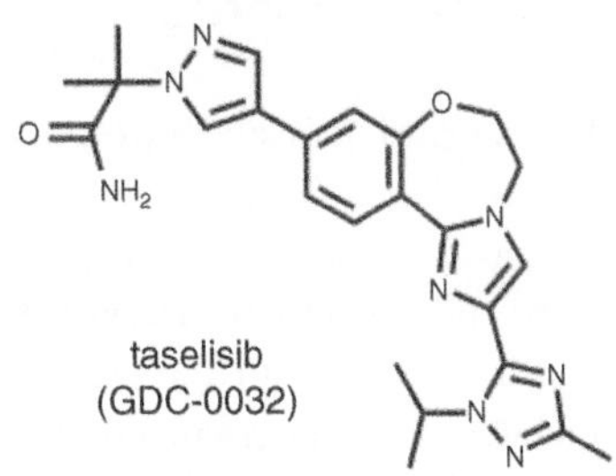

MW = 460; TPSA = 118
Log $D_{7.4}$ = 1.98

Cellular PD	IC50
pAkt:MCF7-neo/HER2 (PIK3CA–E545K)	4 nM
pAkt:PC3 (PTEN -)	31 nM

Cell Viability (Cell Titer Glo)	IC50
MCF7-neo/HER2 (PIK3CA - E545K)	25 nM

Figure 7.7 *In vitro* characterization of taselisib.

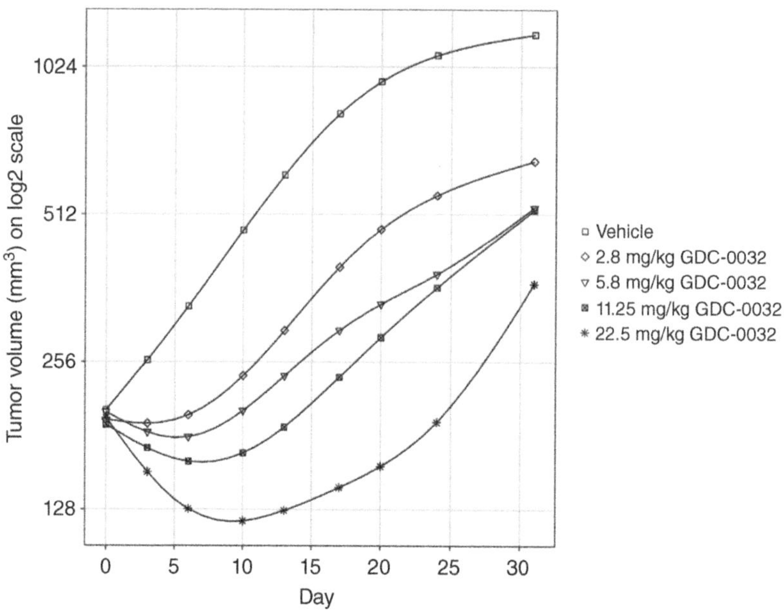

Figure 7.8 Dose–response curves of fitted MCF7-neo/HER2 tumor volumes in response to once daily oral administration of taselisib (GDC-0032). Source: Figure reprinted with permission from [16]. Copyright 2019, American Chemical Society.

7.5 Prediction and Clinical Assessment of Taselisib Human Pharmacokinetics

The ADME properties of taselisib were also evaluated *in vitro* and *in vivo*. Pharmacokinetic studies conducted in preclinical species showed that taselisib had low Cl_p in rodents and monkeys and moderate Cl_p in dogs (Table 7.1). These optimal pharmacokinetic properties in animals appeared very promising and allowed us to pursue the characterization of taselisib in efficacy and toxicology studies. More importantly for the prediction of human pharmacokinetics, the hepatic clearance (Cl_H) extrapolated from *in vitro* metabolic stability assays in liver microsomes and hepatocytes was very consistent with the Cl_p estimated in all preclinical species tested (Table 7.1) [17, 18]. Such *in vitro–in vivo* correlation provided confidence in the relevance and predictive value of the *in vitro* stability assays and suggested that Cl_p in human would be low and correlated with the projected Cl_H, approximately 3 ml/min/kg. Additional experimental parameters necessary for the predictions of pharmacokinetic profiles by physiologically based pharmacokinetic (PBPK) modeling were obtained. These included apparent permeability (Papp) in MDCK cells, plasma protein binding, blood–plasma partitioning as well as physicochemical parameters such as logD, pKa and solubility under various pH and conditions (Table 7.1). PBPK simulations (GastroPlus® v 9.7. SimulationPlus, Lancaster, CA, USA) in preclinical species (rat, monkey, and dog) were conducted to assess the validity of the approach and confirm the experimental input parameters did not lead to major disconnect between simulated and actual pharmacokinetic profiles. Subsequently, simulations of the plasma concentration–time profile in human were performed at the starting dose of 3 mg. Figure 7.9 shows the simulation and actual plasma concentrations measured in patients in the first cohort [19, 20]. While C_{max} appeared relatively well predicted, area under the curve (AUC) and half-life ($t_{1/2}$) did not match the clinical data. At that stage of the program, experimental

Table 7.1 *In vitro* and *in vivo* parameters for taselisib and GastroPlus experimental input parameters.

***In vitro* and *in vivo* parameters for taselisib**				
	Predicted Cl_H[a]			
	Hepatocytes	**Microsomes**	**Cl_p[b]**	***F***
Species	**(ml/min/kg)**		**(ml/min/kg)**	**(%)**
Mouse	5	3	2.73	100
Rat	11	2	1.77 ± 0.07	99.1 ± 6.0
Cyno	8	5	4.15 ± 0.41	41.5 ± 16.1
Dog	11	9	10.1 ± 0.9	66.8 ± 14.7
Human	3	3	–	–
GastroPlus experimental input parameters[c]				
LogD (7.4)			2.5	
pKa			3.2	
Reference solubility			0.022 mg/ml (pH 6.5)	
P_{app} MDCK cells; human P_{eff}			6.44×10^{-6} cm/s; 2.66×10^{-4} cm/s	
Blood/plasma			1.1	
Fraction unbound in plasma			0.24	
Cl_H (=$Cl_{systemic}$)			12.6 l/h (3 ml/min/kg, for 70 kg human)	
Cl_H optimized (=$Cl_{systemic}$)			5 l/h (1.2 ml/min/kg, for 70 kg human)	

[a] Cl_H calculated as described in [14] and incorporating plasma protein and microsomal protein binding [17].
[b] Mean ± SD of 3 animals, except in mice where a pooled sample analysis was performed.
[c] Other inputs were set to the GastroPlus default parameters.
F, oral bioavailability.

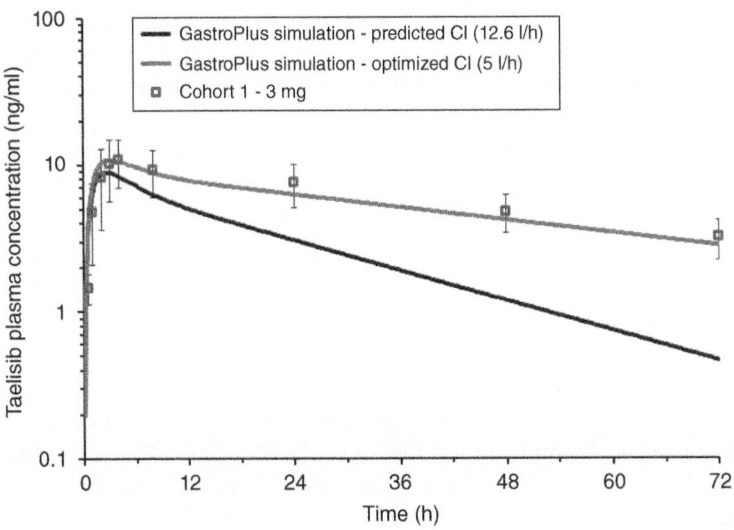

Figure 7.9 Simulations of plasma concentration–time profiles of taselisib in human following a 3 mg dose, using Cl predicted from *in vitro* studies or optimized based on Cohort 1 data.

parameters had been tested in preclinical simulations. The discrepancy between terminal slopes in the simulated and experimental profiles in humans, combined with the known uncertainty associated with Cl_p, prompted us to start the optimization of the model with that parameter. Lowering (i.e. optimizing) Cl_H, which was the only Cl term included in the model and thus corresponded also to systemic Cl (Cl_p), generated the optimized profile presented in Figure 7.9. The optimized simulation matched the clinical data, and implied that the *in vitro* assays overpredicted Cl_H (and Cl_p). Human pharmacokinetic parameters were considered very favorable, with a $t_{1/2}$ of approximately 40 hours and a likely Cl_p of 1.2 ml/min/kg, based on simulations (Figure 7.9; Table 7.1) [18]. Absolute bioavailability studies conducted as the program advanced confirmed that the actual average Cl_p (5.28 l/h ≈ 1.25 ml/min/kg) was similar to the simulated optimized Cl_p [21].

7.6 Conclusion

Through a combination of structure-guided drug design, reaction to structure–activity relationships of isoform selectivity, and conscious optimization of physicochemical properties, taselisib was identified as a PI3Kα inhibitor for clinical development. The evaluation of taselisib in the clinical setting confirmed a desirable human pharmacokinetic profile and began the evaluation of its pharmacology and safety [18, 19, 22]. Additionally, clinical insights from the study of taselisib, as well as other clinical PI3K inhibitors, inspired the pursuit of a next generation PI3Kα inhibitor. Taselisib also laid a solid foundation for the discovery of GDC-0077 which is currently in phase III studies.

References

1 Liu, P., Cheng, H., Roberts, T.M., and Zhou, J.J. (2009). Targeting the phosphoinositide 3-kinase pathway in cancer. *Nat. Rev. Drug Discovery* 8: 627.

2 Bauer, T.M., Patel, M.R., and Infante, J.R. (2015). Targeting PI3 kinase in cancer. *Pharmacol. Ther.* 146: 53e60.

3 (a) Saal, L.H., Holm, K., Maurer, M. et al. (2005). PIK3CA mutations correlate with hormone receptors, node metastasis, and ERBB2, and are mutually exclusive with PTEN loss in human breast carcinoma. *Cancer Res.* 65: 2554–2559.(b) Stemke-Hale, K., Gonzalez-Angulo, A.M., Lluch, A. et al. (2008). An integrative genomic and proteomic analysis of PIK3CA, PTEN, and AKT mutations in breast cancer. *Cancer Res.* 68: 6084–6091.

4 (a) Ihle, N.T. and Powis, G. (2009). Take your PIK: phosphatidylinositol 3-kinase inhibitors race through the clinic and toward cancer therapy. *Mol. Cancer Ther.* 8: 1–9.b Fruman, D.A. and Rommel, C. (2014). PI3K and cancer: lessons, challenges and opportunities. *Nat. Rev. Drug Discovery* 13: 140–156.c Garces, A.E. and Stocks, M.J. (2019). Class 1 PI3K clinical candidates and recent inhibitor design strategies: a medicinal chemistry perspective. *J. Med. Chem.* 62: 4815–4850.d Elmenier, F.M., Lasheen, D.S., and Abouzid, K.A.M. (2019). Phosphatidylinositol 3 kinase (PI3K) inhibitors as new weapon to combat cancer. *Eur. J. Med. Chem.* 183: 111718.

5 Markham, A. (2019). Alpelisib: first global approval. *Drugs* 79: 1249–1253.

6 Folkes, A.J., Ahmadi, K., Alderton, W.K. et al. (2008). The identification of 2-(1*H*-indazol-4-yl)-6-(4-methanesulfonyl-piperazin-1-ylmethyl)-4-morpholin-4-yl-thieno[3,2-*d*]pyrimidine (GDC-0941) as a potent, selective, orally bioavailable inhibitor of class I PI3 kinase for the treatment of cancer. *J. Med. Chem.* 51: 5522–5532.

7 Sutherlin, D.P., Sampath, D., Berry, M. et al. (2010). Discovery of (thienopyrimidin-2-yl) aminopyrimidines as potent, selective, and orally available pan-PI3-kinase and dual pan-PI3-kinase/mTOR inhibitors for the treatment of cancer. *J. Med. Chem.* 53: 1086–1097.

8 Ndubaku, C.O., Heffron, T.P., Staben, S.T. et al. (2013). Discovery of 2-{3-[2-(1-isopropyl-3-methyl-1*H*-1,2–4-triazol-5-yl)-5,6-dihydrobenzo[f]imidazo[1,2-*d*][1,4]oxazepin-9-yl]-1*H*-pyrazol-1-yl}-2-methylpropanamide (GDC-0032): a β-sparing phosphoinositide 3-kinase inhibitor with high unbound exposure and robust in vivo antitumor activity. *J. Med. Chem.* 56: 4597–4610.

9 Staben, S.T., Siu, M., Goldsmith, R. et al. (2011). Structure-based design of thienobenzoxepin inhibitors of PI3-kinase. *Bioorg. Med. Chem. Lett.* 21: 4054–4058.

10 Staben, S.T., Blaquiere, N., Tsui, V. et al. (2013). *Cis*-amide isosteric replacement in thienobenzoxepin inhibitors of PI3-kinase. *Bioorg. Med. Chem. Lett.* 23: 897–901.

11 Heffron, T.P., Wei, B., Olivero, A. et al. (2011). The rational design of PI3 kinase inhibitors exhibiting selectivity over the PI3K-β isoform. *J. Med. Chem.* 54: 7815–7833.

12 Huang, C.-H., Mandeker, D., Schmidt-Kittler, O. et al. (2007). The structure of a human p110alpha/p85alpha complex elucidates the effects of oncogenic PI3Kalpha mutations. *Science* 318: 1744.

13 Zhang, X., Vadas, O., Perisic, O. et al. (2011). Structure of lipid kinase p110β/p85β elucidates an unusual SH2-domain-mediated inhibitory mechanism. *Mol. Cell* 41: 567–578.

14 Berndt, A., Miller, S., Williams, O. et al. (2010). The p110δ crystal structure uncovers mechanisms for selectivity and potency of novel PI3K inhibitors. *Nat. Chem. Biol.* 6: 117.

15 Staben, S.T., Ndubaku, C., Blaquiere, N. et al. (2013). Discovery of thiazolobenzoxepin PI3-kinase inhibitors that spare the PI3-kinase β isoform. *Bioorg. Med. Chem. Lett.* 23: 2606–2613.

16 Angelaud, R., Staben, S., Heffron, T. et al. (2019). From discovery to market readiness: the research and development of the B-sparing phosphatidylinositol 3-kinase inhibitor taselisib. In: *Complete Accounts of Integrated Drug Discovery and Development: Recent Examples from the Pharmaceutical Industry*, vol. 2 (eds. J.A. Pesti, A.F. Abdel-Magid and R. Vaidyanathan), 61–83. Washington, D.C.: American Chemical Society.

17 Obach, R.S., Baxter, J.G., Liston, T.E. et al. (1997). The prediction of human pharmacokinetic parameters from preclinical and *in vitro* metabolism data. *J. Pharmacol. Exp. Ther.* 283: 46–58.

18 Obach, R.S. (1999). Prediction of human clearance of twenty-nine drugs from hepatic microsomal intrinsic clearance data: an examination of in vitro half-life approach and nonspecific binding to microsomes. *Drug Metab. Dispos.* 27: 1350–1359.

19 Juric, D., Krop, I., Ramanathan, R.K. et al. (2017). Phase I dose-escalation study of Taselisb, an oral PI3K inhibitor, in patients with advanced solid tumors. *Cancer Discov.* 7: 704–715.

20 Pang, J., Baumgardner, M., Cheong, J. et al. (2014). Preclinical evaluation of the β isoform-sparing PI3K inhibitor GDC-0032 and prediction of its human pharmacokinetics. *Drug Metab. Rev.* 45 (Suppl. 1): 1–286.

21 Sahasranaman, S., Ma, S., Hsu, J. et al. (2015). Assessment of absolute bioavailability and mass balance of the PI3K inhibitor taselisib (GDC-0032) in healthy subjects. *Clin. Pharmacol. Ther.* 97 (S60-S96): PII-075.

22 Abramson, V.G., Oliveira, M., Cervantes, A. et al. (2019). A phase Ib, open-label, dose-escalation study of the safety and pharmacology of taselisib (GDC-0032) in combination with either docetaxel or paclitaxel in patients with HER2-negative, locally advanced, or metastatic breast cancer. *Breast Cancer Res. Treat.* 178: 121–133.b Baird, R.D., Gao, M., Garcia-Corbacho, J. et al. (2019). POSEIDON trial phase 1b results: safety, efficacy and circulating tumor DNA response of the beta isoform-sparing PI3K inhibitor taselisib (GDC-0032) combined with tamoxifen in hormone receptor positive metastatic breast cancer patients. *Clin. Cancer Res.* 25: 6598–6605.c Tamura, K., Kodaira, M., Shimizu, C. et al. (2018). Phase I study of taselisib in Japanese patients with advanced solid tumors or hormone receptor-positive advanced breast cancer. *Cancer Sci.* 109: 1592–1601.d Langer, C.J., Redman, M.W., Wade, J.L. et al. (2019). SWOG S1400B (NCT02785913), a phase II study of GDC-0032 (taselisib) for previously treated PI3K-positive patients with stage IV squamous cell lung cancer (lung-MAP sub-study). *J. Thorac. Oncol.* 14: 1839–1846.e Dickler, M.N., Saura, C., Richards, D. et al. (2018). Phase II study of taselisib (GDC-0032) in combination with fulvestrant in patients with HER2-negative, hormone receptor-positive advanced breast cancer. *Clin. Cancer Res.* 18: 4380–4387.f Saura, C., Hlauschek, D., Oliveira, M. et al. (2019). Neoadjuvant letrozole plus taselisib versus letrozole plus placebo in postmenopausal women with oestrogen receptor-positive, HER2-negative, early-stage breast cancer (LORELEI): a multicentre, randomised, double-blind, placebo-controlled, phase 2 trial. *Lancet Oncol.* 20: 1226–1238.

8

Drug Discovery with DNA-Encoded Library Technology

Inhibitor of Soluble Epoxide Hydrolase to Clinical Candidate

Yun Ding and Sarah K. Scott

Encoded Library Technologies/NCE Molecular Discovery, GSK, 200 Cambridge Park Drive, MA 02140, Cambridge, MA, USA

The discovery of small molecule ligands that bind to biological targets of interest is fundamental to discovering new therapeutics. DNA-encoded library (DEL) technology links the powers of genetics and chemical synthesis, using DNA tags to encode individual organic molecules in a combinatorial library. It allows the construction and screening of combinatorial libraries with unprecedented size and provides access to much broader chemical diversity, thus facilitating the discovery of ligands to many different biological targets. Over the last 15 years, this emerging technology has been proven to be an effective approach for hit identification and target validation.

8.1 Background of DNA-Encoded Library Technology

The concept of DEL technology was first proposed by Lerner and Brenner at Scripps in 1992 [1] and was shortly expanded in 1993 by Janda and colleagues at Scripps [2], and researchers at Affymax [3] who reported the synthesis of DNA-encoded peptide libraries on a solid support. They envisaged that combinatorial syntheses could be performed in which oligonucleotides on one end of a linker could be used to encode the chemical synthesis performed on the other end of the linker. A split-and-pool combinatorial approach could be applied to library synthesis in which each individual component would have a unique DNA barcode. This could provide several advantages over traditional methods of both library synthesis as well as library screening due to the ability to amplify and sequence the DNA tags [2, 3]. Encoding a combinatorial synthetic library with oligonucleotides provides a mechanism for addressing the major limitations of ambiguity and sensitivity encountered in the direct structural analysis of ligands isolated from large combinatorial libraries. The high capacity of DNA for information storage can be exploited to archive the precise details of the library constructions. Inherently, an oligonucleotide-based coding scheme can achieve tremendous levels of amplification through quantitative polymerase chain reaction (qPCR).

8.1.1 Development of Encoding Strategies

Since the initial reports, a variety of methods for associating DNA with a chemotype have been developed. The Liu group developed DNA-templated synthesis (DTS), in which a DNA oligonucleotide directs bond formation by bringing DNA-appended reactants into proximity via

Contemporary Accounts in Drug Discovery and Development, First Edition. Edited by Xianhai Huang, Robert G. Aslanian, and Wayne H. Tang.

Watson–Crick base pairing (Figure 8.1a) [5]. The Liu group went on to apply the method of DTS to the synthesis of a fumaramide macrocycle library from which a known binder was enriched during selection, serving as a proof of concept for both the use of DTS in the synthesis of a chemical library and the ability to use DTS-produced libraries in selection experiments to elucidate binding interactions [6]. Shortly after the Liu group first disclosed DTS, scientists at Vipergen applied the concept to a unique system, termed "YoctoReactor" (yR), which overcomes some of the challenges associated with directing chemical reactions using DTS (Figure 8.1b). Namely, because of inconsistent folding which can occur within regions of the single-stranded DNA (ssDNA), as well as variation in environment at the reaction site, efficiencies of chemical reactions can vary [7]. The yR overcomes these challenges by formation of three-dimensional DNA junctions to bring the reactants into proximity in a confined environment [8]. However, the biggest challenge for DTS has been the need to separate synthesis and purification of DNA-linked building blocks before the library synthesis. Harbury reported a solution to this problem [9] with DNA-controlled routing of starting materials to separate reaction vessels. Complementarity to resin-bound ssDNA separates DNA-tagged templates into various compartments where solid-phase synthesis can be performed (Figure 8.1c). This method allows off-the-shelf reagents to be utilized directly.

All three of the aforementioned techniques use DNA to control the synthesis of the library members. An alternative strategy is to use DNA sequences which are not involved in directing the synthesis but are assembled beforehand as the "recipe" for each library synthesis step. Parallel synthesis of encoding DNA and library molecules like Lerner, Brenner, and Janda proposed can be achieved via a split-and-pool strategy. The most popular method employed for DNA synthesis is enzymatic ligation with T4 ligase, though chemical ligation has also been reported [10]. Neri and co-workers introduced the concept of DNA-encoded self-assembling chemical libraries which were generated by the combinatorial hybridization of two complementary single-stranded sub-libraries, each containing a set of compounds individually linked to distinctive oligonucleotides [11]. Large DNA-encoded libraries can be constructed by the combinatorial self-assembly of relatively small complementary sub-libraries. The resulting DNA fragments can carry two neighboring pharmacophores and adapt them onto the target protein surface.

Instead of using dual-pharmacophores, a convenient strategy to construct large libraries via split-and-pool synthesis with a single pharmacophore was developed by GSK [12] (formerly Praecis) and Nuevolution (now Amgen). The synthesis of DNA-encoded libraries using this split-and-pool method (detailed in Figure 8.2) offers a unique advantage over traditional combinatorial chemistry because of the ability to pool all the products from a single synthetic step and perform subsequent chemical transformations on the pooled material. Not only does this make handling simpler, as individual components do not need to be treated separately, but this also enables the synthesis of libraries of enormous size (containing millions or billions of members if desired), as the libraries grow exponentially with each cycle of chemistry.

8.1.2 The Encoding Strategy at GSK

Within GSK, we use a short, covalently linked DNA duplex, which we refer to as the "headpiece". We believe this provides advantages over other DNA encoding techniques utilizing single-stranded coding oligonucleotide, as the duplex structure may provide some level of protection for the heterocyclic bases positioned within the helix, limiting undesired reactivity of the DNA during library synthesis. We commonly use a form of this DNA duplex in which a PEG linker is installed using Fmoc-15-amino-4,7,10,13-tetraoxapentadecanoic acid (AOP), which upon Fmoc deprotection reveals a free amine on which small molecule synthesis can be initiated (Figure 8.3). We believe

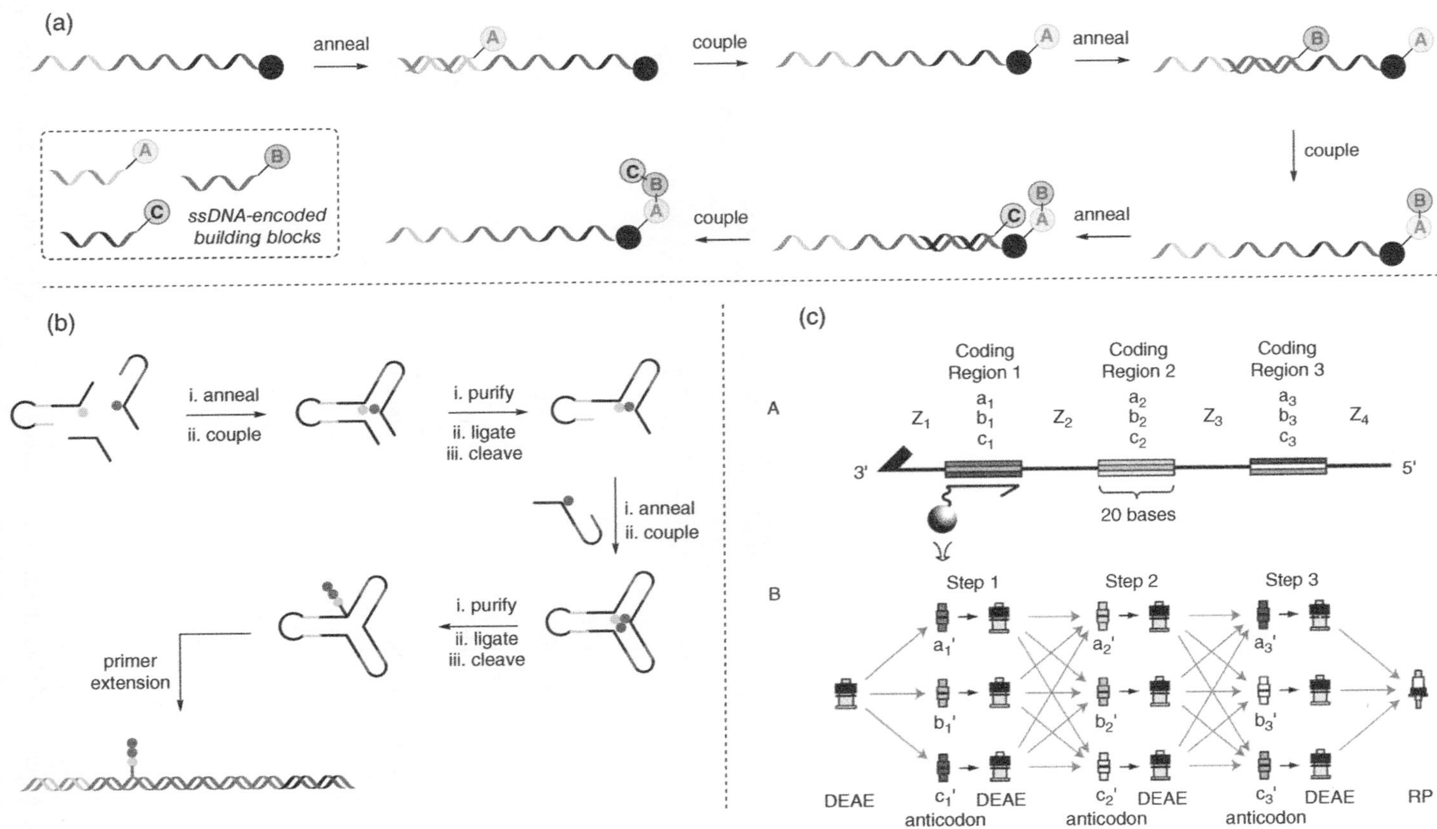

Figure 8.1 Encoding strategies: (a) DNA-templated synthesis; (b) Vipergen YoctoReactor; and (c) DNA-routed synthesis [4]. *Source:* (c) Adapted with permission under Creative Commons Attribution License (CC BY).

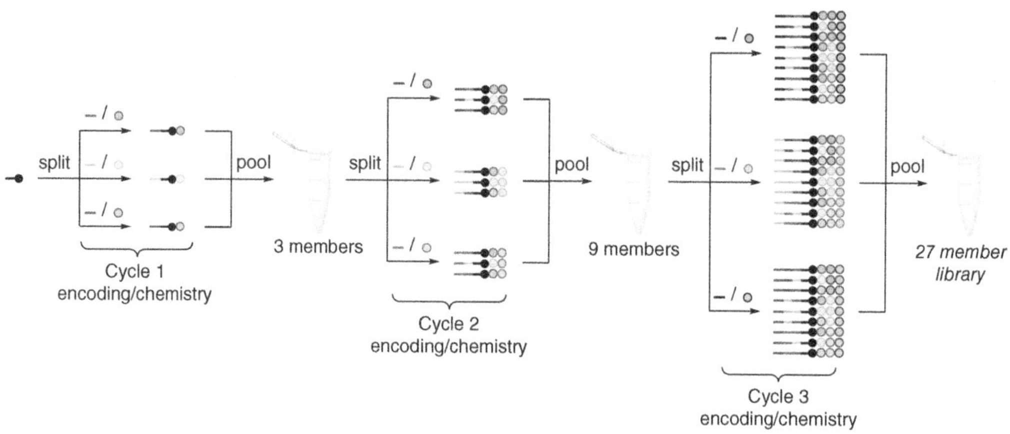

Figure 8.2 Split-and-pool synthesis of DELs.

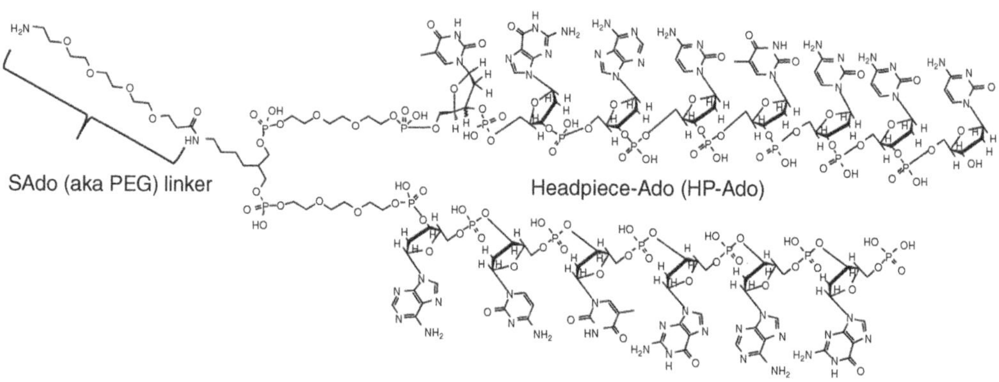

Figure 8.3 Structure of DNA for DEL synthesis.

use of this variation of the headpiece can be advantageous as the PEG linker serves to increase the distance between the oligonucleotides and the position in which the small molecule synthesis occurs. The two-base 3′-overhang on the right-hand portion of the duplex, as depicted in Figure 8.3, provides a substrate for subsequent ligation to coding tags for the first cycle of chemistry. With each cycle of chemistry, short double-stranded DNA coding tags are installed bearing a 7-base variable region and a 3′-overhang which is held constant for all tags in the given cycle. The overhangs continue to serve as handles for subsequent ligation at each cycle of chemistry and varies between cycles to prevent ligation to truncated sequences. With each cycle liquid chromatography–mass spectrometry (LCMS) analysis, or gel analysis if the molecular weight (MW) is too large for LCMS, can be used to confirm quantitative ligation.

Library synthesis typically commences with ligation of the headpiece in an arrayed format using 96-well plates to install the cycle 1 DNA tags using T4 ligase. Upon completion, the samples are subjected to ethanol precipitation and the resulting pellets are redissolved in the appropriate buffer and cycle 1 chemistry is performed. Reaction progress is typically monitored by LCMS in a subset of representative wells, and once the desired reactivity has been achieved, the wells are pooled and cleaned up as appropriate, often using reverse-phase high pressure liquid chromatography (HPLC) or 3k-cutoff membrane filtration. The pooled material is then split again into 96-well plates for the next cycle of ligation and chemistry. If each cycle has 96 building blocks, after 3 cycles of split-and-pool, a library of 0.8 million unique compounds, each with a unique DNA barcode will be generated. Following the last cycle of ligation and synthesis, a ligation on the final pooled material using a unique 30 base-pair primer sequence will be used to close out the library. This primer can be used as an identifier for the library later and provides a control for polymerase chain reaction (PCR) artifacts that may occur during sequencing.

It is worth noting that as libraries progress beyond the first cycle, it becomes difficult to quantify conversions, yields, and byproduct formation with each reaction due to the diversity and high MWs of members in the pooled libraries. Thus, success of each step is typically determined by seeing a gross shift in MW of the library and emphasis is put on developing a fully encompassing protocol during initial validation experiments to ensure the highest probability of success during library synthesis. Although these fail-safes alone cannot guarantee success for all library members, the impact of individual failed reactions is minimized because deep sampling and affinity selection of libraries often gives rise to families of related structures.

This method has now been broadly adapted by pharmaceutical industries and academic groups. In this chapter, the application of this format of DEL technology in drug discovery will be described in detail, along with the identification of the first clinical candidate from DEL technology.

8.1.3 Development of DNA-Compatible Chemistry

At the time Lerner and Brenner first proposed the concept of DELs in 1992, the libraries encoded with genetic material were limited to peptides. One of the key advancements necessary for the widespread success of DELs has been the development of chemistry that is compatible with the oligonucleotide tags. Some of the most obvious limitations that preclude the use of more conventional synthetic methods include the requirement for aqueous conditions, low concentrations, and the avoidance of low pH and high temperatures. Additionally, the chemistries used must present orthogonal reactivity to oligonucleotide reactivity. Initially, only a handful of reactions were able to be performed in the presence of the DNA tags, but over the last decade quite a few more have been developed. Some of the most common and widely used reactions for DELs include S_NAr reactions, amine capping reactions, formation of amide bonds, reductive amination, transition metal-catalyzed cross-coupling reactions, and simple deprotection reactions (such as Fmoc and Boc) [13–19]. More recently the field has seen increasingly complex reactions being developed to be compatible with DELs, including cross-coupling strategies to incorporate higher C-sp^3 content [20]. Additionally, a new strategy involving reversible adsorption of DNA to a solid support so the reactions can be performed in organic solvents opens the opportunity to employ previously inaccessible chemical reactivities in DEL [21–23].

8.1.4 Methods for *in vitro* Selection of DNA-Encoded Libraries

With the development of effective DNA-compatible synthetic methods, large scale libraries can be produced and subjected to biological targets of interest to discover functional ligands. The discovery of functional ligands to biological targets is fundamental to understanding living systems and discovering new therapeutics. There are various methods available to identify such small molecule ligands. One of the most popular methods is high-throughput screening (HTS), which has been widely adapted by the pharmaceutical industry. Traditional HTS procedures require the storage and handling of hundreds of thousands of chemical compounds and the cost related to HTS infrastructure, maintenance, and bulk reagents for the screenings is tremendous which has limited its application in small drug discovery companies and academic groups. Various assay formats used in HTS solely rely on the detection of the biochemical activity of the targets. Proteins with unknown biochemical function cannot be readily screened with HTS, which limits its application in the validation of therapeutic targets. Certain targets that do not have appropriate functional assays to adapt to HTS will be challenging as well.

DEL selection is an affinity-based selection which presents several important advantages over HTS. First, the time, cost, and material consumed are greatly reduced. Secondly, affinity-based screening provides the opportunity for the validation of targets without appropriate functional assays. Thus, the traditionally separated phases in drug discovery, target validation and hit identification, can be merged to give a shorter drug discovery timeline. Thirdly, affinity screening of DELs enables a variety of parallel screening to be conducted, such as the competition with known ligands, cross-selection with selectivity targets, addition or subtraction of co-factors, substrates, or binding partners, and titration of target concentrations. Such selection output can generate hypotheses about selectivity, binding sites, dissociation constants, and biological function before the compounds are synthesized off-DNA and assayed. Fourthly, the selection output from the large chemical collections creates large sets of data that are very valuable for the off-DNA hit follow-up. If closely related structures are selected, it provides confidence that a structurally driven binding event is occurring. By examining the commonalities and diversities among the family members, one can surmise what the key pharmacophore is, and what regions can be further optimized or truncated. Overall, the data generated from the DEL selection provides rich information that is very valuable for the later stage of hit-to-lead optimization. This will be well demonstrated in this chapter with soluble epoxide hydrolase (sEH) which delivered the first drug candidate from DEL technology.

Affinity-based selection of DELs serves to separate or select bound from unbound compounds providing enrichment for some "binders". To succeed it is very critical to develop the right conditions to keep the target protein in its native state. It is also very important to identify the optimal affinity matrix to immobilize the protein and capture the DNA-tagged small molecule binders. There are a number of options available to immobilize proteins with affinity tags, and a number of different conditions can be tested side by side to optimize selections for a given target. One of the most popular tags is the His tag, which can be captured with IMAC resins. The second way to capture protein is through the biotin–streptavidin interaction. The binding between biotin and streptavidin is one of the strongest non-covalent biological interactions. Both His and biotin are preferred tags for the protein since they can be captured instantly. Labeling protein with FLAG-tag has also been broadly used to immobilize the target protein. FLAG-tag, a polypeptide protein tag, is one of the most specific tags that is recognized by specific, high affinity monoclonal antibodies. Compared with the previous two labeling methods, antibody labeling requires some time to capture the protein. The strategy of using multiple tags at different ends of the protein provides different capture methods, yielding additional selection output for the target. No matter what type of matrix, it is important to run pre-selection protein construct assessment to make sure the protein can be captured on the matrix and maintain functional activity. The purity of the protein will directly affect the output of the DEL screening.

Affinity selection is often performed using an immobilized target that separates active library members bound to the target from non-binding library members still in solution (Figure 8.4). In practice, two selection methods are typically used. In Method A, an aliquot of library is incubated with protein. Blocking agents such as sheared salmon sperm DNA and bovine serum albumin are included in the buffer to reduce non-specific binding. The protein is immobilized on resin and the non-binders washed away. In Method B, the protein is first immobilized on the resin at a concentration sufficient to saturate the matrix. The library aliquot is then added to the immobilized

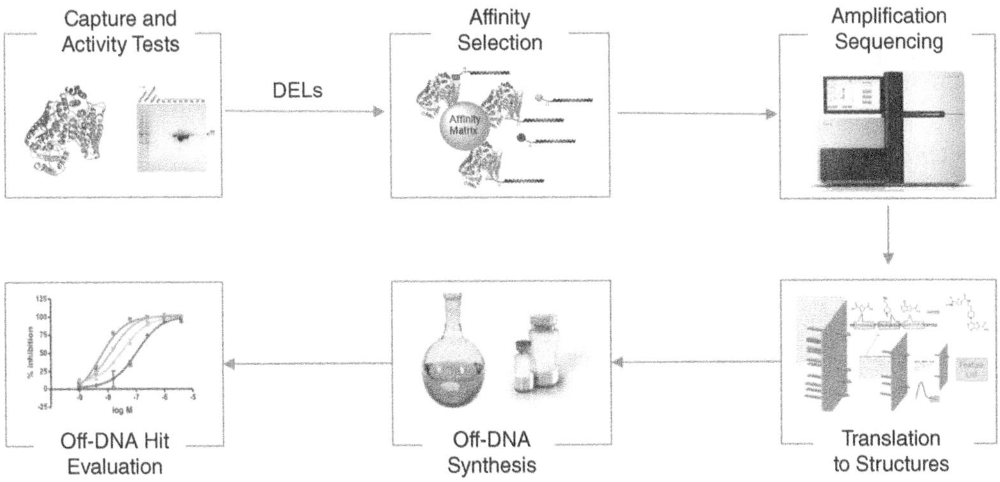

Figure 8.4 Summary of the ELT selection process.

protein target. After incubation, the non-binders are rinsed away. In both methods, the protein is then denatured, and the binding population recovered. This eluant is then incubated with fresh protein to start the next round of selection. Typically, three rounds of selection are performed. In order to rule out molecules whose enrichment was due to binding to the resin matrix, parallel selections are conducted in an identical fashion, but without the addition of protein target (referred to as "no target control"). Library members which are observed to be enriched in the no target control are omitted from the analysis of target-based selection experiments.

8.1.5 Decoding, Data Analysis and off-DNA Hit Follow Up

After selection, the collected binders are PCR-amplified followed by next generation sequencing (NGS) to determine the DNA sequence. In translating these sequences to identify molecules that bind the target, false positives and false negatives can arise for numerous reasons. Sequence bias originating from DNA specific protein binding or from imperfect DNA amplification reactions can produce artifacts that are some of the causes of false positives. DNA contamination from other experiments is another potential positive error driver. False negatives may also arise if the contamination triggers informatics filters against target promiscuity or matrix binding. Data visualization can be integrated into the TIBCO Spotfire platform, allowing scientists to view summaries of the large data sets. Often important trends can be discovered during analysis of the binders, and commonalities between binders give confidence in the pharmacophores that are most promising for follow-up synthesis off DNA. This analysis usually leads to highly informed decisions on which compounds should be synthesized using traditional synthetic methods. Once the molecules have been made off DNA, they can be subjected to the relevant assays to determine the validity of the chosen binders. In some cases, binders that are followed-up on show no desired activity or binding, and this can be due to either a false-positive result in the original screening, or could indicate that the "recipe" used in the synthesis of the DEL produced an undesired byproduct which is responsible for the binding observed. To address the latter, it is common to also follow-up on byproducts which are known or anticipated for the synthesis used in construction of the DEL. The other possible cause of no activity is that the compound only binds to the target. To address this, a binding assay such as SPR can be used to determine whether the compound is the binder.

8.2 Application of DNA-Encoded Library Technology in Small Molecule Drug Discovery

The first hit from a DEL was reported by Harbury and colleagues in 2007. They identified polypeptoid binders against the SH3 domain of the proto-oncogene Crk with affinities similar to those of natural SH3 ligand peptides [24]. Since then, application of DEL technology has produced ligands against various soluble proteins, including kinases such as Src, [25] MAPKAPKK2, [25] Akt3, [25] RIPK1, [26] Aurora A kinase, [12] p38 mitogen-activated protein (MAP) kinase, [12, 27] phosphoinositide 3-kinase a (PI3Ka), [28] glycogen synthase kinase-3 (GSK-3β), [29] and Bruton's tyrosine kinase (BTK) [30]. Many of the inhibitors were found to bind with novel binding modes instead of the traditional ATP binding pocket. Allosteric binding provides the possibility that the hits are more selective against other kinases. Among these reported kinase inhibitors, the benzoxazepinone inhibitor series of RIPK1 was the most successful kinase hit identified from DEL. It resides further in the ATP binding pocket and no interactions with the hinge residues are observed. Though it is not strictly allosteric, part of molecule does occupy the allosteric lipophilic pocket at the back of the ATP binding site. The unique binding conformation drives its high kinase selectivity, along with high potency and good pharmacokinetic (PK) profiles.

DNA-encoded libraries were not only successful in finding potent and selective kinase inhibitors, but proved to be successful across other enzyme families as well. Discovery of matrix metalloproteinase 3 (MMP3) inhibitors represented the first enzyme inhibitors isolated from a DNA-encoded chemical library [31]. Inhibitors against the other metalloproteases were also identified, such as ADAMTS-4 [32] and ADAMTS-5 [33]. In contrast to most MMP inhibitors, the molecules discovered using DEL lack the common Zn-chelating groups, such as carboxylate and hydroxamate. Inhibitors of other enzymes were reported, such as poly(ADP-ribose) polymerase tankyrase 1 (TNKS1), [34] arginine deiminase 4 (PAD4), [35] p53-induced phosphatase 1 (Wip1), [36] insulin degrading enzyme (IDE), [37] sEH, [10, 38] ATPase family AAA-domain containing protein 2 (ATAD2), ADP-ribosyltransferases (PARPs and SIRT6), oxacillinase-48 (OXA-48), [39] bromodomain and extraterminal (BET) family, [40] indoleamine 2,3-dioxygenase-1 (IDO1), [41] undecaprenyl pyrophosphate synthase (Upps), [42] phophodiesterase 12 (PDE12), [43] branched chain aminotransferase mitochondrial (BCATm), [44] autotaxin, [45] Mcl-1, [46] NADH-dependent enoyl-acyl-carrier protein (ACP) reductase, InhA, hepatitis C virus NS4B, [47] etc.

DELs have delivered a number of compounds capable of inhibiting protein–protein interactions. For example, B-cell lymphoma-extra large (Bcl-xL) antagonist, [48] the interleukin 2 (IL-2) inhibitor, [49] and the lymphocyte function-associated antigen-1 (LFA-1) antagonist [50] were identified from DNA-encoded libraries.

DELs have also been screened against membrane proteins, which is more technically demanding compared with soluble proteins. To overcome the lack of stability of isolated transmembrane proteins, one strategy is to overexpress a target protein and perform the selection with protein-overexpressing cell lines. Several potent tachykinin receptors (NK3) antagonists were identified from several different DELs [51]. Another strategy is to stabilize the transmembrane receptors by strategic mutations. Selection of encoded libraries on a stabilized protease activated receptor 2 (PAR2) discovered both agonists and antagonists of PAR2 [52]. The first β2-adrenoceptor (β2AR) small-molecule positive allosteric modulators (PAMs) were isolated by panning DELs against the inactive receptor in which the orthosteric site was unoccupied [53]. This is one of the many examples in which DEL screening discovered active ligands binding at an allosteric site. However, for the G protein-coupled receptors (GPCRs), controlling the discovery of antagonist vs. agonist from DEL screening remains unsolved.

Most applications of DEL are to identify non-covalent inhibitors, but technically it is also suitable for the discovery of covalent inhibitors. Zhu and colleagues at GSK validated the screening method to identify irreversible (covalent) ligands using 3C protease and on-DNA irreversible tool compounds spiked into a library at the same concentration as individual members of that library [54]. Clark and colleagues reported potent irreversible covalent inhibitors of Bruton tyrosine kinase (BTK) from a 27 million acrylamide DEL. Liu and colleagues confirmed the known small molecule–protein interactions and revealed ethacrynic acid, a known ligand for glutathione *S*-transferase, is a novel covalent inhibitor of MAP2K6 kinase [55]. Zimmermann and colleagues also reported a specific and covalent JNK-1 inhibitor from an encoded self-assembling chemical library [56].

Since DEL screening is an affinity-based selection, there is always a risk that identified binders do not have the desired biochemical activity, especially if the protein is unstable and the structure is compromised in any way during the affinity binding process. However, even if the identified binders do not have the desired activity, they can be applied in PROTACs, which uses small molecules to control the cellular levels of drug targets through their precise degradation rather than the inhibition of their activity [57]. There are active researchers employing DELs for identifying binders/inhibitors of deubiquitylating enzymes (DUBs) and other ubiquitin-specific proteases such as E3 ligase and E2 ligase. However, the ultimate outcome remains unknown since no data has been published yet.

8.3 Discovery of Soluble Epoxide Hydrolase Inhibitors Via DNA-Encoded Library Technology

The case study presented in this chapter will look at the application of DELs in the development of novel triazine-based inhibitors of sEH, an enzyme in the α,β-hydrolase fold family of epoxide hydrolases that is responsible for the metabolism of epoxy-lipids to the corresponding diols [58]. For example, sEH is responsible for converting epoxyeicosatrienes (EETs) to their corresponding less active dihydroxyeicosatrienoic acids (DHETs). Inhibiting sEH and, thus, metabolism of EETs is of interest because EETs are known to modulate endothelial cell function in the control of vasomotor tone [59, 60] and in the proliferation and survival of endothelial cells [61, 62]. It has been proposed that sEH inhibitors which block the metabolism of EETs can act as therapeutic agents for several diseases such as diabetes, metabolic syndrome, cardiovascular disease, respiratory disease, and neuropathic pain [62]. Additionally, increasing the half-life of EETs using sEH inhibitors has potential application in the treatment of chronic obstructive pulmonary disease (COPD [63]) by protecting against damage to endothelial cells and in preservation of pulmonary vasculature and alveolar structure in patients with emphysema [63, 64]. Additionally, it has been reported that EETs mediate vasorelaxation (EDHF) and stimulate secretion of vascular endothelial growth factor (VEGF) by endothelial cells with the potential to restore vascular tone and reduce loss of endothelial cells [61, 63, 64]. Thus, discovery of selective and potent sEH inhibitors is attractive for potential application in treating this disease.

Several classes of sEH inhibitors have been identified including urea [65], amide [66, 67], hydrazone derivatives [68], and others [69]. Our goal was to discover novel inhibitors of sEH through affinity selection against DELs. In this chapter, we will describe the discovery of unique triaminotriazine derivatives as potent inhibitors of sEH and the follow up structure–activity relationship (SAR) study that led to the expansion of the hit series. One of the hit series led to a drug candidate GSK2256294 for the treatment of COPD after minimal lead optimization [70, 71].

8.3.1 DELs for sEH Screening

Two related triazine-based DELs were synthesized and screened against sEH. As shown in Figure 8.5a, both libraries commenced with acylation of the headpiece (Figure 8.3) using 192 Fmoc-amino acids. After Fmoc-deprotection, the triazine scaffold was installed using cyanuric chloride. Cycles 2 and 3 both employed S_NAr reactions to substitute the remaining two chlorides, with 32 bifunctional acids installed at cycle 2 and 340 amines installed at cycle 3. The libraries diverged at the final cycle, in which amidation was performed using two sets of building blocks. DEL-A was produced using 384 amines, giving a library containing 802 160 640 compounds and DEL-B was produced using 48 amino esters. Final hydrolysis of DEL-B gave a library containing 100 270 080 members. Nearly all building blocks used were commercially available and each was tested in representative validation experiments using a DNA-linked substrate prior to synthesis of the libraries (>400 Fmoc-amino acids for cycle 1 and >1000 amines for cycles 2–4). Only those building blocks which gave high conversions (>70–80%) were used in the synthesis of DELs A and B. In addition, "on-DNA AUDA" was synthesized by coupling the known sEH inhibitor AUDA to the headpiece appended with a unique DNA tag (Figure 8.5b). On-DNA AUDA showed a similar IC_{50} to the parent AUDA inhibitor, and thus served as a tool compound for validation of the selection protocol.

8.3.2 sEH ELT Selection

Initial validation of the selection protocol was done using DEL-A spiked with on-DNA AUDA at a ratio of 1000:1. A single round of selection was simulated using 1 μM sEH protein and binding of library compounds and on-DNA AUDA was quantified by qPCR. The results of the spike-in experiment are summarized in Figure 8.6, comparing the initial input to the results of the selection in the

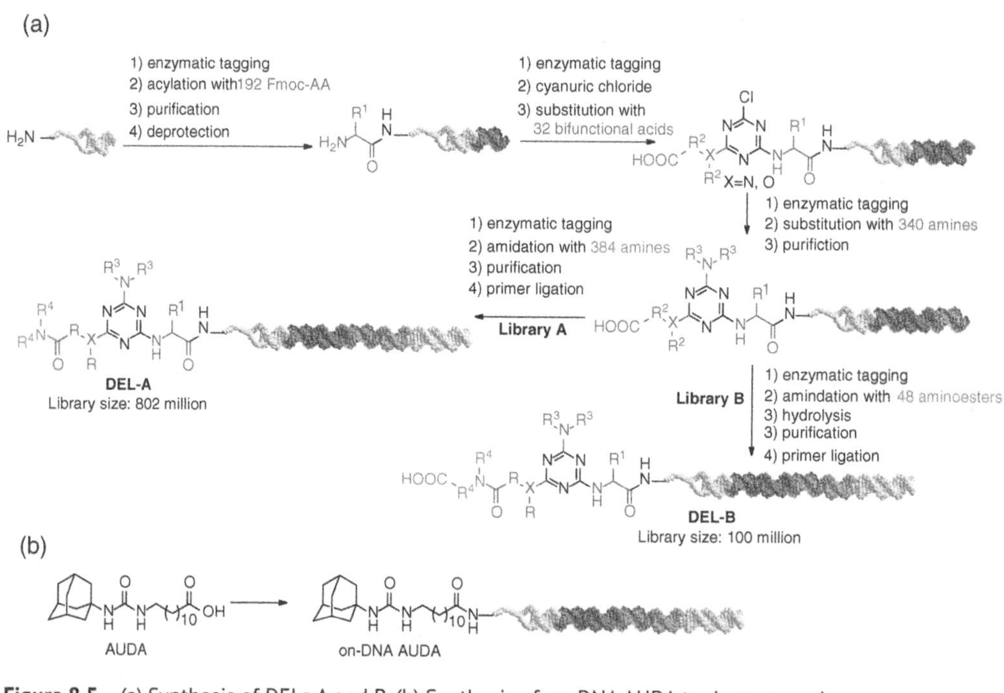

Figure 8.5 (a) Synthesis of DELs A and B. (b) Synthesis of on-DNA AUDA tool compound.

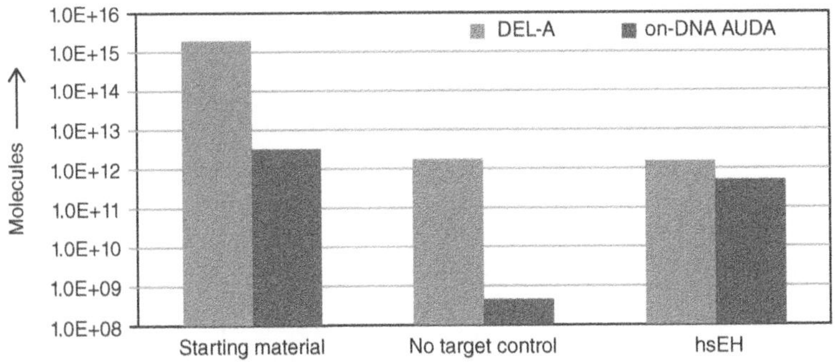

Figure 8.6 qPCR analysis measuring the yield of DEL-A and on-DNA AUDA in the absence (no target control) and presence of target protein (hsEH).

absence of the target protein (no target control) and in the presence of human sEH (hsEH). Although the yield from DEL-A is similar under both conditions, the yield of on-DNA AUDA in the presence of target protein was more than 100 times higher compared with no target control. The same trend was also observed when the concentration of on-DNA AUDA was lowered to that of an individual member of the library. Given the observed dependence of on-DNA AUDA enrichment on the presence of sEH protein, this experiment validated not only quality of the protein for affinity based selection but also the methods used for performing and analyzing the selection.

Upon validating the protein quality and methods, selections were carried out using the protocol previously described [12], incubating the library with either 1 mM protein (to capture all binders), 1 mM protein with AUDA (to elucidate competitive binding against a known inhibitor), or with buffer alone (to control for binding unrelated to the target protein). Streptavidin affinity matrix was then used to capture the protein with bound library members, while the unbound members were washed away. The protein was then denatured and the binders eluted to undergo two more rounds of selection with fresh aliquots of protein, and only after the third round of selection were the binders quantified using qPCR. In analyzing the results from the qPCR, the number of binders observed in the conditions employing sEH alone was approximately 15 times higher than in the conditions employing sEH + on-DNA AUDA inhibitor, indicating competitive binding of the known inhibitor over some of the sEH binders identified. Further, a large number of library members were observed in the conditions employing sEH alone and sEH + on-DNA AUDA relative to the no target control conditions. Even after removing those binders observed in the no target control conditions, a large number of binders remained, most of which were represented with fewer than two copies, making it difficult to narrow down which binders to follow up on using conventional synthesis.

In a similar fashion, DEL-B was screened against sEH; however, based on the results with DEL-A, it was predicted that lowering the protein concentration may help distinguish more potent binders. Thus, selections were performed using protein concentrations of 100 and 10 nM in addition to the original protein concentration of 1 mM and no target control conditions. After removing all binders with only a single copy, the selection output data was visualized using cubic scatter plots (Figure 8.7b). As shown in these plots, the number of binding events decreases with protein concentration, although there is consistency in the binders observed at low concentrations versus high, giving increased confidence in following up on these binders. Therefore, binders observed at 10 nM concentrations of sEH were prioritized for follow-up synthesis off-DNA. Only a single synthon, 4-(aminomethyl)-benzoic acid, was selected at cycle 4 with at least two copies, and

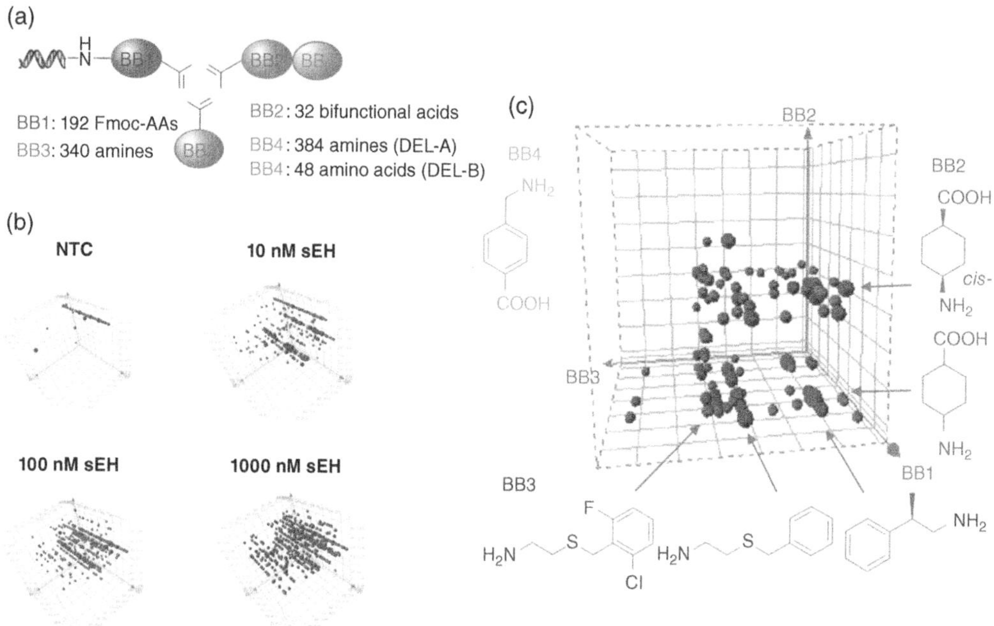

Figure 8.7 (a) DELs A and B. (b) Selection output with DEL-B under different concentrations of sEH. (c) Cubic scatter plot corresponding to 4-(aminomethyl) benzoic acid at cycle 4. The copy number is indicated by the shade and size of the points (darker and larger points indicate higher copy number). The structures shown represent the building blocks defined by a plane or line along the BB axis (arrow).

preferences for the remaining cycles are shown in the cubic scatter plot in Figure 8.7c. Two highly populated planes were observed corresponding to *cis*-4-aminocyclohexanecarboxylic acid and its racemic form at cycle 2, giving high confidence in this motif for follow-up synthesis. Within both planes, several lines were observed which correspond to selected cycle 3 building blocks, further helping to narrow down the optimal structures for follow-up synthesis. These lines also indicate that the choice of building block at cycle 1 likely has little influence on the binding observed.

8.3.3 ELT Hit Confirmation, SAR and Hit-To-Lead Optimization

Follow-up on the selected binders began with the synthesis of a small, representative set of compounds (Table 8.1). The cis isomer of the selected cycle 2 building block was used for each of the initial follow-up compounds and cycles 1 and 3 building blocks were varied. Given that the carboxylic acid functional group in cycle 1 amino acids was used as the attachment point for DNA, the decarboxylic analogues of these building blocks were used in the off-DNA synthesis of compounds **1–3** and **6**. The biochemical assay of these four compounds had IC_{50} values ranging from 25 to 39 nM. Removing the cycle 1 synthon altogether (compound **4**), resulted in a dramatic loss in activity; however, the use of simple methylamine resulted in no loss of activity (compound **5**). This result confirmed the prediction based on the selection results that cycle 1 may not significantly influence activity. Cycle 1 was also used as the attachment point for DNA, which may indicate that this is the position exposed to solvent.

Although there were not many obvious features pulled from the selection results of DEL-A, there were some highly enriched cycle 4 amines, including (2-((trifluoromethyl)thiol) phenyl) methanamine and (2-(trifluoromethoxyl)phenyl) methanamine. These amines, as well as a few other benzylamine derivatives, and one aniline derivative were used to explore the SAR of cycle 4

Table 8.1 Off-DNA hit confirmation.

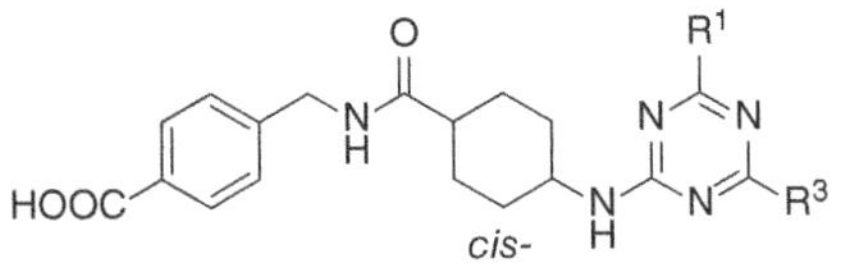

Cmpd	R^1	R^3	IC_{50} (nM)	Cmpd	R^1	R^3	IC_{50} (nM)
1	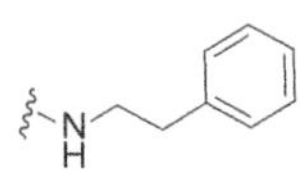	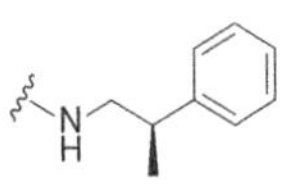	25	4	H	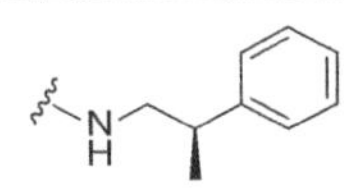	>5000
2	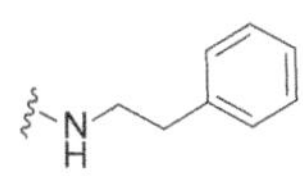	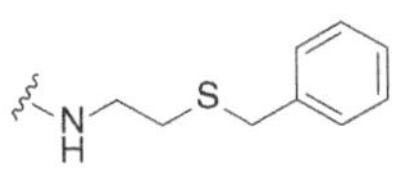	27	5	NHMe	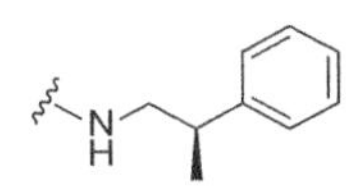	24
3	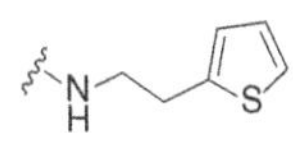	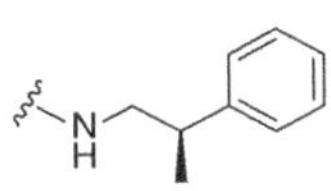	39	6	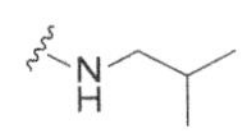	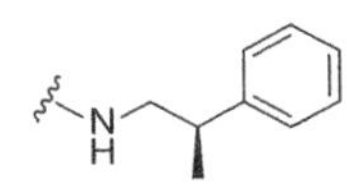	38

Table 8.2 SAR at cycle 4.

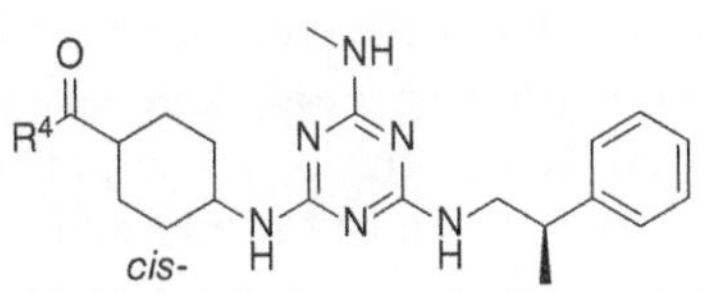

Cmpd	R^4	IC_{50} (nM)	Cmpd	R^4	IC_{50} (nM)
7	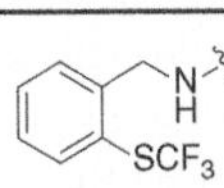	38	10	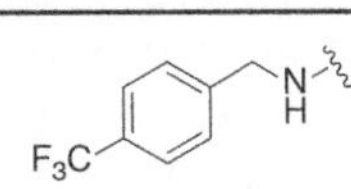	63
8	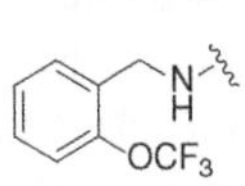	29	11	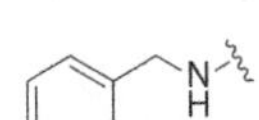	273
9	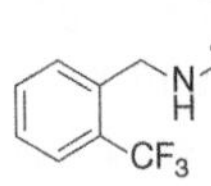	39	12	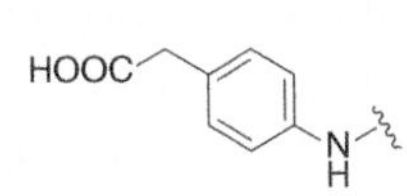	19 000

(Table 8.2). Compounds **7** and **8**, using the enriched cycle 4 amines, both showed similar activity to those discussed in Table 8.1. The use of an *ortho*-CF_3 group (compound **9**) also showed similar activity. However, incorporating the CF_3 group at the para position (compound **10**) led to a twofold decrease in activity. Removing all additional substitution on the benzene ring (compound **11**) led to an even more dramatic loss in activity. Despite the inclusion of many aromatic amines at cycle 4 in the synthesis of both libraries, none were observed in the selection results. To probe this, the final compound (**12**) incorporated an aniline derivative. Consistent with the selection results, a nearly 700-fold drop in activity was observed.

Table 8.3 SAR at cycle 3.

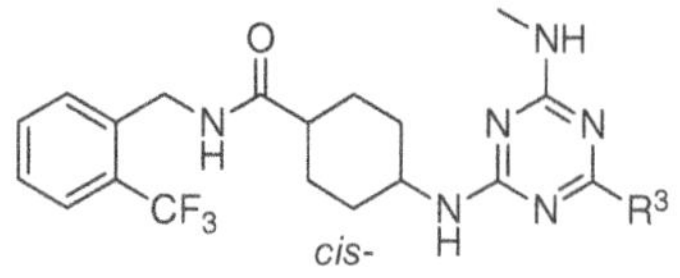

Cmpd	R^3	IC_{50} (nM)	Cmpd	R^3	IC_{50} (nM)
13	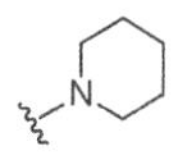	31	**16**	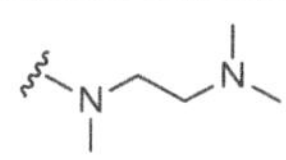	16
14	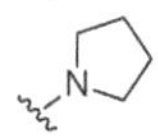	19	**17**	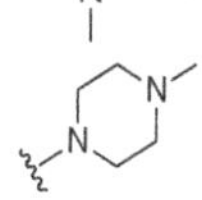	9
15	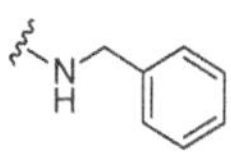	36			

The influence of the cycle 3 building block was then explored, emphasizing the use of truncated synthons (Table 8.3) given the cycle 3 diversity observed in the selection data. Initially piperidine, pyrrolidine, and benzylamine were explored as cycle 3 building blocks giving compounds **13–15**, all of which maintained similar activity. As with the cycle 1 building blocks, the data indicated that the cycle 3 synthon has little effect on the activity of the compound, and thus may be a good handle to influence the physical and PK properties. This was explored by installing trimethylethane-1,2-diamine (compound **16**) and 1-methylpiperazine (compound **17**) to increase solubility, both of which showed increased potency. Compound **17** displayed the highest potency with an IC_{50} of 9 nM and the piperazine synthon was therefore carried through for the remaining SAR studies.

The final set of compounds synthesized went back to explore the effect of the cycle 2 building block. The enrichment of both racemic and *cis*-4-aminocyclohexanecarboxylic acid led to use of the cis isomer in the initial SAR studies. However, the trans isomer of compound 5 was also prepared for comparison and no significant change in potency was observed, indicating that the size of the binding pocket may allow for further modification at cycle 2. This also presented the opportunity to explore synthons which lacked one or both of the stereocenters on the cycle 2 building block (Table 8.4). The aniline derivative (compound **18**) was first prepared, which showed a >4-fold drop in activity relative to compound **17**, while the piperidine analogue (compound **19**) showed similar activity to that of compound **17**. However, the use of azetidine (compound **20**) in place of the cyclohexyl group led to a dramatic drop in activity. Additional motifs at cycle 2 were explored including benzylamine, 1,3-cyclohexylamine, furanylmethanamine, tetrahydroquinoline, and phenol (compounds **21–25**), all of which showed improved activity in the single digit nanomolar range.

CYP450 inhibition was also evaluated for compounds **16–19**. In general, the compounds showed a good CYP450 inhibition profile (Table 8.5). Activity was in the micromolar range across all targets tested, showing potency >1000× lower for compounds **16**, **17**, and **19** and >250× lower for compound **18** against CYP450s compared with sEH.

Rat PK studies were also performed for lead compounds **17–19** and **25** (Table 8.6) with generally moderate to high plasma clearance levels, large distributional volume, an i.v. half-life of around two hours, and good oral bioavailability.

Table 8.4 SAR at cycle 2.

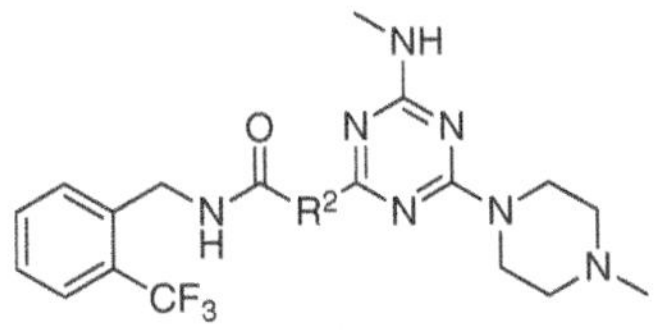

Cmpd	R^2	IC_{50} (nM)	Cmpd	R^2	IC_{50} (nM)
18	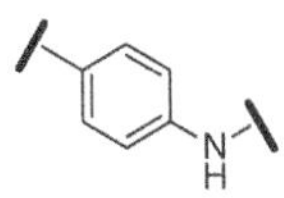	41	**22**	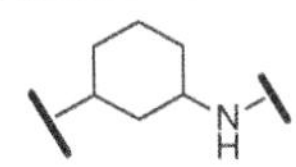	3
19	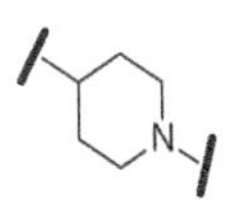	11	**23**	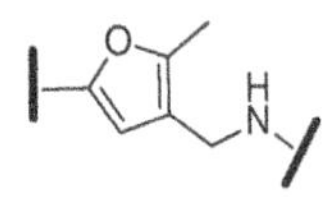	3
20	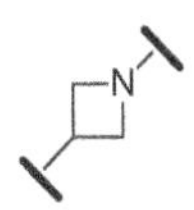	5000	**24**	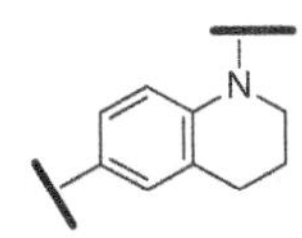	6
21	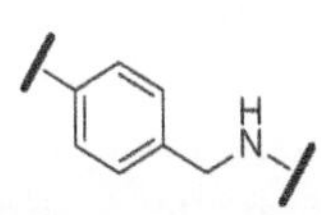	0.7	**25**	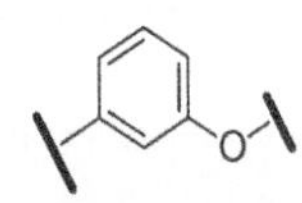	1

Table 8.5 P450 inhibition summary.

	IC_{50} (μM)					
Cmpd	**1A2**	**2C9**	**2C19**	**2D6**	**3A4 (DEF)**	**3A4 (7BQ)**
16	>33	32	>33	33	22	>33
17	>33	32	>33	>33	22	>33
18	>33	12	21	23	19	>33
19	>33	>33	>33	>33	>33	>33

Table 8.6 Rat PK data summary for selected compounds.

Cmpd	CL (ml/min/kg)	V_{dss} (l/kg)	i.v. $t_{1/2}$ (h)	*F* (%)
17	110	8.3	2.4	26
18	60	6.8	1.7	82
19	46	3.6	1.8	81
25	160	5.0	0.7	83

CL, clearance; *F*, bioavailability; $t_{1/2}$, half-life; Vdss, steady state volume of distribution.

8.3.4 Lead Optimization, Preclinical and Clinical Development: GSK2256294 as a Clinical Asset

The use of DELs and extensive SAR studies during the hit-to-lead optimization led to a series of potent, triazine-based sEH inhibitors with cycle 2 diversity and flexibility embedded in the cycle 1 and cycle 3 positions which were used to improve oral bioavailability and physiochemical properties. Attempts were also made to replace the triazine core with pyrimidine, pyridine, or phenyl [72] given the limited knowledge and application of triazine scaffolds in developed drugs [73, 74]. However, these alternative scaffolds proved to be either less potent or displayed poor CYP450 inhibition profiles relative to the triazine scaffold. Among the triazine-based compounds explored, compound **22** containing a 1,3-aminocyclohexanecarboxylate core gave rise to the lead compound **26** (GSK2256294) which was progressed to preclinical and clinical studies (Figure 8.8).

The lead compound GSK2256294 (**26**) was found to be a potent inhibitor of recombinant human sEH (IC_{50} = 27 pmol) with high selectivity. Additionally, **26** displays sEH inhibition activity in human whole blood with an IC_{50} = 6.83 nM [70]. The lead compound also shows high selectivity as no significant inhibition was observed for microsomal epoxide hydrolase (mEH, EPXHI), the phosphatase activity of EPHX2, and other enzymes, receptors, and ion channels. Further, **26** exhibits a very good aqueous solubility profile and good oral bioavailability in rat and dog (94 and ~100%, respectively), with moderate lipophilicity (clogP = 3.7) and plasma protein binding (12.8% free fraction in human blood) [70].

Studies also showed that GSK2256294 may provide an appropriate treatment for diseases such as COPD and cardiovascular diseases where cigarette smoke inhalation is a risk factor. In short-term murine studies, it was shown that administration of GSK2256294 during a 10-day period in which the mice were exposed to cigarette smoke inhibited both initiation and maintenance of pulmonary inflammation. Additionally, administration following 10 days of exposure resulted in a reduction of pulmonary inflammation [70]. Given the efficacy shown in these mice studies coupled with the observed selective inhibition of sEH and the preclinical safety profile, **26** was ultimately progressed to human clinical trials.

8.3.5 Clinical Trials with GSK2256294

In clinical study, the candidate GSK2256294 has been generally well tolerated with the most frequently reported adverse events being headache and contact dermatitis and is effective regardless of age, gender, and food intake [71, 75–77]. In the first-time-in-human (FTIH) study single escalating doses (2–20 mg) were administered to both healthy, non-smoking males and overweight,

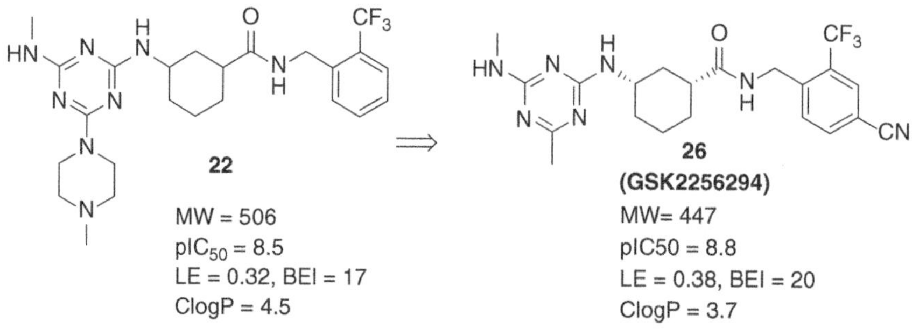

Figure 8.8 Structure of ELT-delivered asset **22** and GSK2256294, a candidate molecule for preclinical and clinical studies of sEH inhibition.

otherwise healthy, male smokers while repeat dosing was administered to two groups of overweight male smokers (receiving either 6 or 18 mg/dose). Overweight male smokers were chosen to represent a population in which the endothelial dysfunction of interest would be present. The study also indicated rapid absorption of GSK2256294 and a half-life consistent with daily dosing, with dose-dependent inhibition levels sustained for up to 24 hours after repeat dosing [75]. A separate, but related report also showed that repeat dosing over two weeks in overweight smokers did in fact impact vasodilation mediated by EET in human resistance arteries [76]. It is worth noting that some concerns exist around using sEH inhibitors to regulate CYP-derived EETs, as studies in perfused mouse lungs showed that these EETs are involved in hypoxic pulmonary vasoconstriction and it is speculated that sEH inhibitors may promote the development of pulmonary hypertension [78]. Clinical studies around GSK2256294 showed no changes in pulmonary artery pressures in healthy patients who received dosing of 15 mg once daily for seven days. Additionally it did not augment the normal hypoxic pulmonary vasoconstrictive response [79]. These reports provide exciting evidence for continued studies of sEH inhibitor GSK2256294 in the treatment of diseases, which are characterized by endothelial dysfunction or abnormal tissue repair.

8.4 Summary

In summary, DELs are a powerful tool which combine genetic encoding, combinatorial synthesis, and affinity-based screening and, as such, are proving to be an effective platform for drug discovery. This has been showcased herein using the case study in which the DEL technology at GSK led to the clinical candidate GSK2256294 for the treatment of COPD. To our knowledge, this represents the first example in which DEL technology has been instrumental in delivering a candidate which has entered human clinical testing. It also serves as a testament to how powerful this technology can be with the ability to efficiently deliver quality compounds for drug discovery efforts. Additional GSK DELs, both triazine and non-triazine derived, have also yielded binders with structurally unrelated chemotypes to those described here and testing of all these binders that were synthesized off DNA were found to be potent sEH inhibitors. Scientists at X-chem have also used a similar technology to elucidate a potent sEH inhibitor of a chemotype unrelated to that of GSK2256294, which may indicate that sEH is a highly compatible target for encoded library technologies [10].

Following the work with GSK2256294, a potent and selective RIP1 kinase [26] inhibitor was also identified using DEL technology and has since led to clinical candidates in multiple therapeutic areas [80, 81]. GSK2982772, a RIP1 clinical candidate for the treatment of inflammatory diseases, was identified with only a two atom modification of the original ELT hit. Further modification led to the second RIP1 clinical candidate, GSK3145095, for the treatment of pancreatic cancer [80]. It is also worth noting that, while certain limitations certainly preclude the exclusive use of DELs in drug discovery, the technology has become an important tool which is not only capable of standing on its own but also working alongside existing drug discovery operations, such as fragment-based drug discovery and HTS, to drive more efficient discovery efforts. Additionally, even when binders identified by DEL technology cannot be progressed as a candidate due to poor pharmacological properties, they can in some cases be repurposed to elucidate important information about a target. In 2016, GSK scientists reported an effort supported by DELs which led to the identification and characterization of GSK583, a potent and selective RIP2 kinase inhibitor. Although the compound was ultimately not progressed as a drug candidate, it proved to be a valuable tool compound for elucidating key functions of RIP2 kinase [82].

Though DEL technology has been proven to be highly useful for hit identification and target validation, it has certain limitations that the current format of encoded libraries cannot overcome. One of the biggest limitations is that it is not practical for cell-based screening, especially for phenotypic screening. To overcome such limitations and broaden the application of DNA-encoded technology in drug discovery, the next generation of encoded technology would enable functional screening along with cell-based screening. Paegel's group demonstrated the feasibility of bead-based DELs for biochemical screening [83, 84]. Though it is still too early to predict its industrialization in drug discovery, it opens the opportunity for biochemical screening, cell-based screening, and ultimately phenotypic screening.

As our knowledge and understanding of DEL technology continues to expand and advancements continue to emerge, we anticipate considerable improvements in the efficiency and quality of DEL-based screening efforts. Advancements of supporting technologies such as DNA sequencing and informatics have been and will continue to be an important feature of promoting the field. Additionally, development of new methods for synthesis of these libraries is critical for providing libraries that cover a wide breadth of chemical space. We look forward to seeing the ways in which the various DNA-encoded chemical library platforms progress and how they are applied to the drug discovery process. We are optimistic that as improvements continue to be made, the ability for DELs to transform drug discovery will be realized.

References

1 Brenner, S. and Lerner, R.A. (1992). Encoded combinatorial chemistry. *Proc. Natl. Acad. Sci. USA* 89 (12): 5381–5383.

2 Nielsen, J., Brenner, S., and Janda, K.D. (1993). Synthetic methods for the implementation of encoded combinatorial chemistry. *J. Am. Chem. Soc.* 115 (21): 9812–9813.

3 Needels, M.C., Jones, D.G., Tate, E.H. et al. (1993). Generation and screening of an oligonucleotide-encoded synthetic peptide library. *Proc. Natl. Acad. Sci. USA* 90 (22): 10700–10704.

4 Halpin, D.R., Harbury, P.B., and DNA Display I. (2004). Sequence-encoded routing of DNA populations. *PLoS Biol.* 2 (7): e173.

5 Gartner, Z.J. and Liu, D.R. (2001). The generality of DNA-templated synthesis as a basis for evolving non-natural small molecules. *J. Am. Chem. Soc.* 123 (28): 6961–6963.

6 Gartner, Z.J., Tse, B.N., Grubina, R. et al. (2004). DNA-Templated organic synthesis and selection of a library of macrocycles. *Science* 305 (5690): 1601–1605.

7 Snyder, T.M., Tse, B.N., and Liu, D.R. (2008). Effects of template sequence and secondary structure on DNA-Templated reactivity. *J. Am. Chem. Soc.* 130 (4): 1392–1401.

8 Hansen, M.H., Blakskjaer, P., Petersen, L.K. et al. (2009). A Yoctoliter-scale DNA reactor for small-molecule evolution. *J. Am. Chem. Soc.* 131 (3): 1322–1327.

9 Halpin, D.R., Lee, J.A., Wrenn, S.J., and Harbury, P.B. (2004). DNA display III. Solid-phase organic synthesis on unprotected DNA. *PLoS Biol. 2* (7): E175.

10 Litovchick, A., Dumelin, C.E., Habeshian, S. et al. (2015). Encoded library synthesis using chemical ligation and the discovery of sEH inhibitors from a 334-million member library. *Sci. Rep.* 5: 10916.

11 Melkko, S., Scheuermann, J., Dumelin, C.E., and Neri, D. (2004). Encoded self-assembling chemical libraries. *Nat. Biotechnol.* 22 (5): 568–574.

12 Clark, M.A., Acharya, R.A., Arico-Muendel, C.C. et al. (2009). Design, synthesis and selection of DNA-encoded small-molecule libraries. *Nat. Chem. Biol. 5* (9): 647–654.

13 Gerry, C.J., Wawer, M.J., Clemons, P.A., and Schreiber, S.L. (2019). DNA barcoding a complete matrix of stereoisomeric small molecules. *J. Am. Chem. Soc.* 141 (26): 10225–10235.

14 Goodnow, R.A., Dumelin, C.E., and Keefe, A.D. (2017). DNA-encoded chemistry: enabling the deeper sampling of chemical space. *Nat. Rev. Drug Discov.* 16 (2): 131–147.

15 Satz, A.L., Cai, J., Chen, Y. et al. (2015). DNA compatible multistep synthesis and applications to DNA encoded libraries. *Bioconjug. Chem.* 26 (8): 1623–1632.

16 Eidam, O. and Satz, A.L. (2016). Analysis of the productivity of DNA encoded libraries. *Med. Chem. Commun.* 7 (7): 1323–1331.

17 Flood, D.T., Kingston, C., Vantourout, J.C. et al. (2020). DNA encoded libraries: a visitor's guide. *Israel J. Chem.* 60 (3–4): 268–280.

18 Malone, M.L. and Paegel, B.M. (2016). What is a "DNA-compatible" reaction? *ACS Comb. Sci.* 18 (4): 182–187.

19 Kunig, V., Potowski, M., Gohla, A., and Brunschweiger, A. (2018). DNA-encoded libraries – an efficient small molecule discovery technology for the biomedical sciences. *Biol. Chem.* 399 (7): 691–710.

20 Patel, S., Badir, S.O., and Molander, G.A. (2021). Developments in photoredox-mediated alkylation for DNA-encoded libraries. *Trends Chem.* 3 (3): 161–175.

21 Flood, D.T., Asai, S., Zhang, X. et al. (2019). Expanding reactivity in DNA-encoded library synthesis via reversible binding of DNA to an inert quaternary ammonium support. *J. Am. Chem. Soc.* 141 (25): 9998–10006.

22 Flood, D.T., Zhang, X., Fu, X. et al. (2020). RASS-enabled S/P-C and S-N bond formation for DEL synthesis. *Angew. Chem. Int. Ed.* 59 (19): 7377–7383.

23 Ruff, Y., Martinez, R., Pellé, X. et al. (2020). An Amphiphilic polymer-supported strategy enables chemical transformations under anhydrous conditions for DNA-encoded library synthesis. *ACS Comb. Sci.* 22 (3): 120–128.

24 Doyon, J.B., Snyder, T.M., and Liu, D.R. (2003). Highly sensitive in vitro selections for DNA-linked synthetic small molecules with protein binding affinity and specificity. *J. Am. Chem. Soc.* 125 (41): 12372–12373.

25 Kleiner, R.E., Dumelin, C.E., Tiu, G.C. et al. (2010). in vitro selection of a DNA-templated small-molecule library reveals a class of macrocyclic kinase inhibitors. *J. Am. Chem. Soc.* 132 (33): 11779–11791.

26 Harris, P.A., King, B.W., Bandyopadhyay, D. et al. (2016). DNA-encoded library screening identifies benzo[b][1,4]oxazepin-4-ones as highly potent and monoselective receptor interacting protein 1 kinase inhibitors. *J. Med. Chem. 59* (5): 2163–2178.

27 Petersen, L.K., Blakskjaer, P., Chaikuad, A. et al. (2016). Novel p38α MAP kinase inhibitors identified from yoctoReactor DNA-encoded small molecule library. *Med. Chem. Commun.* 7 (7): 1332–1339.

28 Yang, H., Medeiros, P.F., Raha, K. et al. (2015). Discovery of a potent class of PI3Kα inhibitors with unique binding mode via encoded library technology (ELT). *ACS Med. Chem. Lett.* 6 (5): 531–536.

29 Gentile, G., Merlo, G., Pozzan, A. et al. (2012). 5-Aryl-4-carboxamide-1,3-oxazoles: potent and selective GSK-3 inhibitors. *Bioorg. Med. Chem. Lett.* 22 (5): 1989–1994.

30 Cuozzo, J.W., Centrella, P.A., Gikunju, D. et al. (2017). Discovery of a potent BTK inhibitor with a novel binding mode by using parallel selections with a DNA-encoded chemical library. *Chembiochem* 18 (9): 864–871.

31 Scheuermann, J., Dumelin, C.E., Melkko, S. et al. (2008). DNA-encoded chemical libraries for the discovery of MMP-3 inhibitors. *Bioconjug. Chem.* 19 (3): 778–785.

32 Ding, Y., O'Keefe, H., DeLorey, J.L. et al. (2015). Discovery of potent and selective inhibitors for ADAMTS-4 through DNA-encoded library technology (ELT). *ACS Med. Chem. Lett.* 6 (8): 888–893.

33 Deng, H., O'Keefe, H., Davie, C.P. et al. (2012). Discovery of highly potent and selective small molecule ADAMTS-5 inhibitors that inhibit human cartilage degradation via encoded library technology (ELT). *J. Med. Chem. 55* (16): 7061–7079.

34 Samain, F., Ekblad, T., Mikutis, G. et al. (2015). Tankyrase 1 inhibitors with drug-like properties identified by screening a DNA-encoded chemical library. *J. Med. Chem.* 58 (12): 5143–5149.

35 Lewis, H.D., Liddle, J., Coote, J.E. et al. (2015). Inhibition of PAD4 activity is sufficient to disrupt mouse and human NET formation. *Nat. Chem. Biol. 11* (3): 189–191.

36 Gilmartin, A.G., Faitg, T.H., Richter, M. et al. (2014). Allosteric Wip1 phosphatase inhibition through flap-subdomain interaction. *Nat. Chem. Biol.* 10 (3): 181–187.

37 Maianti, J.P., McFedries, A., Kleiner, R.E. et al. (2014). Anti-diabetic activity of insulin-degrading enzyme inhibitors mediated by multiple hormones. *Nature* 511 (7507): 94–98.

38 Belyanskaya, S.L., Ding, Y., Callahan, J.F. et al. (2017). Discovering drugs with DNA-encoded library technology: from concept to clinic with an inhibitor of soluble epoxide hydrolase. *ChemBioChem* 18 (9): 837–842.

39 Taylor, D.M., Hu, L., Prasad, B.V.V. et al. (2020). Identifying oxacillinase-48 carbapenemase inhibitors using DNA-encoded chemical libraries. *ACS Infect. Dis. 6* (5): 1214–1227.

40 Wellaway, C.R., Amans, D., Bamborough, P. et al. (2020). Discovery of a bromodomain and extraterminal inhibitor with a low predicted human dose through synergistic use of encoded library technology and fragment screening. *J. Med. Chem. 63* (2): 714–746.

41 Kazmierski, W.M., Xia, B., Miller, J. et al. (2020). DNA encoded library technology-based discovery, lead optimization and prodrug strategy toward structurally-unique indoleamine 2,3-dioxygenase (IDO1) inhibitors. *J. Med. Chem.* 63 (7): 3552–3562.

42 Concha, N., Huang, J., Bai, X. et al. (2016). Discovery and characterization of a class of pyrazole inhibitors of bacterial undecaprenyl pyrophosphate synthase. *J. Med. Chem.* 59 (15): 7299–7304.

43 Wood, E.R., Bledsoe, R., Chai, J. et al. (2015). The role of phosphodiesterase 12 (PDE12) as a negative regulator of the innate immune response and the discovery of antiviral inhibitors. *J. Biol. Chem.* 290 (32): 19681–19696.

44 Deng, H., Zhou, J., Sundersingh, F.S. et al. (2015). Discovery, SAR, and X-ray binding mode study of BCATm inhibitors from a novel DNA-encoded library. *ACS Med. Chem. Lett. 6* (8): 919–924.

45 Cuozzo, J.W., Clark, M.A., Keefe, A.D. et al. (2020). Novel autotaxin inhibitor for the treatment of idiopathic pulmonary fibrosis: a clinical candidate discovered using DNA-encoded chemistry. *J. Med. Chem.* 63 (14): 7840–7856.

46 Johannes, J.W., Bates, S., Beigie, C. et al. (2017). Structure based design of non-natural peptidic macrocyclic Mcl-1 inhibitors. *ACS Med. Chem. Lett. 8* (2): 239–244.

47 Arico-Muendel, C., Zhu, Z., Dickson, H. et al. (2015). Encoded library technology screening of hepatitis C virus NS4B yields a small-molecule compound series with in vitro replicon activity. *Antimicrob. Agents Chemother.* 59 (6): 3450–3459.

48 Melkko, S., Mannocci, L., Dumelin, C.E. et al. (2010). Isolation of a small-molecule inhibitor of the antiapoptotic protein Bcl-xL from a DNA-encoded chemical library. *Chem. Med. Chem.* 5 (4): 584–590.

49 Leimbacher, M., Zhang, Y., Mannocci, L. et al. (2012). Discovery of small-molecule interleukin-2 inhibitors from a DNA-encoded chemical library. *Chem. A Eur. J.* 18 (25): 7729–7737.

50 Kollmann, C.S., Bai, X., Tsai, C.-H. et al. (2014). Application of encoded library technology (ELT) to a protein-protein interaction target: discovery of a potent class of integrin lymphocyte function-associated antigen 1 (LFA-1) antagonists. *Bioorg. Med. Chem.* 22 (7): 2353–2365.

51 Wu, Z., Graybill, T.L., Zeng, X. et al. (2015). Cell-based selection expands the utility of DNA-encoded small-molecule library technology to cell surface drug targets: identification of novel antagonists of the NK3 tachykinin receptor. *ACS Comb. Sci. 17* (12): 722–731.

52 Brown, D.G., Brown, G.A., Centrella, P. et al. (2018). Agonists and antagonists of protease-activated receptor 2 discovered within a DNA-encoded chemical library using mutational stabilization of the target. *SLAS Discov. 23* (5): 429–436.

53 Ahn, S., Pani, B., Kahsai, A.W. et al. (2018). Small-molecule positive allosteric modulators of the β2-adrenoceptor isolated from DNA-encoded libraries. *Mol. Pharmacol.* 94 (2): 850–861.

54 Zhu, Z., Grady, L.C., Ding, Y. et al. (2019). Development of a selection method for discovering irreversible (covalent) binders from a DNA-encoded library. *SLAS Discov.* 24 (2): 169–174.

55 Chan, A.I., McGregor, L.M., Jain, T., and Liu, D.R. (2017). Discovery of a covalent kinase inhibitor from a DNA-encoded small-molecule library × protein library selection. *J. Am. Chem. Soc.* 139 (30): 10192–10195.

56 Zimmermann, G., Rieder, U., Bajic, D. et al. (2017). A specific and covalent JNK-1 ligand selected from an encoded self-assembling chemical library. *Chem. A Eur. J.* 23 (34): 8152–8155.

57 Chamberlain, P.P. and Hamann, L.G. (2019). Development of targeted protein degradation therapeutics. *Nat. Chem. Biol.* 15 (10): 937–944.

58 Harris, T.R. and Hammock, B.D. (2013). Soluble epoxide hydrolase: gene structure, expression and deletion. *Gene* 526 (2): 61–74.

59 Morisseau, C. and Hammock, B.D. (2005). Epoxide hydrolases: mechanisms, inhibitor designs, and biological roles. *Annu. Rev. Pharmacol. Toxicol. 45*: 311–333. 3 plates.

60 Spector, A.A. and Norris, A.W. (2007). Action of epoxyeicosatrienoic acids on cellular function. *Am. J. Physiol. Cell Physiol.* 292 (3): C996–C1012.

61 Campbell, W.B., Gebremedhin, D., Pratt, P.F., and Harder, D.R. (1996). Identification of epoxyeicosatrienoic acids as endothelium-derived hyperpolarizing factors. *Circ. Res.* 78 (3): 415–423.

62 Morisseau, C. and Hammock, B.D. (2013). Impact of soluble epoxide hydrolase and epoxyeicosanoids on human health. *Annu. Rev. Pharmacol. Toxicol.* 53: 37–58.

63 Peinado, V.I., Ramirez, J., Roca, J. et al. (2006). Identification of vascular progenitor cells in pulmonary arteries of patients with chronic obstructive pulmonary disease. *Am. J. Respir. Cell Mol. Biol.* 34 (3): 257–263.

64 Kasahara, Y., Tuder, R.M., Cool, C.D. et al. (2001). Endothelial cell death and decreased expression of vascular endothelial growth factor and vascular endothelial growth factor receptor 2 in emphysema. *Am. J. Respir. Crit. Care Med.* 163 (3 Pt 1): 737–744.

65 Kim, I.-H., Morisseau, C., Watanabe, T., and Hammock, B.D. (2004). Design, synthesis, and biological activity of 1,3-disubstituted ureas as potent inhibitors of the soluble epoxide hydrolase of increased water solubility. *J. Med. Chem.* 47 (8): 2110–2122.

66 Barnych, B., Singh, N., Negrel, S. et al. (2020). Development of potent inhibitors of the human microsomal epoxide hydrolase. *Eur. J. Med. Chem.* 193: 112206.

67 Eldrup, A.B., Soleymanzadeh, F., Taylor, S.J. et al. (2009). Structure-based optimization of arylamides as inhibitors of soluble epoxide hydrolase. *J. Med. Chem.* 52 (19): 5880–5895.

68 Cardozo, M.G. and Ingraham, R.H. (2006). Acyl hydrazones as sEH inhibitors for treatment of metabolic and cardiovascular diseases. WO2006121684A2.

69 Ota, T., Takahashi, H., Kakinuma, H., and Busujima, T. (2007). Preparation of benzimidazole-5-carboxamide derivatives as soluble epoxide hydrolase inhibitors. WO2007043653A1.

70 Podolin, P.L., Bolognese, B.J., Foley, J.F. et al. (2013). in vitro and in vivo characterization of a novel soluble epoxide hydrolase inhibitor. *Prostaglandins Other Lipid Mediat.* 104–105: 25–31.

71 Lazaar, A.L., Yang, L., Boardley, R.L. et al. (2016). Pharmacokinetics, pharmacodynamics and adverse event profile of GSK2256294, a novel soluble epoxide hydrolase inhibitor. *Br. J. Clin. Pharmacol.* 81 (5): 971–979.

72 Thalji, R.K., McAtee, J.J., Belyanskaya, S. et al. (2013). Discovery of 1-(1,3,5-triazin-2-yl) piperidine-4-carboxamides as inhibitors of soluble epoxide hydrolase. *Bioorg. Med. Chem. Lett. 23* (12): 3584–3588.

73 Allain, H. and Bentue-Ferrer, D. (1998). Clinical efficacy of almitrine-raubasine: an overview. *Eur. Neurol.* 39 (Suppl. 1): 39–44.

74 Kim, E.S. (2017). Tivozanib: first global approval. *Drugs* 77 (17): 1917–1923.

75 Lazaar, A., Yang, L., Robertson, J. et al. (2015). Safety and pharmacology of a soluble epoxide hydrolase inhibitor. *Eur. Respir. J.* 46: PA2120.

76 Yang, L., Cheriyan, J., Gutterman, D.D. et al. (2017). Mechanisms of vascular dysfunction in COPD and Effects of a novel soluble epoxide hydrolase inhibitor in smokers. *Chest* 151 (3): 555–563.

77 Yang, L., Cheriyan, J., Lazaar, A. et al. (2016). The role of epoxyeicosatrienoic acids in regulating endothelial function and the effects of a novel soluble epoxide hydrolase inhibitor GSK2256294 in humans. *J. Am. Coll. Cardiol.* 67: 2308.

78 Keserue, B., Barbosa-Sicard, E., Popp, R. et al. (2008). Epoxyeicosatrienoic acids and the soluble epoxide hydrolase are determinants of pulmonary artery pressure and the acute hypoxic pulmonary vasoconstrictor response. *FASEB J.* 22 (12): 4306–4315. https://doi.org/10.1096/fj.08-112821.

79 Ding, Y. and Clark, M.A. (2015). Robust Suzuki-Miyaura cross-coupling on DNA-linked substrates. *ACS Comb. Sci.* 17 (1): 1–4.

80 Harris, P.A., Marinis, J.M., Lich, J.D. et al. (2019). Identification of a RIP1 kinase inhibitor clinical candidate (GSK3145095) for the treatment of pancreatic cancer. *ACS Med. Chem. Lett. 10* (6): 857–862.

81 Harris, P.A., Berger, S.B., Jeong, J.U. et al. (2017). Discovery of a first-in-class receptor interacting protein 1 (RIP1) kinase specific clinical candidate (GSK2982772) for the treatment of inflammatory diseases. *J. Med. Chem. 60* (4): 1247–1261.

82 Haile, P.A., Votta, B.J., Marquis, R.W. et al. (2016). The identification and pharmacological characterization of 6-(tert-Butylsulfonyl)-N-(5-fluoro-1H-indazol-3-yl)quinolin-4-amine (GSK583), a highly potent and selective inhibitor of RIP2 kinase. *J. Med. Chem. 59* (10): 4867–4880.

83 MacConnell, A.B., Price, A.K., and Paegel, B.M. (2017). An integrated microfluidic processor for DNA-encoded combinatorial library functional screening. *ACS Comb. Sci.* 19 (3): 181–192.

84 Cochrane, W.G., Malone, M.L., Dang, V.Q. et al. (2019). Activity-based DNA-encoded library screening. *ACS Comb. Sci.* 21 (5): 425–435.

9

Discovery of HTL26119: Family B GPCR Structure-Based Drug Design Is Now a Reality

Andrea Bortolato[1] and Jonathan S. Mason[2]

[1] *Research and Development, Computer-Assisted Drug Design, Molecular Discovery Technologies, Bristol Myers Squibb, PO Box 5400, NJ 08453, Princeton, NJ, USA*
[2] *Sosei Heptares, Steinmetz Building, Granta Park, CB21 6DG, Great Abington, Cambridge, UK*

9.1 Introduction

HTL26119 is a novel allosteric antagonist of the glucagon-like peptide-1 receptor (GLP1R), a member of the Family B G protein-coupled receptors (GPCRs), identified through a structure-based drug design (SBDD) program at Heptares Therapeutics [1]. Under normal physiological conditions the binding of GLP1 to its receptor on pancreatic β-cells results in enhanced glucose-dependent insulin secretion and a subsequent decrease in plasma blood glucose levels. GLP1R agonists have been successfully developed for the management of type 2 diabetes mellitus and obesity [2]. Antagonists have instead the potential to treat an overactive GLP1R system that can result under rare pathophysiological conditions into significantly increased insulin secretion and concomitant hypoglycemia [3].

There are only limited examples of GLP1R small molecule antagonists known, all in preclinical development. The discovery of HTL26119 was possible because of the recent advances in GPCR structural biology. A SBDD approach was used for the initial hit identification and the following lead optimization. The original hit was identified in silico using a virtual screening (VS) protocol. The available structural data enabled an understanding at the molecular level of the negative allosteric modulator (NAM) mode of action of the ligand. In particular, the predicted ligand binding mode using molecular docking combined with homology modeling (HM), binding site analysis and molecular dynamics (MD) simulation allowed the rational design of HTL26119 selectivity against the glucagon receptor.

This chapter is divided into two parts. The first part provides crucial GPCR SBDD information that creates a solid foundation for the second part, which describes in detail how the increased understanding of this family of receptors enabled the rational structure-based design of HTL26119.

9.2 G Protein-Coupled Receptor Structure-Based Drug Discovery

Integral membrane proteins account for approximately 30% of the human genome, but represent almost 50% of the targets of pharmaceutical agents [4, 5]. Among them, GPCRs are one of the largest families of related proteins in the human genome. They are generally considered excellent

Contemporary Accounts in Drug Discovery and Development, First Edition. Edited by Xianhai Huang, Robert G. Aslanian, and Wayne H. Tang.

Table 9.1 Number of unique GPCR complexes.

Class (Family)	Receptors[a]	Receptor–ligand complexes	Receptor–G protein complexes	Active-state receptors[b]
Total	64	220	6	25
Class A (Rhodopsin)	54	192	2	20
Class B (Secretin)	6	13	4	4
Class B2 (Adhesion)	0	0	0	0
Class C (Glutamate)	2	7	0	0
Class F (Frizzled)	2	8	0	1
Taste	0	0	0	0

[a] Receptor-orthologues are only counted once. Updated table available at www.gpcrdb.org.
[b] Active state is defined as agonist-bound and opened intracellular TM bundle.

targets for drug discovery [6]. This view stems partly from the compelling role GPCRs play in regulating the pathophysiology of diverse diseases. To date, there are almost 500 approved drugs that act on more than 100 unique GPCRs [6]. They represent more than one third of all drugs approved by the US Food and Drug Administration (FDA). As many GPCRs have evolved to bind low molecular weight, fragment-like ligands, such as dopamine and acetylcholine, drugs directed at these receptors are invariably analogues of the physiological ligands. Indeed, around 70% of the drugs that target GPCRs are derived from the natural ligand [6]. In the case of the aminergic receptor subfamily, the availability of the low molecular weight native ligand as a starting point provided a strong basis to design low molecular weight drugs with good pharmaceutical properties. Around 15% of GPCRs for which there are drugs on the market were targeted unintentionally; the drug was discovered in phenotypic assays and the mechanism of action only later revealed to involve a GPCR [7].

Despite their popularity as drug targets, membrane proteins have consistently been poorly represented in collections of X-ray structures of proteins, primarily due to difficulties in purifying sufficient quantities of high-quality protein stable enough to form regular crystals. The first breakthrough in GPCR structural biology came in 2000 with the solution of the structure of bovine rhodopsin [8]. This confirmed the presence of a bundle of seven transmembrane helices and it gave a boost for HM of many receptors of therapeutic interest. Despite this exceptional result, it took another seven years for the real GPCR structural biology revolution to start. The first structures reported in the inactive state and bound to small molecules included the adrenergic β_2, β_1, and adenosine A_{2A} receptors [9–11]. Since then, the number of receptors for which structures have been reported has increased dramatically (Table 9.1, Figure 9.1) [12]. They include different activation states of Family A, B, C, and F, in complex with small molecules, fatty acids, peptides, and globular proteins [6, 13].

The initial SBDD revolution for the Family A GPCRs was in 2007. The availability of X-ray structures for diverse ligand-bound GPCRs in multiple conformations enabled rational drug design efforts for this protein family that have historically been applied to soluble enzyme targets. Computational analyses of GPCR binding sites revealed the crucial features required to achieve high affinity binding [7]. Protein druggability proved to be directly linked to the combination of key physicochemical properties, such as lipophilicity and hydrogen bonding propensities [14]. In

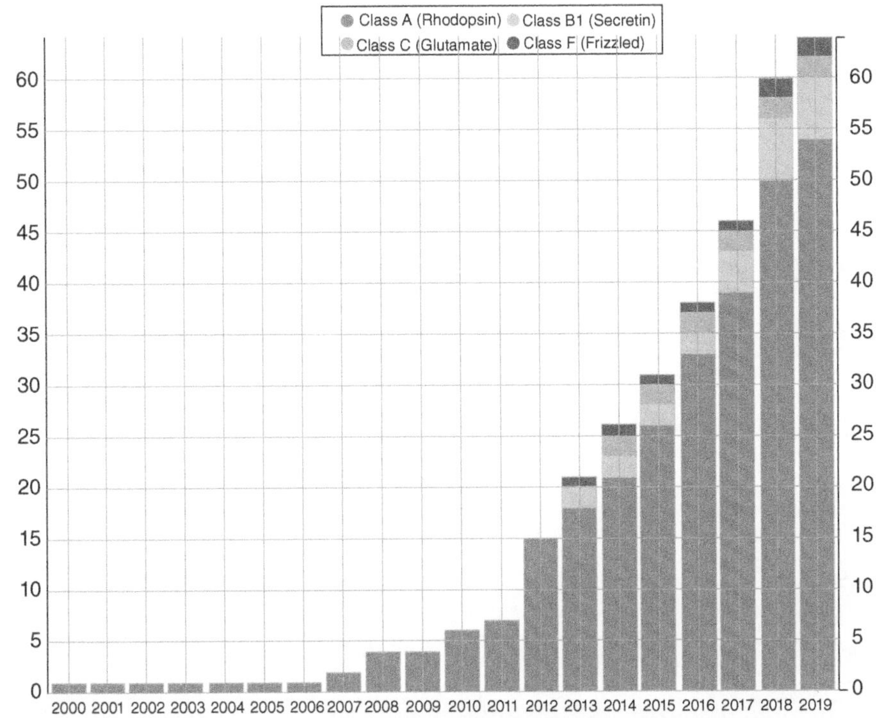

Figure 9.1 Number of total unique receptor complexes per year (www.gpcrdb.org).

parallel, the size and the shape of the binding site combined with the calculated free energies of water in a fully hydrated binding pocket highlighted the most important interactions required for ligand binding (Figure 9.2). Computational methods to evaluate the role of water in ligand binding enable the prediction of the position and relative free energy of the water molecules in the protein active site and the analysis of the perturbation of an explicit water network as a consequence of ligand binding [18, 20]. Water molecules in the GPCR orthosteric binding pocket are very important in determining not only ligand binding affinity, but also kinetics. One of the first studies showing correlation between high energy trapped solvent molecules and ligand residence time focused on related triazine adenosine A_{2A} antagonists [21], including AZD4635, now in phase II clinical trials. For this receptor, conformational changes can also play an important role in controlling ligand binding kinetics. The understanding of the detailed molecular mechanism was possible through the combination of several high-resolution crystal structures combined with metadynamics simulation [22]. X-ray data highlighted how related ligands were able to interact in different ways with a salt bridge on the extracellular side of the receptor. Mutagenesis experiments confirmed the role of this salt bridge in controlling the dissociation kinetics of the ligands from the receptor, while metadynamics simulations demonstrated the ability of ligands to modulate salt bridge stability. These results shed light on a structural determinant of ligand dissociation kinetics, defined as the rate at which a ligand dissociates from a protein and identified the means by which this property may be optimized.

While GPCR X-ray structures enabled SBDD, druggability analysis, ligand–protein binding affinity and kinetics predictions for Family A, the start of the same revolution for Family B GPCRs did not occur for another six years.

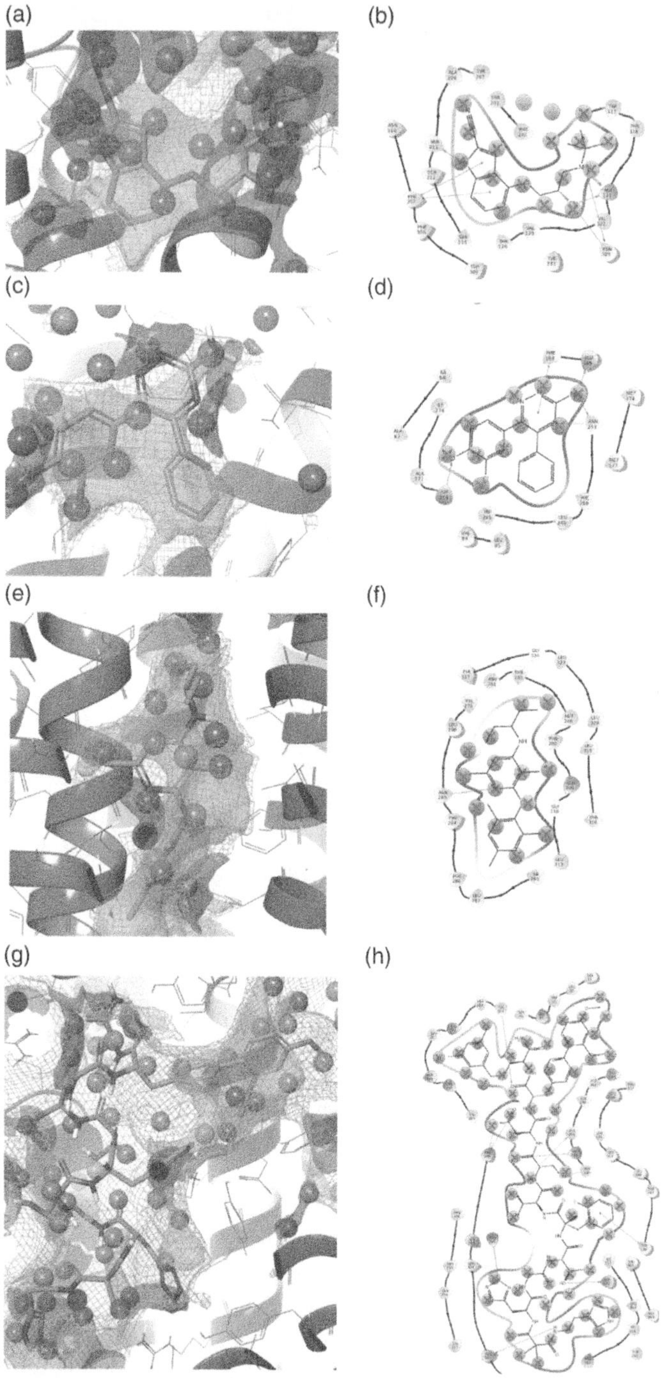

Figure 9.2 GPCR binding site computational analysis. Comparison of two Family A (a–d) and two Family B (e–h) GPCRs. The following receptors are included: (a,b) β_1 adrenergic receptor (PDB ID 2VT4 [10]); (c,d) adenosine A_{2A} receptor (PDB ID 5OM4 [15]); (e,f) corticotropin-releasing hormone receptor 1 (CRF1 receptor, PDB ID 4K5Y [16]); and (g,h) GLP1 receptor (PDB ID 5NX2 [17]). Left panels include the binding site region: the ligand is represented as a stick model. The protein secondary structure is shown with the residues at less than 5Å from the ligand. Predicted Apo hydration sites from WaterMap a physic-based computational chemistry method to map the locations and thermodynamic properties of water molecules that solvate protein binding sites [18]. The different solvation sites are colored based on their free energy from dark gray (high, unstable water molecules) to light gray (low, stable water molecules). The predicted binding pocket physical–chemical properties from SiteMap [19] are included as semitransparent surfaces. SiteMap is a fast computational approach to evaluate the interaction energies between chemical probes and the binding site amino acids. Based on the analysis results, hydrophobic hotspots are included with H-bond donor and acceptor hotspots in different shades of gray. The SiteMap shape of the pocket is included as a grey mesh. Right panels include a two-dimensional depiction of the ligand interactions with the protein and WaterMap waters displaced upon ligand binding. Displaced waters are shown as circles with crosses, H-bonds as arrows, and π–π stacking interactions as lines.

9.3 The Beginning of the Family B GPCR Structural Biology Revolution

Family B GPCRs, also referred to as the secretin family of GPCRs, include receptors for 15 peptide hormones, which can be grouped into five subfamilies based on their physiological role [23]. These receptors are important drug targets in many human diseases, including diabetes, osteoporosis, obesity, cancer, neurodegeneration, cardiovascular disease, headache, and psychiatric disorders. However, the identification of small-molecule oral drugs for this family has proved extremely challenging [24]. Extensive high-throughput screening by the pharmaceutical industry has yielded few small drug-like molecules, generally with bad pharmacokinetic properties, that mimic or block the large peptide ligands.

Until 2013 the lack of any crystal structures of Family B GPCRs limited the use of SBDD to only homology models. This computational method creates an atomic-resolution model of the protein of interest from its amino acid sequence and an experimental three-dimensional structure of a related homologous protein (template) [25]. It was known that structurally, Class B GPCRs consist of a large N-terminal extracellular domain (ECD) and a transmembrane domain (TMD) making up the machinery for signal transduction across the cell membrane [26]. However, with only Family A GPCRs available as a template for HM, the sequence alignment for the TMD was extremely challenging due to the low sequence identity between Class A and B GPCRs. This resulted in errors in the models reducing dramatically their impact in SBDD. For every family the conserved TMD scaffold allows for the alignment of sequences or structures to identify the corresponding residues, which are indexed with a generic residue number [27]. This is why the CRF1 X-ray was extremely important: it afforded a generic template for Family B GPCRs for comparison of mutant effects, ligand interactions, and structural features across receptor subtypes, species orthologs and receptor subfamilies. These conserved positions are common placeholders, while receptor sequences vary and so compose the unique mosaic responsible for ligand affinity and selectivity.

Using a mutagenesis approach called "conformational thermostabilization" [28], a crystallizable version of the TMD of the human CRF1 receptor was generated. This technique allows dramatic increase in protein conformational stabilization including specific point mutations resulting in thermal stability improvement. A CRF1R biased toward the inactive conformation was generated, and its structure was determined by X-ray crystallography in complex with CP-376395 (Figure 9.2e, f), a small-molecule NAM [16]. Mutational screening in the presence of the ligand resulted in the identification of a construct with 12 point mutations that exhibited significantly increased thermal stability in a range of detergents. The structure was studied using an explicit-solvent all-atom MD simulation of the CRF1R in a detergent micelle. This computer simulation method allowed the analysis of the physical movements of atoms and molecules. All atoms of the system are part of the calculation, including solvent water molecules, ions and the detergent forming the micelle. The results showed that the conformation of the helical bundle of the mutated receptor is more stable than the wild type, in agreement with the experimental data. The simulations suggested that TM5 and TM6 are in particular stabilized and the mutations help to prevent the insertion of a detergent molecule close to the ligand binding site [28].

Unlike Class A GPCRs, the CRF1R X-ray showed the helical bundle has a pronounced V-shape (Figure 9.3), presenting a large cavity where the orthosteric peptide binds the receptor in the extracellular side. There are striking differences from Family A GPCRs in the arrangement of TM7 and smaller, but substantial, differences in TM6. In both receptors TM7 assumes a similar, sharply kinked structure. In CRF1R, however, its extracellular end is located approximately 10 Å further

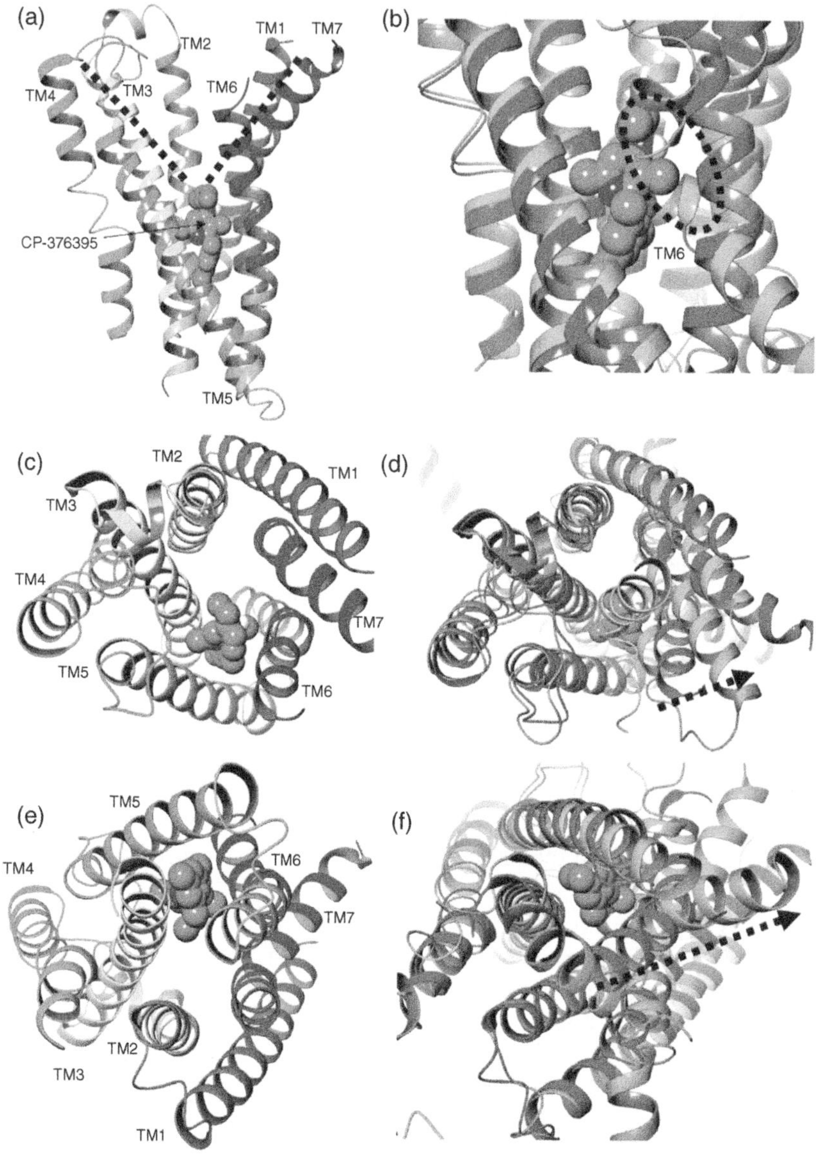

Figure 9.3 CRF1R structures. In the left panels, the CRF1R inactive conformation in complex with CP-376395 is shown (PDB ID 4K5Y [16]). The ligand is always included as a space-filling (van der Waals) representation with the carbon atoms colored green. In (a) the "V" shape of the orthosteric site between TM3 and TM7 is highlighted by a black dotted line. The panels on the right compare the inactive state of CRF1R to the active state structure (PDB ID 6P9X [29]) bound to a peptide agonist. (b) shows how the active conformation of TM6 clashes with the small molecule ligand in the inactive state, the overlapping region is highlighted by a dotted black oval line. (c,d) include an extracellular view of the receptors. The arrow in (d) shows the movement of TM6 upon receptor activation. (e,f) show an intracellular view of CRF1R. (e) shows how the small molecule interacts only with TM3, TM6, and TM5. The intracellular portion of TM6 undergoes a large conformational change during activation (shown by the dotted black arrow in (f)) as a consequence of G protein binding. In particular, an α-helix of the G protein displaces TM6.

away from the axis of the helical bundle. However, the intracellular side is very similar across GPCR families since they all signal through the same effector proteins, which interact with the cytoplasmic side of the receptors.

CP-376395 unexpectedly binds in a pocket located in the cytoplasmic half of the receptor, approximately 18 Å away from the center of the large orthosteric cavity and 13–23 Å away from the center of mass of the small-molecule ligands in the structures of Class A GPCRs. The ligand was located in a predominantly hydrophobic pocket defined by residues of TM3, TM5, and TM6. The side chain of the highly conserved Asn283 formed an essential hydrogen bond with the pyridine nitrogen. The mutation of that residue to alanine resulted in a complete loss of ligand binding [16]. This ligand binding site is an allosteric site distinct from the orthosteric peptide-binding pocket.

The large size and hydrophilic features of the Family B orthosteric site explain why big peptide-like molecules are required to achieve high affinity binding, such as the peptidomimetic ligand crystallized in complex with GLP1R (Figure 9.2g, h). In contrast, the allosteric CRF1R binding site (Figure 9.2e, f) is similar to the druggable orthosteric Family A GPCRs (Figure 9.2a–d). These pockets are small with a good balance of hydrophobic and hydrophilic regions allowing a high affinity ligand to have drug-like properties. In addition these druggable pockets are predicted to contain several unstable waters that when displaced upon ligand binding allow the creation of very productive apolar and polar interactions between the ligand and the target.

In conclusion, the CRF1R X-ray highlighted two key lessons for medicinal chemists trying to target Family B GPCRs: (i) it will be very challenging to reach a high level of activity with a small molecule if binding to the orthosteric site; and (ii) there are important alternative opportunities to affect the activity of this family of receptors in other regions of the helical bundle. These lessons were crucial for the discovery of HTL26119.

9.4 Lessons Learned from the Corticotropin-Releasing Factor Receptor Type 1 Crystal Structure

The X-ray crystal structures for the glucagon receptor and corticotropin-releasing factor receptor published in 2013 allowed for the first time a structural comparison of Family B GPCRs to Class A and F receptors [30]. It particular, it was possible to evaluate similarities in the GPCR activation mechanisms. The CRF1 X-ray suggested the ligand CP-376395 was able to keep the receptor in an inactive conformation by tethering the cytoplasmic half of TM6 to TM3 and TM5 (Figure 9.3). This proposition is supported by the fact that CP-376395 acts as an inverse agonist reducing basal signaling [16]. TM3 is at the center of the contact network and is a structural and functional hub for the GPCR family. In particular, in Class A receptors TM3 creates important hydrophobic interactions with TM6 involving particular residues with generic numbers [27] 3.43, 6.44, 6.40, and 6.41. These contacts represent the central hydrophobic core of the receptor. They create a hydrophobic hindering mechanism (HHM) stabilizing the inactive state hampering the channel of water seen in available crystal structures that facilitates the formation of the receptor active conformation [31]. The role of TM3 in protein function is linked to the rearrangement of the HHM residues seen in all active state structures allowing the upward movement of TM3 along its axis and the rotation of TM6 [31]. From the comparison of Class A, B, and F, 3.43 creates topologically equivalent contacts with TM5 (residue 5.54) while the nearby residues 3.42 and 3.46 are in contact with TM2 and TM4. The interaction between the HHM residue 3.43 and TM6 is very similar, but not identical among the different subfamilies: in Class B, 3.43 is interacting with 6.44 and 6.40, while in Class F the contact is created between 3.43 and 6.41. Despite this, the role of the HHM is shared among subfamilies and possibly linked to the receptor function. Extensive mutagenesis data supports this role for Class A [31]. For Class B and F, no experimental data are available; however, the intracellular resemblance of the overall folds of Class A, B, and F suggests a similar movement of TM6 upon activation, requiring rearrangement of the hydrophobic core, to allow G protein interaction.

One unique feature of the Family B GPCR is the presence of the ECD (also known as N-terminus domain). This region has been shown by X-ray crystallography to interact with the peptidic physiological agonists of the receptors and its role has been proven to be important for protein activation [29]. In particular the ECD of CRF1R has been shown previously to act as an intrinsic negative regulator of receptor activity, with extracellular loop 3 (ECL3) of the TMD playing an important role in mediating crosstalk with the ECD [32, 33]. Enhanced sampling MD, mutagenesis, photoaffinity labeling and cysteine trapping studies provided experimental evidence that the extracellular loops and the juxtamembranous (JM) region connecting TM1 to the ECD, played an important role in the structure, function, and context of full-length multidomain Class B receptors [34, 35]. MD simulation suggested electrostatic interactions between ECL3, the ECD, and the JM region as the major determinants in the relative positioning of the two receptor domains. These results were in agreement with previous studies on Family B GPCRs [32, 36, 37]. Taken together, these experimental and computational data suggest that following initial binding of the 41 amino acid corticotropin releasing hormone to the ECD [38, 39], the peptide agonist relieves receptor inactivation through affecting ECD interactions with ECL3 before subsequently binding to the TMD to induce receptor activation.

The novel X-ray allowed extensive studies to try to understand the ligand binding event to the receptor. MD simulation using an apo CRF1R model strongly suggests that the allosteric pocket (found deep toward the intracellular side of the receptor) is induced by ligand binding (Figure 9.3a). Steered MD supported two possible access routes for the ligand to the induced-fit site in CRF1R: one from the putative orthosteric site in the extracellular side of the receptor and a second from within the membrane, in a location close to TM5 and TM6 [35]. Metadynamics analysis of the free energy surface of the ligand dissociation suggested that the lowest energy escape route for the ligand from the induced-fit pocket is between TM5 and TM6 toward the membrane. During ligand dissociation the small-molecule can exit from the CRF1R-TMD in a region close to G3246 and P3216. These predicted conformational changes required for ligand binding and dissociation also provide a potential rationale for the selectivity of CP-376395 to the human CRF1R subtype. The functional antagonist activity of CP-376395 for CRF1R is 12 nM, while it is greater than 10 000 nM for CRF2R [40]. Sequence identity within the helical bundle between the receptor subtypes is high (78%) and residues participating in direct interactions with CP-376395 in CRF1R are completely conserved in CRF2R. Across the second shell of residues in proximity to CP-376395 (less than 5 Å), only two residues differ between CRF1R and CRF2R, these being H199 and M276 corresponding to V1953 and I272, respectively. Mutation of these residues in CRF1R to the corresponding amino acids in CRF2R reduced the binding affinity of standard non-peptide antagonists in CRF1R while concomitantly having little effect on peptide ligand binding [41, 42]. From the MD simulation, M276 and H199 appeared to play a crucial role in modulating the conformational changes required for CP-376395 binding/dissociation. In CRF2R the corresponding V195 and I272 may potentially result in a receptor conformation incompatible with CP-376395 gaining access to the induced-fit allosteric site.

The discovery of an allosteric binding site in the CRF1 receptor gave an unprecedented opportunity to understand at a molecular level the difference in ligand residence time. The residence time, typically expressed as the inverse of the dissociation rate, is generally in a range that can vary from seconds to hours. The lifetime of the complex controls the effectiveness and duration of the

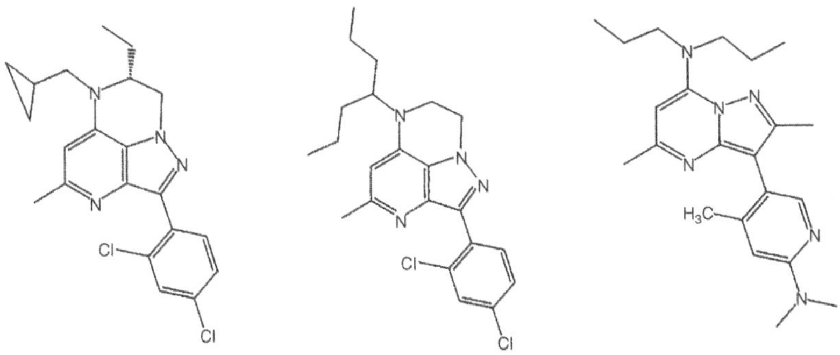

NBI35965
Ki = 1.1 nM
$t_{1/2}$ *in vitro* = 16 min
$t_{1/2}$ *in vivo* < 2 h

NBI34041
Ki = 1.3 nM
$t_{1/2}$ *in vitro* = 53 min
$t_{1/2}$ *in vivo* = 4 h

NBI30775
Ki = 2.6 nM
$t_{1/2}$ *in vitro* = 130 min
$t_{1/2}$ *in vivo* < 6 h

Figure 9.4 Importance of residence time for CRF1R negative allosteric modulators. These three CRF1R NAMs have similar nanomolar binding affinity for the protein target, similar fast pharmacokinetics, but different *in vitro* residence time ($t_{1/2}$), consistent with their sustained suppression of adrenocorticotropin *in vivo* in adrenalectomized rats ($t_{1/2}$ *in vivo*) [44].

biological action much more than the simple affinity of the ligand for its receptor [43]. CRF1R NAMs with similar nanomolar binding affinity can have very different residence time, spanning from less than two minutes to more than seven hours [44]. In particular, the ranking of dissociation half-lives of three of the ligands (NBI35965, $t_{1/2}$ = 16 minutes; NBI34041, $t_{1/2}$ = 53 minutes; NB30775, $t_{1/2}$ = 130 minutes) is consistent with the ranking of their sustained suppression of adrenocorticotropin *in vivo* in adrenalectomized rats (NBI35965, <2 hours; NBI34041, 4 hours; NBI30775, 6 hours; Figure 9.4). The fact that these three ligands have similar fast pharmacokinetics confirms that the sustained adrenocorticotropin suppression shown uniquely by NBI30775 is the result of its prolonged CRF1R occupancy (Figure 9.3). Using a MD-based computational protocol, it was possible to evaluate the energy necessary to break the ligand–protein interactions to start the unbinding event. This physics-based, fully flexible, and pose-dependent ligand scoring function evaluates the maximum energy required to move the ligand from the starting energy basin to the next. In an initial test, this approach was able to predict differences in residence time for a set of 12 CRF1R NAMs. Additionally, it was able to highlight key differences in the degree of accessibility to water molecules during the predicted ligand unbinding to better understand the role of the solvent in determining the small molecule residence time providing opportunities to rationally optimize ligand binding kinetics.

9.5 Structural Understanding of Glucagon and GLP1 Receptor Activation

The CRF1R X-ray structure was the first milestone that allowed the application of SBDD to Family B GPCRs and paved the way for the final discovery of HTL26119. However, our structural understanding of this family of receptors was still limited. The real breakthrough highlighting the new SBDD opportunity was the crystal structure of the glucagon receptor (GCGR) in complex with a small molecule NAM, MK-0893 (Figure 9.5c) [45]. Glucagon is a 29-amino-acid peptide released

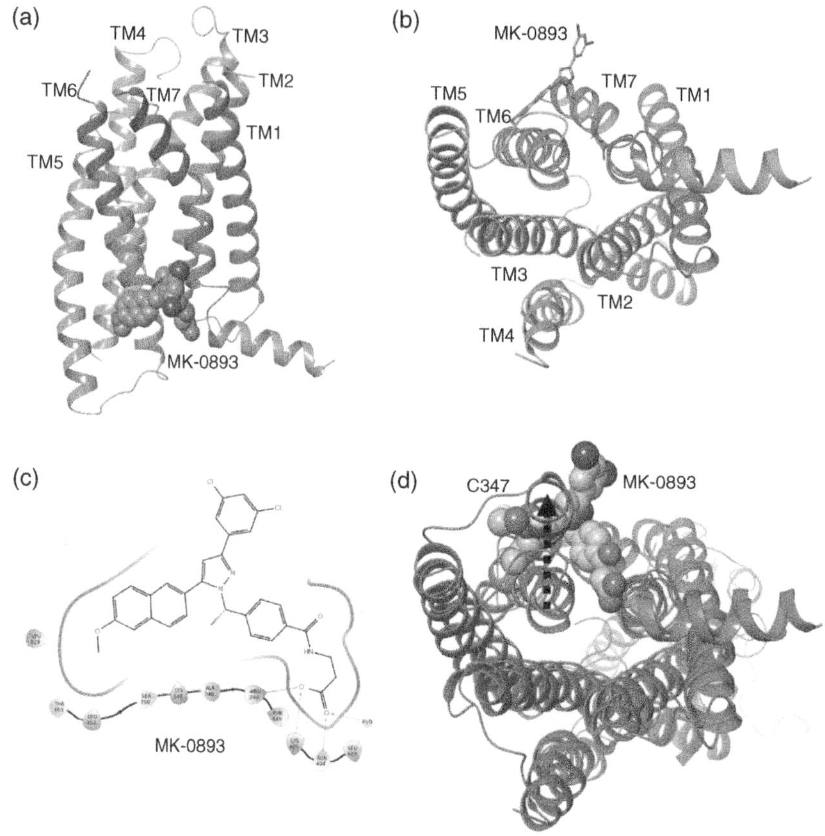

Figure 9.5 Crystal structure of the GCGR bound to MK-0893 (PDB ID 5EE7) [45]. (a,b) show the protein as cartoon representation and the NAM as space filling. A side view is shown in (a) and a view from the intracellular side in (b). The ligand two-dimensional structure and its interactions with GCGR are included in (c). (d) shows the intracellular side view comparison between the X-ray of the inactive state of the GCGR in complex with MK (CPK representation) and GLP1R bound to a peptide agonist (hidden) [17]. GLP1R C347 is shown in CPK. This residue is known to create a covalent bond with PAMs like compound **2**. A black arrow highlights the intracellular movement of TM6 upon activation and how the conformational change is stopped by the negative allosteric modulator MK.

from the α-cells of the islet of Langerhans, which has a key role in glucose homeostasis [46]. Glucagon action is transduced by the GCGR, which is located on liver, kidney, intestinal smooth muscle, brain, adipose tissue, heart and pancreas cells, and this receptor has been considered an important drug target in the treatment of diabetes. Administration of recently identified small-molecule GCGR antagonists in patients with type 2 diabetes results in a substantial reduction of fasting and postprandial glucose concentrations [47].

As for CRF1R, the crystal structure revealed the NAM MK-0893 binding location was unexpected: outside the helical bundle, straddling TM6 from within the lipid bilayer (Figure 9.5). TM6 divided the binding site into two distinct regions: a hydrophobic interface with TM5, and a polar cleft toward TM7. The different physicochemical properties of this bipartite antagonist pocket corresponded to the dual hydrophilic/hydrophobic nature of the ligand. The apolar methoxynaphthalene moiety of the small molecule made hydrophobic contacts in the TM5–TM6 interface. On the opposite site, within the TM6–TM7 cleft, MK-0893 participated in a network of polar contacts. The phenylethylpyrazole core of the molecule created further interactions with TM6 and the position of the pyrazole moiety parallel to the membrane provided two vectors toward the bipartite NAM subpockets. A third vector starting from the ligand pyrazole ring led to the dichlorophenyl group, which made a crystal contact to TM4 of another crystallographic asymmetric unit, defined as the smallest portion of a crystal that, when duplicated and moved by crystal symmetry operations, can produce the unit cell of the crystal. MD simulations of MK-0893 binding to the wild-type receptor demonstrated that the receptor–ligand interaction is stable, and MK-0893 remained at the cytoplasmic membrane boundary. Sequence conservation analysis demonstrated that most of the residues in the binding site showed good conservation across other members of human Class B receptors. The MK-0893 binding mode suggested that the ligand acted as a clamp holding TM6 in the inactive state and hampering receptor conformational changes required for G-protein coupling. In general, GPCR activation results in a rotation of TM6 in response to agonist binding, which is transmitted through a rigid-body movement amplified along TM6, thereby altering the interface between TM5 and TM6 and leading to an outward movement of the cytoplasmic end of TM6 to enable G-protein binding. In conclusion, the ability of MK-0893 to block signaling of GCGR was explained by the novel X-ray and directly linked to the ligand interaction with the inactive conformation of TM5 and TM6.

The overall sequence identity between the GCGR and the GLP1R is 31%. In the helical bundle region it reaches 55%, while in the NAM binding site 4 residues are different out of 12 between the two related GPCRs. The glucagon X-ray suggested for the first time the possibility this binding site could be present and druggable also in the GLP1R. At the same time the 4 residue difference in the pocket could explain the weak activity of MK-0893 versus GLP1R. GLP1 is a hormone released primarily from the gastrointestinal tract in response to food intake and controls the secretion of insulin from pancreatic islets in a glucose-dependent manner [48]. Several GLP1 agonists are now approved for the treatment of type 2 diabetes and, in addition to their insulinotropic action, also have beneficial effects on body weight and cardiovascular parameters [49]. Under rare pathophysiological conditions in which the response to GLP-1 is significantly increased, the patient may display increased insulin secretion and concomitant hypoglycemia. Currently, no small-molecule GLP-1R antagonist is available that can be used to further understand clinically the consequences of an overactive GLP-1 system.

The X-ray structure of GLP1R in complex with an agonist unveiled the remarkably complex network of interactions between a peptidomimetic ligand and receptor protein [17]. The structure increased the understanding of the mechanism of activation of the Class B secretin group of GPCRs. The intracellular portion of TM6 swings away from the central axis of the TMD by over 10 Å. When TM6 and TM7 are excluded from the superposition of GLP1R with GCGR and CRF1R, the structures are in much closer agreement, demonstrating the substantial rearrangements of TM6 and TM7 relative to TM1–TM5 between agonist-bound GLP1R and antagonist-bound GCGR and CRF1R. This “opening” of the intracellular GLP1R surface probably facilitates G-protein coupling and β-arrestin signaling, in a manner similar to that observed for the inactive to active receptor state transitions in Class A GPCR structures, indicating a common activation mechanism across Class A and B receptors [17].

In GLP1R, the NAM binding site includes C347$^{6.36b}$, identified as the site for covalent interaction with electrophilic groups of the agonist positive allosteric modulators (PAMs) 6,7-dichloro-3-meth anesulfonyl-2-*tert*-butylamino-quinoxaline (compound **2**) and 4-(3-benzyloxyphenyl)-2-ethylsulfi nyl-6-(trifluoromethyl)pyrimidine (BETP) (Figure 9.6) [50, 51]. Compound **2** was proposed to create interactions with TM5, TM6 and TM7 [52]. The ligand is inactive on GCGR where C347 is a

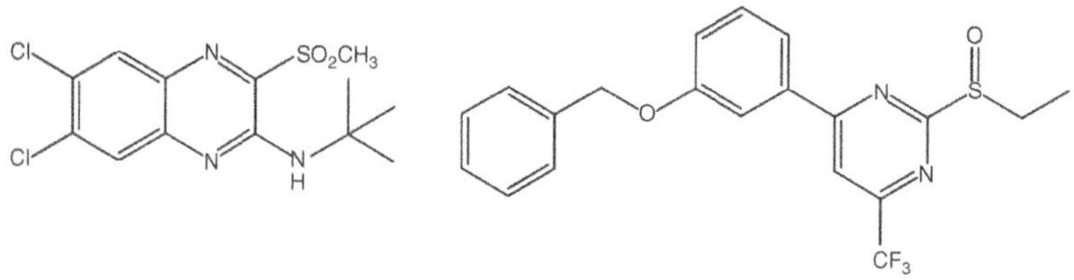

Figure 9.6 Molecular structures of agonist positive allosteric modulators compound **2** and BETP.

phenylalanine. The C347$^{6.36b}$F mutant abolished the potency of compound **2**, in line with the covalent binding mode hypothesis. This is also consistent with structure–activity relationship (SAR) studies showing the importance of the electrophilic character of the dichloroquinoxaline moiety of compound **2** [51]. The comparison between GCGR and GLP1R in this binding site and specific mutation effects on NAMs and agonist PAM provided a reliable framework for the SBDD of novel NAMs of GLP1R.

9.6 Hyperinsulinemic Hypoglycemia

Glucose homeostasis requires appropriate and synchronous coordination of metabolic events and hormonal activities to keep plasma glucose concentrations in a narrow range of 3.5–5.5 mmol/l [53]. Hyperinsulinemic hypoglycemia (HH) refers to a genetically and phenotypically heterogeneous disease. It is characterized by excessive insulin secretion from pancreatic β-cells [3]. The resulting low plasma concentration causes hypoketotic, hypofattyacidemic and hyperinsulinemic hypoglycemia. It is the most frequent cause of persistent hypoglycemia in neonates and infants. Patients with HH have increased risk of permanent brain injury secondary to the metabolic actions of insulin. It is linked to high risk for neurological complications, such as epilepsy, cerebral palsy and cognitive impairment [54]. HH usually presents soon after birth. It is essential to make a rapid diagnosis to prevent potentially associated neurological complications. Symptoms of hypoglycaemia can be mild and unspecific, like lethargy and irritability, or more severe like apnoea, seizures or even coma.

At a molecular level, mutations in nine different genes can lead to the dysregulation of insulin secretion and cause this disorder. Hyperinsulinism may be transient due to risk factors, such as intrauterine growth retardation, perinatal asphyxia, maternal diabetes mellitus or in the setting of syndromes, such as Beckwith–Wiedemann [55]. In adults, HH accounts for 0.5–5.0% of cases of hypoglycemia and can be due either to β-cell tumors (insulinomas) or β-cell hyperplasia. Hypoglycemia typically occurs after a period of fasting, exercise or occasionally during the course of an illness, when there are increased metabolic demands [56]. The incidence of the condition is estimated as 1 in 40000–50000 in the general population to 1 in 2500 in populations with high rates of consanguinity [57].

Advances in the field of HH include use of rapid molecular genetic testing for the disease, application of novel imaging techniques like 6-[fluoride-18]fluoro-levodopa [18F-DOPA] PET-CT and GLP1R imaging. Research into new treatments includes surgical therapies like laparoscopic surgery and evaluation of long-acting octreotide formulations. Drug discovery is focusing on mTOR inhibitors, and GLP1R antagonists [3]. Currently, no small-molecule GLP1R antagonist is

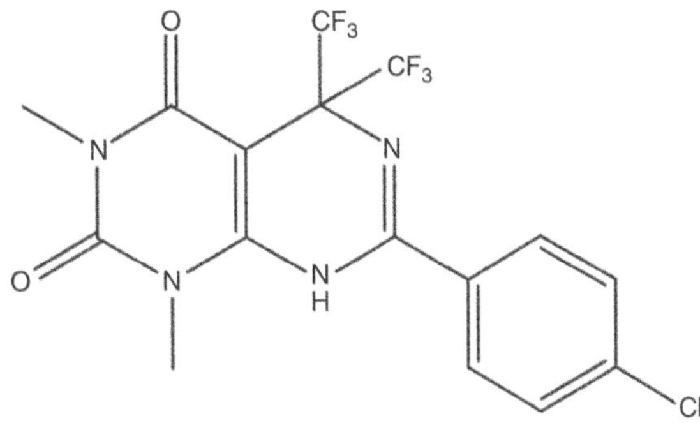

Figure 9.7 Example of known GLP1R antagonist from Vanderbilt University [58, 59].

available that can be used to further understand clinically the potential of this strategy. There are limited examples of GLP1R small molecule antagonists known. For example, the group at Vanderbilt University reported the non-competitive GLP1R antagonist which exhibits insulin and glucose lowering effects *in vivo* (Figure 9.7) [58, 60].

9.7 GLP1 Receptor Negative Allosteric Modulator Lead Identification

VS is widely used in drug discovery to identify novel hit molecules. Structure-based VS uses the protein three-dimensional coordinates to predict a small molecule binding mode and its interactions with the target. Based on these predictions, molecular docking can be applied to evaluate quickly in silico the probability of a molecule having a desired biological activity. Before 2007, GPCR VS was mainly ligand based or used homology models derived from the rhodopsin X-ray data. Recent breakthroughs in membrane protein crystallography greatly increased its application to GPCRs. In addition SBDD methods, relying on the coordinates of GPCR targets, can be further expanded by using HM techniques [61]. Specifically, the discovery of HTL26119 (Figure 9.8) started with a VS campaign based on a GLP1R HM, created using the GCGR X-ray in complex with MK-0893 as a template [1, 45].

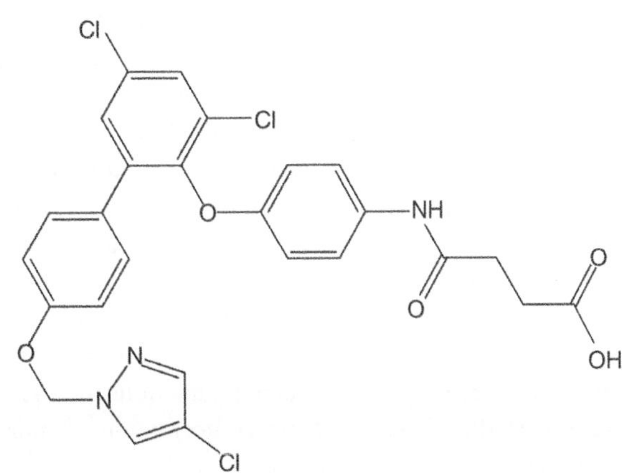

Figure 9.8 Molecular structure of HTL26119.

Even if it is possible to screen millions of molecules in silico, only a small percent of the results can be visually analyzed, typically a few hundred of the top-ranked ligands. Molecular docking software aims to find the best compromise between speed and accuracy. These methods use efficient algorithms to explore the conformational space accessible to the ligand in the binding site and then evaluate the poses using simple scoring functions estimating the ligand–protein interactions. The so-called enrichment is generally linked to the ability to detect and give a low score to a ligand not able to bind the protein target. The top-scoring ligands will include a high number of false positives that an expert user can often identify and discard. The selection of compounds from a three-dimensional analysis of hits in a binding site is of course highly subjective and biased, but it represents a crucial part of a VS campaign.

Structure-based VS of GPCRs is now routine and often a very fruitful method for the discovery of new ligands [61]. The use of retrospective benchmarking sets of ligands and decoys to compare docking methods is well established [62–64]. VS results for a broad range of GPCR targets have high variability in success depending on the computational protocol used and the specific protein target. The key challenge for some targets remains the need to effectively include waters at the docking stage and the need for protein flexibility [62].

Another important aspect often oversimplified in the literature is the preparation of the database used for the VS. Accurate preparation can dramatically reduce the number of false positives that would need to be discarded during the visual inspection step. The GLP1R NAM VS took advantage of a small molecule database that had been optimized several times over years using information from numerous VS campaigns on structurally diverse GPCRs. A schematic summary of the protocol used to prepare and update the Heptares small molecule database is illustrated in Figure 9.9. Key aspects included: (i) ad-hoc selection of suppliers known to be reliable; (ii) a large number of custom filters to remove unwanted chemistry, manually refined several times with iterations of visual inspection of the ligands in the database; (iii) pre-generation of ionic/tautomeric states, three-dimensional conformations of the ligands including several saturated ring puckering coordinates that are known to be challenging to explore by the docking conformational search algorithms; and (iv) division of the database in three distinct groups (small fragments, templates, and analogues) to allow easy custom database shaping depending on the project needs and the unique features of the target GPCR binding pocket.

Two different docking software programs were used for the VS campaign: Glide [66] and Hybrid [67]. Hybrid allows the use of a reference molecule to focus the conformational search of new ligand poses resulting in high similarity in shape and polar properties to a provided reference compound. Since the coordinates of the GLP1R HM were in the same frame of reference of the GCGR template, it was easy to use the crystallographic pose of MK-0893 as the reference for Hybrid. Due to the similarity between the two binding sites, we wanted to evaluate specifically ligands with similar shape and polarity to known GCGR NAMs. The β-alanine *N*-benzamide is a common structural motif in GCGR antagonists that resulted from the VS results to be important also for GLP1R small molecules. The other key requirement was the ability of the ligand to adopt a conformation able to clamp TM6 stopping its conformational change required for activation. Among the VS hits, the most promising had the β-alanine *N*-benzamide group and specific aromatic rings able to fold around TM6 (Figure 9.10). The two hits had low affinity GLP1R antagonist activity, with GLP1R pKi equal to 5.0 (**1**) and 5.5 (**2**), respectively, and no significant GCGR binding. The ortho phenyl ring in **1** was predicted to create favorable interactions with TM6. The dichloro moieties in **2** were predicted to stabilize the bioactive ligand conformation. In particular, MD simulation showed the chloro groups were also able to stabilize the NAM position at the interface between the membrane and the intracellular water molecules. Structural overlay of these two compounds suggested a hybrid

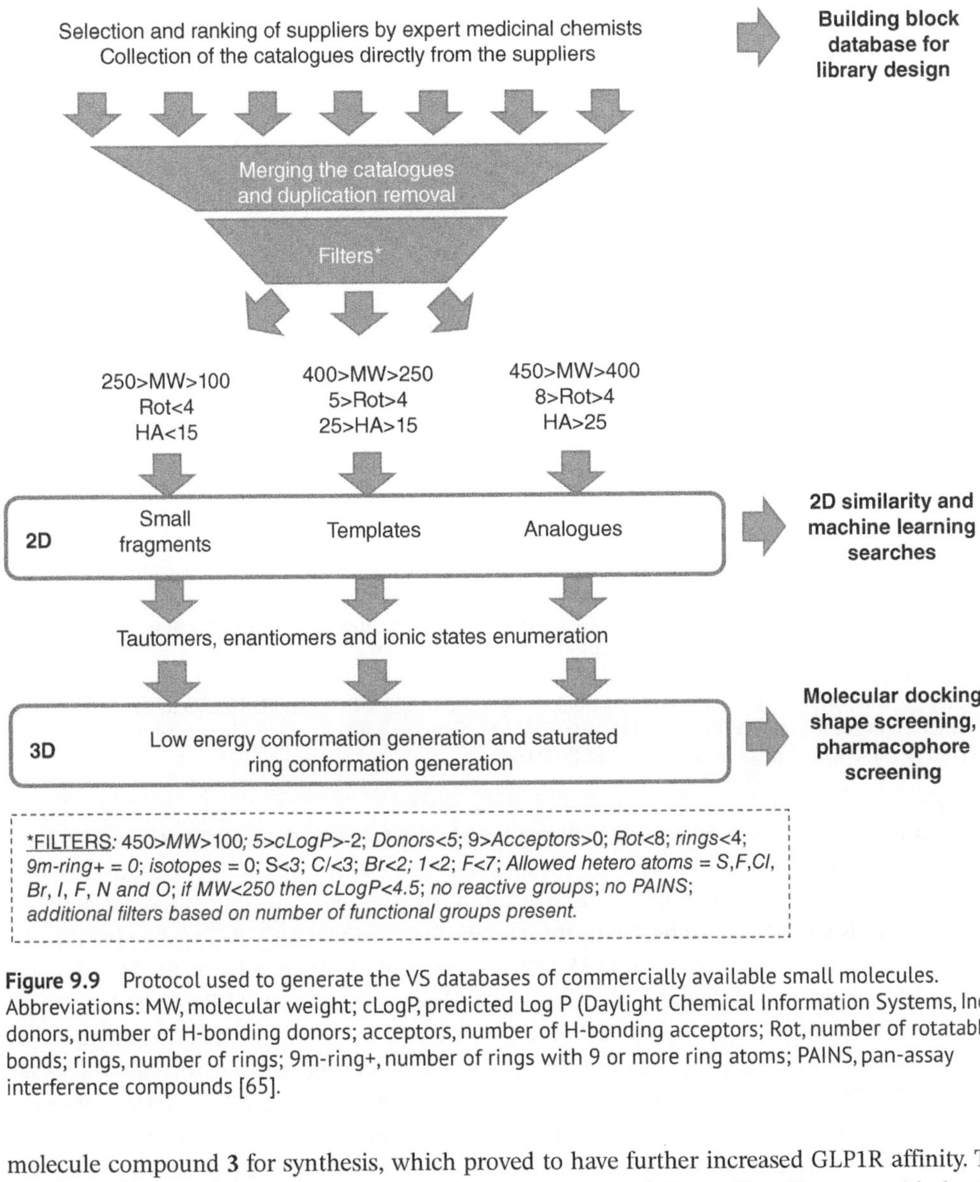

Figure 9.9 Protocol used to generate the VS databases of commercially available small molecules. Abbreviations: MW, molecular weight; cLogP, predicted Log P (Daylight Chemical Information Systems, Inc.); donors, number of H-bonding donors; acceptors, number of H-bonding acceptors; Rot, number of rotatable bonds; rings, number of rings; 9m-ring+, number of rings with 9 or more ring atoms; PAINS, pan-assay interference compounds [65].

molecule compound **3** for synthesis, which proved to have further increased GLP1R affinity. The predicted poses highlighted that these ligands were not able to fully reach and interact with the cleft between TM5 and TM6. Further docking studies identified a long hydrophobic para substitution to be promising: experimental data showed ligand **4** was able to reach promising nanomolar activity.

9.8 GLP1 Receptor Negative Allosteric Modulator Lead Optimization

HTL26119 was rationally designed using the GLP1R HM in which molecular docking and MD were applied extensively. Based on the results of these methods, one important aspect was the optimization of the interactions between the NAM and the TM5/TM6 interface. This region is also

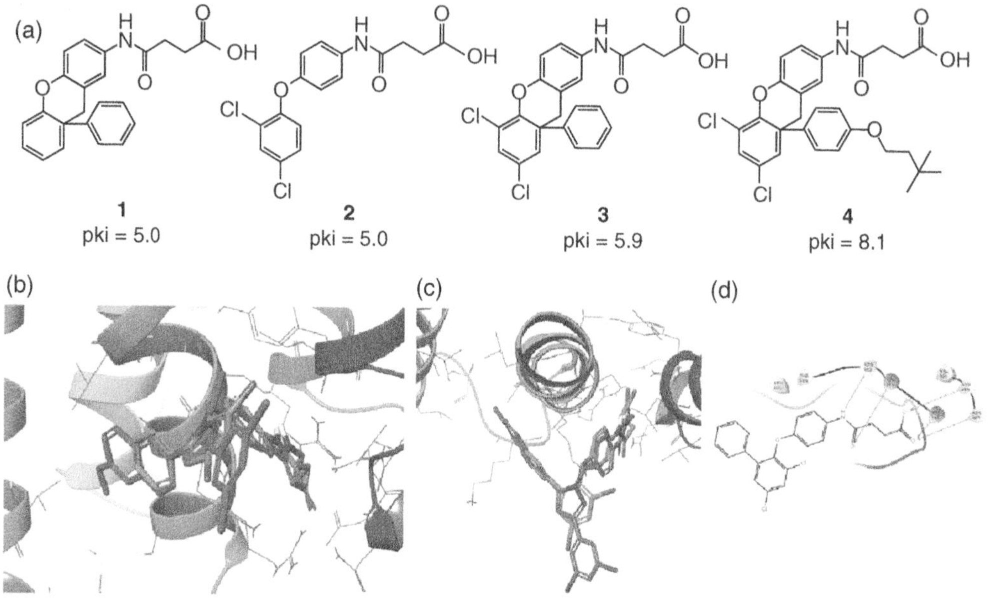

Figure 9.10 VS hits and initial lead. (a) The two-dimensional structures of the VS hits **1** and **2** are included. **3** was designed as a hybrid of the original hits, and **4** was the lead compound rationally designed based on the predicted binding pose in the GLP1R HM. (b) (side view) and (c) (top view, from the extracellular side) show a comparison of the docked ligand **3** to GLP1R HM and MK-0893 in the conformation present in the GCGR X-ray [45]. (d) Two-dimensional representation of the interactions predicted in the docking pose of **3** to the GLP1R HM.

important for the selectivity versus GCGR (Figure 9.11a, b). C347 (generic number 6.36b [27]) on the intracellular side of TM6 is replaced by a phenylalanine (F345) in GCGR. It is the most relevant difference resulting in key variations in the shape of the binding site. A second difference between the two GPCRs is V331. This residue facing the cysteine is a leucine in GCGR. To take advantage of these differences, exploration of the ligand SAR has been focused in particular on the ligand vector toward both these amino acids that create a mainly hydrophobic subpocket (Figure 9.11c). This is in contrast with the interface between TM6 and TM7 where the ligand is predicted to create essentially polar interactions with S352, K351, R348, N406, and N407. The balance of polar and apolar regions of the NAM molecular surface was essential to stabilize the ligand at the interface of a very complex environment composed of the receptor, the intracellular solvent, and the membrane. To better understand these aspects, it was essential to analyze the different ligand docking poses in a fully explicit MD simulation in the presence of the membrane (Figure 9.11d). The results showed how the bilayer adapted the location of the lipidic hydrophobic tails and hydrophilic head groups to the different features of the intracellular side of the receptor near the NAM binding region. A water pocket was present at the N-terminal part of TM6 and TM7, solvating in part the charged residues and the ligand acid group. The NAM was at the very interface between the two environments with its intracellular face in contact with water molecules and on the opposite side was touching lipid lipophilic tails. The chloropyrazole moiety in HTL26119 was perfect in creating optimal apolar interactions with C347 and V331, while the polar nitrogen atom in the ring was able to be fully solvated at the interface with the water molecules.

Reducing the logP of these lipophilic acid ligands was challenging due to the binding site location in the membrane. Efforts to increase the ligand polarity were restricted to regions of the NAM

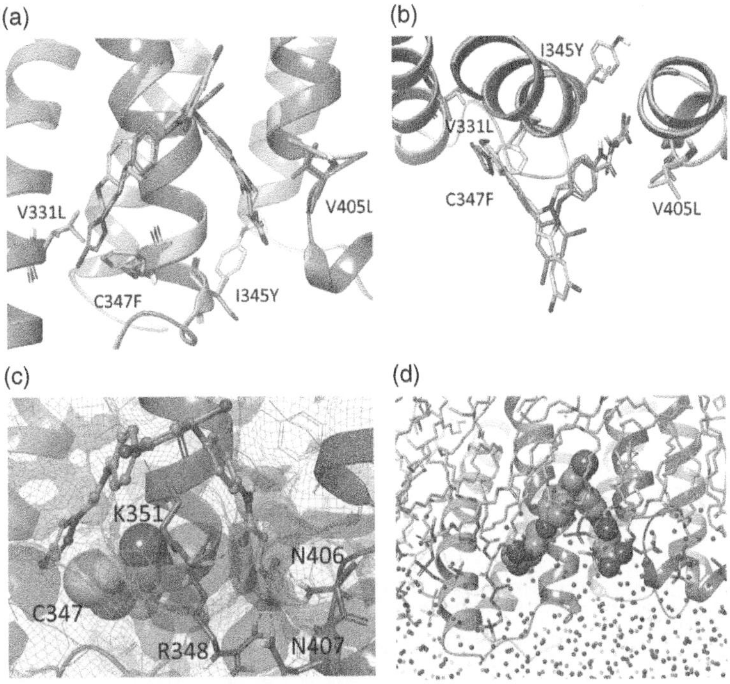

Figure 9.11 (a) (side view) and (b) (extracellular view) include a comparison between HTL26119 docked (stick representation) to the GLP1R HM and the X-ray of MK-0893 (stick representation) in complex with GCGR. Residues of the allosteric binding site differing between the two Family B GPCRs are shown in stick representation with a label including the GLP1R residue type, followed by the GLP1R residue number and at the end the corresponding GCGR residue type. (c) Predicted binding pose of HTL26119 to the GLP1R-HM. Pocket hotspot regions are shown as in Figure 9.2 and residues creating polar interactions with the ligand are included in stick representation with labels. Residue C347, known to be important for the selectivity versus the GCGR, is included as CPK. (d) MD snapshot of HTL26119 (space filling representation) in complex with GLP1R HM (ribbon representation). Small spheres represent water molecules of the intracellular side, while lipid 1-palmitoyl-2-oleoyl-sn-glycero-3-phosphocholine (POPC) carbons are in the stick representation.

pointing toward the solvent. One successful attempt was the introduction of a pyridine nitrogen in the aromatic ring connected to the succinic acid amide head group (Figure 9.12). In parallel, SAR exploration of the para position of the 2-phenoxy-3-benzene showed increasing activity versus GLP1R growing from the methoxy to bigger linear groups optimizing the interactions with TM5 and TM6. Selectivity versus GCGR was only slightly improved suggesting that the phenylalanine replacing C347 in GCGR could adapt its rotamer state to the ligand. Another possibility was that if the ligand group was not rigid enough, it could assume a conformation compatible with the GCGR's bulkier F345. This hypothesis is supported by the increase in activity and selectivity of a linear and rigid dimethyl-2-pentyne with a pKi = 8.7 versus GLP1R and good 14-fold selectivity versus GCGR. An increase in both activity and selectivity was also possible with a more flexible chain and a pyrazole moiety that was predicted to be locked in a position near C347 by interactions with the solvent at the intracellular end of TM6. This strategy resulted in HTL26119 with an optimal 4-chloro-pyrazole group resulting in promising activity and selectivity.

The GLP-1R functional potency of HTL26119 (pKb = 6.8) was lower compared with its binding affinity (pKi = 7.8). This may be related to a low coupling efficiency between the allosteric and orthosteric binding sites and was observed for other compounds in the series. It is an important

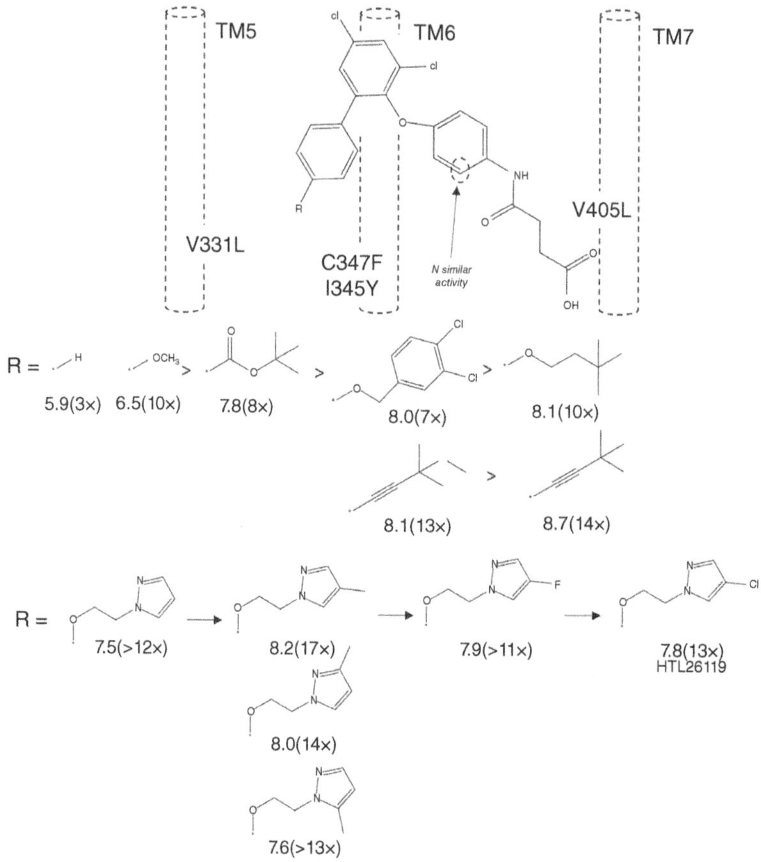

Figure 9.12 SAR highlights that resulted in the discovery of HTL26119. In the upper part, a schematic two-dimensional depiction of the NAM location relative to the transmembrane helices is shown. Residues different in the binding site between GLP1R and GCGR and their relative location are included: the first letter corresponds to the GLP1R amino acid, followed by the GLP1R residue number and the corresponding residue in GCGR. In the lower part, different R groups are shown with their binding pKi to the GLP1R. The fold selectivity versus GCGR is included in parentheses (calculated as GLP1R pKi – GCGR pKi).

aspect that can play a role also on the final NAM selectivity. This effect can be more pronounced for GLP1R over GCGR, resulting in a more marked decrease in the functional activity compared with the binding affinity for GLP1R over GCGR (unpublished data). It is possible an extra 10-fold binding selectivity and affinity for GLP-1R over GCGR in the binding assay may be needed to achieve the desired functional selectivity in this series.

9.9 Conclusion

In the last 10 years, Family B GPCR SBDD has been transformed. Heptares Therapeutics played a crucial role in this revolution that made SBDD for this subfamily a reality. Their three X-ray structures originally published in *Nature* and described in this chapter have been essential milestones in our understanding of every aspect of this complex receptor class. The CRF1R X-ray with a completely unexpected negative allosteric binding site provided a first glimpse of their structural complexity [16]. The agonist bound GLP1R X-ray highlighted the challenges presented by their orthosteric site for small molecule ligand design [17]. In addition, this X-ray started to explain the role of the ECD and the conformational changes happening during receptor activation. Finally, the GCGR X-ray presented new opportunities for the design of allosteric modulators, both positive and negative [45]. These structures together with advanced computational chemistry methods provided structural understanding, at a molecular level, of Family B GPCR activation, modulation, ligand binding kinetics, and selectivity.

HTL26119 represents just an example of what is now possible through this novel structural understanding of Family B GPCRs. This is just the beginning of an exciting new era for rational structure-based Class B GPCR drug design.

References

1. O'Brien, A., Andrews, S.P., Baig, A.H. et al. (2019). Identification of a novel allosteric GLP-1R antagonist HTL26119 using structure-based drug design. *Bioorg. Med. Chem. Lett.* 29 (20): 126611.
2. Andersen, A., Lund, A., Knop, F.K., and Vilsbøll, T. (2018). Glucagon-like peptide 1 in health and disease. *Nat. Rev. Endocrinol.* 14 (7): 390–403.
3. Shah, P., Rahman, S.A., Demirbilek, H. et al. (2017). Hyperinsulinaemic hypoglycaemia in children and adults. *Lancet Diabetes Endocrinol.* 5 (9): 729–742.
4. Overington, J.P., Al-Lazikani, B., and Hopkins, A.L. (2006). How many drug targets are there? *Nat. Rev. Drug Discov.* 5 (12): 993–996.
5. Bakheet, T.M. and Doig, A.J. (2009). Properties and identification of human protein drug targets. *Bioinformatics* 25 (4): 451–457.
6. Hauser, A.S., Attwood, M.M., Rask-Andersen, M. et al. (2017). Trends in GPCR drug discovery: new agents, targets and indications. *Nat. Rev. Drug Discov.* 16 (12): 829–842.
7. Mason, J.S., Bortolato, A., Congreve, M., and Marshall, F.H. (2012). New insights from structural biology into the druggability of G protein-coupled receptors. *Trends Pharmacol. Sci.* 33 (5): 249–260.
8. Palczewski, K. (2000). Crystal structure of rhodopsin: a G protein-coupled receptor. *Science* 289 (5480): 739–745.
9. Cherezov, V., Rosenbaum, D.M., Hanson, M.A. et al. (2007). High-resolution crystal structure of an engineered human 2-adrenergic G protein-coupled receptor. *Science* 318 (5854): 1258–1265.
10. Warne, T., Serrano-Vega, M.J., Baker, J.G. et al. (2008). Structure of a β1-adrenergic G-protein-coupled receptor. *Nature* 454 (7203): 486–491.
11. Jaakola, V.-P., Griffith, M.T., Hanson, M.A. et al. (2008). The 2.6 angstrom crystal structure of a human A2A adenosine receptor bound to an antagonist. *Science* 322 (5905): 1211–1217.
12. Isberg, V., Mordalski, S., Munk, C. et al. (2016). GPCRdb: an information system for G protein-coupled receptors. *Nucleic Acids Res.* 44 (D1): D356–D364.
13. Cooke, R.M., Brown, A.J.H.H., Marshall, F.H., and Mason, J.S. (2015). Structures of G protein-coupled receptors reveal new opportunities for drug discovery. *Drug Discov. Today* 20 (11): 1355–1364.
14. Goodford, P.J. (1985). A computational procedure for determining energetically favorable binding sites on biologically important macromolecules. *J. Med. Chem.* 28 (7): 849–857.
15. Rucktooa, P., Cheng, R.K.Y., Segala, E. et al. (2018). Towards high throughput GPCR crystallography: in meso soaking of adenosine A2A receptor crystals. *Sci. Rep.* 8 (1): 41.
16. Hollenstein, K., Kean, J., Bortolato, A. et al. (2013). Structure of class B GPCR corticotropin-releasing factor receptor 1. *Nature* 499 (7459): 438–443.

17 Jazayeri, A., Rappas, M., Brown, A.J.H. et al. (2017). Crystal structure of the GLP-1 receptor bound to a peptide agonist. *Nature* 546 (7657): 254–258.

18 Abel, R., Young, T., Farid, R. et al. (2008). Role of the active-site solvent in the thermodynamics of factor Xa ligand binding. *J. Am. Chem. Soc.* 130 (9): 2817–2831.

19 Halgren, T.A. (2009). Identifying and characterizing binding sites and assessing druggability. *J. Chem. Inf. Model.* 49 (2): 377–389.

20 Barillari, C., Taylor, J., Viner, R., and Essex, J.W. (2007). Classification of water molecules in protein binding sites. *J. Am. Chem. Soc.* 129 (9): 2577–2587.

21 Bortolato, A., Tehan, B.G., Bodnarchuk, M.S. et al. (2013). Water network perturbation in ligand binding: adenosine A2A antagonists as a case study. *J. Chem. Inf. Model.* 53 (7): 1700–1713.

22 Segala, E., Guo, D., Cheng, R.K.Y. et al. (2016). Controlling the dissociation of ligands from the adenosine A2A receptor through modulation of salt bridge strength. *J. Med. Chem.* 59 (13): 6470–6479.

23 Lagerström, M.C. and Schiöth, H.B. (2008). Structural diversity of G protein-coupled receptors and significance for drug discovery. *Nat. Rev. Drug Discov.* 7 (4): 339–357.

24 Hollenstein, K., De Graaf, C., Bortolato, A. et al. (2014). Insights into the structure of class B GPCRs. *Trends Pharmacol. Sci.* 35 (1): 12–22.

25 Sánchez, R. and Šali, A. (1997). Evaluation of comparative protein structure modeling by MODELLER-3. *Proteins Struct. Funct. Genet.* 1 (1): 50–58.

26 Hoare, S. (2005). Mechanisms of peptide and nonpeptide ligand binding to class B G-protein-coupled receptors. *Drug Discov. Today* 10 (6): 417–427.

27 Isberg, V., De Graaf, C., Bortolato, A. et al. (2015). Generic GPCR residue numbers – aligning topology maps while minding the gaps. *Trends Pharmacol. Sci.* 36 (1): 22–31.

28 Kean, J., Bortolato, A., Hollenstein, K. et al. (2015). Conformational thermostabilisation of corticotropin releasing factor receptor 1. *Sci. Rep.* 5 (1): 11954.

29 Liang, Y.-L., Belousoff, M.J., Zhao, P. et al. (2020). Toward a structural understanding of class B GPCR peptide binding and activation. *Mol. Cell* 77 (3): 656–668.e5.

30 Bortolato, A., Doré, A.S., Hollenstein, K. et al. (2014). Structure of class B GPCRs: new horizons for drug discovery. *Br. J. Pharmacol.* 171 (13): 3132–3145.

31 Tehan, B.G., Bortolato, A., Blaney, F.E. et al. (2014). Unifying family a GPCR theories of activation. *Pharmacol. Ther.* 143 (1): 51–60.

32 Koth, C.M., Murray, J.M., Mukund, S. et al. (2012). Molecular basis for negative regulation of the glucagon receptor. *Proc. Natl. Acad. Sci. USA* 109 (36): 14393–14398.

33 Sydow, S., Flaccus, A., Fischer, A., and Spiess, J. (1999). The role of the fourth extracellular domain of the rat corticotropin-releasing factor receptor type 1 in ligand binding. *Eur. J. Biochem.* 259 (1–2): 55–62.

34 Dong, M., Koole, C., Wootten, D. et al. (2014). Structural and functional insights into the juxtamembranous amino-terminal tail and extracellular loop regions of class B GPCRs. *Br. J. Pharmacol.* 171 (5): 1085–1101.

35 Dore, A.S., Bortolato, A., Hollenstein, K. et al. (2017). Decoding corticotropin-releasing factor receptor type 1 crystal structures. *Curr. Mol. Pharmacol.* 10 (4): 334–344.

36 Barwell, J., Conner, A., and Poyner, D.R. (2011). Extracellular loops 1 and 3 and their associated transmembrane regions of the calcitonin receptor-like receptor are needed for CGRP receptor function. *Biochim. Biophys. Acta – Mol. Cell Res.* 1813 (10): 1906–1916.

37 Wille, S., Sydow, S., Palchaudhuri, M.R. et al. (1999). Identification of amino acids in the N-terminal domain of corticotropin-releasing factor receptor 1 that are important determinants of high-affinity ligand binding. *J. Neurochem.* 72 (1): 388–395.

38 Dong, M., Pinon, D.I., Cox, R.F., and Miller, L.J. (2004). Importance of the amino terminus in secretin family G protein-coupled receptors. *J. Biol. Chem.* 279 (2): 1167–1175.

39 Parthier, C., Reedtz-Runge, S., Rudolph, R., and Stubbs, M.T. (2009). Passing the baton in class B GPCRs: peptide hormone activation via helix induction? *Trends Biochem. Sci.* 34 (6): 303–310.

40 Chen, Y.L., Obach, R.S., Braselton, J. et al. (2008). 2-Aryloxy-4-alkylaminopyridines: discovery of novel corticotropin-releasing factor 1 antagonists. *J. Med. Chem.* 51 (5): 1385–1392.

41 Liaw, C.W., Grigoriadis, D.E., Lorang, M.T. et al. (1997). Localization of agonist- and antagonist-binding domains of human corticotropin-releasing factor receptors. *Mol. Endocrinol.* 11 (13): 2048–2053.

42 Hoare, S.R.J., Brown, B.T., Santos, M.A. et al. (2006). Single amino acid residue determinants of non-peptide antagonist binding to the corticotropin-releasing factor1 (CRF1) receptor. *Biochem. Pharmacol.* 72 (2): 244–255.

43 Tummino, P.J. and Copeland, R.A. (2008). Residence time of receptor–ligand complexes and its effect on biological function. *Biochemistry* 47 (20): 5481–5492.

44 Fleck, B.A., Hoare, S.R.J.J., Pick, R.R. et al. (2012). Binding kinetics redefine the antagonist pharmacology of the corticotropin-releasing factor type 1 receptor. *J. Pharmacol. Exp. Ther.* 341 (2): 518–531.

45 Jazayeri, A., Doré, A.S., Lamb, D. et al. (2016). Extra-helical binding site of a glucagon receptor antagonist. *Nature* 533 (7602): 274–277.

46 Ahrén, B. (2015). Glucagon – early breakthroughs and recent discoveries. *Peptides* 67: 74–81.

47 Bagger, J.I., Knop, F.K., Holst, J.J., and Vilsbøll, T. (2011). Glucagon antagonism as a potential therapeutic target in type 2 diabete. *Diabetes. Obes. Metab.* 13 (11): 965–971.

48 Todd, J.F. and Bloom, S.R. (2007). Incretins and other peptides in the treatment of diabetes. *Diabet. Med.* 24 (3): 223–232.

49 Chatterjee, S., Ghosal, S., and Chatterjee, S. (2016). Glucagon-like peptide-1 receptor agonists favorably address all components of metabolic syndrome. *World J. Diabetes* 7 (18): 441.

50 Bueno, A.B., Showalter, A.D., Wainscott, D.B. et al. (2016). Positive allosteric modulation of the glucagon-like peptide-1 receptor by diverse electrophiles. *J. Biol. Chem.* 291 (20): 10700–10715.

51 Nolte, W.M., Fortin, J.-P., Stevens, B.D. et al. (2014). A potentiator of orthosteric ligand activity at GLP-1R acts via covalent modification. *Nat. Chem. Biol.* 10 (8): 629–631.

52 Song, G., Yang, D., Wang, Y. et al. (2017). Human GLP-1 receptor transmembrane domain structure in complex with allosteric modulators. *Nature* 546 (7657): 312–315.

53 Demirbilek, H., Rahman, S.A., Buyukyilmaz, G.G., and Hussain, K. (2017). Diagnosis and treatment of hyperinsulinaemic hypoglycaemia and its implications for paediatric endocrinology. *Int. J. Pediatr. Endocrinol.* 2017: 9.

54 Carlier, Y., Sosa-Estani, S., Luquetti, A.O., and Buekens, P. (2015). Congenital Chagas disease: an update. *Mem. Inst. Oswaldo Cruz* 110 (3): 363–368.

55 Hussain, K., Thornton, P.S., Otonkoski, T., and Aynsley-Green, A. (2004). Severe transient neonatal hyperinsulinism associated with hyperlactataemia in non-asphyxiated infants. *J. Pediatr. Endocrinol. Metab.* 17 (2): 203–209.

56 Kostopoulou, E. and Shah, P. (2019). Hyperinsulinaemic hypoglycaemia – an overview of a complex clinical condition. *Eur. J. Pediatr.* 178 (8): 1151–1160.

57 Fournet, J.-C. and Junien, C. (2003). The genetics of neonatal hyperinsulinism. *Horm. Res. Paediatr.* 59 (1): 30–34.

58 Nance, K.D., Days, E.L., Weaver, C.D. et al. (2017). Discovery of a novel series of orally bioavailable and CNS penetrant glucagon-like peptide-1 receptor (GLP-1R) noncompetitive antagonists based on a 1,3-disubstituted-7-aryl-5,5-bis(trifluoromethyl)-5,8-dihydropyrimido[4,5-d]pyrimidine-2,4(1H,3H)-dione core. *J. Med. Chem.* 60 (4): 1611–1616.

59 Ghosal, S., Packard, A.E.B., Mahbod, P. et al. (2017). Disruption of glucagon-like peptide 1 signaling in Sim1 neurons reduces physiological and behavioral reactivity to acute and chronic stress. *J. Neurosci.* 37 (1): 184–193.

60 Ward, S.E., Harries, M., Aldegheri, L. et al. (2010). Discovery of N-[(2 S)-5-(6-fluoro-3-pyridinyl)-2,3-dihydro-1H-inden-2-yl]-2-propanesulfonamide, a novel clinical AMPA receptor positive modulator. *J. Med. Chem.* 53 (15): 5801–5812.

61 Congreve, M., Bortolato, A., Brown, G., and Cooke, R.M. (2017). Modeling and design for membrane protein targets. In: *Comprehensive Medicinal Chemistry III*, vol. 3–8 (eds. S. Chackalamannil, D. Rotella and S. Ward), 145–188. Elsevier.

62 Weiss, D.R., Bortolato, A., Tehan, B., and Mason, J.S. (2016). GPCR-bench: a benchmarking set and practitioners' guide for G protein-coupled receptor docking. *J. Chem. Inf. Model.* 56 (4): 642–651.

63 Lagarde, N., Zagury, J.-F., and Montes, M. (2015). Benchmarking data sets for the evaluation of virtual ligand screening methods: review and perspectives. *J. Chem. Inf. Model.* 55 (7): 1297–1307.

64 Mysinger, M.M., Carchia, M., Irwin, J.J., and Shoichet, B.K. (2012). Directory of useful decoys, enhanced (DUD-E): better ligands and decoys for better benchmarking. *J. Med. Chem.* 55 (14): 6582–6594.

65 Baell, J.B. and Holloway, G.A. (2010). New substructure filters for removal of pan assay interference compounds (PAINS) from screening libraries and for their exclusion in bioassays. *J. Med. Chem.* 53 (7): 2719–2740.

66 Repasky, M.P., Shelley, M., and Friesner, R.A. (2007). Flexible ligand docking with glide. *Curr. Protoc. Bioinform.* https://doi.org/10.1002/0471250953.bi0812s18.

67 McGann, M. (2012). FRED and HYBRID docking performance on standardized datasets. *J. Comput. Aided Mol. Des.* 26 (8): 897–906.

10

Discovery and Potential Application of [^{11}C]MK-6884

A Positron Emission Tomography Imaging Agent for the Study of M4 Muscarinic Receptor Positive Allosteric Modulators in Neurodegenerative Diseases

Ling Tong[1] *and Wenping Li*[2]

[1]*Discovery Chemistry, Merck Research Laboratories, 2000 Galloping Hill Road, NJ 07033, Kenilworth, NJ, USA*
[2]*Translational Imaging, Merck Research Laboratories, 770 Sumneytown Pike, PA 19486, West Point, PA, USA*

10.1 Introduction

10.1.1 Positron Emission Tomography

There are significant unmet medical needs in central nervous system (CNS) disorders, such as Alzheimer's disease (AD), schizophrenia, and Parkinson's disease (PD). For example, at least 50 million people globally are believed to be living with AD [1]. In the United States, the burden of AD-related dementia in 2014 was an estimated 5.0 million and by 2060, 13.9 million Americans are projected to have the disease [2]. These CNS disorders are among some of the most debilitating diseases. However, CNS drug discovery has met with significant challenges and setbacks due to the complex disease biology of human brain and the lack of predictive preclinical efficacy models and disease-relevant biomarkers. The research carries great inherent risk of failure as prospective treatments generally lack clinical validation [3]. Improvement in the ability to identify molecules with insufficient efficacy or safety issues prior to clinical development is necessary to reduce the costs and increase the success rate for novel therapeutics [4]. Strategies for early decision making are therefore critical to improve efficiency and increase the odds of success from laboratory concept to commercial product. One possible approach is the use of biomarkers, which can objectively measure target engagement of a therapeutic intervention and evaluate the normal biological, pathogenic, or pharmacologic responses [5]. Noninvasive imaging can provide information in preclinical studies with regard to target engagement, treatment response, safety, or mechanism of action that in turn has a critical impact on internal decision making to help improve the rate of success for drugs taken into the clinic [4].

Positron emission tomography (PET) is a nuclear imaging technique that measures the distribution of radiotracer in living subjects. PET allows the assessment of biochemical and physiological processes in a fully quantitative manner, providing information about the distribution of a drug target and its interaction with drug molecules, and changes over time upon therapeutic intervention [6, 7]. The approach consists of labeling a tracer with a short-lived positron-emitting radionuclide, such as ^{11}C, ^{18}F, and ^{68}Ga, and the administration of such tracer to the subject. This is

Contemporary Accounts in Drug Discovery and Development, First Edition. Edited by Xianhai Huang, Robert G. Aslanian, and Wayne H. Tang.

followed by the acquisition of dynamic or static imaging in a PET scanner following the tracer injection, and the quantitative determination of the regional concentration of the PET tracer (in μCi/cm^3) in the organ studied by means of a suitable mathematical model [8]. Drug discovery and development for targets in the CNS have relied on target engagement for increasing confidence, guiding dose selection, and reducing risk at key stages of the discovery and development process. PET is the key enabling method that allows direct assessment of human CNS biology and pharmacology, providing information on target engagement [9]. In the early discovery stage, radiotracers help establish target (e.g. receptor, enzyme, ion channel) distribution and thus validate the feasibility of therapeutic approach in animal models. During the process of drug discovery and development, imaging of the brain with PET has become an essential tool to determine target engagement and establish a dose- or exposure-to-occupancy relationship as well as correlation with therapeutic efficacy in animal models. PET imaging facilitates translation from preclinical species into humans. The information obtained can be easily interpreted in pharmacological terms, and readily incorporated into the pharmacodynamics process to set exposure targets and make go/no-go decisions with confidence [10]. The process was illustrated by [^{11}C]MK-4232, a CGRP-R-specific PET tracer [11].

10.1.2 Muscarinic Acetylcholine Receptor 4 Positive Allosteric Modulator

Muscarinic acetylcholine receptors (mAChRs) are members of the G protein-coupled receptor super family that mediate the actions of the neurotransmitter acetylcholine (ACh) in both the central and peripheral nervous system. There are a total of five distinct muscarinic receptors (M1–M5) that have been identified in mammals. The mAChR subtypes modulate multiple functions within the CNS. Among these, M1 and M4 are mostly expressed in the CNS and are implicated in regulation of cognitive functions and thus present as potential therapeutic targets for neurodegenerative diseases [12]. In particular, M4 is implicated in the regulation of psychotic-like behaviors. Preclinical experiments identified the M4 receptor as the mAChR subtype responsible for antipsychotic-like efficacy as well as modest cognitive enhancement [13]. However, developing a highly selective cholinergic agent has been extremely challenging due to the highly conserved nature of the mAChR orthosteric binding site among the subtypes. Activation of peripheral mAChRs M2 and M3 are thought to be responsible for the adverse effects of non-selective cholinergic agents [14]. Clinical efficacy via activation of the muscarinic cholinergic system to alleviate psychosis and behavioral disturbances in AD and schizophrenia patients has been demonstrated by xanomeline, an M1/M4-preferring agonist [15–20]. However, gastrointestinal side effects led to a high drop-out rate in clinical trials, presumably due to its lack of subtype selectivity. Eventually xanomeline was discontinued [21, 22].

Over the past decade, there has been a significant effort to develop selective cholinergic agents via a new molecular mechanism, activating the receptors through topographically distinct allosteric sites. There are potential advantages for this approach. The allosteric sites are distinct from the highly conserved binding site of the endogenous neurotransmitter and allow more opportunities to further optimize subtype selectivity. This feature should translate to an improved clinical safety and tolerability profile [13, 23–30]. In addition, allosteric modulation may offer advantages such as maintaining both spatial and temporal neurotransmission through modulation of physiologically relevant receptor-mediated neural signaling pathways [31]. As a result, several selective M4 positive allosteric modulators (PAMs) have been reported, including LY-2033298 [32], VU-0152100 [33], and VU-0152099 [33]. However, none of them has reached clinical evaluation.

10.2 Discovery of a Selective PET Tracer for M4 PAM

10.2.1 Criteria for a PET Tracer

We launched a discovery program aiming to identify a suitable M4 PAM for clinical development. The lack of an available receptor-relevant functional biomarker or selective M4 PAM PET tracer presented challenges to the goal. Therefore, we also initiated a concurrent PET tracer discovery effort for the *in vivo* imaging of the M4 allosteric binding site in brain. At the time this work commenced, there were no reported successful M4 PAM PET tracers suitable for human use. In this chapter, we describe our research leading to the identification of an M4 PAM PET tracer, [^{11}C]MK-6884 [34], and its potential applications in CNS drug discovery. As mentioned previously, noninvasive imaging of M4 PAMs using PET allows quantification of the distribution, expression, and modulation of this target under physiological and pathological conditions. With a validated PET tracer for the M4 receptor, PET imaging can assess receptor occupancy (RO) of a drug candidate at the same binding sites, thereby defining its pharmacodynamic profile and identifying therapeutically relevant dose ranges.

The challenges in developing a PET tracer for imaging are different from developing a therapeutic agent. To enable PET imaging, it is critical to develop a high quality M4 PAM PET ligand that meets a specific set of criteria. The ligand should have high affinity and selectivity (no significant imaging signal generated by off-target binding) for the M4 PAM binding site, have moderate lipophilicity, be a nonsubstrate for the human P-glycoprotein transporter, and afford good blood–brain barrier penetrability. With regard to qualification, the PET tracer should provide high specific signal (total/non-specific ≥1.5:1) to allow the quantitative mapping of the target of interest. It should exhibit *in vivo* displaceable binding suitable to determine pharmacokinetics/RO relationships in preclinical species and humans. The ratio of the concentration of binding sites in a tissue (B_{max}) to the radiotracer binding affinity (K_d) derived from *in vitro* saturation binding studies is proportional to the tracer's non-displaceable binding potential (BPND). In order to provide a specific *in vivo* signal, a B_{max}/K_d ratio >1 is preferred, and a high affinity compound would be required. In order to achieve a useful specific signal, the radiotracer must have low nondisplaceable binding, which is often related to the lipophilicity of the tracer. The ideal ^{18}F- or ^{11}C-labeled M4 PAM PET must have good brain permeability ($P_{app} > 20 \times 10^{-6}$ cm/s in using a monolayer of LLC-PK1 [epithelial-like pig kidney cell line] cells transfected with human MDR1), low P-glycoprotein (P-gp) efflux (transport ratio <2 across the cell membrane monolayer), and low to moderate lipophilicity (shake-flask Log D, sFLog D < 3.5) [35, 36]. Considering the short half-lives of PET radionuclides (20 minutes for ^{11}C and 110 minutes for ^{18}F), the chemical structure of the tracer should be amendable for late-stage radiolabel incorporation using simple chemistry that can be reliably executed at clinical PET centers. Additionally, the formation of brain-penetrant radioactive metabolites of a PET tracer should be minimal because they can interfere with the detection of the M4 specific binding signal.

10.2.2 PET Feasibility Study

The first step in developing a PET tracer is to assess feasibility. For any CNS target of unknown target concentration B_{max} and distribution in brain, one should first identify a tool compound suitable for ^{3}H, ^{125}I or ^{18}F labeling to enable *in vitro* binding studies in brain tissues for the determination of B_{max}. Compound **1** was discovered from a high-throughput screening (HTS) campaign for the program [37]. It was further evolved with iterative library synthesis that identified potency-enhancing heterocycles at C-2 and *N*-alkyl pyrazole substitutions to afford compound **2**, a potent human M4 PAM. Additional characterization of this compound revealed reasonable physiochemical properties that can serve as a starting point to assess PET tracer feasibility (Figure 10.1).

Our efforts began with [^{3}H]**2** to map M4 receptor distribution and binding specificity in mouse brains by autoradiography (ARG). The ARG images in brain sections from wild type (WT) mice indicated that M4 PAM has the highest binding density in the striatum region (Figure 10.2a). The ARG images also show that the M4 PAM binding is dependent on the presence of the orthosteric agonist carbachol (Figure 10.2a vs. b). Negligible displaceable binding was observed in the brain sections of M4 receptor knockout (KO) mice, demonstrating specific binding of [^{3}H]**2** to M4 receptors (Figure 10.2d) [38]. Similar results were obtained with monkey and human brain tissues.

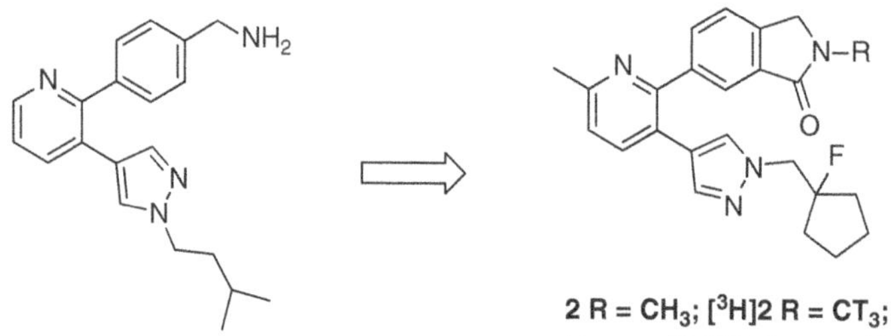

1

Human M4 FLIPR IP = 248 nM
Rat M4 FLIPR IP = 1233 nM

2 R = CH_3; [^{3}H]2 R = CT_3; [^{11}C]1 R = $^{11}CH_3$
Human FLIPR IP = 16 nM
K_i = 3.1 nM
sFLog D = 2.5
P-gp (h/r) = 1.2/7.2
Papp = 29.5 × 10^{-6} cm/s

Figure 10.1 Profiles of compounds **1** and **2**.

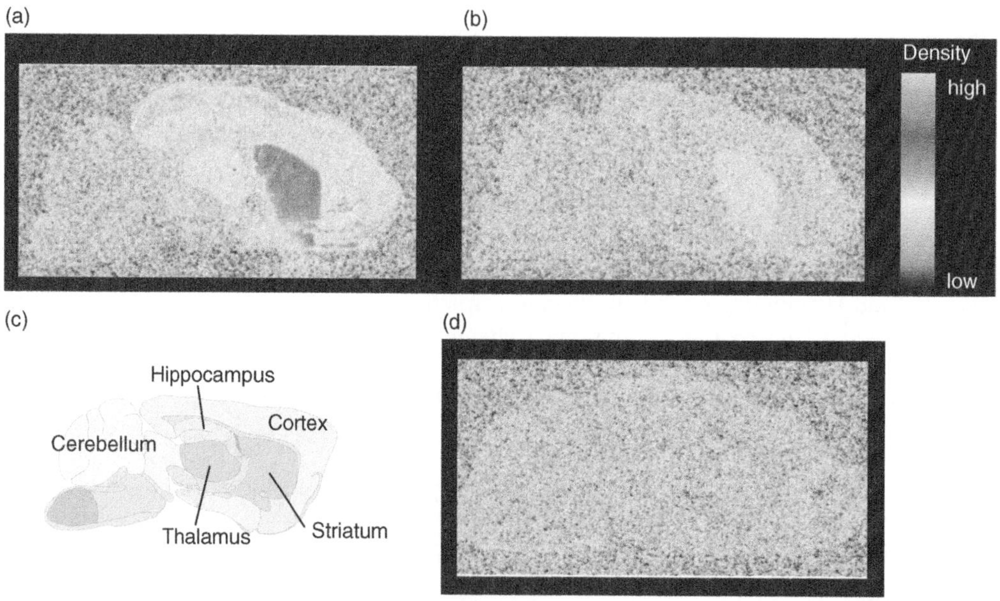

Figure 10.2 *In vitro* autoradiography with [^{3}H]**2** in mouse brain slices. (a)Total binding of [^{3}H]**2** in wild type mouse brain in the presence of 10 μM carbachol. (b) Total binding of [^{3}H]**2** in wild type mouse brain in the absence of 10 μM carbachol. (c) Mouse brain anatomy. (d) Total binding of [^{3}H]**2** in knock out mouse brain in the presence of 10 μM carbachol.

The striatum homogenates of monkey and human were then used to determine B_{max} and the regional K_d by *in vitro* saturation binding studies. From this study, the B_{max} was measured to be 7.7 nM in rhesus and 7.2 nM in human, a modest to low target density for the M4 receptor [38]. When receptor B_{max} is low (<10 nM), PET tracer discovery becomes increasingly challenging because K_d has to be even lower to meet the $B_{max}/K_d > 1$ requirement. A successful PET tracer to image M4 receptors would thus require the tracer having subnanomolar binding affinity. Compound **2** has similar K_d in both monkey (13 nM) and human (18 nM) striatum homogenates.

10.2.3 PET Specific Signal Is Driven by an Increase in Binding Affinity

For a given target (fixed B_{max}), compounds with higher binding affinity (smaller K_d) will increase the ratio of B_{max}/K_d, which is a good *in vitro* estimate of *in vivo* binding potential (BP_{ND}). Here we used binding affinity K_i as a surrogate to predict K_d. To improve binding affinity of M4 PAMs, we found that the alkyl tail group off the pyrazole substituent can influence binding affinity. As we often observed poor correlation between the functional assay and competition binding assay, several earlier compounds with potent functional activity in the FLIPR assay were evaluated in a tissue-based competitive binding assay using tool compound [^{3}H]**2** [38]. Initial studies showed that heterocycle replacements of the lactam moiety at the 2-position of the pyridine core can improve K_i (compound **2**, K_i = 3.1 nM), as exemplified by compound **3** (K_i = 0.6 nM). However, we also found that the improved binding affinity of **3** comes with increased lipophilicity, as measured by HPLC Log D. HPLC Log D is a high throughput measurement of polarity of the compound. This value corrolated well with sFLog D in this chemical series [38]. In a parallel structure–activity relationship (SAR) study, we found that a cyclopentyl methyl group off the pyrazole ring exhibited an improved K_i compared with other types of alkyl groups, as in compound **2** (cf. **6**, **7**, **8**). Further study identified compound **5** (Figure 10.3) with a methyl group replacing a fluorine atom at the tertiary position on the cyclopentyl ring, which afforded moderately improved binding affinity (K_i = 1.0 nM). Both **2** and **5** possess reasonable physicochemical properties and were radiolabeled with ^{11}C for the evaluation of brain distribution and specific binding to the M4 PAM binding site in rhesus monkeys.

The *in vivo* study of [^{11}C]**2** in monkey showed that the compound was brain penetrant but did not show any differentiated PET signals between the striatum and cerebellum curves, resulting in

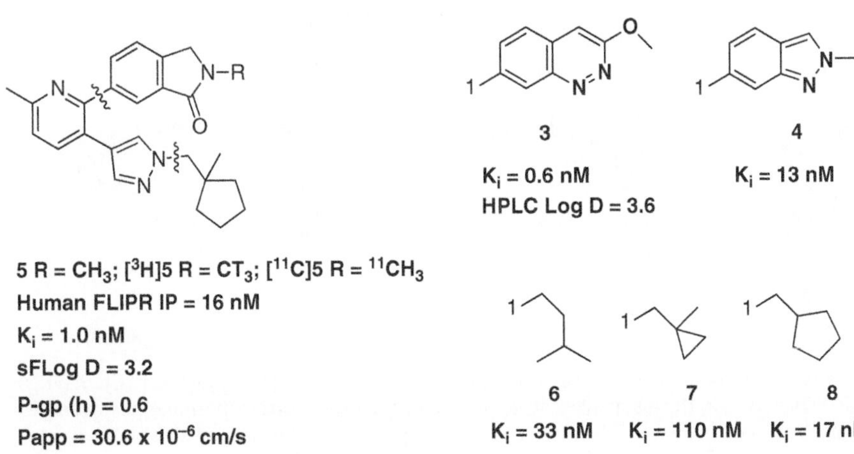

Figure 10.3 Profile of compound **5** and SAR on the tail piece and northeastern R2.

BP_{ND} ~ 0 for the striatum region (Figure 10.4a). On the other hand, [^{11}C]**5** provided a striatum BP_{ND} value of 0.29. In addition, the [^{11}C]**5** baseline PET image (0–90 minutes) in monkey revealed an improved brain uptake compared with [^{11}C]**2** in striatum (Figure 10.4b). These results warranted additional improvement of binding affinity as a way to increase *in vivo* PET specific signal.

We investigated additional vectors to further improve binding affinity on the molecule. A separate SAR study of the core revealed that 4- and 5-substitutions on the central pyridine were detrimental to binding affinity (Table 10.1, compound **11** and **12**), while 6-substituted and 5,6-bis substituted pyridyl were well tolerated. In particular, compound **10,** with a cyano group at the 6-position, and compound **14**, a 5,6-fused cyclopentanone, displayed subnanomolar binding affinity. However, further profiling of compound **14** revealed it to be a P-gp substrate (P-gp ratio = 9.4), which precluded it as a potential candidate for *in vivo* evaluations.

Compound **10** displayed acceptable overall physical properties in addition to its subnanomolar binding affinity (Figure 10.4). It has high permeability (P_{app} = 28.8 × 10^{-6} cm/s) and low P-gp susceptibility in human MDR-1 monolayer (P-gp ratio = 0.7). The lipophilicity of compound **10** was also in line with a CNS PET tracer (sFLog D = 2.7, Figure 10.5). It was radiolabeled with ^{11}C and characterized in preclinical species to project its utility in clinical applications. We were encouraged to find that a baseline PET scan with [^{11}C]**10** in rhesus monkey showed rapid, high tracer distribution and the expected regional pattern of tracer uptake in the brain, consistent with the *in vitro* ARG results of [^{3}H]**2**. The highest uptake was in the striatum. A baseline BP_{ND} of 0.7 was measured in the striatum in this monkey, adequate for the quantification of M4 RO in monkeys (Figure 10.4c). Blockade with a M4 PAM therapeutic compound resulted in substantial reduction of [^{11}C]**10** uptake in striatum [38]. To optimize the clinical utility of a PET tracer, it is preferred to have a tracer with BP_{ND} greater or close to 1. We then devoted efforts to further improve the *in vivo* BP_{ND} by optimizing the *in vitro* properties of PET tracer candidates. We hypothesized that if we

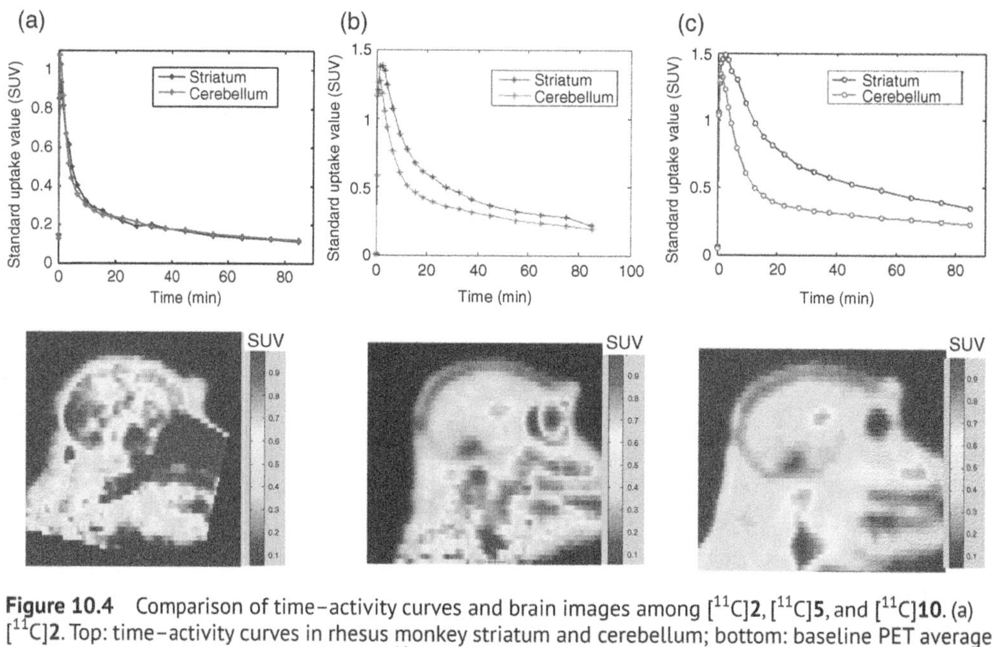

Figure 10.4 Comparison of time–activity curves and brain images among [^{11}C]**2**, [^{11}C]**5**, and [^{11}C]**10**. (a) [^{11}C]**2**. Top: time–activity curves in rhesus monkey striatum and cerebellum; bottom: baseline PET average image from first 3 minutes in monkey. (b) [^{11}C]**5**. Top: time–activity curves in rhesus monkey striatum and cerebellum; bottom: baseline PET image (0–90 minutes) in monkey. (c) [^{11}C]**10**. Top: time–activity curves in rhesus monkey striatum and cerebellum; bottom: baseline PET image (0–90 minutes) in monkey.

Table 10.1 SAR on core substitution.

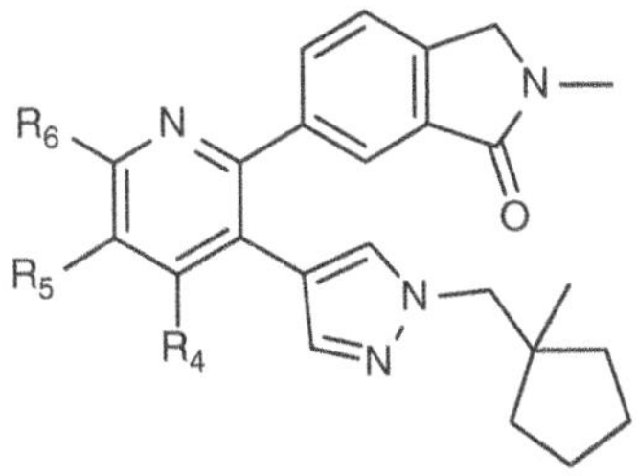

Compound	R_6	R_5	R_4	K_i (nM)
9	H	H	H	10
10	CN	H	H	0.19
11	H	F	H	7.5
12	H	H	CN	22
13	Me	Me	H	1.0
14	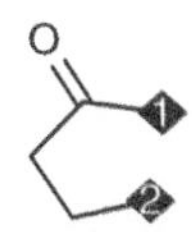		H	0.16

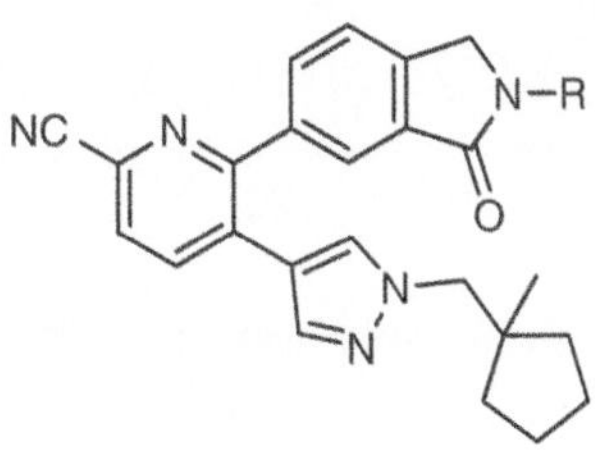

10 R = CH_3; [^{11}C]10 R = $^{11}CH_3$
Human FLIPR IP = 20.4 nM
K_i = 0.19 nM
sFLog D = 2.7
P-gp (h) = 0.7
Papp = 28.8 × 10^{-6} cm/s

Figure 10.5 Chemical structure and *in vitro* properties of compound **10**.

could further improve binding affinity while maintaining favorable physical properties, we would likely achieve this goal.

10.2.4 The Implication of Lipophilicity and Free Fraction on *in vivo* BP_{ND}

We were encouraged by the promising *in vivo* data obtained for [^{11}C]**10**. The cyano group at the 6-position of the pyridine ring not only boosted the binding affinity but also offered a handle for ^{11}C radiolabeling. In this case, one could introduce [^{11}C]CN using [^{11}C]HCN generated by [^{11}C]CO_2

Table 10.2 Variation of R_2 with fixed R_3 and R_6.

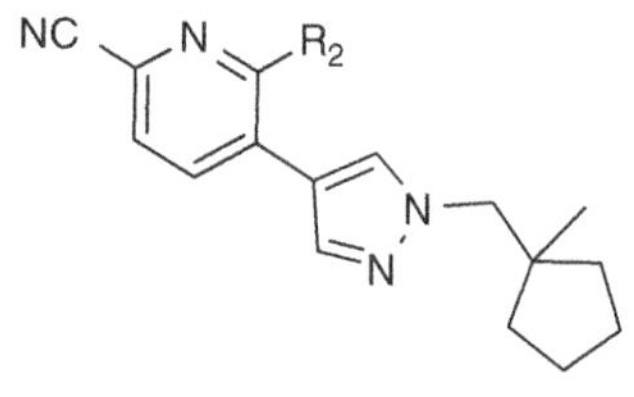

Compound	R_2	K_i (nm)	P-gp ratio (h)	HPLC Log D
15	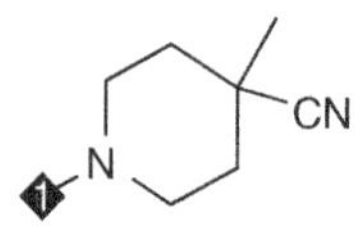	1.9	ND	4.3
16	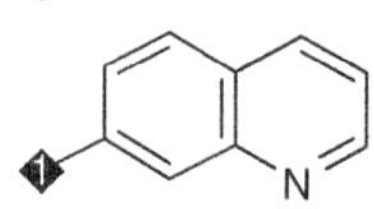	0.40	0.6	3.6
17	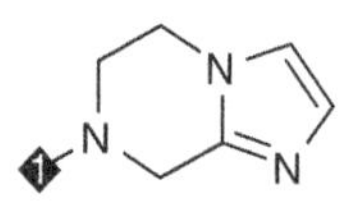	0.20	0.6	3.0
18	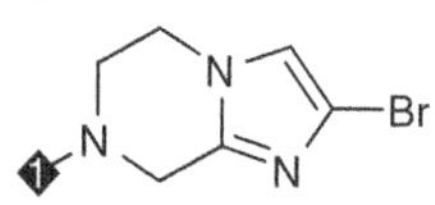	0.12	0.8	3.8
19	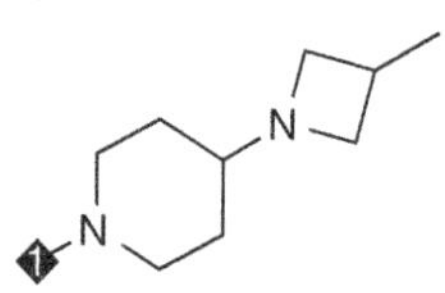	0.24	3.4	3.3
20	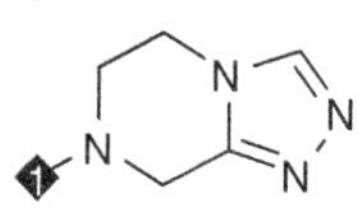	0.62	8.4	2.3

ND, not determined.

via mixing of [^{11}C]CH_4 with anhydrous ammonia at high temperature [39]. This alternative radiolabeling handle opened up additional opportunities to further drive down binding affinity. We proceeded with invariant R_6 and R_3 groups and evaluated a wider variety of R_2 groups. Table 10.2 summarizes the data from several selected monocyclic and bicyclic R_2. Promisingly, several 6,6 and 6,5 fused bicyclic ring systems as well as one 6,4-bicyclic ring system exhibited subnanomolar binding affinity. However, when the P-gp ratio and HPLC LogD of these compounds were considered, only compound **17** was selected for *in vivo* evaluation as it showed no human P-gp liability and is under our Log D cutoff (3.5).

Besides the R_2 group, we looked for additional modifications on the molecule for enhancing binding affinity and maintaining the required physicochemical properties. For example, we found that a methyl substitution on the R_5 position on the pyridine core provided compound **21** with excellent binding affinity and acceptable physicochemical properties. Other core modifications were less desirable. For example, the 5-fluoro-6-cyano core showed stability issues, while the 5-chloro-6-cyano or 5-methyl-6-fluoro core is too lipophilic (Figure 10.6).

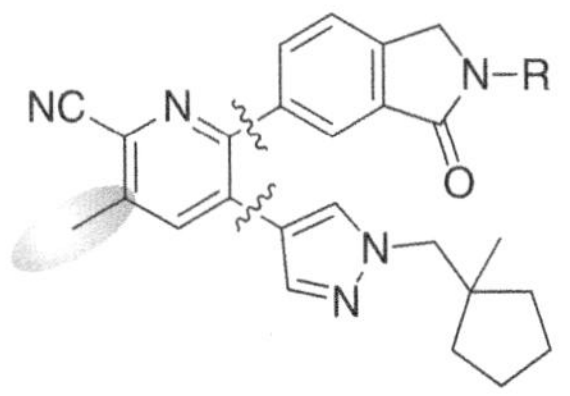

21 R = CH_3; [^{11}C]21 R = $^{11}CH_3$
K_i = 0.26 nM
sFLog D = 3.1
P-gp (h) = 1.0
Papp = 25 × 10^{-6} cm/s

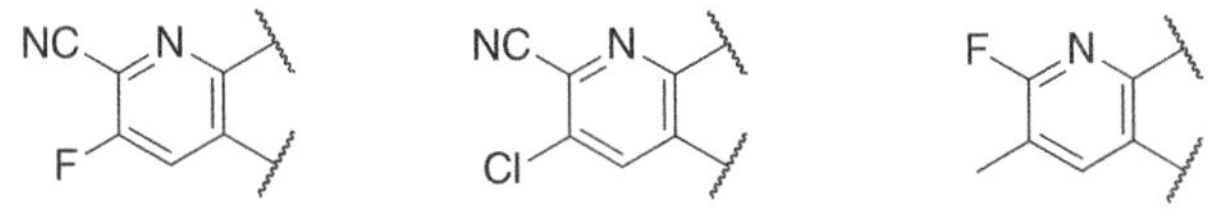

stability issue | high HPLC LogD | high HPLC LogD

Figure 10.6 Modifications on the core.

Substitution on the 5-position of the pyrazole was also evaluated. A fluorine atom at this position provided a compound with exquisite binding affinity as demonstrated by compound **22**. Substitutions such as Cl or CN at the same position were inferior to fluorine with regard to the binding affinity. Chlorine and cyano groups also increased the lipophilicity of the molecule beyond acceptable HPLC Log D values.

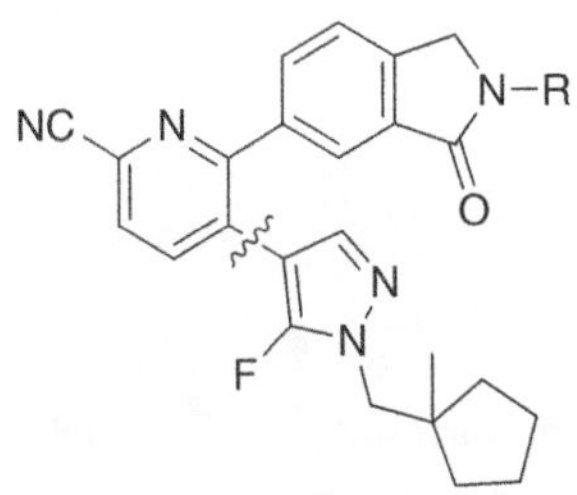

22 R = CH_3; [^{11}C]22 R = $^{11}CH_3$
K_i = 0.08 nM
sFLog D = 3.2
P-gp (h) = 0.7
Papp = 27 × 10^{-6} cm/s

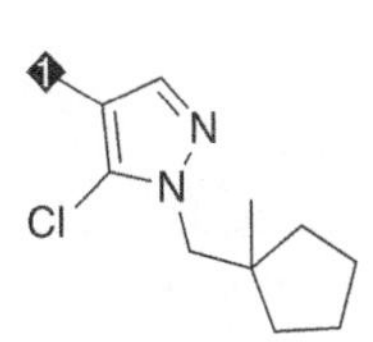

23
K_i = 1.7 nM
HPLC LogD = 3.7

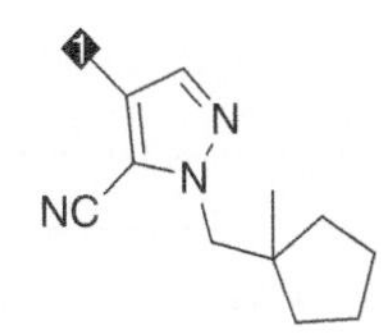

24
K_i = 2.0 nM
HPLC LogD = 3.4

Along with compound **17**, compounds **21** and **22** were radiolabeled with ^{11}C and studied *in vivo* in rhesus monkey brains. To our surprise, despite similar or better binding affinity, none of the new analogs improved *in vivo* BP_{ND} compared with [^{11}C]**10** (Table 10.3).

A comparison of time–activity curves of [^{11}C]**10** and [^{11}C]**22** is shown in Figure 10.7. The brain uptake in striatum peaked with a higher standard uptake value (SUV) and slightly different pattern for [^{11}C]**22** compared with [^{11}C]**10.** However, [^{11}C]**22** also displayed higher radioactivity concentration measured by uptake in the cerebellum, which we used as a reference region as it has few or no M4 receptors for imaging quantification. The positive effect of a higher under-the-curve radioactivity in striatum by [^{11}C]**22** was canceled out by a higher non-specific binding and resulted in a lower overall *in vivo* BP_{ND} than [^{11}C]**10.**

Exactly what causes the improved profile of [^{11}C]**10** as compared with other tracers ([^{11}C]**17**, [^{11}C]**21**, [^{11}C]**22**), which seemingly exhibited similar physicochemical properties and similar-or-better binding affinity? A close examination of these molecules revealed that the difference may lie

Table 10.3 BP_{ND} of [^{11}C]**17**, [^{11}C]**21** and [^{11}C]**22** compared with [^{11}C]**10**.

Compound	10	17	21	22
BP_{ND}	0.7	0.2	0.37	0.42

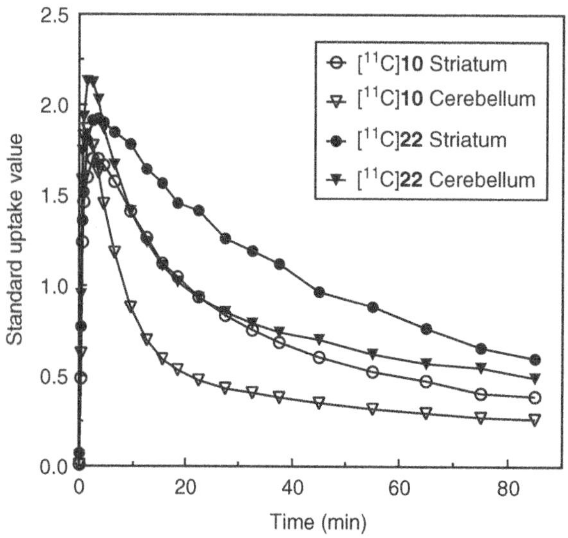

Figure 10.7 Time–activity curves for [^{11}C]**10** and [^{11}C]**22** in monkey brain.

Table 10.4 Key parameters for tracer candidates.

Compound	BP_{ND}	K_i (nM)	sFLog D	f_u% plasma (monkey)	f_u% brain (rat)
2	0	3.1	2.5	7.2	16.6
5	0.2	1.0	2.8	9.6	4.1
10	0.70	0.19	2.7	17	4.3
17	0.20	0.20	2.7	5.1	3.3
21	0.37	0.26	3.1	11.4	3.2
22	0.42	0.08	3.2	8.4	2.7

f_u, free fraction.

in the lipophilicity and free fractions (Table 10.4. Each of these compounds were assessed for lipophilicity by measuring the sFLog D. All of the tracers met the cutoff criteria of <3.5, with compounds **10** and **17** under 3.0, and compounds **21** and **22** between 3.0 and 3.5. Furthermore, compound **10** exhibited the highest unbound fraction measured in monkey plasma (17%) compared with 11.4% for compound **21** and single digit percentages for compounds **17** and **22**. The rat brain free fraction values using rat brain homogenate were relatively close for all of these tracers with the exception of compound **2**. But within the rest of the tracer candidates, compound **10** was still moderately higher in rat brain free fraction. The *in vivo* results of these tracer candidates indicated that the tracer's physical properties including free fraction and lipophilicity played important roles in the *in vivo* specific signal.

10.2.5 Fluorine-18 Labeling Opportunity

Compared with ^{11}C, ^{18}F has a longer half-life of 109.8 minutes. The time between administration of the PET tracer and scanning is an important parameter for a proper transfer of activity from blood to tissues, and the longer half-life of ^{18}F may offer an advantage over ^{11}C on metabolite analysis. In addition, the radiochemistry to incorporate ^{18}F is different from ^{11}C and can expand the scope of compounds that can be radiolabeled [40]. Furthermore, fluorine is also a favorable atom in drug development due to its physical properties including small van der Waals radius (1.47 Å), high electronegativity and ability to form a strong bond with carbon (C—F energy bond of 112 kcal/mol) which is thermally stable and resistant to oxidation [40]. ^{18}F tracers may also offer different *in vivo* profiles compared with their ^{11}C counterparts. We have recently described the development of ^{18}F PET tracers [^{18}F]MK-6240 and [^{18}F]MK-8553 [41, 42].

We have also explored the possibility of a ^{18}F-labeled PET tracer for M4 PAM. This required us to find a handle to introduce fluorine at the last synthetic step in a way that did not alter the affinity of the molecule. In this chemical series, we examined the opportunity to install a fluorine atom at various positions which allow late-stage labeling. As some of the examples shown in Table 10.5, we were able to obtain compounds with high *in vitro* binding affinity that are amenable to ^{18}F labeling.

Table 10.5 Examples of fluorine-containing PET tracer candidates.

Compound	Structure	K_i (nM)	HPLC Log D	P-gp	P_{app}	Free fraction in monkey (%)
23	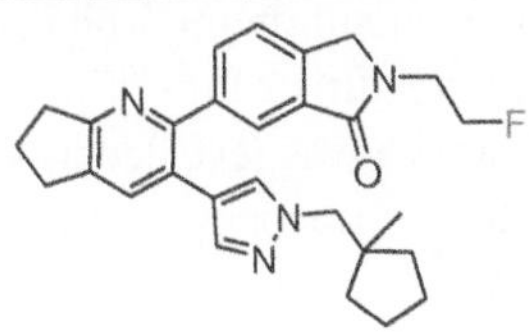	0.31	3.2	1.0	26	4.9
24	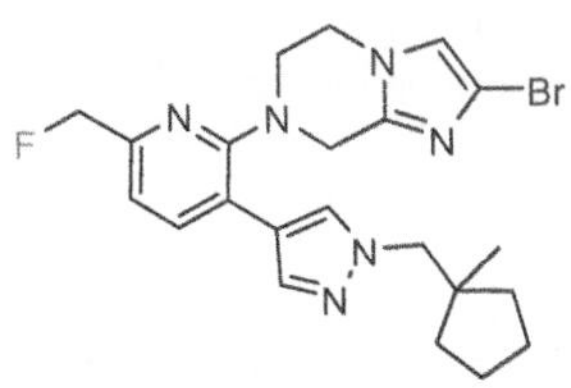	0.40	4.1	0.5	14	ND
25	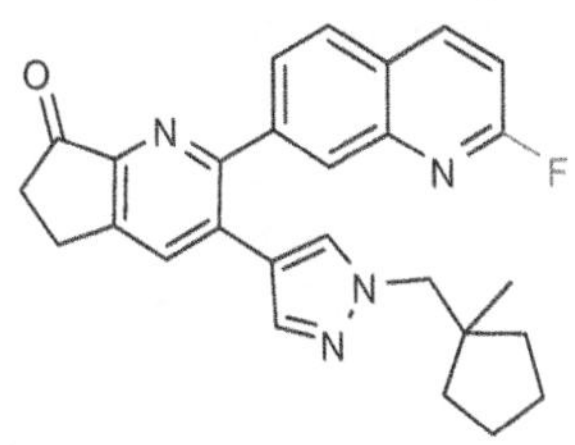	0.13	3.2	0.4	21	1.6
26	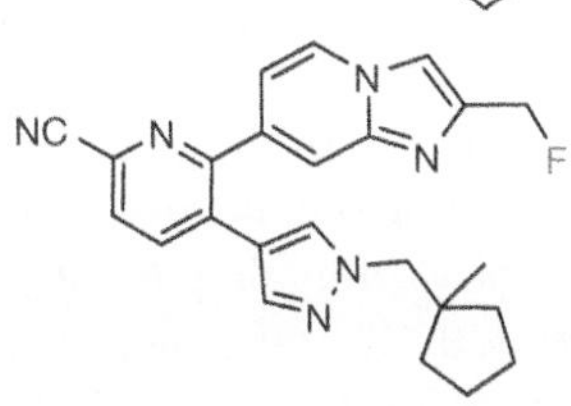	0.67	3.0	0.6	32	6.4

ND, not determined.

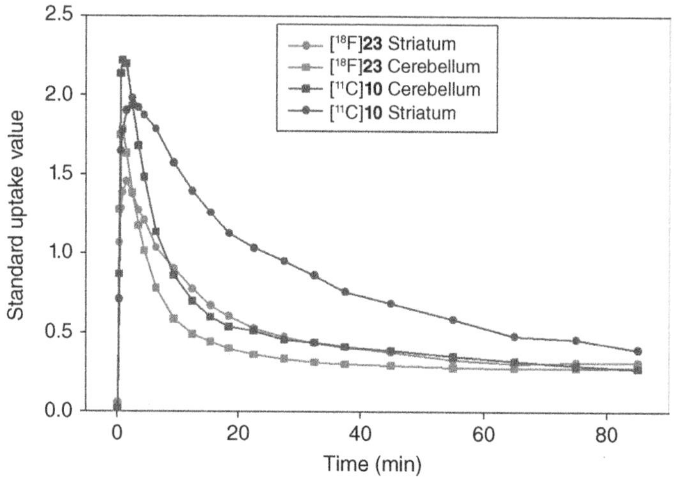

Figure 10.8 Time–activity curves for [^{11}C]**10** and [^{18}F]**23** in monkey brain.

However, they tend to have higher lipophilicity than desired (HPLC Log D ≥3.2), suggesting greater non-specific binding *in vivo* (compounds **23–25** greater or equal to 3.2). For compound **26**, which showed a HPLC Log D of 3.0, the measured free fraction in monkey was 6.4%. Two of the compounds which met all the PET tracer criteria were selected to be radiolabeled with ^{18}F: compounds **23** and **26**. PET baseline studies with [^{18}F]**23** and [^{18}F]**26** in rhesus monkeys showed that the molecules were distributed as expected as [^{11}C]**10** but showed suboptimal pharmacokinetic properties and lower specific signal in striatum. The time–activity curves for [^{18}F]**23** and [^{11}C]**10** are shown in Figure 10.8. Some defluorination was observed, which was revealed as skull uptake during the PET scan period. Overall, for M4 PAM imaging, the ^{18}F tracers tested in comparison with [^{11}C]**10** displayed less desirable physicochemical (e.g. lower free fraction, and higher HPLC Log D) properties and were less likely to be successful PET tracers *in vivo*.

10.3 A PET Tracer that Images M4 in Rat

Small animal PET imaging is an important tool for preclinical drug research as it facilitates the study of rodent models of a variety of human brain diseases. Our research project utilized the rat amphetamine-induced hyperactivity model as an *in vivo* tool to evaluate M4 PAM therapeutic candidates. A rat M4 PET tracer provides information of treatment effects longitudinally and allows us to measure physiological processes, biochemical pathways and neurotransmitters *in vivo* in preclinical rat model studies. It greatly facilitates the investigation of chronic diseases and the evaluation of neuropharmacological interventions.

To identify a compound suitable for the imaging of M4 receptors in rat, we first evaluated our existing PET tracer candidates. We noticed that for this chemical series, there is a greater P-gp susceptibility in rat than in human and monkey. For example, tracer **10** showed adequate *in vivo* BP_{ND} in monkey brain, but its rat P-gp efflux ratio is 4.4, suggesting that it would likely have low level of brain penetration in rats. It was also observed that *in vitro* binding affinity right-shifts between species; tracer **10** showed a K_i = 0.19 nM in monkey vs. 0.74 in rat. Among the tracer candidates evaluated in rat, compound **22** displayed the lowest rat P-gp efflux ratio of 2.0. To our surprise, it showed low brain penetration in wild-type rats despite high binding affinity and lower

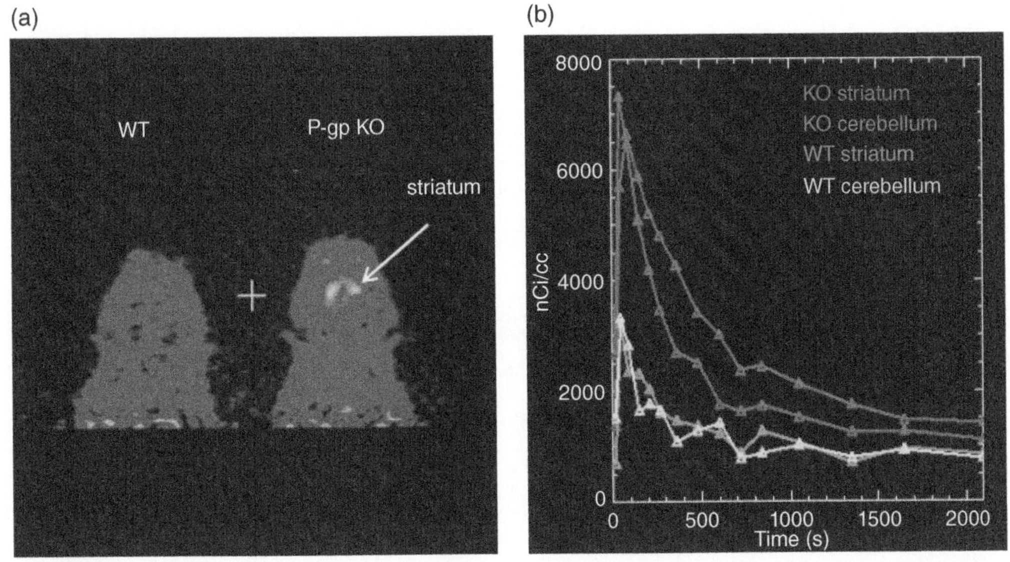

Figure 10.9 (a) Snapshot of [^{11}C]**22** brain imaging in WT and P-gp KO rat. (b) [^{11}C]**22** time–activity curves in WT and P-gp KO rat.

Table 10.6 P-gp ratios in human, monkey, and rat at 1 and 0.1 μM.

Compound	P-gp, h 1 μM /0.1 μM	P-gp, m 1 μM /0.1 μM	P-gp, r 1 μM /0.1 μM
10	1.7/1.8	1.9/2.3	4.4/6.2
22	1.1/1.3	1.3/1.7	2.0/4.2

h, human; m, monkey; r, rat.

P-gp liability. In the MDR1 (P-gp) KO rat, radioactivity of [^{11}C]**22** was observed in the brain (Figure 10.9). This prompted us to further investigate the P-gp liability. The current P-gp ratio was measured at 1 μM compound concentration. A more relevant concentration in the context of a PET tracer under physiological condition would be 0.1 μM. Further investigation revealed that compounds in this chemical series were likely both P-gp substrates and P-gp inhibitors in rat which was consistent with the fact that at lower dose, the P-gp ratio was higher. Table 10.6 summarizes the P-gp ratio measured at both 1 and 0.1 μM for compounds **10** and **22**. In human and monkey, the ratio measured in the two concentrations varied slightly. However, in rat, the P-gp ratio went up significantly under lower compound concentration. This explained the lack of brain uptake for tracer **22** in rat PET baseline studies.

Tracer **22** has the lowest rat P-gp efflux ratio among the viable candidates that we evaluated. To improve the level of brain penetration, tracer **22** was further evaluated in P-gp knock out rats. In this study, moderate specific signal was obtained in rat brain (Figure 10.9), suggesting that tracer **22** may be useful in evaluating RO in P-gp KO rat studies.

The P-gp issue posted a significant challenge in developing a PET tracer for rat in this chemical series. In an attempt to reduce P-gp substrate activity, we looked into possible correlations of various physicochemical properties. However, no apparent relationships could be identified in PSA and HPLC Log D with P-gp, as shown by the scatter plots in Figure 10.10. Other parameters, such

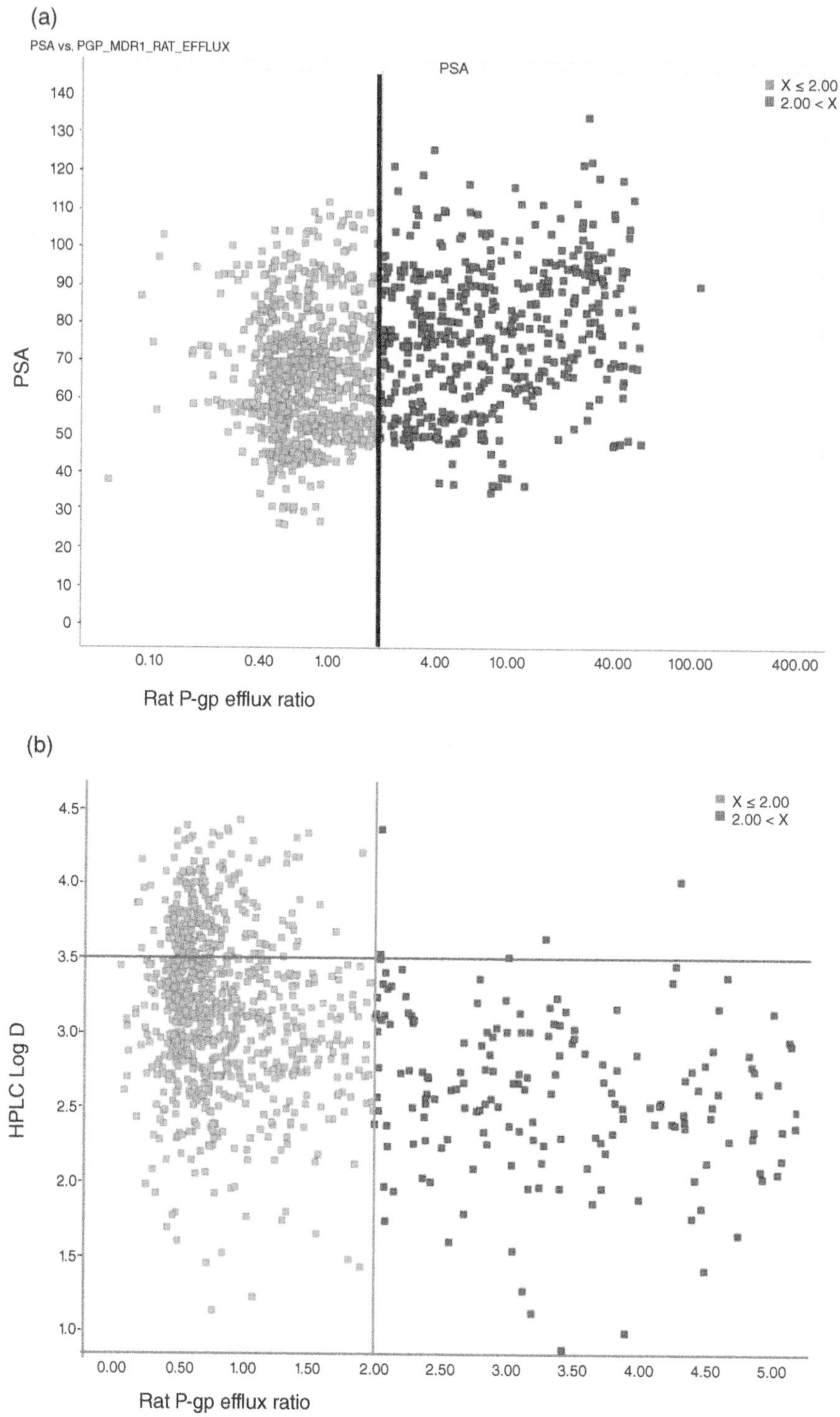

Figure 10.10 (a) Scatter plot of PSA vs. rat P-gp efflux ratio. Vertical line: X = 2.00. (b) Scatter plot of HPLC Log D vs. rat P-gp. Vertical line: X = 2.00; horizontal line: Y = 3.5.

as pKa, also did not reveal any additional information. We resorted to an empirical approach to try to solve this problem. Some SAR trends that we observed including adding fluorine to aromatic rings seemed to be beneficial in dialing out the rat P-gp issue. An example is shown by compound **27**.

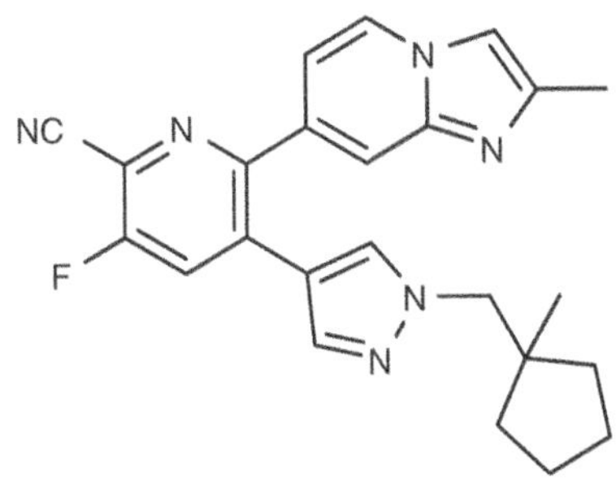

K_i (r) = 0.7 nM

HPLC Log D = 3.3

P-gp (r) = 1.2

Papp(r) = 23 × 10^{-6} cm/s

Compound **27**

A key parameter that we learned from the evaluation of tracer **10** is plasma free fraction (f_u), which is higher for tracer **10** than for some other candidates such as **22**. We believed the same higher free fraction would be needed for rat brain tissue. Therefore, we looked for structural features that increased plasma free fractions in rat. From the scatter plot of rat plasma protein binding (PPB) vs. HPLC Log D, we determined an empirical rule that HPLC Log D should be less than 3 to afford compounds with rat $f_u > 10\%$ (Figure 10.11). Combining with the observation that a lower Log D would increase the likelihood of P-gp liability (Figure 10.10b), the lipophilicity requirement would be even more difficult to achieve.

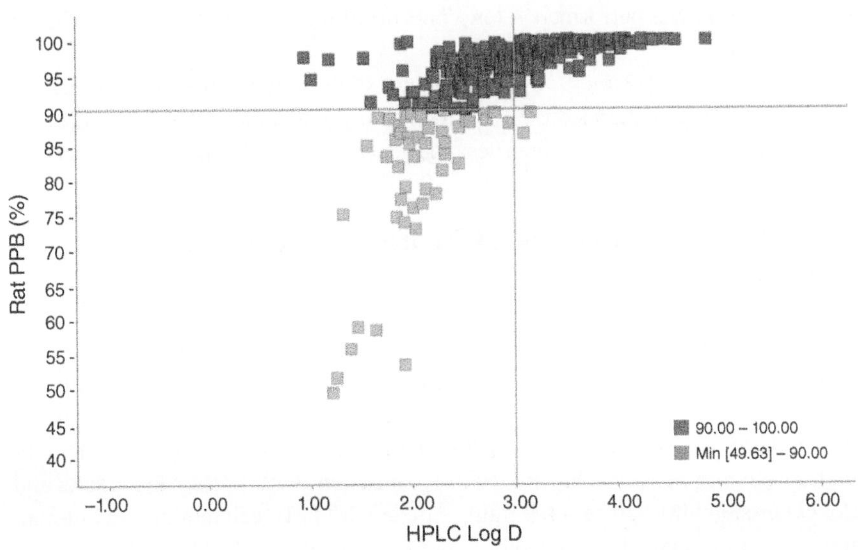

Figure 10.11 Scatter plot of rat PPB vs. HPLC Log D. Vertical line: X = 3.00; horizontal line: Y = 90.

Lipophilicity as measured by sFLog D gave good predictability for the free fraction in this series. Lower lipophilicity also predicted lower non-specific binding. For example, we noticed that certain heterobicycles on the 2 position of the pyridine ring can increase plasma free fraction, exemplified by compounds **28, 29,** and **30**. However, these compounds also have high P-gp efflux ratios, illustrating the challenge of finding the right balance of lipophilicity.

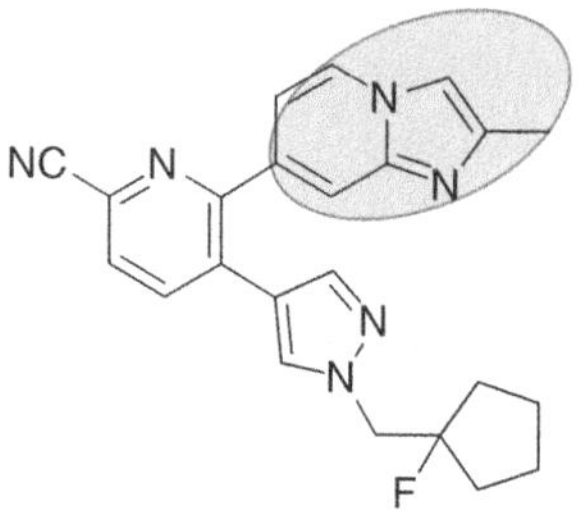

28

K_i = 0.8 nM

FF%,r = 11

HPLC Log D = 2.4

Pgp (h/r) = 0.9/ 2.3

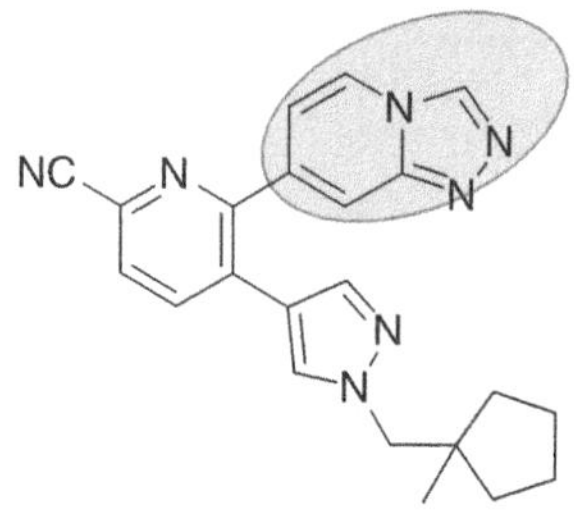

29

K_i = 0.8 nM

FF%, m, r = 36, 28

HPLC Log D = 2.2

Pgp (h/r) = 4.2/20.8

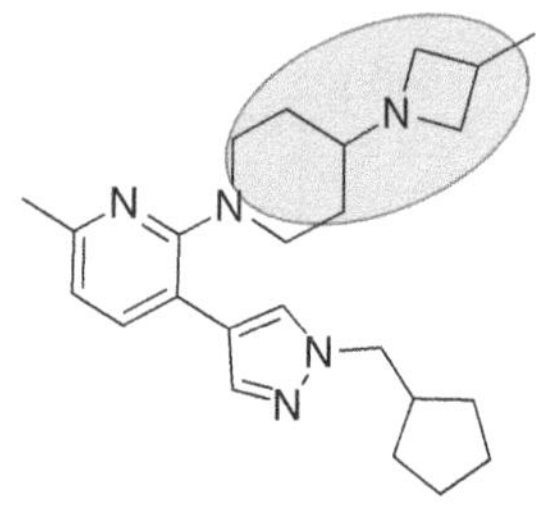

30

K_i = 2.8 nM

FF%,r = 10.5

HPLC Log D = 2.8

Pgp (h/r) = 2.7/12.8

Another SAR effort is shown in Figure 10.12a. Starting with compound **31**, where the aza-lactam increased the free fraction in rat plasma to 46% but also contributed to relatively poor binding affinity (K_i ~ 9 nM) and P-gp liability, we sought to work on other parts of the molecule to balance these properties. We knew that fluorine on the pyrazole as well as a tertiary methyl group on the cyclopentyl tail could improve the binding affinity, so we synthesized compound **32**. These two modifications improved the binding affinity significantly while lowering the P-gp liability. Another modification was to elongate the alkyl chain off the lactam nitrogen to a cyclopropyl methyl, which improved the binding affinity further while maintaining high free fraction in rat plasma and brain (**33**). Most importantly, **33** showed a BA/AB ratio of 1.8, which was indicative of a non-P-gp substrate when the P-gp evaluation was performed at low concentration. The radiolabeling handle for **33** would be the cyano group.

Baseline studies with tracers [^{11}C]**32** and [^{11}C]**33** candidates showed that neither compound is a viable PET tracer. This series demonstrates the difficulty in identifying an M4 PAM PET tracer for rat that balances potency, physicochemical properties, and especially P-gp liability (Figure 10.12).

10.4 Characterization of [^{11}C]10 as a PET Tracer Preclinical Candidate for Human Use

As further optimization of the [^{11}C]**10** *in vivo* specific signal was unsuccessful, we then focused our resources to qualify [^{11}C]**10** as a PET tracer preclinical candidate (PCC). Our goal was to qualify [^{11}C]**10** for determining target engagement of M4 PAMs and measuring changes in cholinergic tone in the striatum. *in vitro* ARG studies with [^{3}H]**10** in monkey and human brain slices demonstrated that the highest binding density is in the striatum, consistent with [^{3}H]**2** ARG results and known M4 receptor expression [43]. Saturation binding studies with [^{3}H]**10** in monkey and human brain striatum homogenates indicated the receptor concentration (B_{max}) of M4 is similar across species, suggesting [^{11}C]**10** would provide a signal of similar magnitude in monkey and human

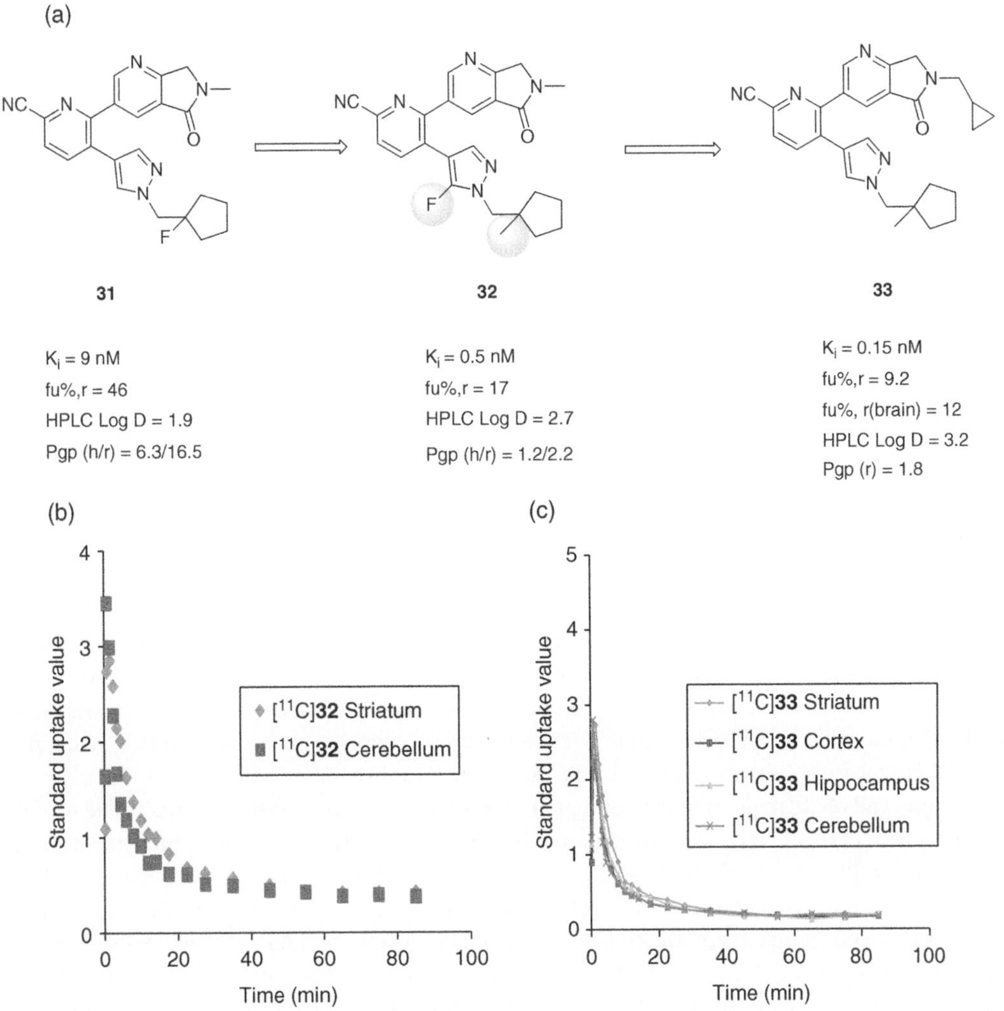

Figure 10.12 (a) Structures and *in vitro* properties of **31, 32,** and **33.** (b) PET baseline time-activity curves of [^{11}C]**32** in P-gp KO rat brain. (c) PET baseline time-activity curves of [^{11}C]**33** in P-gp KO rat brain.

(Table 10.7). Tracer **10** showed weak binding ($IC_{50} > 1\,\mu M$) to selected ion channels, mAChR subtypes, and the standard survey of receptors and enzymes. Therefore off-target binding is not expected to be a concern for the safety or imaging characteristics of [^{11}C]**10** [38].

Baseline PET scans with [^{11}C]**10** in a translational preclinical rhesus monkey model demonstrated rapid, high tracer distribution and the expected regional pattern of tracer uptake in the brain, consistent with the known distribution of M4 receptors (Figure 10.13a, left panel). M4 receptor binding was determined using a reference-tissue model and expressed as BP_{ND}. The highest uptake was in the striatum, consistent with the *in vitro* ARG results. An average baseline BP_{ND} of 0.83 ± 0.17 ($n = 5$ rhesus monkeys) was measured in the striatum, which was determined to be adequate for quantification of M4 RO. Repeat baseline PET studies were conducted in five different monkeys. Comparison of BP_{ND} determinations in five different monkeys resulted in an average test–retest (T-RT) variability of 8% (Table 10.8). Less than 10% T-RT variability is necessary and ideal for a PET tracer to measure occupancy in clinical applications.

Table 10.7 B_{max} and K_d values for [^{3}H]**10** in rhesus monkey and human striatum.

	K_d (nM)[a]	B_{max} (nM)[a]	B_{max}/K_d
Rhesus monkey striatum	0.9 ± 0.2	13.1 ± 1.8	15
Human striatum	1.2 ± 0.1	7.0 ± 0.9	6

B_{max} has been corrected with the assay dilution factor to express B_{max} relative to wet tissue weight.
[a] Wet tissue weight; mean ± SD; *n* = 3 for monkey and human.

Blockade of M4 receptors with the M4 PAM MK-4710 resulted in substantial reduction of [^{11}C]**10** uptake in striatum (Figure 10.13a, right panel) [44]. *in vivo* occupancy studies in rhesus monkey demonstrated the utility of [^{11}C]**10** to determine plasma concentration vs. RO relationships for therapeutic lead candidate MK-4710. In addition, a "chase" study was performed in which an intravenous (IV) bolus of compound **34** (1 mg/kg) was dosed 30 minutes after administration of [^{11}C]**10**. Time–activity curves show that there was a faster clearance of [^{11}C]**10** from striatum after drug injection, while there is no change in the trajectory of the cerebellum (Figure 10.13b, c). The distinct increase in tracer clearance in the striatum indicates reversible binding of [^{11}C]**10** to the M4 receptors.

Additional studies were performed, including whole-body distribution and radiation dosimetry at tracer level in monkeys to estimate human radiation dosimetry. A seven-day repeat dose IV toxicity study in rats was also necessary to qualify [^{11}C]**10** to be a PET tracer PCC. *in vivo* distribution studies conducted at tracer level with [^{11}C]**10** in rhesus monkeys (2 male and 1 female) using the PET/CT Discovery ST scanner showed that the tracer was distributed into brain, liver, kidneys, testes, lung, gall bladder, heart and a gut region representing the gastrointestinal tract. The overall effective dose (ED) of [^{11}C]**10** was estimated to be 3.8 ± 0.2 μSv/MBq, which is in the typical range for ^{11}C-labeled PET tracers (3–16 μSv/MBq). Organs found to have a high absorbed radiation dose included gallbladder (0.034 ± 0.020 mSv/MBq), small intestine (0.016 ± 0.016 mSv/MBq), and liver (0.014 ± 0.007 mSv/MBq). Using the non-human primate dosimetry data as an estimate, administration of one or two 370 MBq (10 mCi) doses of [^{11}C]**10** would result in a human ED of 1.4 or 2.8 mSv, respectively. Thus, the radiation dose to human in the planned studies, after also accounting for concomitant computerized tomography (CT) scanning, is predicted to be within acceptable yearly limits (ED < 10 mSv) for healthy research volunteers. Human radiation exposure was assessed in the whole body and individual organs in the first human PET study.

In accordance with ICH M3(R2) (International Conference on Harmonization Guidance M3(R2)), Nonclinical Safety Studies for the Conduct of Human Clinical Trials and Marketing Authorization for Pharmaceuticals, the nonclinical safety evaluation to support microdose trials of [^{11}C]**10** included a seven-day IV toxicity study in rats at doses of 0, 4, 8, and 80 μg/kg/day. Dose selection was based upon margins of 50×, 100×, and 1000× relative to the projected highest clinical dose on a μg/kg basis. Compound **10** and its precursor were well tolerated in a seven-day IV toxicity study in rats at doses of 0, 4, 8, and 80 μg/kg/day. There were no test article-related histomorphologic findings at doses up to 80 μg/kg/day. The no-adverse-effect-level (NOAEL) of ≥80 μg/kg/day provides a margin of ≈1000-fold over the maximum anticipated clinical dose of 4.9 μg per scan (based on a human equivalent dose in a 60 kg adult).

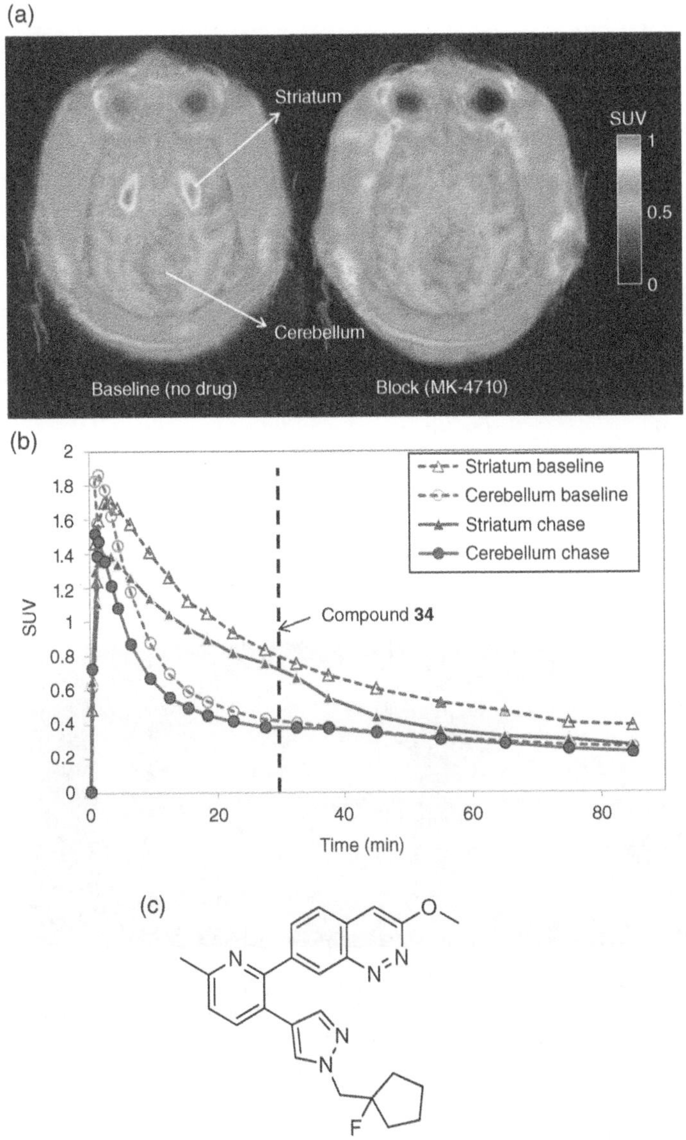

34

Human M4 FLIPR IP = 17 nM

Rat M4 FLIPR IP = 29 nM

M1/M4 IP ratio = 134

M2/M4 IP ratio >1700

M3/M4 IP ratio > 1700

M5/M4 IP ratio > 1700

Figure 10.13 (a) Summed image (0–90 minutes) of [^{11}C]**10** in rhesus monkey highlighting tracer uptake in the striatum. The left panel is [^{11}C]**10** baseline, and the right panel includes blockade with **MK-4710**. The scale is shown in SUV, which is normalized for the injected dose and mass of the monkey. (b) Time–activity curves overlaid for two different PET studies with [^{11}C]**10** in rhesus monkey at baseline (open symbols) and chase study of M4 PAM **34** (filled symbols). No change of tracer binding in cerebellum indicates the cerebellum is a valid reference region for quantification of [^{11}C]**10**. (c) Structure and *in vitro* properties of compound **34**.

Table 10.8 Test–retest variability for [^{11}C]**10** in rhesus monkey[a].

Monkey	BP_{ND}	T-RT (%)
A	Study 1: 0.89	16
	Study 2: 1.04	
B	Study 1: 0.69	5.9
	Study 2: 0.65	
C	Study 1: 0.58	11.4
	Study 2: 0.65	
D	Study 1: 1.0	1.0
	Study 2: 0.99	
E	Study 1: 0.91	4.5
	Study 2: 0.87	
Average	0.83 ± 0.17	7.6 ± 5.8

[a] Test–retest (T-RT) was determined by the equation: BP_{ND}(study 1)-BP_{ND}(study 2)/[BP_{ND}(study 1) + BP_{ND}(study 2)]/2.

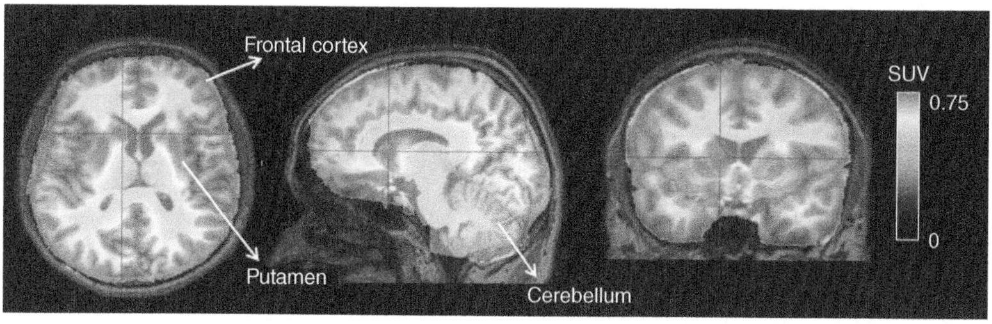

Figure 10.14 PET image of [^{11}C]MK-6884 distribution in the brain of a healthy adult. PET images of [^{11}C]MK-6884 uptake into the brain under baseline conditions, taken 90 minutes after IV administration of 300 mBq of [^{11}C]MK-6884. PET images are superimposed upon magnetic resonance imaging scans from the same subject and presented in coronal, sagittal, and transverse planes.

10.5 Development of [^{11}C]MK-6884

[^{11}C]**10** was qualified as a PCC and was subsequently renamed as [^{11}C]MK-6884 in clinical validation studies. It was evaluated in an open-label, three-part phase I study (determine radiation dosimetry (n = 3), brain regional distributions, and T-RT variability (n = 6), regional receptor distribution of [^{11}C]MK-6884 in patients with AD (n = 10)). These studies characterized the safety, dosimetry, T-RT variability and pharmacokinetics of [^{11}C]MK-6884 as a PET imaging agent for quantifying muscarinic 4 (M4) PAM receptor density in brain regions of interest in healthy subjects and AD patients [45]. Up to two IV doses of ≈ 370 MBq [^{11}C]MK-6884 were generally well-tolerated by all participants in the study. [^{11}C]MK-6884 was rapidly cleared from the body by hepatic and renal routes. The ED for [^{11}C]MK-6884 was 7.20 ± 1.27 μSv/MBq, which is typical of ^{11}C-labeled tracers. The kinetics of [^{11}C]MK-6884 distribution were investigated in six elderly healthy subjects. The rate of brain uptake was moderate, peaking very early (<3 minutes) after injection, with an average peak SUV) of 2.5. The brain region rank order of tracer uptake was putamen > cortex > cerebellum (Figure 10.14). The tracer was rapidly cleared from all brain regions, with a rank order of clearance of cerebellum > cortex > putamen. The region with the highest measured BP_{ND} was the putamen, with a value of 0.93 ± 0.28 as determined using transient equilibrium tissue ratio (TE-TR) methods, using the cerebellum as the reference region. The intra-individual T-RT variability for the BP_{ND} was ≈10%, an average T-RT variability of <20% is acceptable for a PET tracer. In AD patients, high levels of [^{11}C]MK-6884 were taken up in a relatively symmetrical and homogeneous way by the striatum receiving AChEI treatment (either donepezil or rivastigmine), comparable with the striatal BP_{ND} observed in healthy elderly participants under baseline conditions. This supports the potential for M4 PAMs to treat neuropsychiatric symptoms (NPSs) in AD patients, as the M4 PAM site in the striatum, which is believed to be most relevant for mediating the behavioral effects of M4 PAM therapeutics, appears to be unaffected in AD patients. In contrast, [^{11}C]MK-6884 uptake by the neocortex was regionally heterogeneous, with the BP_{ND} in the frontal cortex of AD patients trending lower than healthy elderly participants (0.92 ± 0.15 vs. 0.74 ± 0.20, healthy vs. AD, respectively, $P = 0.07$) This heterogeneity in cortical uptake was observed in all AD patients scanned. [^{11}C]MK-6884 showed relatively high uptake in striatum of AD patients compared with healthy subjects. Not only was [^{11}C]MK-6884 binding regionally heterogeneous in the cortex, it was lower on average in the frontal cortex of AD patients than healthy elderly subjects. The regional decreases in cortical BP_{ND} in AD patients deserves further assessment, as it may be related to various characteristics of cognitive impairment, including deficits in motor coordination, short-/long-term memory, recognition, and communication.

In an early study presented at the American Society for Clinical Pharmacology and Therapeutics (ASCPT) 2019 Annual Meeting, our scientists reported that cholinergic tone alters M4 positive allosteric modulation (PAM) and PET imaging with [^{11}C]MK-6884 was able to measure cholinergic tone change [46]. [^{11}C]MK-6884 has a high specificity with favorable physicochemical properties and *in vivo* pharmacokinetics that warrant further clinical investigation as a potential PET neuroimaging agent.

On 15 March 2019, Enigma Biomedical Group announced an exclusive license agreement with Merck & Co., Inc., for the global development and commercialization of MK-6884, which has potential applications as an imaging agent across multiple neurodegenerative diseases [47]. This deal allows Enigma and its partner, Cerveau Technologies Inc., to continue to execute the collective strategy of providing biomarkers to industry and academia to accelerate research in the field of neurodegenerative diseases.

Acknowledgement

The authors would like to thank Xiaolei Gao, Michael Lo, Robert Mazzola, Thomas Graham, Yuchuan Wang, Kerry Riffel, Talakad Lohith, Hyking Haely, Eric Hostetler, Fiona Thompson, John Morrow, David Tellers, Michael Rudd, Jenny Wai, Tjerk Bueters, Frederick Monsma, and Brian Magilaro.

References

1 BrightFocus Foundation. 2019. Alzheimer's disease: Facts & figures. https://www.brightfocus.org/alzheimers/article/alzheimers-disease-facts-figures (accessed 28 December 2019).

2 Matthews, K.A., Xu, W., Gaglioti, A.H. et al. (2018). Racial and ethnic estimates of Alzheimer's disease and related dementias in the United States (2015–2060) in adults aged≥ 65 years. *Alzheimers Dement.* 15 (1): 17–24.

3 Gribkoff, V. and Kaczmarek, L. (2017). The need for new approaches in CNS drug discovery: why drugs have failed, and what can be done to improve outcomes. *Neuropharmacology* 120: 11–19.

4 Campbell, B., Gonzalez Trotter, D., Hines, C. et al. (2016). in vivo imaging in pharmaceutical development and its impact on the 3Rs. *ILAR J.* 57 (2): 212–220.

5 Biomarkers Definitions Working Group (2001). Biomarkers and surrogate endpoints: preferred definitions and conceptual framework. *Clin. Pharmacol. Therapeutics* 69 (3): 89–95.

6 Jaszczak, R. and Coleman, R. (1985). Single photon emission computed tomography (SPECT) principles and instrumentation. *Invest. Radiol.* 20 (9): 897–910.

7 Wahl, R. and Beanlands, R. (2012). *Principles and Practice of PET and PET/CT*. Philadelphia: Wolters Kluwer Health.

8 Reivich, M. (2009). Positron emission tomography (PET). In: *Encyclopedia of Neuroscience* (ed. L. Squire), 771–782. Amsterdam: Elsevier.

9 Honer, M., Gobbi, L., Martarello, L., and Comley, R. (2014). Radioligand development for molecular imaging of the central nervous system with positron emission tomography. *Drug Discov. Today* 19 (12): 1936–1944.

10 Matthews, P., Rabiner, E., Passchier, J., and Gunn, R. (2012). Positron emission tomography molecular imaging for drug development. *Br. J. Clin. Pharmacol.* 73 (2): 175–186.

11 Hostetler, E., Joshi, A.D., Sanabria-Bohórquez, S. et al. (2013). in vivo quantification of calcitonin gene-related peptide receptor occupancy by telcagepant in rhesus monkey and human brain using the positron emission tomography tracer [^{11}C]MK-4232. *J. Pharmacol. Exp. Ther.* 347 (2): 478–486.

12 Lebois, E., Thorn, C., Edgerton, J. et al. (2018). Muscarinic receptor subtype distribution in the central nervous system and relevance to aging and Alzheimer's disease. *Neuropharmacology* 136: 362–373.

13 Wood, M., Noetzel, M., Melancon, B. et al. (2016). Discovery of VU0467485/AZ13713945: an M4 PAM evaluated as a preclinical candidate for the treatment of schizophrenia. *ACS Med. Chem. Lett.* 8 (2): 233–238.

14 Bymaster, F., Carter, P., Yamada, M. et al. (2003). Role of specific muscarinic receptor subtypes in cholinergic parasympathomimetic responses,in vivo phosphoinositide hydrolysis, and pilocarpine-induced seizure activity. *Eur. J. Neurosci.* 17 (7): 1403–1410.

15 Bodick, N., Offen, W., Levey, A. et al. (1997). Effects of xanomeline, a selective muscarinic receptor agonist, on cognitive function and behavioral symptoms in Alzheimer disease. *Arch. Neurol.* 54 (4): 465–473.

16 Shekhar, A., Potter, W., Lightfoot, J. et al. (2008). Selective muscarinic receptor agonist xanomeline as a novel treatment approach for schizophrenia. *Am. J. Psychiatry* 165 (8): 1033–1039.

17 Raedler, T., Bymaster, F., Tandon, R. et al. (2006). Towards a muscarinic hypothesis of schizophrenia. *Mol. Psychiatry* 12 (3): 232–246.

18 Bridges, T., LeBois, E., Hopkins, C. et al. (2010). The antipsychotic potential of muscarinic allosteric modulation. *Drug News Perspect.* 23 (4): 229.

19 Conn, P., Lindsley, C., Meiler, J., and Niswender, C. (2014). Opportunities and challenges in the discovery of allosteric modulators of GPCRs for treating CNS disorders. *Nat. Rev. Drug Discov.* 13 (9): 692–708.

20 Menniti, F.S., Lindsley, C.W., Conn, P.J. et al. (2013). Allosteric modulators for the treatment of schizophrenia: targeting glutamatergic networks. *Curr. Top. Med. Chem.* 13 (1): 26–54.

21 Messer, W. (2002). The utility of muscarinic agonists in the treatment of Alzheimer's disease. *J. Mol. Neurosci.* 19 (1–2): 187–193.

22 Mirza, N., Peters, D., and Sparks, R. (2006). Xanomeline and the antipsychotic potential of muscarinic receptor subtype selective agonists. *CNS Drug Rev.* 9 (2): 159–186.

23 Tarr, J., Wood, M., Noetzel, M. et al. (2017). Challenges in the development of an M4 PAM preclinical candidate: the discovery, SAR, and in vivo characterization of a series of 3-aminoazetidine-derived amides. *Bioorg. Med. Chem. Lett.* 27 (13): 2990–2995.

24 Long, M., Engers, J., Chang, S. et al. (2017). Discovery of a novel 2,4-dimethylquinoline-6-carboxamide M 4 positive allosteric modulator (PAM) chemotype via scaffold hopping. *Bioorg. Med. Chem. Lett.* 27 (22): 4999–5001.

25 Melancon, B., Wood, M., Noetzel, M. et al. (2017). Optimization of M 4 positive allosteric modulators (PAMs): the discovery of VU0476406, a non-human primate in vivo tool compound for translational pharmacology. *Bioorg. Med. Chem. Lett.* 27 (11): 2296–2301.

26 Salovich, J., Vinson, P., Sheffler, D. et al. (2012). Discovery of N-(4-methoxy-7-methylbenzo[d]thiazol-2-yl)isonicatinamide, ML293, as a novel, selective and brain penetrant positive allosteric modulator of the muscarinic 4 (M4) receptor. *Bioorg. Med. Chem. Lett.* 22 (15): 5084–5088.

27 Wood, M., Noetzel, M., Poslusney, M. et al. (2017). Challenges in the development of an M_4 PAM *in vivo* tool compound: the discovery of VU0467154 and unexpected DMPK profiles of close analogs. *Bioorg. Med. Chem. Lett.* 27 (2): 171–175.

28 Bubser, M., Bridges, T., Dencker, D. et al. (2014). Selective activation of M4 muscarinic acetylcholine receptors reverses MK-801-induced behavioral impairments and enhances associative learning in rodents. *ACS Chem. Nerosci.* 5 (10): 920–942.

29 Birdsall, N.J., Farries, T., Gharagozloo, P. et al. (1999). Subtype-selective positive cooperative interactions between brucine analogs and acetylcholine at muscarinic receptors: functional studies. *Mol. Pharmacol.* 55 (4): 778–786.

30 Christopoulos, A. (2002). G protein-coupled receptor allosterism and complexing. *Pharmacol. Rev.* 54 (2): 323–374.

31 Jeffrey Conn, P., Christopoulos, A., and Lindsley, C. (2009). Allosteric modulators of GPCRs: a novel approach for the treatment of CNS disorders. *Nat. Rev. Drug Discov.* 8 (1): 41–54.

32 Chan, W., McKinzie, D., Bose, S. et al. (2008). Allosteric modulation of the muscarinic M4 receptor as an approach to treating schizophrenia. *PNAS* 105 (31): 10978–10983.

33 Brady, A., Jones, C., Bridges, T. et al. (2008). Centrally active allosteric potentiators of the M4 muscarinic acetylcholine receptor reverse amphetamine-induced hyperlocomotor activity in rats. *J. Pharmacol. Exp. Ther.* 327 (3): 941–953.

34 Acton, J., Bao, J., Egbertson, M. et al. (2017). 3-(1H-Pyrazol-4-yl) pyridines as allosteric modulators of M4 muscarinic acetylcholine receptor and their preparation. PCT Int. Application, WO2017107089A1.

35 Eckelman, W.C., Gibson, R.E., Rzeszotarski, W. et al. (1979). The design of receptor binding radiotracers. In: *Principles of Radiopharmacology*, vol. 1 (ed. L. Colombetti), 251. New York: CRC Press.

36 Waterhouse, R. (2003). Determination of lipophilicity and its use as a predictor of blood–brain barrier penetration of molecular imaging agents. *Mol. Imaging Biol.* 5 (6): 376–389.

37 Schubert, J., Harrison, S., Mulhearn, J. et al. (2019). Discovery, optimization, and biological characterization of 2,3,6-trisubstituted pyridine-containing M4 positive allosteric modulators. *ChemMedChem* 14 (9): 943–951.

38 Tong, L., Li, W., Lo, M. et al. (2020). Discovery of [^{11}C]MK-6884: a positron emission tomography (PET) imaging agent for the study of M4 muscarinic receptor positive allosteric modulators (PAMs) in neurodegenerative diseases. *J. Med. Chem.* 63 (5): 2411–2425.

39 Miller, P., Long, N., Vilar, R., and Gee, A. (2008). Synthesis of ^{11}C,^{18}F,^{15}O, and ^{13}N radiolabels for positron emission tomography. *Angew. Chem. Int. Ed.* 47 (47): 8998–9033.

40 Jacobson, O., Kiesewetter, D., and Chen, X. (2014). Fluorine-18 radiochemistry, labeling strategies and synthetic routes. *Bioconjug. Chem.* 26 (1): 1–18.

41 Hostetler, E., Walji, A., Zeng, Z. et al. (2016). Preclinical characterization of ^{18}F-MK-6240, a promising PET tracer for in vivo quantification of human neurofibrillary tangles. *J. Nucl. Med.* 57 (10): 1599–1606.

42 Li, W., Salinas, C., Riffel, K. et al. (2016). The discovery and characterization of [18F]MK-8553, a novel PET tracer for imaging O-GlcNAcase (OGA). International Symposium on Function NeuroReceptor Mapping of the Living Brain (NRM2016), Boston.

43 Li, W., Wang, Y., Lohith, T. G. et al. Translation of the M4 PAM specific PET tracer [^{11}C]MK-6884 from rhesus monkeys to Alzheimer's disease patients. *Science Translational Medicine* 14 (627). DOI: 10.1126/scitranslmed.abg3684

44 Mazzola, R., Lee, S., Lo, M. et al. (2020). Discovery of MK-4710: A potent, selective, and bioavailable M4 positive allosteric modulator (PAM). Fall ACS Virtual Meeting & Expo.

45 ClinicalTrials.gov (2015). [^{11}C]MK-6884 positron emission tomography (PET) tracer validation trial (MK-6884-001). https://clinicaltrials.gov/ct2/show/study/NCT02621606 (accessed 3 March 2020).

46 Li, W., Wang, Y., Zeng, Z. et al. (2019). Discovery and first-in-human evaluation of M4 PAM PET tracer [^{11}C]MK-6884. *J. Label. Compd. Radiopharm.* 62: S87–S88.

47 StreetInsider.com (2019). Enigma Biomedical Group signs license agreement with Merck for novel MK-6884 imaging agent. https://www.streetinsider.com/Press+Releases/Enigma+Biomedical+Group+Signs+License+Agreement+with+Merck+for+Novel+MK-6884+Imaging+Agent/15263685.html (accessed 3 Mar ch 2020).

11

Targeted Protein Degradation by Proteolysis Targeting Chimeras: A Revolution in Small Molecule Drug Discovery

Wu Du

Hinova Pharmaceuticals Inc., No. 2-3, 4th Floor, RongYao Tower 1, 5 South Keyuan Road, Chengdu, 610041, People's Republic of China

11.1 The Concept of Targeted Protein Degradation

11.1.1 Introduction

Proteins are at the frontline of modulating physiological processes, and they are the biological targets of many drugs. Traditional drug discovery efforts have been focused mainly on modulating protein function directly using antagonists or agonists to inhibit or promote a protein's function, respectively. The agents for directly modulating a protein's function can be small or large molecules. There have been tremendous advances and achievements in the discovery of new drugs modulating protein function. However, many challenges still remain. A protein inhibitor must bind to a protein to achieve inhibition. Large molecules such as peptides and antibodies can interact with proteins on a cell surface, but they are usually unable to cross the cell membrane to interact with intracellular protein targets. In comparison, small molecules are usually capable of crossing cell membranes, but some proteins are considered "undruggable" for having shallow surfaces that do not have a binding pocket deep enough to accommodate small molecule ligands that bind tightly enough to achieve bioactivity. About 85% of the therapeutically relevant proteins remain untargeted. New approaches to target these untargeted proteins are desirable and remain both a big challenge and a big opportunity. As our understanding of the life cycle of a protein advances, new technologies are emerging to address different aspects of it, and are accordingly leading to more efficacious and safer therapeutics.

A protein's life cycle starts from DNA. DNA is transcribed to mRNA, which upon translation produces a protein. An antagonist or agonist interacts with the protein and modulates its function. The protein does not last forever. It is eventually degraded to small peptides and amino acids, mainly through the ubiquitin–proteasome system (UPS). As shown in Figure 11.1, all phases of this life cycle have been targeted for the development of new therapeutics. While CRISPR modulates DNA, and RNAi modulates RNA, targeted protein degradation (TPD) modulates the protein degradation process.

Regulating the degradation of a protein offers unique advantages compared with directly modulating a protein's function by antagonists or agonists. Promoting the degradation of a specific protein will lead to lower protein levels and accordingly achieve the biological effect similar to

Contemporary Accounts in Drug Discovery and Development, First Edition. Edited by Xianhai Huang, Robert G. Aslanian, and Wayne H. Tang.

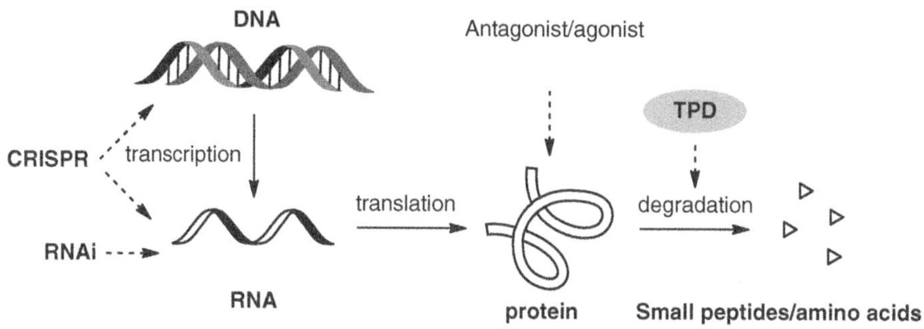

Figure 11.1 Regulation of protein drug targets by different approaches.

antagonism. Conversely, suppressing the degradation of a protein will lead to higher protein levels and achieve the biological effect that is similar to agonism. Furthermore, an antagonist needs to bind and remain bound to the target protein forming a ligand-protein complex to achieve sustained efficacy. When the ligand dissociates from the target protein, the antagonism effect is lost. This requires the compound to have an off rate and a pharmacokinetic (PK) profile sufficient to provide continuous coverage. In contrast, in the case of TPD, once a protein is degraded, its biological function is eliminated regardless of whether the degrader compound is still present or not. Thus, presumably a degrader is catalytic in nature and does not require an off rate and a PK profile to support continuous coverage. The biological effect achieved by a degrader is largely controlled by the protein's degradation rate versus its re-synthesis rate.

Since the first publication in 2001, TPD, especially the proteolysis targeting chimeras (PROTACs), has emerged as a new frontier for small molecule drug discovery. The enthusiasm from both academia and industry is well demonstrated by the explosive number of publications in this field, and the number of emerging biotech companies focusing on this technology. Over the last few years, a number of well-written reviews have been published [1].

11.1.2 The Ubiquitin-Proteasome System

The UPS is the major protein degradation machinery that is responsible for degradation of 80% of intracellular proteins in eukaryotes while extracellular proteins are degraded by lysosomes [2]. This process has a profound impact on all aspects of physiological processes. The importance of this discovery was recognized in 2004 when Drs. Aaron Ciechanover, Avram Hershko, and Irwin Rose were awarded the Nobel Prize in Chemistry for the discovery of the ubiquitin mediated protein degradation mechanism.

Ubiquitin is a small (8.6 kDa) regulatory protein composed of 76 amino acid residues. It was identified by Dr. Gideon Goldstein in 1975. Four genes in the human genome code for this protein: *UBB, UBC, UBA5*, and *RPS27A*. Its structure features a C-terminal carboxylic acid tail and 7 lysine residues. This protein is highly conserved during evolution. The ubiquitins of different species share high sequence homology (>95%).

As "ubiquitous" as its name indicates, ubiquitin is found in all eukaryotic cells. During the protein degradation process, ubiquitin is first covalently attached to a protein substrate marking this protein to be degraded in the proteasome. This process is known as ubiquitination (also known as ubiquitylation or ubiquitinylation) [3]. As shown in Figure 11.2, in the presence of ATP, a ubiquitin protein is first attached to the cystine residue in a ubiquitin activating enzyme E1 ligase by formation of a thioester bond. Then the activated ubiquitin is transferred to a cystine residue in the

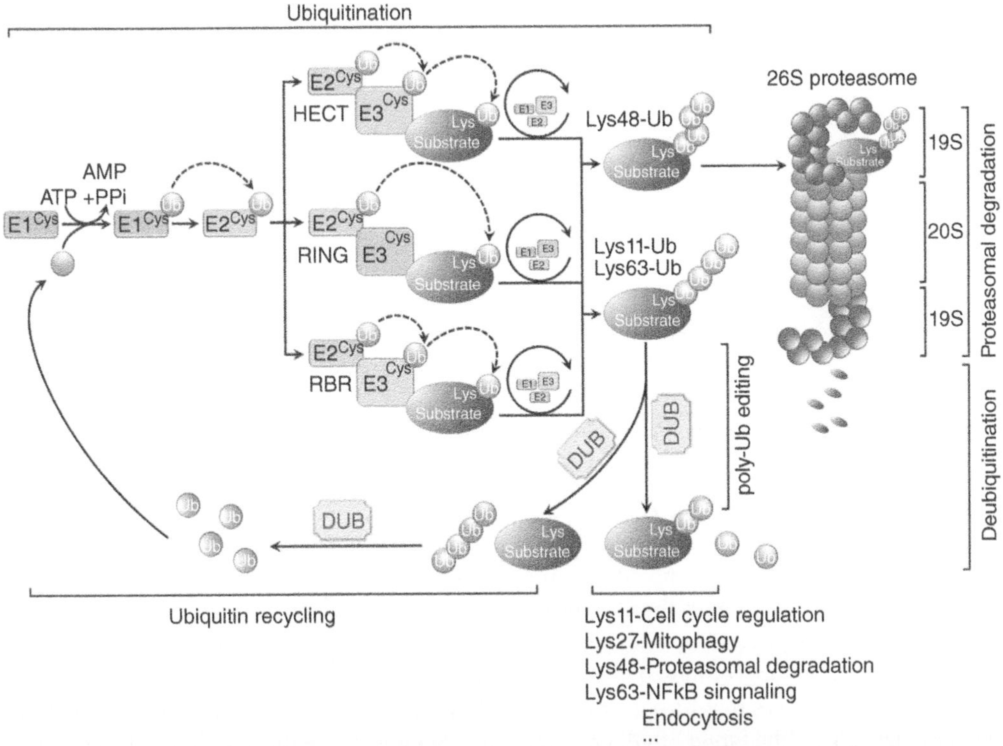

Figure 11.2 Ubiquitination. *Source:* The image is copied from [3] according to Creative Commons License.

conjugating enzyme E2 ligase. Then catalyzed by E3 ligase, the ubiquitin is transferred to a lysine residue in the substrate protein to give a mono-ubiquitinated substrate protein. The specificity of the ubiquitination is determined by the E3 ligase. There are three classes of E3 ligases: homologous to the E6AP C-terminus (HECT) domain-containing E3 ligases, Really Interesting New Gene (RING) or U-box domain-containing E3 ligases, and RING-between-RING (RBR) domain-containing E3 ligases [4]. When HECT and RBR E3 ligases are involved in ubiquitination, the ubiquitin is transferred from the E2 ligase to the E3 ligase and then to the protein substrate. When RING E3 ligase is involved, the ubiquitin is transferred directly from the E2 ligase to the substrate protein.

This ubiquitination process is further repeated to give a polyubiquitin substrate protein for degradation in the proteasome. A minimum attachment of four ubiquitin proteins is required in order for a protein to be recognized by the proteasome. The connection between two ubiquitins is made by formation of an amide bond between the amino group in the lysine residue from the first ubiquitin and the glycine residue from the C-terminus of the second ubiquitin.

There is also a reverse process that removes ubiquitin from a substrate protein that is known as deubiquitination (DUB). This process is catalyzed by deubiqitinase. Inhibition of the deubiqutinase will block removal of ubiquitin from the target protein and promote degradation of this protein [5].

Proteasomes are the protein machinery that are responsible for decomposing an undesired protein by proteolysis [6]. They are located in both the nucleus and cytoplasm in all eukaryotes. This rather large protein complex has a cylindrical barrel shape with a hollow internal chamber that is the site of degradation of the protein, and two gates that control the entrance of the protein to be

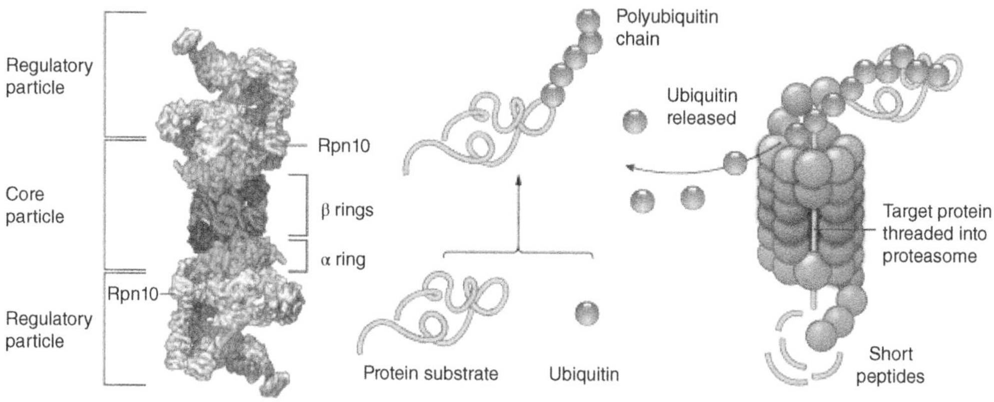

Figure 11.3 Degradation of polyubiquitinated protein by the proteasome. *Source:* This image is used with permission from Steven Graepel. Copyright: Steven Graepel.

degraded and export of the short peptide fragments and amino acid products after a protein is degraded. This machinery is composed of four stacked rings. The two internal rings are composed of seven β-subunit proteins with three to seven protease active sites in each protein facing toward the inside of the "barrel" and are responsible for cleaving the peptide bond in the target protein. The two external rings are composed of seven α-subunits that function as the gate keeper by binding to regulatory particles that recognizes the polyubiquitin tag on the protein to be degraded. As shown in Figure 11.3, the target protein bearing the polyubiquitin chain is recognized by the proteasome and is threaded through the proteasome central cavity where it is cleaved into amino acids and short peptides.

Proteasomes play a pivotal role in physiological activity. In addition to degradation of undesired proteins, they are also involved in cell cycle control, apoptosis, and immune response. As such, they are targets for new drug development. The proteasome inhibitors Bortezomib, Carfilzomib, and Ixazomib are FDA approved anticancer drugs, and several more are in clinical development.

11.1.3 Targeted Protein Degradation by Proteolysis Targeting Chimeras

Since the magnitude of a protein's biological effect is usually related to the amount of this protein, regulating protein levels would be a novel approach to modulate protein function. By selectively promoting degradation of a target protein, this strategy results in a significantly reduced level of the target protein, and accordingly, a suppressed biological effect. This revolutionary approach was realized using PROTACs and was first demonstrated in 2001 by the Crews group. This approach has been intensively pursued by academia and industry scientists with explosive interest in recent years. Since the key step in the degradation of a protein is the formation of the substrate-E3 ligase binary complex for ubiquitination, a logical approach is to artificially select and modulate this process. As shown in Figure 11.4 [1e], a PROTAC molecule is designed to include three moieties: the target protein binding moiety, the E3 ligase binding moiety, and a linker connecting these two. By binding to both the target protein and the E3 ligase, the PROTAC molecule brings the target protein and E3 ligase into proximity and facilitates subsequent ubiquitination, and accordingly subsequent protein degradation through the UPS. Unlike traditional small molecule inhibitors, the unique mechanism of action of the PROTAC approach provides features such as high potency, high selectivity, and is catalytic in nature. Furthermore, the three-moiety feature of PROTACs

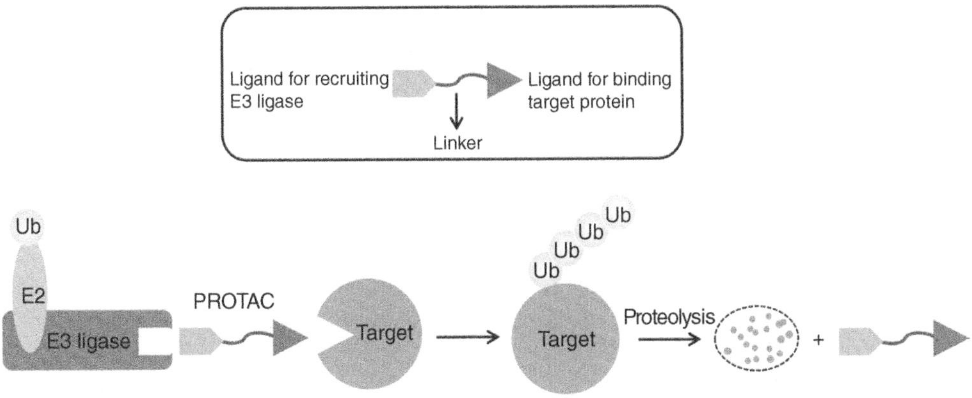

Figure 11.4 The mechanism of action of PROTACs. *Source:* The image is copied from [1e] according to STM permission guidelines.

Table 11.1 Comparison of PROTACs with other therapeutics.

	PROTACs	Small molecules	Antibodies	siRNA	CRISPR
Intracellular target	√	√	X	√	√
Tissue distribution	√	√	Poor	Poor	√
Systemic delivery	√	√	√	X	√
Oral dosing	√	√	X	X	X
Catalytic nature	√	X	X	√	√
High selectivity	√	Moderate	√	√	√
Long-lasting effect	√	Moderate	X	√	√
Effect reversible	√	√	√	X	X

√, yes; X, no.

makes it possible to design PROTACs in rational ways. Compared with inhibition by traditional small molecules, the distinct mechanism of action of the PROTAC approach allows the potential for exceptionally high potency and selectivity, high efficacy due to event driven nature, and low dose due to its catalytic nature. PROTACs are becoming a new tool to tackle targets that were once "undruggable."

The unique mechanism of action of PROTACs imbues this approach with special advantages. A side-by-side comparison of PROTACs with traditional small molecules, antibodies, small interfering RNA (siRNA), and CRISPR is shown in Table 11.1. PROTACs are believed to have the attributes of large molecules and traditional small molecules, but without their drawbacks. They are relatively small molecules, and can thus incorporate features characteristic of traditional small molecules such as oral dosing and tissue distribution. Meanwhile, they can also offer the advantages of large molecules, such as being catalytic, having high potency, high selectivity, and long-lasting effects.

The proteasome is mainly located in the cytoplasm in cells. Thus, PROTACs have been applied to intracellular protein targets in most cases. PROTACs can still be considered as small molecules, but they are at the high end of the molecular weight (MW) range. By joining the target binding moiety, linker and the E3 ligase binding moiety, the MW of a PROTAC is usually close to a thousand daltons. Despite the high MW, it has already been demonstrated that PROTAC molecules can achieve systemic exposure and permeate into tissues through oral dosing, which will be addressed in detail in later sections. In the PROTAC facilitated ubiquitination, once the polyubiquitination is completed, the protein-PROTAC-ligase ternary complex dissociates and releases the PROTAC molecule for the next round of catalysis. The catalytic nature of PROTAC makes it possible to achieve high potency and a low dose. While the biological effect of traditional small molecules is "occupancy driven," that of a PROTAC is "event driven." The efficacy of a single dose of PROTAC depends more on the rate of re-synthesis of the protein, and less on the clearance rate of the PROTAC molecule. This also makes it possible to achieve low dose and less frequent dosing. Unlike CRISPR, the biological effect of PROTAC is reversible. A single dose of PROTAC promotes degradation of a protein resulting in a biological effect. Then upon clearance of the PROTAC and re-synthesis of the protein, this biological effect can be restored. PROTAC is thus a reversible protein knock down strategy in addition to the RNA/DNA strategy.

11.2 Advances with PROTACs

11.2.1 Proof of Concept and Early Peptide Based PROTACs

The first PROTAC was published in 2001 by Sakamoto and co-workers [7]. In this historical proof of concept study, PROTAC-1 (Figure 11.5) was prepared by linking the IκBα phosphopeptide (IPP) to Ovalicin through a linker. The IPP was used as the ligase binding moiety to be recognized by the F-box protein β-TRCP in the ligase complex, the Skp1-Cullin-F box complex (SCF). Ovalicin was used as the target protein binding moiety to recruit methionine aminopeptidase-2 (MetAP-2). It was demonstrated that the target protein can be tethered to the ligase complex, ubiquitinated and degraded in a PROTAC-1 dependent manner.

Further exploration in this field continued using a peptide based E3 ligase binding moiety. As shown in Figure 11.6, an androgen receptor (AR) targeting PROTAC was achieved by linking dihydrotestosterone (DHT) with IκBα phosphopeptide. An estrogen receptor (ER) targeting PROTAC was achieved by linking GA-1 with IκBα phosphopeptide. These PROTACs were shown to enable the ubiquitination and degradation of AR and ER *in vitro* in a proteasome-dependent manner respectively. These PROTACs were delivered into cells by microinjection [8].

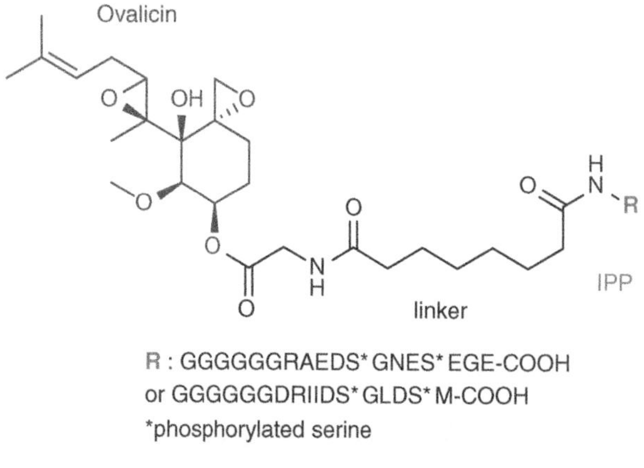

Figure 11.5 PROTAC-1.

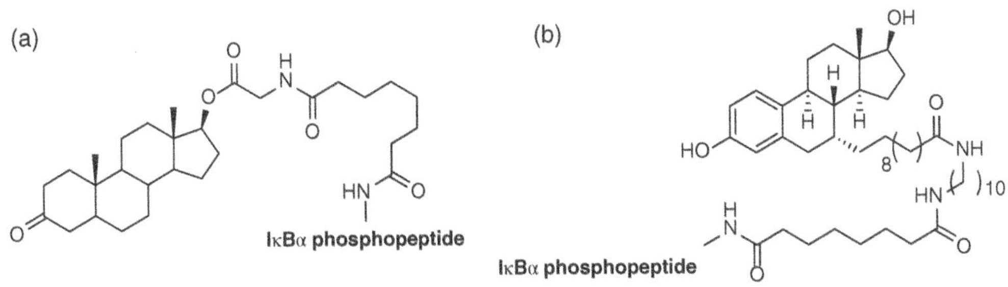

Figure 11.6 Peptide based PROTACs targeting AR (a) and ER (b).

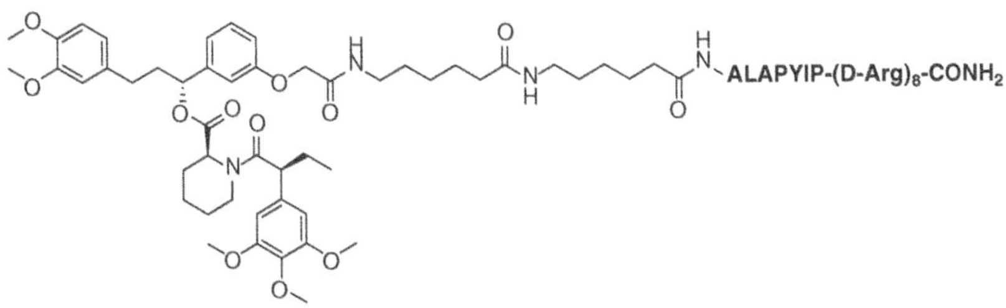

Figure 11.7 HIF-1α based PROTAC.

Subsequently, additional peptide based PROTAC molecules were discovered and microinjection was no longer needed. As shown in Figure 11.7, the HIF-1α peptide carries a cell penetrating sequence that can recruit Von Hippel–Lindau (VHL) protein E3 ligase. It was applied in a FK506 binding protein (FKBP) targeting PROTAC [9]. Later, a shorter fragment of HIF-1α peptide was applied in an aryl hydrocarbon receptor nuclear translocator (ARNT) targeting PROTAC [10].

The early generation of peptide based PROTACs was successful in degrading the targeted protein, however with low degradation activity. They were usually active at low μM concentrations. This is apparently due to poor cell permeability caused by the peptide structure. To achieve PROTACs with therapeutic potential, a breakthrough in the E3 ligase moiety was needed.

11.2.2 Small Molecule Based PROTACs: The Discovery of VHL and CRBN E3 Ligands

The first PROTAC consisting entirely of small molecules tethered together was reported in 2008 [11]. Nutlin, a MDM2-P53 PPI inhibitor, was applied to recruit MDM2 E3 ligase (Figure 11.8). By linking with the AR targeting moiety, the resulting PROTAC succeeded in degrading the AR. Although the potency was still at micromolar concentration, it demonstrated that a cell permeable PROTAC was feasible and stimulated further interest in this field.

Encouraged by the small molecule based PROTAC, further efforts were undertaken to identify E3 ligase moieties that would lead to higher degradation activity and better cell permeability. The identification of E3 ligase ligand VHL in 2012 further fueled interest in this field [12]. Ciulli's lab studied the structure–activity relationship of the VHL scaffold and optimized this ligand to achieve improved activity and physicochemical properties [13]. On another front, the thalidomide-class molecules (thalidomide, pomalidomide, lenalidomide) were found to recruit cereblon (CRBN) E3 ligase and represent the other major class of E3 ligand that is broadly applied to current PROTAC research [14]. The PROTAC targeting receptor-interacting protein kinase 2 (RIPK2) with

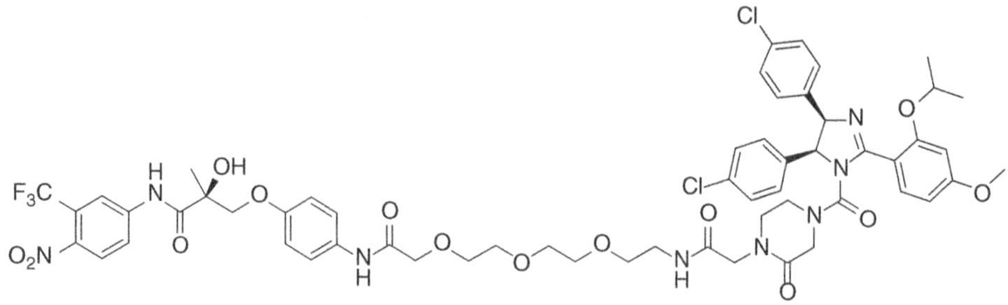

Figure 11.8 The first small molecule based PROTAC.

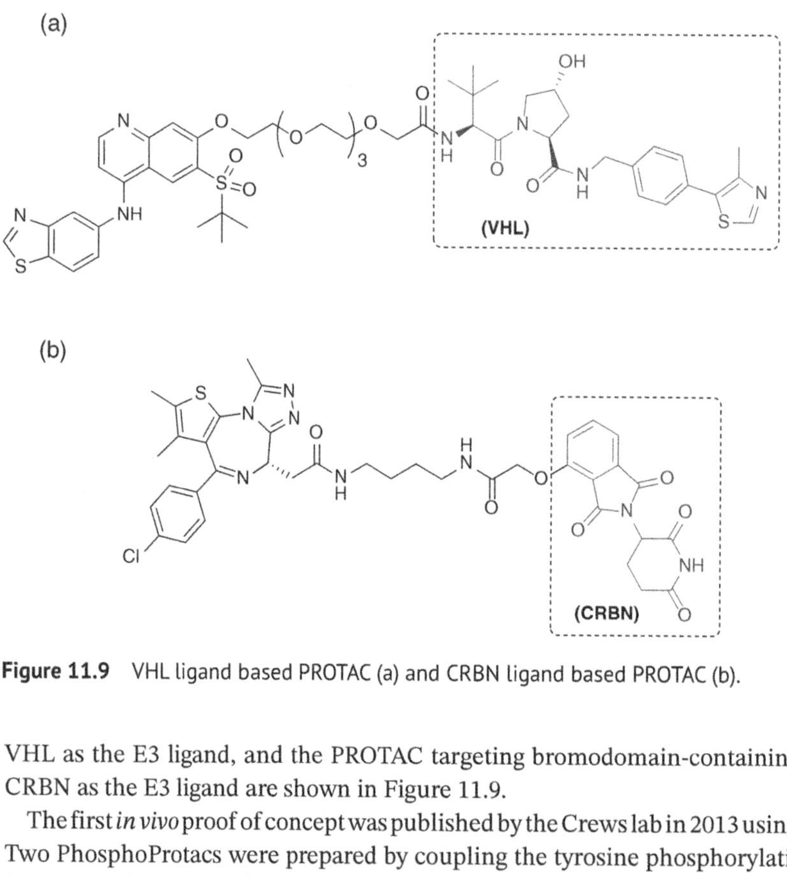

Figure 11.9 VHL ligand based PROTAC (a) and CRBN ligand based PROTAC (b).

VHL as the E3 ligand, and the PROTAC targeting bromodomain-containing protein (BRD) with CRBN as the E3 ligand are shown in Figure 11.9.

The first *in vivo* proof of concept was published by the Crews lab in 2013 using PhosphoProtac [15]. Two PhosphoProtacs were prepared by coupling the tyrosine phosphorylation sequence of TrkA (tropomyosin receptor kinase A), or the neuregulin receptor, ErbB3 (erythroblastosis oncogene B3) with a peptide ligand for the E3 ligase VHL protein. As shown in Figure 11.10a, TrkA PhosphoProtac has an aminohexanoic amide linker. These PROTACs were able to suppress the short- and long-term effects of their respective activating receptor tyrosine kinase pathways both *in vitro* and *in vivo*. As shown in Figure 11.10b, the ErbB3 PhosphoProtac was dosed at 10 mg/kg for 47 days by i.p. injection to mice implanted with the human ovarian carcinoma cell line OVCAR8. The tumor mass reduction compared with the vehicle group, although moderate, was statistically significant.

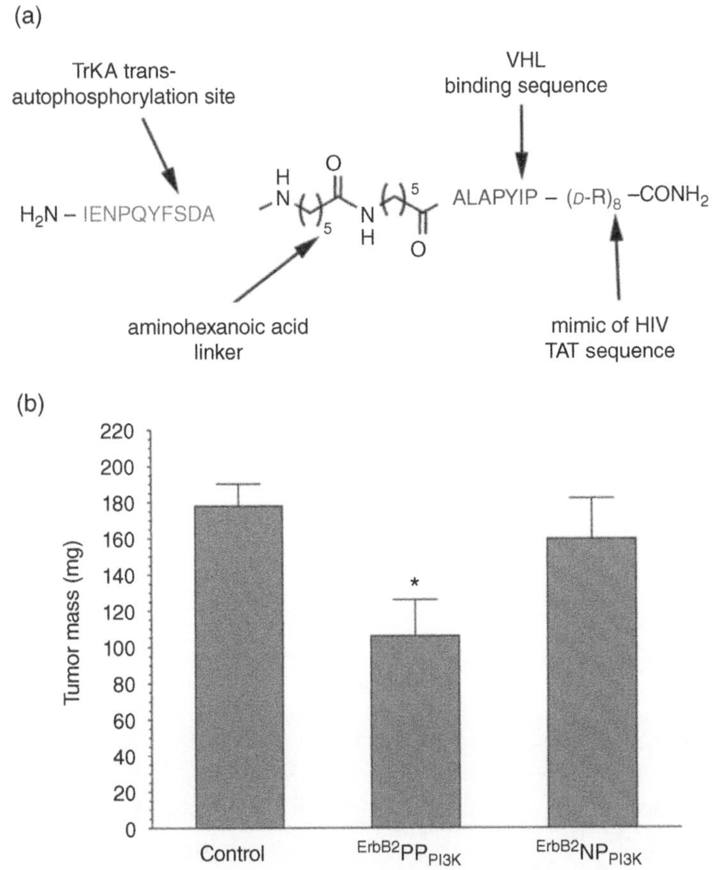

Figure 11.10 (a) Structure of TrkA PhosphoProtac. (b) PhosphoPROTAC $^{ErbB2}PP_{PI3K}$ significantly reduced growth of implanted OVCAR8 ovarian cell tumors *in vivo* in nude mice (asterisk denotes significant difference from control mice as assessed; $p<0.05$). *Source:* The images are copied from [15] with permission.

11.2.3 Ligands for E3 Ligase

E3 ligase is one of the three major components of the ternary complex and is crucial for the proteasome mediated targeted degradation process to proceed. There are hundreds of different E3 ligases known to date, but only a few of them were found to have small molecule ligands [16]. Some of these small molecule ligands are summarized in Figure 11.11.

Thalidomide is infamous in the history of drug safety. It was notoriously known for causing birth defects when given as a treatment for morning sickness due to pregnancy in the 1950s. It was later approved as a cancer treatment. Further studies of its mechanism of action found that thalidomide, along with its analogs lenalidomide and pomalidomide, interacts with the E3 ligase CRBN and promotes ubiquitination of the endogenous substrates of CRBN E3 ligase [14b]. This was confirmed by crystal structures of the damage specific DNA binding protein 1 (DDB1)-CRBN complex bound to thalidomide and lenalidomide [14c, h]. To date, the thalidomide family of CRBN E3 ligase ligands represents one of the major classes of E3 ligase ligands studied and applied in PROTAC based TPD. Among all the known E3 ligands, thalidomide and analogs have the smallest MW, which has the potential advantage of providing better physicochemical properties and pharmacological profiles.

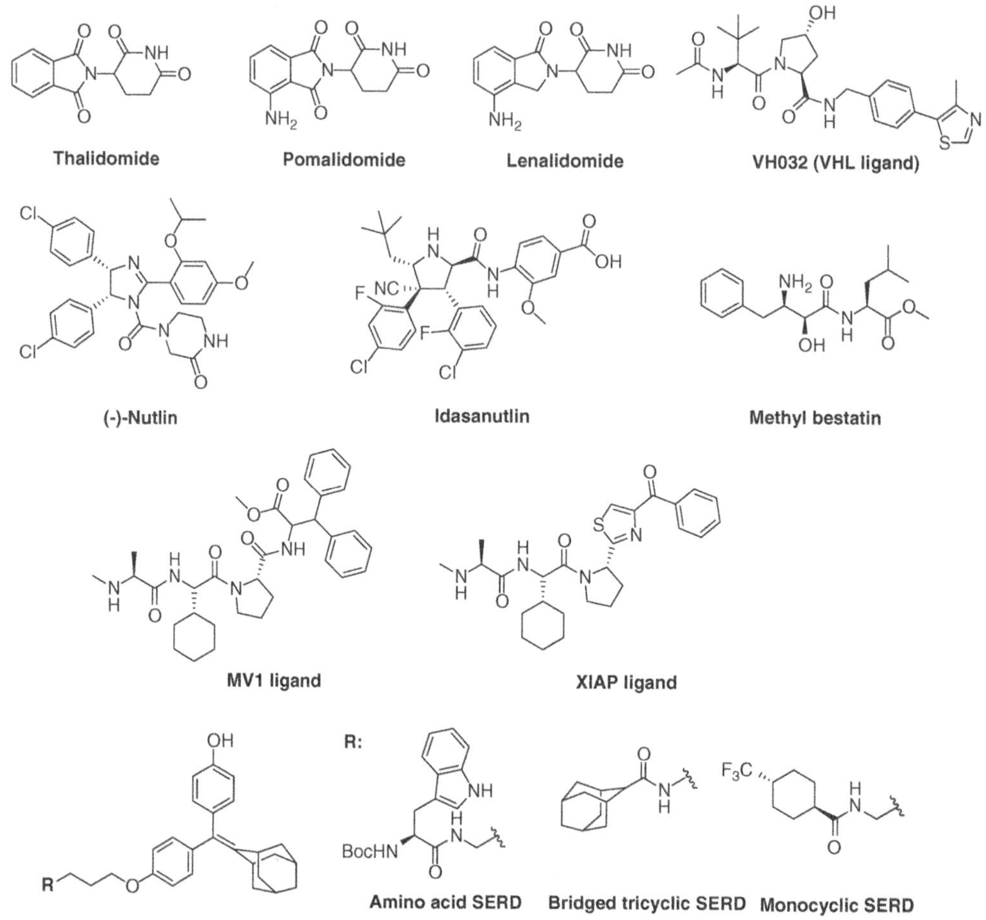

Figure 11.11 Small molecule E3 ligase ligands.

VHL E3 ligase was utilized in PROTAC studies early on by using a peptide sequence to recruit it. Further investigation into small-molecule-like VHL ligands led to the discovery of the potent ligand VH032. Additional screening and optimization identified other VHL ligands of this scaffold family. Judging by the number of publications, VHL ligands represent another major family of E3 ligands broadly applied in PROTAC research.

Nutlin is a ligand for MDM2 E3 ligase by binding to its p53 pocket. It was applied to PROTACs that degrade the AR to generate the first cell permeable all-small molecule-PROTAC (see Section 11.2.5). A structurally similar ligand, idasanutlin, was also applied to the construction of PROTAC molecules [17].

Other E3 ligands include methyl bestatin which recruits cellular inhibitor of apoptosis protein 1 (cIAP1) [18]. MV1 is a pan ligand of cIAP1/cIAP2/XIAP [19]. XIAP ligand belongs to the same scaffold as MV1, but has higher affinity and preferentially recruits X-linked IAP (XIAP) rather than cellular IAP1 [20]. Sharma's group reported a new class of selective estrogen receptor degraders (SERDs) that applied a bisphenolic-adamantyl moiety that binds to the ER and an optimized lipophilic amide moiety to recruit E3 ligase. The representative examples of the E3 ligand moiety include a Boc-protected lipophilic amino acid, bridged tricyclic moiety, and trifluoromethyl-substituted monocyclic ring moiety. The details on E3 ligase recruiting were not fully elucidated [21].

11.2.4 Mechanistic Considerations: The Ternary Complex and the Kinetics

Quite distinct from the traditional target–inhibitor interaction mediated by formation of the target-inhibitor binary complex, PROTACs recognize both the protein of interest (POI) and the E3 ligase at the same time by formation of a target-Protac-E3 ligase ternary complex. The formation of the ternary complex is the critical step in promoting subsequent ubiquitination, polyubiquitination, and degradation. The Ciulli group reported the first crystal structure of the ternary complex formed between the bromodomain BRD4, the PROTAC molecule, and the VHL E3 ligase [22]. Figure 11.12 shows this ternary complex in ribbon representation. In this paper, a mathematical model was applied to the ternary complex and predicted a U-shaped PROTAC concentration dependent formation of the ternary complex.

The Hook effect is a phenomenon observed for PROTAC, but not for small molecule inhibitors. When the concentration of the PROTAC is high enough to overwhelm the concentration of target protein and E3 ligase, there will be formation of more POI-PROTAC binary complex and PROTAC-E3 binary complex. This results in reduction of the concentration of the POI-PROTAC-E3 ternary complex and accordingly, reduction of the degradation activity. Such counter-intuitive higher drug concentration leading to lower efficacy (activity) phenomena is known as the Hook Effect and was previously known in the immunology field when the formation of multi-component immune complexes involving antibodies or antigens is required. The hook effect is illustrated in Figure 11.13. The interaction between POI and E3 ligase can be favorable (due to protein–protein interactions that stabilize the ternary complex) or repulsive (due to charge repulsion and/or steric clash). The former has been shown to reduce the hook effect by enhancing the formation of the ternary complex.

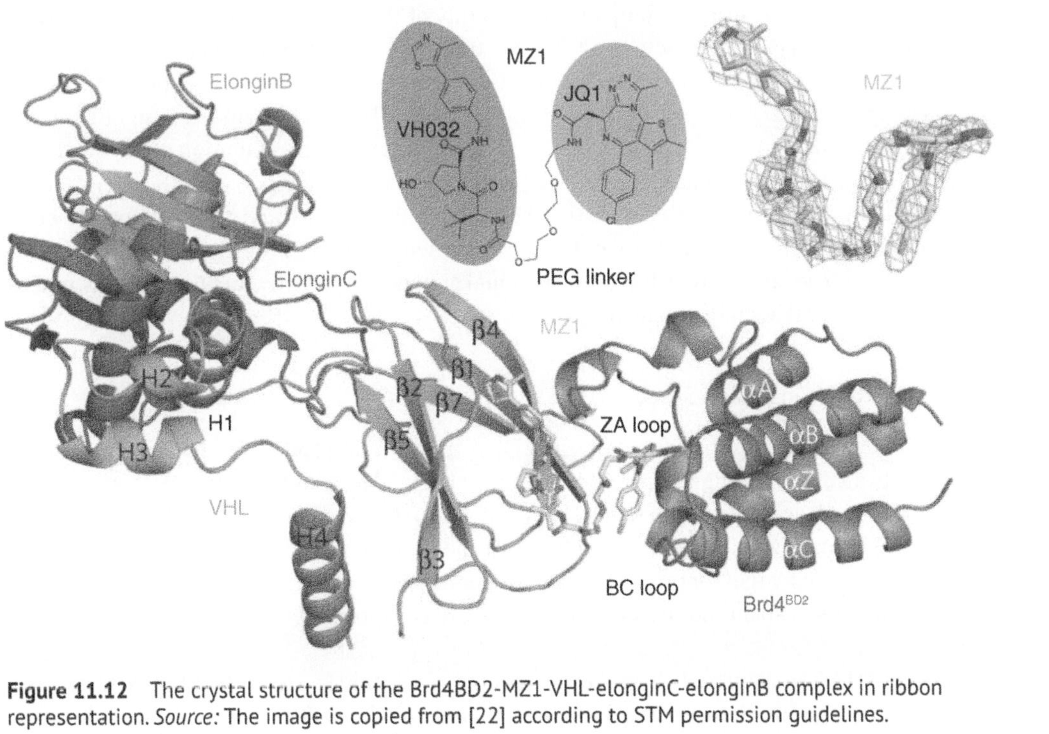

Figure 11.12 The crystal structure of the Brd4BD2-MZ1-VHL-elonginC-elonginB complex in ribbon representation. *Source:* The image is copied from [22] according to STM permission guidelines.

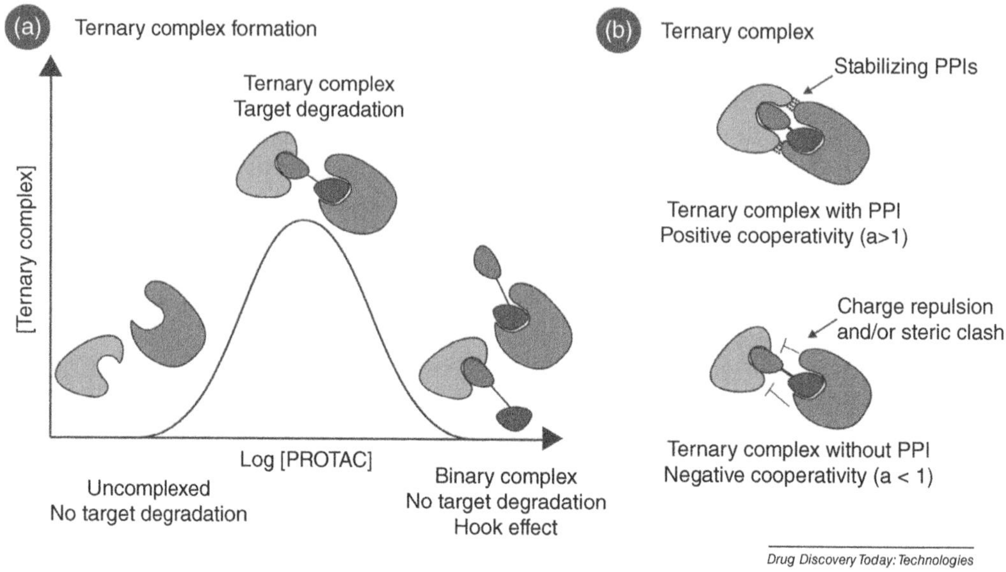

Figure 11.13 Hook effect. (a) PROTAC-mediated ternary complex formation and the hook effect as a function of PROTAC concentration. PROTAC compound: POI ligand, E3 ligand connected by a linker, POI, and E3. (b) Top: positive cooperativity is obtained when stabilizing PPIs are obtained between POI and E3. Bottom: Negative cooperativity is observed when charge repulsion and/or steric clashes are observed between POI and E3. *Source:* The images are copied from [1a] according to STM permission guidelines.

As we further consider the ternary complex, apparently the linker's role is not just as simple as bringing the target protein and the E3 ligase into proximity, but also plays an important recognition role in stabilizing the ternary complex. The stability of the ternary complex not only relies on how strong the POI ligand moiety in the PROTAC molecule binds to the target protein and how strong the E3 ligase ligand binds to the E3 ligase, but also how much the linker contributes to stabilizing the whole ternary complex. Thus, a PROTAC molecule is more than the sum of its parts, and should be considered as a whole new entity instead of a simply a combination of the POI binding moiety, linker and the E3 ligase ligand. This is demonstrated by a report from Fischer's lab on an integrated study of degraders comprising the BRD4 inhibitor JQ1 and the thalidomide class CRBN E3 ligase ligand [23]. It was shown that the linker length and attaching position could affect degradation selectivity and activity by conformational change. Apparently, such complexity makes optimization of PROTACs more challenging.

11.2.5 AR PROTACs: A Case Study

The AR is a nuclear receptor that functions as a DNA transcription factor to regulate gene expression. It is activated by binding androgenic hormones such as testosterone and DHT and plays a critical role in the development and maintenance of the male sexual phenotype. However, activation of the AR signaling pathway is associated with the development and progression of prostate cancer. It is a validated target for prostate cancer treatment. Several AR antagonists, such as bicalutamide, enzalutamide, apalutamide, and darolutamide, were approved by the FDA to treat prostate cancers.

The first PROTAC that entered clinical trials was the AR degrader ARV-110 that was discovered and developed by the New Heaven based PROTAC pioneer Arvinas. The first small molecule based

PROTAC was also the AR targeting PROTAC that was reported by Crews, the founder of Arvinas. After selection of the AR as the POI, it took about 15 years to evolve from a conceptual PROTAC to a clinical compound. Thus, a closer look into the AR PROTAC may provide a better understanding of this field.

The early peptide based PROTACs targeting AR and ER published in 2004 were important in validating the concept. However, due to their peptide nature, these early PROTACs were unable to cross the cell membrane to engage proteasome mediated protein degradation. Thus, they had little potential practical application. Nevertheless continued efforts eventually delivered AR and ER PROTACs as the first two PROTACs that entered clinical trials, and more importantly, as orally bioavailable PRTOACs.

The first small molecule based AR targeting PROTAC dates back to 2008 when the Crews group reported a PROTAC (compound **14** in Figure 11.14) employing an amide scaffold non-steroidal AR ligand similar to the AR antagonist bicalutamide, a PEG based linker and a triphenyl imidazoline nutlin type ligand of MDM2 E3 ligase [11]. A Hela cell line that expresses the AR was used to test AR degradation. Upon compound treatment at a rather high concentration of 10 μM, Western blot analysis with an AR antibody showed significant AR degradation. The proteasome inhibitor epoxomicin (10 μM) was also applied and showed inhibition of AR degradation, thus verifying the degradation was proteasome mediated. This study showed that cell permeability can be achieved by PROTACs, but significant improvement was still needed to have a clinically relevant PROTAC.

Advances in prostate cancer treatment by suppressing the AR signaling pathway led to the discovery of more potent AR antagonists as AR ligands. Enzalutamide was approved as a second generation nonsteroidal antiandrogen to treat metastatic castration-resistant prostate cancer (mCRPC). Thus, the diaryl thioxoimidazolidinone scaffold was explored as an AR ligand for PROTAC. In 2015, the Crews group reported a new class of selective AR PROTAC degraders named SARD279 and SARD033 [24]. These compounds used the AR antagonist RU59063 as the AR ligand and a highly lipophilic adamantane moiety to recruit E3 ligase. Both AR degraders showed AR degradation activity at low μM concentrations and were comparable with enzalutamide (MDV3100), but much better than their warhead molecule RU59063 in the inhibition of lymph node carcinoma of the prostate (LNCaP) cell (an AR expressing prostate cancer cell line) proliferation (Figure 11.15).

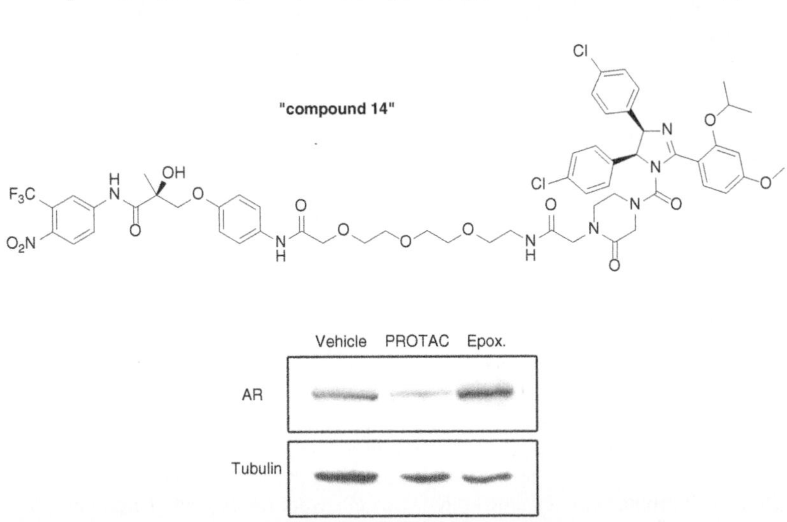

Figure 11.14 The first all small molecule based PROTAC and verification of proteasome mediated AR degradation. *Source:* The image is copied from [11] according to STM permission guidelines.

Later, a more potent AR PROTAC was discovered in the Crews lab by applying the VHL ligand to recruit the E3 ligase [25]. DC_{50} is the concentration of a compound required to achieve 50% degradation. D_{max} is the maximum degradation achieved by a compound. DC_{50} and D_{max} values were calculated using quantified band intensities from western blots or from the degradation curve of an ELISA assay. These two are frequently used to evaluate the degradation activity of a degrader. While the previous compound SARD279 had a DC_{50} of 1.1 μM and a D_{max} of 69%, ARCC-4 turned out to be much more potent, with a DC_{50} of 5 nM and D_{max} of 98%. ARCC-4 was also potent in inhibiting proliferation of AR expressing prostate cancer cell lines such as LNCaP, VCaP (vertebral cancer of the prostate), LNCaP/AR, but not PC-3 cells that are deficient in AR expression.

More importantly, as shown in Figure 11.16b, ARCC-4 had not only degradation activity against wild type AR, but also degradation activity against a number of clinically relevant AR mutants such as AR-7877A, AR-H874Y, AR-F876L, AR-L702H, and AR-M896V. AR mutations are observed in samples from patients with metastatic prostate cancer after drug treatment. These mutants

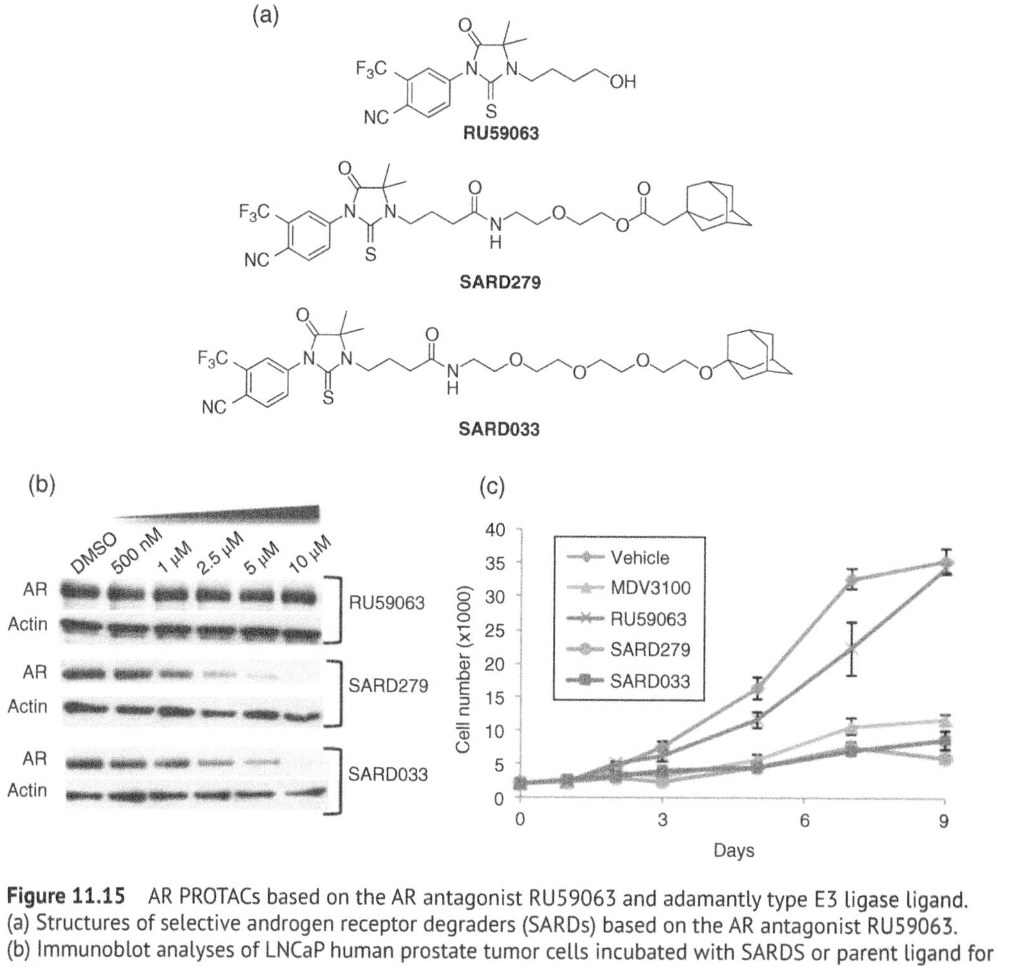

Figure 11.15 AR PROTACs based on the AR antagonist RU59063 and adamantly type E3 ligase ligand. (a) Structures of selective androgen receptor degraders (SARDs) based on the AR antagonist RU59063. (b) Immunoblot analyses of LNCaP human prostate tumor cells incubated with SARDS or parent ligand for 24 hours. (c) Inhibition of LNCaP cell proliferation by SARD PROTACs. Cells were treated with compounds at 3 μM and in the presence of 10% fetal bovine serum (FBS) for nine days. Viable cells were counted at each of the indicated timepoints. *Source:* The images are copied from [24] according to STM permission guidelines.

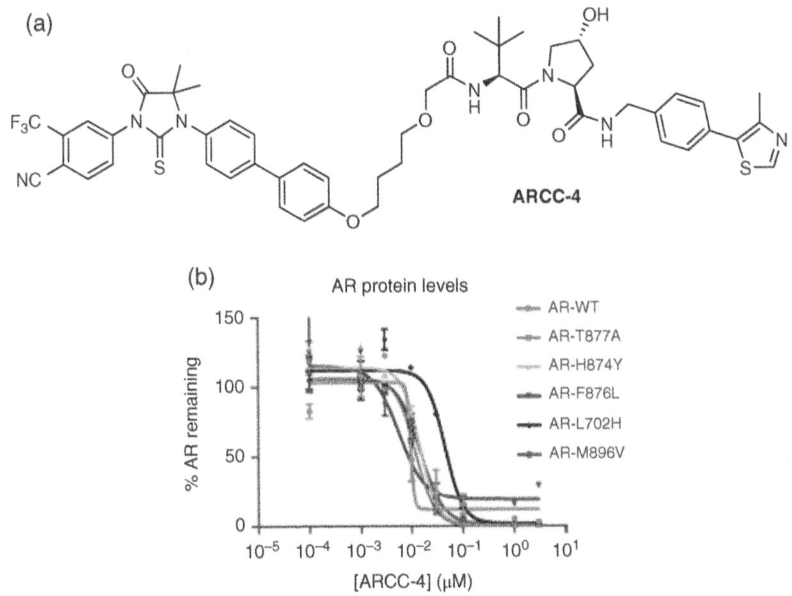

Figure 11.16 AR PROTAC ARCC-4. (a) Structure of ARCC-4. (b) Degradation curve of ARCC-4 on AR-wt and AR mutants. *Source:* The images are copied from [25] according to STM permission guidelines.

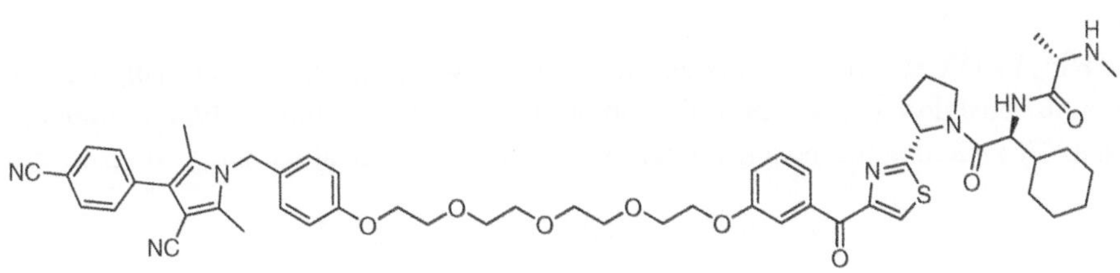

Figure 11.17 AR degrader SNIPER(AR)-51.

induce drug resistance by various mechanisms. For example, F876L and T877A mutants convert AR antagonists into agonists. Some mutants allow other hormones to activate the AR signaling pathway. Successful degradation of these mutants using the PROTAC approach suggests a potential pathway for treating drug resistant prostate cancer.

More AR PROTACs have been reported by other labs. Naito and co-workers from Takeda reported a class of AR degraders prepared by conjugation of AR ligands and inhibitor of apoptosis protein (IAP) ligands [26]. They were named specific and nongenetic inhibitors of apoptosis protein dependent protein erasers (SNIPERs). These AR degraders recruit cellular inhibitor of apoptosis protein 1 (cIAP1) ubiquitin ligases to degrade target proteins. After optimization on the linker and the ligase ligand moiety, SNIPER(AR)-51 (Figure 11.17) was identified. This compound effectively facilitated AR degradation, downregulated AR-dependent gene expression, and inhibited proliferation of AR-positive prostate cancer cells. It was well noted that "the ligand for the target protein, ligand for E3 ligase, linker composition, and linker length are all important factors that affect the efficacy of target protein degradation." Unlike other PROTACs, SNIPERs can induce the degradation of both the target protein and cIAP1. On the one hand, this feature may limit the full potential and duration of target protein degradation due to a gradual decrease in the E3 ligase cIAP1, but on the other hand, it may enhance the antitumor efficacy by promoting cancer cell apoptosis since cIAP1 is an antiapoptotic protein frequently overexpressed in human tumor cells and is believed to be involved in drug resistance. Therefore, the SNIPER strategy remains a promising tool in the protein degrader family of antitumor molecules.

Wang's lab recently reported a highly potent AR degrader ARD-69 (Figure 11.18) for the treatment of prostate cancer [27]. This family of AR PROTACs was based on the AR antagonist tetramethyl cyclobutylbenzamide scaffold and VHL E3 ligase ligand. Since the molecule is composed of three moieties, AR antagonist, linker and VHL ligand, the optimization was performed on each moiety individually and was repeated over several rounds to achieve collaborative optimization. This extensive medicinal chemistry campaign eventually led to the discovery of ARD-69, which achieved DC_{50} values of 0.86, 0.76, and 10.4 nM and D_{max} > 95% for the full length androgen receptor (AR-fl) in LNCaP, VCaP, and 22Rv1 AR+ prostate cancer cell lines, respectively. ARD-69 was also highly potent in inhibiting tumor cell proliferation with an IC_{50} of 0.25 nM for LNCaP cells, 0.34 nM for VCaP cells and 183 nM for 22RV1 cells. Overall, ARD-69 is about 100-fold more potent than enzalutamide in these cell lines. This compound was dosed in a VCaP Xenograft mouse model by IP injection at 50 mg/kg. Western blot analysis of the tumor tissue showed long-lasting AR degradation up to 48 hours.

The *in vivo* efficacy of a similar AR PROTAC, ARD-61 (Figure 11.19a), was reported by Kregel et al. [28]. This AR PROTAC differed from ARD-69 only on the VHL ligand moiety. It was highly potent in degrading the AR-fl and resulted in higher inhibition of tumor cell proliferation compared with AR antagonist drugs that are on the market. AR variants, particularly AR-v7, are believed to be responsible for most drug resistance. Although ARD-61 did not directly degrade AR-v7, it inhibited tumor cell growth in models overexpressing AR-v7. The authors also developed a number of isogenic prostate cell line models in which AR-v7 was highly expressed. ARD-61 maintained its anti-proliferation activity. In a VCaP Xenograft model, ARD-61 was given by IP injection at 25 mg/kg and 50 mg/kg. It demonstrated dose dependent inhibition of tumor growth (Figure 11.19b). In another castration resistant VCaP Xenograft model, animals were treated with enzalutamide (10 mg/kg) for

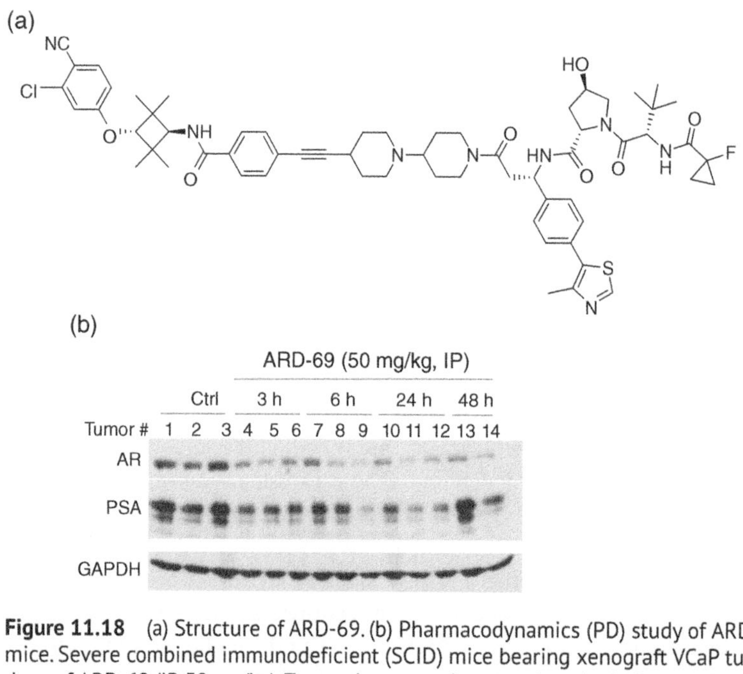

Figure 11.18 (a) Structure of ARD-69. (b) Pharmacodynamics (PD) study of ARD-69 in VCaP tumor tissue in mice. Severe combined immunodeficient (SCID) mice bearing xenograft VCaP tumors were given a single dose of ARD-69 (IP, 50 mg/kg). Tumor tissue was harvested at the indicated time points for western blot. *Source:* The image is copied from [27] according to STM permission guidelines.

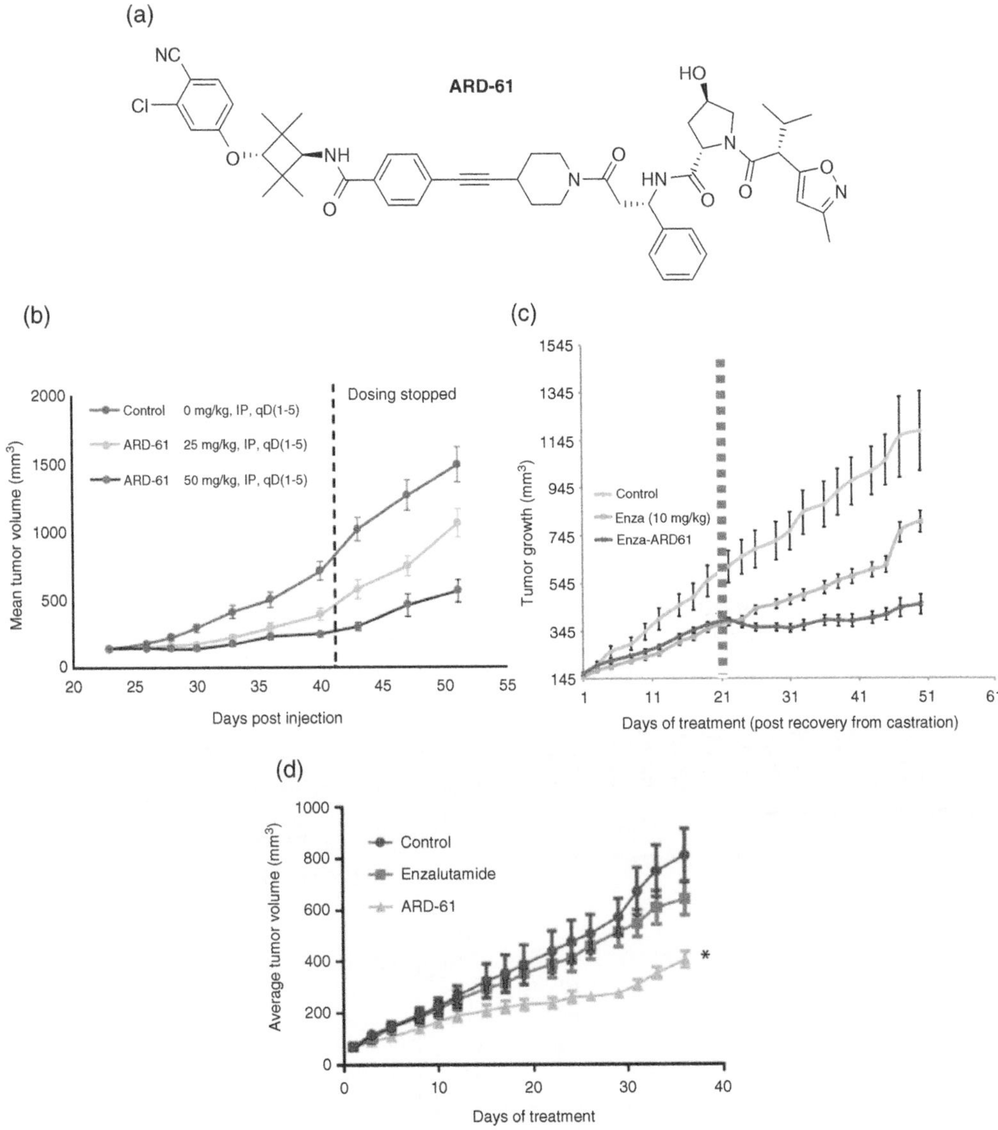

Figure 11.19 Androgen receptor targeting PROTAC ARD-61. (a) Structure of ARD-61. (b) Intact mouse VCaP xenograft model. The dashed line indicates cessation of dosing. (c) Castration resistant VCaP mouse model. Mice were injected with VCaP cells subcutaneously, and when the tumors reached 200 mm^3 in size, the mice were castrated. Once the tumor grew back to the pre-castration size, the animals were treated with control (n = 10) or enzalutamide (10 mg/kg). After three weeks (21 days indicated by a dashed line), half of the mice (n = 8) treated with enzalutamide were switched to ARD-61 (50 mg/kg). (d) CWR-R1 Enz^R xenograft model. Castrated mice were injected subcutaneously with human prostate cancer CWR-R1 Enz^R cells, and when the tumors reached 100 mm^3 in size, the mice were treated with control (n = 10), enzalutamide (50 mg/kg, n = 10), or ARD-61 (IP, 50 mg/kg, n = 10) QD (once daily), 5 times a week throughout the duration of the experiment. *Source:* The image is copied from [28] according to STM permission guidelines.

three weeks and then were switched to ARD-61 (IP, 50 mg/kg) for continued treatment. The ARD-61 treated group showed more tumor growth inhibition compared with the enzalutamide treated group and thus demonstrated that the AR degrader ARD-61 is superior to the AR antagonist enzalutamide. An enzalutamide resistant CWR-R1 Enz^R model was established on castrated mice. In this model, enzalutamide showed no tumor inhibition, but ARD-61 did (Figure 11.19d). The AR PROTAC, although only targeting AR-fl, demonstrated the potential to treat drug resistant prostate cancers. Furthermore, the fact that AR degraders like ARD-61/69 showed no degradation of AR-v7, but some antiproliferation activity in AR-v7 expressing cancer cells is intriguing. Whether this is clinically relevant warrants further studies.

As previously mentioned, ARV-110 is the first orally bioavailable AR PROTAC which entered phase I clinical trials in 2019. Shortly thereafter, ARV-471, an orally bioavailable ER PROTAC, entered phase I trials in the same year. The structures of these oral PROTACs have not been disclosed by Arvinas, but some data have been released [29].

As shown in Figure 11.20, ARV-110 is a highly potent AR degrader with a DC_{50} ~1 nM. It robustly degraded the AR in AR expressing cell lines and was highly selective for the AR as demonstrated by proteomics analysis. It also degraded AR point mutations such as F876L, T877A, M896A, and H874Y. Through degradation of the AR, it suppressed the expression of the PSA and inhibited AR-dependent cell proliferation. When tested in mouse xenograft models, ARV110 achieved >90% AR degradation in tumor tissue at 1 mg/kg QD oral dosing. It was efficacious in the castrated VCaP

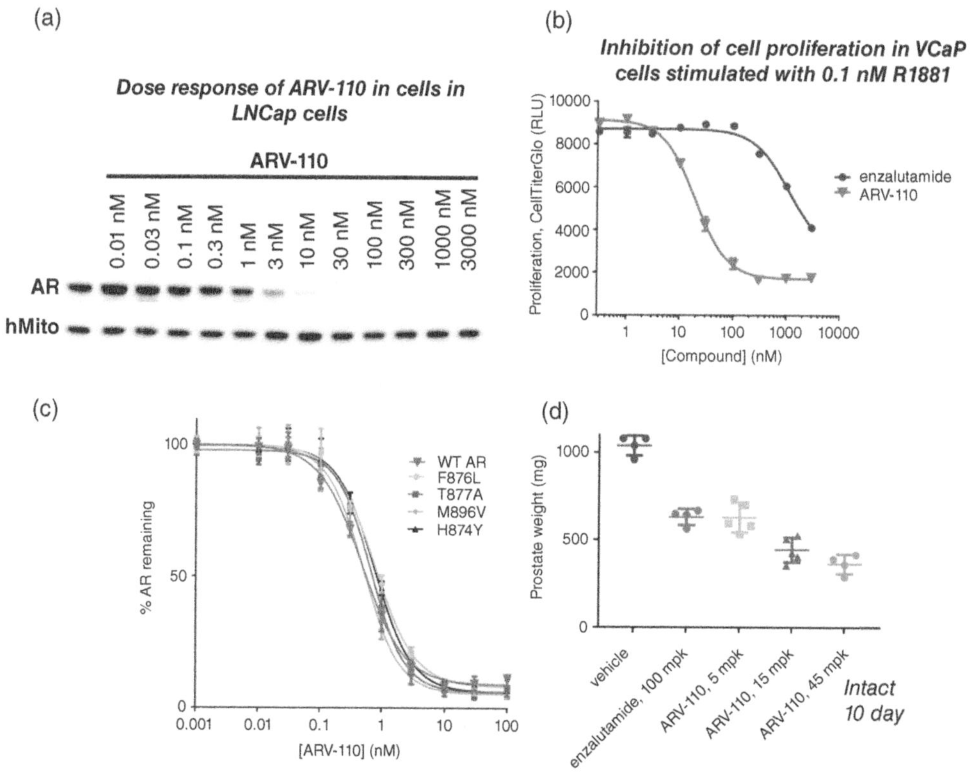

Figure 11.20 (a) Western blot analysis of ARV-110 in LNCaP cells showing potent dose dependent degradation of the AR. (b) Antiproliferation activity of ARV-110 in VCaP cells compared with enzalutamide. (c) Degradation of AR mutants by ARV-110. (d) ARV-110 demonstrated involution of rat prostate. (e,f) ARV-110 demonstrates (e) tumor growth inhibition and (f) prostate specific antigen (PSA) reductions in an AR expressing prostate patient derived xenograft model TM00298. (g,h) ARV-110 demonstrates tumor growth inhibition in (g) castrated and (h) intact VCaP models. (i) AR amplified, TMPRSS2-ERG translocation positive VCaP tumors were passaged in castrated, enzalutamide treated (20 mpk) mice for three years. In this enzalutamide resistant model, ARV-110 retains efficacy. *Source:* PROTAC is a registered trademark of Arvinas Operations, Inc. and is used with permission from Arvinas Operations, Inc. The image is copied from [29] with permission from Arvinas Operations, Inc.

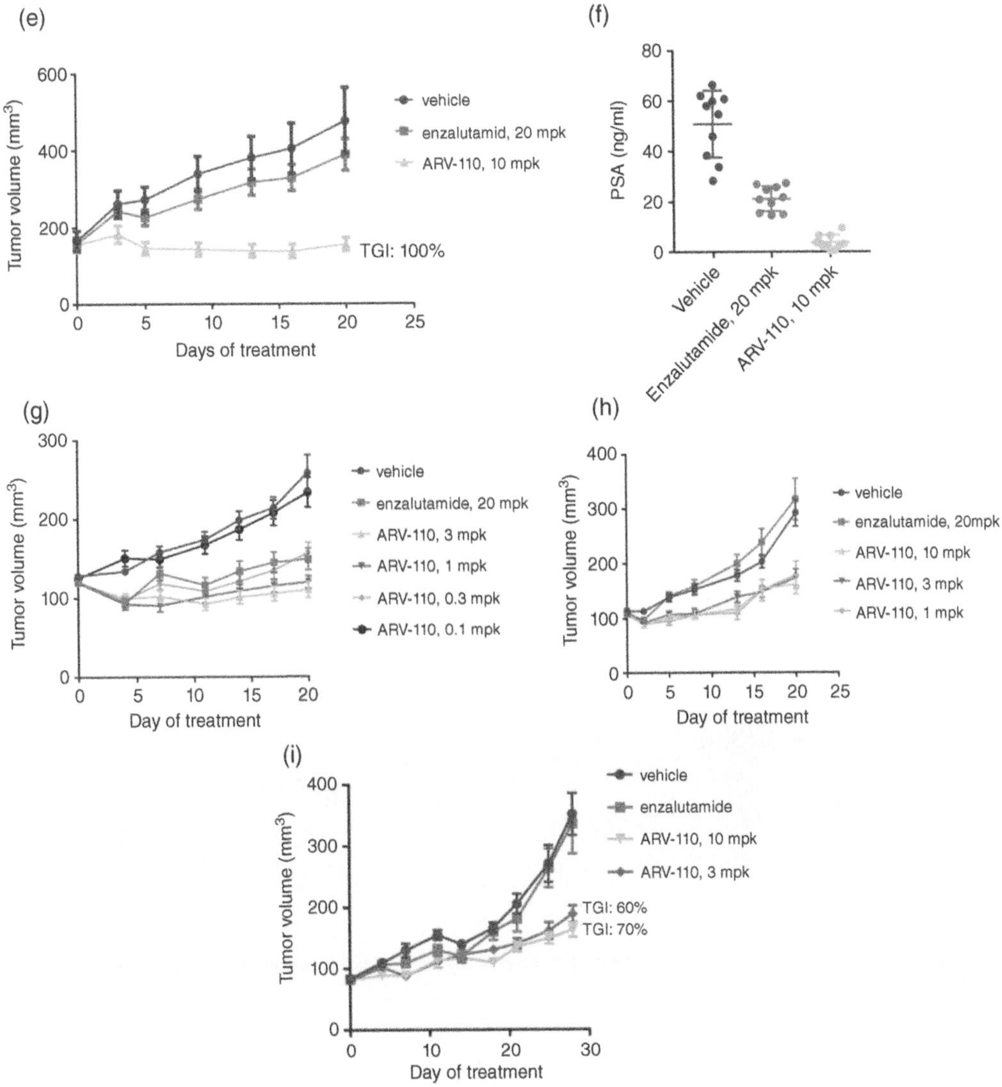

Figure 11.20 (Continued)

model at 0.3 mg/kg. It was also efficacious in the intact VCaP model at 1 mg/kg where enzalutamide was not efficacious at 20 mg/kg. In an intact rat, prostate involution was confirmed after a QD dose of 45 mg/kg for 10 days. ARV-110 demonstrated tumor growth inhibition and PSA reduction in the AR expressing prostate patient derived xenograft (PDX) mouse model TM00298. ARV-110 was also efficacious in an enzalutamide resistant mouse model at 3 mg/kg. The physicochemical properties and PK profiles in preclinical setting have not been fully disclosed.

11.2.6 Novel PROTACs: Self-Assembled Click-Formed PROTACs, Photochemically Controlled PROTACs, Antibody-PROTAC Conjugates

The high MW of PROTAC molecules makes it challenging to achieve the desired drug-like properties. To address this challenge, the Heightman group reported a conceptually novel bio-orthogonal strategy using a relatively low MW POI ligand fragment and the E3 ligand fragment to assemble the linker *in vivo* using click chemistry to yield the PROTAC molecule for degradation (Figure 11.21) [30]. Trans-alkenes and teterazines are known click cycloaddition partners giving a smooth Diels–Alder reaction at room temperature. Thus, cells were treated sequentially with a fragment bearing both the ligand to BRD4 and a trans-cyclooctene moiety, and the other fragment bearing both thalidomide as CRBN ligand and a tetrazine moiety. The intracellular formation of the PROTAC compound was successful and subsequent degradation of BRD4 was confirmed by Western blots.

Another example of this strategy is shown in Figure 11.22. A covalent ERK 1/2 inhibitor was tethered to a trans-cyclooctene and it was allowed to react with a CRBN ligand tethered to a tetrazine moiety in cells to generate the ERK-CLIPTAC compound. Degradation of ERK was confirmed by Western blot.

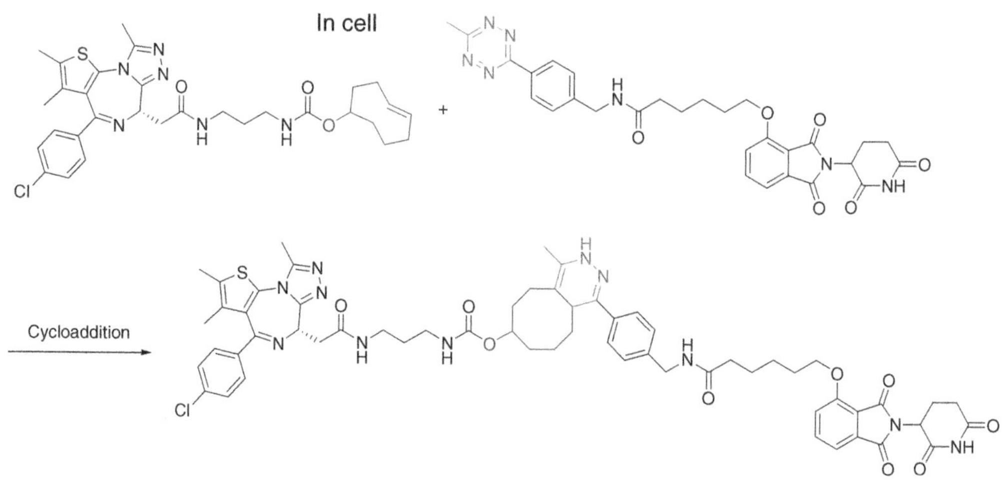

Figure 11.21 Bio-orthogonal strategy using click chemistry to assemble the linker *in vivo* to yield the PROTAC molecule.

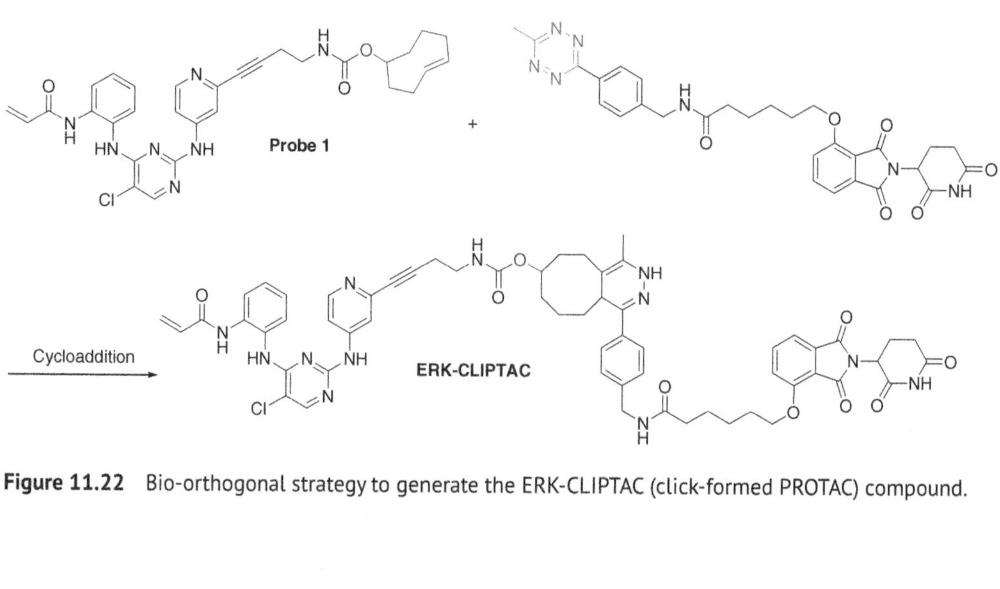

Figure 11.22 Bio-orthogonal strategy to generate the ERK-CLIPTAC (click-formed PROTAC) compound.

The photochemically controlled PROTAC (PHOTAC) reported by the Trauner group demonstrated how a photochemically induced linker conformation change affected degradation activity [31]. This concept was demonstrated by a BET degrader and a FKBP12 degrader. As shown in Figure 11.23, PHOTAC-I-3 used the JQ1 class of BET ligand and the CRBN ligand. A diazo group was used in its linker as the photo-switch. Normally, the diazo group exists in the lower energy trans-conformation which is inactive. Irradiating with 390 nm light causes the molecule to adopt the higher energy cis-conformation that demonstrates degradation activity. The ultraviolet–visible (UV–vis) spectrum of PHOTAC-I-3 indicates how the conformation changes with different wavelengths of light. The conformational change led to different degrees of degradation toward BRD4 and BRD2. When in the dark, the molecule showed no degradation activity, while under UV light with wavelength of 390 nm, the compound showed the maximum degradation. A similar FKBP123 targeting PHOTAC is shown in Figure 11.24.

Drugs need to be delivered to the desired tissue for therapeutic effects and to avoid side effects. To achieve tissue specificity, an antibody-PROTAC conjugate strategy was reported by Tate's group [32]. Since the E3 ligase involved in protein degradation is widely expressed in various tissues, it is difficult to modulate a PROTAC molecule's distribution into a specific tissue or cell type. Analogous to antibody-drug conjugates (ADCs), an antibody-PROTAC conjugate was designed as an alternative approach to selectively deliver a broad spectrum of PROTAC molecules into a

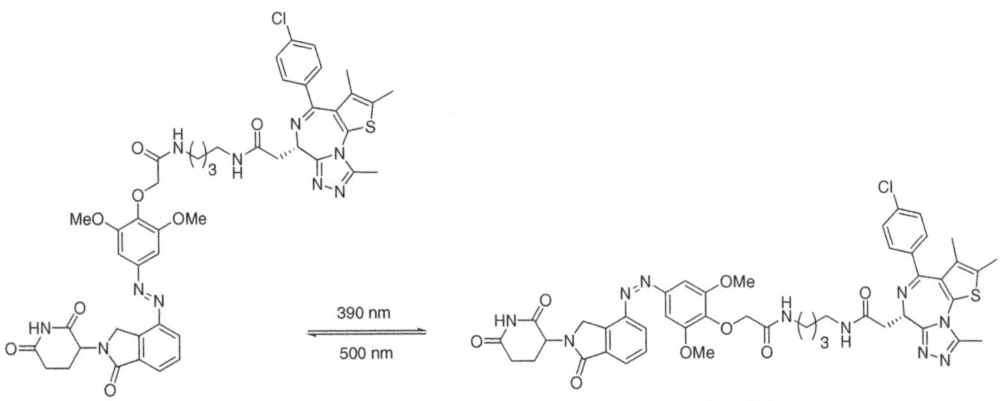

trans-PHOTAC-I-3 cis-PHOTAC-I-3

Figure 11.23 BET targeting PHOTAC-I-3 [31].

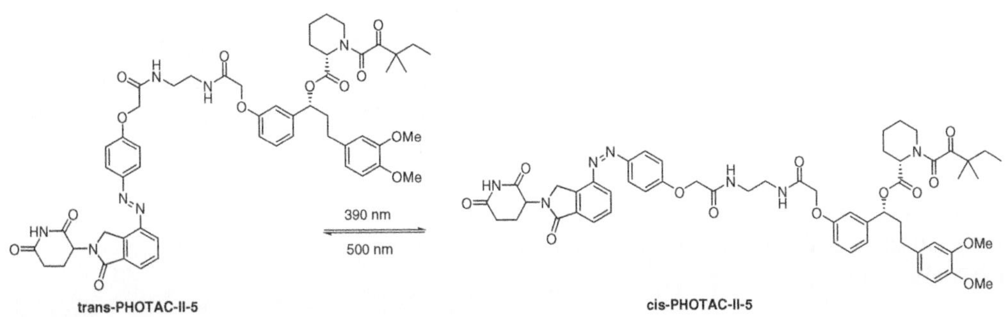

trans-PHOTAC-II-5 cis-PHOTAC-II-5

Figure 11.24 FKBP123 targeting PHOTAC-II-5.

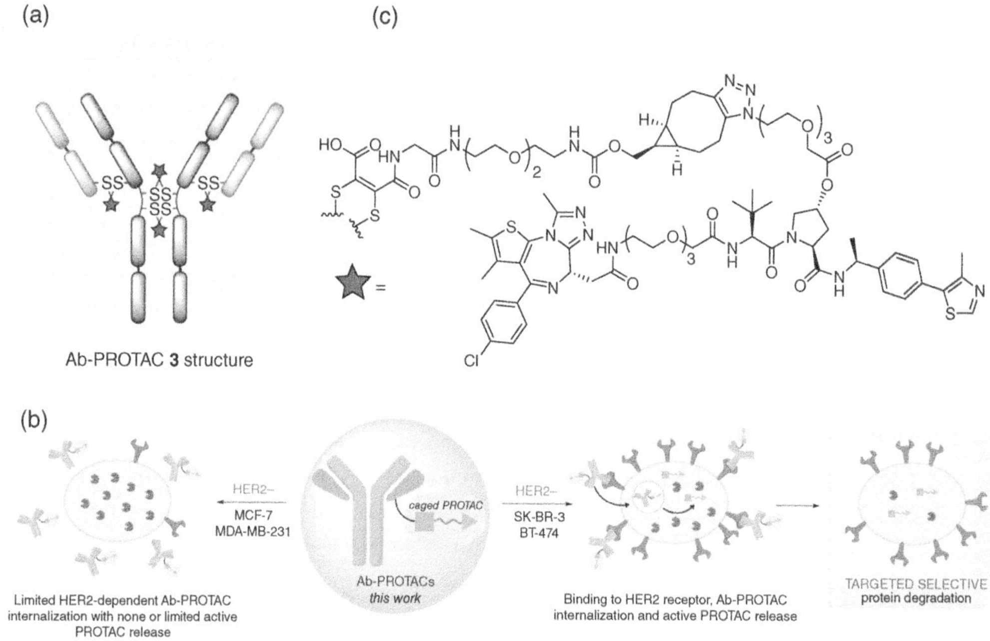

Figure 11.25 Antibody-PROTAC conjugate. *Source:* The image is copied from [32] according to STM permission guidelines.

specific cell type. As shown in Figure 11.25a, Ab-PROTAC 3 was prepared by linking the HER2 inhibitor antibody trastuzumab with a BRD4 PROTAC through a linker constructed by click chemistry. This conjugate bound to the HER2 receptor and was then subjected to endosomal internalization and lysosomal release of the active BRD4 PROTAC by hydrolysis of the ester bond in the linker between the antibody and the PROTAC moiety (Figure 11.25b). It was shown that this conjugate only degraded BRD4 in HER2 positive breast cancer cell lines, but not HER2 negative cell lines. It made this study a proof of concept for tissue specific BRD4 degradation and demonstrated significant therapeutic potential.

11.2.7 Examples of Small Molecule Based PROTACs

It has been more than a decade since the discovery of the first all small molecule based AR PROTAC in 2008. The revolutionary concept and continued advances in this field are inspiring more and more scientists, from both academia and industry, to dedicate efforts in this field. More targets are now degraded by diverse PROTAC compounds. For targets with high therapeutic importance, different PROTAC compounds are emerging to achieve better drug-like properties. Some previous reviews provided a timely summary of PROTACs [1]. An updated list of PROTACs is given in Table 11.2.

To date, the *in vitro* degradation of target proteins and subsequent antitumor activity have been well demonstrated. To advance into *in vivo* efficacy studies, one has to face the DMPK challenge of PROTACs.

Table 11.2 Examples of PROTACs.

Entry	Target (compound ID)	Structure	References
1	AR (compound **14**)	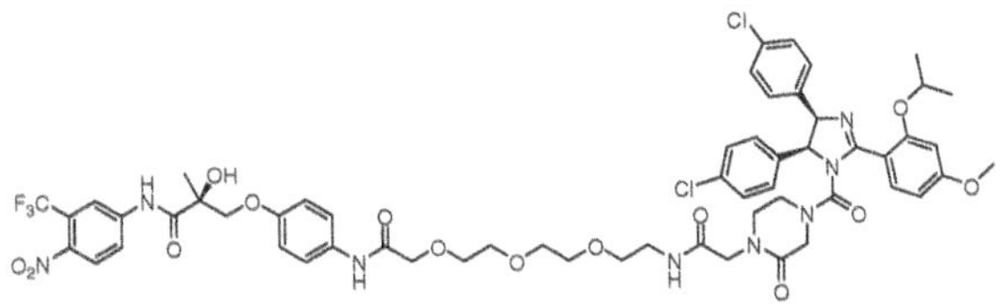	[11]
2	BRD4 (A1874)	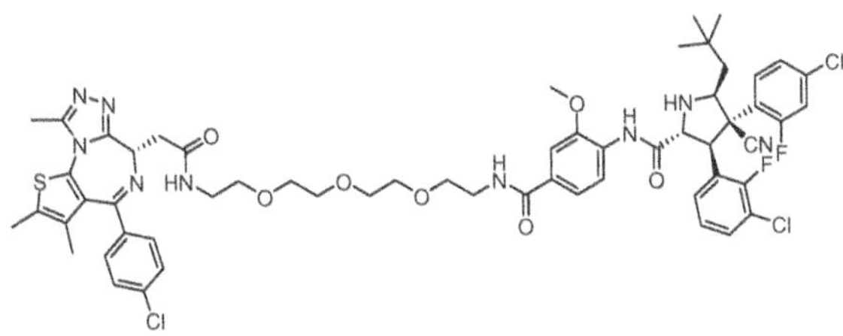	[33]
3	BRD4 (SNIPER(BRD4)-1)	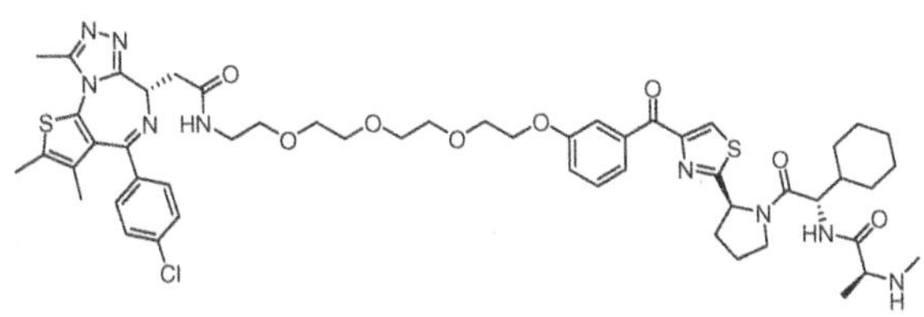	[18]
4	ABL (SNIPER(ABL)-39)	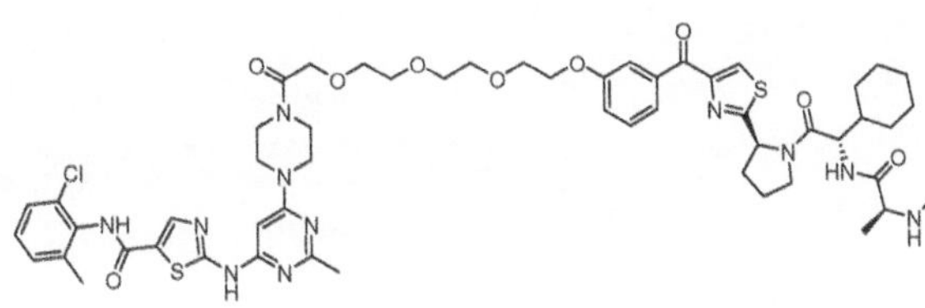	[20]
5	ERα (SNIPER(AR)-51)	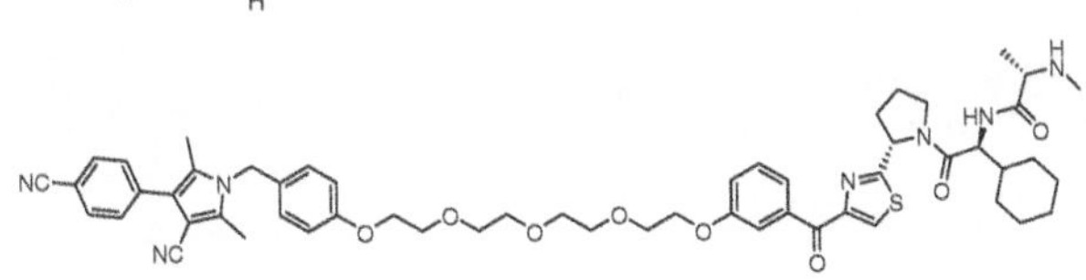	[26]
6	ERα (SNIPER(ER)-110)	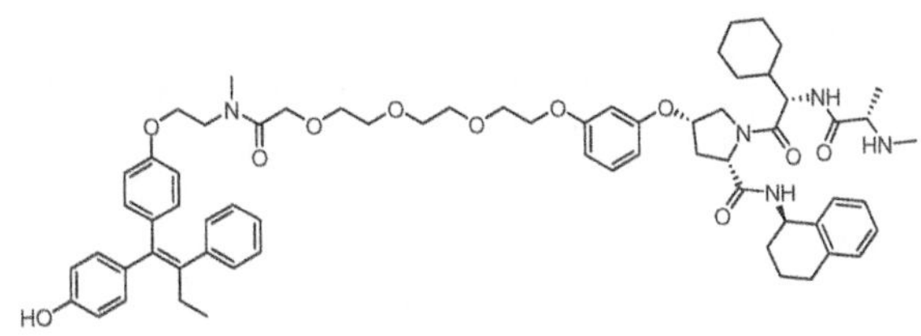	[34]
7	BRD4 (MZ1)	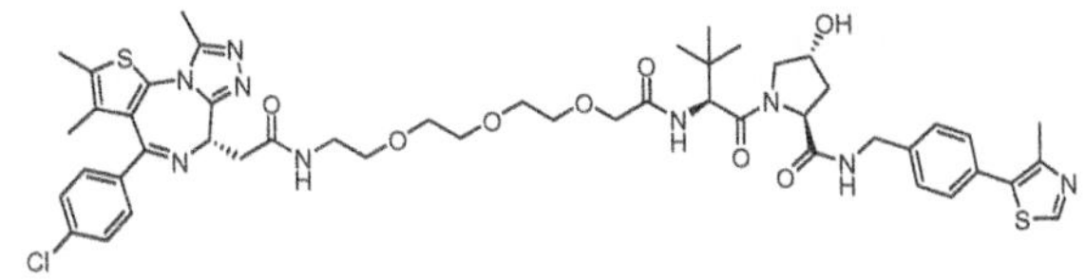	[35]

(Continued)

Table 11.2 (Continued)

Entry	Target (compound ID)	Structure	References
8	BRD4 (**12b**)	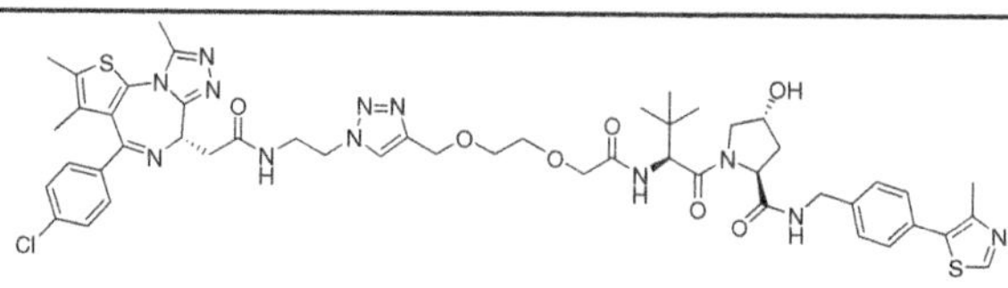	[36]
9	BRD4 (ARV-771)	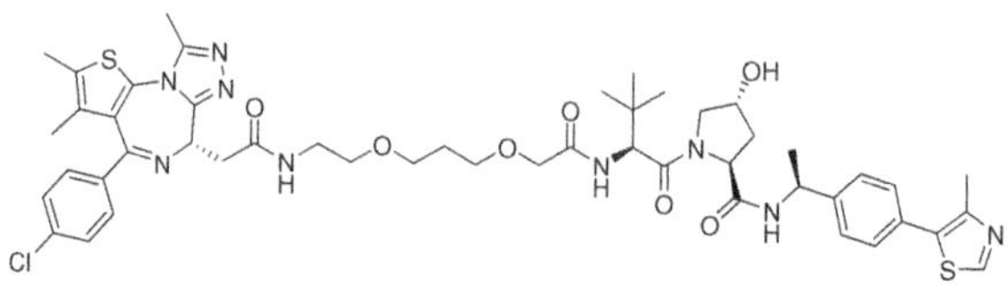	[37]
10	BET (AT1)	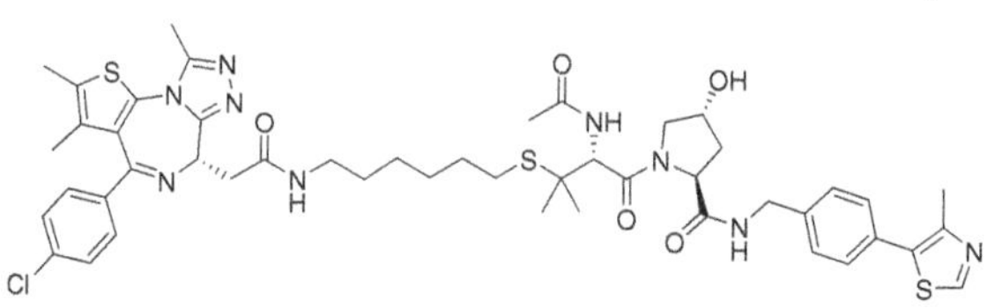	[22]
11	BET (MZT54)	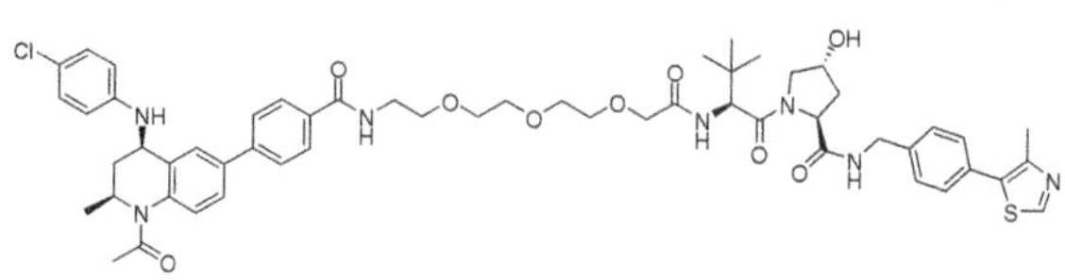	[38]
12	ERRα (PROTAC_ ERRα(1))	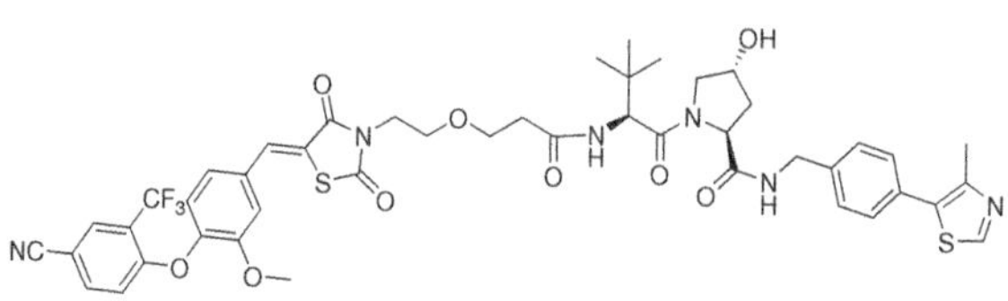	[39]
13	RIPK2 (PROTAC_RIPK2)	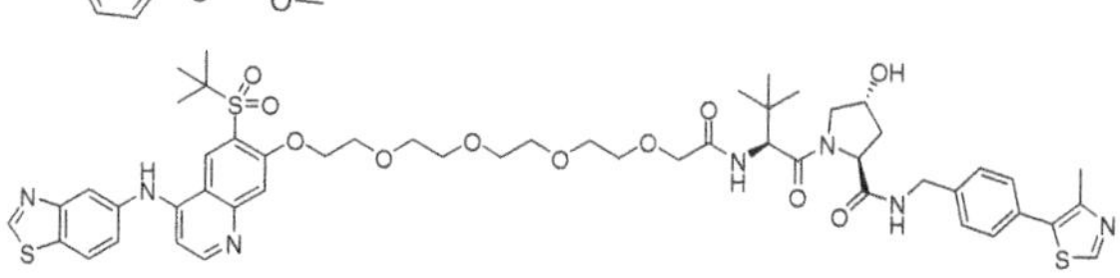	[39]
14	ABL (DAS-VHL)	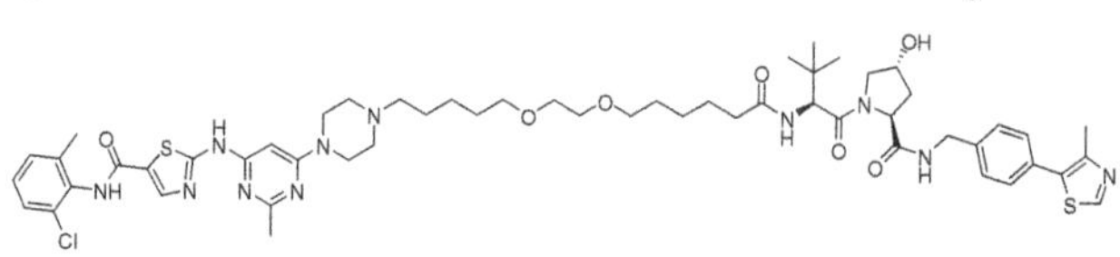	[40]
15	Multiple kinases (TL12-186)	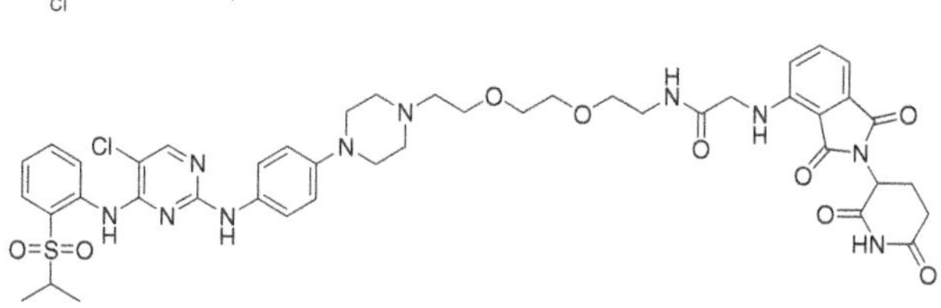	[41]
	FLT3 (FLT3 PROTAC)	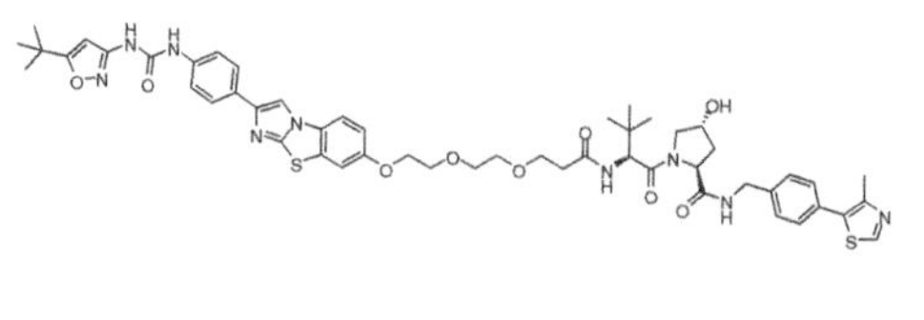	

Table 11.2 (Continued)

Entry	Target (compound ID)	Structure	References
16	FAK (PROTAC 3)	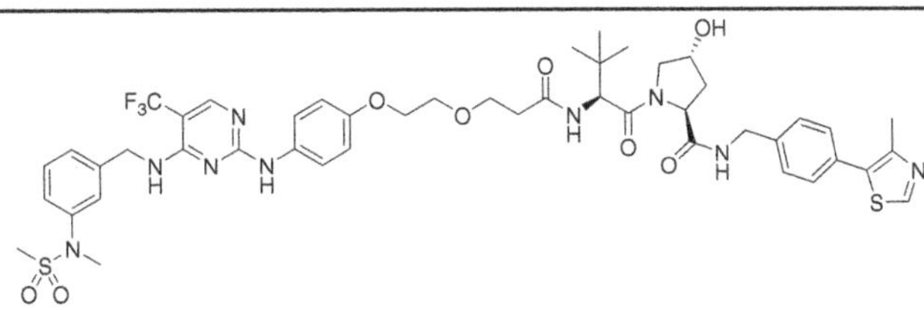	[42]
17	TBK1	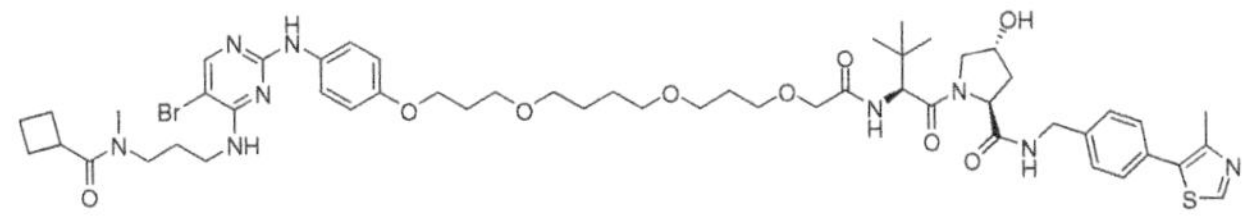	[43]
18	BRD7/9 (VZ185)	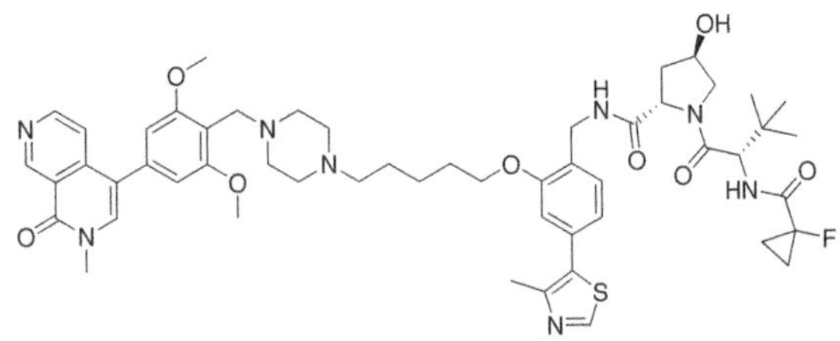	[44]
19	AR (ARD-69)	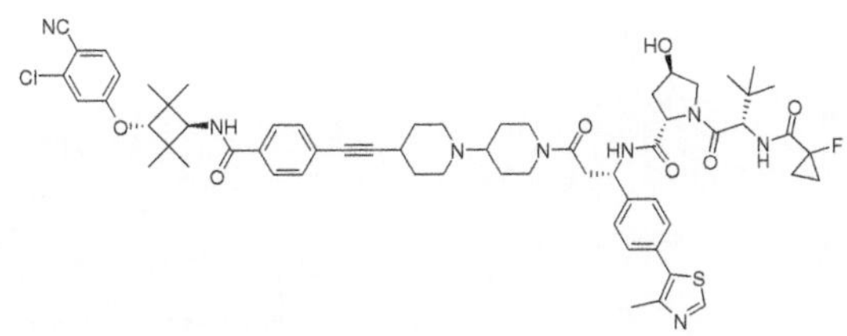	[27]
20	BRD4 (dBET1)	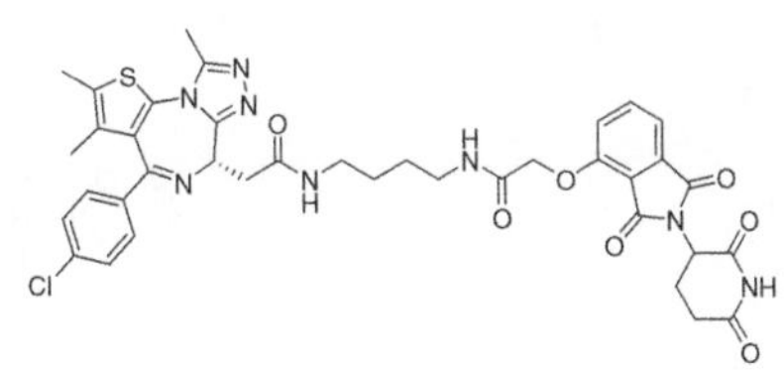	[45]
21	BRD4 (ARV-825)	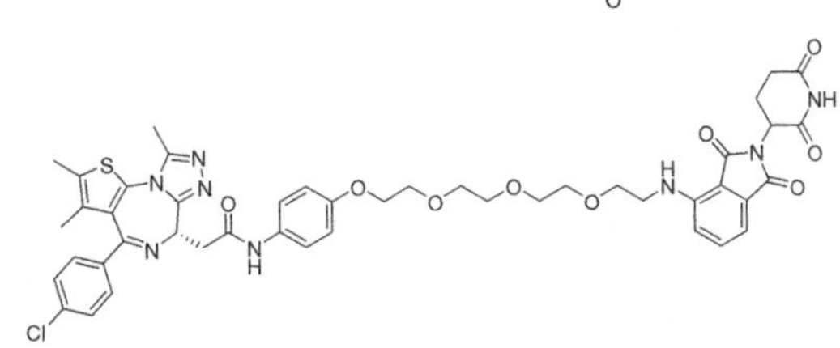	[46]
22	BET (BETd-260)	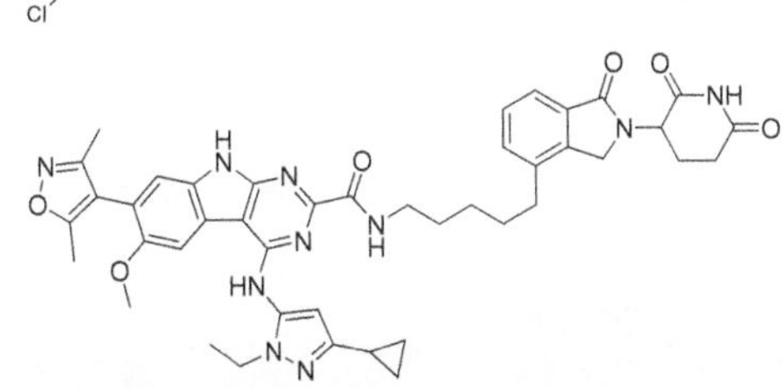	[47]

(Continued)

Table 11.2 (Continued)

Entry	Target (compound ID)	Structure	References
23	BET (QCA-570)	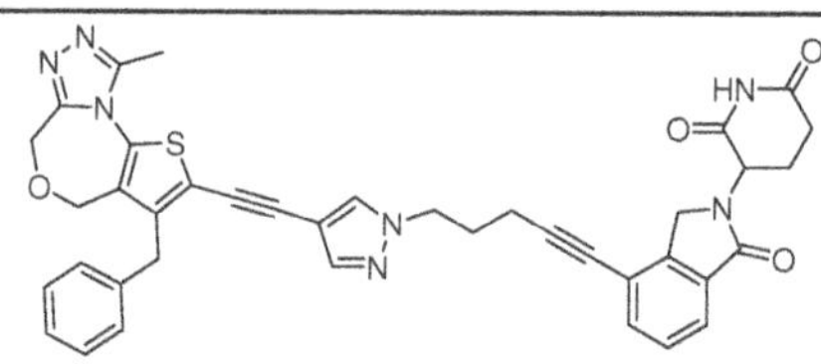	[48]
24	BCL6 (compound **15**)	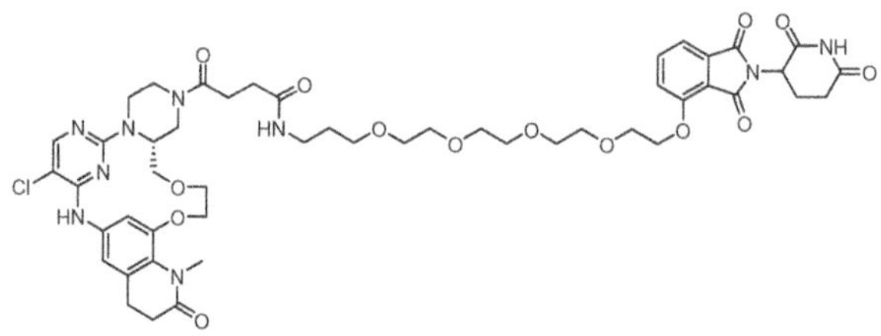	[49]
25	CDK8 (JH-XI-10)	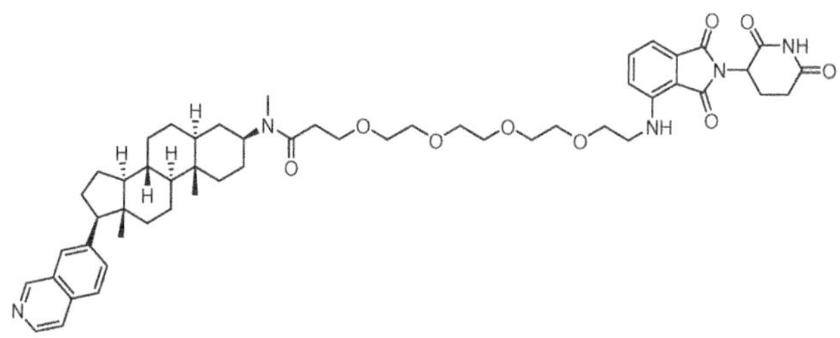	[50]
26	CDK9 (THAL-SNS-032)	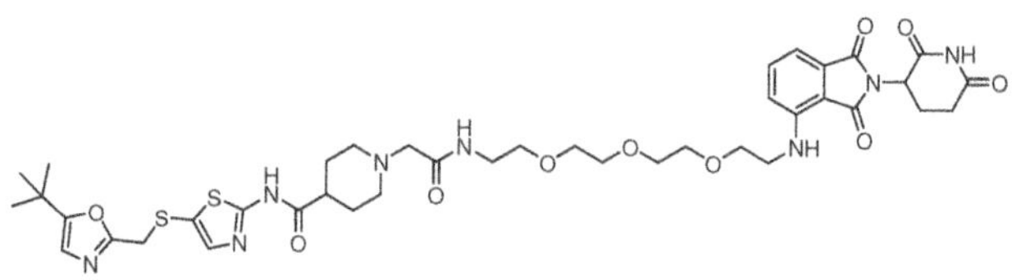	[51]
27	CDK9 (compound **3**)	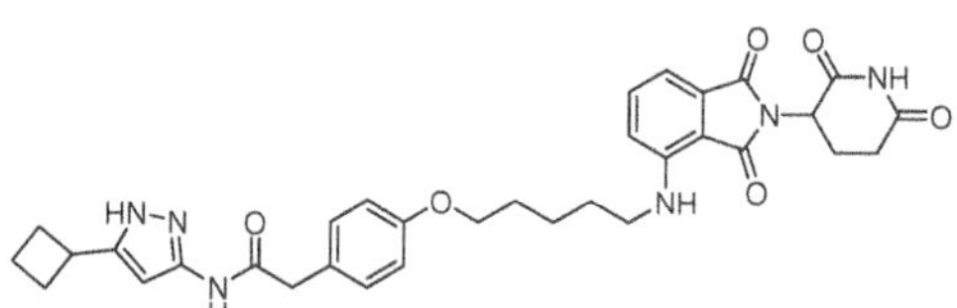	[52]
28	PI3K (compound **D**)	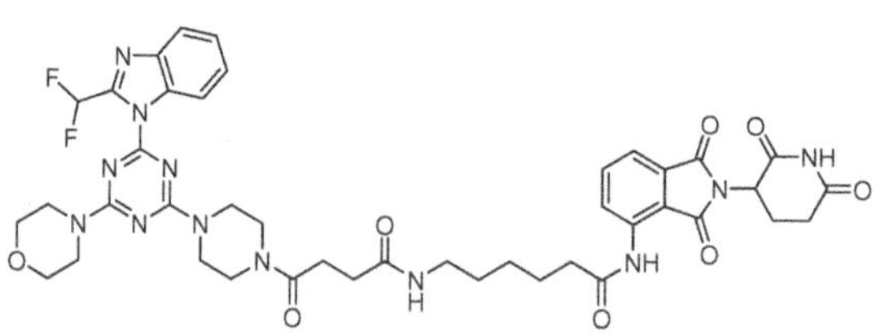	[53]
29	BTK (MT-802)	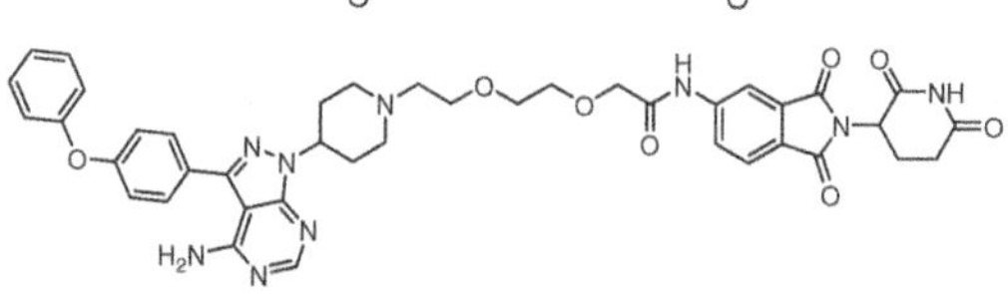	[54]

Table 11.2 (Continued)

Entry	Target (compound ID)	Structure	References
30	BTK (P13I)	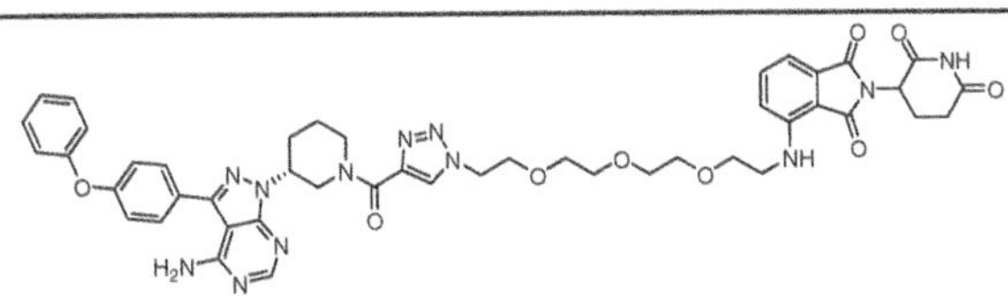	[55]
31	BTK (compound **10**)	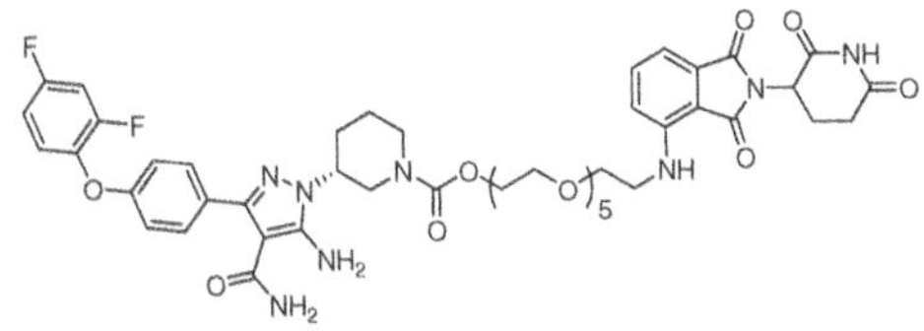	[56]
32	ALK (TL13-112)	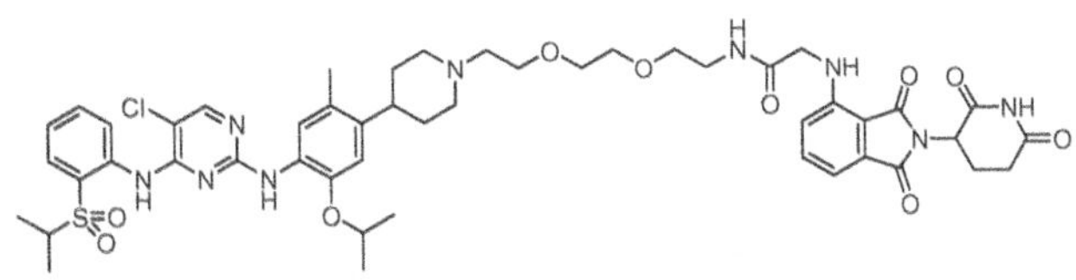	[57]
33	HDCA6 (compound **9C**)	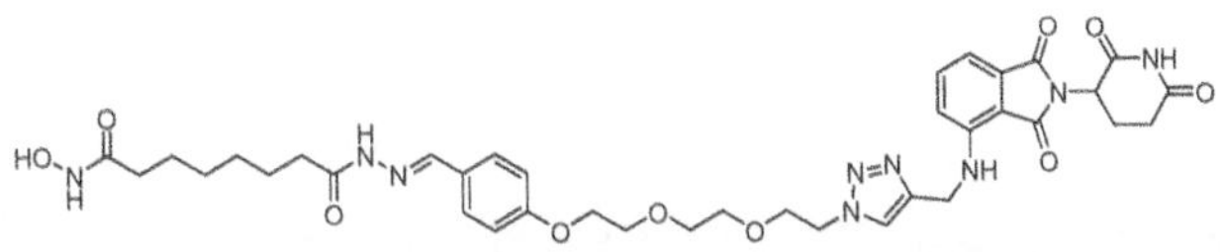	[58]
34	Sirtuin (compound **12**)	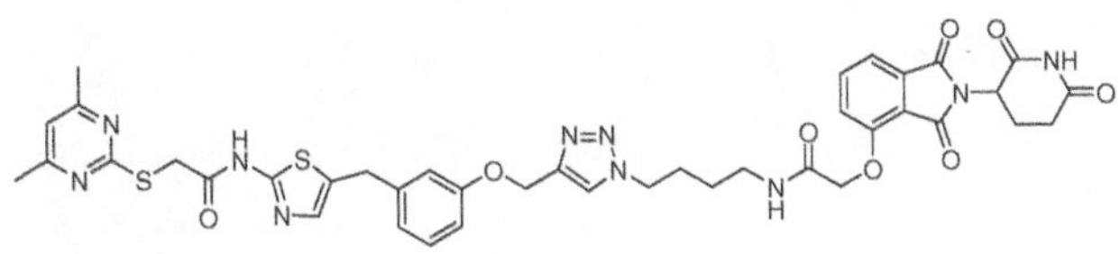	[59]
35	PCAF/GCN5 (GSK983)	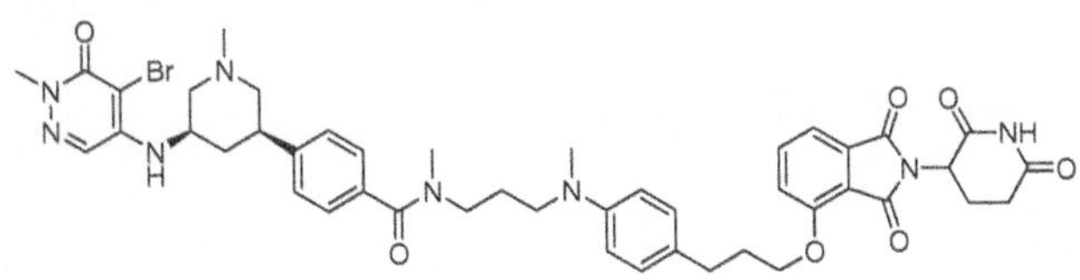	[60]
36	Pirin (CCT367776)	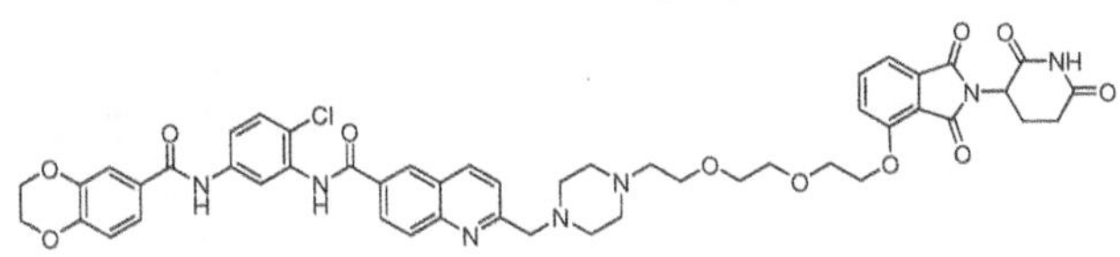	[61]
37	$FKBP12^{F36V}$ (dTAG13)	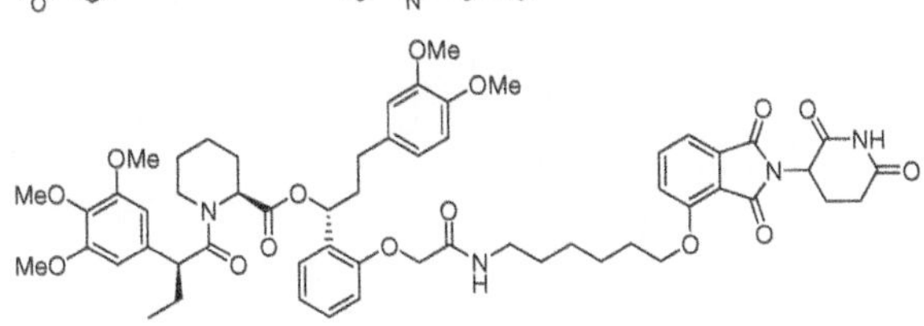	[62]

(Continued)

Table 11.2 (Continued)

Entry	Target (compound ID)	Structure	References
38	FKBP12 (RC32)	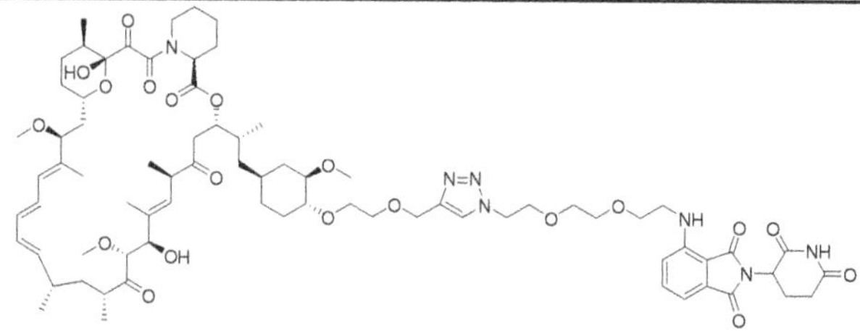	[63]
39	BCL-XL (XZ424)	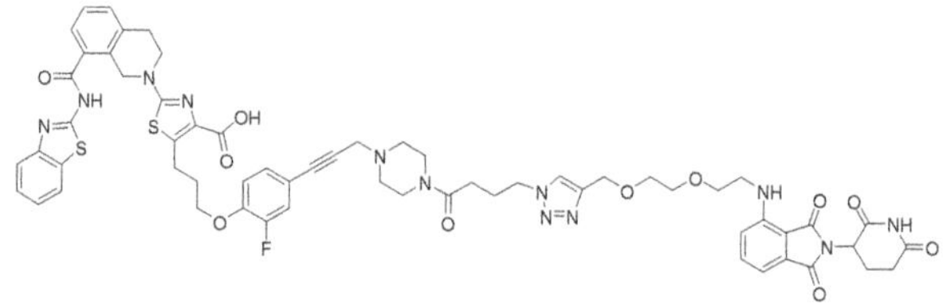	[64]
40	TrKC (IY-IY-pom)	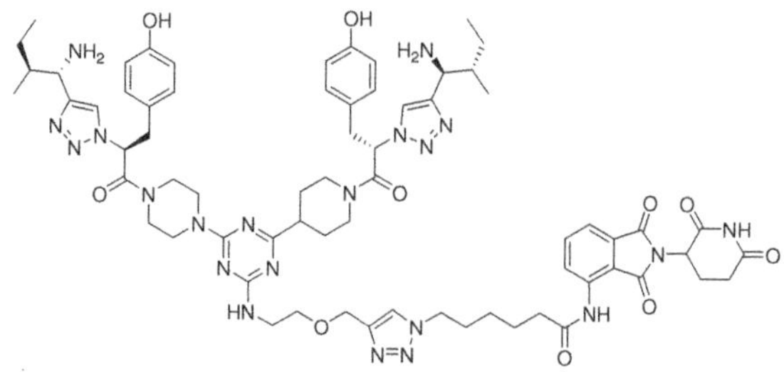	[65]
41	IRAK4 (compound **9**)	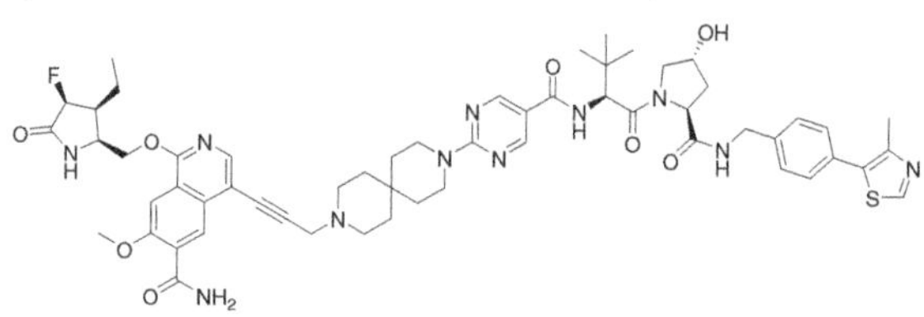	[66]
42	BCR-ABL1 (GMB-475)	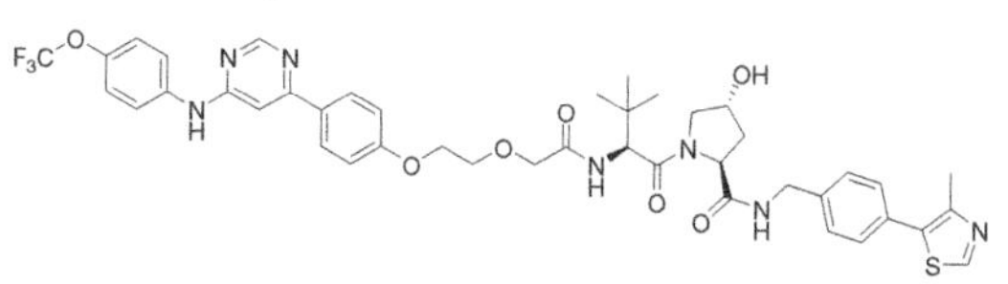	[67]
43	MDM2 (MD-224)	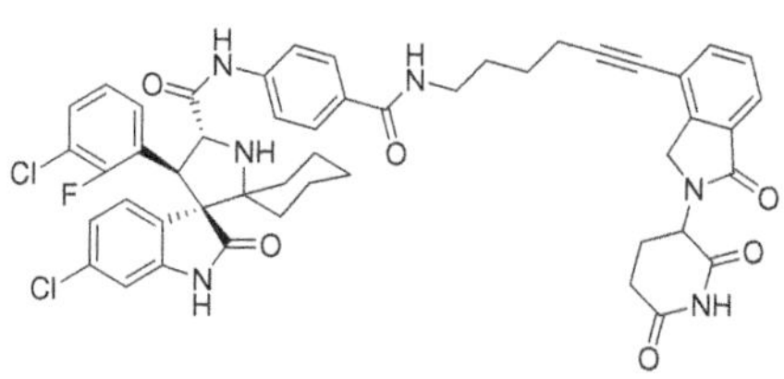	[68]
44	STAT3 (SD-36)	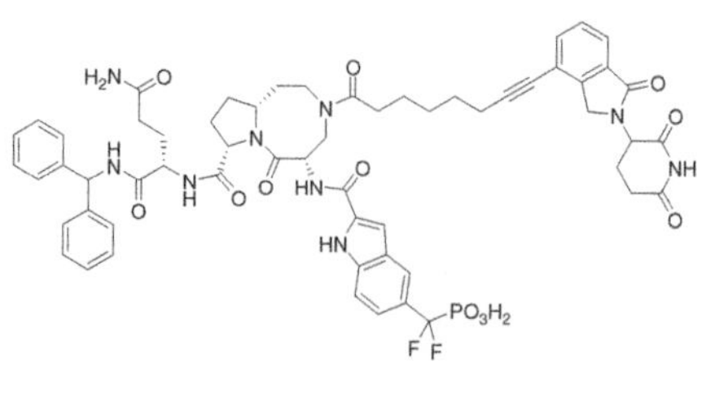	[69]

Table 11.2 (Continued)

Entry	Target (compound ID)	Structure	References
45	CDK6 (compound **6**)	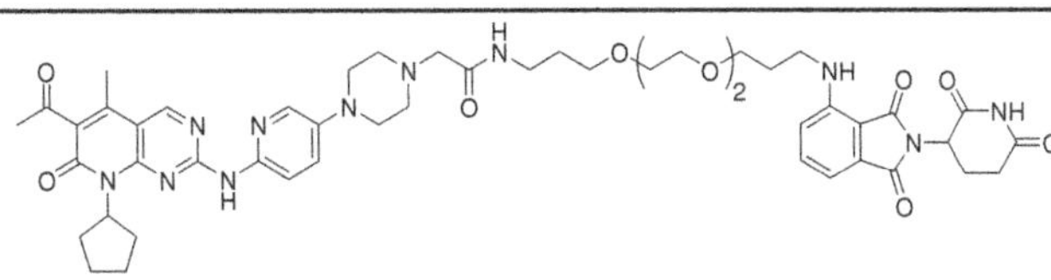	[70]
46	CDK4/6 (Pal-pom, rib-pom)	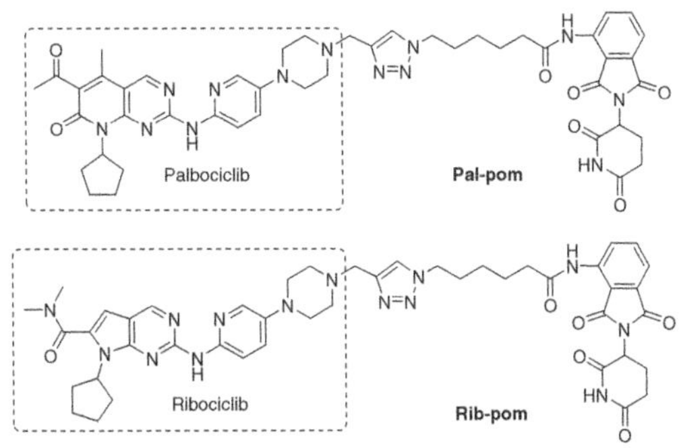	[71]
47	CDK6 (CP-10)	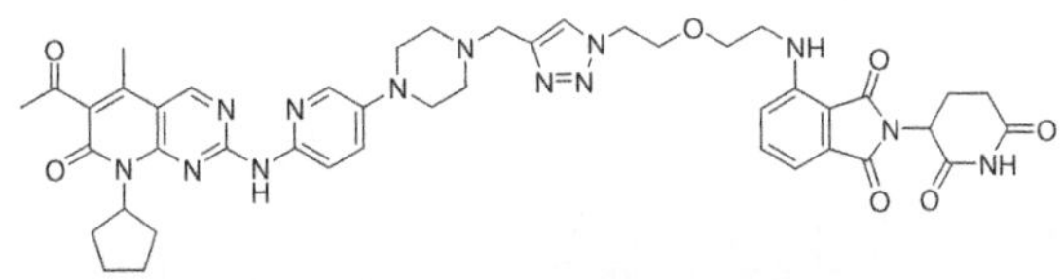	[72]
48	HDAC6	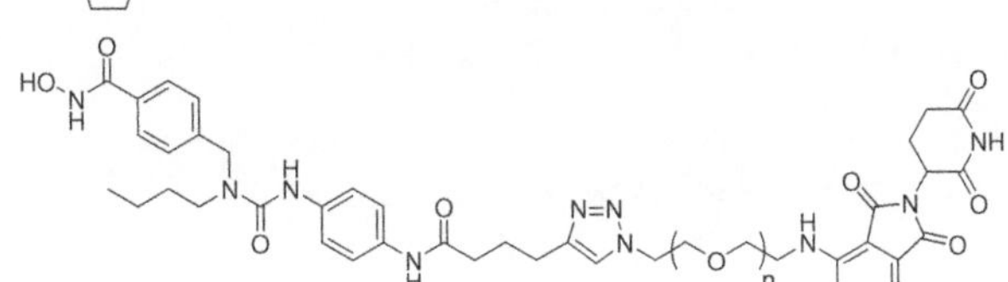	[73]
49	PARP1 (compound **3**)	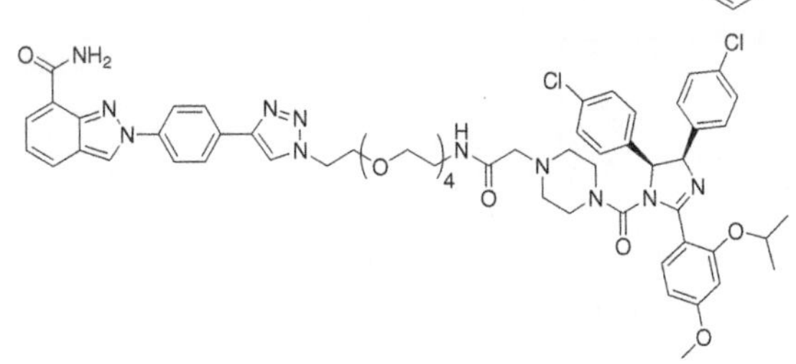	[74]
50	ERRα	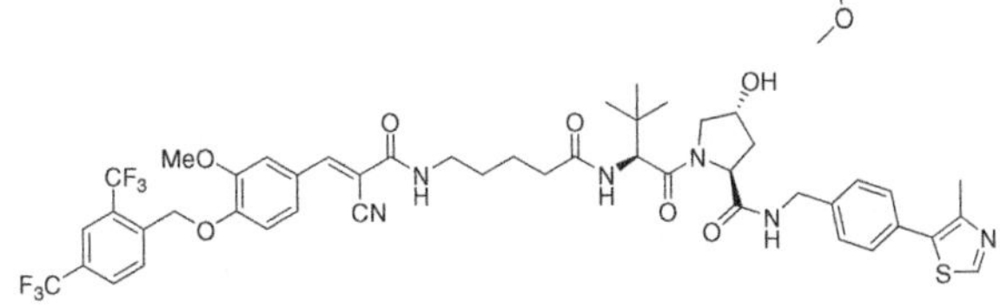	[75]
51	MCL1 (dMCL1-2)	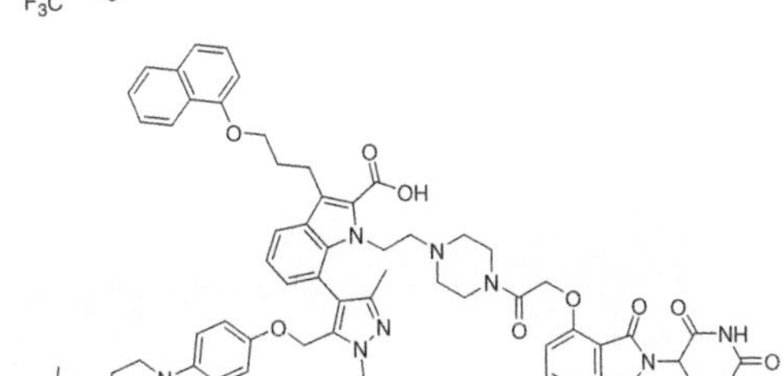	[76]

(Continued)

Table 11.2 (Continued)

Entry	Target (compound ID)	Structure	References
52	BCR-ABL (SIAIS178)	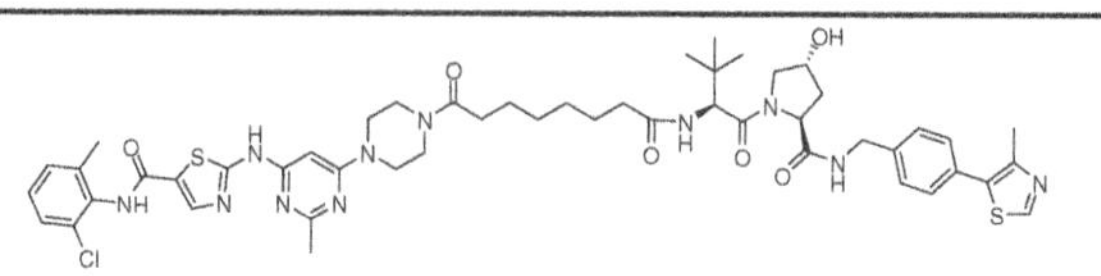	[77]
53	AR (ARD-266)	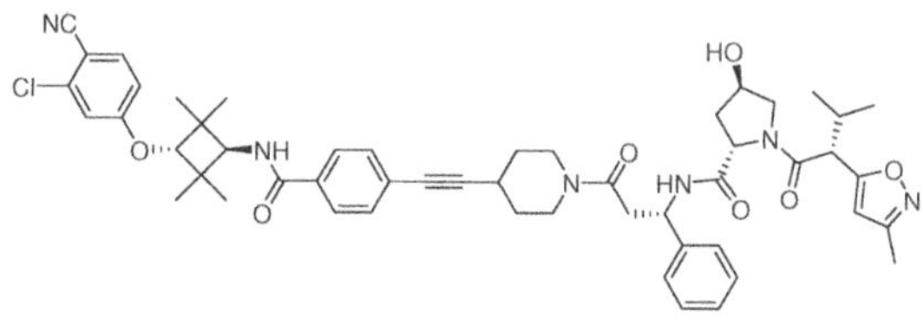	[78]
54	ER (ERD-308)	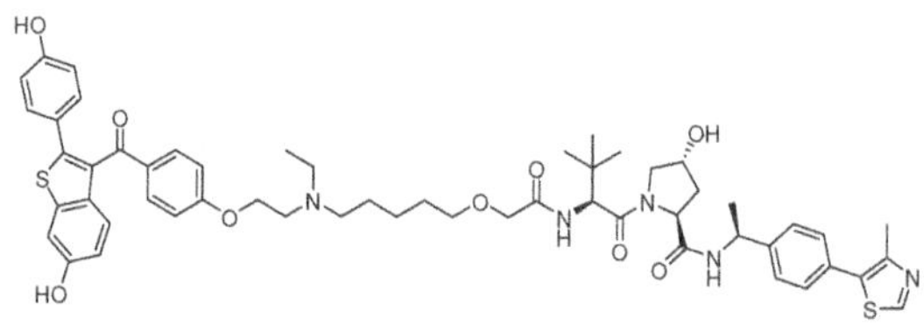	[79]
55	HDAC6	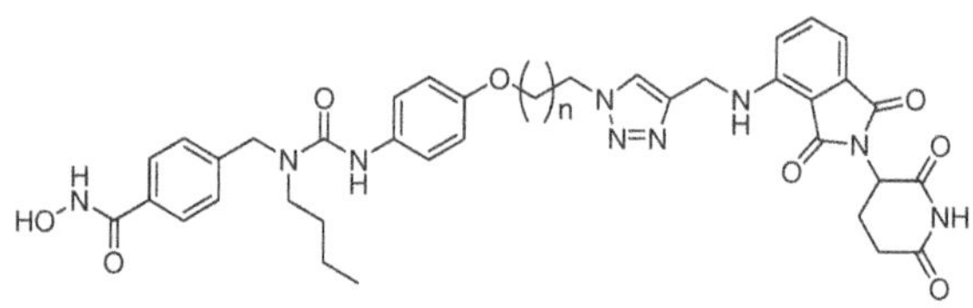	[80]
56	FAK (FC-11)	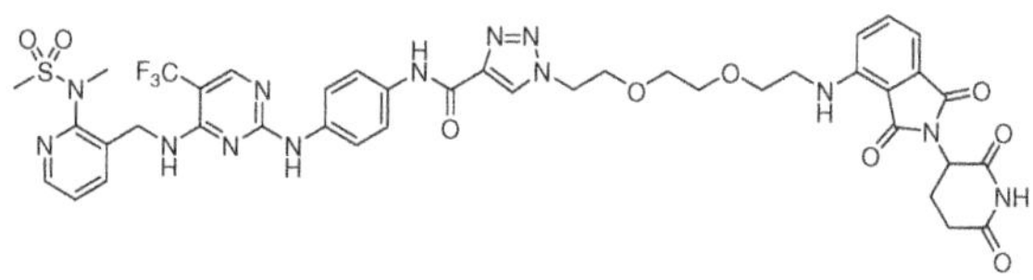	[81]
57	BTK (SJF620)	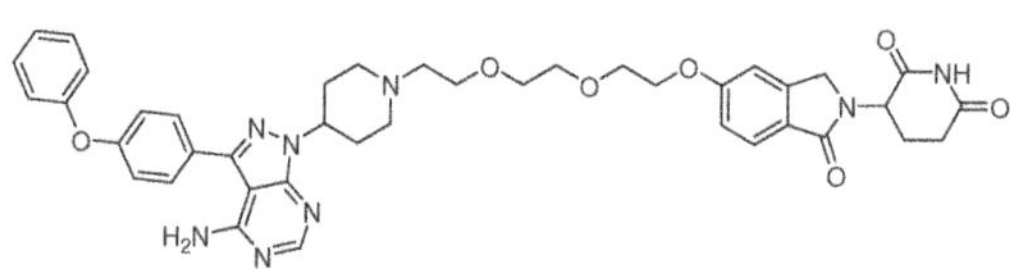	[82]
58	SGK3 (SGK3-PROTAC1)	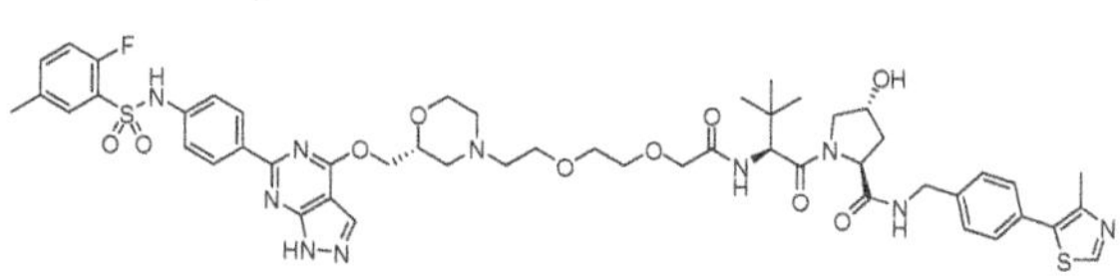	[83]
59	BET (SIAIS213110)	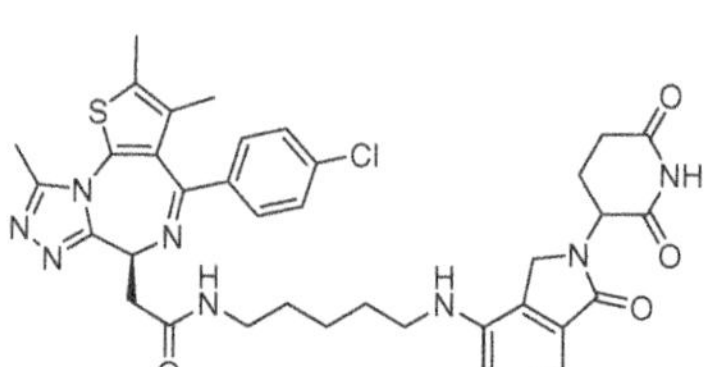	[84]
60	VHL/CRBN (**14a**)	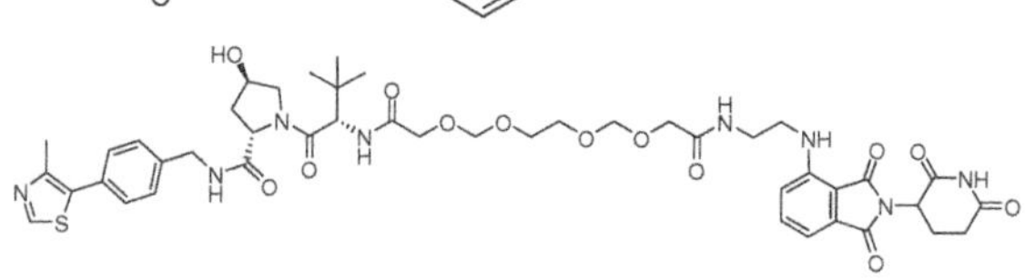	[85]

Table 11.2 (Continued)

Entry	Target (compound ID)	Structure	References
61	SMARCA2, SMARCA4 (PROTAC 1)	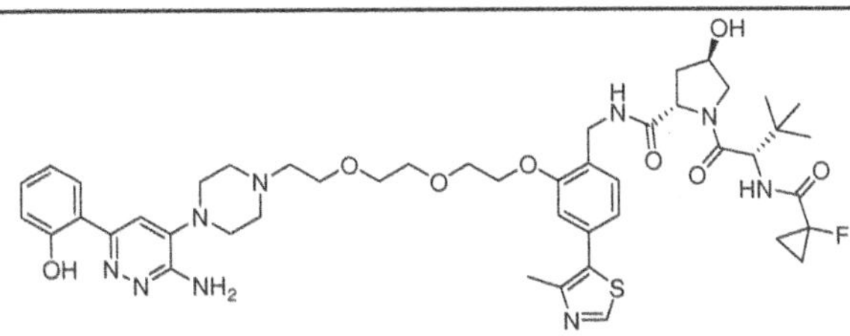	[86]
62	BRD4 (TD-230, TD-207)	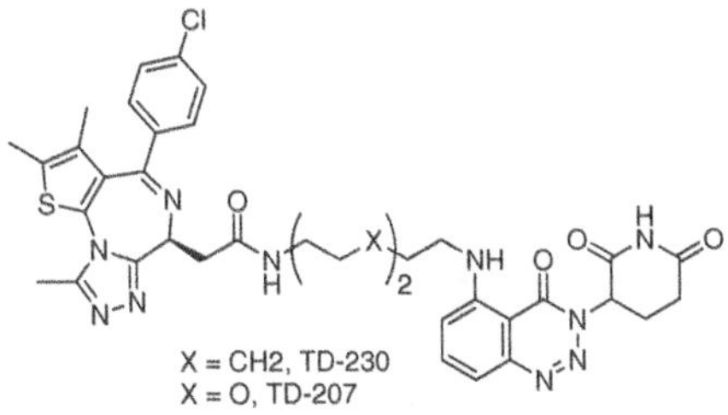	[87]
63	CDK2/9 (F3)	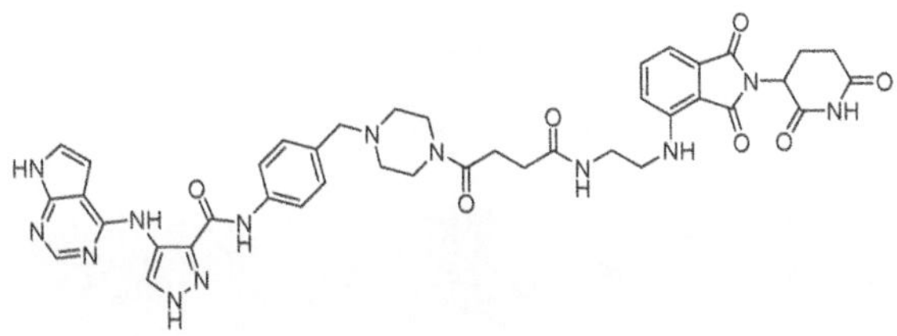	[88]
64	BRD4	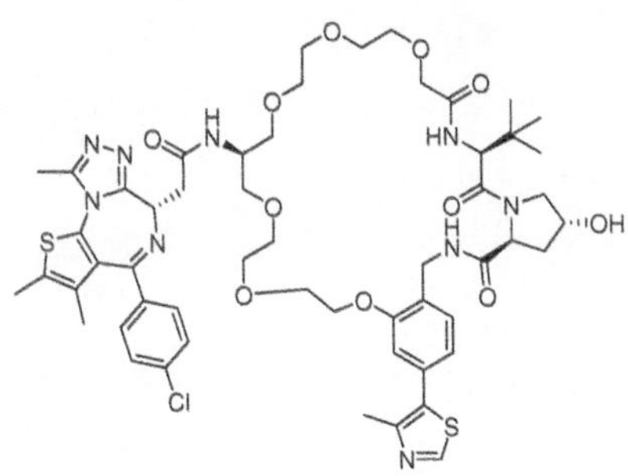	[89]
65	BRD4 (compound **21**)	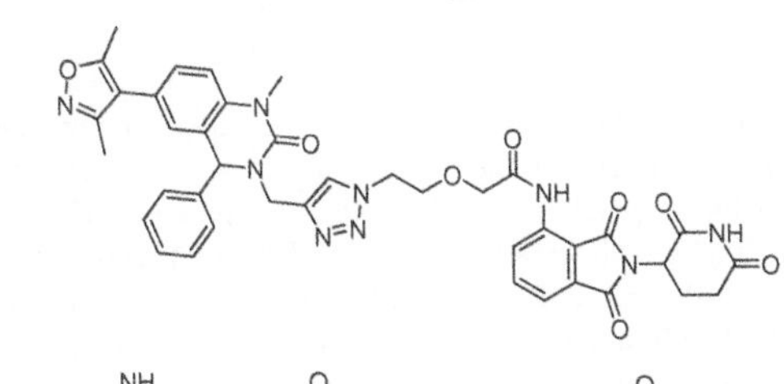	[90]
66	ERRα (Arg-$T^{ERR}\alpha$)	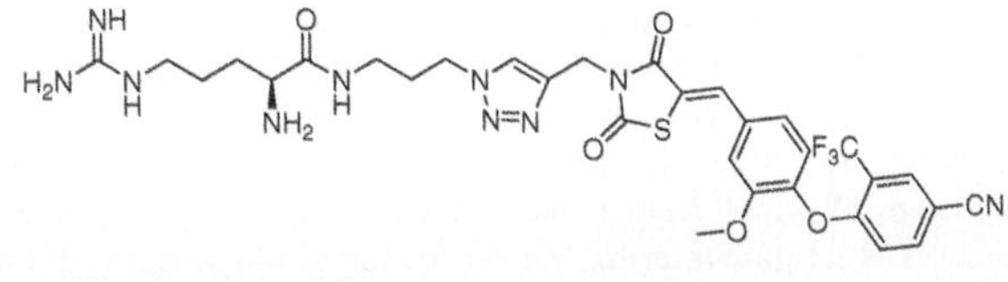	[91]

(Continued)

Table 11.2 (Continued)

Entry	Target (compound ID)	Structure	References
67	ERRα (His-T^{ERR}α)	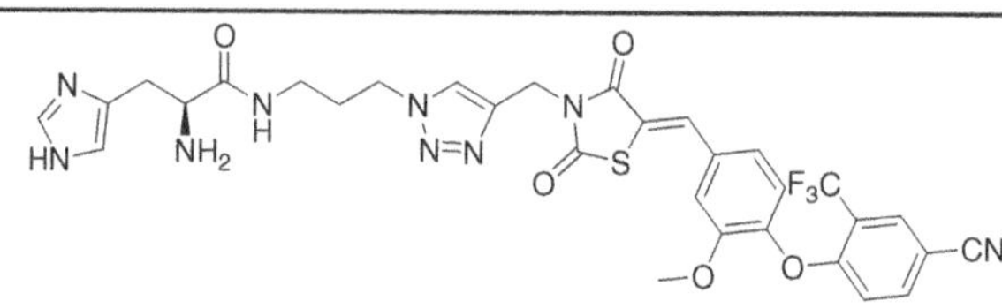	[91]
68	BCL-2/BCL-X_L (DT2216)	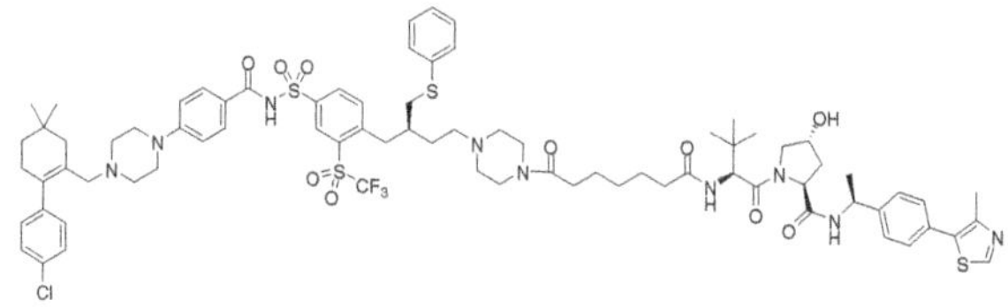	[92]
69	BRD4/PLK1 (HBL-4)	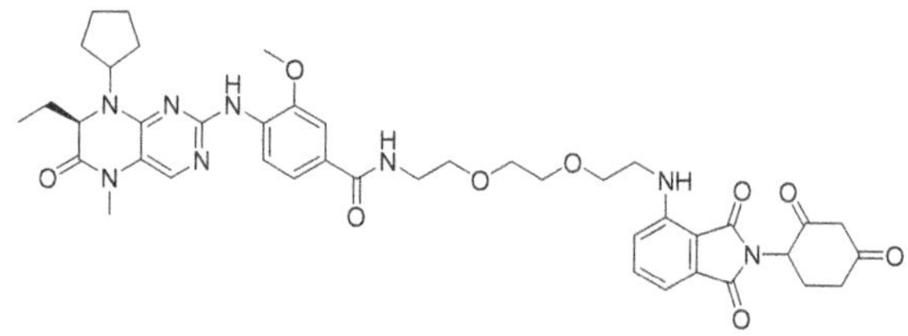	[93]
70	BTK/BLK (PROTAC 7)	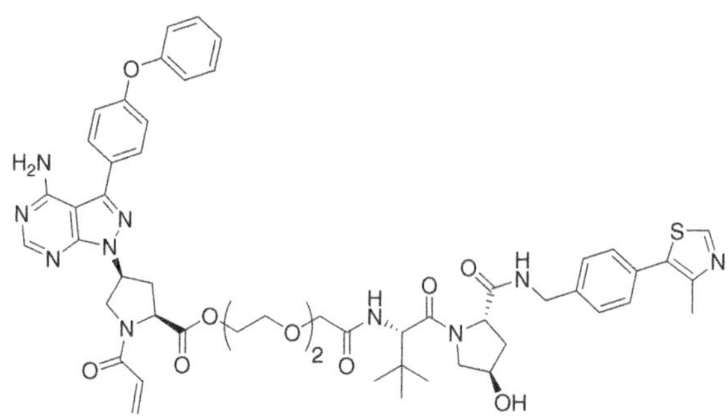	[94]
71	BTK (DD-03-171)	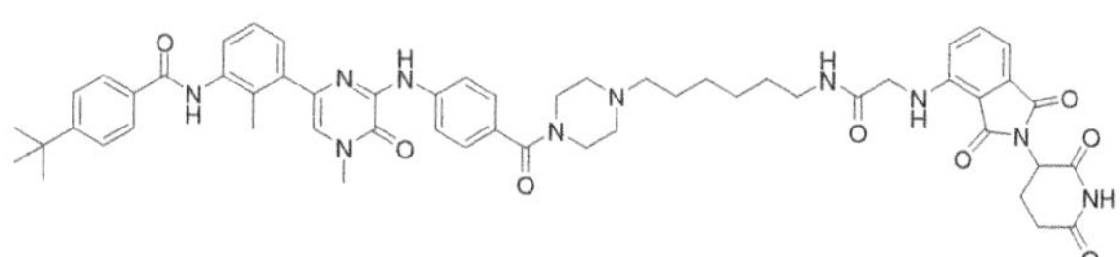	[95]

ABL, Abelson tyrosine-protein kinase coded by Abelson murine leukemia viral oncogene homolog 1; ALK, anaplastic lymphoma kinase; BCL6, B-cell lymphoma 6 protein; BCL-XL, B-cell lymphoma-extra large protein; BCR-ABL1, a kinase protein coded for by the abnormal BCR-ABL1 fusion gene; BET, bromodomain and extraterminal domain proteins; BLK, B lymphocyte kinase; BRD4, bromodomain-containing protein 4; BTK, Bruton tyrosine kinase; CDK8, cyclin-dependent kinase 8; ERRα, estrogen receptor-related receptor-α; FAK, focal adhesion kinase; FKBP12F36V, FK506 binding protein with F36V mutation; FLT3, FMS-like tyrosine kinase 3; GCN5, general control of amino acid synthesis protein 5; HDCA6, histone deacetylase 6; IRAK4, interleukin-1 receptor-associated kinase 4; MCL1, myeloid cell leukemia 1; MDM2, mouse double minute 2 homolog; PARP1, poly polymerase 1; PCAF, P300/CBP-associated factor; PI3K, phosphoinositide 3-kinase; PLK1, polo-like kinase 1; RIPK2, receptor-interacting serine/threonine-protein kinase 2; SGK3, serine/threonine-protein kinase 3; SMARCA2, protein coded by SWI/SNF related, matrix associated, actin dependent regulator of chromatin, subfamily A, member 2; STAT3, signal transducer and activator of transcription 3; TBK1, TANK binding kinase 1; TrKC, tropomyosin receptor kinase C.

11.3 Pharmacokinetics and Oral Absorption Challenge

Achieving a drug-like DMPK profile is one of the core challenges for PROTACs [96]. Due to its tripartate nature, PROTACs usually have a MW at the high end for a small molecule, usually ~1000 Da. This chemical space is not favored in the traditional view of oral medicines. The high MW makes permeability and absorption particularly challenging. It appears that Arvinas has already overcome this hurdle by delivery of two oral PROTACs into clinical trials: the AR PROTAC ARV-110 and ER PROTAC ARV-471. BMS's AR degrader CC-94676 became the third oral PROTAC to enter the clinic. But the structures, physicochemical properties and preclinical DMPK profiles have not been fully disclosed. Limited human PK data from the phase I trial of the Arvinas compound have been presented at conferences. In this highly competitive field, additional oral PROTACs beyond the AR and ER from a few PROTAC focused companies are believed to be moving toward Investigational New Drug (IND) filing soon, not surprisingly without any publications on the profiles of their compounds. From peer reviewed publications, IP and SC injections are still the major routes of administration of PROTAC compounds for *in vivo* studies. As shown in Table 11.3, IP or SC injection gave poor to moderate drug exposure while still demonstrating efficacy. Entries 7 and 8 make an interesting comparison as they demonstrate how a small change in the PROTAC structure can cause a significant change in the PK profile. While oral drugs are available for many drug targets, oral PROTACs shall make this new modality truly a practical and competitive therapeutic approach.

PROTACs obviously are outliers of Lipinski's "rule of five" and Veber's rotatable bond/polar surface area guidelines, however there was a recent discussion on the oral absorption challenge of the "beyond rule of five" chemical space [1j]. While each PROTAC molecule contains a different target binding moiety and linker, the selection of the E3 ligand moiety remains rather limited. When evaluated using parameters relevant to oral absorption such as MW, cLogP, hydrogen bond donors (HBDs), hydrogen bond acceptors (HBAs), polar surface area, number of rotatable bonds (nRotB), and aromatic rings, the CRBN class of E3 ligands overwhelms other E3 ligands as having a better chance to lead to oral bioavailability due to their lower MWs and more drug-like properties. However, every PROTAC is unique and one cannot completely rule out other E3 ligands that may lead to oral bioavailability. We will have a better understanding of this issue when data related to current oral drug candidates such as ARV-110 and ARV-471 are fully disclosed. Although very challenging, the current clinical candidates have proven that even for high MW compounds like PROTACs, oral PK is still achievable.

11.4 PROTACs in Clinical Development

There are only three PROTACs in clinical development so far, the AR PROTAC ARV-110 and the ER PROTAC ARV-471 by Arvinas, and AR PROTAC CC-94676 by BMS.

11.4.1 Androgen Receptor Targeting PROTAC ARV-110

The phase I trial of ARV-110 was an open-label, dose escalation clinical trial to evaluate the safety, tolerability, PK, and PD of ARV-110 in patients with metastatic mCRPC who have progressed after at least two prior systemic therapies. At least one of these two therapies needs to be abiraterone or enzalutamide. ARV-110 was given as a tablet once daily (QD) with food in a 28-day cycle. The trial

Table 11.3 Pharmacokinetic data for exemplary PROTACs.

Entry	Target (compound) and references	Structure	PK profile
1	BET (dBET1) [45]	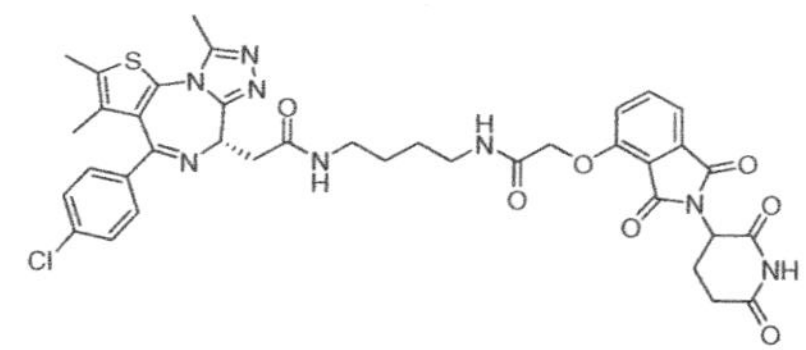	50 mg/kg IP: $C_{max} = 12\,394$ ng/ml $T_{max} = 0.292$ h $AUC_{inf} = 15\,972$ h.ng/ml With >100 ng/ml at 4 h postdose
2	BET (BETd-260) [47]	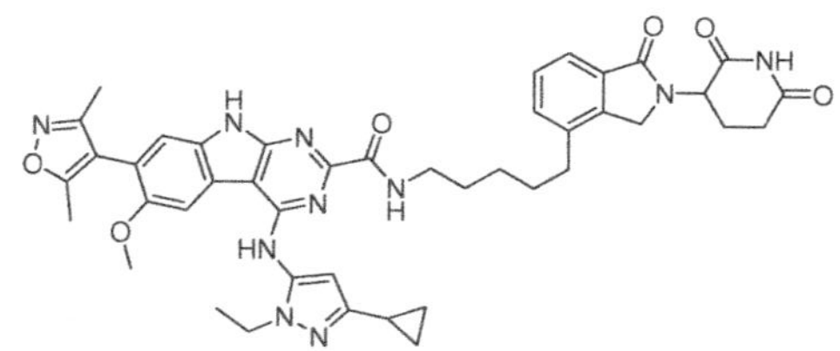	5 mg/kg IV: $C_{1\,h} = 392$ ng/ml $C_{6\,h} = 6.4$ ng/ml $C_{24\,h} < 1$ ng/ml
3	BET (ARV771) [37]	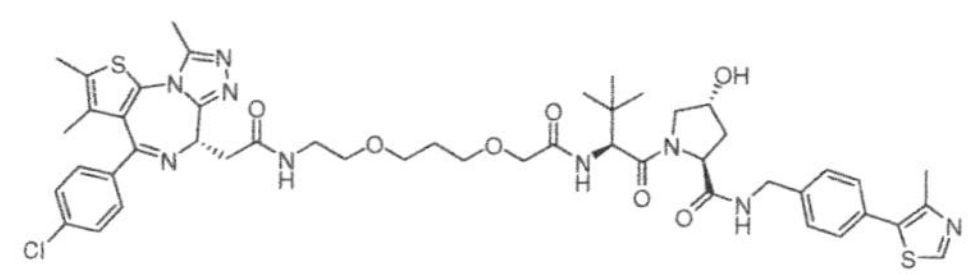	10 mg/kg SC: $C_{max} = 1.7$ μM $T_{max} = 1$ h $AUC = 7.3$ μM.h $C_{24\,h} = 8$ nM
4	BET (QCA570) [48]	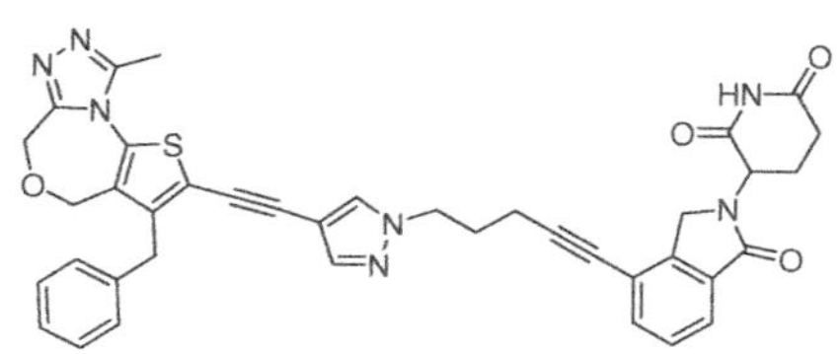	5 mg/kg IV: $C_{1\,h} = 1968$ ng/ml $C_{3\,h} = 155$ ng/ml $C_{6\,h}$ = BLQ
5	BTK (DD-03-171) [95]	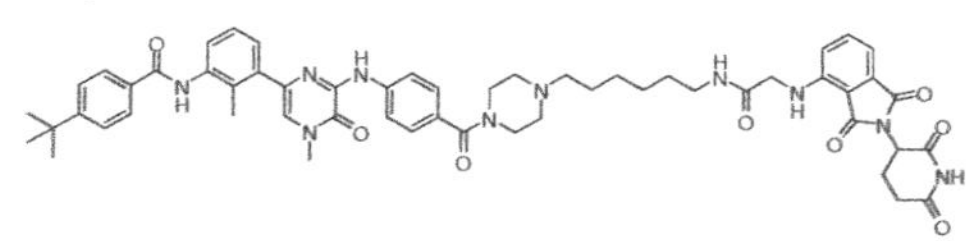	50 mg/kg BID $C_{max} = 192$ ng/ml $AUC = 22\,238$ ng.h/ml

6	FLT-3 [97]	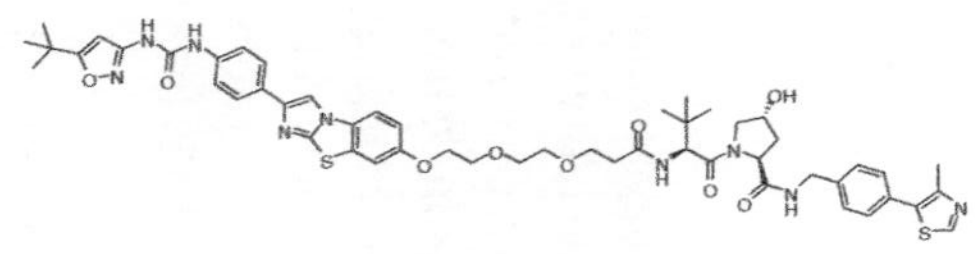	10 mg/kg IP: $C_{max} = 1069$ ng/ml $T_{max} = 2$ h $T_{1/2} = 2.3$ h $C_{22h} > 5$ nM
7	BTK (MT802) [54, 82]	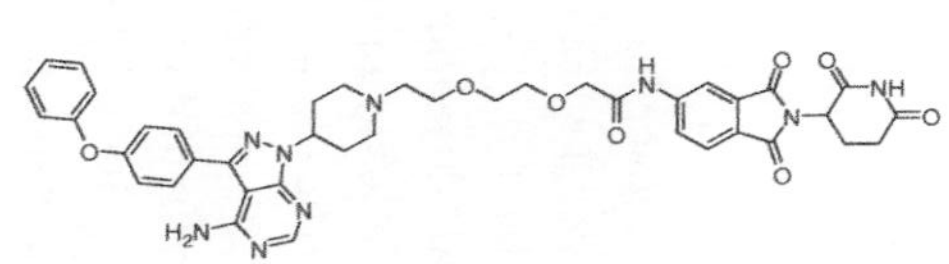	1 mg/kg IV: $C_{max} = 0.07$ μg/ml $T_{1/2} = 0.12$ h $AUC_{last} = 10.2$ min.ng/ml
8	BTK (SJF620) [82]	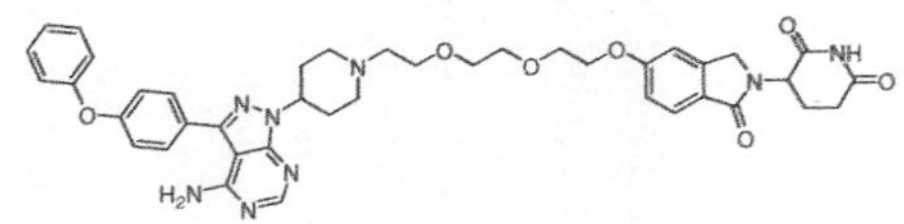	1 mg/kg IV: $C_{max} = 2.1$ μg/ml $T_{1/2} = 1.64$ h $AUC_{last} = 405$ min.ng/ml

Table 11.4 Phase I PK data for ARV-110.

Dose (PO, QD)	Day 1 AUC_{0-24} (ng·h/ml) mean	Day 1 C_{max} (ng/ml) mean	Day 15 AUC_{0-24} (ng·h/ml) mean[a]	Day 15 C_{max} (ng/ml) mean
35 mg	160.5	11.1	1701	83
70 mg	300	19.6	2538	141
140 mg[b]	865	54	5023	353

[a] Accumulation occurs between day 1 and day 15.
[b] Exposure at 140 mg entered the preclinical efficacious range associated with tumor growth inhibition in a mouse model.

used a conventional 3 + 3 design with a starting dose of 35 mg, escalating to 70, 140, and 280 mg. The phase I human PK data were presented at the second Targeted Protein Degradation Summit in Boston in October 2019 (Table 11.4), and also the American Society of Clinical Oncology (ASCO) meeting in May 2020 (Figure 11.26). For comparison purposes, preclinical mouse data are presented in Table 11.5. The human PK showed good dose linearity. At 140 mg, the single dose did not reach efficacious drug exposure. After continued dosing for 15 days, the drug exposure exceeded the efficacious drug exposure with an $AUC_{0\text{-}24h}$ = 5023 ng.h/ml, and C_{max} = 354 ng/ml. The T_{max} is four to eight hours, and the $T_{1/2}$ is ~three to seven days. The AUC is still dose proportional at the 280 mg dose.

The overall safety profile of ARV-110 is satisfactory. In preclinical toxicology studies, animals were dosed orally once a day for 28 days. The animals were allowed to recover for 14 days before moving to a higher dose. Male rats were dosed at 20, 60, and 120 mg/kg per day. Female rats were dosed at 20, 40, and 80 mg/kg per day. Overall, ARV-110 was well tolerated. The no observed adverse effect level (NOAEL) was found to be 40 mg/kg for female rats and 120 mg/kg for male rats. Dogs were dosed at 3, 10, and 30 mg/kg. The NOAEL was found to be 10 mg/kg, and the 30 mg/kg dose exceeded the maximum tolerated dose (MTD). The dog was found to be the species most sensitive to ARV-110. Dose limiting toxicity (DLT) was found to be gastrointestinal alteration. All dogs were found to have loose/discolored stools at all dose levels. In some mid/high dose animals, liver function enzyme elevation was found, but it was reversible. Decreased prostate weights were found for all male rats and male dogs. This agrees with the mechanism of suppressing the AR signaling pathway and is further evidence indicating that the drug is engaging the target.

ARV-110 was generally well tolerated in the phase I trial at doses of 35, 70, 140, and 280 mg. Increases in ALT/AST was the only treatment-emergent adverse event (TEAE) found. There was one case of a grade >3 TEAE found in the 70 mg group (N = 4), 140 mg group (N = 8), and 280 mg group (N = 7), respectively. One of the 22 patients had DLT with ALT/AST grade 3/4 and renal failure at the 280 mg dose. Another patient showed a grade 3 ALT/AST increase. Both patients received concomitant rosuvastatin (Crestor™). This suggested a drug–drug interaction (DDI) of ARV-110 with rosuvastatin. PK data showed rosuvastatin concentrations increased in both patients. *in vitro* transport pump studies confirmed that ARV-110 inhibited the BCRP transporter, which is responsible for clearing rosuvastatin. No DDI was found with other medicines. Six patients on other statins (including atorvastatin) showed no ALT/AST adverse effects. The one patient who was initially taking rosuvastatin continued the ARV-110 treatment but stopped rosuvastatin. No further toxicity was found for this patient.

ARV-110 showed on-target AR degradation in human tumor tissue and reduction of PSA. Patients were pretreated at baseline with either chemotherapy and/or a second generation AR treatment

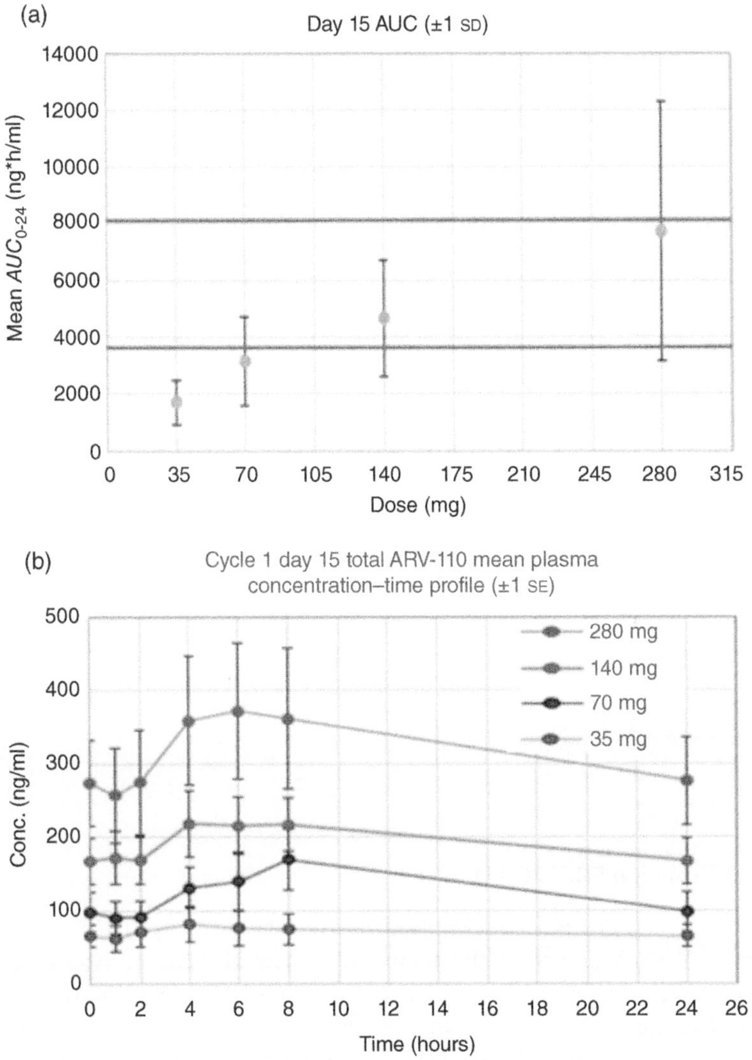

Figure 11.26 ARV-110 showed dose proportional human PK. (a) AUC_{0-24} of the four doses administered in the phase I human clinical trials. The upper horizontal line is the minimum efficacious exposures for tumor growth inhibition based on enzalutamide-resistant VCaP model. The lower horizontal line is based on the castrated and non-castrated VCaP model. (b) Human PK data for the four doses administered in the phase I clinical trial. $T_{1/2}$ = 110 hours. *Source:* PROTAC is a registered trademark of Arvinas Operations, Inc. and is used with permission from Arvinas Operations, Inc. The image is copied from [98] with permission from Arvinas Operations, Inc.

Table 11.5 Preclinical efficacious exposure range of ARV-110 in mouse.

Dose (PO, QD)	AUC_{0-24} (ng·h/ml)	C_{max} (ng/ml)
1 mpk	3628	224
3 mpk	8106	507

including abiraterone and/or enzalutamide. A tumor sample from a patient who followed six weeks of ARV-110 dosing (280 mg) showed AR degradation. Two patients with AR mutants T878A, H875Y that are resistant to abiraterone or enzalutamide showed >50% PSA reduction on 140 mg ARV-110 treatment. One of the two patients confirmed a partial response according to the Response Evaluation Criteria in Solid Tumors (RECIST). Although the number of responders is not as high as many had expected, the confirmed response certainly is a clinical proof of concept for this revolutionary therapy.

The favorable safety profile and initial success in biomarker read out are encouraging. The trial of ARV-110 is ongoing at a dose of 420 mg. Meanwhile, since certain AR mutations responded better to ARV-110, future patient selection will be important. This may lead to an accelerated approval path.

A news release by Arvinas in December 2020 reported that a dose escalation to 700 mg was well-tolerated. Based on the phase I results, a phase II ARDENT trial has initiated with a once daily, oral dose of 420 mg design. The phase II trial is expected to confirm the potential for accelerated approval in a molecularly defined population and to explore the potential for approval for early-line mCRPC treatment.

The structure of ARV-110 was finally released by Arvinas as shown in Figure 11.27 [98d].

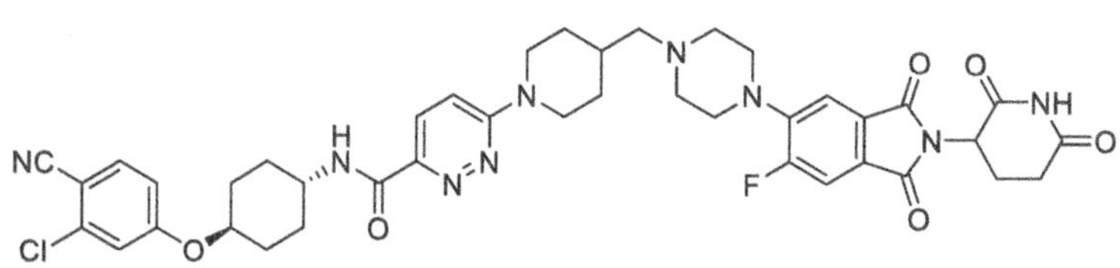

Figure 11.27 Structure of ARV-110.

11.4.2 Estrogen Receptor Targeting PROTAC ARV-471

The phase I trial for ER PROTAC ARV-471 was an open-label, dose escalation study that assessed the safety and tolerability of ARV-471 in patients with ER+/HER2- locally advanced or metastatic breast cancer, who have received prior hormonal therapy and chemotherapy. In preclinical studies, ARV471 showed inhibition of tumor growth by 99% when orally dosed at 10 mg/kg, QD, in the estrogen receptor 1 (ESR1) mutant PDX model. The $AUC_{0\text{–}24h}$ for efficacious drug exposure is 2538 ng·h/ml, with C_{max} = 312 ng/ml (Table 11.6). The human PK after 15 days of QD oral dosing of 30 mg of ARV471 gave $AUC_{0\text{–}24h}$ = 4100 ng·h/ml, with C_{max} = 224 ng/ml, exceeding the preclinical efficacious drug exposure (Table 11.7 and Figure 11.28). The T_{max} is ~4 hours, and the $T_{1/2}$ is ~24 hours. No treatment-related AEs or DLTs were observed. The trial is currently ongoing.

A news release by Arvinas in December 2020 reported that rising doses of 30, 60, 120, 180, and 360 mg were well tolerated. The PK exposure is dose proportional and far exceed preclinical

Table 11.6 Preclinical efficacious exposure range of ARV-471 in mouse.

Dose (PO, QD)	Mean $AUC_{0\text{–}24}$ (ng·h/ml)	Mean C_{max} (ng/ml)
3 mpk	658	84
10 mpk	2538	312
30 mpk	5717	962

Table 11.7 Phase I PK data for ARV-471.

Dose (PO, QD)	Day 1 AUC_{0-24} (ng·h/ml) mean	Day 1 C_{max} (ng/ml) mean	Day 15 AUC_{0-24} (ng·h/ml) mean	Day 15 C_{max} (ng/ml) mean
30 mg	1690	109	4100	224

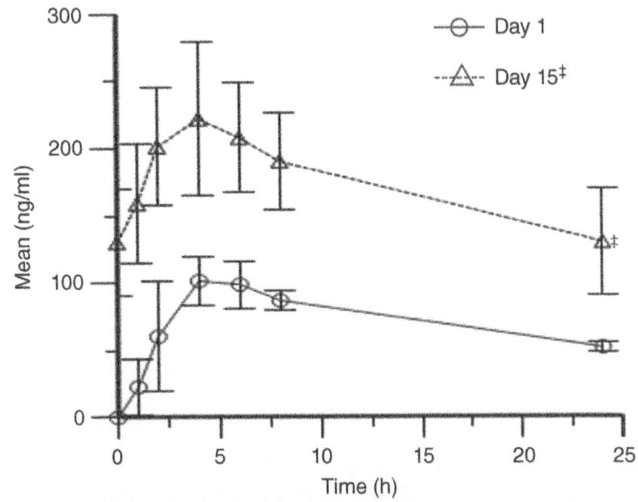

Figure 11.28 Human PK of ARV-471. PO 30 mg, QD. (Source: The image is copied from [98] with permission from Arvinas Operations, Inc.)

efficacy thresholds. The effective $T_{1/2}$ is about 28 hours. At the 180 mg dose, a high benefit rate of 42% was achieved in a heavily pretreated patient population. ARV-471 is promising to become the best-in-class ER targeting therapy. Additional clinical trials of ARV-471 are ongoing.

The structure of ARV-471 was finally released by Arvinas as shown in Figure 11.29 [98e].

Figure 11.29 Structure of ARV-471.

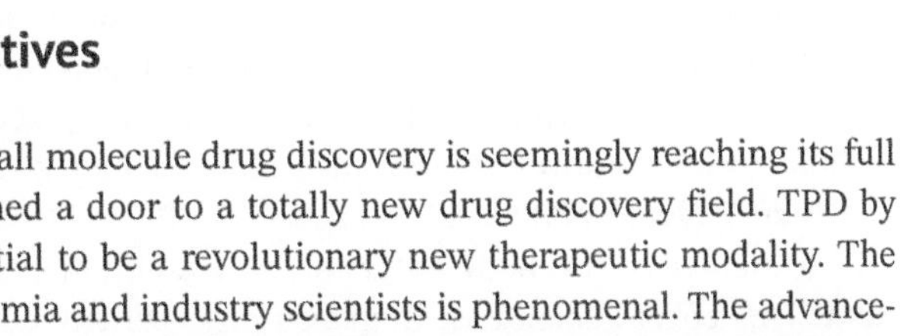

11.5 Challenges and Perspectives

Emerging at a time when traditional small molecule drug discovery is seemingly reaching its full potential, TPD by small molecules opened a door to a totally new drug discovery field. TPD by PROTACs undoubtedly has great potential to be a revolutionary new therapeutic modality. The enthusiasm demonstrated by both academia and industry scientists is phenomenal. The advancement of ARV-110 and ARV-471 into clinical trials has demonstrated that, first, oral PK of high MW compounds like PROTACs is achievable; secondly, target engagement and down regulation of biomarkers by PROTACs are confirmed in humans. This will certainly encourage more participation in the field.

On the other hand, there are still many challenges. Despite the success of ARV-110 and ARV-471, and another AR degrader CC-94676, the oral PK still remains a hurdle for the PROTAC class of therapeutics. So far, no other oral PROTACs have been officially announced, despite several under intense preclinical development. Dosing remains either via IP, SC or IV. Even for ARV-110, a dose of 140 mg in humans was required to reach the efficacious drug exposure although it had a dose of 1 mg/kg as the efficacious dose in the mouse Xenograft model. This may suggest low oral bioavailability in humans. The predicted low dose due to the high potency and the catalytic nature of PROTACs is not achieved as expected in the AR case. Apparently improving human oral PK is still a hurdle to be overcome.

Potential drug resistance is another challenge that many may have overlooked. PROTACs were expected to be a solution that would overcome drug resistance by degradation of the mutant target protein. However, how the body responds to down regulated protein levels is not fully understood. There is already a reported case of PROTAC induced drug resistance [99]. Unlike the drug resistance that we often see with small molecule inhibitors, the acquired resistance to the VHL and CRBN based BET PROTACs reported in this case was not caused by mutation of the target protein. Instead, the acquired resistance was mainly caused by alternation of the E3 ligase. This report reminded us that the possibility for drug resistance induced by the PROTAC approach still remains.

Optimization of PROTACs remains a more labor intensive and time-consuming process compared with traditional small molecule drugs, due to the more complicated structure of PROTACs. The optimization of a PROTAC is usually divided into optimization of the three components individually followed by the combination of the optimized pieces. There were examples showing that a weak target protein ligand can still lead to a potent degrader due to a contribution of the linker moiety to binding. This highlighted the need for global optimization instead of local optimization. The complex structures of PROTACs largely increased the workload for medicinal chemists. Fortunately, computational chemistry demonstrated some success in helping chemists design PROTACs [100]. Nevertheless, a PROTAC program is considered to be more resource-consuming compared with a traditional small molecule program.

Much is still unknown about the Chemistry Manufacturing and Controls (CMC) development of PROTACs due to the lack of information disclosed on the development of the current clinical compounds. Presumably, the larger and more complex structure of a PROTAC is likely to contain more metabolically labile groups and to generate a more complex metabolic profile compared with traditional small molecules. Thus, toxicity studies may be more time and resource consuming. The formulation of PROTACs is also a new topic that is largely unexplored. The oral PK hurdles of PROTACs puts additional burdens on formulations to improve bioavailability. There is only one reported formulation study of a PROTAC [101]. Pre-formulation studies were performed on the BRD4 degrader ARV-825. This PROTAC compound has poor aqueous solubility and compromised stability due to pH dependent hydrolytic degradation. ARV-825 loaded self-nanoemulsifying preconcentrate (ARV-SNEP) was prepared using dimethyl acetamide, medium chain triglycerides and Kolliphor ELP. It showed significantly enhanced aqueous solubility and reasonable pH and fed state independent stability.

The potential *in vivo* hook effect added an additional challenge to the PD and clinical development. This effect generates a U-shaped dose–activity curve. As such, a higher dose may give a lower PD readout. The *in vitro* hook effect is well documented. On the other hand, there is no report of an *in vivo* hook effect yet. This may be due to limited reports on *in vivo* efficacy studies, and less likely that there is not an *in vivo* Hook effect. Thus, this adds difficulty to find the optimum dose range to achieve the best efficacy.

The clinical development of ARV-110, ARV-471 and CC-94676 is very exciting. With the enormous advantages offered by this revolutionary approach, TPD by PROTACs will continue to be an exciting and dynamic field with more and more attention and effort. The value of degraders eventually will depend on the advantages they can offer over traditional small molecule inhibitors and agonists. There are already many publications on PROTAC degraders for various targets. As our understanding of this field increases, the expectation for it certainly rises. The research in this field will likely move from "identify a degrader" toward "identify a degrader with more drug-like properties." Future publications are expected to address the more challenging topics such as DMPK, CMC development, pharmacology, and clinical development. More drug candidates are expected to enter clinical development, and eventually new drugs from the PROTAC class are expected to reach the market in the near future.

Acknowledgement

The author wishes to thank Drs. Xinghai Li, Chengzhi Zhang, and Jiang Yu for their suggestions.

References

1 (a) Pettersson, M. and Crews, C.M. (2019). PROteolysis TArgeting Chimeras (PROTACs) – past, present and future. *Drug Discov. Today Technol.* 31: 15. (b) Konstantinidou, M., Li, J., Zhang, B. et al. (2019). PROTACs – a game-changing technology. *Expert Opin. Drug Discovery* 14: 1255. (c)Neklesa, T.K., Winkler, J.D., and Crews, C.M. (2017). Targeted protein degradation by PROTACs. *Pharmacol. Ther.* 174: 138. (d) Ottis, P. and Crews, C.M. (2017). Proteolysis-targeting chimeras: induced protein degradation as a therapeutic strategy. *ACS Chem. Biol.* 12: 892. (e) Gu, S., Cui, D., Chen, X. et al. (2018). PROTACs: an emerging targeting technique for protein degradation in drug discovery. *Bioessays* 1700247. (f) Wang, P. and Zhou, J. (2018). Proteolysis Targeting Chimera (PROTAC): a paradigm-shifting approach in small molecule drug discovery. *Curr. Top. Med. Chem.* 18: 1. (g) Sun, X., Gao, H., Yang, Y. et al. (2019). PROTACs: great opportunities for academia and industry. *Signal Transduct. Target. Ther.* 4: 64. (h) Zou, Y., Ma, D., and Wang, Y. (2019). The PROTAC technology in drug development. *Cell Biochem. Funct.* 37: 21. (i) Pei, H., Peng, Y., Zhao, Q., and Chen, Y. (2019). Small molecule PROTACs: an emerging technology for targeted therapy in drug discovery. *RSC Adv.* 9: 16967. (j) Edmondson, S.D., Yang, B., and Fallan, C. (2019). Proteolysis Targeting Chimeras (PROTACs) in 'Beyond Rule-of-Five' chemical space: recent Progress and future challenges. *Bioorg. Med. Chem. Lett.* 29: 1555.

2 Pickart, C.M. (2001). Mechanisms underlying ubiquitination. *Annu. Rev. Biochem.* 70: 503.

3 Zheng, Q., Huang, T., Zhang, L. et al. (2016). Dysregulation of ubiquitin-proteasome system in neurodegenerative diseases. *Front. Aging Neurosci.* 8: 303.

4 (a)Atkin, G. and Paulson, H. (2014). Ubiquitin pathways in neurodegenerative disease. *Front. Mol. Neurosci.* 2014 (7): 63. (b)Morreale, F.E. and Walden, H. (2016). Types of ubiquitin ligases. *Cell* 165: e241.

5 He, M., Zhou, Z., Shah, A.A. et al. (2016). The emerging role of deubiquitinating enzymes in genomic integrity, diseases, and therapeutics. *Cell Biosci.* 6: 62.

6 (a) Thibaudeau, T.A. and Smith, D.M. (2019). A practical review of proteasome pharmacology. *Pharmacol. Rev.* 71: 170. (b) Hines, J. and Crews, C. M. (2017). Infographic: proteasome basics, the structure and function of the cell's protein-degrading machine.*The Scientist* (30 April).

7 Sakamoto, K.M., Kim, K.B., Kumagai, A. et al. (2001). Protacs: chimeric molecules that target proteins to the Skp1-Cullin-F box complex for ubiquitination and degradation. *J. Proc. Natl. Acad. Sci. USA*. 98: 8554.

8 Sakamoto, K.M., Kim, K.B., Verma, R. et al. (2003). Development of protacs to target cancer-promoting proteins for ubiquitination and degradation. *Mol. Cell. Proteomics* 2: 1350.

9 Schneekloth, J.S., Fonseca, F.N., Koldobskiy, M. et al. (2004). Chemical genetic control of protein levels: selective in vivo targeted degradation. *J. Am. Chem. Soc.* 126: 748.

10 Lee, H., Puppala, D., Choi, E.-Y. et al. (2007). Targeted degradation of the aryl hydrocarbon receptor by the PROTAC approach: a useful chemical genetic tool. *ChemBioChem* 2058: 8.

11 Schneekloth, A.R., Pucheault, M., Tae, H.S., and Crews, C.M. (2008). Targeted intracellular protein degradation induced by a small molecule: en route to chemical proteomics. *Bioorg. Med. Chem. Lett.* 18: 5904.

12 (a) Buckley, D.L., Gustafson, J.L., Van Molle, I. et al. (2012). Small-molecule inhibitors of the interaction between the E3 ligase VHL and HIF1a. *Angew. Chem. Int. Ed.* 51: 11463. (b) Buckley, D.L., Van Molle, I., Gareiss, P.C. et al. (2012). Targeting the von Hippel–Lindau E3 ubiquitin ligase using small molecules to disrupt the VHL/HIF-1α interaction. *J. Am. Chem. Soc.* 134: 4465. (c) Van Molle, I., Thomann, A., Buckley, D.L. et al. (2012). Dissecting fragment-based lead discovery at the Von Hippel-Lindau protein: hypoxia inducible factor 1a protein–protein interface. *Chem. Biol.* 19: 1300.

13 (a) Testa, A., Lucas, X., Castro, G.V. et al. (2018). 3-Fluoro-4-hydroxyprolines: synthesis, conformational analysis, and stereoselective recognition by the VHL E3 ubiquitin ligase for targeted protein degradation. *J. Am. Chem. Soc.* 140: 9299. (b) Soares, P., Lucas, X., and Ciulli, A. (2018). Thioamide substitution to probe the hydroxyproline recognition of VHL ligands. *Bioorg. Med. Chem.* 26: 2992. (c) Soares, P., Gadd, M.S., Frost, J. et al. (2018). Group based optimization of potent and cell-active inhibitors of the von Hippel–Lindau (VHL) E3 ubiquitin ligase: structure–activity relationships leading to the chemical probe (2S,4R)-1-((S)-2-(1- cyanocyclopropanecarboxa mido)-3,3-dimethylbutanoyl)-4-hydroxy-N-(4-(4-methylthiazol-5-yl)benzyl)pyrrolidine-2-carboxamide (VH298). *J. Med. Chem.* 61: 599. (d) Frost, J., Galdeano, C., Soares, P. et al. (2016). Potent and selective chemical probe of hypoxic signaling downstream of HIF-a hydroxylation via VHL inhibition. *Nat. Commun.* 7: 13312.

14 (a) Ito, T., Ando, H., Suzuki, T. et al. (2010). Identification of a primary target of thalidomide teratogenicity. *Science* 327: 1345. (b) Lopez-Girona, A., Mendy, D., Ito, T. et al. (2012). Cereblon is a direct protein target for immunomodulatory and antiproliferative activities of lenalidomide and pomalidomide. *Leukemia* 26: 2326. (c) Fischer, E.S., Böhm, K., Lydeard, J.R. et al. (2014). Structure of the DDB1–CRBN E3 ubiquitin ligase in complex with thalidomide. *Nature* 512: 49. (d) Krönke, J., Udeshi, N.D., Narla, A. et al. (2014). Lenalidomide causes selective degradation of IKZF1 and IKZF3 in multiple myeloma cells. *Science* 343: 301. (e) Petzold, G., Fischer, E.S., and Thoma, N.H. (2016). Structural basis of lenalidomide induced CK1a degradation by the CRL4CRBN ubiquitin ligase. *Nature* 532: 127. (f) Krönke, J., Fink, E.C., Hollenbach, P.W. et al. (2015). Lenalidomide induces ubiquitination and degradation of CK1a in del(5q) MDS. *Nature* 523: 183. (g) Gandhi, A.K., Kang, J., Havens, C.G. et al. (2014). Immunomodulatory agents lenalidomide and pomalidomide costimulate T cells by inducing degradation of T cell repressors Ikaros and Aiolos via modulation of the E3 ubiquitin ligase complex CRL4CRBN. *Br. J. Haematol.* 164: 811. (h) Chamberlain, P.P., Lopez-Girona, A., Miller, K. et al. (2014). Structure of the human cereblon-DDB1-lenalidomide complex reveals basis for responsiveness to thalidomide analogs. *Nat. Struct. Mol. Biol.* 21: 803.

15 Hines, J., Gough, J.D., Corson, T.W., and Crews, C.M. (2013). Posttranslational protein knockdown coupled to receptor tyrosine kinase activation with phosphoPROTACs. *Proc. Natl. Acad. Sci. USA* 110: 8942.

16 Zheng, N. and Shabek, N. (2017). Ubiquitin ligases: structure, function, and regulation. *Annu. Rev. Biochem.* 86: 129.

17 Hines, J., Lartigue, S., Dong, H. et al. (2019). MDM2-recruiting PROTAC offers superior, synergistic antiproliferative activity via s simultaneous degradation of BRD4 and stabilization of p53. *Cancer Res.* 79: 251.

18 Itoh, Y., Ishikawa, M., Naito, M. et al. (2010). Protein knockdown using methyl bestatin-ligand hybrid molecules: design and synthesis of inducers of ubiquitination-mediated degradation of cellular retinoic acid-binding proteins. *J. Am. Chem. Soc.* 132: 5820.

19 Itoh, Y., Ishikawa, M., Kitaguchi, R. et al. (2012). Double protein knockdown of cIAP1 and CRABP-II using a hybrid molecule consisting of ATRA and IAPs antagonist. *Bioorg. Med. Chem. Lett.* 22: 4453.

20 Ohoka, N., Okuhira, K., Ito, M. et al. (2017). *in vivo* Knockdown of pathogenic proteins via specific and nongenetic inhibitor of apoptosis protein (IAP)-dependent protein erasers (SNIPERs). *J. Biol. Chem.* 292: 4556.

21 Wang, L., Guillen, V.S., Sharma, N. et al. (2018). New class of selective estrogen receptor degraders (SERDs): expanding the toolbox of PROTAC degrons. *ACS Med. Chem. Lett.* 9: 803.

22 Gadd, M.S., Testa, A., Lucas, X. et al. (2017). Structural basis of PROTAC cooperative recognition for selective protein degradation. *Nat. Chem. Biol.* 13: 514.

23 Nowak, R.P., DeAngelo, S.L., Buckley, D. et al. (2018). Plasticity in binding confers selectivity in ligand induced protein degradation. *Nat. Chem. Biol.* 14: 706.

24 Gustafson, J.L., Neklesa, T.K., Cox, C.S. et al. (2015). Small-molecule-mediated degradation of the androgen receptor through hydrophobic tagging. *Angew. Chem. Int. Ed. Engl.* 54: 9659.

25 Salami, J., Alabi, S., Willard, R.R. et al. (2018). Androgen receptor degradation by the proteolysis targeting chimera ARCC-4 outperforms enzalutamide in cellular models of prostate cancer drug resistance. *Commun. Biol.* 1: 100.

26 Shibata, N., Nagai, K., Morita, Y. et al. (2018). Development of protein degradation inducers of androgen receptor by conjugation of androgen receptor ligands and inhibitor of apoptosis protein ligands. *J. Med. Chem.* 61: 543.

27 Han, X., Wang, C., Qin, C. et al. (2019). Discovery of ARD-69 as a highly potent proteolysis targeting chimera (PROTAC) degrader of androgen receptor (AR) for the treatment of prostate cancer. *J. Med. Chem.* 62: 941.

28 Kregel, S., Wang, C., Han, X. et al. (2020). Androgen receptor degraders overcome common resistance mechanisms developed during prostate cancer treatment. *Neoplasia* 22: 111.

29 Neklesa, T.K., Snyder, L.B., Willard, R.R. et al. (2019). ARV-110: an oral androgen receptor PROTAC degrader for prostate cancer. *J. Clin. Oncol.* 37 (7 suppl): 259.

30 Lebraud, H., Wright, D.J., Johnson, C.N., and Heightman, T.D. (2016). Protein degradation by in-cell self-assembly of proteolysis targeting chimeras. *ACS Cent. Sci.* 2: 927.

31 Reynders, M., Matsuura, B.S., Bérouti, M. et al. (2020). PHOTACs enable optical control of protein degradation. *Sci. Adv.* 6: eaay5064.

32 Maneiro, M., Forte, N., Shchepinova, M.M. et al. (2020). Antibody-PROTAC conjugates enable HER2 dependent targeted protein degradation of BRD4. *ACS Chem. Biol.* 15: 1306–1312.

33 Hines, J., Lartigue, S., Dong, H. et al. (2019). MDM2-recruiting PROTAC offers superior, synergistic antiproliferative activity via simultaneous degradation of BRD4 and stabilization of p53. *Cancer Res.* 79: 251.

34 Ohoka, N., Morita, Y., Nagai, K. et al. (2018). Derivatization of inhibitor of apoptosis protein (IAP) ligands yields improved inducers of estrogen receptor alpha degradation. *J. Biol. Chem.* 293: 6776.

35 Zengerle, M., Chan, K.H., and Ciulli, A. (2015). Selective small molecule induced degradation of the BET bromodomain protein BRD4. *ACS Chem. Biol.* 10: 1770.
36 Wurz, R.P., Dellamaggiore, K., Dou, H. et al. (2018). A "click chemistry platform" for the rapid synthesis of bispecific molecules for inducing protein degradation. *J. Med. Chem.* 61: 453.
37 Raina, K., Lu, J., Qian, Y. et al. (2016). PROTAC-induced BET protein degradation as a therapy for castration-resistant prostate cancer. *Proc. Natl. Acad. Sci. USA* 113: 7124.
38 Chan, K.H., Zengerle, M., Testa, A., and Ciulli, A. (2018). Impact of target warhead and linkage vector on inducing protein degradation: comparison of bromodomain and extra-terminal (BET) degraders derived from triazolodiazepine (JQ1) and tetrahydroquinoline (I-BET726) BET inhibitor scaffolds. *J. Med. Chem.* 61: 504.
39 Bondeson, D.P., Mares, A., Smith, I.E. et al. (2015). Catalytic in vivo protein knockdown by small-molecule PROTACs. *Nat. Chem. Biol.* 11: 611.
40 Lai, A.C., Toure, M., Hellerschmied, D. et al. (2016). Modular PROTAC design for the degradation of oncogenic BCR-ABL. *Angew. Chem. Int. Ed. Engl.* 55: 807.
41 Huang, H.T., Dobrovolsky, D., Paulk, J. et al. (2018). A chemoproteomic approach to query the degradable kinome using a multi-kinase degrader. *Cell Chem. Biol.* 25: 88–99 e86.
42 Cromm, P.M., Samarasinghe, K.T.G., Hines, J., and Crews, C.M. (2018). Addressing kinase-independent functions of FAK via PROTAC-mediated degradation. *J. Am. Chem. Soc.* 140: 17019.
43 Crew, A.P., Raina, K., Dong, H. et al. (2018). Identification and characterization of von Hippel-Lindau-recruiting proteolysis targeting chimeras (PROTACs) of TANK-binding kinase 1. *J. Med. Chem.* 61: 583.
44 Zoppi, V., Hughes, S.J., Maniaci, C. et al. (2019). Iterative design and optimization of initially inactive proteolysis targeting chimeras (PROTACs) identify VZ185 as a potent, fast and selective von Hippel-Lindau (VHL)-based dual degrader probe of BRD9 and BRD7. *J. Med. Chem.* 62: 699.
45 Winter, G.E., Buckley, D.L., Paulk, J. et al. (2015). Drug development. Phthalimide conjugation as a strategy for in vivo target protein degradation. *Science* 348: 1376.
46 Lu, J., Qian, Y., Altieri, M. et al. (2015). Hijacking the E3 ubiquitin ligase Cereblon to efficiently target BRD4. *Chem. Biol.* 22: 755.
47 Zhou, B., Hu, J., Xu, F. et al. (2018). Discovery of a small-molecule degrader of bromodomain and extra-terminal (BET) proteins with picomolar cellular potencies and capable of achieving tumor regression. *J. Med. Chem.* 61: 462.
48 Qin, C., Hu, Y., Zhou, B. et al. (2018). Discovery of QCA570 as an exceptionally potent and efficacious proteolysis targeting chimera (PROTAC) degrader of the bromodomain and extra-terminal (BET) proteins capable of inducing complete and durable tumor regression. *J. Med. Chem.* 61: 6685.
49 McCoull, W., Cheung, T., Anderson, E. et al. (2018). Development of a novel B-cell lymphoma 6 (BCL6) PROTAC to provide insight into small molecule targeting of BCL6. *ACS Chem. Biol.* 13: 3131.
50 Hatcher, J.M., Wang, E.S., Johannessen, L. et al. (2018). Development of highly potent and selective steroidal inhibitors and degraders of CDK8. *ACS Med. Chem. Lett.* 9: 540.
51 Olson, C.M., Jiang, B., Erb, M.A. et al. (2018). Pharmacological perturbation of CDK9 using selective CDK9 inhibition or degradation. *Nat. Chem. Biol.* 14: 163.
52 Robb, C.M., Contreras, J.I., Kour, S. et al. (2017). Chemically induced degradation of CDK9 by a proteolysis targeting chimera (PROTAC). *Chem. Commun. (Camb.)* 53: 7577.
53 Li, W., Gao, C., Zhao, L. et al. (2018). Phthalimide conjugations for the degradation of oncogenic PI3K. *Eur. J. Med. Chem.* 151: 237.

54 Buhimschi, A.D., Armstrong, H.A., Toure, M. et al. (2018). Targeting the C481S Ibrutinib-resistance mutation in Bruton's tyrosine kinase using PROTAC-mediated degradation. *Biochemistry* 57: 3564.

55 Sun, Y., Zhao, X., Ding, N. et al. (2018). PROTAC-induced BTK degradation as a novel therapy for mutated BTK C481S induced ibrutinibresistant B-cell malignancies. *Cell Res.* 28: 779.

56 Zorba, A., Nguyen, C., Xu, Y. et al. (2018). Delineating the role of cooperativity in the design of potent PROTACs for BTK. *Proc. Natl. Acad. Sci. USA* 115: E7285.

57 Powell, C.E., Gao, Y., Tan, L. et al. (2018). Chemically induced degradation of anaplastic lymphoma kinase (ALK). *J. Med. Chem.* 61: 4249.

58 Yang, K., Song, Y., Xie, H. et al. (2018). Development of the first small molecule histone deacetylase 6 (HDAC6) degraders. *Bioorg. Med. Chem. Lett.* 28: 2493.

59 Schiedel, M., Herp, D., Hammelmann, S. et al. (2018). Chemically induced degradation of Sirtuin 2 (Sirt2) by a proteolysis targeting chimera (PROTAC) based on Sirtuin rearranging ligands (SirReals). *J. Med. Chem.* 61: 482.

60 Bassi, Z.I., Fillmore, M.C., Miah, A.H. et al. (2018). Modulating PCAF/GCN5 immune cell function through a PROTAC approach. *ACS Chem. Biol.* 13: 2862.

61 Chessum, N.E.A., Sharp, S.Y., Caldwell, J.J. et al. (2018). Demonstrating in-cell target engagement using a Pirin protein degradation probe (CCT367766). *J. Med. Chem.* 61: 918.

62 Nabet, B., Roberts, J.M., Buckley, D.L. et al. (2018). The dTAG system for immediate and target-specific protein degradation. *Nat. Chem. Biol.* 14: 431.

63 Sun, X., Wang, J., Yao, X. et al. (2019). A chemical approach for global protein knockdown from mice to non-human primates. *Cell Discov.* 5: 10.

64 Zhang, X., Thummuri, D., He, Y. et al. (2019). Utilizing PROTAC technology to address the on-target platelet toxicity associated with inhibition of BCL-XL. *Chem. Commun.* 55: 14765.

65 Zhao, B. and Burgess, K. (2019). TrkC-targeted kinase inhibitors and PROTACs. *Mol. Pharm.* 16: 4313.

66 Nunes, J., McGonagle, G.A., Eden, J. et al. (2019). Targeting IRAK4 for degradation with PROTACs. *ACS Med. Chem. Lett.* 10: 1081.

67 Burslem, G.M., Schultz, A.R., Bondeson, D.P. et al. (2019). Targeting BCR-ABL1 in chronic myeloid leukemia by PROTAC-mediated targeted protein degradation. *Cancer Res.* 79: 4744.

68 Wurz, R.P. and Cee, V.J. (2019). Targeted degradation of MDM2 as a new approach to improve the efficacy of MDM2-p53 inhibitors. *J. Med. Chem.* 62: 445.

69 Zhou, H., Bai, L., Xu, R. et al. (2019). Structure-based discovery of SD-36 as a potent, selective, and efficacious PROTAC degrader of STAT3 protein. *J. Med. Chem.* 62: 11280.

70 Rana, S., Bendjennat, M., Kour, S. et al. (2019). Selective degradation of CDK6 by a palbociclib based PROTAC. *Bioorg. Med. Chem. Lett.* 29: 1375.

71 Zhao, B. and Burgess, K. (2019). PROTACs suppression of CDK4/6, crucial kinases for cell cycle regulation in cancer. *Chem. Commun.* 55: 2704.

72 Su, S., Yang, Z., Gao, H. et al. (2019). Potent and preferential degradation of CDK6 via proteolysis targeting chimera degraders. *J. Med. Chem.* 62: 7575.

73 Yang, H., Lv, W., He, M. et al. (2019). Plasticity in designing PROTACs for selective and potent degradation of HDAC6. *Chem. Commun. (Camb.)* 55: 14848.

74 Zhao, Q., Lan, T., Su, S., and Rao, Y. (2019). Induction of apoptosis in MDA-MB-231 breast cancer cells by a PARP1-targeting PROTAC small molecule. *Chem. Commun.* 55: 369.

75 Peng, L., Zhang, Z., Lei, C. et al. (2019). Identification of new small-molecule inducers of estrogen-related receptor α (ERRa) degradation *ACS med. Chem. Lett.* 10: 767.

76 Papatzimas, J.W., Gorobets, E., Maity, R. et al. (2019). From inhibition to degradation: targeting the antiapoptotic protein myeloid cell leukemia 1 (MCL1). *J. Med. Chem.* 62: 5522.
77 Zhao, Q., Ren, C., Liu, L. et al. (2019). Discovery of SIAIS178 as an effective BCR-ABL degrader by recruiting Von Hippel-Lindau (VHL) E3 ubiquitin ligase. *J. Med. Chem.* 62: 9281.
78 Han, X., Zhao, L., Xiang, W. et al. (2019). Discovery of highly potent and efficient PROTAC degraders of androgen receptor (AR) by employing weak binding affinity VHL E3 ligase ligands. *J. Med. Chem.* 62: 11218.
79 Hu, J., Hu, B., Wang, M. et al. (2019). Discovery of ERD-308 as a highly potent proteolysis targeting chimera (PROTAC) degrader of estrogen receptor (ER). *J. Med. Chem.* 62: 1420.
80 Wu, H., Yang, K., Zhang, Z. et al. (2019). Development of multifunctional histone deacetylase 6 degraders with potent antimyeloma activity. *J. Med. Chem.* 62: 7042.
81 Gao, H., Wu, Y., Sun, Y. et al. (2019). Design, synthesis, and evaluation of highly potent FAK-targeting PROTACs. *ACS Med. Chem. Lett.* 11 (10): 1855–1862.
82 Jaime-Figueroa, S., Buhimschi, A.D., Toure, M. et al. (2019). Design, synthesis and biological evaluation of proteolysis targeting chimeras (PROTACs) as a BTK degraders with improved pharmacokinetic properties. *Bioorg. Med. Chem. Lett.*: 126877.
83 Tovell, H., Testa, A., Zhou, H. et al. (2019). Design and characterization of SGK3-PROTAC1, an isoform specific SGK3 kinase PROTAC degrader. *ACS Chem. Biol.* 14: 2024.
84 Qiu, X., Sun, N., Kong, Y. et al. (2019). Chemoselective synthesis of Lenalidomide-based PROTAC library using alkylation reaction. *Org. Lett.* 21: 3838.
85 Girardini, M., Maniaci, C., Hughes, S.J. et al. (2019). Cereblon versus VHL: hijacking E3 ligases against each other using PROTACs. *Bioorg. Med. Chem.* 27: 2466.
86 Farnaby, W., Koegl, M., Roy, M.J. et al. (2019). BAF complex vulnerabilities in cancer demonstrated *via* structure-based PROTAC design. *Nat. Chem. Biol.* 15: 672.
87 Kim, S.A., Go, A., Jo, S. et al. (2019). A novel cereblon modulator for targeted protein degradation. *Eur. J. Med. Chem.* 166: 65.
88 Zhou, F., Chen, L., Cao, C. et al. (2020). Development of selective mono or dual PROTAC degrader probe of CDK isoforms. *Eur. J. Med. Chem.* 187: 111952.
89 Testa, A., Hughes, S.J., Lucas, X. et al. (2020). Structure-based design of a macrocyclic PROTAC. *Angew. Chem. Int. Ed.* 59: 1727.
90 Zhang, F., Wu, Z., Chen, P. et al. (2020). Discovery of a new class of PROTAC BRD4 degraders based on a dihydroquinazolinone derivative and lenalidomide/pomalidomide. *Bioorg. Med. Chem.* 28: 115228.
91 Shanmugasundaram, K., Shao, P., Chen, H. et al. (2019). A modular PROTAC design for target destruction using a degradation signal based on a single amino acid. *J. Biol. Chem.* 294: 15172.
92 Khan, S., Zhang, X., Lv, D. et al. (2019). A selective BCL-XL PROTAC degrader achieves safe and potent antitumor activity. *Nat. Med.* 25: 1938.
93 Mu, X., Bai, L., Xu, Y. et al. (2020). Protein targeting chimeric molecules specific for dual bromodomain 4 (BRD4) and polo-like kinase 1 (PLK1) proteins in acute myeloid leukemia cells. *Biochem. Biophys. Res. Commun.* 521: 833.
94 Xue, G., Chen, J., Liu, L. et al. (2020). Protein degradation through covalent inhibitor-based PROTACs. *Chem. Commun.* 56: 1251.
95 Dobrovolsky, D., Wang, E.S., Morrow, S. et al. (2019). Bruton's tyrosine kinase degradation as a therapeutic strategy for cancer. *Blood* 133: 952.
96 Cantrill, C., Chaturvedi, P., Rynn, C. et al. (2020). Fundamental aspects of DMPK optimization of targeted protein degraders. *Drug Discov. Today* 25 (6): 969–982.

97 Burslem, G.M., Song, J., Chen, X. et al. (2018). Enhancing antiproliferative activity and selectivity of a FLT-3 inhibitor by proteolysis targeting chimera conversion. *J. Am. Chem. Soc.* 140: 16428.

98 (a) Taylor, I. (2019). Moving PROTAC protein degraders from the laboratory to the clinic. https://ir.arvinas.com/static-files/ce10f577-753f-4063-af5e-403f11f75128 (accessed 26 September 2021; (b) Arvinas (2020). ARV-110 phase 1/2 dose escalation interim update. https://ir.arvinas.com/static-files/8f6f9c9b-e738-4ea5-8655-0f90a2fb5faa (accessed 26 September 2021); (c) Arvinas (2020). Leading the way in targeted protein degradation therapeutics. https://ir.arvinas.com/static-files/0fb6b5b5-be22-4bd4-b656-e1fdd238c1c6 (accessed 26 September 2021; (d) Snyder, L.B. (2021). Discovery of ARV-110, a first in class androgen receptor degrading PROTAC® for the treatment of men with metastatic castration resistant prostate cancer https://ir.arvinas.com/static-files/99279395-e91e-44d1-8ad9-88db2024c1fa (accessed 26 September 2021; (e) Snyder, L.B. (2021). The discovery of ARV-471, an orally bioavailable estrogen receptor degrading PROTAC® for the treatment of patients with breast cancer. https://ir.arvinas.com/static-files/7a4db470-3d7f-4d4b-98a4-481b8c573169 (accessed 26 September 2021).

99 Zhang, L., Riley-Gillis, B., Vijay, P., and Shen, Y. (2019). Acquired resistance to BET-PROTACs (proteolysis-targeting chimeras) caused by genomic alterations in Core components of E3 ligase complexes. *Mol. Cancer Ther.* 18: 1302.

100 a Imrie, F., Bradley, A.R., van der Schaar, M., and Deane, C.M. (2020). Deep Generative Models for 3D Linker Design. *J. Chem. Inf. Model.* 60: 1983.b Drummond, M.L. and Williams, C.I. (2019). In silico modeling of PROTAC-mediated ternary complexes: validation and application. *J. Chem. Inf. Model.* 59: 1634.

101 Rathod, D., Fu, Y., and Patel, K. (2019). BRD4 PROTAC as a novel therapeutic approach for the treatment of vemurafenib resistant melanoma: Preformulation studies, formulation development and in vitro evaluation. *Eur. J. Pharm. Sci.* 138: 105039.

12

Entrepreneurial Drug Hunter

Macrocyclic Peptide Modalities

Tomi Sawyer

Maestro Therapeutics, MA 01772, Southborough, MA, USA

12.1 Introduction

As I reflect upon my career, which has interwoven academic, biotech and pharma drug discovery, I have witnessed many "breakthroughs" in terms of novel molecules advanced for varying disease targets. Perhaps at this point in my scientific work of now more than four decades, I've joined the ranks of many who may be described as entrepreneurial drug hunters. Of course, this embraces the disciplines of chemistry, biology, structural biology, biophysical chemistry, computational chemistry, and other related fields of specialization. What has amazed me throughout the years has been the ambition and imagination of many individuals that I have had the privilege to collaborate with or, in some cases, competed with, but hopefully in the spirit of ultimately providing hope a reality for those whom we strive to create new medicines.

In this chapter, I am sharing a perspective of macrocyclic peptide modalities. Albeit it is not intended to be comprehensive or describing exemplary work in technical detail, the objective of which I am aspiring to achieve is to highlight "breakthroughs" throughout more than a century of peptide science, technology and medicine dating back to the synthesis of the first and simplest peptide Gly-Gly and discovery of the antibiotic cyclic dipeptide penicillin. Nowadays, powerful tools to leverage diversity, design and delivery of peptides has greatly impacted the development of peptide drugs. One of the most well-recognized opportunities within the academic, biotech and pharma community in this endeavor has been macrocyclic peptide modalities that leverage unique conformational properties, molecular size, and structural complexity. Collectively, this will be the framework of offering a perspective of the past, present, and future of macrocyclic peptide modalities.

12.2 Macrocyclic Peptide Modalities in Retrospect

As highlighted in Figure 12.1, perhaps it is fitting that the first peptide synthesized, Gly-Gly, was in fact the simplest in terms of chemical structure. However, the significance of such an achievement in 1901 should neither be underestimated, nor the scientist, Emil Fischer (Nobel Laureate), who is credited for this work. Albeit a greater understanding of both its structure and biology required

Contemporary Accounts in Drug Discovery and Development, First Edition. Edited by Xianhai Huang, Robert G. Aslanian, and Wayne H. Tang.

subsequent research for many years, the early discovery of insulin [1] well-represents one of the most spectacular of structurally complex macrocyclic peptides (see Sections 12.3 and 12.4) by virtue of its size and three intramolecular S—S bonds. In fact, the discovery of insulin by Frederick Banting and Charles Best as a pancreatic hormone preceded its chemical nature being determined relative to its constituent 51 amino acid and heterodimeric structure by way of sequencing (Frederick Sanger, Nobel Laureate) and crystallography (Dorothy Hodgkin, Nobel Laureate). Also, and especially noteworthy in the middle part of the past century, the discovery of oxytocin [2], including its structural determination, chemical synthesis, and biological characterization by the team of Vincent DuVigneaud (Nobel Laureate) exemplifies an extraordinary achievement in the history of macrocyclic peptide modalities. Furthermore, the synthesis of RNase-A [3] and RNA-S [4] using solid-phase methodology by the team of Robert Merrifield (Nobel Laureate) and solution-phase methodology by the team of Ralph Hirschmann (National Medal of Science), respectively, illustrate hallmark breakthroughs in complex macrocyclic peptide (protein) chemistry to make an enzyme! Such pioneering science and technology established the foundation for what would next transpire in terms of propelling peptide drug discovery toward a myriad of therapeutic applications.

12.3 Receptor and Extracellularly Targeted Macrocyclic Peptides

Insulin's complex macrocyclic structure is achieved by proteolytic processing of a single chain precursor peptide having intramolecular S—S bonding between three pairs of Cys residues to form the final heterodimeric form as shown in Figure 12.2. In fact, intramolecular S—S bonding via Cys residues is a common "chemistry" known for many endogenous peptides of varying size and

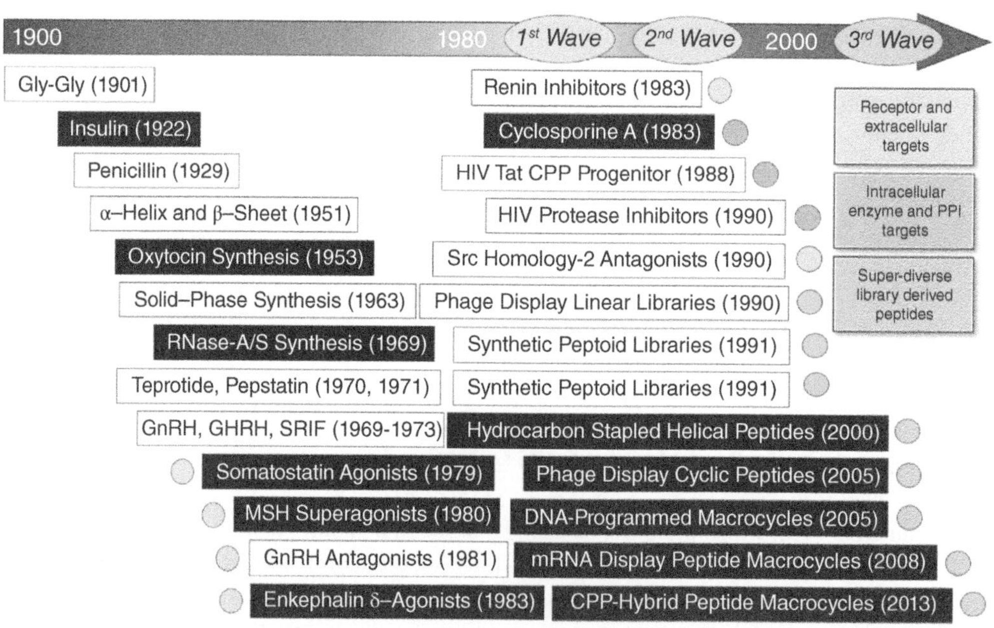

Figure 12.1 Peptide science milestones. Macrocyclic peptides are highlighted (dark shaded boxes). See text for details.

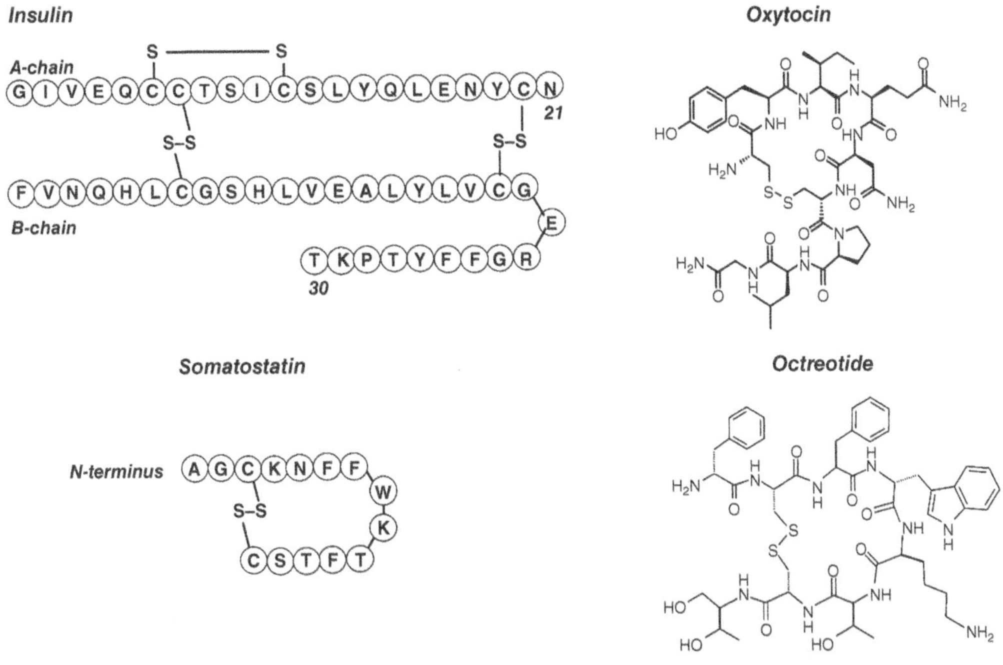

Figure 12.2 Structures of macrocyclic peptides insulin, oxytocin, somatostatin, and octreotide. See text for details.

shape, including the macrocyclic peptide oxytocin (see Section 12.2) as shown in Figure 12.2. Synthetic modifications of the macrocyclic peptide somatostatin to create simpler analogs and redesigning the architecture of the Cys-to-Cys placement led to the discovery of octreotide [5] (Figure 12.2) as well as next-generation analogs leveraging N-to-C-terminus cyclization [6].

In contrast to insulin or oxytocin, design strategies to convert endogenous linear peptides to macrocyclic peptides were first demonstrated [7] by $[Cys^{4,10}]\alpha$-melanocyte-stimulating hormone ($[Cys^{4,10}]\alpha$-MSH) and, most recently, by setmelanotide [8] (FDA-approved 2020) as shown in Figure 12.3. The first X-ray structure of a chemically related, macrocyclic analog, Ac-$[Nle^4$, *cyclo*$(Asp^5$, D-Nal$[2']^7$, $Lys^{10})]\alpha$-MSH_{4-10}-NH_2 (SHU-9119), complexed to the melanocortin-4 receptor (MC4R) has been recently achieved [9] and it supported previously hypothesized conformations (e.g. β-turn) for α-MSH or macrocyclic peptide analogs thereof relative to their binding at specific melanocortin receptors (i.e. MC1R, MC3R, MC4R, and MC5R).

Beyond the accounts of early G protein-coupled receptor (GPCR) targeted peptide drug discovery (e.g. somatostatin and α-MSH) a few other macrocyclic peptide modalities are noteworthy to mention, including linaclotide [10], Zilucoplan [11], and the PCSK9 inhibitor [12] shown in Figure 12.4. The macrocyclic peptide linaclotide (FDA-approved 2012) is highly conformationally constrained by virtue of its three intramolecular S—S bonding architecture, and such contributes to its selectivity at its targeted guanylate cyclase-C receptor. Zilucoplan (FDA-approved 2020) is a relatively complex peptide having a macrocyclic (*cyclo*$[Lys^1$, $Asp^6]$ substructure and a carboxy-terminally extended linear peptide sequence that also contains a terminal ε-Lys-polyethylene glycol-γ-Glu-conjugated C-18 acid moiety as shown in Figure 12.4. Zilucoplan targets complement component C5 to prevent its cleavage into downstream complement components C5a and C5b. Recently, Merck disclosed [12] novel macrocyclic peptide inhibitors targeting PCSK9, including

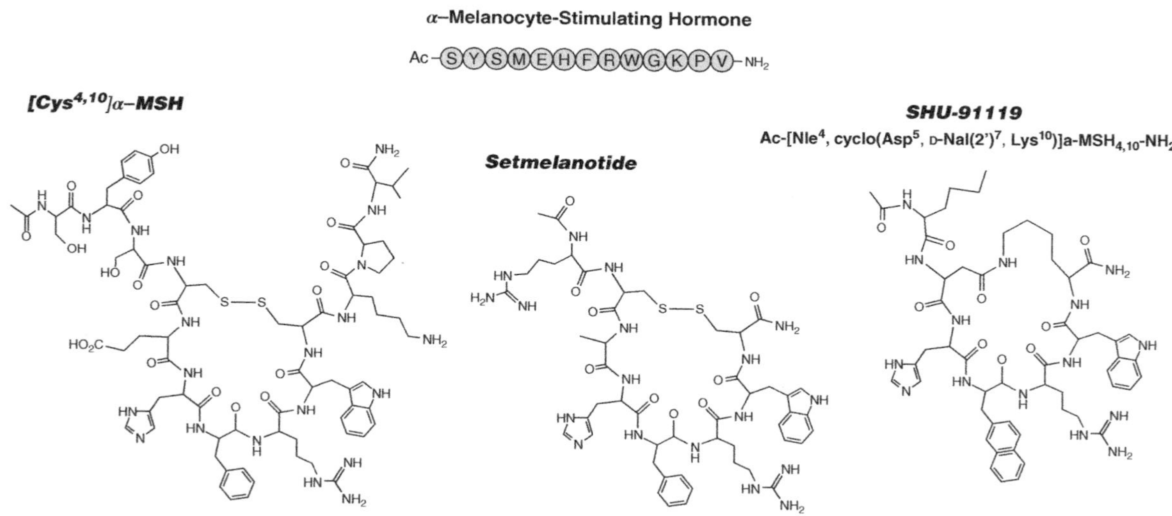

Figure 12.3 Structures of α-melanocyte-stimulating hormone (α-MSH) and macrocyclic analogs [Cys^{4,10}]α-MSH, setmelanotide, and Ac-[Nle^4, *cyclo*(Asp^5, D-Nal[2′]^7, Lys^{10})]α-MSH$_{4-10}$-NH$_2$ (SHU-9119). See text for details.

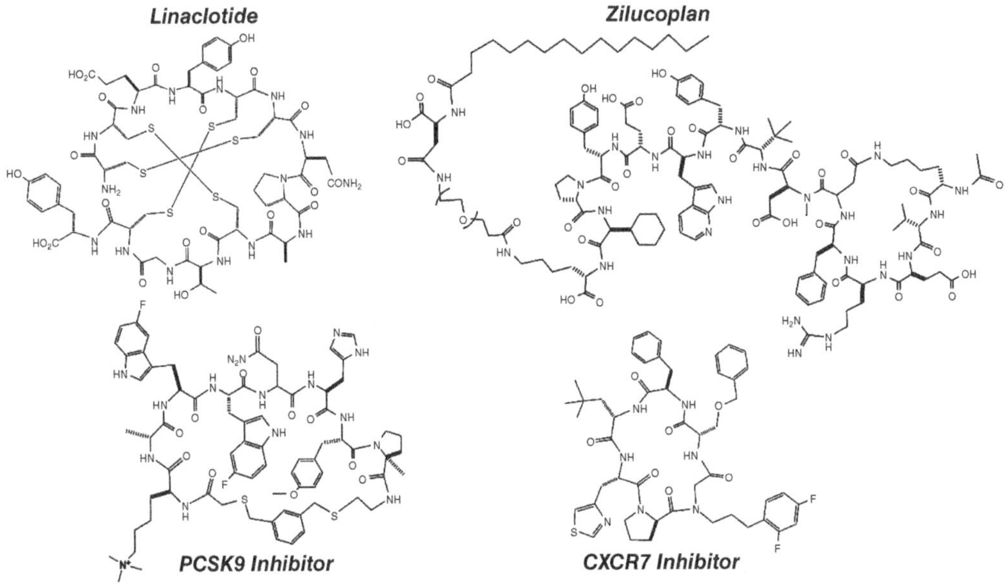

Figure 12.4 Structures of macrocyclic peptides linaclotide, zilucoplan, a PCSK9 inhibitor, and a CXCR7 inhibitor. See text for details.

the preclinical lead molecule shown in Figure 12.4. This macrocyclic peptide leverages two thioethers bridging its N- and C-termini via a *m*-xylene moiety. The relatively low molecular weight of this molecule was especially desired for future analog optimization and to achieve oral bioavailability. Noteworthy is that both Zilucoplan and the PCSK9 inhibitor were each advanced by iterative design and structure–activity studies originating from progenitor macrocyclic peptides generated via mRNA display screening methodologies. The CXCR7 inhibitor [13] shown in Figure 12.4 is an orally bioavailable macrocyclic peptide exemplifying backbone N-alkylation and optimized hydrophobicity in a design strategy reminiscent of cyclosporine A (see Section 12.4).

A rather stunning example of designing and synthesizing a trimeric macrocyclic peptide is BCY12491 [14] (Figure 12.5) having high affinity and specificity to both an immune costimulatory receptor (i.e. CD137) and tumor antigen (i.e. EphA2) by virtue of two CD137-binding substructures and one EphA2-binding substructure connected via a tris-PEG-linker. Such a trimeric macrocyclic peptide exemplifies a significant and compelling option to antibodies. BCY12491 represents a novel macrocyclic peptide modality that has its origin using phage display in terms of varying length and chemical cyclization (i.e. between three Cys residues and reactive linker moieties to create intramolecular tris-thioether bridged, bicyclic peptides) on the bacteriophage surface.

12.4 Intracellular Protein–Protein Interaction Targeted Macrocyclic Peptides

The design of peptides having cell permeability properties and ability to modulate intracellular protein–protein interaction targets has likely been the most challenging, with the exception of that of peptide oral bioavailability, in terms of expanded drug development beyond receptor and extracellular target space. Yet, significant progress has been achieved relative to the first

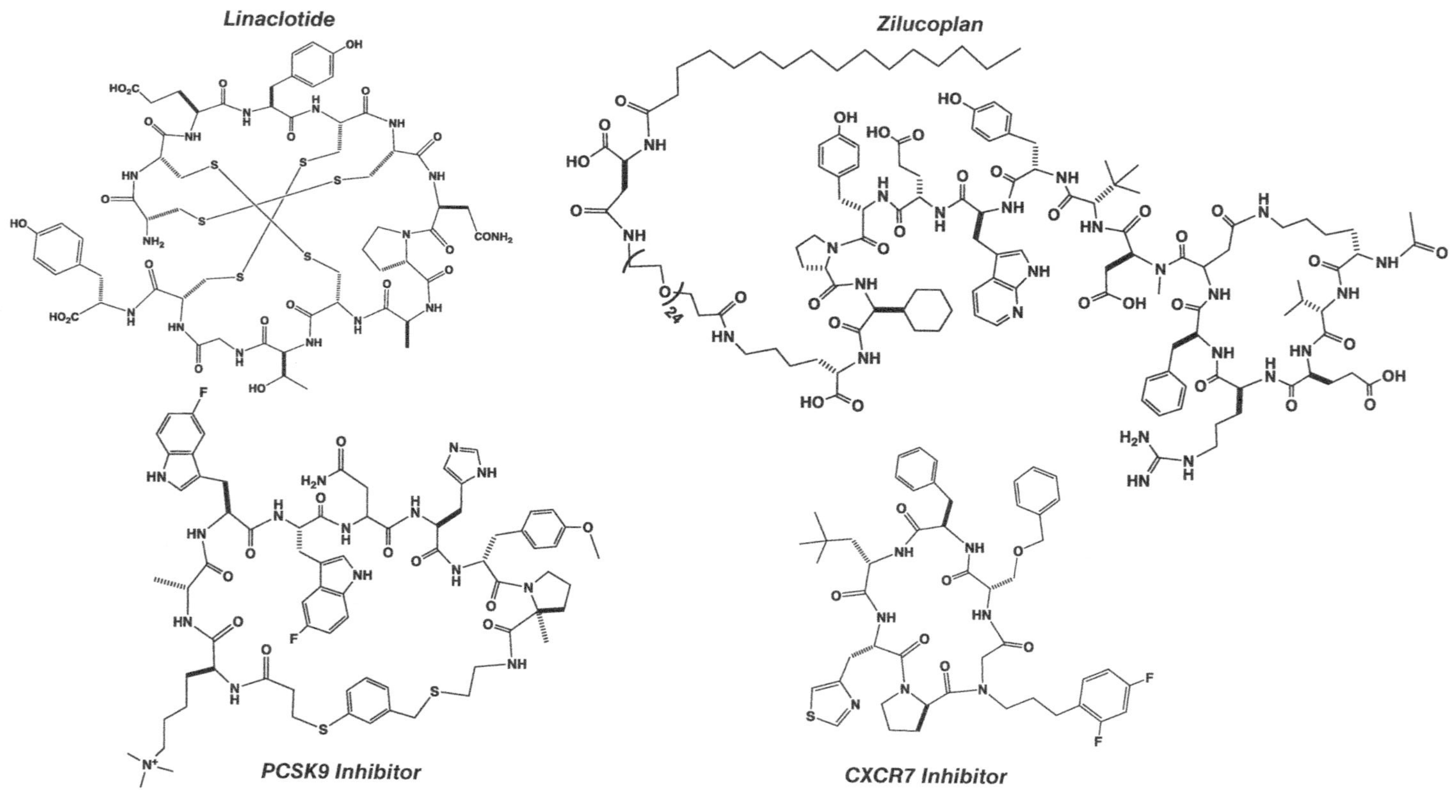

Figure 12.5 Structure of macrocyclic peptide BCY12491 highlighting its CD137-binding and EphA2-binding bicyclic domains connected via a tris-PEG-linker in a trimeric architecture. See text for details.

identification of the natural product cyclosporine A [15] (CsA; Figure 12.6) which has several attributes to understand its structure–activity and structure–permeability relationships. The N-to-C macrocyclic structure of CsA includes several N-methylated amino acids and one D-amino acid. It is known to exhibit passive permeability, despite its higher molecular weight, and such has been attributed to its conformational flexibility and ability to leverage intramolecular H-bonding for favorable lipid membrane interaction as well as intermolecular H-bonding with water. The impact of CsA to the field of peptide drug discovery targeting intracellular protein–protein interactions with respect to its intrinsic cell permeability properties has been extraordinary. In many ways, CsA is a key benchmark macrocyclic peptide for passive transport (e.g. cell permeability and/or oral bioavailability) for ongoing macrocyclic peptide drug discovery efforts [16–21]. Perhaps the most powerful technology that is contributing to such efforts nowadays is that of leveraging super-diverse macrocyclic peptide libraries derived from synthetic (e.g. one-bead, one-compound) or biologic (e.g. phage-, mRNA- or DNA-display) methodologies [22–26]. Exemplary macrocyclic peptides (i.e. XIAP-BIR3 inhibitor [27], KRAS inhibitor [28] and NEMO inhibitor [29]) are shown in Figure 12.6. The XIAP-BIR3 inhibitor is a homodimer that is cyclized using Click chemistry via modifications of two amino acid side chains of each of its monomeric precursors. The KRAS inhibitor KS-58 is bicyclic by virtue of N-to-C cyclization and a dithioether bridge between two Cys residues to create a conformationally constrained analog of previously disclosed macrocyclic peptides. The NEMO inhibitor is also bicyclic by virtue of three amino acids linked by amides to a 1,3,5-tricarboxy-benzene moiety. The NEMO inhibitor's cell permeability properties are specifically related to its cationic substructure (three Arg residues) as well as its hydrophobicity and conformational constraints which confer partitioning into lipid membranes and triggering of endocytosis to drive cellular uptake.

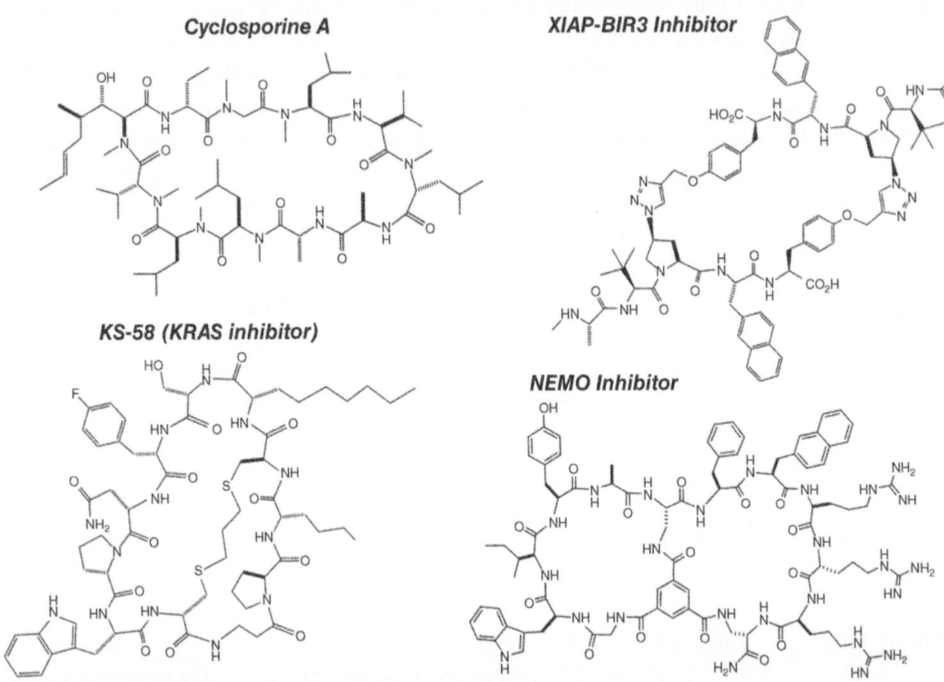

Figure 12.6 Structures of macrocyclic peptides cyclosporin A, a XIAP-BIR3 inhibitor, a KRAS inhibitor (KS-58), and a NEMO inhibitor. See text for details.

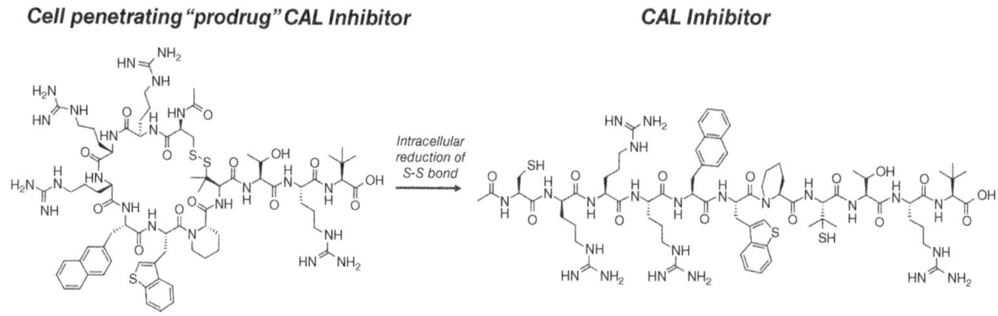

Figure 12.7 Structures of cell-permeable macrocyclic peptide "prodrug" of CAL inhibitor. See text for details.

An innovative design of a cell-penetrating macrocyclic peptide "prodrug" of CFTR-associated ligand (CAL) inhibitor [30] is shown in Figure 12.7. Upon cell uptake, the macrocyclic peptide will undergo ring opening (i.e. reduction of the disulfide) to form the linear peptide that is a potent inhibitor of CAL binding to CFTR. This work provides incentive to further design strategies to create cell-permeable macrocyclic peptide "prodrugs" exploiting intracellular S—S ring opening to enable target binding properties.

Interestingly, the precedence of macrocyclic peptides from natural products (e.g. CsA) as exemplified exclude a fundamental and canonical secondary structure, namely the α-helix. Hence, perhaps it is not surprising that macrocyclic α-helical peptides have raised intriguing therapeutic promise in terms of modulating receptor and other extracellular as well as intracellular protein–protein interaction targets. The macrocyclic α-helical peptide ATSP-7041 [31–33] (Figure 12.8) exemplifies a key benchmark for this modality class as related to both its structural features (e.g. α-methyl amino acids and cyclization by hydrocarbon stapling) and cell uptake properties (e.g. lipophilic partitioning and translocation from membrane to the cytosolic compartment). Beyond what has been significantly expanding macrocyclization chemistries (e.g. Click, thioether, amide, and other linkages) of such α-helical peptides [34–37], a noteworthy strategy to design all-D-macrocyclic peptides has been recently exemplified by ^{d}PMI-d(1–5-12) stitched all-D-peptide MDM2/X antagonist [38] as shown in Figure 12.8. This bicyclic peptide leverages a gem-dialkylated amino acid having olefinic side chains to then enable ring-closing metastasis with partner amino acids at the N- and C-terminus. The inspiration of this series of all-D-macrocyclic peptides was based upon previously described [39] linear all-D-peptides having dual MDM2/X high affinity binding and derived from mirror image display (i.e. all-D-MDM2 screening with phage display libraries to identify subsequent linear L-peptides to then provide a basis to synthesize their all-D-enantiomeric analogs).

12.5 Macrocyclic Peptide Advancement to Clinical Development and FDA Approval

It is unquestionable that macrocyclic peptide modalities will be an increasingly major part of future peptide therapeutics and no longer limited to receptor and extracellular target space. The potential for macrocyclic peptides designed for intracellular targets, such as protein–protein interactions, has been realized with respect to many ongoing drug discovery efforts, including

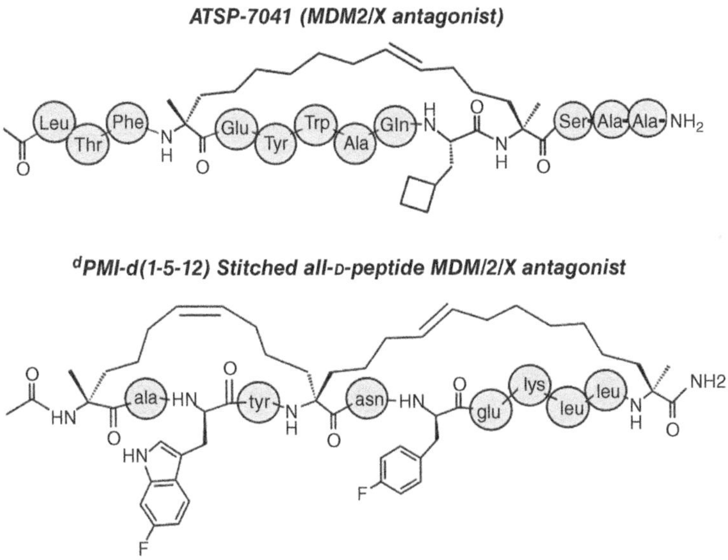

Figure 12.8 Structures of macrocyclic peptides ATSP-7041 and dPMI-d(1–5–12) stitched all-D-peptide MDM2/X antagonist. See text for details.

optimization of those preclinical lead molecules exemplified in this chapter (e.g. the XIAP-BIR3 inhibitor, KRAS inhibitor and NEMO inhibitor ATSP-7041 and the dPMI-d [1-5-12] stitched all-D-peptide MDM2/X antagonist).

Several macrocyclic peptides have been successfully advanced into clinical trials and FDA-approved over recent years (Table 12.1) and they include Somatuline (a somatostatin receptor agonist developed by Ipsen), Signifor (a somatostatin receptor agonist developed by Novartis), Linzess (a guanylate cyclase-A receptor agonist developed by Ironwood/Forest), Orbactiv (an antibiotic developed by the Med Company), Dalvance (an antibiotic developed by Vicuron), Cubicin (an antibiotic developed by Cubist [now MSD]), Grazoprevir (a HCV protease inhibitor developed by MSD), Trulance (a guanylate cyclase-A receptor agonist developed by Synergy), and Vyleesi (a melanocortin receptor agonist developed by AMAG) [40–48]. Of the four exemplary FDA-approved inhibitors of intracellular proteases, Grazoprevir [46] exemplifies a macrocyclic peptidomimetic. With respect to intracellular protein–protein interaction target space, there are many macrocyclic peptides advancing with respect to preclinical research and development, and ALRN-6924 (an analog of ATSP-7041) exemplifies a promising molecule that is currently in phase 2 clinical studies [49–51]. If it successfully proceeds through further clinical testing, it will exemplify a significant advancement for macrocyclic peptide modalities leveraging the α-helix. Importantly, for intracellular protein–protein interaction target space, it is known that α-helical interfaces constitute more than 50% of those known relative to analysis of crystallographic studies [52] (e.g. Protein Data Bank). Lastly, as indicated in Table 12.1 the predominant route of administration of most peptide drugs that have been advanced into clinical trials or have been FDA-approved to date is subcutaneous injection (SC), including long-acting formulations and implants, or intravenous (IV). However, there are some examples of peptide/peptidomimetic drugs given via the per oral (PO) route of administration, including a recently FDA-approved oral version of semaglutide that is marketed as Rybelsus [53].

Table 12.1 Some FDA approved peptide drugs (2005–2019) exemplifying their therapeutic targets, disease indications, routes of administration, and recommended doses. See text for details.

Year	Name[a]	Company	Target[b]	Indication	Route[c]	Dose
2005	**Byetta**	Amylin	GPCR	Diabetes	SC	10 μg
2005	**Symlin**	Amylin	GPCR	Diabetes	SC	15 μg
2007	**Somatuline** ←	Ipsen	GPCR	Agromegaly	IM	30 mg
2008	**Firmagon**	Ferring	GPCR	Prostate Cancer	SC	120 mg
2009	**Victoza**	Novo Nordisk	GPCR	Diabetes	SC	1.2 mg
2010	**Egrifta**	Theratech.	GPCR	Lipodystrophy	SC	2 mg
2012	**Signifor** ←	Novartis	GPCR	Cushing's	SC	20 mg
2012	**Kyprolis**	Onyx	Protease ←	Multiple Myeloma	IV	1 mg
2012	**Linzess** ←	Ironwood/Forest	Enzyme	IBS; idiopathic constipation	PO ←	266 μg
2012	**Gattex**	NPS	GPCR	Short bowl syndrome	SC	5 mg
2013	**Lyxumia**	Sanofi	GPCR	Diabetes	SC	20 μg
2014	**Orbactiv** ←	The Med. Co.	Cell membrane	Antibacterial	IV	1.2 g
2014	**Dalvance** ←	Vicuron	Cell membrane	Antibacterial	IV	1.0 g
2015	**Tanzeum**	GSK	GPCR	Diabetes	SC	30 mg qw
2015	**Trulicity**	Lilly	GPCR	Diabetes	SC	1.5 mg qw
2015	**Natpara**	NPS	GPCR	Hypocalcemia	IM	100 μg
2015	**Ninlaro**	Takeda	Protease ←	Multiple Myeloma	PO ←	3 mg
2015	**Cubicin** ←	Cubist/MSD	Cell membrane	Antibacterial	IV	~300 mg
2016	**Epslusa**	Gilead	Protease ←	Antiviral	PO ←	100 mg
2016	**Grazoprevir** ←	MSD	Protease ←	Antiviral	PO ←	50 mg
2017	**Macrilen**	Aeterna Zentaris	GPCR	Growth hormone deficiency	PO ←	5 mg
2017	**Trulance** ←	Synergy	Enzyme	Chronic idiopathic constipation	PO ←	3 mg
2017	**Ozempic**	Novo Nordisk	GPCR	Diabetes	SC	0.25 mg qw
2019	**Scenesse**	Clinuvel	GPCR	Skin damage and pain	SC	16 mg q2m
2019	**Vyleesi** ←	AMAG	GPCR	Hypoactive sexual desire	SC	1.75 mg
2019	**Rybelsus**	Novo Nordisk	GPCR	Diabetes	PO ←	3, 7, 14 mg

[a] Macrocyclic peptides are indicated by arrows.
[b] Intracellular targets are indicated by arrows.
[c] Oral delivery is indicated by arrows.

12.6 Macrocyclic Peptide Drug Discovery Paradigm and Future Directions

Macrocyclic peptide drug design strategies to tackle cell permeability and oral bioavailability are expected to become yet more sophisticated by virtue of increasing empirical data from cellular and *in vivo* models that provide more quantitative assessment of such drug delivery properties. Key physicochemical and biophysical parameters of macrocyclic peptides may be systematically analyzed relative to compelling lead molecules having target binding and functional activities. Such parameters include molecular weight (typically in the 500–2000 Da range), lipophilicity (experimental and/or calculated LogP and typically in the 2–5 range), H-bond donor and acceptors (typically seeking lower H-bond donors, in particular, via intramolecular H-bonding and/or masking by N-methylation), and polar surface area (PSA, typically seeking lower PSA). As aforementioned, it is recommended that lead optimization studies include benchmarks such as CsA and/or other well-characterized macrocyclic peptides, possibly selecting from those that have been determined to exhibit cell permeability by varying mechanisms [54–58]. Chemistry-wise, some potential ways to explore structure–permeability relationships include the "engineering" of macrocyclic peptide modalities by way of incorporating unnatural amino acids (e.g. N-methylated or cyclic amino acids, Cα-methylated amino acids and possibly β-amino acids) and/or enhanced conformational constraints relative to ring size, bicyclization and other innovative modifications that may impact overall physicochemical and biophysical properties. In the case of macrocyclic α-helical peptides (e.g. ATSP-7041), there has been significant learning from systematic structure–activity (biochemical and cellular) and structure–permeability relationship studies [31–33]. Interestingly, ATSP-7041 lacks classic passive permeability as is otherwise well-established for CsA, but ATSP-7041 and analogs thereof have been determined to possess "lipophilic partitioning" and subsequent translocation into the cytosolic compartment of cells in what is an apparent unidirectional cell uptake process. In terms of physicochemical and biophysical parameters, ATSP-7041 is a relatively lipophilic peptide and is negatively charged by virtue of a Glu substitution. In rather extreme contrast, Arg-rich macrocyclic peptides [22, 58–61] have shown significant cell permeability and functional potency by leveraging "cationic partitioning" to cell membranes and subsequent endocytic mechanisms, including efficient endosomal escape, to enter into the cytosolic compartment. Albeit the collective learnings from such passive permeability, lipophilic partitioning and cationic partitioning for optimizing macrocyclic peptide lead molecules to modulate intracellular targets is gaining momentum, at this point in time the translation of such knowledge to oral bioavailability remains yet somewhat limited to passive permeability properties. Such is not surprising, however, since passive permeability is the predominant mechanism underlying oral bioavailability for a plethora of small molecule drugs. Importantly, there is emerging a number of robust screening tools (i.e. computational, biophysical and biological) which may enable both predictive design rules and experimental permeability analysis [55, 62–77] as summarized in Table 12.2. Indeed, a tremendous opportunity exists to expand our understanding of peptide oral drug delivery for macrocyclic peptide modalities beyond the realm of those limited to passive permeability properties. In future, the potential to exploit to the highly diverse chemical space of the superclass of macrocyclic peptides by leveraging formulations with permeability enhancers (for transcellular and/or paracellular transport) to achieve oral bioavailability [78–85] will likely be the ultimate tool for drug hunters in their efforts to create the next generation of peptide therapeutics.

As a concluding thought to share in this perspective of macrocyclic peptide modalities, the "classic" multidisciplinary paradigm that has been used for many years in drug discovery still prevails

Table 12.2 Some computational, biophysical and biological screening tools used to investigate permeability properties and structure–permeability relationships of macrocyclic peptide modalities. See text for details.

Permeability screening tool	Some key features and properties evaluated
EPSA	Experimental EPSA values are determined using supercritical fluid chromatography
	Low EPSA values have been shown to correlate with high passive permeability and predicted oral bioavailability
ΔG(insertion)	ΔG(insertion) is a calculated value that refers to the solvation free difference for transferring peptide in a low-dielectric conformation from water to a low-dielectric environment (membrane-like) and is predictive of passive permeability
PAMPA	PAMPA uses mixtures of phospholipids infused into lipophilic microfilters with net negative charge (surrogate model system to correlate with experimental oral bioavailability)
Lipid:water phase partitioning	Both octanol:water phase partitioning and recent modification to incorporate fatty acids and pH gradient as shown for guanidinium-rich peptides to be predictive of energy-independent translocation
Cell monolayer transcytosis	Caco-2 cells, or other cell types, are used to measure passive permeability from donor to acceptor compartments to correlate with *in vivo* oral bioavailability
Cell-based, target agnostic penetration assay (CAPA, NanoClick)	CAPA and NanoClick measure cytosolic delivery of specifically tagged-peptides (chloroalkyl and azide, respectively) into cells that stably express a haloenzyme-reporter fusion protein (i.e. fusion with GFP and luciferase, respectively)
Radius of gyration	This is calculated as the root-mean-square distance between a peptide's atoms and its center of gravity
	This is an alternative property for MW for beyond-rule-of-five molecules
NMR analysis of intramolecular versus solvent hydrogen bonding	Solution NMR studies using hydrogen/deuterium (H/D) exchange to determine peptide backbone amide temperature coefficients and intramolecular hydrogen bonding
Label-free mass spectrometry and fluorescently tagged cell uptake analysis	Direct measurement methods for cell uptake of peptides using mass spectrometry methods and/or imaging studies (e.g. fluorescence correlation microscopy) to quantitate intracellular exposure

EPSA, exposed polar surface area; GFP, green fluorescent protein; PAMPA, parallel artificial membrane permeability assay.

for lead optimization to a drug candidate as illustrated in Figure 12.9. This typically integrates structure–activity relationships established from iterative design, synthesis and testing of molecules. Such generally requires target binding as well as cellular and *in vivo* functional properties. Biophysical, pharmacokinetic (PK) and absorption, distribution, metabolism, and excretion (ADME) properties are also extremely important for translating preclinical lead molecules to clinical candidates. Such properties may generally be "target agnostic" and have more to do with optimizing solubility, permeability, stability, and exposure levels *in vivo* to enable possible correlation between pharmacological efficacy and PK/ADME. It is anticipated that macrocyclic peptide modalities as highlighted in this chapter will greatly expand our working knowledge of peptide

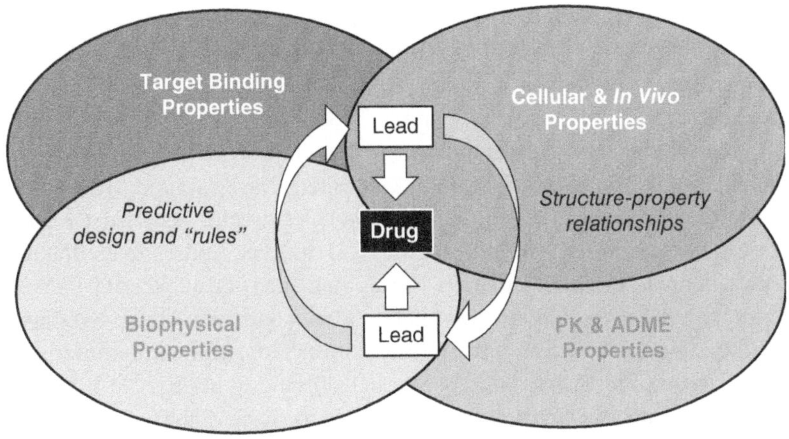

Figure 12.9 Multidisciplinary paradigm for macrocyclic peptide drug discovery. See text for details.

drug development, especially for those designed to modulate intracellular targets. Relative to receptor- and extracellular target space, the example of the synthetic trimeric macrocyclic peptide BCY12491 [14] (see Section 12.3) illustrates the potential for macrocyclic peptide designs to create novel peptide–peptide conjugates capable of interacting with two different receptors and expanding upon recent progress involving yet other types of peptide (albeit linear in many cases) conjugation with small molecule drugs, albumin-binding ligands, antibodies, proteins, oligonucleotides, carbohydrates and other modalities [86–98]. In retrospect, the pioneering work of many individuals has contributed to our current macrocyclic peptide chemical space [16–26, 32–36, 52–54, 57, 60–62, 64–66, 84, 97–131], and it is unquestionable that there yet exists significant opportunity for innovation to advance this superclass therapeutic modality (e.g. computational design, cyclization chemistry, drug delivery, and peptide conjugation). Hopefully, such will inspire the imagination of future peptide drug hunters.

Acknowledgements

I would like to acknowledge the following peptide drug hunters (and their colleagues) for the outstanding scientific contributions to some of the work that I have highlighted in this chapter as well as their collaboration in varying ways over my career (listed alphabetically): Emel Adaligil, Gregory Adams, Fahad Al-Obeidi, Juan Alvarez, Paramjit Arora, Per Artursson, Joseph Audie, Alexander Bayden, Annette Beck-Sickinger, Elisabetta Bianchi, Kaustav Biswas, Nicolas Boyer, David Brayden, Thomas Britton, Christopher Brown, Zachary Brown, Minying Cai, Yong Chang, Jean Chmielewski, Wayne Cody, Roger Cone, David Craik, Christian Cunningham, Suzanne D'Addio, Krzysztof Darlak, Paul Darman, Phil Dawson, Charles Deber, Qiaolin Deng, Tanvi Desai, Patrick Dougherty, Ruchia Duggal, Scott Edmonson, Carl Elkin, Mark Embry, Wayne Fairbrother, David Fairlie, Paul Feldman, Fernando Jose Ferrer Gago, Eric Feyfant, Nader Fotouhi, Roger Freidinger, John Frost, Prajakta Gadgil, Pranoti Gangurde, Robert Garbaccio, Michael Garrigou, Victor Garsky, Sam Gellman, Lila Gierasch, Jenifer Gifford, Marian Gindy, Meir Glick, Bradford Graves, Vincent Guerlavais, Sookhee Ha, Mac Hadley, Carrie Haskell-Luevano, Jessica Hawes, Christian Heinis, Jennifer Hickey, Ralph Hirschmann, Jerome Hochman, Robert Hodges, Brett Hopkins, Victor Hruby, Chunhui Huang, Charles Johannes, Jennifer Johnston, Srinivasaraghavan Kannan, Noriyuki

Kawahata, Wieslaw (Wes) Kazmierski, Nicholas Keen, Stephen Kent, Angela Kerekes, Kamala Kesavan, Horst Kessler, Hai-Young (Anne) Kim, James Knittel, Thomas Kodadek, Joshua Kritzer, Shane Krska, Krishna Kumar, Peter Kutchukian, Kit Lam, David Lane, Hyelee Lee, Scott Lentini, Immanuel Lerner, Charles Lesberg, Songnian Lin, Yu-Shan Lin, Scott Lokey, Derek Maclean, Rumit Maini, Claudio Mapelli, Alexey Makarov, Santiago Esteban Martin, Kevin McDonnell, Amit Michaeli, Francesca Milletti, Lauren Monovich, Henry (Hank) Mosberg, Ravi Nargund, Huw Nash, Matthew Naylor, Laura Nevola, Simon Ng, Justin Noehre, Daniel Obrecht, Federica Orvieto, Anandan Palani, Craig Parish, Emma Parmee, Anthony Partridge, Dehua Pei, Andrea Peier, Bradley Pentelute, Antonello Pessi, Nicholas Pierson, Cameron Pye, Ziqing (Leo) Qian, Ronald Raines, Patrick Reid, Alonso Ricardo, Matt Robers, Michael Rosenblatt, Manoj Samant, Solimar Santiago, Berengere Sauvagnat, Jonathon Sawyer, Thomas Sawyer, Joshua Schwochert, Natarajan Sethuraman, Aurash Shahripour, Shubh Sharma, Ed Sherer, Woody Sherman, Brandon Smith, Clark Smith, Zhanna Sobol, Ved Srivastava, Douglas Staples, Yongchao Su, Hiro Suga, Chengzao Sun, Jon Swanson, David Tellers, Nick Terrett, Thomas Tucker, Punit Upadhyaya, Lex Van der Ploeg, Daniel Veber, Greg Verdine, Chandra Verma, Abbas Walji, Ann Weber, Blair Wood, Chengwei Wu, Yusheng Xiong, Andrei Yudin, Doris Zane, and Hongjun (David) Zhang.

References

1 Mayer, J., Zhang, F., and DiMarchi, R.D. (2007). Insulin structure and function. Biopolymers 88: 687–713.

2 Wisneiwski, K. (2001). Design of oxytocin analogs. *Methods Mol. Biol.* 2019: 235–271.

3 Gutte, B. and Merrifield, R.B. (1969). Total synthesis of an enzyme with ribonuclease a activity. *J. Am. Chem. Soc.* 92 (2): 501–502.

4 Hirschmann, R., Nutt, R.F., Veber, D.F. et al. (1969). Total synthesis of an enzyme. V. Preparation of enzymatically active material. *J. Am. Chem. Soc.* 91 (2): 507–508.

5 Rosenberg, J. and Octreotide, M. (1988). A synthetic analog of somatostatin. *Drug Intell. Clin. Pharm.* 22 (10): 748–754.

6 Gomes-Porras, M., Cardenas-Sales, J., and Alvarez-Escola, C. (2020). Somatostatin analogs in clinical practice: a review. 21 (5): 1682.

7 Sawyer, T.K., Hruby, V.J., Darman, P.S., and Hadley, M.E. (1982). [Half-Cys^4, half-Cys^{10}]alpha-melanocyte-stimulating hormone: a cyclic alpha-melanotropin exhibiting superagonist biological activity. *Proc. Natl. Acad. Sci. USA* 79 (6): 1751–1755.

8 Sharma, S., Garfield, A.S., Shah, B. et al. (2019). Current mechanistic and pharmacodynamic understanding of melanocortin-4 receptor activation. *Molecules* 24 (10): 1892.

9 Yu, J., Gimenez, L.E., Hernandez, C.C. et al. (2020). Determination of the melanocortin-4 receptor structure identifies Ca^{2+} as a cofactor for ligand binding. *Science* 368 (6489): 428–433.

10 Rothstein, R.D. and Friedenberg, F.K. (2013). Linaclotide: a novel compound for the treatement of irritable bowel syndrome with constipation. *Expert Opin. Pharmacother.* 14 (15): 2125–2132.

11 Gorman, D.M., Lee, J., Payne, C.D. et al. (2021). Chemical synthesis and characterization of the complement C5 inhibitory peptide zilucoplan. *Amino Acids* 53: 143–147.

12 Alleyne, C., Amin, R.P., Bhatt, B. et al. (2020). Series of novel and highly potent cyclic peptide PCSK9 inhibitors derived from an mRNA display screen and optimized via structure-based design. *J. Med. Chem.* 63 (22): 13796–13824.

13 Boehm, M., Beaumont, K., Jones, R. et al. (2017). Discovery of potent and orally bioavailable macrocyclic peptide-peptoid hybrid CXCR7 modulators. *J. Med. Chem.* 60 (23): 9653–9663.

14 Upadhyaya, P., Lahdenranta, J., Hurov, K. et al. (2021). Anticancer immunity induced by a synthetic-targeted CD137 agonist. *J. Immunother. Cancer* 9 (1): e0017762.

15 Stahelin, H.F. (1996). The history of cyclosporin a (Sandimmune) revisited: another point of view. *Experientia* 52: 5–13.

16 Nielsen, D.S., Shepherd, N.E., Xu, W. et al. (2017). Orally absorbed cyclic peptides. *Chem. Rev.* 117 (12): 8094–8128.

17 Naylor, M.R., Bockus, A.T., Blanco, M.-J., and Lokey, R.S. (2017). Cyclic peptide natural products chart the frontier of oral bioavailability in the pursuit of undruggable targets. *Curr. Opin. Chem. Biol.* 38: 141–147.

18 Pye, C.R., Hewitt, W.M., Schwochert, J. et al. (2017). Nonclassical size dependence of permeation defines bounds for passive adsorption of large drug molecules. *J. Med. Chem.* 60 (5): 1665–1672.

19 Chatterjee, J., Laufer, B., and Kessler, H. (2012). Synthesis of N-methylated cyclic peptides. *Nat. Protoc.* 7 (3): 432–333.

20 Kelly, C.N., Townsend, C.E., Jain, A.N. et al. (2021). Geometrically diverse lariat peptide scaffolds reveal an uptapped chemical space of high membrane permeability. *J. Am. Chem. Soc.* 143 (2): 705–714.

21 Naylor, M.R., Ly, A.M., Handford, M.J. et al. (2018). Lipophilic permeability efficiency reconciles the opposoing roles of lipophilicity in membrane permeability and aqueous solublity. *J. Med. Chem.* 61 (24): 11169–11182.

22 Qian, Z., Upadhyaya, P., and Pei, D. (2015). Synthesis and screening of one-bead-one-compound cyclic libraries. *Methods Mol. Biol.* 1248: 39–53.

23 Dougherty, P.G., Qian, Z., and Pei, D. (2017). Macrocycles as protein-protein interaction inhibitors. *Biochem. J.* 474 (7): 1109–1125.

24 Chen, S. and Heinis, C. (2015). Phage selection of bicyclic peptides based on two disulfide bridges. *Methods Mol. Biol.* 1248: 119–137.

25 Bashiruddin, N.K. and Suga, H. (2015). Construction and screening of vast libraries of natural product-like macrocyclic peptides using in vitro display technologies. *Curr. Opin. Chem. Biol.* 24: 131–138.

26 Zhu, Z., Shaginian, A., Grady, L.C. et al. (2018). Design and application of a DNA-encoded macrocyclic peptide library. *ACS Chem. Biol.* 13 (1): 53–59.

27 Seigal, B.A., Connors, W.H., Fraley, A. et al. (2015). The discovery of macrocyclic XIAP antagonists from a DNA programmed chemistry library, and their optimization to give lead compounds with in vivo antitumor activity. *J. Med. Chem.* 58 (6): 2855–2861.

28 Sakamoto, K., Mautani, T., and Hirokawa, T. (2020). Generation of KS-58 as the first K-Ras(G12D)-inhibitory peptide presenting anti-cancer activity in vivo. *Sci. Rep.* 10 (1): 21671.

29 Rhodes, C.A., Dougherty, P.G., Cooper, J.K. et al. (2018). Cell-permeable bicyclic peptidyl inhibitors against NEMO-IκB kinase interaction directly from a combinatorial library. *J. Am. Chem. Soc.* 140 (38): 12102–12110.

30 Dougherty, P.G., Wellmerling, J.H., Koley, A. et al. (2020). Cyclic peptidyl inhibitors against CAL/CFTR interaction for treatment of cystic fibrosis. *J. Med. Chem.* 63 (24): 15773–15784.

31 Chang, Y.S., Graves, B., Guerlavais, V. et al. (2013). Stapled α-helical peptide drug development: a potent dual inhibitor of MDM2 and MDMX for p53-dependent cancer therapy. *Proc. Natl. Acad. Sci. USA* 110 (36): E3445–E3454.

32 Sawyer, T.K., Partridge, A.W., Kaan, H.Y.K. et al. (2018). Macrocyclic α-helical peptide therapeutic modality: a perspective of learnings and challenges. *Bioorg. Med. Chem.* 26 (10): 2807–2815.

33 Partridge, A.W., Kaan, H.Y.K., Juang, Y.C. et al. (2019). Incorporation of putative helix-breaking amino acids in the design of novel stapled peptides: exploring biophysical and cellular permeability properties. *Molecules* 24 (12): 2292.

34 Wu, Y., Kaur, A., Fowler, E. et al. (2019). Toolbox of diverse linkers for navigating the cellular efficacy landscape of stapled peptides. *ACS Chem. Biol.* 14 (3): 526–533.

35 Spokoyny, A.M., Zou, Y., Ling, J.J. et al. (2013). A perfluoroaryl-cysteine S(N)Ar chemistry approach to unprotected peptide stapling. *J. Am. Chem. Soc.* 135 (16): 5946–5949.

36 Bernal, F. and Katz, S.G. (2014). Synthesis of stabilized alpha-helical peptides. *Methods Mol. Biol.* 1176: 107–114.

37 Tiwari, G. and Verma, C.S. (2017). Toward understanding the molecular recognition of albumin by p53-activating stapled peptide ATSP-7041. *J. Phys. Chem. B* 121 (4): 657–670.

38 Kannan, S., Aronica, P.G.A., Ng, S. et al. (2020). Macrocyclization of an all-d linear α-helical peptide imparts cellular permeability. *Chem. Sci.* 11 (21): 5577–5591.

39 Zhan, C., Zhao, L., Wei, X. et al. (2012). An ultrahigh affinity d-peptide antagonist of MDM2. *J. Med. Chem.* 55 (13): 6237–6241.

40 Haris, B., Saraswathi, S., and Hussain, K. (2020). Somatostatin analogues for the treatment of hyperinsulinaemic hypoglycaemia. *Ther. Adv. Endocrinol. Metab.* 11: 1–23.

41 Cuevas-Ramos, D. and Fleseriu, M. (2016). Pasireotide: a novel treatment for patients with acromegly. *Drug Des. Devel. Ther.* 10: 227–239.

42 Love, B.L., Johnson, A., and Smith, L.S. (2014). Linaclotide: a novel agent for chronic constipation and irritable bowel syndrome. *Am. J. Health Syst. Pharm.* 71 (13): 1081–1091.

43 Brade, K.D., Rybak, J.M., and Rybak, M.J. (2016). Oritavancin: a new lipoglycopeptide antibiotic in the treatment of gram-positive infections. *Infect. Dis. Ther.* 5 (1): 1–15.

44 Roecker, A.M. and Pope, S.D. (2008). Dalbavancin: a lipoglycopeptide antibacterial for gram-positive infections. *Expert Opin. Pharmacother.* 9 (10): 1745–1754.

45 Steenbergen, J.N., Alder, J., Thorne, G.M., and Tally, F.P. (2005). Daptomycin: a lipopeptide antibiotic for the treatment of serious gram-positive infections. *J. Antimicrob. Chemother.* 55 (3): 282–288.

46 Bell, A.M., Wagner, J.L., Barber, K.E., and Stover, K.R. (2016). Elbasvir/Grazoprevir: a review of the latest agent in the fight against hepatitis C. *Int. J. Hepatol.* 2016: 3852126.

47 Miner, P.B. (2020). Plecanatide for the treatment of constipation-predominant irritable bowel syndrome. *Expert Rev. Gastroenterol. Hepatol.* 14 (2): 71–84.

48 Dhillon, S. and Keam, S.J. (2019). Bremelanotide: first approval. *Drugs* 79 (14): 1599–1606.

49 Eskandari, M., Shi, Y., Liu, J. et al. (2021). The expression of MDM2, MDM4, p53 and p21 in myeloid neoplasms and the effect of MDM2/MDM4 dual inhibitor. *Leuk. Lymphoma* 61 (1): 167–175.

50 Ng, S.Y., Yoshida, N., Christie, A.L. et al. (2018, 2024). Targetable vulnerabilities in T- and NK-cell lymphomas identified through preclinical models. *Nat. Commun.* 9 (1).

51 Carvajal, L.A., Neriah, D.B., Senecal, A. et al. (2018). Dual inhibition of MDMX and MDM2 as a therapeutic strategy in leukemia. *Sci. Transl. Med.* 10: 436, eaao3004.

52 Watkins, A.M., Wuo, M.G., and Arora, P.S. (2015). Protein-protein interactions mediated by helical tertiary structure motifs. *J. Am. Chem. Soc.* 137 (36): 11622–11630.

53 Davies, M., Pieber, T.R., Hartoft-Nielsen, M.-L. et al. (2017). Effect of oral semaglutide compared with placebo and subcutaneous semaglutide on glycemic control in patients with type 2 diabetes: a randomized clinical trial. *JAMA* 318 (15): 1460–1470.

54 Hochman, J., Sawyer, T.K., and Duggai, R. (2021). Overcoming cellular and systemic barriers to design the next wave of peptide therapeutics. In: *Quantitative Analysis of Cellular Drug Transport, Disposition, and Delivery* (eds. G. Rosani, G. Thurber and R.K. Keswani), 201–227. Springer Nature.

55 Peier, A., Ge, L., Boyer, N. et al. (2021). NanoClick assay: a high-throughput, target-agnostic peptide cell permeability assay. *ACS Chem. Biol.* 16: 293–309.

56 Peraro, L., Deprey, K.L., Moser, M.K. et al. (2018). Cell penetration profiling using the chloroalkane penetration assay. *J. Am. Chem. Soc.* 140 (36): 11360–11369.

57 Furukawa, A., Schwochert, J., Pye, C.R. et al. (2020). Drug-like properties in macrocycles above MW 1000: backbone rigidity versus side-chain lipophilicity. *Angew. Chem. Int. Ed. Engl.* 59 (48): 21571–21577.

58 Sahni, A., Qian, Z., and Pei, D. (2020). Cell-penetrating peptides escape the endosome by inducing vesicle budding and collapse. *ACS Chem. Biol.* 15 (9): 2485–2492.

59 Qian, Z., LaRochelle, J.R., Jiang, B. et al. (2014). Early endosomal escape of a cyclic cell-penetrating peptide allows effective cytosolic cargo delivery. *Biochemistry* 53 (24): 4034–4046.

60 Appiah Kubi, G., Dougherty, P.G., and Pei, D. (2001). Designing cell-permeable macrocyclic peptides. *Methods Mol. Biol.* 2019: 41–59.

61 Dougherty, P.G., Sahni, A., and Pei, D. (2019). Understanding cell penetration of cyclic peptides. *Chem. Rev.* 119 (17): 10241–10287.

62 Furukawa, A., Townsend, C.E., Schwochert, J.A. et al. (2016). Passive membrane permeability in cyclic peptomer scaffolds is robust to extensive variation in sidechain functionality and backbone geometry. *J. Med. Chem.* 59: 9503–9512.

63 Schwochert, J., Lao, Y., Pye, C.R. et al. (2016). Stereochemistry balances cell permeability and solubility in the naturally derived phepropeptin cyclic peptides. *ACS Med. Chem. Lett.* 7: 757–761.

64 Ahlbach, C.L., Lexa, K.W., Bockus, A.T. et al. (2015). Beyond cyclosporine a: conformation-dependent passive membrane permeabilities of cyclic peptide natural products. *Future Med. Chem.* 7: 2121–2130.

65 Bockus, A.T., Lexa, K.W., Pye, C.R. et al. (2015). Probing the physicochemical boundaries of cell permeability and oral bioavailability in lipophilic macrocycles inspired by natural products. *J. Med. Chem.* 58: 4581–4589.

66 Hewitt, W.M., Leung, S.S.F., Pye, C.R. et al. (2015). Cell-permeable cyclic peptides from synthetic libraries inspired by natural products. *J. Am. Chem. Soc.* 137: 715–721.

67 Goetz, G.H., Philippe, L., and Shapiro, M.J. (2014). EPSA: a novel supercritical fluid chromatography technique enabling the design of permeable cyclic peptides. *ACS Med. Chem. Lett.* 5: 1167–1172.

68 Nielsen, D.S., Lohman, R.-J., Hoang, H.N. et al. (2015). Flexibility versus rigidity for orally bioavailable cyclic Hexapeptides. *Chembiochem* 16: 2289–2293.

69 Rafi, S.B., Hearn, B.R., Vedantham, P. et al. (2012). Predicting and improving the membrane permeability of peptidic small molecules. *J. Med. Chem.* 55 (7): 3163–3169.

70 Guimarães, C.R., Mathiowetz, A.M., Shalaeva, M. et al. (2012). Use of 3D properties to characterize beyond rule-of-5 property space for passive permeation. *J. Chem. Inf. Model.* 52: 882–890.

71 Wang, C.K., Northfield, S.E., Swedberg, J.E. et al. (2015). Exploring experimental and computational markers of cyclic peptides: charting islands of permeability. *Eur. J. Med. Chem.* 97: 202–213.

72 Herce, H.D., Garcia, A.E., and Cardoso, M.C. (2014). Fundamental molecular mechanism for the cellular uptake of guanidinium-rich molecules. *J. Am. Chem. Soc.* 136 (50): 17459–17467.

73 Rezgui, R., Blumer, K., Yeoh-Tan, G. et al. (2016). Precise quantification of cellular uptake of cell-penetrating peptides using fluorescence-activated cell sorting and fluorescence correlation spectroscopy. *Biochim. Biophys. Acta* 1858 (7 Pt A): 1499–1506.

74 LaRochelle, J.R., Cobb, G.B., Steinauer, A. et al. (2015). Fluorescence correlation spectroscopy reveals highly efficient cytosolic delivery of certain penta-arg proteins and stapled peptides. *J. Am. Chem. Soc.* 137: 2536–2541.

75 Holm, T., Andaloussi, S.E., and Langel, U. (2011). Comparison of CPP uptake methods. *Methods Mol. Biol.* 683: 207–217.

76 Aubry, S., Aussedat, B., Delaroche, D. et al. (1798). MALDI-TOF mass spectrometry: a powerful tool to study the internalization of cell-penetrating peptides. *Biochim. Biophys. Acta* 2010: 2182–2189.

77 Gordon, L.J., Allen, M., Artursson, P. et al. (2016). Direct measurement of intracellular compound concentration by RapidFire mass spectrometry offers insights into cell permeability. *J. Biomol. Screen.* 21: 156–164.

78 Lemmer, H.J. and Hamman, J.H. (2013). Paracellular drug absorption enhancement through tight junction modulation. *Expert Opin. Drug Deliv.* 10 (1): 103–114.

79 Maher, S., Mrsny, R.J., and Brayden, D.J. (2016). Intestinal permeation enhancers for oral peptide delivery. *Adv. Drug Deliv. Rev.* 106 (Pt B): 277–319.

80 Danielsen, E.M. (1863). Intestinal permeation enhancers: lessons learned from studies using an organ culture model. *Biochim. Biophys. Acta Biomembr.* 2021 (1): 18374.

81 Brayden, D.J., Hill, T.A., Fairlie, D.P. et al. (2020). Systemic delivery of peptides by the oral route: formulation and medicinal chemistry approaches. *Adv. Drug Deliv. Rev.* 157: 2–36.

82 Bucheit, J.D., Pamulapati, L.G., Carter, N. et al. (2020). Oral semaglutide: A review of the first oral glucagon-like peptide 1 receptor agonist. *Diabetes Technol Ther.* 11: 10–18.

83 Di, L. (2015). Strategic approaches to optimizing peptide ADME properties. *AAPS J.* 17 (1): 134–143.

84 Rader, A.F.B.M., Weinmuller, M., Reichart, F. et al. (2018). Orally active peptides: is there a magic bullet? *Angew. Chem. Int. Ed. Engl.* 57 (44): 14414–14438.

85 Zizzari, A.T., Pliatsika, D., Gall, F.M. et al. (2021). New perspectives in oral peptide delivery. *Drug Discov. Today* (26, 4): 1097–1105.

86 Bohme, D. and Beck-Sickinger, A.G. (2015). Controlling toxicity of peptide-drug conjugates by different chemical linker structures. *ChemMedChem* 10: 804–814.

87 He, R., Finan, B., Mayer, J.P., and DiMarchi, R.D. (2019). Peptide conjugates with small molecules design to enhance efficacy and safety. *Molecules* 24: 1855.

88 Hoppenz, P., Elis-Heindl, S., and Beck-Sickinger, A.G. (2020). Peptide-drug conjugates and their targets in advanced cancer therapies. *Front. Chem.* 8: 571.

89 Balogh, B., Ivanczi, M., Nizami, B. et al. (2021). ConjuPepDB: a database of peptide-drug conjugates. *Nucleic Acids Res.* 49: D1102–D1112.

90 Lear, S., Pfimlin, E., Zhou, Z. et al. (2021). Engineering of a potent, long-acting NPY2R agonist for combination with GLP-1R as a multi-hormonal treatment for obesity. *J. Med. Chem.* 63: 9660–9671.

91 Zhou, X., Smith, Q.R., and Liu, X. (2021). Brain penetrating peptides and peptide–drug conjugates to overcome the blood-brain barrier and target CNS diseases. *Wiley Interdiscip. Rev. Nanomed. Nanobiotechnol.* 13 (4): e1695.

92 Parrasia, S., Rossa, A., Varanita, T. et al. (2021). An angiopep2-PAPTP construct overcomes the blood-brain barrier. New perspectives against brain tumors. *Pharmaceuticals (Basel)* 14 (2): 129.

93 Hua, D., Tang, L., Wang, W. et al. Improved angioblastoma activity and BBB permeability by conjugation of paclitaxel to cell-penetrative MMP-2-cleavable peptide. *Adv. Sci. (Weinh)* **2020**, 8 (2): 2001960.

94 Duskey, J.T., Ottonelli, I., Da Ros, F. et al. (2020). Novel peptide-conjugated nanomedicines for brain targeting: in vivo evidence. *Nanomed. Nanotechnol. Biol. Med.* 28: 102226.

95 Kurnkoff, K., Vunk, B., and Langel, U. (2020). Status update in the use of cell-penetrating peptides for the delivery of macromolecular therapeutics. *Expert Opin. Biol. Ther.* 21 (3): 361–370.

96 Kaarsholm, N.C., Lin, S., Yan, L. et al. (2018). Engineering glucose responsiveness into insulin. *Diabetes* 67 (2): 299–308.

97 Bhat, M., Jatyan, R., Mittal, A. et al. (2021). Opportunities and challenges of fatty acid conjugated therapeutics. *Chem. Phys. Lipids* 236: 105053.

98 Zhang, C., Vingradova, E.V., Spokoyny, A.M. et al. (2019). Arylation chemistry for bioconjugation. *Angew. Chem. Int. Ed. Engl.* 58 (15): 4810–4839.

99 Philippe, G.J.B., Craik, D.J., and Henriques, S.T. (2021). Converting peptides into drugs targeting intracellular protein–protein interactions. *Drug Discov. Today* 26 (6): 1521–1531.
100 Gonzalez-Muniz, R., Bonache, M.A., and de Vega, M.J.P. (2021). Modulating protein–protein interactions by cyclic and macrocyclic peptides: prominent strategies and examples. *Molecules* 26 (2): 445.
101 Gonzalez-Castro, R., Gomez-Lim, M.A., and Plisson, F. (2021). Cysteine-rich peptides: Hyperstable scaffolds for protein engineering. *Chembiochem* 22 (6): 961–973.
102 Buckton, L.K., Rahimi, M.N., and McAlpine, S.R. (2021). Cyclic peptides as drugs for intracellular targets: the next frontier in peptide therapeutic development. *Chemistry* 27 (5): 1487–1513.
103 Wills, R., Adebomi, V., and Raj, M. (2021). Site-selective peptide macrocyclization. *Chembiochem* 22 (1): 52–62.
104 De Veer, S.J., Kan, M.-W., and Craik, D.J. (2019). Cyclotides: from structure to function. *Chem. Rev.* 119 (24): 12375–12421.
105 Malins, L.R., de Gruyter, J.N., Robbins, K.J. et al. (2017). Peptide macrocyclization inspired by non-ribosomal imine natural products. *J. Am. Chem. Soc.* 139 (14): 5233–5241.
106 Atangcho, L., Navaratna, T., and Thurber, G.M. (2019). Hitting undruggable targets: viewing stabilized peptide development through the lens of quantitative systems pharmacology. *Trends Biochem. Sci.* 44 (3): 241–257.
107 Lau, Y.H., de Andrade, P., Wu, Y., and Spring, D.R. (2015). Peptide stapling techniques based on different macrocyclisation chemistries. *Chem. Soc. Rev.* 44 (1): 91–102.
108 White, C.J. and Yudin, A.K. (2011). Contemporary strategies for peptide macrocyclization. *Nat. Chem.* 3 (7): 509–524.
109 Zorzi, A., Deyle, K., and Heinis, C. (2017). Cyclic peptide therapeutics: past, present and future. *Curr. Opin. Chem. Biol.* 38: 24–29.
110 Vinogradov, A.A., Yin, Y., and Suga, H. (2019). Macrocyclic peptides as drug candidates: recent progress and remaining challenges. *J. Am. Chem. Soc.* 141 (10): 4167–4181.
111 Rivera, D.G., Ojeda-Carralero, G.M., Reguera, L., and Van der Eycken, E.V. (2020). Peptide macrocyclization by transition metal catalysis. *Chem. Soc. Rev.* 49 (7): 2039–2059.
112 Tsiamantas, C., Otero-Ramirez, M.E., and Suga, H. (2001). Discovery of functional macrocyclic peptides by means of the RaPID system. *Methods Mol. Biol.* 2019: 299–315.
113 Ermert, P., Luther, A., Zbinden, P., and Obrecht, D. (2019). Frontier between cyclic peptides and macrocycles. *Methods Mol. Biol.* 2001: 147–202.
114 Maide, A.K., Hill, T., Iyer, A. et al. (2019). Crystal structures of protein-bound cyclic peptides. *Chem. Rev.* 119 (17): 9861–9914.
115 Otero-Ramirez, M.E., Passioura, T., and Suga, H. (2018). Structural features and binding modes of thioether-cyclized peptide ligands. *Biomedicine* 6 (4): 116.
116 Robertson, N.S. and Spring, D.R. (2018). Using peptidomimetics and constrained peptides as valuable tools for inhibiting protein–protein interactions. *Molecules* 23 (4): 959.
117 Michaeli, A., Mezan, S., Kuhbacher, A. et al. (2018). Computationally designed bispecific MD2/CD14 binding peptides show TLR4 agonist activity. *J. Immunol.*: 3383–3391.
118 Ng, S., Juang, Y.-C., Chandramohan, A. et al. (2020). De-risking drug discovery of intracellular targeting peptides: screening strategies to eliminate false-positive hits. *ACS Med. Chem. Lett.* 11: 1993–2001.
119 Diller, D., Swanson, J., Jarasinski, M. et al. (2019). Rigorous retrospective and prospective computational investigations on MDM2/MDMX-targeted stapled and non-stapled peptides. *Molecules* 24 (24): E4586.

120 Gadgil, P., Alleyne, C., Feng, K.-I. et al. (2019). Assessing the utility of in vitro screening tools for predicting bio-performance of oral peptide delivery. *Pharm. Res.* 36: 151.

121 Fan, L., McElroy, W.T., Sawyer, T.K. et al. (2020). The challenge of prosecuting intracellular protein–protein interactions: a multi-modality examination of chemical matter identification, biophysical characterization and membrane permeability. In: *Inhibitors of Protein–Protein Interactions: Small Molecules, Peptides and Macrocycles* (ed. A. Tavassoli), 66–110. London: Royal Society of Chemistry.

122 Sawyer, T.K., Hochman, J., D'Addio, S.M. et al. (2019). Transport, stability and delivery considerations for design of peptide drugs. In: *Peptide Therapeutics: Strategy and Tactics for Chemistry, Manufacturing, and Controls*, Drug Discovery Series No. 72 (ed. V. Srivastava), 421–472. London: Royal Society of Chemistry.

123 Sawyer, T.K. (2017). Renaissance in peptide drug discovery: the third wave. In: *Peptide-Base Drug Discovery: Challenges and Opportunities* (ed. V. Srivastava), 1–34. London: Royal Society of Chemistry.

124 Diller, D., Jarosinski, M., Sawyer, T.K., and Audie, J. (2015). Peptide drug discovery: innovative technologies and transformational medicines. In: *Advances in the Discovery and Development of Peptide Therapeutics* (eds. G. Karger and F. Albericio), 8–27. London: Future Science Publishers.

125 Sawyer, T.K., Guerlavais, V., Darlak, K., and Feyfant, E. (2014). Macrocyclic α-helical peptide drug discovery. In: *Macrocycles in Drug Discovery* (ed. J. Levin), 339–366. London: Royal Society of Chemistry.

126 Guerlavais, V. and Sawyer, T.K. (2014). Advancements in stapled peptide drug discovery and development. In: *Annual Reports in Medicinal Chemistry*, vol. 49 (ed. P. Berstein), 331–345. Washington, D.C.: American Chemical Society.

127 Zane, D., Feldman, P.L., Sawyer, T.K. et al. (2021). Development and regulatory challenges for peptide therapeutics. *Int. J. Toxicol.* 40 (2): 108–124.

128 Davenport, A.P., Scully, C.C.G., de Graaf, C. et al. (2020). Advances in therapeutic peptides targeting G protein-coupled reeeptors. *Nat. Rev. Drug Discov.*: 389–413.

129 Phyo, P., Zhao, X., Templeton, A.C. et al. (2021). Understanding molecular mechanism of biologics drug delivery and stability from NMR spectroscopy. *Adv. Drug Deliv. Rev.* 174: 1–29.

130 Muttenthaler, M., King, G.D., Adams, D.J., and Alewood, P.F. (2021). Trends in peptide drug discovery. *Nature Rev. Drug Discov.* 20: 309–325.

131 Golosov, A.A., Flyer, A.N., Amin, J. et al. (2021). Design of thioether cyclic peptide scaffolds with passive permeability and oral exposure. *J. Med. Chem.* 64 (5): 2622–2633.

13

Application of Pyrrolobenzodiazepines in Antibody Drug Conjugates

Ning Zou and Amy Han

R&D Chemistry, Therapeutic Proteins, Regeneron Pharmaceuticals, Inc., 777 Old Saw Mill River Road, NY 10591, Tarrytown, NY, USA

13.1 Introduction

The isolation of anthramycin (Figure 13.1a) from *Streptomyces refuineus* in 1965 revealed a new class of molecules that exhibited potent antitumor biological activities [1]. This family of compounds has attracted wide interest in chemical synthesis and drug development due to the presence of a novel pyrrolobenzodiazepine (PBD) core (Figure 13.1b) within their structures. This PBD core consists of three fused rings: (i) a substituted aromatic "A" ring; (ii) a seven-member diazepine "B" ring; and (iii) a pyrrolidine "C" ring. The S-configuration at the C11 position is critical to the activity of PBD [2, 3].

The molecules exist in three interconvertible forms at the C11-N10 position: imine and carbinolamine (Figure 13.1c) with the C10-N11 imine having the ability to form a covalent aminal bond between the C11 carbon and the C2-NH_2 group of a guanine base once the molecule has settled within the minor groove (Figure 13.1c). Formation of a covalent bond between PBD and a guanine base of DNA (Figure 13.2) makes PBD serve as a strong DNA alkylator and causes interference with the DNA replication processing, thus shunting cellular growth [5].

Recent attempts to improve *in vivo* efficacy and/or reduce toxicity to normal cells include: (i) conducting a structure–activity relationship (SAR) study on the A- and C-ring [6], (ii) linking DNA alkylating agents to the PBD skeleton [7], (iii) combining two PBD monomers to form the respective dimers, and (iv) creating unsymmetrical PBD dimers by coupling two different PBD monomers with one PBD monomer containing a type of DNA alkylating agent [8]. A summary of SAR studies on the monomers of PBD, dimers of PBD, and unsymmetrical dimers of PBD is shown in Figure 13.3 [4].

It was found that electron-donating groups in the C-7 and C-8 positions of the A-ring increased electrophilicity of the N-10/C-11 imine moiety, and in turn increased potency. Substituents on the C-9 position of the A-ring caused a decrease in potency, primarily due to steric hindrance of the substituents with the minor groove of DNA. Hydroxyl groups installed at the C-9 position had the potential to produce free radicals and promote cardiotoxicity. S stereochemistry at C-11a was imperative for DNA binding because it allowed the correct conformation relative to the DNA minor groove. Unsaturation of the C-ring, along with C-2 exo and/or C1-2 and C2-3 endo double

Contemporary Accounts in Drug Discovery and Development, First Edition. Edited by Xianhai Huang, Robert G. Aslanian, and Wayne H. Tang.

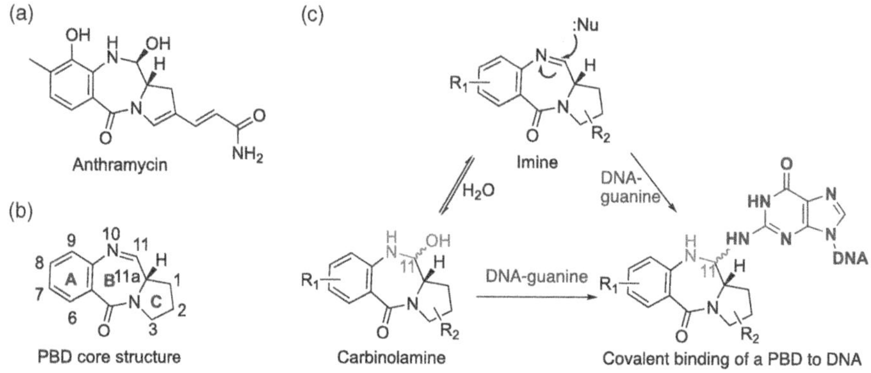

Figure 13.1 (a) Anthramycin (PBD monomer), (b) PBD core structure, and (c) interconvertible forms of PBDs (imine, carbinolamine) and the mechanism of covalent binding of a PBD to DNA.

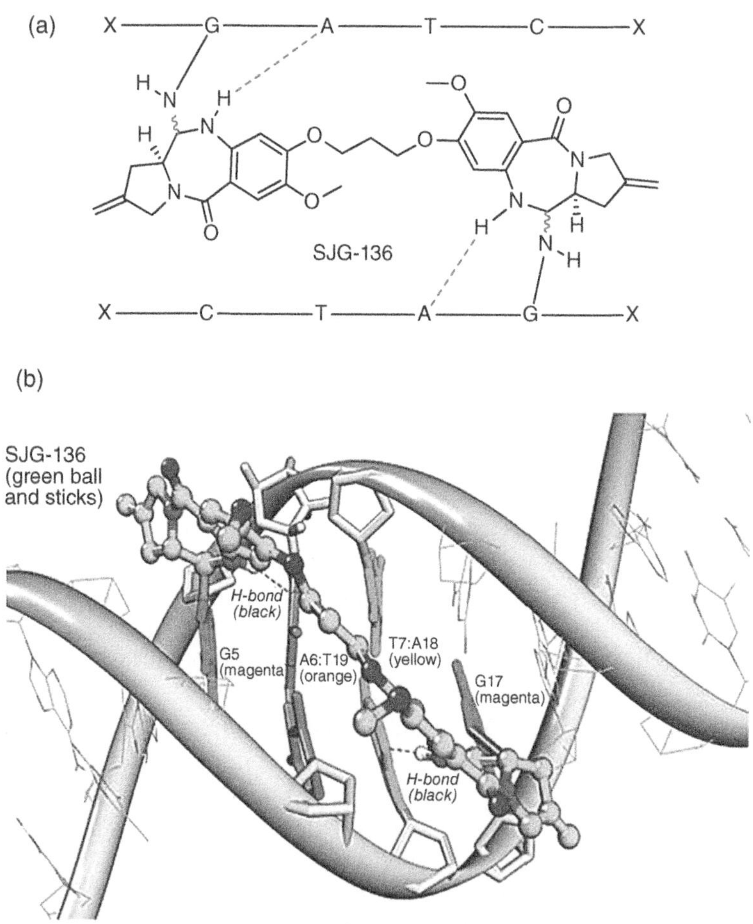

Figure 13.2 (a) Structure of SJG-136 binding with DNA. (b) Interstrand cross-link formation of PBD dimer SJG-136 (SG2000) with the DNA sequence. Source: Adapted with permission from [4]. Copyright (2016) John Wiley and Sons.

bonds, caused "flattening" of the C-ring, which has been proposed as a way to provide optimal van der Waals interactions with the DNA minor groove.

Interestingly, it was found that odd-numbered linker systems between the two PBD monomers were favored over even-numbered linkers [9]. Polyethylene units were installed to increase hydrophilicity and to improve the solubility and toxicology profile of the molecules. Installation of aromatic groups off the C ring (fused or off position C-3) at the R-position in Figure 13.3b improved "flatness" and aided in interactions with the DNA minor groove. To keep molecular weight under control, many PBD dimers had a methyl or methylene group at C2 instead of the aniline group. To increase potency, D-rings in Figure 13.3c can be installed to improve van der Waals interactions with the DNA minor groove

In order to make linkers between the PBD monomers also connect to linkers to antibodies, frameworks such as in Figure 13.3b,c were constructed and studied. Installation of a 3,5-dimethleneoxy aniline unit as a link between two PBD monomers created a PBD

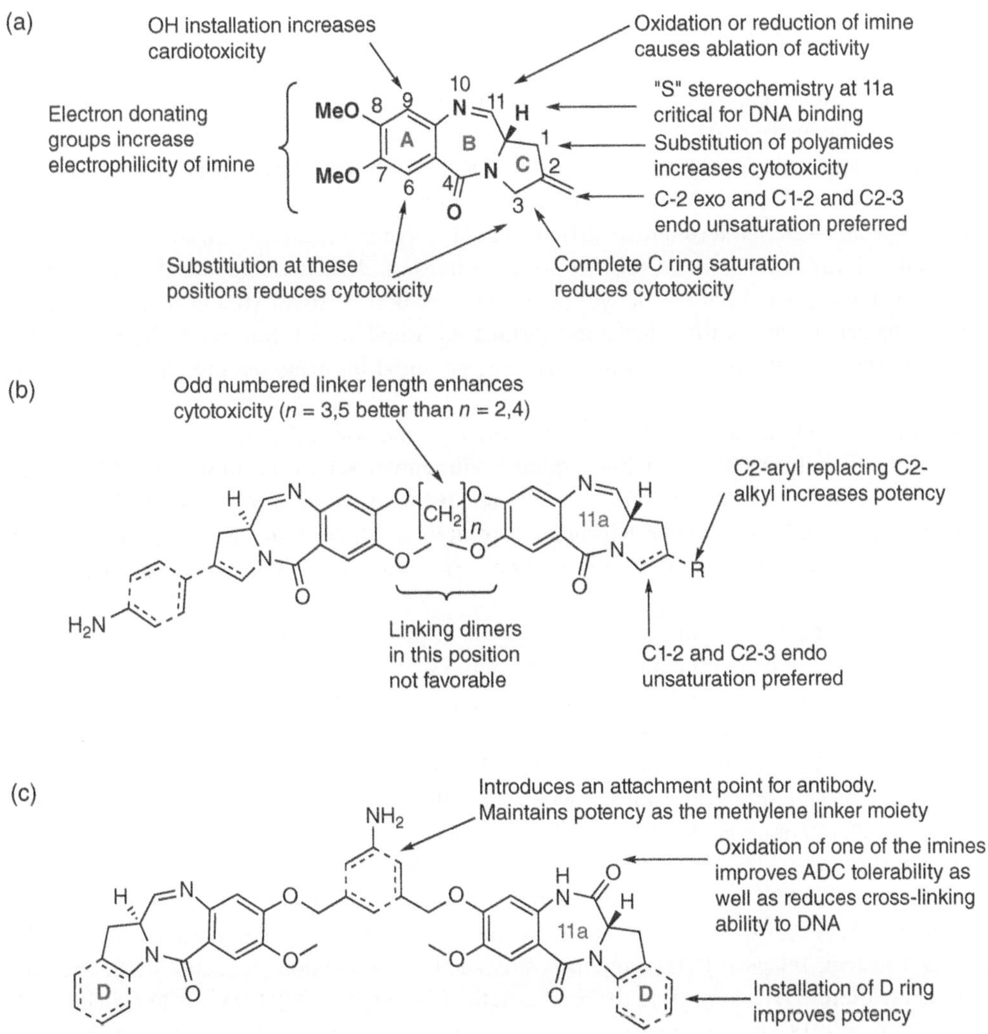

Figure 13.3 The SAR of the PBDs: (a) monomers, (b) dimers, and (c) unsymmetrical dimers [4].

unsymmetrical dimer that possessed a "handle" to link to an antibody for antibody-drug conjugates (ADCs). Further, to modulate the toxicity of the dimer, one imine moiety was changed to an amide group. This improved the toxicity profile while also decreasing cytotoxicity.

Both the naturally occurring monomer anthramycin (Figure 13.1a) and the synthetic dimer SJG-136 (Figure 13.2) have undergone clinical evaluation as standalone agents [10]. SJG-136 displayed significant cytotoxic activity due to the formation of DNA cross-links (Figure 13.2a,b) [11]. SJG-136 showed activity against both solid tumors and hematologic malignancies [12]. SJG-136 had been studied in multiple phase 1 clinical trials and progressed into phase 2 clinical trials [13]. More ADCs utilizing the PBD-payloads are in clinical trials. ADCs with SG-3199 are summarized in Table 13.1, ADCs with SG-1882 are summarized in Table 13.2, and ADCs with indolinobenzodiazepine pseudodimer (IGN)-based payloads are summarized in Table 13.3 [14].

Current non-site-specific PBD-ADCs were conjugated on antibody-cystines or lysine-residues (Figure 13.4). The PBD-ADCs were also site-specifically conjugated on engineered cysteines using THIOMAB technology and on glutamine-residues using enzymatic transglutaminase-based conjugations (Figure 13.4) [15], or on a glycan site using Glycoconnect™ technology.

The lysine-based conjugation is one of the most widely used non-specific conjugation strategies. The lysine ε-amino groups of an antibody react with an amine-reactive group of a linker linked PBD to generate amide bonds. Moreover, site-specific lysine conjugation is also developed based on azetidinone chemistry [16].

Unpaired cysteines are absent from the constant regions of human IgGs and are seldom present in the variable regions. Thus, introduced cysteines can provide exploitable chemical handles to create site-specific ADCs, referred to as THIOMABs [17]. For conjugations, all solvent-accessible (i.e. introduced and interchain) disulfides were reduced, and then interchain disulfides were reoxidized using copper sulfate or dehydroascorbic acid. Both THIOMAB and native antibody-ADCs can be generated via thioether bond formed with maleimide. Further hydrolysis of the succinimide ring, and the open ring minimized the potential for drug loss due to retro-Michael reactions [18].

Microbial transglutaminase (MTG) enzymes catalyze the formation of an isopeptide bond between primary amines and the γ-carboxamide of glutamine. MTG is promiscuous with respect to amine-based substrates but has more restricted recognition of protein-based glutamines. Antibodies with N297Q or N297D aglycosylated mutant [19], or with inserted glutamine tag (LLQG) [20] are preferentially conjugation positions to generate highly homogeneous site-specific ADCs [21].

In the past 30 years, the ADC approach has been established as an important drug delivery technology, not only for oncotherapy but also for immunotherapy [22]. In its simplest form, ADCs consist of three components: antibody, linker, and payload. The targeting antibody must have a specific binding affinity toward a targeted antigen on the cell surface of the targeted cells of interest, as well as have a good internalization rate for cellular processing. The internalized ADCs then traffic through the *endosome–lysosome* pathway where the warhead is released from the ADC via antibody degradation or upon cleavage of a cleavable linker.

There have been 204 ADC Investigational New Drugs (INDs) filed in the past 30 years and only 12 ADCs approved by the Food and Drug Administration (FDA), all of which were developed as oncotherapies (Table 13.4) [14]. The first FDA approved ADC was Gemtuzumab ozogamicin (Mylotarg®) to treat relapsed CD33 protein-positive acute myelogenous leukemia [23]. In June 2010, Pfizer withdrew Mylotarg® from the market at the request of the US FDA due to toxicity [24]. However, in April 2018, Gemtuzumab ozogamicin was reapproved for use in the United States [25] and in the European Union with changes to remove higher drug–antibody ratio (DAR)

Table 13.1 ADCs with SG-3199 (payload) in clinical trials.

Identifier	Phase	Starting date	Sponsor	Disease indication	Drug names	Target
NCT03700294	1	Dec. 2018	ADC Therapeutics S.A.	Advanced solid tumors	ADCT-601	Axl
NCT03489525	1	May 2018	MedImmune LLC	Myeloma/multiple myeloma	MEDI2228	BCMA
NCT03859752	1	Aug. 2019	Tanabe Research Laboratories	cMet positive solid tumors	MT-8633	c-MET
NCT03685344	1	Dec. 2018	ADC Therapeutics S.A.	DLBCL, mantle cell lymphoma, follicular lymphoma	ADCT-402	CD19
NCT02669017	1	Mar. 2016	ADC Therapeutics S.A.	Relapsed or refractory B-NHL	ADCT-402	CD19
NCT02669264	1	Mar. 2016	ADC Therapeutics S.A.	Relapsed or refractory B-ALL	ADCT-402	CD19
NCT03684694	1/2	Dec. 2018	ADC Therapeutics S.A.	DLBCL, mantle cell lymphoma	ADCT-402	CD19
NCT03589469	2	Aug. 2018	ADC Therapeutics S.A.	Relapsed or refractory DLBCL	ADCT-402	CD19
NCT04384484	3	Sep. 2020	ADC Therapeutics S.A.	DLBCL, high grade B-cell lymphoma	ADCT-402	CD19
NCT03698552	1/2	Aug. 2018	M.D. Anderson Cancer	ALL, B-ALL	ADCT-602	CD22
NCT03621982	1	Nov. 2018	ADC Therapeutics S.A.	Selected advanced solid tumors	ADCT-301	CD25
NCT04052997	2	Sep. 2019	ADC Therapeutics S.A.	Hodgkin lymphoma, classical Hodgkin lymphoma	ADCT-301	CD25
NCT02432235	1	Jun. 2015	ADC Therapeutics S.A.	Relapsed or refractory Hodgkin lymphoma and NHL	ADCT-301	CD25
NCT02588092	1	Feb. 2016	ADC Therapeutics S.A.	Relapsed or refractory AML, ALL	ADCT-301	CD25
NCT01901653	1/2	Jul. 2013	Stemcentrx	SCLC	Rova-T	DLL3
NCT02819999	1	Oct. 2016	AbbVie	SCLC, extensive-stage SCLC	Rova-T	DLL3
NCT02674568	2	Jan. 2016	AbbVie	SCLC	Rova-T	DLL3
NCT03086239	1	Apr. 2017	AbbVie	SCLC, advanced SCLC	Rova-T	DLL3
NCT03061812	3	Apr. 2017	AbbVie	Advanced or metastatic SCLC	Rova-T	DLL3
NCT03334487	3	Mar. 2018	AbbVie	SCLC	Rova-T	DLL3
NCT02874664	1	Sep. 2016	Stemcentrx	SCLC, extensive-stage SCLC	Rova-T	DLL3
NCT03503890	U		Abbvie	SCLC	Rova-T	DLL3
NCT03000257	1	Dec. 2016	Abbvie	Advanced solid tumors	Rova-T	DLL3

(Continued)

Table 13.1 (Continued)

Identifier	Phase	Starting date	Sponsor	Disease indication	Drug names	Target
NCT02709889	1/2	Aug. 2016	Abbvie	Solid tumors, advanced solid tumors	Rova-T	DLL3
NCT03033511	3	Feb. 2017	Abbvie	SCLC, extensive-stage SCLC	Rova-T	DLL3
NCT02500914	1	Jun. 2015	Stemcentrx	SCLC, LCNEC	SC-002	DLL3
NCT03543358	2	Sep. 2018	Abbvie	Cancer indications	Rova-T	DLL3
NCT03026166	1/2	Mar. 2017	Abbvie	SCLC	Rova-T	DLL3
NCT02539719	1	Aug. 2015	Stemcentrx	Platinum resistant/refractory ovarian cancer	Tamrintamab Pamozirine	DPEP3
NCT03125200	1	May 2017	ADC Therapeutics S.A.	Advanced solid tumors	ADCT-502	HER-2
NCT02991911	1	Jan. 2017	MedImmune LLC	Metastatic castration-resistant prostate cancer	ADCT-401	PSMA

ALL, acute lymphoblastic leukemia; AML, acute myeloid leukemia; B-NHL, B-cell lineage non-Hodgkin lymphoma; DLBCL, diffuse large B-cell lymphoma; LCNEC, large-cell neuroendocrine carcinoma; NHL, non-Hodgkin lymphoma; SCLC, small cell lung cancer.

Table 13.2 ADCs with SG-1882 (payload) in clinical trials.

Identifier	Phase	Starting date	Sponsor	Disease indication	Drug names	Target
NCT02614560	1/2	Nov. 2015	Seattle Genetics, Inc.	AML	Vadastuximab talirine	CD33
NCT02706899	1/2	Feb. 2016	Seattle Genetics, Inc.	Myelodysplastic syndrome	Vadastuximab talirine	CD33
NCT02785900	3	May. 2016	Seattle Genetics, Inc.	AML	Vadastuximab talirine	CD33
NCT03145909	1	Jul. 2017	AbbVie	Advanced solid tumors	ABBV-176	PRLR
NCT02954796	1	Dec. 2016	Seattle Genetics, Inc.	Myeloma/multiple myeloma	SGN-CD352A	CD352
NCT02216890	1	Aug. 2014	Seattle Genetics, Inc.	Lymphoma, RCC, metastatic RCC	SGN-CD70A	CD70
NCT02702141	1	Feb. 2016	Seattle Genetics, Inc.	Relapse/refractory aggressive B-NHL	SGN-CD19B	CD19
NCT02848248	1	Aug. 2016	Seattle Genetics, Inc.	AML	SGN-CD123A	CD123
NCT01902329	1	Jul. 2013	Seattle Genetics, Inc.	AM /AML, APL	Vadastuximab talirine	CD33
NCT02326584	1	Dec. 2014	Seattle Genetics, Inc.	AML	Vadastuximab talirine	CD33
NCT03234712	1	Oct. 2017	AbbVie	Advanced solid tumors	ABBV-321	EGFR

AM, acute myelogenous; AML, acute myeloid leukemia; APL, acute promyelocytic leukemia; B-NHL, B-cell lineage non-Hodgkin lymphoma; RCC, renal cell carcinoma.

Table 13.3 ADCs with IGN-based payloads in clinical trials.

Identifier	Phase	Starting date	Sponsor	Disease indication	Drug names	Target
NCT02674763	1	Mar. 2016	ImmunoGen, Inc.	AML	IMGN779	CD33
NCT04086264	1/2	Nov. 2019	ImmunoGen, Inc.	AML	IMGN632	CD123
NCT04086264	1/2	Nov. 2019	ImmunoGen, Inc.	AML	Venclexta	Bcl-2
NCT03386513	1/2	Jan. 2018	ImmunoGen, Inc.	Relapse/refractory AML BPDCN, ALL and other CD123+ Hem malignancies	IMGN632	CD123
NCT03449030	1	Apr. 2018	Millennium Pharmaceuticals, Inc.	Advanced gastrointestinal cancer	TAK-164	GCC

ALL, acute lymphoblastic leukemia; AML, acute myeloid leukemia; BPDCN, blastic plasmacytoid dendritic cell neoplasm.

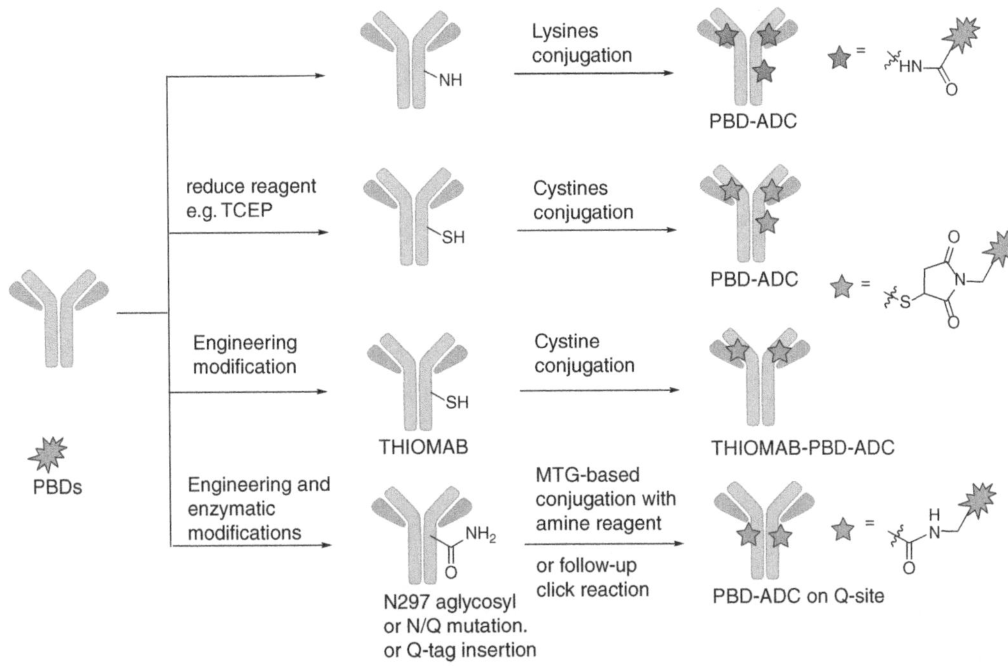

Figure 13.4 Non-site-specific and site-specific PBD-ADCs conjugations.

contaminations [26]. Brentuximab vedotin (Adectris®) [27] targets Hodgkin's lymphoma (HL)-associated CD30 protein (CD30) antigen and was approved by the FDA in 2011 to treat relapsed or refractory (R/R) Hodgkin lymphoma and systemic anaplastic large cell lymphoma. Trastuzumab emtansine (Kadcyla® from Roche) consists of the humanized monoclonal antibody (mAb), trastuzumab (Herceptin) coupled to the cytotoxic agent mertansine (DM1). The antibody trastuzumab binds to cell surfaces that express the human epidermal growth factor receptor-2 (EGFR2). Kadcyla® was approved by the FDA in 2013 as a breast cancer therapy [28]. Inotuzumab ozogamicin (Besponza® from Pfizer) targets the CD22 antigen and was approved by the FDA in 2017 to treat R/R B-cell precursor acute lymphoblastic leukemia (ALL) [29a, b]. In 2018 the FDA approved Moxetumomab pasudotox-tdfk (Lumoxiti® from Agensys and Seattle Genetics), which consists of the binding fragment (Fv) of an anti-CD22 antibody fused to a toxin called PE38; this was approved by the FDA in 2021 to treat hairy cell leukemia. This toxin is a 38 kDa fragment of Pseudomonas exotoxin A [29c]. Polatuzumab vedotin (Polivy® from Roche) was approved by the FDA in July 2019 and is currently used to treat diffuse large B-cell lymphoma (DLBCL) by targeting the antigen CD79b protein [30]. Trastuzumab deruxtecan also known as fam-trastuzumab deruxtecan (Enhertu® from Daiichi Sankyo) is a treatment for adults who have unresectable (unable to be removed with surgery) or metastatic (when cancer cells spread to other parts of the body) HER2-positive breast cancer and have received two or more prior anti-HER2-based dosing regimens in the metastatic setting [31]. In 19 December 2019, the FDA granted accelerated approval of Enfortumab vedotin-ejfv (Padcev® from Astellas Pharma US, Inc.) to treat adult patients with locally advanced or metastatic urothelial cancer [32]. In April 2020, Sacituzumab govitecan-hziy (Trodelvy® from Immunomedics), was approved as the first and only approved ADC for the treatment of advanced triple-negative breast cancer (TNBC) in adult patients [33]. Belantamab mafodotin (Blenrep® from GlaxoSmithKline) is for the treatment of relapsed and refractory multiple

Table 13.4 Twelve FDA approved ADCs.

Drug name (organization)	ADC	Antibody	Linker	Payload	DAR	Indication
ADCETRIS brentuximab vedotin Seattle Genetics	Brentuximab vedotin	Anti-CD30 IgG1	VC-PAB	MMAE	4	Hodgkin lymphoma
Kadcyla trastuzumab emtansine Roche/Genentech	Trastuzumab emtansine	Anti-HER2 IgG1	MCC	DM1	3.5	HER2 positive breast cancer
MYLOTARG Pfizer	Gemtuzumab ozogamicin	Anti-CD33 IgG4	AcBut hydrazone and disulfide	Calicheamicin	2–3	Acute lymphoblastic
BESPONSA inotuzumab ozogamicin Roche/Genentech	Inotuzumab ozogamicin	Anti-CD22 IgG4	AcBut hydrazone and disulfide	Calicheamicin	6	Acute lymphoblastic
LUMOXITI moxetumomab pasudotox-tdfk for injection AstraZeneca	Moxetumomab pasudotox	Fv of Anti-CD22	Fusion	PE38	—	Hairy cell leukemia
POLIVY polatuzumab vedotin-piiq Roche/Genentech	Polatuzumab vedotin	Anti-CD79b IgG1	VC-PAB	MMAE	3.5	B-cell lymphoma

PADCEV enfortumab vedotin-ejfv Agensys and Seattle Genetics	Enfortumab vedotin	Anti-Nectin-4 IgG1	VC-PAB	MMAE	4	Bladder cancer
ENHERTU fam-trastuzumab deruxtecan-nxki AstraZeneca and Daiichi Sankyo	Trastuzumab deruxtecan	Anti-HER2 IgG1	GGFG	Dxd	8	HER2 positive breast cancer
TRODELVY sacituzumab govitecan-hziy Immunomedics	Sacituzumab govitecan	Anti-Trop-2 IgG1	Hydrolyzable linker (CL2A)	SN-38	7–8	Triple-negative breast cancer
BLENREP belantamab mafodotin-blmf GSK	Belantamab mafodotin	Anti-BCMA IgG1	Non-cleavable linker	MMAF	4	Multiple myeloma
Zynlonta loncastuximab tesirine-lpyl ADC Therapeutics	Loncastuximab tesirine-lpyl	Anti-CD19 IgG1	VA-PAB	PBD	2.3	Relapsed or refractory diffuse large B-cell lymphoma
tivdak tisotumab vedotin-tftv Seagen/Genmab	Tisotumab vedotin-tftv	Anti-TF IgG1	VC-PAB	MMAE	4	Recurrent or metastatic cervical cancer

ADC, antibody-drug conjugate; DAR, drug–antibody ratio; GGFG, tetrapeptide protease-cleavable linker; MCC, non-cleavable linker; VA-PAB, VC-PAB, protease-cleavable dipeptide linker.

myeloma (MM) and was approved by the FDA in August 2020 [34a]. In April 2021, the FDA granted accelerated approval of the first PBD-ADC targeting CD19, loncastuximab tesirine-lpyl (Zynlonta®, ADC Therapeutics SA), for adult patients with R/R large B-cell lymphoma [34b]. In September 2021, the FDA approved Tisotumab vedotin-tftv (Tivdak®, Seagen), an ADC consisting of a tissue factor-directed antibody and MMAE, for adult patients with recurrent or metastatic cervical cancer [34c].

Despite the recent success of ADCs in the oncological arena, issues still exist for ADCs. For instance, efficacy is limited with solid tumors due to the large size of ADCs that precludes their penetration into the tumors [35]. In addition, the antibody of the ADC can induce immunogenic responses [36] and the slow clearance rate of ADCs from non-target organs can lead to undesired side effects [37]. In an effort to circumvent the limitations inherent to ADCs, small molecule drug conjugates (SMDCs) have been designed as an alternative approach. SMDCs replace the antibody of the ADC with a small molecular targeting ligand. The targeting ligands can be composed of small protein scaffolds [38], peptides [39], or low molecular-weight non-peptidic ligands [40]. The high selectivity and/or specificity of the ligand is of paramount importance in the design of the SMDC, as it will potentially lower the toxicity of the SMDC to the normal cells. To lower the toxicity, high affinity ligands are sought since the higher affinity of the ligand to the receptor, the lower the concentration of drug necessary to achieve receptor saturation [41]. Some SMDCs utilizing folate and prostate-specific membrane antigen (PSMA) ligands have been studied in clinical trials [41, 42]. To ensure the active payload is released from the SMDC at the desired site and internalized within the tumor cells at a favorable rate, a cleavable linker, such as a disulfide bond linker or a peptide linker, have been utilized. Highly potent cytotoxic PBD agents have served as warheads in the SMDCs. Compared with ADCs, SMDCs have some unique properties, such as nonimmunogenic nature, small size for improved cell penetration in the solid tumors, faster clearance, and lower cost of manufacturing. These advantages make SMDCs an attractive and potentially promising therapeutic approach to tackle solid tumors. However, to date, no SMDC has reached FDA approval for treating solid tumors.

Recent years have seen the development of synthetic cytotoxic PBD dimers as a new class of warhead in the fight against cancer using drug conjugates, including both ADCs and SMDCs [4, 43]. This chapter summarizes modifications of the PBDs for improvement of *in vitro* cell-killing potency and enhancement of PBD-ADCs conjugation ability to widen the therapeutic index (TI) for PBD-ADCs in anticancer therapy. Examples of ADCs and SMDCs that contain PBD dimers (or PBD derivatives as payloads) are discussed, as well as their corresponding synthesis outlined, and their performance in clinical trials.

13.2 Antibody Drug Conjugating with Pyrrolobenzodiazepine Payloads

13.2.1 SG-3199 (Payload), SG-3249 (Linker-Payload), and SG-3199-Based ADCs

SG-3249 (tesirine) is a linker-payload containing SG-3199 (a PBD dimer) as the payload (Figure 13.5). SG-3249 has been conjugated in various ADCs because it possesses attractive properties, such as high potency, good scalability, and low hydrophobicity. SG-3249 contains a linker that is cleaved in the presence of cathepsin B and a maleimide moiety for efficient conjugation to an antibody of interest. SG-3249-ADCs have been extensively studied in preclinical and clinical trials (Figure 13.5).

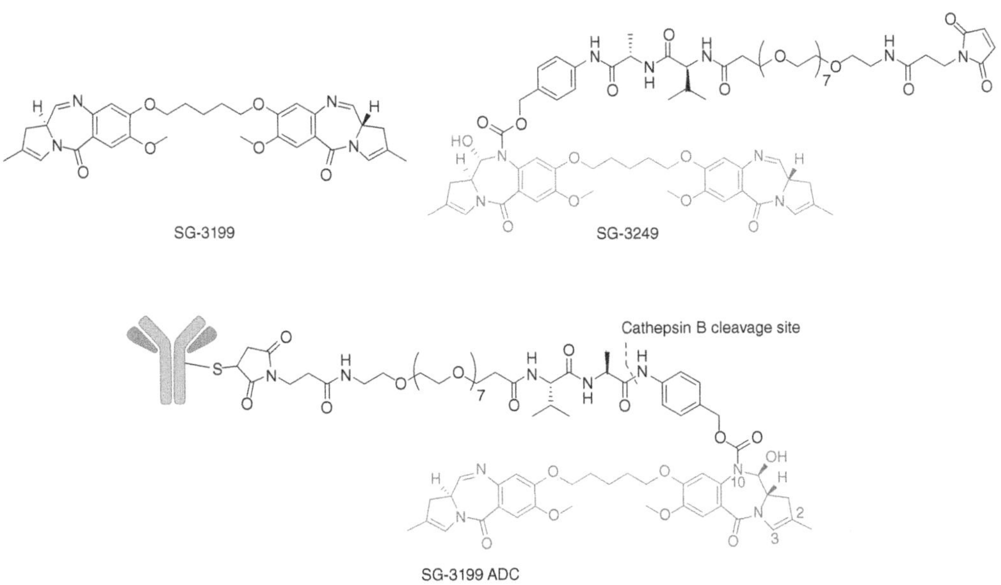

Figure 13.5 SG-3199 (payload), SG-3249 (linker-payload), and SG-3199 ADC.

The PBD dimer payload, SG-3199, has a 5-carbon linker between the two PBD monomers, which imparts additional potency in the molecule than its 3-carbon tethered analog, due to the 5-carbon tether having more flexibility, providing more opportunities to fit in the minor groove of DNA. SG-3199 retained picomolar activity in several *in vitro* cytotoxic assays. For instance, SB-3199 had an EC_{50} value of 150 pM in K562, 20 pM in NCIN87, 1.0 nM in BT474, and 320 pM in SKBR3 cell lines, respectively [44]. To create the ADC containing the PBD dimer payload (SG3199-ADC), the linker from the antibody to the payload is connected to SG-3199 at the N10-position via a carbamate moiety of the one PBD molecule, while the other monomer exists as the imine form at the N10 position (Figure 13.5) [44]. Because the N10-position imine is involved in covalent binding to DNA, the linker attached at the other N10-position, a carbamate moiety, creates a prodrug situation and adds an additional level of safety to the ADC. A PEG8 moiety was introduced between the maleimide and Val-Ala cathepsin B cleavable peptide linker moiety to improve the molecule's overall hydrophilicity and physicochemical properties [45].

The PBD dimer-linker, SG-3249, was synthesized by a convergent synthetic route (Scheme 13.1) [44]. Two monomeric PBD moieties (**1-7** and **1-8**) were prepared from 4-(benzyloxy)-3-methoxybenzaldehyde **1-1**, which served as the A ring of the PBD core. The key intermediate **1-2** was prepared in eight steps, and the amino group of **1-2** was either protected as the alloc group **1-3** for the synthesis of one of the PBD monomers, or as the carbamate through the PABC-Ala-Val-Alloc linker **1-4**, which served as another PBD monomer. Once deprotection of TBS in **1-3** and **1-4**, the primary hydroxyl group is liberated and subjected to oxidation conditions to provide the fused-A, B, C-ring PBD core **1-5** and **1-6**, respectively. TBS protection of the hydroxy in **1-5** and **1-6**, followed by deprotection of triisopropylsilyl provided **1-7** and **1-8**, respectively. Installation of the iodo-C5 chain off the phenolic hydroxy gave **1-9**, which was then coupled with **1-8** to provide a dimer-linker **1-10**. Deprotections of TBS and alloc groups formed **1-11**, which was coupled with maleimido-PEG8-NHS ester to generate SG-3249 (tesirine).

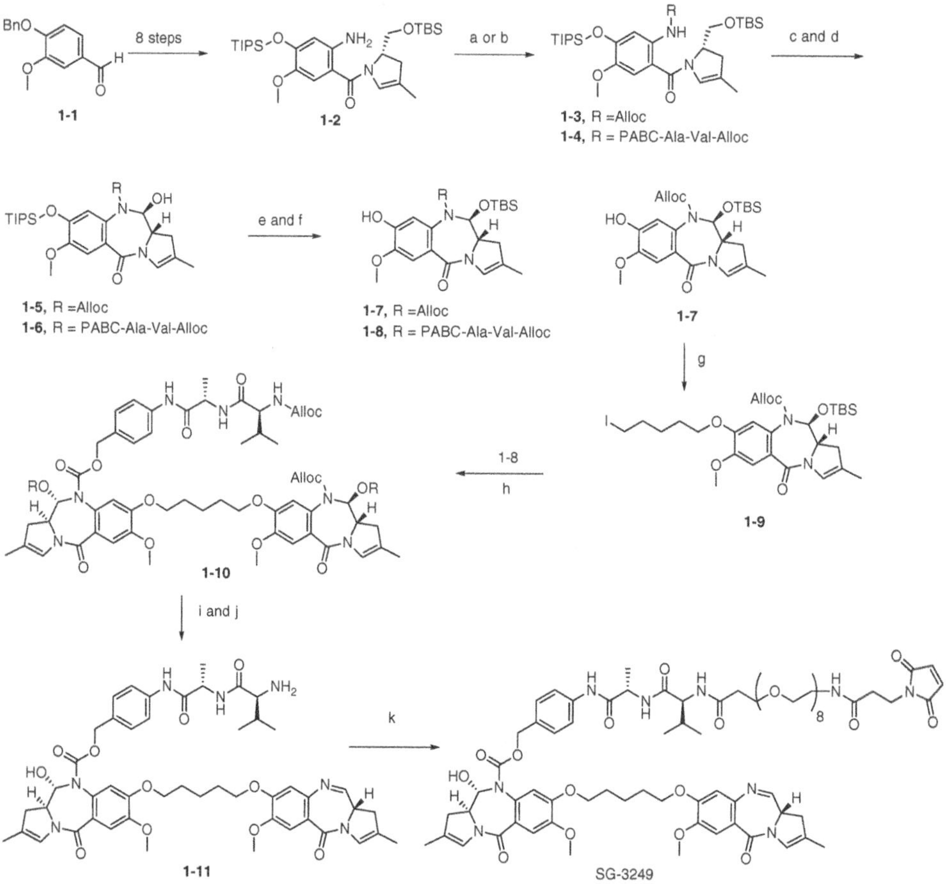

Scheme 13.1 Synthesis of SG-3249. Reagents and conditions: (a) Alloc-Cl, pyridine, DCM, −78 °C to rt, 100%; (b) triphosgene, Et_3N, THF, 5 °C, then Alloc-Val-Ala-PAB-OH, Et_3N, THF, 40 °C, 50%; (c) AcOH/MeOH/THF/water (7/1/1/2), rt; (d) DMSO, $(COCl)_2$, Et_3N, DCM, −78 °C to rt, 66%, **1-5**, 60%, **1-6**; (e) *tert*-Butyldimethylsilyl (TBS)-OTf, 2,6-lutidine, DCM, 0 °C; (f) LiOAc, DMF/water (95/5), rt, 100%, **1-7**, 100%, **1-8**; (g) 1,5-diiodopentane, K_2CO_3, acetone, 60 °C, 90%; (h) **1-8**, K_2CO_3, acetone, 65 °C, 86%; (i) TBAF/AcOH, THF, rt, 80%; (j) $Pd(PPh_3)_4$, pyrrolidine, DCM, rt, 100%; (k) Mal-dPEG8-acid, EDCI, DCM, rt, 73%.

13.2.1.1 ADCT-301

ADCT-301 (Camidanlumab tesirine) is an ADC consisting of the recombinant human IgG1 HuMax-TAC mAb that targets the cell-surface antigen CD25 with SG-3249. The ADC has a DAR of 2.3. IL2R-a (CD25) is overexpressed on many lymphoma and leukemic cells. ADCT-301 binds human CD25 with picomolar affinity and has demonstrated high potency and selective cytotoxicity against a panel of CD25-expressing human lymphoma cell lines [46].

In a phase 1 study comprising 133 patients (NCT02432235) for the treatment of relapsed/refractory lymphoma, ADC dosing resulted in a 70.7% overall response rate (ORR) across all doses with 86.5% ORR at the initial dose. In addition, a 44.0% ORR in heavily pretreated patients with R/R T-cell lymphoma who had received a median of four prior lines of therapy was observed. Overall, ADCT-301 demonstrated significant clinical activity, while maintaining a manageable tolerability profile. ADCT-301 is currently being evaluated in a pivotal phase 2 clinical trial in patients with R/R Hodgkin lymphoma (NCT04052997).

13.2.1.2 ADCT-401

Medimmune's ADCT-401 (MEDI3276) is an ADC consisting of an anti-PSMA-specific IgG1 mAb (J591), engineered to enable site-specific conjugation with SG-3249 (tesirine). The conjugation produces a DAR of 1.8. MEDI3276 displayed potent and specific cytotoxicity in a panel of PSMA positive prostate cancer cell lines [47]. ADCT-401 was tested in a phase 1/1b clinical study to evaluate the safety and efficacy in metastatic castration resistant prostate cancer (NCT02991911). Briefly, 33 patients (median age of 71 years old) with prior regimens of 4 months and median follow-up of 5.4 months, received MEDI3726. Drug-related adverse events (AEs) occurred in 30 patients (90.9%), with 15 patients being grade 3/4 (45.5%), serious in 11 patients (33.3%) and causing discontinuation in 13 patients (39.4%). Composite response rate across all doses was 2/33 (6.1%). Median progression-free survival was 3.9 months and median overall survival (OS) was 10.6 months [48]. A maximum tolerated dose (MTD) was not identified due to drug-related AEs (skin toxicities and effusions) that prevented raising the dose over 300 μg/kg and limited the number of cycles. Responses were seen at higher doses but were not durable as patients discontinued the study due to drug-related AEs. ADCT-401(MEDI3726) was terminated in phase 1 clinical trials due to safety/efficacy issues and subsequently removed from the company's pipeline.

13.2.1.3 ADCT-402

ADCT-402 (loncastuximab tesirine) is composed of a humanized anti-CD19 antibody, stochastically conjugated to SG-3249. The human CD19 antigen is a 95 kDa type I membrane glycoprotein whose expression is limited to the various stages of B-cell development and differentiation and is maintained in the majority of B-cell malignancies, including leukemias and non-Hodgkin lymphomas (NHLs) of B-cell origin. Coupled with its differential and favorable expression profile, CD19 has rapid internalization kinetics and is not shed into the circulation, making it an ideal target for the development of ADCs to treat B-cell malignancies. CD19 has been clinically validated as a therapeutic target by several approaches, including naked and bispecific antibodies, ADCs, and CAR-T cells. ADCT-402 demonstrated highly specific and potent *in vitro* cytotoxicity and bystander killing property in CD-19 expressing human cell lines. In the single cell gel electrophoresis (comet) assay, the released PBD payload formed DNA-crosslinks, reaching a peak at around 12 hours and persisted for 36 hours. Single doses of ADCT-402 showed highly potent and dosedependent in vivo antitumor activity in various human tumor models. ADCT-402 also demonstrated excellent stability and tolerability in rat and cynomolgus monkey pharmacokinetic studies [49].

Since March 2016, ADCT-402 has been studied in patients with R/R B-cell lineage non-Hodgkin lymphoma (B-NHL) (NCT02669017). During dose escalation, 88 patients with R/R B-cell NHL were treated with loncastuximab tesirine at doses of 15–200 μg/kg. Treatment-emergent adverse events (TEAEs) were experienced by 98.7% of the 88 patients. Most common TEAEs (≥20% of patients) were hematologic abnormalities, fatigue, edema, liver test abnormalities, nausea, rash, and dyspnea. TEAEs (≥5% of patients) of grade ≥3 included hematologic abnormalities, liver test abnormalities, fatigue, and dyspnea. ORR at doses ≥120 μg/kg was 59.4% (41 of 69 patients, 40.6% complete response; 18.8% partial response). Median duration of response, progression-free survival, and OS (all doses) were 4.8, 5.5, and 11.6 months, respectively. Drug exposure increased with increasing dose, showing moderate accumulation with multiple doses ≥150 μg/kg. There was no evidence of immunogenicity [50].

A Study of ADCT-402 in patients with R/R B-cell lineage acute lymphoblastic leukemia (B-ALL) commenced in March 2016 (NCT02669264). Three patients with R/R B-ALL achieved a complete response with loncastuximab tesirine. The MTD was not reached. This trial was terminated due to slow accrual [51].

A phase 2 open-label single-arm study to evaluate the efficacy and safety of loncastuximab tesirine in patients with R/R DLBCL was started in August 2018 (NCT03589469). In June 2020, ADC Therapeutics reported that loncastuximab tesirine achieved an ORR of 48.3% (70/145 patients) and a complete response rate (CRR) of 24.1% (35/145 patients), increasing the median duration of response to 10.25 months. The most common grade ≥3 TEAEs in ≥10% of patients were neutropenia (25.5%) with a low incidence of febrile neutropenia (3.4%), thrombocytopenia (17.9%), gamma-glutamyl transferase increases (16.6%), and anemia (10.3%) [52] Based on those results, the FDA granted accelerated approval to loncastuximab tesirine (Zynlonta®) for adult patients with R/R large B-cell lymphoma on 23 April 2021 [34b].

Loncastuximab tesirine has been evaluated in combination with ibrutinib in a phase 1b trial with patients that have relapsed or refractory diffuse large B-cell lymphoma (R/R DLBCL) or mantle cell lymphoma (MCL) (NCT03684694) and another phase 1b trial in combination with durvalumab, for patients with R/R DLBCL, MCL or follicular lymphoma (NCT03685344). These studies are ongoing.

13.2.1.4 ADCT-502

ADCT-502 is an ADC composed of an engineered version of humanized IgG1 trastuzumab, directed against human HER2, site-specifically conjugated to SG-3249 with a DAR of 1.7. ADCT-502 exhibited high *in vitro* potency and targeted cytotoxicity against various solid cancer cell lines. ADCT-502 also demonstrated strong and durable antitumor activity in mouse xenografts with various levels of HER2 expression but was inactive in HER2-negative xenografts [46, 53]. ADCT-502 is stable, well tolerated and has a favorable pharmacokinetic (PK) profile both in rat and cynomolgus monkey [46, 53]. Moreover, ADCT-502 showed an indirect bystander effect in HER2-negative expressed MDA-MB-468 cells that were incubated with conditioned medium from ADCT-502-treated HER2+ SK-BR-3 cells. ADCT-502 also displayed superior *in vivo* antitumor activity compared with market ADC, Kadcyla (T-DM1), in various tumor xenografts, including those with low HER2 expression levels [53].

A phase 1 trial of ADCT-502 in patients with advanced solid tumors with HER2 expression was opened in May 2017 (NCT03125200). Unfortunately, ADC Therapeutics reported (2018 Press Release) that safety, tolerability, pharmacokinetics, and efficacy of ADCT-502 did not demonstrate sufficient patient benefit in HER2 expressing tumors and terminated the trial due to safety reasons [54].

13.2.1.5 ADCT-602

ADCT-602 is an ADC composed of the humanized anti-human CD22 mAb epratuzumab (hLL2), stochastically conjugated to SG-3249 with a DAR of 1.7. CD22 is an attractive and clinically validated ADC target, that is highly expressed on most malignant B-cells, including expression in greater than 90% of patients with B-ALL [55]. ADCT-602 is being evaluated in a phase I/II clinical trial in patients with R/R B-cell ALL (NCT03698552). The trial is being led by The University of Texas MD Anderson Cancer Center.

13.2.1.6 Rovalpituzumab Tesirine (Rova-T)

Rovalpituzumab tesirine (SC16LD6.5 or Rova-T) is a delta-like protein 3 (DLL3)-targeted ADC consisting of the humanized DLL3-specific IgG1 mAb SC16, the DNA cross-linking agent SG-3199 (D6.5), and a protease-cleavable linker that covalently links SG-3199 to SC16 [56, 57]. The cancer-stem cell-associated DLL3 is expressed in more than 80% of small-cell lung cancer (SCLC) patient tumors, and is prevalent on tumor cells, including cancer stem cells, but not present in healthy tissue [56].

In July 2013, Abbvie and Stemcentrx conducted a phase I/II study of SC16LD6.5 as a single agent in patients with recurrent SCLC (NCT01901653). Among 60 evaluable subjects, active dose levels resulted in a confirmed ORR of 18% and a confirmed clinical benefit rate (CBR criteria: stable disease or better) of 68%. Among 26 evaluable subjects with DLL3 expression in at least 50% of tumor cells (DLL3-high), confirmed ORR and CBR were 39 and 89%, respectively. Median duration of response was 5.6 months. One-year survival rates among all and DLL3-high subjects were 18 and 32%, respectively. Among primary sensitive relapse patients, confirmed ORR and CBR among all subjects were 24% (8/33) and 67% (22/33), respectively. Among DLL3-high subjects, ORR and CBR were 53% (8/15) and 100% (15/15), respectively, with one-year survival rates of 17 and 33%, respectively. Among primary resistant/refractory relapse patients, confirmed ORR and CBR among all subjects were 12 (3/25) and 72% (18/25), respectively; and among DLL3-high subjects were 18 (2/11) and 73% (8/11), respectively, with one-year survival rates of 21 and 29%, respectively. The most common grade 3 or higher toxicities included thrombocytopenia (12%), serosal effusions (11%), and skin reactions (8%). ADC pharmacokinetics were linear with a terminal half-life of 10–14 days and no sign of anti-therapeutic antibodies developing [58]. Overall, Rova-T showed encouraging single agent antitumor activity with a manageable safety profile.

A phase 2 study of Rova-T coined "Trinity" was initiated in January 2016 (NCT02674568). The purpose of this study was to determine the efficacy of Rova-T as a "third-line and later" treatment for participants with R/R DLL3 expressing SCLC. Among 339 patients enrolled, 261 (77%) had two prior lines of therapy and 78 (23%) had ≥3. DLL3-high and DLL3-positive tumors, determined by rabbit mAb immunohistochemical analysis, were seen in 238 (70%) and 287 (85%) patients, respectively. The remaining 52 patients (15%) were DLL3-negative only by rabbit mAb immunohistochemical analysis or had missing results. ORR was 12.4, 14.3, and 13.2% in all, DLL3-high, and DLL3-positive patients, respectively. Median OS was 5.6 months in all patients and 5.7 months in DLL3-high patients. The most common AEs were fatigue, photosensitivity reaction, and pleural effusion. Grade 3–5 AEs were seen in 213 (63%) patients. Rova-T showed modest clinical activity with associated toxicities [57].

A phase 3 study to evaluate the efficacy of Rova-T as maintenance therapy following first-line platinum-based chemotherapy (MERU) was carried out in January 2017 (NCT03033511). An Independent Data Monitoring Committee (IDMC) recommended terminating MERU due to lack of survival benefit for patients receiving Rova-T compared with the placebo control arm based on results at a pre-planned interim analysis. The phase 3 trial demonstrated no survival benefit at a pre-planned interim analysis for patients receiving Rova-T as compared with placebo. The overall safety profile was generally consistent with that observed in previous studies. The MERU trial was terminated [59].

In 2017, AbbVie also started a phase 3 clinical trial, called TAHOE. The purpose of this study was to assess the efficacy, safety, and tolerability of Rova-T versus topotecan in participants with advanced or metastatic SCLC with high levels of DLL3 and who have first disease progression during or following front-line platinum-based chemotherapy (NCT03061812). In December 2018, an IDMC recommended stopping enrollment in TAHOE due to shorter OS in the Rova-T arm compared with the topotecan control arm [60].

13.2.1.7 ADCT-601

ADCT-601 is an ADC composed of a humanized mAb that binds to human AXL (from BerGenBio) and conjugated using Glycoconnect technology (from Synaffix BV) to a linker that is attached to SG-3199 (Figure 13.6).

Once bound to an AXL-expressing cell, ADCT-601 is internalized into the cell where enzymes release the PBD-based warhead. AXL, a member of the TAM family of receptor tyrosine kinases, is overexpressed in various cancers and overexpression is linked to metastasis, poor survival, and drug

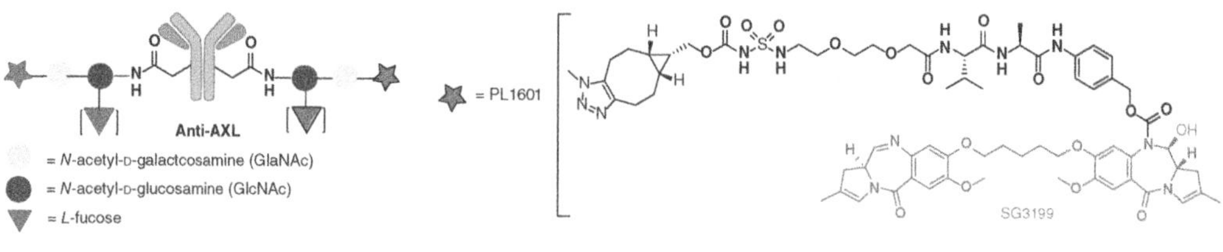

ADCT-601

Figure 13.6 The structures of ADCT-601 and its ADCs.

resistance [61]. AXL is an ideal target for an ADC approach, as it is highly overexpressed in many solid tumors (e.g. lung, breast, prostate, pancreas, glioma, and esophageal) and hematological malignancies (e.g. acute and chronic myeloid leukemia). ADCT-601 is being evaluated in a phase 1 clinical trial in patients with selected advanced solid tumors (NCT03700294).

13.2.1.8 MEDI2228

MEDI2228 is an ADC composed of an engineered version of anti-B-cell maturation antigen (BCMA) antibody site-specifically conjugated to SG-3199 via a cleavable valine–alanine dipeptide linker. BCMA (TNFRSF17) is a suitable therapeutic target for the treatment of MM due to its restricted expression on normal plasma cells and universal expression in myeloma cells. MEDI2228 exhibits potent antitumor activity in preclinical models of MM [62]. MEDI2228 has been evaluated in a phase 1 clinical trial in patients with R/R MM since 2018 (NCT03489525).

13.2.1.9 TR1801-ADC (MT-8633)

TR1801-ADC is a site-specific ADC which consists of a non-agonistic cMet antibody with SG-3249. It has picomolar activity in cancer cell lines derived from different solid tumors including lung, colorectal, and gastric cancers [63]. Tanabe Research Laboratories started a phase 1 study of TR1801-ADC (MT-8633) in patients with advanced solid tumors that express c-Met in February 2019 (NCT03859752).

13.2.2 SGD-1882 (Payload), SGD-1910 (Linker-Payload), and SGD-1882-Based ADCs

SGD-1882 is an unsymmetrical PBD dimer developed by Spirogen (Figure 13.7). Published data with monomeric PBDs indicate that the most potent forms contain unsaturated C-ring and 2-aryl substituents [64]. SGD-1882 contains a C2-*para*-aniline residue, which connects via an amide bond with a cathepsin B labile, maleimidocaproyl-valine-alanine linker to form the linker payload SGD-1910 (talirine).

SGD-1882 was prepared via a convergent synthetic route from commercially available starting materials (Scheme 13.2) [65]. The PBD monomer **2-2** was prepared from methyl vanillate **2-1**. Acetylation and enol triflation of **2-2** produced **2-4**. Suzuki coupling of **2-4** with two different substituted phenyl boronic acids gave **2-5** and **2-8**. Alkylation of **2-7** and **2-9** in the presence of Cs_2CO_3 and dimethylformamide (DMF) yielded non-symmetric dimer **2-10**. SGD-1882 was obtained after deprotection of **2-10**. Coupling of the maleimidocaproyl-valine-alanine acid **2-11** to SGD-1882 using the standard peptide coupling agent EEDQ, provided the linker payload SGD-1910 [66].

13.2.2.1 SGN-CD33A (Vadastuximab Talirine)

SGN-CD33A (vadastuximab talirine) consists of a humanized anti-CD33 antibody with bioengineered cysteines, that are conjugated to SGD-1882 (Figure 13.7). With antibody engineering,

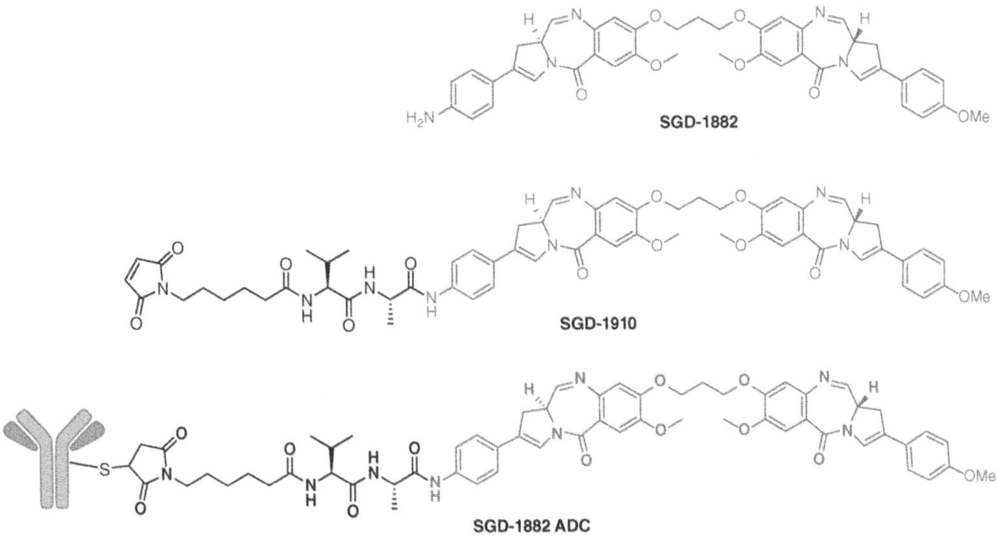

Figure 13.7 SGD-1882 payload, the payload-linker SGD-1910 (Talirine), and SGD-1882 ADC.

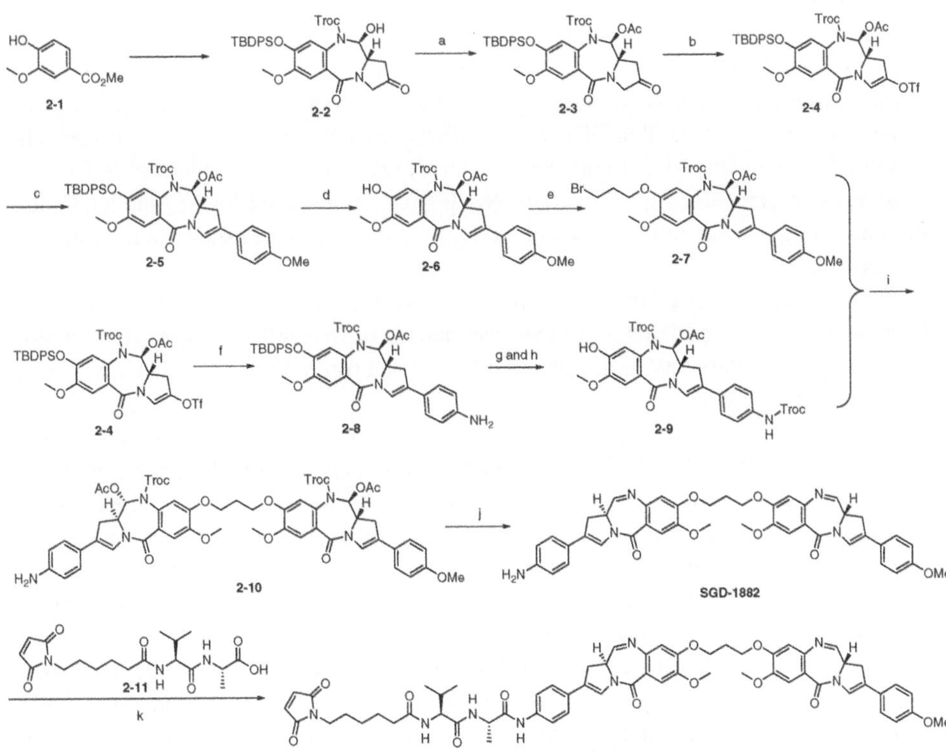

Scheme 13.2 Synthesis of SG-1882. Reagents and conditions: (a) Ac_2O, pyridine (5.0 equiv.), rt, 84%; (b) Tf_2O (3.0 equiv.), 2,6-lutidine (5.0 equiv.), −45 °C, 94%; (c) p-$MeO(C_6H_4)B(OH)_2$, $Pd(PPh_3)_4$ (0.01 equiv.), rt, K_2CO_3, H_2O/EtOH/toluene (10/1/2), 93%; (d) TBAF (1.2 equiv.), THF, rt, 91%; (e) 1,3-dibromopropane, K_2CO_3, DMF; (f) $NH_2(C_6H_4)B$-pinacol ester, $Pd(PPh_3)_4$, (0.01 equiv.), K_2CO_3, H_2O/EtOH/toluene (10/1/2), rt, 84%; (g) TROC-Cl (equiv. 1.1), DCM, pyridine, rt, 97%; (h) TBAF, THF, 93%; (i) **2-7** (1.3 equiv.), Cs_2CO_3, DMF, 80%; (j) Pb–Cd (3 equiv.), THF/NH_4OAc (1 M, aq) (1/1), 58%; (k) Mal-dPEG8-acid, EDCI, DCM, rt, 73%.

very homogenous drug loading is accomplished, averaging 1.9 PBD moieties per antibody [67]. CD33 is highly expressed in most of the acute myeloid leukemia (AML) patients. It is detectable on at least a subset of leukemic blasts in nearly all patients [68]. The protease cleavable linker is highly stable in the circulation and releases the cytotoxic payload only upon internalization into CD33-expressing cells, mitigating the potential for off-target toxicity due to unconjugated drug. After drug internalization, the payload induces DNA damage, cell cycle arrest at the G2/M phase, and apoptotic cell death.

SGN-CD33A has shown robust activity *in vitro* against a broad panel of AML cell lines and primary patient samples as well as *in vivo* using preclinical AML models [67]. SGN-CD33A has been tested alone and with other antileukemia agents in several clinical trials. In July 2013, SGN-33A was evaluated in a phase 1 study to ascertain safety, pharmacokinetics, and preliminary activity of SGN-CD33A. These studies determined the recommended monotherapy dose in patients with R/R AML. Additional expansion cohorts tested SGN-CD33A in specific subpopulations of relapsed AML, and in a cohort of older, treatment-naive patients (NCT01902329). Patients received intravenous (IV) SGN-CD33A on day 1 (5–60 μg/kg) or on days 1 and 4 (20 μg/kg) of 21-day cycles. A total of 131 patients (median age, 73 years in a range of 26–89 years) had intermediate I–II (48%) or adverse (34%) risk (as listed by European Leukemia Net classification) and 50% of patients had underlying myelodysplasia. Two dose-limiting toxicities (grade 2 pulmonary embolism and grade 4 hypocellular marrow) occurred during the dose finding study. Most AEs were consistent with myelosuppression; nonhematologic AEs included fatigue, nausea, and diarrhea. The 30-day mortality was 8%. At the recommended monotherapy dose of 40 μg/kg, the complete remission (CR) with incomplete count recovery (CRi) rate was 28% (5 of 18 patients); 50% of patients who responded achieved minimal residual disease negativity. In patients across dose levels who achieved CR or CRi, the median time to full count recovery was 6.4 weeks for neutrophils (≥1000/ml) and 10.6 weeks for platelets ($\geq 100 \times 10^9$/l). SGN-CD33A demonstrated activity and a tolerable safety profile as a single agent in patients with AML. The recommended monotherapy dose of SGN-CD33A is 40 μg/kg [69].

SGN-CD33A had also been tested in other clinical trials. A safety study of SGN-33A in combination with standard-of-care in patients with AML was started in December 2014 (NCT02326584). This study examined the safety profile of SGN-CD33A by itself (monotherapy) or in combination with other standard treatments.

A phase 1/2 study of SGN-CD33A administered in sequence with allogeneic hematopoietic stem cell transplant in patients with R/R AML was conducted. This study examined the safety and antileukemic profile of SGN-CD33A in patients with relapsed chemoresistant AML, who were given SGN-CD33A in sequence with standard treatments before a planned stem cell transplant, or as maintenance therapy after a stem cell transplant. In 2018, results reported that 6 AML patients in the Pre-allo (before stem cell transplant) arm demonstrated a 66.70% CR and 11-week (median) OS rate (NCT02614560). The study was terminated in January 2017.

A phase 1/2 study began in February 2016 to evaluate the combination of SGN-CD33A and azacitidine in subjects with previously untreated International Prognostic Scoring System (IPSS) intermediate-2 or high-risk myelodysplastic syndrome (MDS). The trial was terminated in January 2018 (NCT02706899) with no results reported.

A phase 3 CASCADE study evaluated SGN-CD33A combined with azacitidine or decitabine in older patients who had newly diagnosed AML. This trial was terminated due to safety; specifically, a higher rate of deaths, including fatal infections, in the SGN-CD33A arm versus the control arm were observed (NCT02785900).

13.2.2.2 SGN-CD70A

SGN-CD70A is composed of a CD70-directed engineered cysteine mAb, h1F6ec, conjugated with SGD-1882. CD70 is a member of the tumor necrosis factor superfamily that is aberrantly expressed in several solid tumors and hematologic malignancies, including clear cell and papillary renal cell carcinoma (RCC) and NHL. Normal expression of CD70 is limited to stromal cells of the thymic medulla, mature dendritic cells, and activated B and T lymphocytes. Thus, CD70 is an attractive target for ADC-based therapy [70].

A phase 1 trial of SGN-CD70A enrolled 20 patients who had R/R CD70-positive NHL including DLBCL, MCL, and grade 3b follicular lymphoma (FL3b) (NCT02216890) [71]. The MTD of SGN-CD70A was 30 μg/kg on the six-week schedule. The most common AEs were thrombocytopenia (75%), nausea (55%), anemia (50%), and fatigue (50%). The onset for treatment-related thrombocytopenia typically occurred during cycle 1. Most of the treatment-related events of thrombocytopenia were grade 3 or greater. Antitumor activity in patients included 1 CR and 3 partial remissions (PRs), 2 of which were ongoing for at least 42.9 weeks. SGN-CD70A exposures were approximately dose proportional, with a mean terminal half-life of three to five days. Modest single-agent activity was observed in heavily pretreated NHL patients, but the applicability of SGN-CD70A is limited by the frequency and severity of thrombocytopenia.

SGN-CD70A was also evaluated in patients with CD70-positive metastatic RCC. The MTD of SGN-CD70A was 30 μg/kg due to thrombocytopenia, as the dose limited toxicity. The most common AEs were fatigue (67%), anemia (61%), and thrombocytopenia (56%). Of 18 enrolled patients, 1 achieved a partial response and 13 achieved stable disease, for a CBR of 78%. Modest single-agent antitumor activity did not support the further development of SGN-CD70A in metastatic RCC [71].

13.2.2.3 SGN-CD19B

SGN-CD19B is composed of the humanized antibody hBU12ec conjugated to SGD-1882 [72]. SGN-CD19B rapidly internalized into CD19-expressing tumor cells and the released PBD drug induced DNA damage, resulting in G2-M cell cycle arrest and cell death. SGN-CD19B demonstrated activity against a broad panel of malignant B-cell lines and induced durable regressions in mice bearing xenografts derived from these B-cell malignancies [73]. SGN-CD19B was evaluated in phase 1 study to examine the safety profile of SGN-CD19B as a single agent in patients with B-NHL. The study was terminated in 2018 with no results reported.

13.2.2.4 SGN-CD123A

SGN-CD123A consists of a humanized anti-CD123 antibody conjugated to SGD-1882. SGN-123A demonstrated effective anti-leukemic activity in multiple preclinical models of AML [74]. The phase 1 study of SGN-123A in patients with AML began in August 2016 and was completed in 2018 without published data.

13.2.2.5 SGN-CD352A

SGN-CD352A consists of a humanized anti-CD352 engineered cysteine (ec) mAb, h20F3ec, conjugated with SGD-1882. CD352 is a tumor antigen expressed on B cell malignancies such as MM, chronic lymphocytic leukemia (CLL) and NHL. CD352 expression was detected by flow cytometry on the surface of malignant plasma cells in 87% (13/15) of human MM patient samples. SGN-CD352A, upon binding CD352 at the MM cell surface, undergoes rapid clathrin-dependent endocytosis (<2 hours) and traffics to lysosomal vesicles, inducing a dose dependent DNA damage signaling response in MM cells, activating ATM and ATR kinases, and caspase 3/7 dependent apoptotic cell death results within 48 hours. SGN-CD352A demonstrated potent cytotoxic activity (EC_{50} values of 6.6–52.6 pM) against a panel of human myeloma and lymphoma cell lines, with efficient cancer cell killing observed at CD352 expression levels as low as 3500 receptors per cell. SGN-CD352A showed strong *in vivo* antitumor effect in MM cell line mouse xenograft models [75].

Seattle Genetics initiated a phase 1 trial of SGN-CD352A for patients with R/R MM in December 2016, and the trial was terminated in July 2019 (NCT02954796).

13.2.2.6 ABBV-176

ABBV-176 is an ADC composed of the humanized antibody h16f (PR-1594804) conjugated to SGD-1882. The antibody targets the prolactin receptor (PRLR), which is overexpressed in several solid tumor types. This phase 1 study (NCT03145909) was performed to evaluate ABBV-176 in patients with advanced solid tumors likely to exhibit elevated levels of PRLR. This study was terminated after the dosing of 19 patients due to lack of objective responses in patients [76].

13.2.2.7 ABBV-321

ABBV-321 (serclutamab talirine) is an ADC composed of EGFR-targeted antibody conjugated with SGD-1882. EGFR is a well-validated oncology target and has been used to treat various cancers, such as lung, head and neck, colon, and pancreatic cancer [77]. ABBV-321 exhibits potent antitumor activity in cellular and *in vivo* assays including xenograft cell-line and patient derived xenografts (PDXs) of glioblastoma (GBM), colorectal, lung, and head and neck tumor models [78].

A phase 1 study to evaluate ABBV-321 in patients with advanced solid tumors associated with overexpression of the EGFR commenced in October 2017 (NCT03234712).

13.2.3 IGN Payload-Based ADCs

IGNs are monoalkylating DNA-interactive payloads, which contain only one imine moiety to eliminate cross-linking with DNA and incorporate an indolino group to lower the toxicity [79].

Immumogen developed several IGN-ADCs (IMGN632, IMGN779, and TAK164) that are currently in clinical trials. IGN-ADCs have unique structural features that are distinct from PBDs. First, the pyrrolidine group in PBD was replaced with an indolino moiety, which increased the binding to DNA. Secondly, a substituted aniline spacer was inserted between two PBD monomers, resulting in increased potency and providing a site for linker attachment [80]. The IGN payload for IMGN779 is DGN462 (Scheme 13.3) and for IMGN632 and TAK164 (DGN549) is DGN549 (Scheme 13.4). Those payloads were attached through different linkers to the desired antibodies.

The synthesis of DGN462 and the conjugations with IMGN779 is described in the following. The PBD monomer was prepared in seven steps from the commercially available starting material of 3-methoxy-4-hydroxy benzoic acid (**3-1**, Scheme 13.3). After protection and nitration, coupling with (*S*)-methyl-indoline-2-carboxylate HCl salt (**3-3**) provided **3-4**. The methyl ester of **3-4** was reduced with DIBAL to form an aldehyde, followed by reduction of the nitro group, induced cyclization to form the "B" ring imine. Subsequent deprotection of the Bn-group gave the key monomer **3-5**. The central spacer unit was prepared from aniline **3-6**. After installation of the PEG moiety, the disulfide side chain was introduced by reductive amination. The diol of **3-9** was mesylated to **3-10**, which reacted with the monomer **3-5** to form the diimine dimers **3-11**. The dimer **3-11** was partially reduced with triacetoxyborohydride to produce the unsymmetric dimer **3-12**. Cleavage of the dithio-group with TECP gave the monoimmine DGN462 [81]. DGN462 was treated with sodium bisulfite to form the sulfonated-DGN462, which was then treated in situ with sulfo-SPDB (**3-13**) to

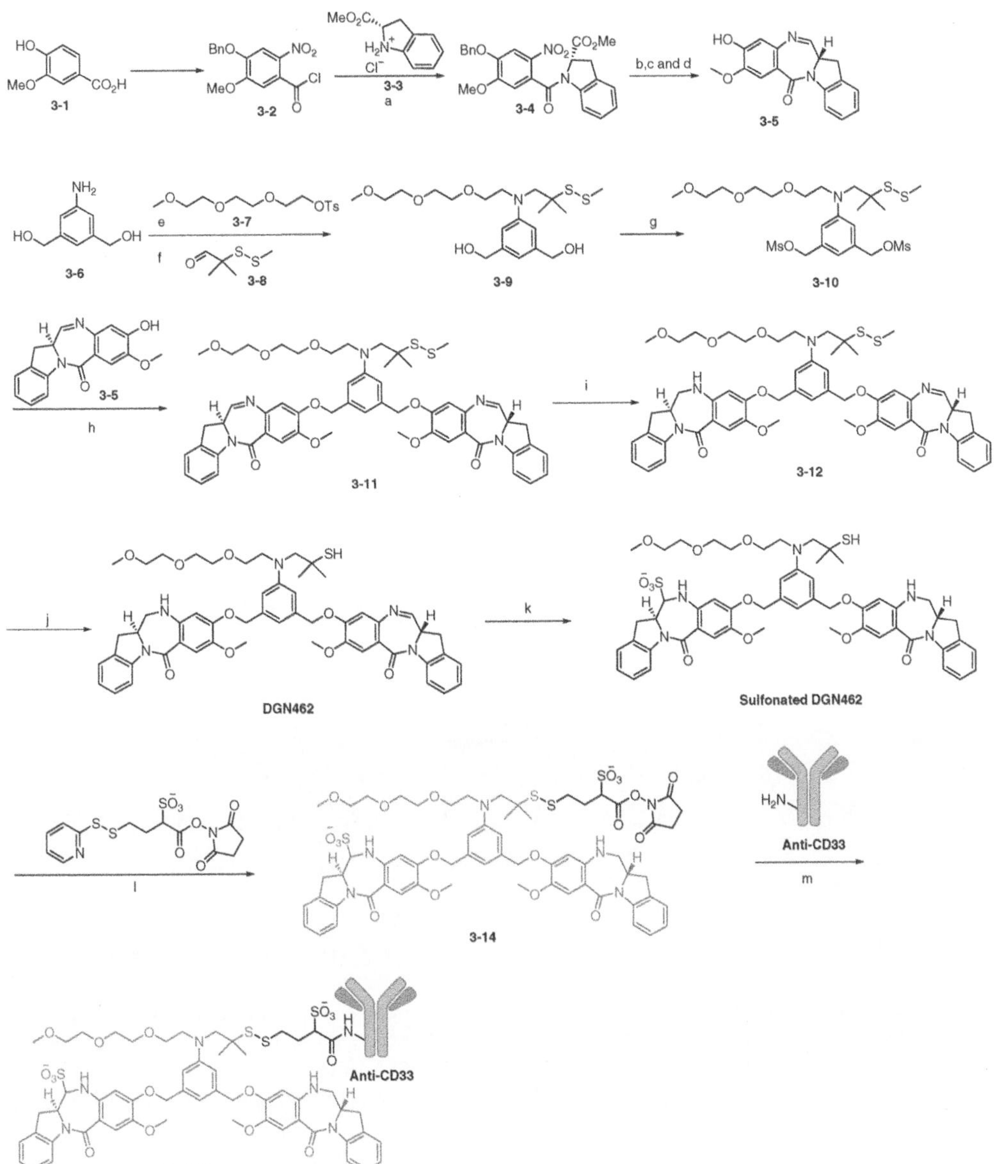

Scheme 13.3 Synthesis of IGN462 and IMGN779. Reagents and conditions: (a) **3-3**, TEA, THF, 94%; (b) DIBAL, DCM, 69%; (c) $Na_2S_2O_4$, MeOH/AcCl, 73%; (d) $MeSO_3H$, DCM, 95%; (e) **3-7**, K_2CO_3, DMF, 48%; (f) **3-8**, $ZnCl_2$, $MgSO_4$, $NaBH(OAc)_3$, 1,2-dichloroethane, 22%; (g) MsCl, TEA, DCM; (h) **3-5**, K_2CO_3, KI, DMF, 56%; (i) $NaBH(OAc)_3$, 1,2-dichloroethane, 31%; (j) TECP, ACN, 100%; (k) $NaHSO_3$, IPA, H_2O, 65%; (l) i-Pr_2NEt, DMA; (m) anti-CD33 antibody, LEPPS buffer, pH = 8.5, DMA.

form the activated eater **3-14**. Conjugation with the anti-CD33A antibody cysteines generated IMGN779.

IMGN632 used a Val-Val-hexanoic acid linker (**4-1**) to conjugate to cysteine residues of the antibody, while TAK164 utilized the Val-Val-hexanoic acid linker to conjugate to lysine residues of the antibody (Scheme 13.4). The payload of DGN549 was prepared from the dipeptide linker **4-1** and

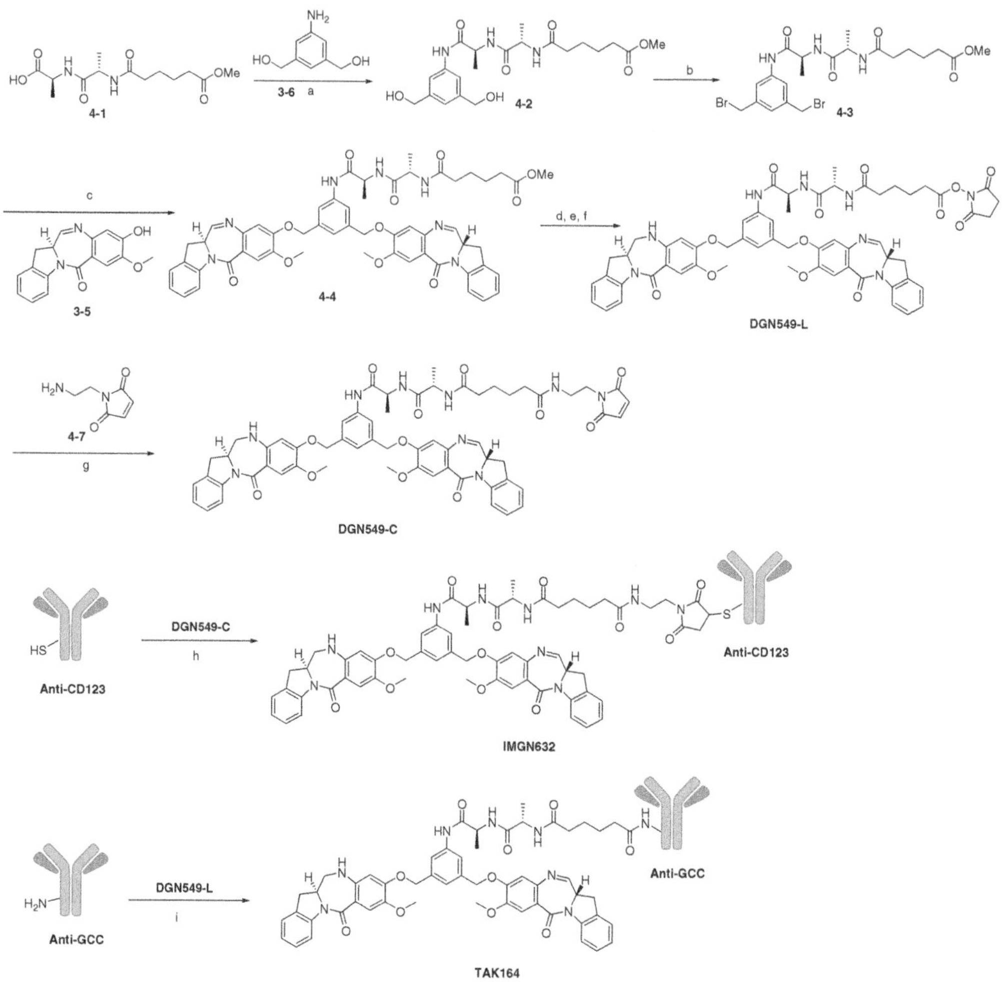

Scheme 13.4 Synthesis of IGN549, IMGN632, and TAK164. Reagents and conditions: (a) **3-6**, EEDQ, DCM, MeOH, rt, 36%; (b) CBr_4, PPh_3, DMF, rt, 39%; (c) **3-5**, K_2CO_3, DMF, rt, 74%; (d) $NaBH(OAc)_3$, DCE, rt, 25%; (e) Me_3SnOH, DCE, 80 °C, 80%; (f) EDC, NHS, DCM, rt, 30%; (g) **4-7**, DIPEA, DCM, rt, 21%; (h) anti-CD123 antibody, 2% v/v DMA and 38% v/v propylene glycol in PBS, 5 mM EDTA, pH 6.0; (i) HEPES buffer, pH = 8.5, DMA.

aniline **3-6**. The diol compound **4-2** was transferred to dibromo compound **4-3** and then coupled with the key monomer **3-5** to form the dimer **4-4**. **4-4** was converted to either DGN549-L for coupling to lysine residues, or DGN549-C for coupling to cysteine residues of antibodies [82, 83].

13.2.3.1 IMGN779

IMGN779 is a humanized anti-CD33 antibody conjugated to DGN462 via a cleavable disulfide linker (Figure 13.8 and Scheme 13.3) [81a].

CD33 is a sialic acid binding receptor expressed on the surface of a majority of AML blasts. It is an established ADC target for AML, and the target of the FDA approved ADC, gemtuzumab ozogamicin (Mylotarg). DGN462 is a novel DNA-alkylating payload, which has potent preclinical anti-leukemia activity. The payload has a monoimine moiety that lacks cross-linking ability giving rise to a better efficacy/toxicity profile as compared with PBD [83]. IMGN779 demonstrated potent

Figure 13.8 IMGN779 and its ADCs.

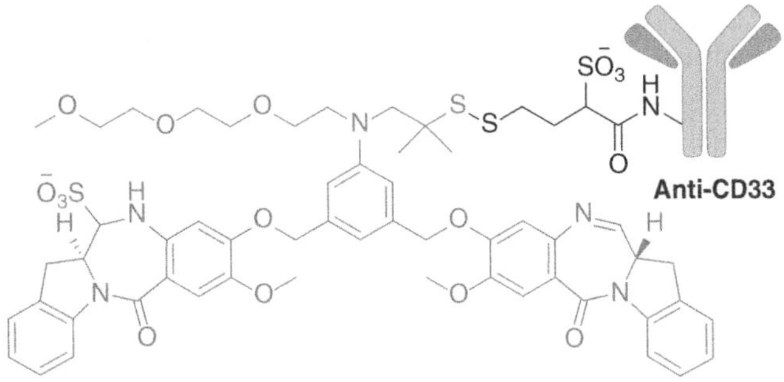

IMGN779

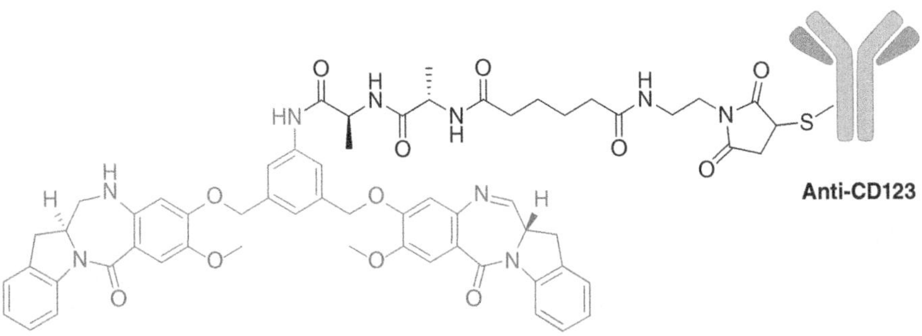

IMGN632

Figure 13.9 IMGN632 and its ADCs.

antitumor activity against AML cell-lines and primary patient samples *in vitro*, and robust efficacy in multiple *in vivo* animal models of AML, such as nude mice bearing EOL-1 xenografts and severe combined immunodeficient (SCID) mice bearing MV4-11 xenografts [77, 84].

A phase 1 study of IMGN779 in adult patients with R/R CD-33-positive AML started in March 2016 (NCT02674763). IMGN779 displayed tolerability with repeat dosing across a wide range of doses in patients with relapsed AML. IMGN779 demonstrated antileukemia activity in 41% (12 of 29) of patients with evaluable bone marrows (≥0.39 mg/kg) [85].

13.2.3.2 IMGN632

IMGN632 is a novel CD123 targeting ADC comprising of an anti-CD123 antibody engineered with site-specific cysteines conjugated with DGN549-C (Figure 13.9 and Scheme 13.4) [76, 83].

IMGN632 was highly cytotoxic to B-ALL/lymphoma cell lines, with half maximal inhibitory concentrations (IC_{50}) between 0.6 and 20 pM. CD123 expression is prevalent across ALL/lymphoma subtypes, and IMGN632 demonstrated promising selective activity in preclinical models of B-ALL/lymphoma [86].

IMGN632 was evaluated in a phase I/II clinical study to treat patients with R/R AML, blastic plasmacytoid dendritic cell neoplasm (BPDCN), ALL, and other CD123+ diseases (NCT03386513). IMGN632 demonstrated an ORR of 18% (13/71) and 33% (3/9) in AML and BPDCN patients, respectively. The dose and schedule of 45 μg/kg given on day 1 every three weeks has been selected as the monotherapy recommended dose for phase 2 studies of IMGN632 [87]. IMGN632 is also being tested as a monotherapy in patients with AML and in combination with azacitidine and venetoclax (NCT04086264).

13.2.3.3 TAK-164

TAK-164 is an ADC comprising a full-length, fully human IgG1 mAb conjugated with DGN549-L (also known as IGN-P1) (Figure 13.10 and Scheme 13.4). The antibody targets the extracellular domain of guanylyl cyclase C (GCC).

TAK-164 demonstrated efficient binding to antigen-expressing cells and GCC-dependent uptake, resulting in cytotoxicity *in vitro*. TAK-164 displayed low plasma clearance in female SCID mice bearing GCC-positive xenografts, after a single administration. TAK-164 also displayed a dose- and time-dependent increase in DNA damage as measured by the pharmacodynamic marker, phospho-histone H2A family member X (pH2A.X) in GCC-positive tumor-bearing xenografts. TAK-164 resulted in durable antitumor activity in multiple GCC-positive PDX models at a single IV dose. Furthermore, TAK-164 preferentially accumulates in GCC-positive tumors monitored in a preclinical imaging study using fluorodeoxyglucose-positron emission tomography. These promising preclinical data warrant advancement of this ADC for clinical evaluation [88, 89].

A phase 1 study of TAK-164 in patients with advanced gastrointestinal cancer expressing GCC was started in April 2018 (NCT03449030). No data have been released to date.

13.2.4 Other PBD-Based Payload ADCs

13.2.4.1 PBD-MA

DHES0815A consists of an anti-human HER-2 antibody linked to a DNA minor groove crosslinking agent, pyrrolo[2,1-*c*] [1, 5] benzodiazepine monoamide (PBD-MA) [90]. DHES0815A is in a phase 1 clinical trial for breast cancer with primary end-points due in April 2022 (NCT03451162) [91].

13.2.4.2 Pyrridinobenzodiazepines

Pyrridinobenzodiazepines (PDDs) are a class of DNA-interactive ADC payloads that form mono-alkylated adducts instead of the cross-linked adducts formed by PBD dimers. PDDs contain a DNA covalent binding PDD moiety and a DNA non-covalent binding sequence-selective component (Figure 13.11) [92, 93].

This DNA-targeting sequence selective ADC payload has a dual mechanism of action, involving DNA alkylation and transcription factor inhibition. These adducts are easier for healthy cells to repair compared with DNA adducts formed with PBD dimers, providing less off-target toxicity. Due to their DNA sequence-selectivity recognition properties, PDD-based payloads specifically bind to transcription factor binding sites and are envisioned to provide another level of

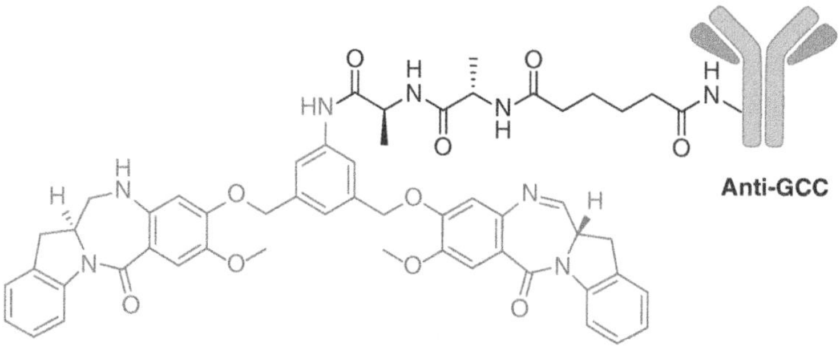

Figure 13.10 TK-164 and its ADCs.

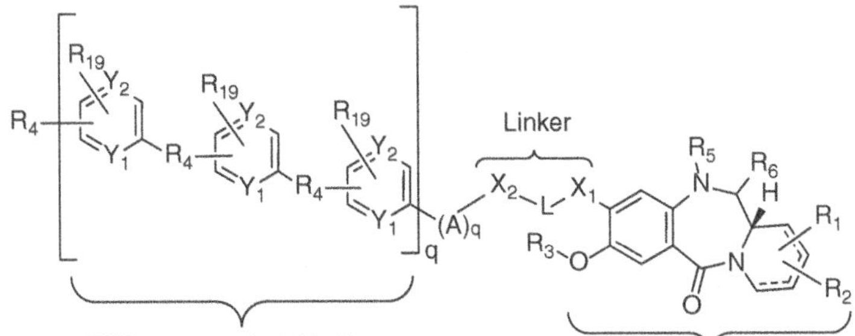

Figure 13.11 Pyrridinobenzodiazepines.

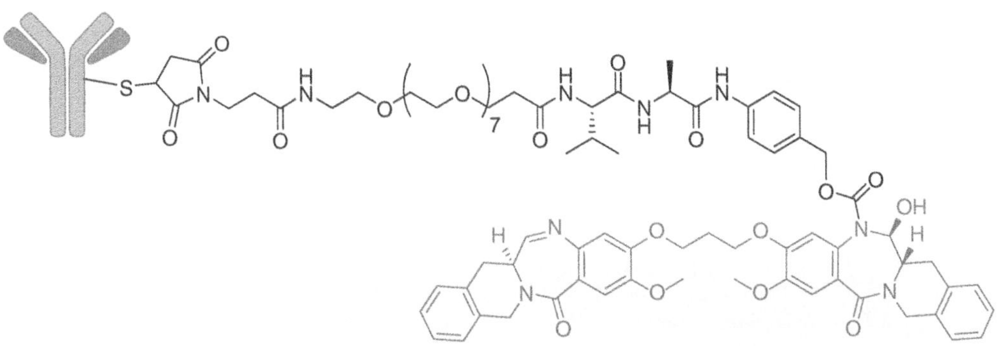

Figure 13.12 D212 and its ADCs.

selectivity [92]. PDD payloads are also designed to have reduced hydrophobicity which can improve conjugation efficiencies to an antibody [92, 93]. Together, these properties can lead to highly targeted cytotoxicity toward selected tumor cells. Anti-CSPG4-(PDD) ADC showed good activity in *in vitro* and *in vivo* human melanoma xenografts [94].

13.2.4.3 Isoquinolidinobenzodiazepine Dimers

Isoquinolidinobenzodiazepines (IQBs) are PBD analogs with the pyrrolidine ring of PBD replaced with benzo-fused rings [95]. D212 is a benzo-fused piperidine ring (IQB) dimer linked via a three-carbon atom length spacer. D212 showed comparable *in vitro* potency with that of SGD-1882 in AML cell lines. Anti-CD33 D212-ADCs are homogeneous ADCs exhibiting favorable physicochemical properties and displaying a robust target-specific potency profile on AML cell lines (Figure 13.12). Preliminary *in vitro* characterization of anti-CD33 ADC using D212 suggested that the IQB-payload exhibited similar potency to the well-characterized SGN-CD33A using PBD as the payload. These results demonstrate that the D212 linker-payload can be utilized for developing novel ADC therapeutics for targeted cancer therapy [96].

13.2.4.4 PBD-Duocarmycin Dimers

The PBD and duocarmaycin heterodimers are DNA-interactive agents that covalently bind to guanine and adenine bases within the minor groove of DNA. Duocarmaycin is a family of cyclopropyl pyrroloindole and cyclopropyl benzoindole (CPI and CBI) derivatives. It was anticipated that the PBD-duocarmycin heterodimer could form cross links between guanine and adenine bases [97], a similar binding model to that of the PBD dimer shown in Figure 13.2 [98].

Genentech developed novel heterodimer payloads containing a mono PBD unit and a CBI unit connected by a methylene linker and applied them as new PBD ADC payloads. Protease cleavable peptide linkers and disulfide linkers were attached to the phenol of CBI or the N-10 position of PBD to generate PBD-CBI linker payloads, respectively. Anti-HER2 and anti-CD22 THIOMAB ADCs were formed by conjugating, via antibody-cystines, with the maleimide-linker-payload using the seco-CBI-PBD dimer, generating ADC1-2 (Figure 13.13) with DARs in a range of 1.9 to 1.6 [99].

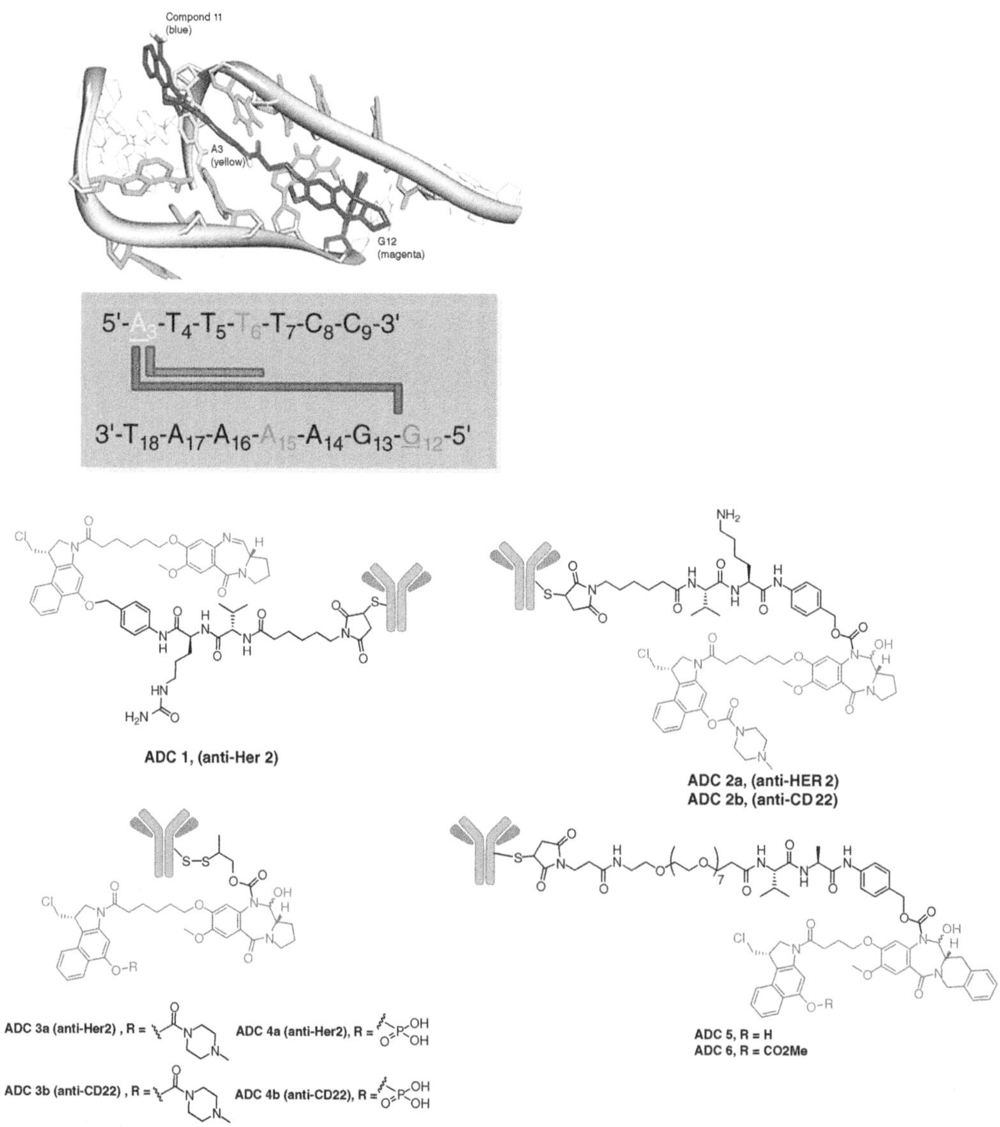

Figure 13.13 PBD-duocarmycin dimer ADCs.

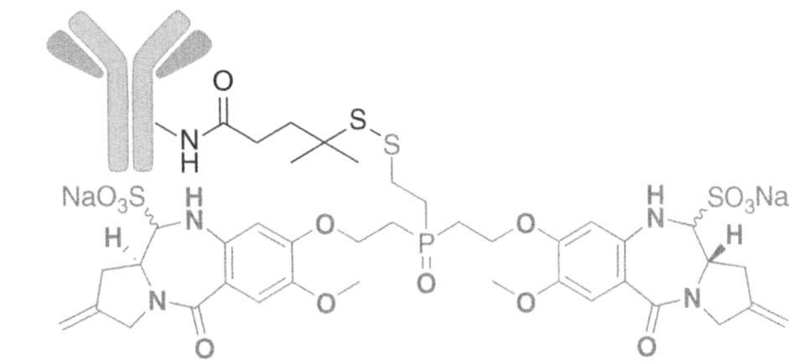

Figure 13.14 PBD dimer with thio-oxophosphane ADC.

ADC3a (anti-Her2), ADC3b (anti-CD22), ADC4a (anti-Her2), and ADC4b (anti-CD22) THIOMAB antibodies were conjugated with a self immolative disulfide linker attached to N-10 of PBD.

ADC1 showed 50% growth inhibition on Her2 positive SKBR3 cell line *in vitro*. ADC2a showed tumor stasis or regression in a Her2 Fo5 mammary transgenic tumor model after 26 days with a single IV dose of 3 or 10 mg/kg [99]. ADC3a and ADC4a exhibited complete response after 43 days in a Fo5 mammary transgenic tumor model at a single IV dose of 5 mg/kg [100].

The IQB-CBI dimer linker payload containing a maleimide PEG8-peptide linker was conjugated with an anti-interleukin-1 receptor accessory protein antibody to generate ADC5 and ADC6 (Figure 13.13) [101]. The ADCs displayed antigen-dependent cytotoxicity with IC_{50}s in the low ng/ml ranges. Duocarmycin-PBD and IQB-CBI ADCs show good preclinical antitumor activity [100, 101].

13.2.4.5 PBD Dimer with Thio-Oxophosphane Moiety

Hangzhou DAC Biotech Ltd. developed a novel series of PBD dimer containing ADCs having a thio-oxophosphane moiety in the middle of the linker between the two mono PBD units. The disulfide linker is "non-cleavable" and imparts improved stability to the ADC linkage (Figure 13.14). The addition of the phosphate group and sulfonic acid moiety at the C-11 position of the PBD monomers improves aqueous solubility to the ADCs. The "non-cleavable" linker was conjugated to an anti-CD22 antibody via lysine conjugation. Various linkers and conjugation sites to the antibody were also investigated [102].

13.3 Small Molecule Drug Conjugates with Pro-Pyrrolobenzodiazepine Payloads

Endocyte developed novel SMDCs targeting several cancer antigens, in particular with folate and PSMA receptors. These SMDC antigens have been conjugated with pro-PBD payloads to operate under various mechanisms of actions [4, 43, 103].

13.3.1 N-Substituted 1,3-Oxazolidine pro-PBD

The first generation of pro-PBD payload SMDCs employed a N-substituted 1,3-oxazolidine as a masked aldehyde, that after cleavage of the disulfide bond, formed a Schiff base. The neighboring amine cyclizes with the aldehyde forming the imine moiety of the PBD core (Figure 13.15).

To apply the concept of oxazoline pro-PBD, a model folate pro-PBD conjugate was prepared, and a release study performed with dithiothreitol (DTT). The formation of the PBD construct and its DTT-adduct demonstrated the efficiency of the pro-PBD approach (Figure 13.16) [104].

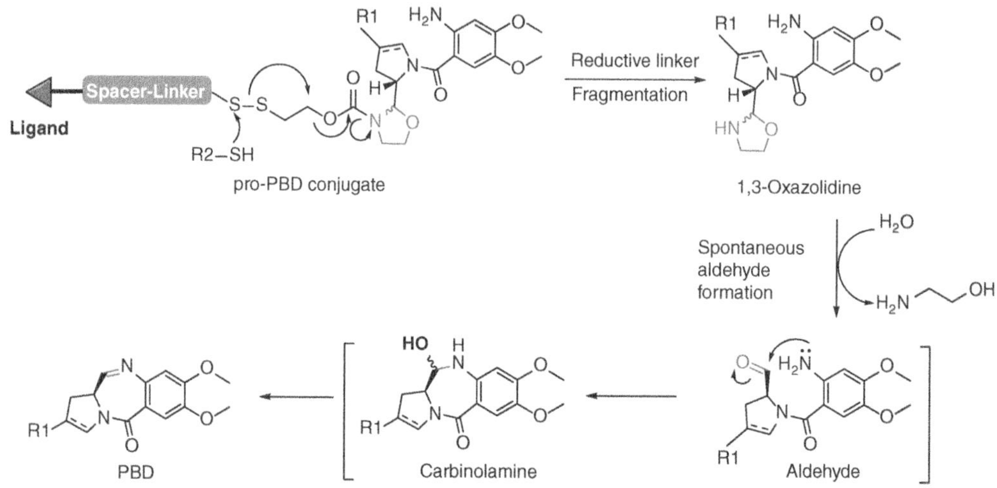

Figure 13.15 Oxazolidine pro-PBD as latent payload [91].

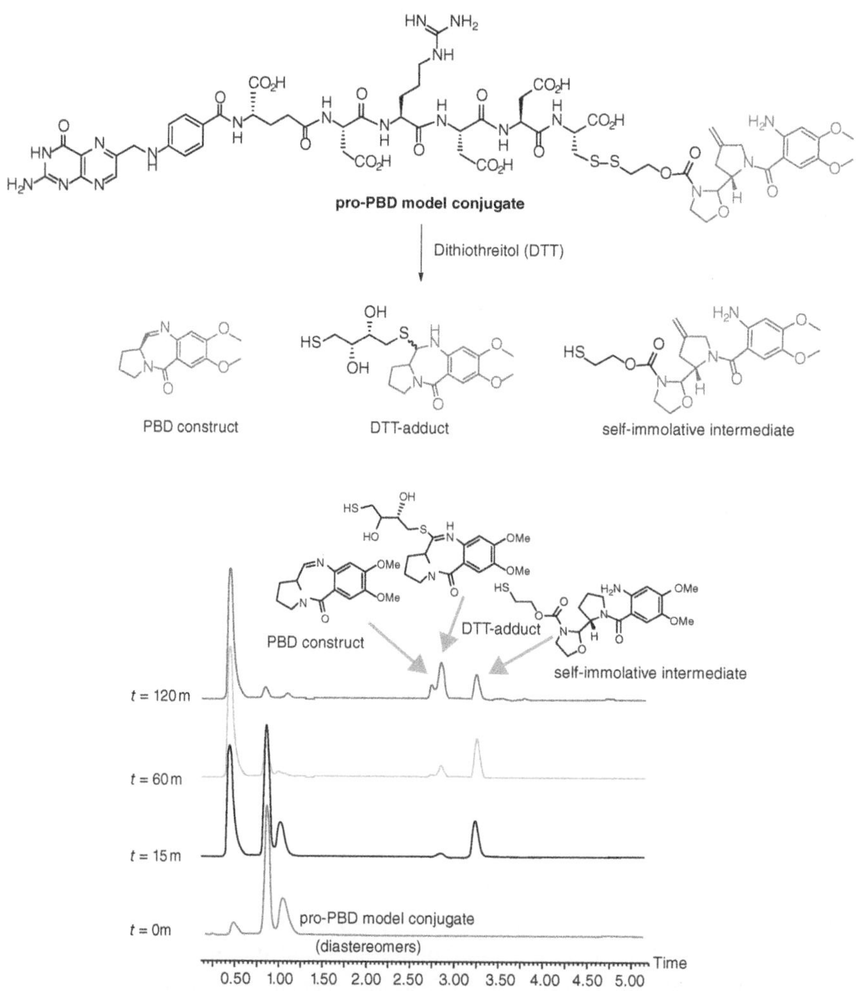

Figure 13.16 The model release study of oxazolidine pro-PBD. *Source:* Adapted with permission from [104]. Copyright (2017) American Chemical Society.

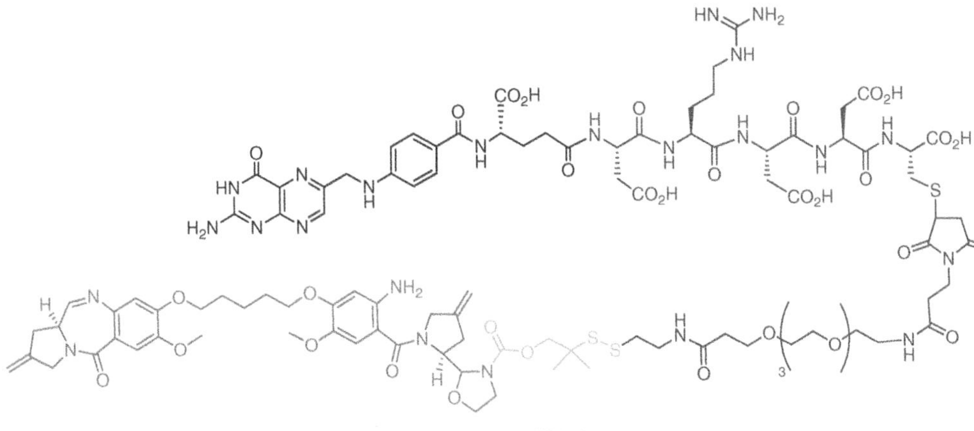

EC2629

Figure 13.17 Pro-PBD SMDC (EC2629).

EC2629 is a SMDC consisting of folate receptor (FR) ligand, a hydrophilic peptide PEG space, a self immolative disulfide linker system, and a pro-PBD dimer with a 1,3-oxazolidine ring (Figure 13.17).

The synthesis of EC2629 began with the construction of the linker and PBD C ring precursor (Scheme 13.5). Prolinal **5-2** was treated with ethanolamine to form the 1,3-oxazolidine ring and then coupled to **5-1** to form **5-3**. Using the same strategy, the other C ring precursor **5-6** was built. Precursor **5-6** was coupled with the bis-A ring precursor **5-5** to yield **5-7**. **5-7** was then coupled with **5-4** to yield **5-8**. The Fmoc protecting groups on **5-8** were removed with diethylamine, resulting in the cyclization of one PBD unit. After installation of the maleimido PEG4 fragment to form **5-9**, conjugation with the folate spacer **5-10** yielded EC2629. EC2629 contains a unique hydrophilic peptide spacer (L-Asp-L-Glu-L-Cys-PEG4) between the disulfide linker and the folate ligand to improve the overall hydrophilicity of the SMDC [105].

EC2629 showed excellent *in vitro* and *in vivo* activity. Treatment once a week to nude mice bearing FR positive human xenografts led to cures in 100% of the mice at very low doses (300 nmol/kg). The observed activity was not accompanied by any noticeable weight loss (up to 20 weeks post end of dosing) or major organ tissue degeneration. Complete responses were also observed against FR-positive paclitaxel (KB-PR) and cisplatin (KB-CR) resistant models. When evaluated against FR-positive PDX models of ovarian (ST070), endometrial (ST040) and TNBCs (ST502, ST738), EC2629 demonstrated significantly greater antitumor activity as compared with their corresponding standard of care treatments [105]. An IND was filed for EC2629 in 2017.

13.3.2 Oxime Ether pro-PBD

Second generation pro-PBD folate conjugates have employed an oxime ether moiety for the C ring precursor [103a, c, 106]. The self-immolative linker was attached at the N-10 position of the aniline while the C11 aldehyde was shielded in oxime ether form. The pro-PBD conjugate possessed good stability and preserved an sp^2 electrophilic center [103c]. When the linker was cleaved under endosomal/lysosomal conditions, the resulting aniline spontaneously condensed with the C-11 oxime ether to form the PBD core (Figure 13.18). SMDC2 was designed based on this concept; the formation of the desired PBD dimer and its DTT-adduct proved the success of oxime ether pro-PBD strategy [103c].

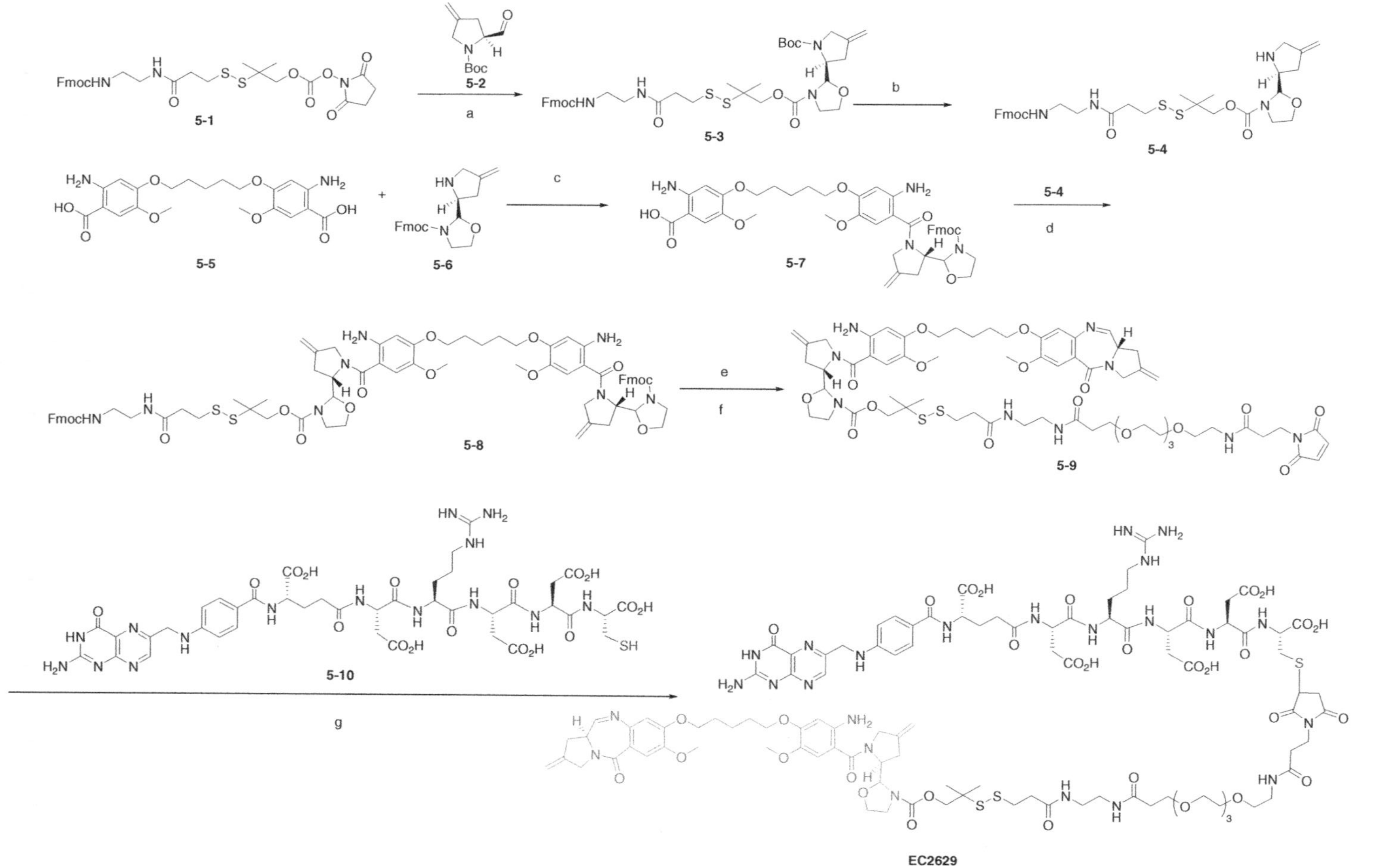

Scheme 13.5 Synthesis of EC2629. Reagents and conditions: (a) ethanolamine, $MgSO_4$, Fmoc-Cl, TEA, DCM, rt, 46%; (b) 50% TFA/DCM; (c) PyBop, DIPEA, DCM, DMF, rt, 60%; (d) **5-4**, PyBop, DIPEA, DMF, 0 °C, 84.9%; (e) Et_2NH, DCM, rt; (f) Mal-PEG4-NHS, TEA, DCM, rt; (g) **5-10**, pH 7 buffer, rt, 24% over three steps.

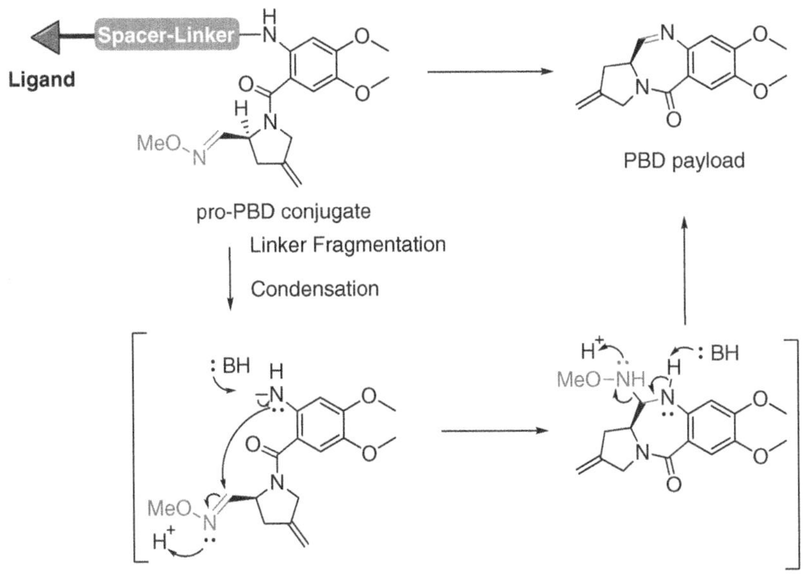

Figure 13.18 Oxime ether pro-PBD as latent payload.

SMDC2 was prepared using this strategy (Scheme 13.6). The pyridyl disulfide linker (**6-1**) was introduced on the amine of the asymmetric bis-A ring intermediate **5-7** by carbamate formation. **6-2** was coupled with the oxime C-ring precursor **6-3** to yield **6-4**. After deprotection of the Fmoc protecting groups on **6-4**, the mono PBD **6-5** was formed and then linked with the folate spacer **6-6** to give SMDC2.

SMDC2 showed excellent *in vitro* and *in vivo* activity. The IC_{50} of SMDC2 in the FR(+)-KB cell line was 56 pM. The *in vivo* efficacy of SMDC2 was evaluated in female nude mice bearing FR(+)-KB tumor xenografts. Treatment with 0.3 μmol/kg of SMDC2 once a week for two weeks resulted in complete responses with acceptable weight loss [103a].

13.4 Discussion

With traditional chemotherapy, the cytotoxic drug is generally dosed IV or orally (PO) for full-body and tissue exposure. Since cytotoxic drugs are indiscriminate in terms of which cells they kill, there is no selectivity observed. This lack of selectivity translates to the lack of a "therapeutic window." That is, toxicity and efficacy are directly proportional to the amount of compound dosed. This usually relates to a narrow "window" of opportunity/efficacy for the drug to exert its antitumor effect versus a general toxicity observation. The minimum dose or concentration of the drug in the body to produce a biological effect is known as the minimum efficacious dose (MED). The MTD is the dose of drug that produces an acceptable level of toxicity such that if that drug level (dose/concentration) is exceeded, it would place a patient at an unacceptable risk for experiencing toxic effects. The therapeutic index (TI) is defined as the difference in dose level between the MED and MTD (Figure 13.19a). This window is the area where efficacy at the safest level of toxicity is observed.

ADCs, on the other hand, are designed to selectively deliver a highly potent cytotoxic agent to tumor cells by utilizing the inherent specificity of a mAb that is designed for a particular antigen

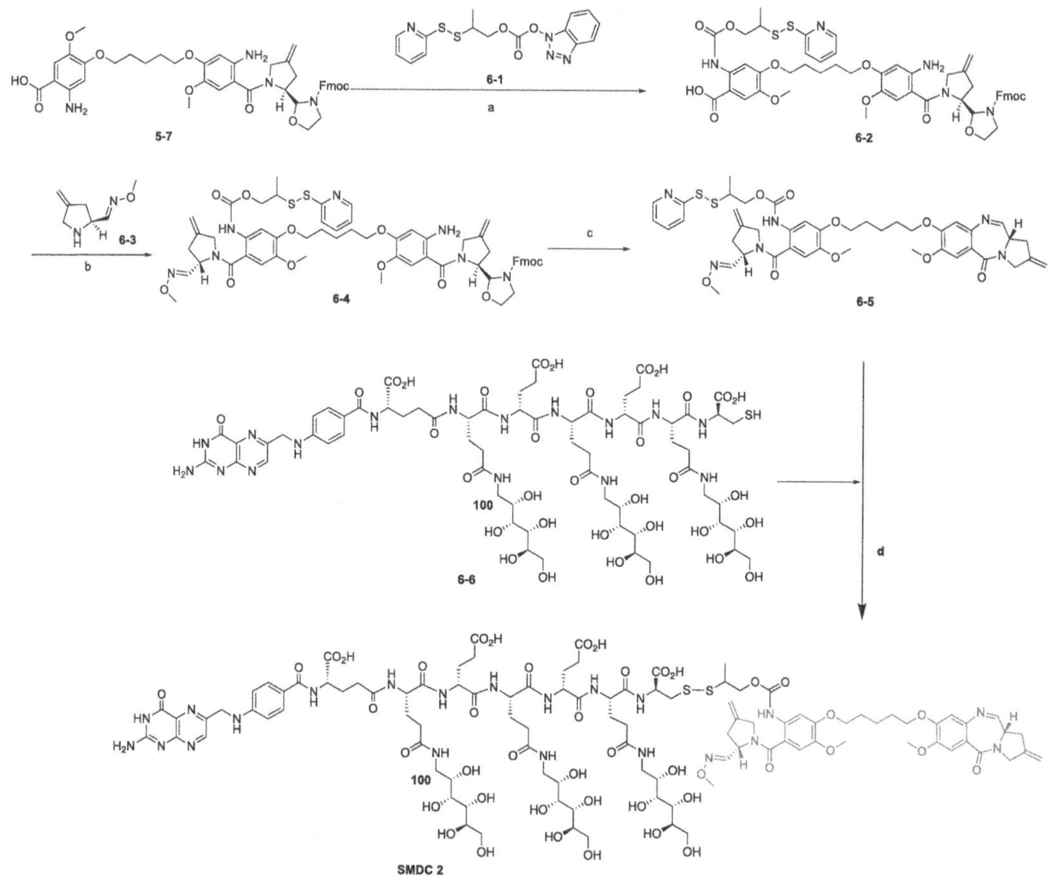

Scheme 13.6 Synthesis of SMDC2. Reagents and conditions: (a) **6-1**, iPr_2NEt, DCM, rt, 85%; (b) **6-3**, PyBop, DIPEA, DCM, rt, 72.2%; (c) Et_2NH, DCM; (d) **6-6**, DMSO, H_2O, Et_3N, 27% over two steps.

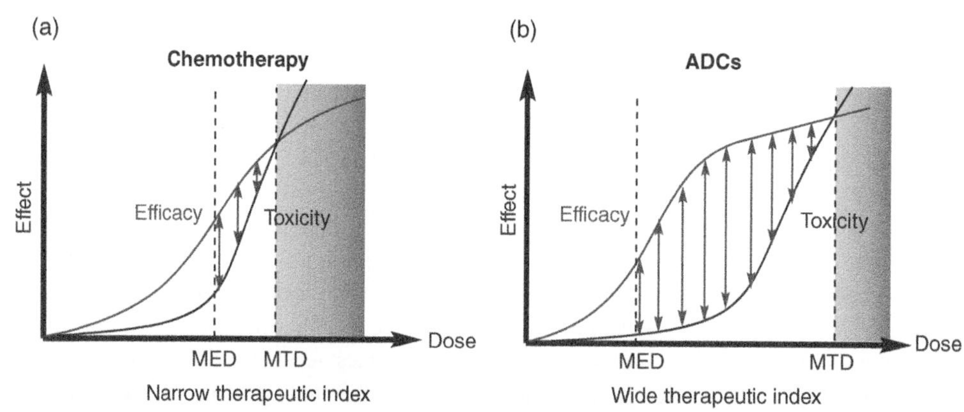

Figure 13.19 Increasing the therapeutic index from (a) by cytotoxic drugs to (b) by ADCs.

expressed on the surface of the target cell. By harnessing this specificity, the cytotoxic agent can be selectively delivered to the cancerous cells. In this way, general toxicity can be lowered as compared with traditional chemotherapy regimens. Further, the dose of the cytotoxic agent can be much lower since it is not released systemically, and thus does not expose all tissues and cells. In other words, it has the potential for substantially lowering toxicity and increasing the therapeutic window as compared with standard chemotherapy protocols (Figure 13.19b).

The ADC approach, however, is not without its issues. Since ADCs are complex in molecular form and metabolic function, a major challenge in developing them has been how to balance increased ADC efficacy in target cells while reducing the toxicity in non-targeted cells. This is compounded by the fact that there are many pathways causing non-specific uptake of ADCs, which leads to overall toxicity of the ADC due to systemic toxic payload (or its metabolite and/or catabolite) exposure. The non-specific uptake routes include ADC tissue distribution, non-specific pinocytosis/phagocytosis, and FcgR-driven and/or FcRn-driven cellular internalization. Therefore, payloads that are extremely toxic (i.e. highly potent) may still not have a broad enough therapeutic window for clinical application.

As discussed, although in concept the ADC approach should be beneficial in widening the therapeutic window of the cytotoxic agent, thus rendering it a viable drug for anticancer therapy, in reality only a few ADCs have shown a transformational difference as compared with standards of care (SoC). In fact, toxicity continues to outpace efficacy by dosage due to the fact that approximately 90% of the ADC never reaches the tumor [107], which implies that a great deal is taken up in normal cells. Thus, the payload and accompanying metabolites that do reach and penetrate the tumor must be highly toxic and efficacious. Couple relatively low antigen level with the low amount of ADC penetrating the tumor (~1%) and it is understandable why the TI of the ADC is in the low to unacceptable range. More than 98 ADC INDs have been filed in the past 10 years. These studies cover a broad range of targets (>63 antigens), and over 35 candidates have been terminated in the clinic due to toxicity issues. Of the 98 ADC INDs filed, approximately 63 (65%) remain active in phase 1A or 1B. Again, the main issue with failure in the clinic is the narrow TI observed in humans.

Indeed, high response rates of efficacy are directly related to a high toxicity profile in the majority of studies. As mentioned previously, the observed toxicity can be due to many factors, one of which is that animal models predominately underpredict the toxicity profile of the ADC in humans. Another key issue is the chemistry/construction of the ADC. The linker between the antibody and the cytotoxic agent may be cleaved in plasma (during systemic circulation), thus causing premature release of the payload or its precursor in healthy tissue, giving rise to non-specific toxic events. One approach to better predict the therapeutic window of an ADC slated for the clinic is to compare the anticipated TI with other ADCs that have well-defined TIs, such as ADCs that are clinically active or FDA-approved. To accomplish this, the quantitative ratio of the exposure levels of the experimental ADC as compared with a reference ADC is considered. The ratio is derived from the exposure level of the ADC at a chosen safety endpoint to the exposure level of that ADC at the chosen efficacy endpoint. This exposure-based method of calculating TI is based on the C_{avg} or $C_{steady\ state}$ levels of the ADCs and is depicted by the equation shown in the caption for Figure 13.20 [108]. C_{avg} at HNSTD stands for the concentration average for human standard (FDA-approved ADC used as standard) and is estimated using PK data from FDA-approved, marketed ADCs.

Overall, as the potency of the cytotoxic agent increases, the more difficult it is to obtain a favorable TI. The inherent selectivity afforded by the ADC approach allows for the use of less potent cytotoxic agents, while still being efficacious. This in turn allows the TI to be increased and the

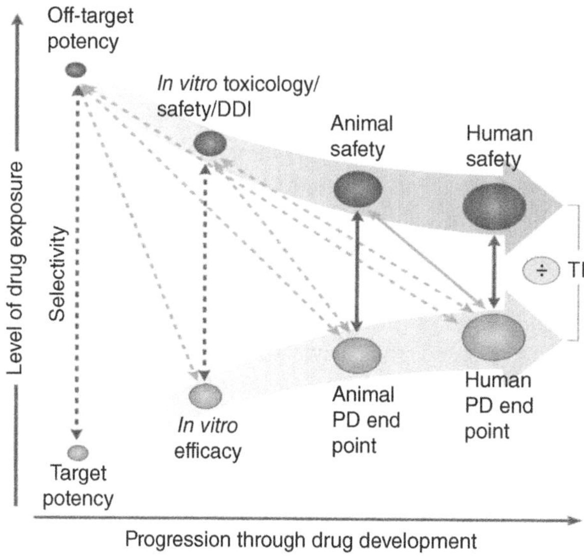

Figure 13.20 An exposure-centric approach to TI determination in humans [108]. TI = C_{avg} at HSNTD/C_{avg} at tumor responsive dose. *Source:* Adapted with permission from [108]. Copyright (2012) Nature Publishing Group.

ADC to exert a more acceptable safety profile. If the payload is too toxic, the TI has a good chance of being too narrow. This has been the case with employing PBD payloads in ADCs to date. PBDs have potencies in the subnanomolar to picomolar range, which makes them extremely potent. It becomes a challenge to balance the efficacy (positives) with the toxicity (negatives). One approach is to design prodrugs to increase the aqueous solubility of the PBD-containing ADCs in an effort to decrease the formation of insoluble aggregates and also to selectively cleave the payload within the tumor microenvironment. It was anticipated that these properties would lower the toxicity profile of the PBD-containing ADCs. For example, it was shown that by employing β-glucuronide as a prodrug (cleaved by β-glucuronidase) to improve the solubility of the ADC, the safety profile and TI could be improved (Figure 13.21). This was attributed to a decrease in off-target internalization of the ADC due to increasing hydrophilicity, and also increased enzymatic activity in tumor cells to cleave β-glucoronide due to the tumor microenvironment being more acidic than normal tissue. The PBD dimer ADC containing the β-glucuronidase cleavable cap also displayed improved conjugation efficiency with less aggregation. Unfortunately, these factors were not enough to dramatically improve the overall TI [109].

It is important to note that the narrow TI of PBD-containing ADCs is not only related to the toxicity of the PBD payload but can also be derived from the normal cells expressing the targeted antigen. Normal cell expression of the antigen can result in significant off-target toxicity due to a highly potent PBD payload. In addition, the bystander effect of PBD-ADCs is necessary to kill neighboring tumor cells, but if "too permeable" the payload can penetrate or carry over to healthy tissue. Further, a stable linker system between the mAb and the payload is crucial for safety [110]. To reduce the off-target toxicity of PBD-ADCs, careful selection of the ADC target is paramount. It should have a high tumor-to-normal expression ratio along with good internalization kinetics. By combining all scenarios described above, a favorable TI could be within reach for PBD-based ADCs.

SMDCs also suffer from problematic PK properties. Although small molecules penetrate tumors more efficiently than antibodies, the small molecules are also cleared more quickly via renal

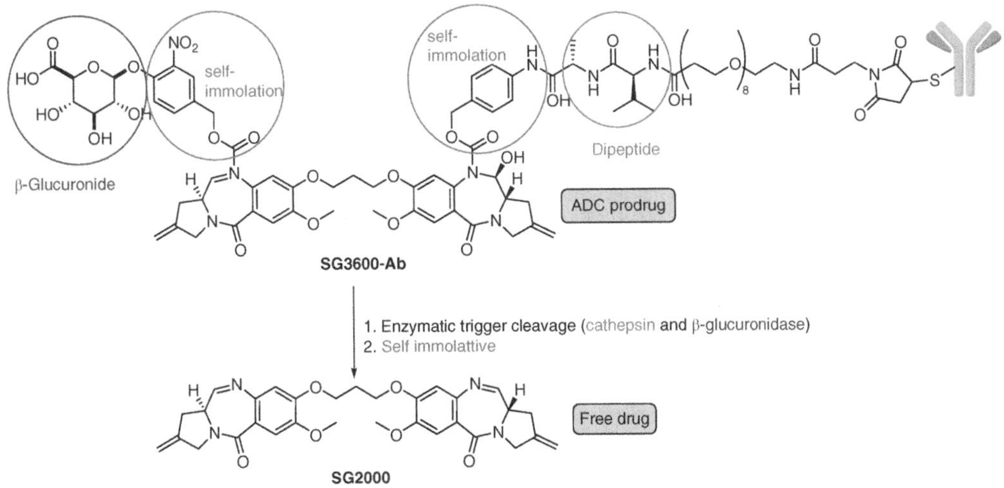

Figure 13.21 β-Glucoronide prodrug of PBD dimer ADCs (SG3600 ADC).

filtration. Thus, half-life extension strategies are needed to lower the clearance rate and allow for improved accumulation levels within tumors.

SMDCs as an alternative for ADCs have their unique structures and requirements for development. The selection of a tumor-specific receptor is very important for the SMDC. The receptor must be overexpressed on the tumor cell relative to a normal cell, and it must be able to sufficiently deliver the therapeutic quantities of drug into the tumor cell [41b]. The ligands with high affinity will not only deliver the drug into the tumor cell but also could attack the normal cell with the receptor expression and cause off- target toxicity of the SMDC. Compared with the ADCs, the size of the ligand is much smaller than that of the antibody and it could not provide protection for linkers and payloads as the antibody did in an ADC. Since a SMDC is accessible to chemical and enzymatic triggers, the stability of each component is crucial to the SMDC. To reduce plasma-protein binding and limit nonspecific adsorption, the linker, spacer, and payload must be carefully designed. Modification of each component of a SMDC to improve the hydrophilicity will dramatically affect the pharmacokinetics/pharmacodynamics of the whole conjugate and its efficacy, respectively.

13.5 Conclusion

From the discovery of anthramycin and its antitumor properties in 1965, the first PBD-ADC, Zynlonta® (loncastuximab tesirine), was approved by the FDA in 2021 for oncological therapies as a single-agent treatment for adult patients with R/R DLBCL.

In the past several decades, several efforts have focused on attenuating the toxicity profiles and increasing the selectivity of these molecules by manipulating the PBD core to create wider therapeutic windows of efficacy, all with the aim of enhancing their potential as viable oncological drugs. In an effort to improve the selectivity of PBD-containing antitumor molecules, PBD-based molecules were attached to antibodies (ADCs) and receptor ligands (SMDCs) to enhance the targeted delivery of these highly potent entities to tumor cells and to widen their therapeutic windows. Although the application of PBD-based payloads in ADCs and SMDCs demonstrated favorable *in vitro* and *in vivo* activities and performed well in some phase 1 and 2 clinical trials using PBD-ADCs, most clinical candidates of PBD-ADCs were a failure or discontinued primarily due to their indiscriminate high toxicity.

Because ADCs and SMDCs are complex molecular constructs, improvements must be made not only to the PBD payloads, but also to the linker, antibody, and targeting ligand systems. Additional SAR studies of PBD molecules need to be conducted in order to design and screen potent PBD analogs with lower potential for toxicity. That said, lower potency could be tolerated if the toxicity profile were improved. One approach to accomplishing this is by increasing the hydrophilicity of the payload, as well as the overall hydrophilicity of the ADC or SMDC. Selecting ligands and antibodies with high affinity and specificity, along with consistent DAR levels, would also improve the efficiency of drug delivery. Engineering new antibodies and developing innovative conjugation methods could provide more options for ADC construction as well and could potentially impact the efficacies and therapeutic windows of the ADCs. Furthermore, the design of novel linker systems imparting excellent stability could reduce off-target toxicity and improve the pharmacokinetics/pharmacodynamics of the conjugates. The development of successful targeting medicines is a challenge, and hinges on choosing the correct combination of payload, linker, and antibody/small ligand systems. Selecting wisely will go a long way toward addressing the narrow therapeutic window issues currently observed and will be crucial for the success of ADCs and SMDCs in the clinic. Overall, for all ADC and SMDC constructs, the following parameters may impact a drug's efficacy: (a) tumor penetration, (b) antigen level expression, (c) internalization rates, (d) stability of payload/metabolites formed, (e) diffusion rates, (f) mutations of targets, and (g) expression of multidrug resistance pump proteins.

Another effort to improve PBD-based ADCs entails using them in combination therapies. To that end, the combination of PBD-ADCs with immune checkpoint inhibitors, T-cell agonists, and other agents has been explored [111]. For example, coupling the unique DNA-alkylating mechanism of PBD-ADCs with other drugs that operate under alternative antitumor mechanisms may provide favorable antitumor synergistic effects and allow a lower dose of the PBD-ADC to be employed than if used as a monotherapy. Further investigation of the synergistic effects of PBD-ADCs with immune checkpoint inhibitors has the potential to provide opportunities to widen the application of PBD-ADCs, as well as improve their TI.

References

1 Leimgruber, W., Stefanovic, V., Schenker, F. et al. (1965). Isolation and characterization of anthramycin, a new antitumor antibiotic. *J. Am. Chem. Soc.* 87: 5791–5793.

2 Kopka, M.L., Goodsell, D.S., Baikalov, I. et al. (1994). Crystal structure of a covalent DNA-drug adduct: anthramycin bound to C-C-A-A-C-G-T-T-G-G and a molecular explanation of specificity. *Biochemistry* 33: 13593–13610.

3 Boyd, F.L., Stewart, D., Remers, W.A. et al. (1990). Characterization of a unique tomaymycin-d(CICGAATTCICG)2 adduct containing two drug molecules per duplex by NMR, fluorescence, and molecular modeling studies. *Biochemistry* 29: 2387–2403.

4 Mantaj, J., Jackson, P.J.M., Rahman, K.M., and Thurston, D.E. (2017). From anthramycin to pyrrolobenzodiazepine (PBD)-containing antibody–drug conjugates (ADCs). *Angew. Chem. Int. Ed.* 56: 462.

5 Kohn, K.W., Glaubiger, D., and Spears, C.L. (1974). The reaction of anthramycin with DNA: II. Studies of kinetics and mechanism. *Biochim. Biophys. Acta, Nucleic Acids Protein Synth.* 361 (3): 288–302.

6 (a) Antonow, D. and Thurston, D.E. (2011). Synthesis of DNA-interactive pyrrolo[2,1-c][1,4]benzodiazepines (PBDs). *Chem. Rev.* 111: 2815–2864. (b) Cipolla, L., Araújo, A.C., Airoldi, C., and Bini, D. (2009). Pyrrolo[2,1-c][1,4]benzodiazepine as a scaffold for the design and synthesis of anti-tumour drugs. *Anticancer Agents Med. Chem.* 9 (1): 1–31. (c) Gerratana, B. (2012). Biosynthesis, synthesis, and biological activities of pyrrolobenzodiazepines. *Med. Res. Rev.* 32 (2): 254–293. (d) Thurston, D.E. and Bose, D.S. (1994). Synthesis of DNA-interactive pyrrolo[2,1-c][1,4]benzodiazepines. *Chem. Rev.* 94 (2): 433–465.

7 (a) Zhou, Q., Duan, W., Simmons, D. et al. (2001). Design and synthesis of a novel DNA–DNA Interstrand adenine–guanine cross-linking agent. *J. Am. Chem. Soc.* 123 (20): 4865–4866. (b) Tercel, M., Stribbling, S.M., Sheppard, H. et al. (2003). Unsymmetrical DNA cross-linking agents: combination of the CBI and PBD pharmacophores. *J. Med. Chem.* 46 (11): 2132–2151. (c) Giddens, A.C., Lee, H.H., Lu, G.-L. et al. (2016). Analogues of DNA minor groove cross-linking agents incorporating aminoCBI, an amino derivative of the duocarmycins: synthesis, cytotoxicity, and potential as payloads for antibody–drug conjugates. *Bioorg. Med. Chem.* 24 (22): 6075–6081.

8 (a) Gregson, S.J., Howard, P.W., and Thurston, D.E. (2003). Synthesis of the first examples of A-C8/C-C2 amide-linked pyrrolo[2,1-c][1,4]benzodiazepine dimers. *Bioorg. Med. Chem. Lett.* 13: 2277–2280. (b) Reddy, B.S., Damayanthi, Y., and Lown, J.W. (2000). Design, synthesis and in vitro cytotoxicity studies of novel pyrrolo[2,1-c][1,4]benzodiazepine (PBD)–polymade conjugates and 2,2'-PBD dimers. *Anti-Cancer Drug Des.* 15: 225–238. (c) Kamal, A., Ramulu, P., Srinivas, O., and Ramesh, G. (2003). Synthesis and DNA-binding affinity of A-C8/C-C2 alkoxyamido-linked pyrrolo[2,1-c][1,4]benzodiazepine dimers. *Bioorg. Med. Chem. Lett.* 13: 3955–3958.

9 (a) Bose, D.S., Thompson, A.S., Smellie, M. et al. (1992). Effect of linker length on DNA-binding affinity, cross-linking efficiency and cytotoxicity of C8-linked pyrrolobenzodiazepine dimers. *J. Chem. Soc., Chem. Commun.* (20): 1518–1520. (b) Smellie, M. et al. (1994). Cellular pharmacology of novel C8-linked anthramycin-based sequence-selective DNA minor groove cross-linking agents. *Br. J. Cancer* 70 (1): 48–53.

10 (a) Janjigian, Y.Y., Lee, W., Kris, M.G. et al. (2010). A phase I trial of SJG-136 (NSC#694501) in advanced solid tumors. *Cancer Chemother. Pharmacol.* 65: 833–838. (b) Hochhauser, D., Meyer, T., Spanswick, V.J. et al. (2009). Phase I study of sequence-selective minor groove DNA binding agent SJG-136 in patients with advanced solid tumors. *Clin. Cancer Res.* 15: 2140–2147. (c) Puzanov, I., Lee, W., Chen, A.P. et al. (2011). Phase I pharmacokinetic and pharmacodynamic study of SJG-136, a novel DNA sequence selective minor groove cross-linking agent, in advanced solid tumors. *Clin. Cancer Res.* 17: 3794–3802.

11 (a) Hartley, J.A. and Hochhauser, D. (2012). Small molecule drugs – optimizing DNA damaging agent-based therapeutics. *Curr. Opin. Pharmacol.* 12: 398–402. (b) Rahman, K.M., Thompson, A.S., James, C.H. et al. (2009). The pyrrolobenzodiazepine dimer SJG-136 forms sequence-dependent intrastrand DNA cross-links and monoalkylated adducts in addition to interstrand cross-links. *J. Am. Chem. Soc.* 131: 13756–13766.

12 Alley, M.C., Hollingshead, M.G., Pacula-Cox, C.M. et al. (2004). SJG-136 (NSC 694501), a novel rationally designed DNA minor groove interstrand cross-linking agent with potent and broad spectrum antitumor activity. *Cancer Res.* 64 (18): 6700–6706.

13 ClinicalTrials.gov (2010). NCT01200797, SJG-136 in treating patients with epithelial ovarian, primary peritoneal, or fallopian tube cancer.

14 Beacon (2021). Beacon Targeted Therapies. https://beacon-intelligence.com (accessed 5 October 2021).

15 Schumacher, D., Hackenberger, C.P.R., Leonhardt, H., and Helma, J. (2016). Current status: site-specific antibody drug conjugates. *J. Clin. Immunol.* 36 (1): 100–107.

16 Redman, E.A., Mellors, J.S., Starkey, J.A., and Ramsey, J.M. (2016). Characterization of intact antibody drug conjugate variants using microfluidic capillary electrophoresis–mass spectrometry. *Anal. Chem.* 88 (4): 2220–2226.

17 Junutula, J.R., Raab, H., Clark, S. et al. (2008). Site-specific conjugation of a cytotoxic drug to an antibody improves the therapeutic index. *Nat. Biotechnol.* 26: 925–932.

18 Shen, B.Q., Xu, K., Liu, L. et al. (2012). Conjugation site modulates the in vivo stability and therapeutic activity of antibody-drug conjugates. *Nat. Biotechnol.* 30: 184–189.

19 Jeger, S., Zimmermann, K., Blanc, A. et al. (2010). Site-specific and stoichiometric modification of antibodies by bacterial transglutaminase. *Angew. Chem.* 49: 9995–9997.

20 Strop, P., Liu, S.H., Dorywalska, M. et al. (2013). Location matters: site of conjugation modulates stability and pharmacokinetics of antibody drug conjugates. *Chem. Biol.* 20: 161–167.

21 Farias, S.E., Strop, P., Delaria, K. et al. (2014). Mass spectrometric characterization of transglutaminase based site-specific antibody-drug conjugates. *Bioconjugate Chem.* 25: 240–250.

22 (a) *ADC Review* (2019).What are antibody-drug conjugates?. https://www.adcreview.com/the-review/antibody-drug-conjugates/what-are-antibody-drug-conjugates/ (accessed 9 March 2020); (b) Gingrich, J. (2020). How the next generation antibody drug conjugates expnads beyond cytotoxic payloads for cancer therapy. https://www.adcreview.com/articles/how-the-next-generation-antibody-drug-conjugates-expands-beyond-cytotoxic-payloads-for-cancer-therapy, (accessed 7 April 2020).

23 Bross, P.F., Beitz, J., Chen, G. et al. (2001). Approval summary: gemtuzumab ozogamicin in relapsed acute myeloid leukemia [published correction appears in *Clin Cancer Res.*, 2002, 8(1), 300]. *Clin. Cancer Res.* 7 (6): 1490–1496.

24 National Cancer Institute (2017). SEER cancer stat facts: acute myeloid leukemia (AML). https://seer.cancer.gov/statfacts/html/amyl.html (accessed 2 October 2021).

25 Selby, C., Yacko, L.R., and Glode, A.E. (2019). Gemtuzumab ozogamicin: back again. *J. Adv. Pract. Oncol.* 10 (1): 68–82.

26 European Medicines Agency (2008). Mylotarg. https://www.ema.europa.eu/en/medicines/human/EPAR/mylotarg (accessed 2 October 2021).

27 Seagen (2021). Brentuximab vedotin. https://www.seattlegenetics.com/pipeline/brentuximab-vedotin (accessed 2 October 2021).

28 US FDA (2013).FDA approves new treatment for late-stage breast cancer. https://web.archive.org/web/20170112023904/https://www.fda.gov/NewsEvents/Newsroom/PressAnnouncements/ucm340704.htm (accessed 2 October 2021).

29 (a) Ricart, A.D. (2011). Antibody-drug conjugates of calicheamicin derivative: gemtuzumab ozogamicin and inotuzumab ozogamicin. *Clin. Cancer Res.* 17 (20): 6417–6427; (b) US FDA (n.d.) Drugs@FDA: FDA-approved drugs. https://www.accessdata.fda.gov/scripts/cder/daf/index.cfm?event=overview.process&ApplNo=761040 (accessed 2 October 2021); (c) US FDA (2018). FDA approves moxetumomab pasudotox-tdfk for hairy cell leukemia. https://www.fda.gov/drugs/resources-information-approved-drugs/fda-approves-moxetumomab-pasudotox-tdfk-hairy-cell-leukemia (accessed 4 October 2021).

30 US FDA (2019).FDA approves first chemoimmunotherapy regimen for patients with relapsed or refractory diffuse large B-cell lymphoma. Press release (10 June).

31 US FDA (2019).FDA approves new treatment option for patients with HER2-positive breast cancer who have progressed on available therapies. Press release (20 December).

32 US FDA (2019). FDA approves new type of therapy to treat advanced urothelial cancer. Press release (18 December).

33 (a)US FDA (2020). FDA approves new therapy for triple negative breast cancer that has spread, not responded to other treatments. Press release (22 April).

34 (a) US FDA (2020).FDA approvals and database. https://www.fda.gov/drugs/drug-approvals-and-databases/fda-granted-accelerated-approval-belantamab-mafodotin-blmf-multiple-myeloma (accessed 2 October 2021); (b) US FDA (2021). FDA grants accelerated approval to loncastuximab tesirine-lpyl for large B-cell lymphoma. https://www.fda.gov/drugs/resources-information-approved-drugs/fda-grants-accelerated-approval-loncastuximab-tesirine-lpyl-large-b-cell-lymphoma (accessed 4 October 2021); (c) US FDA (2021). FDA grants accelerated approval to tisotumab vedotin-tftv for recurrent or metastatic cervical cancer. https://www.fda.gov/drugs/resources-information-approved-drugs/fda-grants-accelerated-approval-tisotumab-vedotin-tftv-recurrent-or-metastatic-cervical-cancer (accessed 4 October 2021).

35 Deonarain, M.P., Yahioglu, G., Stamati, I. et al. (2018). Small-format drug conjugates: a viable alternative to ADCs for solid tumours? *Antibodies* 7: 16.

36 Carrasco-Triguero, M., Yi, J.-H., Dere, R. et al. (2013). Immunogenicity assays for antibody–drug conjugates: case study with ado-trastuzumab emtansine. *Bioanalysis* 5: 1007–1023.

37 Borsi, L., Balza, E., Bestagno, M. et al. (2002). Selective targeting of tumoral vasculature: comparison of different formats of an antibody (L19) to the ED-B domain of fibronectin. *Int. J. Cancer* 102: 75–85.

38 Shahied, L.S., Tang, Y., Alpaugh, R.K. et al. (2004). Bispecific minibodies targeting HER2/neu and CD16 exhibit improved tumor lysis when placed in a divalent tumor antigen binding format. *J. Biol. Chem.* 279: 53907–53914.

39 (a) Kurzrock, R., Gabrail, N., Chandhasin, C. et al. (2012). Safety, pharmacokinetics, and activity of GRN1005, a novel conjugate of angiopep-2, a peptide facilitating brain penetration, and paclitaxel, in patients with advanced solid tumors. *Mol. Cancer Ther.* 11: 308–316. (b) Zhang, X.X., Eden, H.S., and Chen, X. (2012). Peptides in cancer nanomedicine: drug carriers, targeting ligands and protease substrates. *J. Controlled Release* 159: 2–13. (c) Rana, S., Nissen, F., Lindner, T. et al. (2013). Screening of a novel peptide targeting the proteoglycan-like region of human carbonic anhydrase IX. *Mol. Imag.* 12: 497–509. (d) McGuire, M.J., Powell Gray, B., Li, S. et al. (2014). Identification and characterization of a suite of tumor targeting peptides for non-small cell lung cancer. *Sci. Rep.* 4: 4480.

40 (a) Xia, W. and Low, P.S. (2010). Folate-targeted therapies for cancer. *J. Med. Chem.* 53: 6811–6824. (b) Varghese, B., Vlashi, E., Xia, W. et al. (2014). Folate receptor-β in activated macrophages: ligand binding and receptor recycling kinetics. *Mol. Pharmaceutics* 11: 3609–3616. (c) Vaitilingam, B., Chelvam, V., Kularatne, S.A. et al. (2012). A folate receptor-α-specific ligand that targets cancer tissue and not sites of inflammation. *J. Nucl. Med.* 53: 1127–1134. (d) Thomas, M., Kularatne, S.A., Qi, L. et al. (2009). Ligand-targeted delivery of small interfering RNAs to malignant cells and tissues. *Ann. N.Y. Acad. Sci.* 1175: 32–39. (e) Shen, J., Chelvam, V., Cresswell, G., and Low, P.S. (2013). Use of folate-conjugated imaging agents to target alternatively activated macrophages in a murine model of asthma. *Mol. Pharmaceutics* 10: 1918–1927.

41 (a) Trapani, G., Denora, N., Trapani, A., and Laquintana, V. (2012). Recent advances in ligand targeted therapy. *J. Drug Targeting* 20: 1–22. (b) Srinivasarao, M. et al. (2015). Principles in the design of ligand-targeted cancer therapeutics and imaging agents. *Nat. Rev. Drug Discovery* 14 (3): 203–219.

42 (a) Morris, R.T., Joyrich, R.N., Naumann, R.W. et al. (2014). Phase II study of treatment of advanced ovarian cancer with folate-receptor-targeted therapeutic (vintafolide) and companion SPECT-based imaging agent (Tc-99m-etarfolatide). *Ann. Oncol.* 25: 852–858. (b) Harb, W.A., Edelman, M.J., Chambers, S.K. et al. (2017). Phase 1 dose escalation study of the folate receptor-targeted small molecule drug conjugate EC1456. *J. Clin. Oncol.* 35 (15_suppl): 2576–2576. (c) Morris, M.J., Vogelzang, N.J., Sartor, O. et al. (2017). *J. Clin. Oncol.* 35 (15_suppl): 5038–5038.

43 Vlahov, I.R., Leamon, C.P., Qi, L. et al. (2016).Preparation of pyrrolobenzodiazepine prodrugs and their folate conjugates for treating antiproliferative diseases. PCT Patent, WO2016148674A1.

44 (a) Tiberghien, A.C., Levy, J.-N., Masterson, L.A. et al. (2016). Design and synthesis of tesirine, a clinical antibody–drug conjugate pyrrolobenzodiazepine dimer payload. *ACS Med. Chem. Lett.* 7 (11): 983–987. (b) Howard, P.W. (2014). Pyrrolobenzodiazepines and conjugates thereof. WO2014057074.

45 Jeffrey, S.C., Nguyen, M.T., Andreyka, J.B. et al. (2006). Dipeptide-based highly potent doxorubicin antibody conjugates. *Bioorg. Med. Chem. Lett.* 16 (2): 358–362.

46 Flynn, M.J. et al. (2016). ADCT-301, a pyrrolobenzodiazepine (PBD) dimer–containing antibody–drug conjugate (ADC) targeting CD25-expressing hematological malignancies. *Mol. Cancer Ther.* 15 (11): 2709–2721.

47 Cho, S., Zammarchi, F., Williams, D.G. et al. (2018). Antitumor activity of MEDI3726 (ADCT-401), a pyrrolobenzodiazepine antibody–drug conjugate targeting PSMA, in preclinical models of prostate cancer. *Mol. Cancer Ther.* 17 (10): 2176–2186.

48 De Bono, J.S., Fleming, M.T., Sing-Zan Wang, J. et al. (2020). MEDI3726, a prostate-specific membrane antigen (PSMA)-targeted antibody-drug conjugate (ADC) in mCRPC after failure of abiraterone or enzalutamide. *J. Clin. Oncol.* 38 (Suppl 6): 99–99.

49 Zammarchi, F., Corbett, S., Adams, L. et al. (2018). ADCT-402, a PBD dimer–containing antibody drug conjugate targeting CD19-expressing malignancies. *Blood* 131 (10): 1094–1105.

50 Kahl, B.S., Hamadani, M., Radford, J. et al. (2019). A phase I study of ADCT-402 (Loncastuximab tesirine), a novel pyrrolobenzodiazepine-based antibody–drug conjugate, in relapsed/refractory B-cell non-Hodgkin lymphoma. *Clin. Cancer Res.* 25 (23): 6986–6994.

51 Jain, N., Stock, W., Zeidan, A. et al. (2020). Loncastuximab tesirine, an anti-CD19 antibody-drug conjugate, in relapsed/refractory B-cell acute lymphoblastic leukemia. *Blood Adv.* 4 (3): 449–457.

52 Carlo-Stella, C., Luigi Zinzani, P., Kahl, B. et al. (2020). Initial results of a phase 2 study of loncastuximab tesirine, a novel pyrrolobenzodiazepine-based antibody-drug conjugate, in patients with relapsed or refractory diffuse large B-cell lymphoma. EHA Library. 12 June 2020; 295053; S233 [Abstract].

53 Zammarchi, F., Reinert, H.W., Janghra, N. et al. (2017). Abstract 52: mechanistic and benchmarking studies of ADCT-502, a pyrrolobenzodiazepine (PBD) dimer-containing antibody-drug conjugate (ADC) targeting HER2-expressing solid tumors. *Cancer Res.* 77 (13 Supplement): 52.

54 Bloomberg (2018). ADC Therapeutics announces the termination of its ADCT-502 program targeting HER2 expressing solid tumors. https://www.bloomberg.com/press-releases/2018-04-25/adc-therapeutics-announces-the-termination-of-its-adct-502-program-targeting-her2-expressing-solid-tumors (accessed 4 October 2021).

55 ADC Therapeutics (2021) ADCT-602 targeting CD22. https://www.adctherapeutics.com/our-pipeline1-1/adct-602/ (accessed 4 October 2021).

56 Rudin, C.M., Pietanza, M.C., Bauer, T.M. et al. (2017). Rovalpituzumab tesirine, a DLL3-targeted antibody-drug conjugate, in recurrent small-cell lung cancer: a first-in-human, first-in-class, open-label, phase 1 study. *Lancet Oncol.* 18 (1): 42–51.

57 Morgensztern, D., Besse, B., Greillier, L. et al. (2019). Efficacy and safety of rovalpituzumab tesirine in third-line and beyond patients with DLL3-expressing, relapsed/refractory small-cell lung cancer: results from the phase II TRINITY study. *Clin. Cancer Res.* 25 (23): 6958–6966.

58 Spigel, D., Pietanza, M.C., Bauer, T. et al. (2017). OA05.03 single-agent rovalpituzumab tesirine, a delta-like protein 3 (DLL3)-targeted antibody-drug conjugate (ADC), in small-cell lung cancer (SCLC). *J. Thorac. Oncol.* 12: S260–S261.

59 AbbVie (2019). AbbVie discontinues rovalpituzumab tesirine (Rova-T) research and development program. https://news.abbvie.com/news/press-releases/abbvie-discontinues-rovalpituzumab-tesirine-rova-t-research-and-development-program.htm (accessed 2 October 2021).

60 AbbVie (2018). Phase 3 trial of Rova-T as second-line therapy for advanced small-cell lung cancer (TAHOE study) halted. https://news.abbvie.com/news/press-releases/phase-3-trial-rova-t-as-second-line-therapy-for-advanced-small-cell-lung-cancer-tahoe-study-halted.htm (accessed 2 October 2021).

61 Tolcher, A.W., Falchook, G., Bendell, J.C. et al. (2019). A phase 1, open-label, dose-escalation study to evaluate the safety, tolerability, pharmacokinetics, and antitumor activity of ADCT-601 in patients with advanced solid tumors. https://www.adctherapeutics.com/wp-content/uploads/2020/09/ESMO-2019-Poster.pdf (accessed 4 October 2021).

62 (a) Kinneer, K., Meekin, J.H., Varkey, R. et al. (2017). Preclinical evaluation of MEDI2228, a BCMA-targeting pyrrolobenzodiazepine-linked antibody drug conjugate for the treatment of multiple myeloma. *Blood* 130 (Supplement 1): 3153–3153. (b) Tai, Y.-T., Xing, L., Lin, L. et al. (2019). MEDI2228, a novel BCMA pyrrolobenzodiazepine antibody drug conjugate, overcomes drug resistance and synergizes with bortezomib and DNA damage response inhibitors in multiple myeloma. *Clin. Lymphoma Myeloma Leuk.* 19 (10): e154–e155. (c) Kinneer, K., Flynn, M., Thomas, S.B. et al. (2019). Preclinical assessment of an antibody–PBD conjugate that targets BCMA on multiple myeloma and myeloma progenitor cells. *Leukemia* 33 (3): 766–771.

63 Gymnopoulos, M., Betancourt, O., Blot, V. et al. (2020). TR1801-ADC: a highly potent cMet antibody–drug conjugate with high activity in patient-derived xenograft models of solid tumors. *Mol. Oncol.* 14 (1): 54–68.

64 Antonow, D., Kaliszczak, M., Kang, G.-D. et al. (2010). Structure–activity relationships of monomeric C2-aryl pyrrolo[2,1-*c*][1,4]benzodiazepine (PBD) antitumor agents. *J. Med. Chem.* 53 (7): 2927–2941.

65 (a) Kolakowski, R.V., Young, T.D., Howard, P.W. et al. (2015). Synthesis of a C2-aryl-pyrrolo[2,1-c][1,4]benzodiazepine monomer enabling the convergent construction of symmetrical and non-symmetrical dimeric analogs. *Tetrahedron Lett.* 56 (30): 4512–4515. (b) Howard, P.W., Gregson, S.J., and Masterson, L. (2010). Unsymmetrical pyrrolobenzodiazepine-dimers for treatment of proliferative diseases. WO2010043880. (c) Howard, P.W., Jeffrey S., Burke, P., and Senter, P. (2011). Targeted pyrrolobenzodiazapine conjugates. WO2011130613.

66 Jeffrey, S.C., Burke, P.J., Lyon, R.P. et al. (2013). A potent anti-CD70 antibody–drug conjugate combining a dimeric pyrrolobenzodiazepine drug with site-specific conjugation technology. *Bioconjugate Chem.* 24 (7): 1256–1263.

67 Kung Sutherland, M.S., Walter, R.B., Jeffrey, S.C. et al. (2013). SGN-CD33A: a novel CD33-targeting antibody–drug conjugate using a pyrrolobenzodiazepine dimer is active in models of drug-resistant AML. *Blood* 122 (8): 1455–1463. https://doi.org/10.1182/blood-2013-03-491506.

68 (a) Jilani, I., Estey, E., Huh, Y. et al. (2002). Differences in CD33 intensity between various myeloid neoplasms. *Am. J. Clin. Pathol.* 118: 560–566. https://doi.org/10.1309/1WMW-CMXX-4WN4-T55U. (b) Legrand, O., Perrot, J.Y., Baudard, M. et al. (2000). The immunophenotype of 177 adults with acute myeloid leukemia: proposal of a prognostic score. *Blood* 96 (3): 870–877. https://doi.org/10.1182/blood.V96.3.870. (c) Cowan, A.J., Laszlo, G.S., Estey, E.H., and Walter, R.B. (2013). Antibody-based therapy of acute myeloid leukemia with gemtuzumab ozogamicin. *Front. Biosci.* 18: 1311–1334. https://doi.org/10.2741/4181. (d) Walter, R.B., Appelbaum, F.R., Estey, E.H., and Bernstein, I.D. (2012). Acute myeloid leukemia stem cells and CD33-targeted immunotherapy. *Blood* 119 (26): 6198–6208. https://doi.org/10.1182/blood-2011-11-325050.

69 Stein, E.M., Walter, R.B., Erba, H.P. et al. (2018). A phase 1 trial of vadastuximab talirine as monotherapy in patients with CD33-positive acute myeloid leukemia. *Blood* 131 (4): 387–396. https://doi.org/10.1182/blood-2017-06-789800.

70 Sandall, S., Anderson, M.E., Jonas, M. et al. (2014). Abstract 2647: SGN-CD70A, a novel and highly potent anti-CD70 ADC, induces double-strand DNA breaks and is active in models of MDR+ renal cell carcinoma (RCC) and non-Hodgkin lymphoma (NHL). *Cancer Res.* 74 (19 Supplement): 2647. https://doi.org/10.1158/1538-7445.AM2014-2647.

71 Phillips, T., Barr, P.M., Park, S.I. et al. (2019). A phase 1 trial of SGN-CD70A in patients with CD70-positive diffuse large B cell lymphoma and mantle cell lymphoma. *Invest. New Drugs* 37 (2): 297–306. https://doi.org/10.1007/s10637-018-0655-0.

72 Ryan, M.C., Schimpf, B., Anderson, M. et al. (2015). SGN-CD19B, a pyrrolobenzodiazepine (PBD)-based anti-CD19 drug conjugate, demonstrates potent preclinical activity against B-cell malignancies. *Blood* 126 (23): 594. https://doi.org/10.1182/blood.V126.23.594.594.

73 Ryan, M.C., Palanca-Wessels, M.C., Schimpf, B. et al. (2017). Therapeutic potential of SGN-CD19B, a PBD-based anti-CD19 drug conjugate, for treatment of B-cell malignancies. *Blood* 130 (18): 2018–2026. https://doi.org/10.1182/blood-2017-04-779389.

74 Li, F., Sutherland, M.K., Yu, C. et al. (2018). Therapeutic potential of SGN-CD19B, a PBD-based anti-CD19 drug conjugate, for treatment of B-cell malignancies. *Mol. Cancer Ther.* 17 (2): 554–564. https://doi.org/10.1158/1535-7163.MCT-17-0742.

75 Lewis, T., Olson, D., Gordon, K.A. et al. (2016). Abstract 1195: SGN-CD352A: a novel humanized anti-CD352 antibody-drug conjugate for the treatment of multiple myeloma. *Cancer Res.* 76 (14 Supplement): 1195. https://doi.org/10.1158/1538-7445.AM2016-1195.

76 Lemech, C., Woodward, N., Chan, N. et al. (2020). A first-in-human, phase 1, dose-escalation study of ABBV-176, an antibody-drug conjugate targeting the prolactin receptor, in patients with advanced solid tumors. *Invest. New Drugs* https://doi.org/10.1007/s10637-020-00960-z.

77 (a) Saltz, L.B., Meropol, N.J., Loehrer, P.J. Sr. et al. (2004). Phase II trial of cetuximab in patients with refractory colorectal cancer that expresses the epidermal growth factor receptor. *J. Clin. Oncol.* 22 (7): 1201–1208. (b) Cunningham, D., Humblet, Y., Siena, S. et al. (2004). Cetuximab monotherapy and cetuximab plus irinotecan in irinotecan-refractory metastatic colorectal cancer. *N. Engl. J. Med.* 351 (4): 337–345. (c) Van Cutsem, E., Peeters, M., Siena, S. et al. (2007). Open-label phase III trial of panitumumab plus best supportive care compared with best supportive care alone in patients with chemotherapy-refractory metastatic colorectal cancer. *J. Clin. Oncol.* 25 (13): 1658–1664. (d) Jonker, D.J., O'Callaghan, C.J., Karapetis, C.S. et al. (2007). Cetuximab for the treatment of colorectal cancer. *N. Engl. J. Med.* 357 (20): 2040–2048. (e) Vermorken, J.B., Mesia, R., Rivera, F. et al. (2008). Platinum-based chemotherapy plus cetuximab in head and neck cancer. *N. Engl. J. Med.* 359 (11): 1116–1127. (f) Van Cutsem, E., Kohne, C.H., Lang, I. et al. (2011). Cetuximab plus irinotecan, fluorouracil, and leucovorin as first-line treatment for metastatic colorectal cancer: updated analysis of overall survival according to tumor KRAS and BRAF mutation status. *J. Clin. Oncol.* 29 (15): 2011–2019. (g) Shepherd, F.A., Rodrigues Pereira, J., Ciuleanu, T. et al. (2005). Erlotinib in previously treated non-small-cell lung cancer. *N. Engl. J. Med.* 353 (2): 123–132. (h) Moore, M.J., Goldstein, D., Hamm, J. et al. (2007). Erlotinib plus gemcitabine compared with gemcitabine alone in patients with advanced pancreatic cancer: a phase III trial of the National Cancer Institute of Canada Clinical Trials Group. *J. Clin. Oncol.* 25 (15): 1960–1966.

78 Anderson, M.G., Falls, H.D., Mitten, M.J. et al. (2020). Targeting multiple EGFR expressing tumors with a highly potent tumor-selective antibody drug conjugate. *Mol. Cancer Ther.* https://doi.org/10.1158/1535-7163.MCT-20-0149.

79 Hartley, J.A. (2021). Antibody-drug conjugates (ADCs) delivering pyrrolobenzodiazepine (PBD) dimers for cancer therapy. *Expert Opin. Biol. Ther.* 21 (7): 931–943. https://doi.org/10.1080/14712598.2020.1776255.

80 Archer, K.E., Reid, E.E., Shizuka, M. et al. (2019). Synthesis of highly potent N-10 amino-linked DNA-alkylating indolinobenzodiazepine antibody–drug conjugates (ADCs). *ACS Med. Chem. Lett.* 10 (8): 1211–1215. https://doi.org/10.1021/acsmedchemlett.9b00254.

81 (a) Miller, M.L., Fishkin, N.E., Li, W. et al. (2016). A new class of antibody–drug conjugates with potent DNA alkylating activity. *Mol. Cancer Ther.* 15 (8): 1870–1878. https://doi.org/10.1158/1535-7163.MCT-16-0184. (b) Chari Ravi, V.J., Miller, M.L., and Shizuka, M. (2016). Cytotoxic benzodiazepine derivatives. WO2016036801.

82 (a) Bai, C., Reid, E.E., Wilhelm, A. et al. (2020). Site-specific conjugation of the indolinobenzodiazepine DGN549 to antibodies affords antibody–drug conjugates with an improved therapeutic index as compared with lysine conjugation. *Bioconjugate Chem.* 31 (1): 93–103. https://doi.org/10.1021/acs.bioconjchem.9b00777. (b) Reid, E.E., Archer, K.E., and Shizuka, M. (2019). Effect of linker stereochemistry on the activity of indolinobenzodiazepine containing antibody–drug conjugates (ADCs). *ACS Med. Chem. Lett.* 10 (8): 1193–1197. https://doi.org/10.1021/acsmedchemlett.9b00240.

83 Miller, M.L., Shizuka, M., Wilhelm, A. et al. (2018). A DNA-interacting payload designed to eliminate cross-linking improves the therapeutic index of antibody–drug conjugates (ADCs). *Mol. Cancer Ther.* 17 (3): 650–660. https://doi.org/10.1158/1535-7163.MCT-17-0940.

84 Kovtun, Y., Noordhuis, P., Whiteman, K.R. et al. (2018). IMGN779, a novel CD33-targeting antibody–drug conjugate with DNA-alkylating activity, exhibits potent antitumor activity in models of AML. *Mol. Cancer Ther.* 17 (6): 1271–1279. https://doi.org/10.1158/1535-7163.MCT-17-1077.

85 (a) Cortes, J.E., DeAngelo, D.J., Erba, H.P. et al. (2018). Maturing clinical profile of IMGN779, a next-generation CD33-targeting antibody-drug conjugate, in patients with relapsed or refractory acute myeloid leukemia. *Blood* 132 (Supplement 1): 26. https://doi.org/10.1182/blood-2018-99-112891. (b)ImmunoGen (2018). ImmunoGen presents new data highlighting potential of novel ADCs in oral presentations at ASH annual meeting. https://investor.immunogen.com/news-releases/news-release-details/immunogen-presents-new-data-highlighting-potential-novel-adcs (accessed 2 October 2021).

86 (a) Kovtun, Y., Jones, G.E., Adams, S. et al. (2018). A CD123-targeting antibody-drug conjugate, IMGN632, designed to eradicate AML while sparing normal bone marrow cells. *Blood Adv.* 2 (8): 848–858. https://doi.org/10.1182/bloodadvances.2018017517. (b) Angelova, E., Audette, C., Kovtun, Y. et al. (2019). CD123 expression patterns and selective targeting with a CD123-targeted antibody-drug conjugate (IMGN632) in acute lymphoblastic leukemia. *Haematologica* 104 (4): 749–755. https://doi.org/10.3324/haematol.2018.205252.

87 Daver, N.G., Montesinos, P., DeAngelo, D.J. et al. (2019). Clinical profile of IMGN632, a novel CD123-targeting antibody-drug conjugate (ADC), in patients with relapsed/refractory (R/R) acute myeloid leukemia (AML) or blastic plasmacytoid dendritic cell neoplasm (BPDCN). *Blood* 134 (Supplement_1): 734. https://doi.org/10.1182/blood-2019-128648.

88 Abu-Yousif, A.O., Bannerman, B.M., Cvet, D. et al. (2018). Abstract B120: TAK-164, a GCC-targeted antibody-drug conjugate (ADC) for the treatment of colorectal cancers and other GI malignancies. *Mol. Cancer Ther.* 17 (1 Supplement): B120–B120. https://doi.org/10.1158/1535-7163.TARG-17-B120.

89 Koenig, E.M., Li, C., Yang, H. et al. (2018). Abstract 3916: relationship of guanylyl cyclase C (GCC) expression and efficacy of TAK-164, a GCC-targeted antibody-drug conjugate in a panel of 68 subcutaneous HuPrime colorectal cancer PDX models. *Cancer Res.* 78 (13 Supplement): 3916. https://doi.org/10.1158/1538-7445.AM2018-3916.

90 NCI (n.d.). NCI Drug Dictionary. https://www.cancer.gov/publications/dictionaries (accessed 31 January 2019).

91 (a) Rinnerthaler, G., Gampenrieder, S.P., and Greil, R. (2019). HER2 directed antibody-drug-conjugates beyond T-DM1 in breast cancer. *Int. J. Mol. Sci.* 20 (5): 1115. https://doi.org/10.3390/ijms20051115; (b) De Santis, R. (2020). Anti-ErbB2 immunotherapeutics: struggling to make better antibodies for cancer therapy. *mAbs* 12 (1): 1725346. https://doi.org/10.1080/19420862.2020.1725346.

92 *ADC Review* (2019). Study demonstrates superior efficacy and cytotoxicity of femtogenix's sequence-selective DNA-interactive payload. https://www.adcreview.com/news/study-demonstrates-superior-efficacy-and-cytotoxicity-of-femtogenixs-sequence-selective-dna-interactive-payload (accessed 2 October 2021).

93 Jackson, P.J.M., Thurston, D.E., and Rahman, K.M. (2018). PDD compounds. US20180291021A1.

94 Hoffmann, R.M., Crescioli, S., Mele, S. et al. (2020). A novel antibody-drug conjugate (ADC) delivering a DNA mono-alkylating payload to chondroitin sulfate proteoglycan (CSPG4)-expressing melanoma. *Cancers* 12 (4): 1029.

95 Markandeya, N., Shankaraiah, N., Reddy, C.S. et al. (2010). Asymmetric syntheses of piperidino-benzodiazepines through 'cation-pool' host/guest supramolecular approach and their DNA-binding studies. *Tetrahedron: Asymmetry* 21 (21): 2625–2630.

96 Smith, S.W., Jammalamadaka, V., Borkin, D. et al. (2018). Design and synthesis of isoquinolidinobenzodiazepine dimers, a novel class of antibody–drug conjugate payload. *ACS Med. Chem. Lett.* 9 (1): 56–60.

97 Zhou, Q., Duan, W., Simmons, D. et al. (2001). Design and synthesis of a novel DNA—DNA interstrand adenine—guanine cross-linking agent. *J. Am. Chem. Soc.* 123 (20): 4865–4866.

98 Jackson, P.J.M., Rahman, K.M., and Thurston, D.E. (2017). The use of molecular dynamcis simulations to evaluate the DNA sequence-selectivity of G–A cross-linking PBD–duocarmycin dimers. *Bioorg. Med. Chem. Lett.* 27 (1): 102–108.

99 (a) Flygare, J.A., Pillow, T.H., Safina, B. et al. (2015). 1-[Chloromethyl]-2,3-dihydro-1H-benzo[*e*]indole dimer antibody-drug conjugate compounds, and methods of use and treatment. WO2015023355; (b) Zhang, D., Le, H., Cruz-Chuh, J.D. et al. (2018). Immolation of p-aminobenzyl ether linker and payload potency and stability determine the cell-killing activity of antibody-drug conjugates with phenol-containing payloads. *Bioconjugate Chem.* 29 (2): 267–274. (c) Pillow, H.T. and Tercel, M. (2019). Duocarmycin–PBD dimers as antibody–drug conjugate (ADC) payloads. In: *Cytotoxic Payloads for Antibody–Drug Conjugates* (eds. D.E. Thurston and P.J.M. Jackson), 241–258. The Royal Society of Chemistry.

100 Chen, X., Dennis, M., Junutula, J.R. et al. (2016). Anti-HER2 antibodies and immunoconjugates. WO2016040723.

101 (a) Jackson, P.J.M., Thurston, D.E., and Rahman, K.M. (2017). Asymmetric conjugate compounds. WO2017194960; (b) Junutula, J.R., Smith, S.W., Borkin, D., and Ghone, S. (2018). Isoquinolidinobenzodiazepine (iqb)-1(chloromethyl)-2,3-dihydro-1h-benzo[e]indole (cbi) dimers. WO2018071455.

102 Zhao, R.Y. (2015). Novel cytotoxic agents for conjugation of drugs to cell binding molecule. WO2015/028850.

103 (a) Vlahov, I.R., Leamon, C.P., Qi, L. et al. (2017). PBD conjugates for treating diseases. WO2017172930; (b) Vlahov, I.R., Leamon, C.P., Qi, L. et al. (2018). Folate conjugate for use in targeting tumor associated macrophages. WO2018182776; (c) Vlahov, I.R., Kleindl, P.J., Felten, A., and Leamon, C.P. (2019). Pro-pyrrolobenzodiazepines and conjugates as antibody–drug conjugate

(ADC) payloads. In: *Cytotoxic Payloads for Antibody–Drug Conjugates* (eds. D.E. Thurston and P.J.M. Jackson), 279–295. The Royal Society of Chemistry.

104 Vlahov, I.R., Qi, L., Kleindl, P.J. et al. (2017). Latent warheads for targeted cancer therapy: design and synthesis of pro-pyrrolobenzodiazepines and conjugates. *Bioconjugate Chem.* 28 (12): 2921–2931. https://doi.org/10.1021/acs.bioconjchem.7b00476.

105 Reddy, J.A., Nelson, M., Dircksen, C. et al. (2020). Pre-clinical studies of EC2629, a highly potent folate- receptor-targeted DNA crosslinking agent. *Sci. Rep.*: 12772. https://doi.org/10.1038/s41598-020-69682-9.

106 (a)Vlahov, I.R., Felten, A., Zou, N. et al. (2018). MEDI 426, pro-pyrrolobenzodiazepine (pro-PBD) bioconjugates, part 4: Design of novel oxime-based pro-PBD conjugates that release active drug via intramolecular diazepine-ring-closure. 256th ACS National Meeting & Exposition, Boston, MA; (b) Vlahov, I.R., Felten, A.E., Zou, N. et al. (2019) MEDI 104, pro-pyrrolobenzodiazepine (pro-PBD) bioconjugates, part 5: Design and synthesis of bis-pro-PBD conjugates containing a self-immolative linkers that release active drug via intramolecular diazepine-ring-closure. 258th ACS National Meeting & Exposition, San Diego, CA.

107 Polson, A.G., Ho, W.Y., and Ramakrishnan, V. (2011). Investigational antibody-drug conjugates for hematological malignancies. *Expert Opin. Investig. Drugs* 20: 75.

108 Muller, P.Y. and Milton, M.N. (2012). The determination and interpretation of the therapeutic index in drug development. *Nat. Rev. Drug Discovery* 11 (10): 751–761.

109 Gregson, S.J., Barrett, A.M., Patel, N.V. et al. (2019). Synthesis and evaluation of pyrrolobenzodiazepine dimer antibody-drug conjugates with dual β-glucuronide and dipeptide triggers. *Eur. J. Med. Chem.* 179: 591–607.

110 Jackson, P.J.M., Kay, S., Pysz, I., and Thurston, D.E. (2018). Use of pyrrolobenzodiazepines and related covalent-binding DNA-interactive molecules as ADC payloads: is mechanism related to systemic toxicity? *Drug Discov. Today Technol.* 30: 71–83. https://doi.org/10.1016/j.ddtec.2018.10.004.

111 (a) Hamilton, E., O'Malley, D.M., O'Cearbhaill, R. et al. (2020). Tamrintamab pamozirine (SC-003) in patients with platinum-resistant/refractory ovarian cancer: findings of a phase 1 study. *Gynecol. Oncol.* 158 (3): 640–645. (b) Lamber, S., Zhang, C., Englert, S. et al. (2020).Pharmacodynamic and predictive biomarker associated with observed clinic responses on patients with HNSCC and NSCLC. AACR Annual Meeting, Philadelphia, PA; (c) Sutherland, M.S.K., Yu, C., Anderson, M. et al. (2014). SGN-CD33A in combination with cytarabine or hypomethylating agents demonstrates enhanced anti-leukemic activity in preclinical models of AML. *Blood* 124 (21): 3739–3739. (d) Scripture, C., Selvaggi, G., Lakatos, I. et al. (2017). 1538TiP – a phase 1/2 study on safety of rovalpituzumab tesirine in combination with nivolumab or nivolumab + ipilimumab in small cell lung cancer. *Ann. Oncol.* 28: v542. (e) Moskowitz, C.H., Bastos-Oreiro, M., Ungar, D. et al. (2019). Safety and anti-tumor activity study of Loncastuximab tesirine and durvalumab in diffuse large B-cell, mantle cell, or follicular lymphoma. *Blood* 134 (Supplement_1): 2807.

14

Combination Therapy Case Studies in Anticancer and Anti-Infectious Disease Drug Discovery and Development

Xianhai Huang[1] *and David Yu-Kai Chen*[2]

[1] *Discovery Chemistry, InventisBio Co., Ltd., 25A Hanover Road #340, NJ 07932, Florham Park, NJ, USA*
[2] *Department of Chemistry, Seoul National University, 1 Gwanak-ro, 151-742, Gwanak Gu, Seoul, South Korea*

14.1 Introduction

The human body is an extremely complex system consisting of increasing levels of organization from molecules, organelles, cells, tissues, organs, to organ systems. These different levels of organization are interconnected and work together seamlessly to make the body function properly. Even the smallest change at the molecular level may result in dramatic effects at higher levels. Due to this complexity, combined with the differences between individuals at the genetic level, finding cures for human diseases is extraordinarily challenging. The complex interconnectivity at the molecular level of human physiology may require applications of combination therapies in the treatment of life-threatening human diseases since single agent targeted therapies may no longer be sufficient. Combination or multicomponent therapy refers to the simultaneous introduction of two or more therapeutic regiments for the treatment of human diseases. Although concurrent dietary, physical and/or medicinal interventions can all be considered as combination therapies, the discussions in this chapter are focused on the combination of drug products prescribed by health professionals to achieve superior results than their monotherapy counterparts. With the recent advances in human genomics and proteomics, and the understanding of the fundamental biology behind human pathophysiological processes and modes of action (MOA) of drug molecules, combination therapies have expanded the range of therapeutic options in the treatment of many human diseases such as cancer, infectious diseases, metabolic, cardiovascular, autoimmune, and neurological disorders [1–3]. Since a large number of review articles has been published on the general discussions of combination therapy [4–6], this chapter will primarily focus on case studies in anticancer and antibacterial drug discovery. Before proceeding to more in-depth considerations, a few general remarks on combination therapies are outlined to facilitate later discussions:

1) Rationales for combination therapies [7, 8]: (i) to achieve a greater therapeutic effect compared with single drug therapy; (ii) to achieve the same therapeutic effect as single drug therapy, but with fewer dose-limiting toxicities and undesirable side effects; and (iii) to overcome drug resistance toward monotherapies.

Contemporary Accounts in Drug Discovery and Development, First Edition. Edited by Xianhai Huang, Robert G. Aslanian, and Wayne H. Tang.

2) Depending on the therapeutic outcome of the combination treatment, combination therapies can be classified into the following main categories [9]: (i) Synergistic: the combination therapy results in effects greater than the combined effects of the individual drugs. (ii) Additive: the combination therapy results in effects equal to the combined effects of the individual drugs. (iii) Antagonistic: the combination therapy results in effects less than the combined effects of the individual drugs. (iv) Potentiative: the therapeutic effect of one drug is improved by another drug through regulation of its pharmacokinetics (PK) (absorption, distribution, metabolism, excretion, and toxicology [ADMET]). (v) Reductive: the therapeutic effect of one drug is compromised by another drug through regulation of its PKs. (vi) Coalistic: the individual drugs are inactive but render their activities through a combination therapy. Although different types of combinations may provide different pharmacodynamic (PD) effects, in most cases, synergistic drug combinations are most desired in view of improved efficacy and reduced likelihood of developing resistance. On the other hand, while antagonism effects are usually not desirable due to reduced efficacy, this type of combination may provide protective measures against the future development of drug resistance [10].
3) The fundamental design principles for an effective drug combination include but are not limited to the following considerations [11]: (i) Employ drugs with established therapeutic profiles as single agents to improve their efficacies. (ii) Consider drugs with non-overlapping toxicities so that each drug can be administered at near-maximal dose. (iii) Combine agents with different mechanisms of action and minimal cross-resistance in order to reduce the likelihood of developing resistance. In addition, consider different dosing regimens of the combination with a minimal treatment-free period between cycles while allowing the recovery of sensitive target tissues [12, 13].
4) The FDA guidance for the co-development of two or more new investigational drugs for use in combination [14]: (i) The combination should be used to treat a serious disease or condition. (ii) The combination should have a strong biological rationale. (iii) The combination is superior to the individual agents and may provide a significant therapeutic advance over available therapy. (iv) There is a compelling reason why the new investigational drugs cannot be developed independently.
5) Potential disadvantages in drug combination: (i) more side effects and the difficulties to determine the cause of the side effects; (ii) development complications including fixed dose ratio and incompatible PK; and (iii) potential drug–drug interactions.

With numerous drug combinations in clinical trials, combination therapies have become a central drug development strategy and are routinely evaluated for almost every new drug and disease. With literally unlimited possibilities of drug combinations, how to strategically design effective drug combinations has become one of the most important questions in the development of successful life-saving drugs [15]. In this regard, advances in metadata modeling and artificial intelligence may play an important role in selecting the best combinations by bridging the gap between preclinical biology and clinical outcomes. This chapter will use two case studies originating from anticancer and antibacterial drug discovery efforts to showcase how combination therapies can be rationally designed to achieve desirable PD endpoints.

14.1.1 Combination Therapy in Anticancer Drug Discovery and Development

Cancer is the second leading cause of death in the United States. Based on a report from the National Cancer Institute in 2019 [16], about $150.8 billion were spent nationally for cancer care in 2018. The total cost of cancer care in 2020 is projected to be $174 billion. Millions of people are living with cancer or have had cancer, and about one-third of the people in the US will develop cancer during their lifetimes [17]. Cancer as a disease has become an enormous social burden and the need to discover more efficacious and cost-effective treatments has become increasingly more important.

Although impressive progress and extraordinary advances have been made at the fundamental scientific level to identify the best treatments against this devastating disease, challenges remain and only a few have demonstrated efficacy in diverse patient populations. Fortunately, the academic community and pharmaceutical industry are working relentlessly to come up with novel and more effective treatments. A Silicon Valley Bank (SVB) 2020 mid-year report showed that oncology remains as the top investment among all indications [18]. Currently, treatments for cancer patients include surgery, radiation, chemotherapy, hormone therapy, immunotherapy, targeted therapy, gene therapy, and combination therapy. In particular, targeted therapies and immunotherapies have made tremendous progress in recent years, with a few representative examples including the anti-estrogen agent tamoxifen for the treatment of breast cancer [19], the anti-androgen drug bicalutamide and nilutamide for the treatment of prostate cancer [20], the anti-BCR-ABL kinase inhibitor imatinib for the treatment of chronic myelogenous leukemia [21], the anti-HER2 drug afatinib for the treatment of non-small cell lung carcinoma [22], the BRAF inhibitor vemurafenib for the treatment of late stage melanoma [23], and the epidermal growth factor receptor inhibitors such as erlotinib, gefitinib, osimertinib, and cetuximab for the treatment of a variety of cancers [24]. Although the birth of immunotherapy can be traced back to more than 100 years ago, its impact has only been realized recently and it is often perceived as a relatively new technology [25]. Among the specific immunotherapies, PD-1 inhibitors such as pembrolizumab and nivolumab, PD-L1 inhibitors such as atezolizumab, and CTLA4 inhibitors such as ipilimumab have been the most well-studied in recent years. However, the percentage of patients who showed a response to single-agent checkpoint inhibitor is less than 15% despite the application of these immunotherapies to large number of cancer patients of almost all tumor types [26, 27]. There are a number of reasons why most of the current cancer therapies are either transiently effective or not as effective as predicted. Chief among these are tumor heterogeneity or resistance due to genome instability, metabolic changes, and target alterations [28–30]. In addition, most anticancer drugs employ doses close to MTD (maximum tolerated dose) in order to achieve the best efficacy which may lead to unspecific biodistribution leading to unwanted and unacceptable side effects to healthy tissue. For these reasons, combination therapy plays an increasingly important role in cancer treatment because of its potential in tackling a diverse range of cancer evolutionary and resistance pathways, with improved efficacy, reduced side effects and improved patient compliance [31]. With the advances in genomic and computational tools, and a better understanding of the underlying disease mechanisms in recent years, rationally designed drug combinations have emerged as an indispensable strategy to fight cancer [32–34].

The efficacy of an antitumor drug combination can be first evaluated in preclinical settings through *in vitro* cell assays and *in vivo* animal studies. Depending on the outcome of interest, a large panel of cell lines, broad drug combination matrices, and different culture conditions can be included. The concentration and time courses can be precisely selected and the effect of the drug combination on tumor cell growth can be readily measured. After the optimal *in vitro* combination matrix is identified, *in vivo* animal models can be employed to evaluate the safety and efficacy in tumor growth inhibition (TGI) and tumor growth regression (TGR) in preparation for clinical studies. Computational models can play a role in evaluating the effectiveness of anticancer drug combinations in the preclinical setting. Of the six main effects that can be obtained from drug combinations, three are of interest, namely, synergistic, antagonistic, and additive interactions.

Several methods for the quantitative evaluation of combination drugs especially for oncology drugs have been proposed and employed to provide important guidance in the design of *in vivo* studies based on *in vitro* dose level (concentration) and schedule (time course). A number of computational models are available to quantify *in vitro* study results in the prediction of combination effects [35]. These quantification methods can be broadly classified as the following while more are under development: (i) highest single agent (HSA) model, or Gaddum's non-interaction model [36]; (ii) Loewe additivity model (isobologram) [37]; (iii) Bliss independence model [38]; (iv) median effect analysis [39, 40]; (v) zero interaction potency (ZIP) model [35]; (vi) equivalent dose model [41]; (vii) PDE-Hill model [42]; and (i) BRAID model [43]. Although *in vivo* study results are more difficult to quantify and few such models are available, in recent years, a few additional modeling systems and machine learning techniques have been introduced based on the principles of Loewe and Bliss in the prediction of drug combination toxicity and efficacy *in vivo* [44–48]. For all these models, isobologram analysis provides a visual picture of drug combination effects (synergic, additive, or antagonistic) in both two-dimensional (Figure 14.1) and three-dimensional formats (Figure 14.2). For example, SynergyFinder is very useful software for evaluating high-throughput drug combination scores in a three-dimensional fashion, for all the popular synergy scoring models, including HSA, Loewe, Bliss, and ZIP. This software provides the users a convenient tool to examine their drug combination data and interpret the drug combination results [49–52].

Although preclinical evaluation of drug combinations provides extremely useful information on the combination effects, whether a synergic combination will translate into a clinical benefit remains an important question. In fact, the majority of clinical protocols for cancer drug combinations are still derived empirically based on results obtained from analysis of clinical trials instead of preclinical experimental data. Multiple factors may have contributed to this disconnection. How drugs perform in humans in a clinical setting is very different from preclinical *in vitro* assays and *in vivo* animal studies. Tumor heterogeneity, PK profiles, toxicity related adverse events (AEs), and development of resistance can all play a role in reducing the value of preclinical drug combination studies. The translational challenges of drug combinations from promising preclinical results to unpredictable clinical outcomes remain an area of focus in the quest for discovering lifesaving combination drugs.

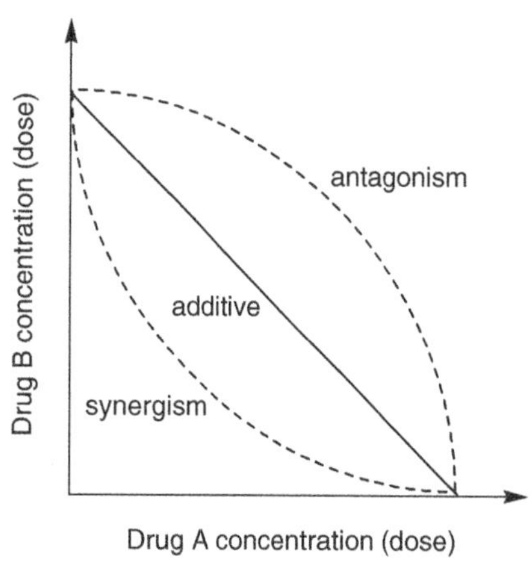

Figure 14.1 Two-dimensional isobologram based on Loewe's additivity model showing additive, synergistic, and antagonistic two-drug combinations. If the combination effect falls within the two dashed lines, the combination is additive; if below the lower isobologram, the combination is synergistic; if above the higher isobologram, the combination is antagonistic. When the drugs' combined effect lies on the linear additive isobole (solid line), they are noninteracting.

14.1.2 Combination Therapy in Antibacterial Drug Discovery and Development

Humanity has a long-standing history in the battle against bacterial infections, and the landmark discovery of penicillin and sulfonamide antibiotics paved the way for antibacterial drug discovery in the early twentieth century. More than 200 antibacterial drugs have been discovered including but not limited to tetracyclines, macrolides, aminoglycosides, and fluoroquinolones. Unfortunately, even with treatment options available from such diverse chemical classes, bacterial

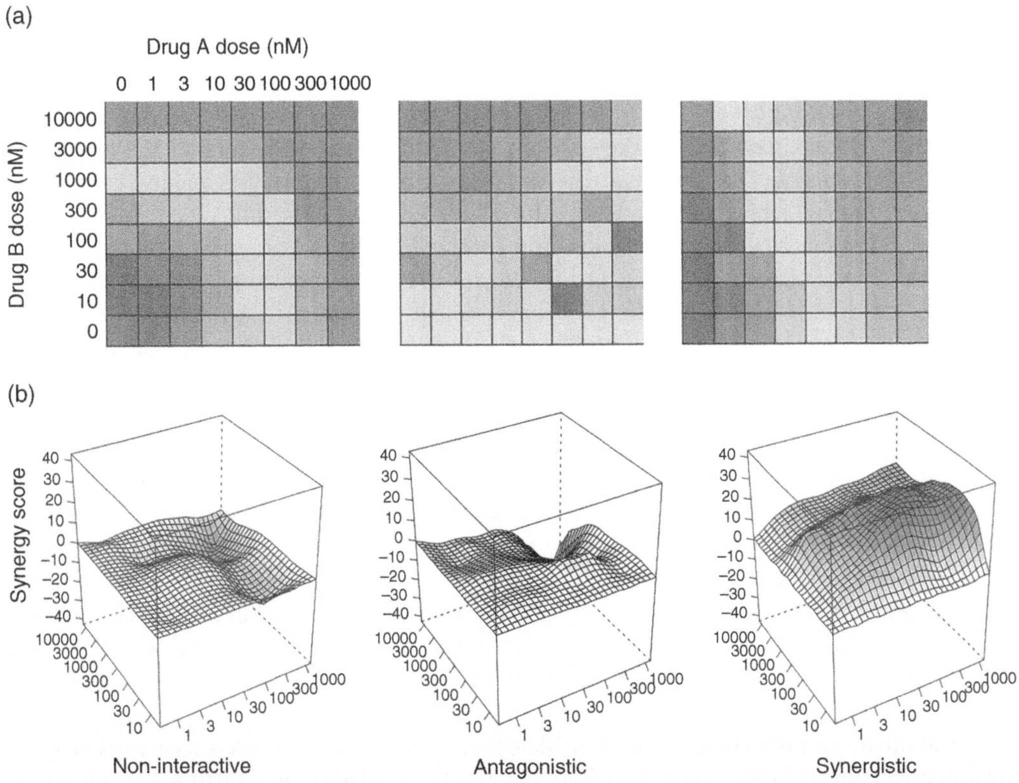

Figure 14.2 Three-dimensional isobologram illustrating drug combination data analysis. (a) A typical high-throughput drug combination screen utilizes a dose–response matrix design where all possible dose combinations for a drug pair can be tested. Shades in the dose–response matrices indicate different levels of phenotypic responses of the cancer cell. (b) Depending on the interaction pattern models derived from the dose–response matrices, a drug combination can be classified as non-interactive, antagonistic, or synergistic. *Source:* Reproduced with permission. Copyright© 2018, Springer Nature [49].

infections remain an ever-increasing threat to human health. The ready availability and often inappropriate use of broad-spectrum antibiotics has led to the emergence of multidrug-resistant Gram-negative organisms (MDRGNs), which represents a serious increased threat of mortality particularly to hospitalized patients. The rampant spread of extended-spectrum β-lactamase (ESBL)-, AmpC β-lactamases-, and carbapenemase-producing Gram-negative bacteria including multidrug resistance bacterial superbugs, i.e. *Enterococcus faecium*, *Staphylococcus aureus*, *Klebsiella pneumoniae*, *Acinetobacter baumanii*, *Pseudomonas aeruginosa*, and *Enterobacter* species (termed as "ESKAPE") have significantly increased the use of broad-spectrum antibiotics and an endless cycle of cause and effect [53, 54]. According to the 2019 AR Threats Report by the US CDC, more than 2.8 million antibiotic-resistant infections occur in the United States each year resulting in the loss of more than 35 000 lives. In 2017, 223 900 cases of *Clostridioides difficile* contributed to at least 12 800 casualties [55]. The annual expenditure associated with the treatment of multidrug resistant (MDR) bacterial infections can reach up to $35 billion in just the US alone, and this enormous social burden places a strong emphasis on discovering new and effective antibiotics. However, nearly 50 years since the golden era of antibiotics (~1940–1960), no new classes of antibacterial drugs were approved against carbapenem-resistant *Enterobacteriaceae*

(CRE). This highlighted the intrinsic difficulties associated with antibacterial drug discovery and its unfavorable return of investment for pharmaceutical companies. In contrast, public and philanthropic investment in the development of antibacterial agents has increased in recent years [56].

Due to the reduced productivity in the discovery of novel antibacterial agents and the substantial possibility of developing resistance associated with single entity antibiotics, combination therapy becomes an essential strategy to combat MDR, extensively drug resistant (XDR), and pan-drug resistant (PDR) bacterial infections. Increased bactericidal activity and suppressing or delaying the emergence of resistance can be attained by the concurrent administration of multiple antibiotics with different profiles, either a broader spectrum of activity or different mechanisms of action, or both [57, 58]. Eight reasons were proposed for antibacterial drug combinations decades ago which remain relevant today [59]: (i) Decreased toxicity without compromising efficacy. (ii) Synergy: For antimicrobial agents, a greater-than-2-log increase in bactericidal activity *in vitro* compared with the bactericidal activity of each agent alone is considered to be synergistic; (iii) "Shotgun therapy" strategy: there is no time to be wrong for initial emergency treatment of seriously ill patients. (iv) To prevent and attack mutant bacteria: the second antibiotic may delay emergence of resistant bacteria and the effect of the active agents can be prolonged. (v) Mixed infection with each microorganism requiring a different drug. (vi) To prevent superinfection by new bacteria. (vii) To attack infection in non-susceptible populations. (viii) To reach otherwise inaccessible organisms. Among these, synergy with less toxicity and overcoming resistance remain the top reasons for antibacterial drug combination. Just like anticancer drug combinations, the three main effects of antibacterial drug combination (synergism, antagonism, and additivity) can be evaluated with models such as Loewe additivity and Bliss independence models [60]. However, these models have not been frequently used since the direct measure of MIC_{50} and MIC_{90} values (where MIC is the minimum inhibitory concentration) could be more indicative of the effect of antibiotic combinations. In addition, there are some key differences between antibacterial drug combinations and other human disease drug combinations due to the faster development of resistance associated with the use of antibacterial agents. Unlike most human diseases, the idea for treating bacterial infections, and in fact all infectious diseases, with a combination of agents is unique in that the drug target is not the host, but evolutionarily unrelated bacteria. Thus, the goal of antibacterial research is to identify drugs and combinations of drugs that have maximum efficacy with a minimum chance for emergence of resistance and minimum human toxicity [61]. In this regard, a few factors need to be considered in order to discover and develop the ideal antibacterial drugs and drug combinations [62]: (i) Antibacterials should target biological processes that are exclusive to the microbe and not the host to minimize the potential for on-target toxicity. The drugs should spare the native gut flora if possible. (ii) The safety profile of the administered drugs should be pristine since high plasma concentrations of antibiotics administered over a short period of time are often required for efficacy. (iii) Unlike drugs targeting human tissue where near or compete inhibition of the target may not be necessary, effective antibacterial compounds should prevent bacterial regrowth in addition to exhibiting rapid and efficient bactericidal or bacteriostatic activity. (iv) Since antibacterial compounds have to penetrate not only the host's tissues, but also the bacterial defenses, the molecules may have to incorporate specific physicochemical characteristics in order to be effective. (v) Drug resistance can develop more rapidly during treatment of an infection due to rapid, random variations that occur during DNA replication of pathogenic bacteria. These considerations have put very stringent chemical, biological and pharmacological prerequisites on the discovery of new classes of effective antibiotics. Due to these attributes, the development of successful antibiotic drug combinations is intrinsically different from traditional drug discovery and development, and this brings additional challenges in the discovery of effective and powerful antibiotic combinations compared with other drug classes. For the same reasons, the translational challenges of antibacterial drug discovery could be significantly higher, and good *in vitro* combination synergy in bactericidal activity might not be translated into good efficacy in humans. This is not to say preclinical combination studies are not needed; in fact, these studies provide valuable information for *in vivo* animal studies and generate important combination matrix and MIC values for clinical considerations [63]. These challenges may be addressed via knowledge-based rationale design of antibacterial combinations that will be more effective at treating life-threatening infections, particularly those found in a health-care setting, than traditional monotherapies. Based on the mode of action of the individual molecules, there are three different potential combinations that can be envisioned: combinations based on drugs in which both components inhibit cell growth of the target organism; combinations wherein at least one component does not have antibiotic activity; combinations of drugs in which the individual components are inactive but show antibacterial activity when combined [64]. The ceftazidime and avibactam combination (Avycaz®) represents such a rationally designed drug combination and is employed as a case study to illustrate how effective combination treatment options can be identified through rational design based on a fundamental understanding of the underlying biology.

14.2 Case Study of Olaparib (Lynparza®) and Bevacizumab (Avastin®) Combination in the Treatment of Advanced Ovarian Cancer

14.2.1 Discovery and Development History of Olaparib and Bevacizumab in the Treatment of Ovarian Cancer

Ovarian cancer (OC) is the fifth most common cancer in women in developed countries worldwide and there is still no cure for advanced OC although medical advancements have helped to improve progression free survival rates dramatically. According to the American Cancer Society Key Statistics for Ovarian Cancer page, it is estimated that about 21 750 American women will be diagnosed with OC and 13 940 women will die from the disease in 2020 [65]. OC is usually diagnosed at more advanced stages since early-stage OC is commonly mistakenly treated as benign ailments due to its mild symptoms. The five-year survival rate of stages 1–3 OC is 90, 70, and 39%, respectively, and only 17% for metastatic stage 4 patients. The four most common types of OCs are serous carcinoma with up to 52% of cases, clear cell carcinoma, mucinous carcinoma, and endometrioid carcinoma, with surgery and chemotherapy as the standard treatment options. Although OC patients are typically very responsive toward chemotherapy, relapse often occurs followed by disease progression within one year of treatment. Even for platinum-sensitive cases, defined as the cancer relapse more than six months after the end of treatment, the disease-free period decreases with each round of treatment [66–68]. Thus, OC treatment remains as an important unmet medical need and the clinical outcomes need to be further improved. To achieve this goal, the pharmaceutical industry has worked relentlessly to discover more efficacious new drugs such as targeted cancer therapies. Among these, two classes of drugs have become standard of care for OC treatment: PARP inhibitors including olaparib (brand name Lynparza) and antiangiogenics such as bevacizumab (brand name Avastin). The combination of these two drugs provides additional options for the treatment of OC and its discovery and development will be discussed in this chapter.

14.2.1.1 Discovery and Development History of Olaparib in the Treatment of Ovarian Cancer

For all living organisms, unwanted lesions can result from deleterious attacks by exogenous agents such as radiation as well as endogenous sources such as reactive oxygen species. One of the most harmful lesions found in genomic DNA is double-strand breaks (DSBs). A complex DNA damage response (DDR) signal transduction network is employed by cells to neutralize DNA lesions and prevent the transmission of incorrect genetic information to daughter cells during cell division. In normal cells, the DDR can coordinate cell cycle arrest, DNA repair and apoptosis through the combined actions of sensors, signal transducers, and effector proteins. In cancer cells, DDR pathways which have been compromised or impaired may lead to erratic growth and tumor formation [69–71].

Based on the damage and/or cell cycle phase, the DDR pathway recruits different proteins. Sensor proteins, like MRE11-RAD50-NSB1 (MRN) or poly(ADP-ribose) polymerase 1 (PARP1), detect the damage first, and these proteins then recruit two kinases, ataxia telangiectasia mutated (ATM) or Rad3-related (ATR). ATM and ATR phosphorylate and activate the DDR effector proteins (e.g. checkpoint kinase 1 [CHK1], checkpoint kinase 2 [CHK2] or p. 53) which leads to repair, cell cycle arrest, or apoptosis. Dysregulation of this process can lead to cancer, so targeting the DDR process promotes cell death and limits the progression of the cancer. As part of the DDR pathway, PARP enzymes have very important functions in DNA repair and contribute to cell survival after DNA damage [69]. Out of the 17-membered PARP protein family, PARP1 and PARP2 are the best studied enzymes that are involved in DNA repair [70]. DNA damage leads to PARP activation and synthesis of poly(ADP)-ribose (pADPr). This is one of the earliest steps in the recognition and subsequent signaling of DNA damage in mammalian cells [71, 72]. In a pADPr-dependent manner, molecules are then recruited to sites of DNA damage. The catalytic function of PARP is blocked by PARP inhibitors which compete for the substrate binding site. Base excision repair (BER) is blocked which leads to the accumulation of unrepaired single-stranded breaks (SSBs). Replication fork collapse causes SSBs to develop into double-stranded breaks (DSBs) in replicating cells (Figure 14.3). There are two mechanisms for repair of DSBs – homologous recombination (HR) and non-homologous end-joining (NHEJ). Unlike HR, which is a precise repair mechanism required by DSBs for cell survival, NHEJ is prone to errors and is a quick fix operating throughout the cell cycle. NHEJ errors can cause translocations and telomere fusion which are hallmarks for cancer cells. A functional HR mechanism is coded for by two genes – BRCA1 and BRCA2 [73]. However, when the HR repair pathway malfunctions due to a mutation, DSBs are repaired by NHEJ which leads to a predisposition for cancer. Interestingly, neither the HR nor BER pathway are necessary for cell viability, but BRCA-deficient cancer cells in the presence of PARP inhibitors are not viable due to a combination of non-functional BER repair and the inability of the cells to undergo normal HR repair. This selective targeting of BRCA-deficient cancer cells by PARP inhibitors represents an important concept of synthetic lethality: when a combination of deficiencies in the expression of two or more genes or proteins leads to cell death, whereas a cell with deficiency in only one of these genes or proteins is viable (Figure 14.3) [74–76].

Up to half of the high-grade serous OCs could be deficient in HR as suggested by The Cancer Genome Atlas (TCGA). Deficiency of either BRCA1 or BRCA2 can be a result of germline mutation, somatic mutation or epigenetic silencing of BRCA1 [77]. Compelling preclinical data demonstrated that BRCA-deficient cells are 1000-times more sensitive to PARP inhibitors than wild type cells. Due to these reasons, PARP inhibition becomes a viable strategy for treating OC patients with BRCA-deficiency [74, 75, 78].

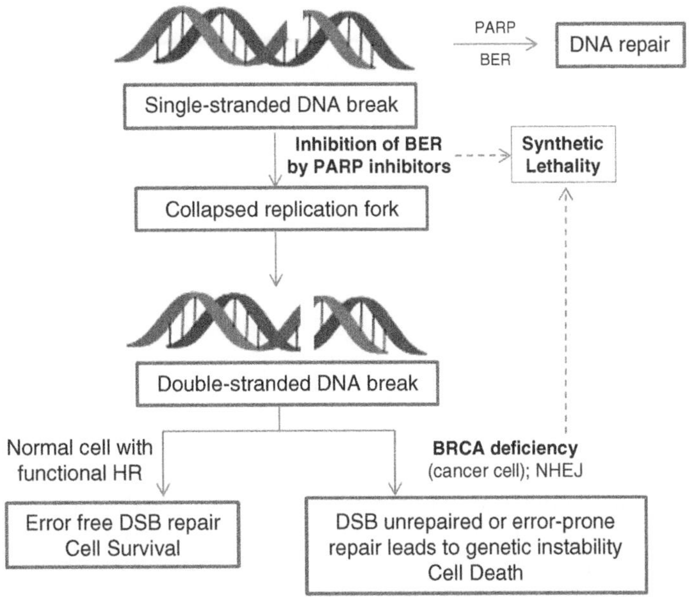

Figure 14.3 Synthetic lethality of PARP inhibitor in BRCA-deficient cells.

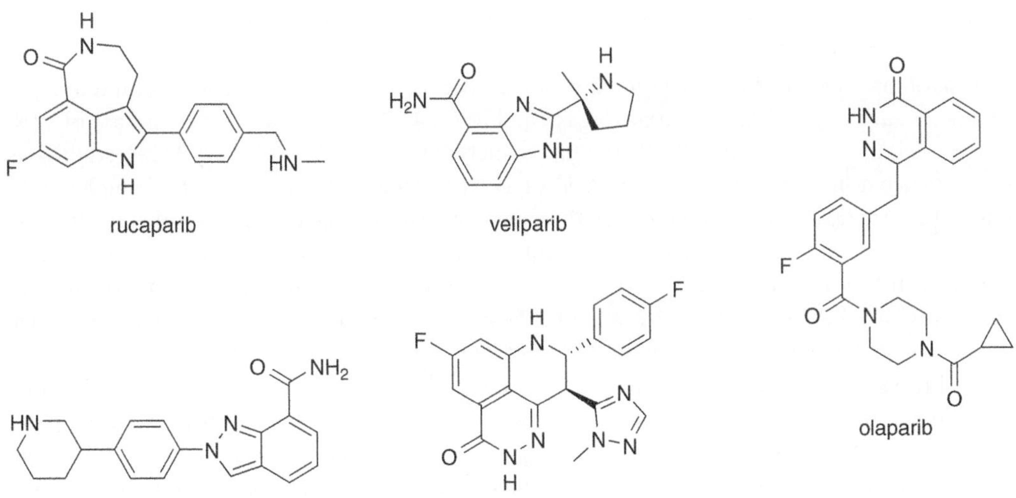

Figure 14.4 Structures of representative PARP inhibitors.

The first set of PARP inhibitors to enter clinical trials (Figure 14.4) included rucaparib (AG014699, PF-01367338) [79], veliparib (ABT-888) [80], olaparib (AZD2281), and niraparib (MK-4827) [81]; more potent second generation PARP inhibitors with picomolar PARP1 IC_{50} values, such as talazoparib (BMN 673) have also been developed [82]. Olaparib is a PARP inhibitor with good *in vitro* binding activity against PARP1 (IC_{50} = 5 nM), PARP2 (IC_{50} = 1 nM), and tankyrase-1 (IC_{50} = 1.5 μM). For cell activity, olaparib showed an inhibitory IC_{50} for PARP-1 of 6 nM and the total inhibition of PARP-1 activity to be at concentrations of 30–100 nM in SW620 cell lysates using a PAR formation assay. Olaparib displayed moderate clearance, volume of distribution, and half-life in both rat and dog PK studies with excellent oral bioavailability of ~100% in both species (Table. 14.1) [83].

Table 14.1 Preclinical PK profiles of olaparib in rat and dog.

Species	Rat[a]	Dog[b]
Cl (ml/min/kg)	49	5.4
V_d (l/kg)	1.18	1.3
$t_{1/2}$ (min)	53	170
$AUC_{0\text{-inf}}$ (μg/ml/min)	304	511
Bioavailability (%)	100	100

[a] i.v. and p.o. at 15 mg/kg in 10% DMSO/10% cyclodextrin in PBS.
[b] i.v. at 2.5 mg/kg (10% DMSO/10% cyclodextrin in PBS) and p.o. at 10 mg/kg in methylcellulose.
AUC, area under the curve; Cl, clearance; $t_{1/2}$, terminal elimination half-life; V_d, volume of distribution.

Olaparib inhibits the growth of selected tumor cell lines *in vitro* and its favorable PK profile was translated into desirable *in vivo* efficacy in animal studies resulting in decreased tumor growth in mouse xenograft models of human cancer. Compared with BRCA1- and BRCA2-proficient cells, BRCA1-deficient cell lines were hypersensitive to olaparib inhibition with some of these BRCA1- and BRCA2-deficient cells up to 1000-fold more sensitive to olaparib inhibition than wild-type cells. Based on these favorable *in vitro* and *in vivo* preclinical results, olaparib was selected as a preclinical candidate (PCC) and moved into clinical development [83].

Olaparib has been in clinical development since 2005 and was evaluated against different types of cancers such as pancreatic, ovarian, breast, and prostate. While the first line treatment of most cancers is still chemotherapy, PARP inhibitors including olaparib are largely being studied as maintenance drugs. Maintenance therapy is a type of treatment that is given after completion of initial standard treatment in order to keep the cancer from returning with the goal of extending the length of time before a new recurrence emerges or with the possibility of turning a temporary remission into a long-term cure. Although PARP inhibitors have worked particularly well in patients with a BRCA1 or BRCA2 mutation, these drugs have been approved for maintenance therapy in women with and without a BRCA mutation [84].

The PK, safety and tolerability, and MTD were established in phase 1 studies [85–87]. These olaparib monotherapy studies in patients with advanced solid tumors demonstrated that olaparib was well-tolerated with generally mild and manageable AEs and established the MTD as 400 mg b.i.d. using a capsule formulation. Clinical unbound steady-state trough concentrations in patients who received 100, 200, and 400 mg of olaparib twice daily were simulated and compared with the estimated IC_{90} values from non-clinical studies [88], and it turned out that only the 400 mg twice daily dose resulted in trough concentrations exceeding the IC_{90} values in all patients. Thus, the 400 mg b.i.d. dose was selected in clinical trials [89, 90].

Subsequently, phase 2 and 3 clinical trials were carried out to evaluate the safety and clinical efficacy of olaparib in treating different tumor types [91, 92]. The detailed clinical trial results are beyond the scope of this chapter and will not be discussed here. To facilitate the discussion of combination therapy of olaparib with bevacizumab in the treatment of OCs in this chapter, a brief summary of the approval history of olaparib in OC related treatments follows:

1) In December 2014, the FDA granted accelerated approval for olaparib (brand name Lynparza) to be used for the treatment of patients with BRCA1/2-associated advanced OC who have received three or more prior lines of chemotherapy. This approval was based on the result of a

phase 2 trial which was further confirmed by recent data from a mandated confirmatory phase 3 SOLO3 trial [84, 92], and represents the first "personalized" therapy for OC. A companion diagnostic test was approved (The Myriad Genetics BRACAnalysis CDx™ test) along with this approval to identify patients eligible for olaparib treatment.

2) In August 2017, based on two randomized trials in patients with recurrent OCs who had a response to platinum-based therapy (SOLO-2 (NCT01874353) and Study 19 (NCT00753545)), olaparib tablets were approved in the United States for the maintenance treatment of adults with recurrent epithelial ovarian, fallopian tube, or primary peritoneal cancer, who are in a complete or partial response to platinum-based chemotherapy [93–95].
3) In December 2018, olaparib was approved based on results from the SOLO-1 (NCT01844986) phase 3 clinical trial for the maintenance treatment of adults who had a complete or partial response to first-line platinum-based chemotherapy but with deleterious or suspected deleterious germline or somatic BRCA-mutated (gBRCAm or sBRCAm) advanced epithelial ovarian, fallopian tube or primary peritoneal cancer [94, 96].

14.2.1.2 Discovery and Development History of Bevacizumab in the Treatment of Ovarian Cancer

In order to support the high metabolic activity associated with tumor growth, cancer cells need to express some specific characteristics. In particular, tumors require significant blood flow to supply nutrients and oxygen and remove waste products. This is accomplished by the generation of new blood vessels which sprout from existing ones, a process known as angiogenesis. Normally this process is tightly controlled, but it can be activated by a tumor through a process known as the "angiogenic switch" [97]. There are numerous factors involved in angiogenesis, but vascular endothelial growth factor A (VEGF-A) has been particularly implicated in tumor cell angiogenesis [98].

Thus, targeting angiogenesis in tumor biology became an important strategy to combat cancer in the past a few decades through the inhibition of tumor angiogenesis and normalization of tumor vasculature. VEGF-A is one such angiogenesis target that was validated in both animal models and human clinical trials [99].

Bevacizumab is a recombinant humanized monoclonal antibody that blocks angiogenesis by binding to VEGF-A isoforms to prevent the interaction of VEGF-A with VEGFR thus inhibiting the activation of VEGF signaling pathways that promote neovascularization. It was first discovered by Genentech scientist Napoleone Ferrara who validated the hypothesis proposed by Judah Folkman in 1971 that inhibiting angiogenesis might be useful in controlling cancer proliferation [99]. Since the expression of VEGF receptors is typically very low or undetectable in most normal tissue but significantly upregulated in the vasculature of many tumors, the inhibition of tumor angiogenesis through neutralization of VEGF by bevacizumab is expected to have high specificity and selectivity in the inhibition of tumor growth and metastasis.

Bevacizumab is a full-length humanized antibody composed of two identical light chains and two heavy chains and was originally derived from a mouse monoclonal antibody generated from mice immunized with recombinant human VEGF [100]. In a bovine capillary endothelial cell proliferation study to assess its potency and efficacy, bevacizumab showed an ED_{50} of 50 ± 5 ng/ml and 90% inhibition at 500 ng/ml [101].

Preclinical PK parameters of bevacizumab were established in animal studies [100, 102]. Table 14.2 summarizes key PK parameters in all species following administration of high doses of bevacizumab. Bevacizumab showed an initial half-life ($t_{1/2}\alpha$) of approximately half a day to one day and a terminal half-life ($t_{1/2}\beta$) of about one to two weeks in mice, rats, and cynomolgus monkeys

Table 14.2 Pharmacokinetic parameters of bevacizumab with intravenous dosing.

Species	Dose (mg/kg)	Cl (ml/kg/day)	V_{ss} (ml/kg)	$t_{1/2}\alpha$ (h)	$t_{1/2}\beta$ (d)	MRT (d)
Mouse	9.3	15.7	152	28.8	6.81	9.69
Rat	10.1	4.83	79.5	6.57	12.3	17.1
Cyno monkey	50	5.78	73.9	19.2	10.3	13.1

Cl, clearance; MRT, mean residence time; $t_{1/2}\alpha$, initial half-life; $t_{1/2}\beta$, terminal half-life; V_{SS}, volume of distribution at steady state.

with i.v. dosing. It had low clearance and a moderately high volume distribution in these species at doses of 10 mg/kg or higher. Bevacizumab had very good absorption with high bioavailability in mouse (>100%), cynomolgus monkey (98%), and rat (69%) with subcutaneous administration.

The preclinical *in vivo* efficacy of bevacizumab was studied in cancer xenograft models in mice and dose-dependent TGI was observed independent of tumor types and route of administration (i.v., i.p., and intratumoral). Moderate to complete inhibition of tumor angiogenesis was achieved in these studies. TGI of 25–>95% was observed in a variety of tumor types including glioblastoma, ovarian carcinoma, prostate carcinoma, colon adenocarcinoma, breast carcinoma, and pancreatic cancer, among others [100, 102].

Clinical development of bevacizumab included angiogenesis driven metastatic colorectal cancer (mCRC), non-small-cell lung cancer (NSCLC), metastatic breast cancer (mBC), glioblastoma multiforme (GBM), OC, and renal cell carcinoma (RCC), and since the first phase 1 clinical trials by Genentech in 1997 a wide range of indications have been studied with bevacizumab in clinical settings. The drug was approved by the FDA in February 2004 based on a phase 3 study (AVF2107g) of bevacizimab administered in combination with bolus IFL (irinotecan, fluorouracil, and leucovorin) chemotherapy as first-line therapy for previously untreated mCRC. This approval represents the first targeted therapy for patients with mCRC and the first approved tumor microenvironment targeting angiogenesis inhibitor [103–105].

Over the years, bevacizumab has been approved by the FDA for the treatment of a number of tumor types. The detailed clinical trial results are beyond the scope of this chapter and will not be discussed here. Instead, a brief summary of the approval history of its application in OC treatments is listed in the following to facilitate the discussion of the application of combination therapy of bevacizumab with olaparib in this cancer type:

1) In November 2014, the FDA approved bevacizumab (Avastin) plus chemotherapy for platinum-resistant recurrent OC based on results from the phase 3 AURELIA study that showed bevacizumab plus chemotherapy reduced the risk of disease worsening or death (progression-free survival, PFS) compared with women who received chemotherapy alone [106]. Bevacizumab was the first targeted therapy approved for treatment of advanced ovarian, fallopian tube and primary peritoneal cancer.
2) In December 2016, the FDA approved bevacizumab (Avastin) plus chemotherapy for a specific type of advanced OC based on results from two phase 3 studies (GOG-0213 and OCEANS) [107, 108], both of which demonstrated a significant improvement in the time patients lived without their disease getting worse as judged by PFS rate.
3) In June 2018, the FDA approved bevacizumab plus chemotherapy including carboplatin and paclitaxel as a treatment for women with stage 3 or stage 4 advanced OC following initial surgery based on data from the pivotal phase 3 (GOG-0218) trial [109].

Bevacizumab is well-tolerated, and its safety profile is well-characterized and AEs are manageable in a diverse range of tumor types and in combination with different chemotherapy regimens. The most frequent AEs include hypertension, fatigue or asthenia, diarrhea, nausea, and abdominal pain; gastrointestinal (GI) perforations, hemorrhage, and arterial thromboembolism are the most frequent serious AEs [102].

14.2.2 Rational Design of Olaparib and Bevacizumab Combination

As discussed in Section 14.1, depending on the strategy drug combinations can be designed based on biological rationales (targeted manner) or through initial effectiveness assessment by *in vitro* screening of a large panel of cell lines in a high throughput fashion to determine the synergy scores (random manner). Since both the PARP inhibitor olaparib and the anti-angiogenesis agent bevacizumab have broad applications in the treatment of OC, the combination of these two drugs in OC treatment was designed based on a strong biological rationale.

As discussed above, bevacizumab is an anti-angiogenesis agent that targets the hallmark angiogenesis of cancer which was thought to be a universal approach for the treatment of solid tumors. However, initial tumor responses are usually followed by disease progression due to adaptive resistance [97]. One of the most common resistance mechanisms to angiogenesis inhibitors is the upregulation of hypoxia inducible factor 1 alpha (HIF1α), which promotes the transcription of over 40 genes related to oxygen delivery, among other functions. *in vitro* studies have shown that deletion of genes encoding for PARP decreases HIF1α accumulation. Therefore, PARP inhibitors may function in overcoming bevacizumab resistance by preventing accumulation and signaling associated with HIF1α thus promoting a hypoxic state [110]. This is an excellent example of "contextual synthetic lethality" due to tumor microenvironment where hypoxia-induced DNA repair-deficient tumor cells can be targeted by disrupting backup pathways [111].

It is clear that concurrent treatment of tumor cells with bevacizumab and PARP inhibitors provides a dual mechanism of action for cancer treatment. Bevacizumab treatment induces a hypoxic state in cells that results in a mutagenic environment with a downregulation of DNA repair. PARP inhibitors work by blocking the secondary DNA repair mechanism and in addition prevent alleviation of the hypoxic state. Therefore, the combination of bevacizumab and olaparib is a logical approach to cancer treatment [112]. With this rationale in mind, the combination of bevacizumab with olaparib was investigated in clinical trials and the combination's therapeutic benefit was evaluated in front-line OC maintenance treatment [113].

14.2.3 Olaparib and Bevacizumab Combination in Clinical Studies

14.2.3.1 Phase 1 Clinical Studies of the Olaparib and Bevacizumab Combination

A phase 1 study was carried out with the olaparib and bevacizumab combination with the primary objectives to establish the safety and tolerability in patients with advanced solid tumors and to select an appropriate dosing regimen for use in future phase 2 and phase 3 efficacy studies [114]. The secondary objectives included the assessment of PK parameters of olaparib when administered alone and in combination with bevacizumab. Continuously increasing doses of twice daily oral olaparib (100, 200 and 400 mg b.i.d. capsule formulation) in combination with intravenous bevacizumab at a fixed dose (10 mg/kg) administered every 14 days (q2w) were employed in this study.

The PK profiles of this combination are summarized in Table 14.3. Overall, the AUC_{ss} and $C_{ss,max}$ for olaparib in combination with bevacizumab was similar to administering olaparib alone. These

Table 14.3 Pharmacokinetic profiles of olaparib dosed twice daily (b.i.d.) in combination and alone.

PK parameter	100 mg	100 mg with bevacizumab	200 mg	200 mg with bevacizumab	400 mg	400 mg with bevacizumab
T_{max} (h)	2	1	1	2	2	2
$C_{ss,max}$ (μg/ml)	2.5	1.8	4.7	5.2	9.1	8.1
AUC_{ss} (μg.h/ml)	NA	NA	25.8	26.6	58.1	50.3
AUC_{ss} ratio	NA	NA	NA	1.0	NA	0.9
$C_{ss,max}$ ratio	NA	0.7	NA	1.1	NA	0.9

AUC_{ss}, area under the plasma concentration–time curve during any dosing interval at steady state; AUC_{ss} ratio, AUC_{ss} of olaparib in combination divided by AUC_{ss} of olaparib dosed alone; $C_{ss,max}$, maximum plasma (peak) concentration in plasma during dosing interval; $C_{ss,max}$ ratio, $C_{ss,max}$ of olaparib in combination divided by $C_{ss,max}$ of olaparib dosed alone; NA, not applicable.

data suggested that olaparib exposure is unaffected by co-administration of bevacizumab, and olaparib 400 mg b.i.d. in capsule formulation combined with bevacizumab 10 mg/kg q2w appeared to be a tolerable regimen in patients with advanced solid tumors based on the results of this study.

In this phase 1 study, no AEs of Common Terminology Criteria for Adverse Events (CTCAE) grade X3, serious adverse events (SAEs) or dose limiting toxicities were observed with olaparib, and other observed AEs such as headache and fatigue have been associated with bevacizumab. Based on the MTD of olaparib at 400 mg b.i.d. previously determined in patients with solid tumors [85], coupled with the profiles from this phase 1 clinical study, the dose level of olaparib 400 mg b.i.d. with bevacizumab 10 mg/kg once biweekly (q2w) was considered for future clinical studies.

14.2.3.2 Phase 2 Clinical Studies of the Olaparib and Bevacizumab Combination

The efficacy of the olaparib and bevacizumab combination was established from phase 2 studies evaluating the efficacy of the combination of an antiangiogenic agent with olaparib or bevacizumab with a PARP inhibitor involving patients with relapsed platinum-sensitive OC. The results showed longer PFS than the use of a PARP inhibitor alone [115–117].

14.2.3.3 Phase 3 Clinical Studies of the Olaparib and Bevacizumab Combination

The randomized phase 3 PAOLA-1 (PAOLA-1/ENGOT-ov25) trial evaluated the efficacy of maintenance therapy with the olaparib and bevacizumab combination as compared with placebo and bevacizumab in patients with newly diagnosed advanced OC who were receiving chemotherapy plus bevacizumab followed by bevacizumab, regardless of BRCA mutation status [118]. The time from randomization until investigator-assessed disease progression or death was the primary endpoint. The secondary endpoints included the global health status-quality of life dimension of the European Organization for Research and Treatment of Cancer (EORTC) Quality of Life Questionnaire (QLQ) in addition to the time from randomization until second disease progression or death, overall survival, the time until the first subsequent therapy or death.

The clinical outcome of this phase 3 trial is summarized in Table 14.4. The trial met its primary objective by showing a significant PFS benefit in the intention-to-treat population. The olaparib and bevacizumab group showed significantly longer duration of investigator-assessed PFS than in the placebo plus bevacizumab group (median of 22.1 months vs. 16.6 months). Results of subgroup

Table 14.4 Summary of clinical outcome of PORLA-1 phase 3 clinical trial [118].

	Median PFS (olaparib and bevacizumab) (mo)	Median PFS (placebo and bevacizumab) (mo)	HR (95% CI)
Primary endpoint	22.1	16.6	0.59 (0.49–0.72)
Patients with BRCA mutant	37.2	21.7	0.31 (0.20–0.47)
Patients without a tumor BRCA mutation	18.9	16.0	0.71 (0.58–0.88)
Patients with HRD-positive tumors	37.2	17.7	0.33 (0.25–0.45)
Patients with HRD-positive tumors without BRCA mutation	28.1	16.6	0.43 (0.28–0.66)
Patients with BRCA mutation-negative tumors	18.9	16.0	0.71 (0.58–0.88)
Patients with tumors with negative or unknown HRD status	16.9	16.0	0.92 (0.72–1.17)
Patients with HRD negative tumors	16.6	16.2	1.00 (0.75–1.35)

CI, confidence interval; HR, hazard ratio for disease progression or death; HRD, homologous recombination deficiency.

analyses of PFS showed a benefit in the majority of predefined subgroups. Olaparib in patients with BRCA-mutated and homologous recombination deficiency (HRD)-positive tumors showed a PFS benefit. Patients with HRD-positive tumors without a BRCA mutation also had a substantial clinical benefit from olaparib. Patients whose tumor HRD status was unknown, also benefited.

The observed improved PFS benefit in patients with HRD-positive tumors that did not have BRCA mutations in the olaparib and bevacizumab arm is especially noteworthy. As discussed previously, an antiangiogenic treatment can induce hypoxia which may contribute to the increase of HRD and bevacizumab may sensitize the activity of olaparib in patients with HRD-positive tumors, especially for patients with HRD-positive tumors without a BRCA mutation. However, it is difficult to draw a conclusion whether the PFS benefit observed with HRD-positive tumors without BRCA mutations was due to a synergistic effect of the combination of olaparib and bevacizumab or whether it was the result of olaparib since there was no comparator olaparib arm in the study. Further exploration might be needed to provide a more conclusive answer.

The AEs of this trial are summarized in Table 14.5. The most common AEs (all grades) that occurred at a higher incidence among patients receiving olaparib–bevacizumab than among those receiving placebo–bevacizumab were fatigue, nausea, and anemia. Anemia was the most common serious AE that occurred at a higher incidence with olaparib–bevacizumab than with placebo–bevacizumab.

Overall this phase 3 trial demonstrated that a significant PFS benefit, especially in those patients with HRD-positive tumors when maintenance olaparib and bevacizumab combination was administered to patients with newly diagnosed advanced OC who were receiving standard treatment including bevacizumab [118]. Based on the positive results from this phase 3 PAOLA-1 trial, a supplemental new drug application (sNDA) for olaparib (Lynparza) in combination with bevacizumab

Table 14.5 The most common adverse events and the incidence of associated grade 3 or higher adverse events for the entire maintenance treatment period.

Events	Olaparib–bevacizumab (all grades) (%)	Placebo–bevacizumab (all grades) (%)	Olaparib–bevacizumab (grade ≥3) (%)	Placebo–bevacizumab (grade ≥3) (%)
Fatigue or asthenia	53	32	5	1
Nausea	53	22	2	1
Hypertension	46	60	19	30
Anemia	41	10	17	<1
Lymphopenia	24	9	7	1
Arthralgia	22	24	1	1
Vomiting	22	11	1	2
Abdominal pain	19	20	1	2
Diarrhea	18	17	2	2
Neutropenia	18	16	6	3
Leukopenia	18	10	2	1
Urinary tract infection	15	10	<1	<1
Headache	14	13	<1	1
Constipation	10	20	0	<1
Thrombocytopenia‖	8	3	2	<1
Proteinuria	6	15	1	<1%
Leading to dose interruption	54	24	NA	NA
Leading to dose reduction	41	7	NA	NA
Leading to discontinuation of intervention	20	6	NA	NA

NA, not available.

(Avastin) was accepted and granted priority review by the FDA in January 2020. Subsequently in May 2020, the FDA approved the combination for the maintenance treatment of adults with advanced epithelial ovarian, fallopian tube or primary peritoneal cancer who have a complete or partial response to first-line platinum-based chemotherapy and whose cancer has been tested to have either a deleterious or suspected deleterious BRCA mutation and/or genomic instability.

14.2.4 Summary of the Olaparib and Bevacuzimab Combination

In recent years, PARP inhibitors have received much attention from the medical community with their potent anticancer activities through a novel concept of synthetic lethality. Since its discovery, olaparib has been extensively studied in clinical trials as both a mono and combination therapy [119]. On the other hand, bevacizumab which targets the hallmarks of cancer angiogenesis has been extensively evaluated in clinical trials in combination therapies with a large number of anticancer drugs [120]. While some of these clinical studies have provided alternative treatment options for cancer patients, the development of effective combination therapies with reduced time and cost remains challenging

As mentioned above, olaparib holds considerable promise in combination therapies, however, there are a few aspects that should be considered before the full potential of such combinations might be realized: (i) a better understanding of the mechanism of action of PARP inhibitors like olaparib so that better combinations can be designed [121, 122]. (ii) Better biomarkers are required so that the appropriate patient populations can be identified. (iii) Resistance to olaparib combination therapy is an emerging problem that must be dealt with, perhaps by development of new combinations with agents that have a different mechanism of action from DNA damaging chemotherapies [113].

Despite considerable efforts, no predictive biomarker, potential biomarkers of resistance, and imaging technical biomarkers have been identified to enable a more targeted use of bevacizumab. Plasma levels of VEGF-A (pVEGFA) and pVEGF in mBC were evaluated as predictive biomarkers for bevacizumab efficacy in clinical settings, but no conclusive results were obtained from these studies [104, 123]. Due to these reasons, bevacizumab is used in general patient populations that were not pre-selected by a biomarker which may have contributed to less than ideal treatment outcomes in certain cases. This remains a challenging area to be addressed in order to realize the full therapeutic potential of bevacizumab in combination therapies.

Notwithstanding the impact that the bevacizumab and olaparib combination has had on OC treatment, there is still no cure for advanced OC and an unmet medical need remains. There is an urgent need for better therapies for patients whose OCs are platinum-resistant or platinum-refractory and for those with normal BRCA genes. Over the past few years, a number of promising novel therapies with different modalities have been making their way through clinical trials including targeted therapies such as the ATR inhibitor AZD6738 and the Wee1 inhibitor adavosertib, antibody-drug conjugates (ADCs) such as mirvetuximab soravtansine, the anti-DLL4/VEGF bispecific antibody navicixizumab, protein therapies such as AVB-S6-500, immunotherapies such as the anti-PD-L1 antibody MEDI4736, gene therapies such as ofranergene obadenovec (VB-111), and vaccines such as DPX-Survivac. The monotherapy or eventual combination therapies of these agents may provide longer-lasting relief for OC patients and hopefully one day this terminal disease will be cured or turned into a chronic illness.

14.3 Case Study of Ceftazidime and Avibactam Combination (Avycaz®) in the Treatment of Complicated Urinary Tract Infections and Intra-abdominal Infections

14.3.1 Brief History of the Discovery of Ceftazidime and Avibactam and the Rational for the Combination of Ceftazidime and Avibactam in the Treatment of Complicated Urinary Tract Infections and Intra-abdominal Infections

β-Lactam antibiotics are some of the most widely prescribed antibacterial drugs that includes penicillins, cephalosporins, and monobactams [124, 125]. To improve effectiveness and overcome drug resistance, several generations of each drug class have been developed. Among these, ceftazidime is a third-generation cephalosporin (Figure. 14.5) introduced into clinical use in the 1980s that exhibits bactericidal activity through binding of the penicillin-binding proteins (PBPs). Among at least eight PBP isoforms, PBP3 is the main target of ceftazidime in the inhibition of bacterial cell-wall synthesis and eventual cell death. It displays broad-spectrum activity including *Pseudomonas*

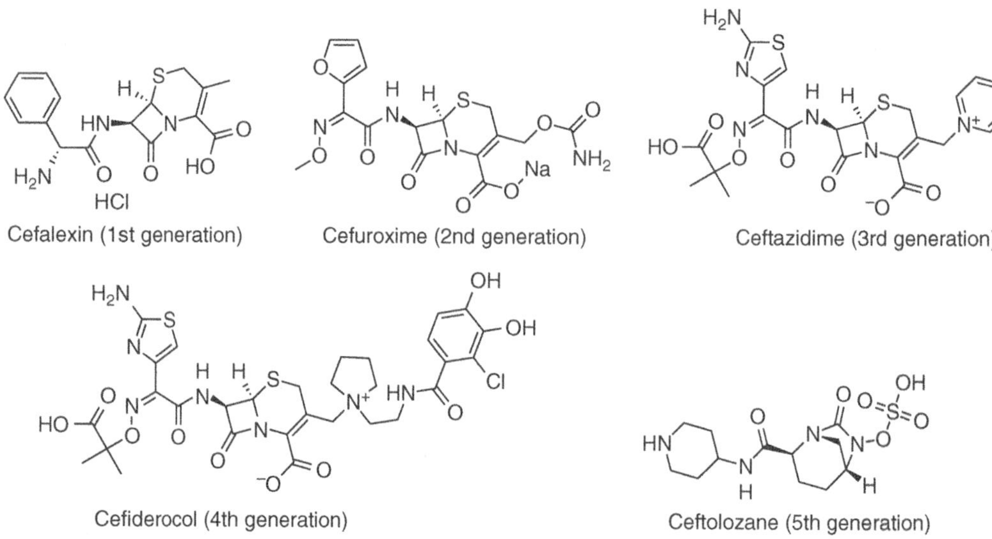

Figure 14.5 Representative drugs from different generations of cephalosporins.

spp. with a good safety profile and well-characterized PK and PD properties. Compared with older antibiotics, ceftazidime is a better treatment option with improved efficacy against MDR bacteria as it is more stable toward extended spectrum β-lactamases (ESBLs), while some other third- and fourth-generation cephalosporins are extensively hydrolyzed by ESBL [126].

Among the antibiotic resistance mechanisms, β-lactamases are one of the major factors contributing to the development of resistance to ceftazidime. β-Lactamases are classified into four groups, A, B, C, and D: classes A, C, and D are serine-derived enzymes which rely on an active-site serine residue to form a transient covalent bond with the antibiotic followed by hydrolysis of the enzyme-linked ester. Common strains in these classes include cefotaximase (CTX-M), sulfhydryl variant β-lactamase (SHV), Temoneira β-lactamase (TEM), *Klebsiella pneumoniae* carbapenemase (KPC), Guiana-extended-spectrum β-lactamase (GES), and *Serratia marcescens* enzymes (SMEs). Class B enzymes are metallo-β-lactamases which rely on Zn^{2+} ion-assisted activation of a water molecule to generate the hydrolytic species to inactivate the antibiotics. New Delhi metallo-β-lactamase (NDM), Veronaimipenemase (VIM), and Imipenemase (IMP) belong to this class of β-lactamases (Figure 14.6). β-Lactamase inhibitors targeting resistance originated from class A, C, and D β-lactamases have been developed, however, currently there are no inhibitors against class B β-lactamases.

Based on an understanding of the β-lactamase resistance mechanism, a number of β-lactamase inhibitors were discovered and developed to overcome resistance to β-lactam antibiotics. These include earlier inhibitors such as clavulanate, sulbactam, tazobactam, recently approved drugs such as avibactam, relbactam, and vaborbactam, and the ones currently in clinical development such as nacubactam and zidebactam (Figure. 14.7).

Avibactam (also known as NXL104 or AVE1330A) is a non-β-lactam β-lactamase inhibitor with a chemical structure different from earlier β-lactamase inhibitors since it does not contain the reactive β-lactam ring. Mechanistically, the urea carbonyl in avibactam acts as the β-lactam carbonyl of β-lactam-based inhibitors to deactivate β-lactamases at the active site. While β-lactam-based β-lactamase inhibitors undergo an irreversible reaction producing intermediates that are hydrolyzed, a recent study found that the release of serine from the avibactam carbamate intermediate proceeds through the regeneration of intact avibactam and not hydrolysis suggesting it is a covalent

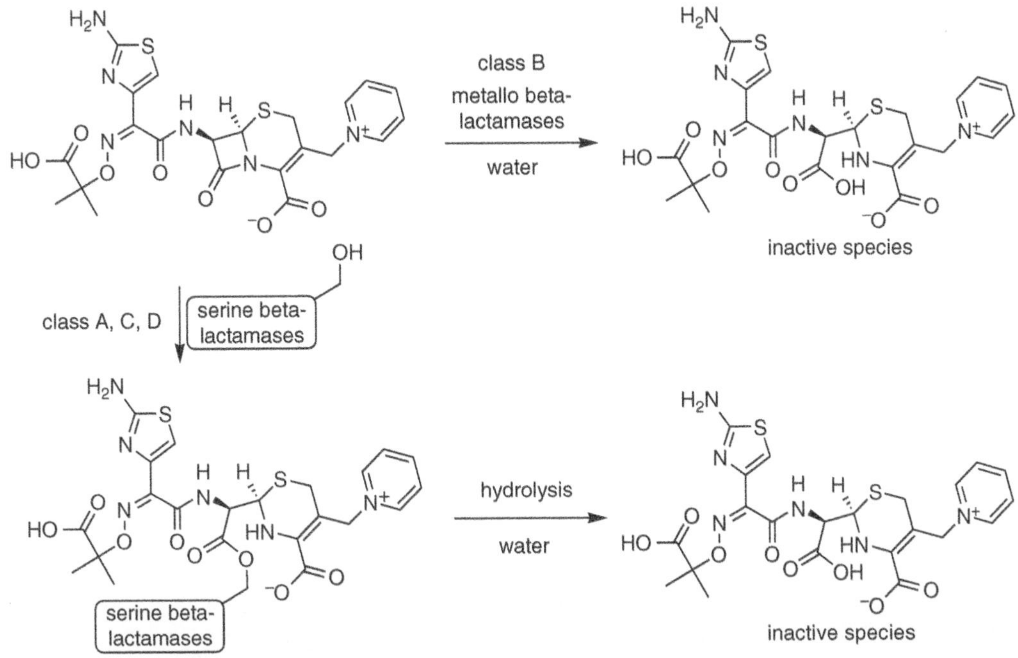

Figure 14.6 The mechanism of class A, B, C, and D β-lactamase catalyzed hydrolysis of β-lactam inhibitors. Serine mediated hydrolysis of the β-lactam ring from class A, C, and D β-lactamases; zinc ion mediated hydrolysis of the β-lactam ring from class B β-lactamases.

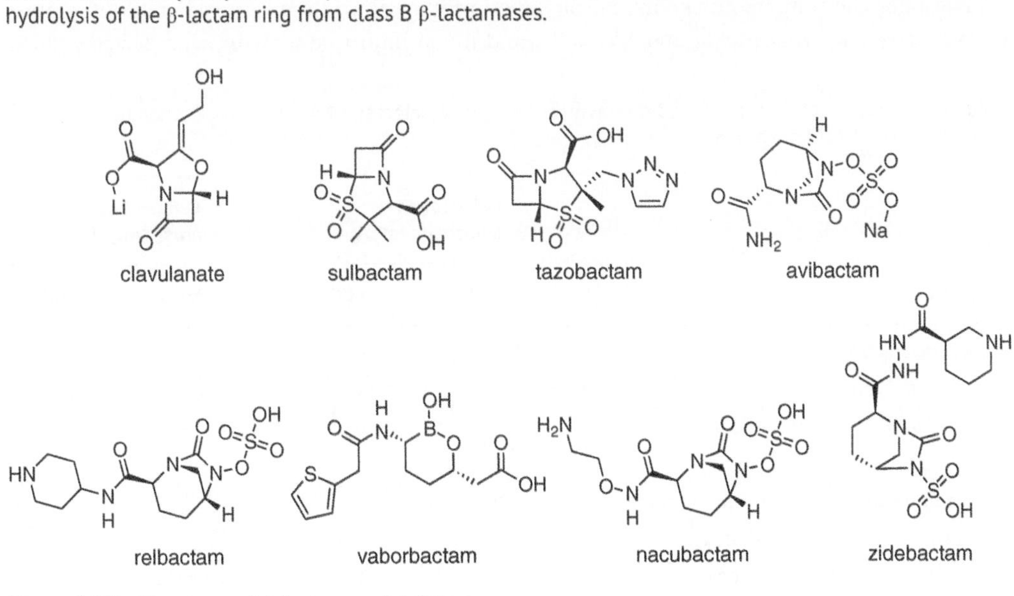

Figure 14.7 Structures of β-lactamase inhibitors.

but slowly reversible inhibitor (Figure 14.8), which is a unique mechanism of inhibition among β-lactamase inhibitors [127]. Avibactam as a single agent has no meaningful antimicrobial activity at therapeutic concentrations, but it exhibits a broader spectrum of β-lactamase inhibition than other available β-lactam-based β-lactamase inhibitors and is active against a large panel of β-lactamase-producing bacteria. Avibactam shows high affinity with class A and class C β-lactamases

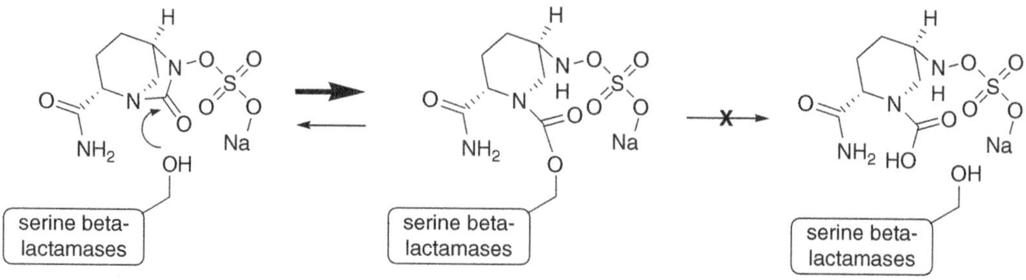

Figure 14.8 The mechanism of inactivation of β-lactamases by avibactam.

and is not active against Ambler's class B metallo β-lactamases due to the obvious reason that class B β-lactamases do not have the active site serine residue as discussed earlier [128, 129].

Although ceftazidime has good efficacy against a broad spectrum of bacterial infections including uncomplicated and complicated urinary tract infections (UTIs) and intra-abdominal infections (IAIs), its effectiveness is increasingly compromised due to the development of resistance from β-lactamase-producing pathogens. Thus, the combination of ceftazidime and avibactam to restore and broaden the coverage of ceftazidime is warranted and the effectiveness of this combination was first assessed in preclinical settings followed by fixed-dose combinations in clinical trials.

The *in vitro* activity of ceftazidime plus avibactam was evaluated against a wide range of organisms in numerous settings to establish the optimal combination options for clinical consideration [130]. The MIC_{50} and MIC_{90} of the combination in selected sensitive pathogens are summarized in Table 14.6. The combination showed clear superiority to ceftazidime alone when tested against *Enterobacteriaceae*, *Escherichia* and *Klebsiella*, and much improved activity when tested against

Table 14.6 *in vitro* activity of ceftazidime–avibactam against selected Gram-negative pathogens in comparison to ceftazidime alone.

	Ceftazidime–avibactam (mg/l)		**Ceftazidime alone (mg/l)**	
Organism	**MIC_{50}**	**MIC_{90}**	**MIC_{50}**	**MIC_{90}**
Enterobacter aerogenes	0.25	0.5	0.5	>32
Enterobacter cloacae	0.25	1	0.5	>32
Escherichia coli (ESBL-producing)	0.12	0.25	16	>32
E. coli (AmpC-hyperproducing)	0.12	0.25	16	>32
Klebsiella pneumoniae (ESBL-producing)	0.5	1	32	>32
K. pneumoniae (OXA-48-producing)	0.25	0.5	256	512
K. pneumoniae (KPC-producing)	0.25	1	>256	>256
K. pneumoniae (Carbapenem-non-susceptible)	0.25	2	>32	>32
Pseudomonas aeruginosa (multidrug-resistant)	8	>16	>16	>16
P. aeruginosa (AmpC-derepressed)	4	8	64	>128
Acinetobacter baumannii	8	>16	8	>32
Bacteroides fragilis (anaerobic)	4	32	0.5	>32
Prevotella/Porphyromonas spp. (anaerobic)	2	4	32	>128

P. aeruginosa. The ceftazidime and avibactam combination had limited activity against *Acinetobacter baumannii* and most *Bacteroides* spp. among Gram-negative anaerobes, therefore more effective antibiotics need to be discovered for treating these infections [131].

The preclinical *in vivo* efficacy of ceftazidime and avibactam was evaluated in a number of animal models infected by bacteria that produce the clinically important class A and C β-lactamases including the ESBL, KPC and AmpC enzymes. The *in vivo* models included but were not limited to murine models of sepsis, septicemia and thigh infection as a result of KPC-producing *K. pneumoniae* infection and with extended spectrum and AmpC-producing Enterobacteriaceae isolates [132, 133], pneumonia caused by an AmpC- and SHV-11-producing *K. pneumoniae* isolates [132], pyelonephritis [134], and kidney infection caused by AmpC- or ESBL-producing Enterobacteriaceae isolates [135], and a rabbit model of meningitis due to ceftazidime-non-susceptible Enterobacteriaceae isolates with an AmpC-producing *K. pneumoniae* infection [136]. The detailed results of these animal studies will not be discussed here but are summarized in Table 14.7 [131]; nonetheless, the data from these studies showed that the combination therapy was very efficacious with significant bacterial killing over ceftazidime alone. These results support a potential role for the combination of ceftazidime and avibactam in a 4:1 ratio in the treatment of infections caused by Gram-negative bacteria especially ceftazidime-non-susceptible pathogens. The PK/PD targets of this combination were established through these animal studies as discussed in the following.

The efficacy of β-lactam antibiotics is best defined as the time the free drug concentration remains above the MIC for the dosing interval ($\%fT > \text{MIC}$) while the efficacy benchmark for avibactam was determined to be the time the free drug concentration remained over a threshold concentration (CT) over the dosing interval ($\%T > \text{CT}$). Based on animal data, a target of 40–50%$T >$ MIC was selected for ceftazidime while a target of 40–50%$T >$ CT was determined for avibactam. This translated into a concentration of 0.5 and 1 mg/l of avibactam combined with ceftazidime versus Enterobacteriaceae and *P. aeruginosa*, respectively [128, 137–143]. Combined with the earlier clinical results, a dosing paradigm of 2.0/0.5 g of ceftazidime–avibactam administered over two hours by infusion was used in the phase 2 trials and compared with ceftazidime alone [133, 144].

Both complicated urinary tract infections (cUTIs) and complicated intra-abdominal infections (cIAIs) caused by Gram-negative bacteria are serious bacterial infections in hospitalized patients that carry a high risk of morbidity and mortality. As mentioned previously, the effectiveness of ceftazidime in treating uncomplicated and complicated UTI and IAI was severely compromised by the spread of β-lactamase-producing pathogens. With the expanded spectrum of activity of avibactam against multiple serine-based β-lactamases, the combination therapy of ceftazidime–avibactam was selected in clinical trials for the treatment of difficult to treat bacterial infections such as cUTI and cIAI in the US. At the European Medicines Agency (EMA), this was also used for hospital-acquired pneumonia (HAP)/ventilator-associated pneumonia (VAP) and severe and deep infections and severe Gram-negative infections with no other therapeutic options available [124, 145]. The ceftazidime and avibactam combination was designated by the FDA as a qualified infectious disease product under the Generating Antibiotics Incentives Now Act to facilitate its development [146].

14.3.2 PK, Safety and Tolerability of the Ceftazidime and Avibactam Combination in Phase 1 Human Clinical Trials

Two phase 1 studies (NXL104-1001 and NXL104-1002) were carried out to assess the safety and PKs of avibactam in healthy subjects when administered alone or with ceftazidime. The PK profile

Table 14.7 Summary of *in vivo* study results of the ceftazidime and vibactam combination in animal models of infections.

Model	Dose	Results
KPC-producing *Klebsiella pneumoniae* strains induced murine lethal septicemia model	Ceftazidime–avibactam weight ratio of 4:1	Combination therapy reduced ED_{50} from 1580 mg/kg (first isolate) and 710 mg/kg (second isolate) to 15.1 mg/kg and 3.8 mg/kg, respectively. ~100% animal survival for ceftazidime-avibactam combination at a dose of 32/8 mg/kg vs. 0% survival with up to 256 mg/kg of ceftazidime
KPC-producing *K. pneumoniae* strains induced neutropenic thigh infection model	Ceftazidime–avibactam at doses of ≥128/32 mg/kg vs. ceftazidime at doses of ≥1024 mg/kg	A > 2 log10 CFU reduction in thighs removed 24 h post-infection in combination therapy vs. no reduction in ceftazidime only arm
Murine septicemia model with extended spectrum and AmpC-producing Enterobacteriaceae isolates	Ceftazidime–avibactam weight ratio of 4:1	ED_{50} of 2–27 mg/kg for ceftazidime–avibactam vs. >60 mg/kg for ceftazidime only arm
AmpC- and SHV-11-producing *K. pneumoniae* isolate induced murine pneumonia model	150/37.5 mg/kg Ceftazidime–avibactam dosed every 8 h starting 16 h post-infection for a duration of 2 d	Mean lung bacterial counts were reduced from 11 to 4.5 log10 CFU with combination vs. 7.9 log10 CFU with ceftazidime only arm
AmpC- or ESBL-producing Enterobacteriaceae isolates induced immunosuppressed mouse model of kidney infection	Ceftazidime–avibactam (ratio of 4:1) at a dose range of 10–25 mg/kg administered subcutaneously 4, 8, 24, and 32 h post-infection	Kidney bacterial burden at 48 h was reduced by 2.6–4.5 log10 CFU with combination therapy compared with ceftazidime only arm
AmpC-producing *K. pneumoniae* isolate induced rabbit meningitis model	Ceftazidime–avibactam dosed at 150/37.5 mg/kg vs. 150 mg/kg ceftazidime alone	5.66 log10 CFU bacterial load was reduced with combination therapy vs. no reduction of the cerebrospinal fluid bacterial load with ceftazidime only arm
P. aeruginosa isolates induced neutropenic and immunocompetent murine thigh infection model	A free ceftazidime–avibactam drug concentration–time profile simulating a ceftazidime–avibactam dose of 2 g/500 mg intravenously every 8 h that seen in humans	A reduction in thigh bacterial density of ≥0.5 log10 CFU for 22 of 27 isolates with the combination vs. 10 of 27 isolates with ceftazidime only arm

CFU, colony-forming unit.

of ceftazidime as a single agent was long established with dose-dependent C_{max} and area under the curve (AUC) [147]. In these more recent phase 1 studies, the PK of avibactam as a single agent demonstrated nearly dose proportional linear PK for both AUC and peak plasma concentrations across the dose range studied (50–2000 mg) for a single i.v. administration, as summarized in Table 14.8 (with 100–1500 mg data shown) [148]. The PK parameters of ceftazidime and avibactam given in combination as a single two-hour infusion and as multiple two-hour infusions every eight hours for 11 days are summarized in Table 14.9 [126]. No clear accumulation of ceftazidime (2000 mg) or avibactam (500 mg) was observed following multiple i.v. infusions in these studies. Key PK parameters had no clear changes either after single or repeated dosing when the two drugs were administered as a combination or alone [143].

Table 14.8 Single rising dose PK profiles of ceftazidime and avibactam alone and in combination.

Avibactam dose (mg)	Ceftazidime dose (mg)	Cl (l/h)	Cl_R (l/h)	V_{ss} (l)	$t_{1/2}$ (h)	C_{max} (mg/l)	AUC_{last} (h.mg/l)
100		12.18	11.12	23.58	1.80	4.89	8.15
250		12.80	15.68	22.81	1.70	11.88	19.44
250	1000	12.03 (7.74)[a]	11.90 (6.63)[a]	21.08 (18.88)[a]	1.79 (1.82)[a]	12.93 (57.35)[a]	20.66 (128.3)[a]
500		13.39	12.68	22.72	1.83	26.15	37.13
500	2000	13.36 (10.07)[a]	12.71 (9.70)[a]	25.43 (28.23)[a]	2.16 (3.07)[a]	23.33 (93.17)[a]	37.15 (195.9)[a]
1000		11.63	11.22	21.99	2.18	48.48	85.75
1500		10.33	8.53	19.39	2.65	99.06	145.10

[a] The data in parentheses are for ceftazidime and all remaining data are for avibactam.
AUC_{last}, area under concentration–time curve from time 0 to time of last measurable plasma concentration; C_{max}, maximum observed concentration; Cl, plasma clearance; Cl_R, renal clearance; $t_{1/2}$, terminal elimination half-life; V_{ss}, volume of distribution at a steady state.

Table 14.9 Single and multiple dose PK profiles of ceftazidime (2000 mg) in combination with avibactam (500 mg) in healthy adult males.

	Avibactam		Ceftazidime	
Parameter	Ceftazidime–avibactam single dose	Ceftazidime–avibactam multiple dose	Ceftazidime–avibactam single dose	Ceftazidime–avibactam multiple dose
Cl (l/h)	11.9	13.1	6.93	6.86
V_{ss} (l)	23.2	22.2	18.1	17
$t_{1/2}$ (h)	2.22	2.71	3.27	2.76
C_{max} (mg/l)	15.2	14.6	88.1	90.4
AUC (mg.h/l)	42.1	38.2	289	291

AUC, area under concentration–time curve (from time 0 to infinity used for single dose infusion and over dosing interval used for multiple dose infusions); C_{max}, maximum observed concentration; Cl, plasma clearance; $t_{1/2}$, terminal elimination half-life; V_{ss}, volume of distribution at a steady state.

These phase 1 studies showed that avibactam (500 mg) administered in the presence or absence of ceftazidime (2000 mg) was generally well tolerated and no serious or severe AEs were reported in either study. There were no observed drug–drug interactions in the drug combination since the PK parameters were not affected by one another. No dosage adjustment was necessary based on age or sex, however, the clearance of avibactam and ceftazidime was reduced in patients with renal impairment which may require dose adjustment for these patients as discussed in later sections. Based on the PK and safety findings in these preliminary phase 1 studies, phase 2 studies were initiated to evaluate the efficacy and safety of ceftazidime–avibactam in patients with cUTIs (NCT00690378), and concomitantly with metronidazole in patients with cIAIs (NCT00752219) which eventually led to phase 3 clinical trials evaluating ceftazidime (2000 mg) in combination with avibactam (500 mg) (NCT01499290, NCT01500239, NCT01595438, NCT01599806, NCT01726023, and NCT01808092).

14.3.3 Clinical Efficacy of the Ceftazidime and Avibactam Combination

14.3.3.1 Ceftazidime and Avibactam Combination Phase 2 Clinical Trials

Two randomized and comparative phase 2 clinical studies were conducted with ceftazidime–avibactam, one in patients with serious cUTI and the other in patients with cIAI [136, 149]. The goal of the cUTI study (NCT00690378) was to compare the efficacy and safety of ceftazidime–avibactam and imipenem–cilastatin (the current treatment of choice in cUTI) in hospitalized adults with serious cUTI due to Gram-negative pathogen infection. The primary efficacy objective was a favorable microbiological response in microbiologically evaluable (ME) patients at the test-of-cure (TOC) visit five to nine days post-therapy. Secondary endpoints included microbiological response four to six weeks post-therapy in the ME population at the end of i.v. therapy and at the late follow-up (LFU) visit.

In the ME population at the TOC visit, a favorable microbiological response was observed in 70.4% patients in the ceftazidime and avibactam arm and 71.4% in the imipenem and cilastatin arm. Favorable microbiological response rates at the end of i.v. therapy were 96.2 and 100% in the ceftazidime plus avibactam and imipenem plus cilastatin arms, respectively, and 57.7 and 60.0% at the LFU visit. In the clinically evaluable (CE) population and modified intent-to-treat (MITT) population, both the ceftazidime plus avibactam and imipenem plus cilastatin arms also achieved similar favorable clinical response as summarized in Table 14.10.

In this study, the ceftazidime plus avibactam arm and the imipenem plus cilastatin arm displayed similar safety and tolerability profiles. Treatment-emergent SAEs were reported in 8.8% of patients in the ceftazidime plus avibactam arm compared with 3.0% in the imipenem plus cilastatin arm.

Table 14.10 Clinical response of ceftazidime and avibactam combination for cUTI in the phase 2 clinical trial.

Population	Ceftazidime and avibactam response rate (%)	Imipenem and cilastatin response rate (%)
ME (TOC)	70.4	71.4
ME (End i.v.)	96.2	100
ME (LFU)	57.7	60.0
CE (TOC)	85.7	80.6
CE (End i.v.)	100	100
CE (LFU)	74.1	66.7
MITT (TOC)	67.4	63.3
MITT (End i.v.)	87.0	91.8
MITT (LFU)	50.0	46.9

CE, clinically evaluable; End i.v., end of i.v. therapy; LFU, late follow-up; ME, microbiologically evaluable; MITT, modified intent-to-treat; TOC, test-of-cure.

The main objective of the cIAI phase 2 trial (NCT00752219) was to evaluate the efficacy, safety and tolerability of ceftazidime–avibactam plus metronidazole versus meropenem in the treatment of cIAIs in hospitalized adults.

The results of this study are summarized in Table 14.11. In the ME population at the TOC visit, a favorable clinical response was observed in 91.2 and 93.4% of ceftazidime–avibactam plus metronidazole and meropenem patients, respectively. At the end of i.v. therapy, a favorable clinical response was observed in the ME population patients, as well as in the CE and the mMITT populations, as summarized in Table 14.11. Among all randomized patients who received a study drug, patients treated with ceftazidime–avibactam plus metronidazole and those treated with meropenem had similar favorable clinical responses at the TOC visit, at the end of i.v. therapy, and at the LFU visit. These findings suggested that the efficacy of ceftazidime–avibactam plus metronidazole is similar to that of meropenem. In addition, a favorable microbiological response was observed with ceftazidime–avibactam plus metronidazole in 96% of patients with pathogens non-susceptible to ceftazidime alone at baseline, which is comparable with that observed in patients with ceftazidime-susceptible pathogens [150].

In this phase 2 study, AEs were mild or moderate in intensity in the majority of cases and AEs occurring in ≥5% patients in either group are summarized in Table 14.12.

Because of the urgent unmet medical need for antibiotics in the treatment of MDR Gram-negative bacterial infections, the FDA allowed for a new regulatory pathway during this trial period. Subsequently, the ceftazidime and avibactam combination was approved in February 2015 for the treatment of adult patients with cUTI, prior to the completion of the phase 3 trials [130]. At that time, ceftazidime–avibactam (brand name Avycaz®) was the only combination therapy available to have activity against KPCs. Consequently, it was recommended that the combination be reserved for those patients with no other treatment options [151].

The PK data from phase 1 and 2 studies were used to develop population PK models for the ceftazidime and avibactam combination and a cIAI population was used to simulate exposures and calculate the PK/PD target attainment (PTA). Using the preclinical animal data for ceftazidime and avibactam as a guide, the joint PTA for ceftazidime and avibactam (joint PTA) was defined as 50%fT > MIC for ceftazidime using the ceftazidime and avibactam MIC and 50%fT > 1.0 mg/l for

Table 14.11 Clinical response of ceftazidime and avibactam combination plus metronidazole for cIAI in the phase 2 clinical trial.

Population	Ceftazidime and avibactam plus metronidazole response rate (%)	Meropenem response rate (%)
ME (TOC)	91.2	93.4
ME (End i.v.)	97.1	97.4
ME (LFU)	Similar to TOC group	Similar to TOC group
CE (TOC)	92.0	94.4
CE (End i.v.)	96.6	97.8
CE (LFU)	91.9	94.4
mMITT (TOC)	82.4	88.8
Ceftazidime-NS	96	94

ME, microbiologically evaluable; CE, clinically evaluable; End i.v., end of i.v. therapy; LFU, late follow-up; ME, microbiologically evaluable; mMITT, microbiologically modified intent-to-treat; NS, nonsusceptible; TOC, test-of-cure.

Table 14.12 Percentages of phase 2 ceftazidime/avibactam treatment-emergent adverse events in ≥5% of patients.

	Complicated urinary tract infection		Complicated intra-abdominal infection	
AE	**Ceftazidime–avibactam (%)**	**Imipenem–cilastatin (%)**	**Ceftazidime–avibactam plus metronidazole (%)**	**Meropenem (%)**
Nausea	0	3.0	9.9	5.9
Vomiting	0	0	13.9	4.9
Constipation	10.3	3.0	4.0	1.0
Diarrhea	8.8	10.4	5.0	4.9
Abdominal pain	14.7	6	7.9	2.9
Headache	19.1	31.3	3.0	2.9
Dizziness	5.9	0	0	2.0
Insomnia	5.9	6.0	0	2.0
Hypertension	5.9	3.0	2.0	2.9
Pyrexia	0	1.5	8.9	10.8
Chest pain	5.9	4.5	1.0	1.0
Pyuria	0	0	5.0	4.9

AE, adverse event.

avibactam. The PTA calculations indicated a breakpoint of 8/4 mg/l for the dose regimen of 2000 mg ceftazidime combined with 500 mg avibactam given as a two-hour i.v. infusion every eight hours. This was the dose employed in the phase 3 cUTI clinical trials and was the dose eventually approved by the FDA. Creatinine clearance was identified as one of the primary predictors of variability in ceftazidime and avibactam PK. The dose regimens based on renal function were simulated for ceftazidime dose adjustments, with the avibactam dose adjusted to maintain the ceftazidime to avibactam ratio at 4:1. Based on the simulation, dosage adjustments are recommended for patients with renal impairment and are summarized in Table 14.13 [128, 143, 151, 152].

14.3.3.2 Ceftazidime and Avibactam Combination Phase 3 Clinical Trials

A number of phase 3 clinical trials were carried out to evaluate the efficacy and safety of the ceftazidime and avibactam combination. Three phase 3 clinical trials (RECAPTURE 1 [NCT01595438],

Table 14.13 Dosage recommendations for the ceftazidime and avibactam combination based on the patients' renal function.

Creatinine clearance (ml/min)	Dose recommendation: ceftazidime–avibactam
>50	2 g/0.5 g every 8 h
31–50	1 g/0.25 g every 8 h
16–30	0.75 g/0.19 g every 12 h
6–15	0.75 g/0.19 g every 24 h
≤5	0.75 g/0.19 g every 48 h

RECAPTURE 2 [NCT01599806] [153], and REPRISE [NCT01644643] [154]) were studied for the treatment of cUTI patients, and another four phase 3 clinical trials (RECLAIM 1 [NCT01499290], RECLAIM 2 [NCT01500239] [155], RECLAIM 3 [NCT01726023] [156], and REPRISE [NCT01644643, with a much smaller number of cIAI patients] [154]) were employed for the treatment of cIAI patients, and an additional phase 3 study (REPROVE, NCT01808092) was carried out to treat hospital-acquired bacterial pneumonia (HABP) (including ventilator-associated bacterial pneumonia [VABP]) [157].

Ceftazidime and Avibactam Combination Phase 3 Clinical Trials in the Treatment of cUTI [153, 154]

The RECAPTURE-1 and RECAPTURE-2 trials evaluated the efficacy and tolerability of ceftazidime–avibactam in the treatment of adults with cUTI (including pyelonephritis) which were two identical phase 3 parallel-group trials designed to incorporate FDA and EMA guidance. Primary endpoints as required by the FDA included the proportion of patients with symptomatic resolution (or return to premorbid state) of UTI-specific symptoms and the proportion of patients who had symptomatic resolution and bacterial eradication. EMA guidance was similar to the FDA, wherein the primary endpoint included the proportion of patients with a favorable per-patient microbiological response at TOC.

As summarized in Table 14.14, the primary efficacy results were generally consistent across the mMITT, extended microbiologically evaluable (eME), and ME populations. Noninferiority of ceftazidime–avibactam versus doripenem was demonstrated for the FDA co-primary endpoints, as well for the EMA primary endpoint. Secondary and exploratory efficacy endpoints were met as well, which are summarized in Table 14.14.

AEs were predominantly mild or moderate in intensity, and generally balanced across groups. No new safety concerns were identified in the combination therapy (Table 14.15).

REPRISE was the first multidrug-resistant Gram-negative pathogen-directed phase 3 study for ceftazidime–avibactam, focusing specifically on the efficacy, safety, and tolerability in patients with cUTI or cIAI due to ceftazidime-resistant Gram-negative pathogens. The primary endpoint was assessment of clinical response in the mMITT population at the TOC visit, 7–10 days after last infusion of study therapy.

The clinical outcome is summarized in Table 14.16. This study achieved both the primary endpoint and key secondary endpoints. Similar clinical cure rates were observed with the ceftazidime and avibactam combination and best available therapy for the treatment of serious ceftazidime resistant Gram-negative cUTI. The number of patients with cIAI in this study for ceftazidime–avibactam was small which limited the interpretation of the findings in this population, but the proportion of patients with a clinical and microbiological response was also similar to that observed with the best available therapy.

In summary, the results from the phase 3 RECAPTURE and REPRISE trials demonstrated the effectiveness of the ceftazidime and avibactam combination including against ceftazidime nonsusceptible pathogens and reaffirmed its potential clinical usage as a carbapenem replacement treatment in patients with resistant Gram-negative infections. The safety and tolerability profile of ceftazidime–avibactam in this phase 3 study is generally similar to the safety profile of ceftazidime alone.

Ceftazidime and Avibactam Combination Phase 3 Clinical Trials in the Treatment of cIAI [154–156]

RECLAIM 1 (NCT01499290) and RECLAIM 2 (NCT01500239) were two identical, prospective, and comparative global studies to evaluate the efficacy and safety of ceftazidime–avibactam plus metronidazole versus meropenem in hospitalized adults with cIAI. A clinical cure at days 28–35

Table 14.14 Primary and secondary efficacy endpoints in mMITT population.

Endpoint	Ceftazidime–avibactam (%)	Doripenem (%)
FDA co-primary endpoints		
Patient-assessed symptomatic resolution at day 5	70.2	66.2
Combined patient-assessed symptomatic resolution and favorable per-patient microbiological response at TOC	71.2	64.5
Per-patient favorable microbiological response at TOC	77.4	71.0
Patient-reported symptomatic resolution at TOC	84.5	86.3
EMA primary endpoint		
Per-patient favorable microbiological response at TOC	77.4	71.0
Secondary endpoints		
Microbiological		
Per-patient favorable microbiological response at End i.v.	95.2	94.7
Per-patient favorable microbiological response at LFU	68.2	60.9
Per-patient favorable microbiological response at TOC in patients with a ceftazidime-nonsusceptible pathogen	62.7	60.7
Per-patient favorable microbiological response at LFU in patients with a ceftazidime-nonsusceptible pathogen	61.3	45.2
Per-patient favorable microbiological response at TOC in patients with a ceftazidime-susceptible pathogen	81.0	73.0
Per-patient favorable microbiological response at LFU in patients with a ceftazidime-susceptible pathogen	69.9	64.1
Clinical		
Investigator-determined clinical cure (End i.v.)	96.2	97.6
Investigator-determined clinical cure (TOC)	90.3	90.4
Investigator-determined clinical cure (LFU)	85.2	83.9
Sustained clinical cure at LFU in patients who were cured at TOC	93.0	91.5
Investigator-determined clinical cure at TOC in patients with a ceftazidime-susceptible pathogen	90.8	90.5
Investigator-determined clinical cure at TOC in patients with a ceftazidime-nonsusceptible pathogen	89.3	89.3

End i.v., end of i.v. therapy; LFU, late follow-up; TOC, test-of-cure.

assessed by noninferiority of ceftazidime–avibactam plus metronidazole to meropenem was the primary endpoint.

The efficacy data of these phase 3 studies is summarized in Table 14.17. This study met its primary endpoint. For the secondary endpoint, a favorable clinical cure rate was also observed for ceftazidime–avibactam plus metronidazole in comparison with meropenem. Based on these favorable clinical response rates, it was clear that ceftazidime–avibactam in combination with metronidazole achieved noninferiority compared with meropenem and was effective in the treatment of cIAI. This combination was also effective in patients with ceftazidime susceptible Gram-negative pathogens with similar clinical cure rates to meropenem. Since an invasive procedure is required

Table 14.15 Safety data in safety population up to late follow-up.

AE	Ceftazidime–avibactam (%)	Doripenem (%)
Any AE	36.2	31
Any serious AE	4.1	2.4
Any AE leading to discontinuation of study drug	1.4	1.2
Any AE of severe intensity	2.0	1.4
AEs reported in ≥2% of patients in either treatment group by system organ class and preferred term		
Headache	7.4	7.9
Nausea	2.9	2.0
Diarrhea	2.7	1.2
Constipation	2.2	1.4

AE, adverse event.

Table 14.16 Clinical response in the REPRISE trial[a] in mMITT population at test-of-cure visit.

	Complicated urinary tract infection		Complicated intra-abdominal infection	
Clinical response	Ceftazidime–avibactam	Best available therapy	Ceftazidime–avibactam	Best available therapy
Clinical cure at TOC visit				
Overall response	91% (cUTI and cIAI combined) vs. 91% (best available therapy)			
	92%	94%	80%	55%
Patients with acute pyelonephritis	91%	90%		
Patients without acute pyelonephritis	92%	99%		
Microbiological response at the TOC visit				
	82%	64%		
Patients with acute pyelonephritis	88%	70%		
Patients without acute pyelonephritis	78%	58%		

[a] In this phase 3 study, the cIAI population is very small, thus some data were not included for comparison.

for intra-abdominal cultures, microbiological outcomes in the mMITT population were presumed to be similar to clinical responses based on clinical outcome.

In conclusion, in these phase 3 studies, ceftazidime–avibactam plus metronidazole met all criteria for the primary endpoint of noninferiority compared with meropenem across all primary analysis populations and the combination therapy was effective in treating hospitalized adults with cIAI with no new safety signals. These results support the use of ceftazidime–avibactam plus metronidazole as a potential alternative to carbapenems in the treatment of cIAI.

Table 14.17 Clinical cure rate in the RECLAIM 1 and 2 clinical trials.

Population	mMITT		MITT		CE at TOC	
	Ceftazidime–avibactam plus metronidazole	Meropenem	Ceftazidime–avibactam plus metronidazole	Meropenem	Ceftazidime–avibactam plus metronidazole	Meropenem
Primary endpoint						
Cure at TOC (%)	81.6	85.1	82.5	84.9	91.7	92.5
Difference (% [95% CI])	−3.5 (−8.64 to 1.58)		−2.4 (−6.90 to 2.10)		−0.8 (−4.61 to 2.89)	
Secondary endpoint						
Cure at EOT (%)	87.4	92.4	88.3	92.2	92.9	95.2
Difference (% [95% CI])	−5.0 (−9.24 to −0.93)		−3.9 (−7.57 to −0.29)		Not reported	
Cure at LFU visit (%)	82.3	84.6	82.5	83.4	90.0	90.4
Difference (% [95% CI])	−2.3 (−7.41 to 2.79)		−0.9 (−5.45 to 3.72)		Not reported	

CE, clinically evaluable; CI, confidence interval; EOT, end-of-treatment; LFU, late follow-up; MITT, modified intention-to-treat; mMITT, microbiologically modified intention-to-treat; TOC, test-of-cure.

RECLAIM 3 (NCT01726023) was a phase 3 comparative study to assess the efficacy and safety of ceftazidime–avibactam plus metronidazole compared with meropenem in patients with cIAI in Asian countries. The primary efficacy endpoint was the clinical cure rate in CE populations at the TOC visit and secondary endpoints including clinical cure rates in a variety of additional infected groups.

The clinical results of this study are summarized in Table 14.18. The data confirmed the non-inferiority of ceftazidime–avibactam plus metronidazole versus meropenem. Secondary endpoints were not formally assessed against a noninferiority margin; however, clinical cure rate at the EOT and LFU visits in the CE population, as well as the EOT, TOC and LFU visits in the ME and eME populations, and the EOT visit in the mMITT population were all similar to the primary analysis result (Table 14.18). Microbiological responses were again presumed from clinical responses due to the involvement of an invasive procedure in intra-abdominal culture. Ceftazidime–avibactam plus metronidazole was effective in treating infections caused by ceftazidime-nonsusceptible Gram-negative pathogens and by ceftazidime-susceptible isolates with similar clinical cure rates.

In summary, ceftazidime–avibactam plus metronidazole was noninferior to meropenem in the treatment of patients with cIAI in Asian countries including China, the Republic of Korea and Vietnam with safety findings for ceftazidime–avibactam in line with other populations and the known safety profiles of ceftazidime. In addition, for ceftazidime-nonsusceptible pathogens, ceftazidime–avibactam plus metronidazole was also effective as shown in this study.

Ceftazidime and Avibactam Combination Phase 3 Clinical Trials in the Treatment of Nosocomial Pneumonia and Ventilator-Associated Pneumonia [157, 158]

REPROVE was a prospective phase III non-inferiority trial in patients with nosocomial pneumonia (NP) commonly associated with antimicrobial resistant Gram-negative pathogens, including

Table 14.18 Clinical cure rate of ceftazidime–avibactam plus metronidazole compared with meropenem by visit and analysis population in the RECLAIM 3 study.

Population	Ceftazidime–avibactam plus metronidazole (%)	Meropenem (%)
CE (EOT)	96.3	95.4
CE (TOC)	93.8	94.0
CE (LFU)	93.5	93.9
mMITT (EOT)	88.1	92.1
mMITT (TOC)	83.2	88.8
mMITT (LFU)	81.1	86.8
eME (EOT)	97.2	96.0
eME (TOC)	93.0	95.0
eME (LFU)	92.8	94.6
ME (EOT)	97.2	95.8
ME (TOC)	92.9	94.7
ME (LFU)	92.7	94.3

CE, clinically evaluable; eME, extended microbiologically evaluable; EOT, end-of-treatment; LFU, late follow-up; ME, microbiologically evaluable; mMITT, microbiologically modified intention-to-treat; TOC, test-of-cure.

VAP. The primary endpoint was the proportion of patients clinically cured in co-primary clinically modified intention-to-treat (cMITT) and CE populations at the TOC visit [158].

The clinical results of this study are summarized in Table 14.19. Non-inferiority of ceftazidime–avibactam to meropenem was demonstrated in both co-primary analysis populations. Similar results were observed in secondary analysis populations (Table 14.20).

The REPROVE phase 3 results confirmed the efficacy and safety of ceftazidime–avibactam and demonstrated non-inferiority of ceftazidime–avibactam versus meropenem for the treatment of NP including VAP caused by ceftazidime-nonsusceptible or ceftazidime-susceptible Gram-negative aerobic pathogens. The data further support that the ceftazidime and avibactam combination can be used as a carbapenem-replacement strategy for serious Gram-negative infections including NP. The safety profile of ceftazidime–avibactam was comparable with that

Table 14.19 Clinical cure rate of ceftazidime–avibactam compared with meropenem by visit and analysis population in the REPROVE study.

Population	Ceftazidime–avibactam (%)	Meropenem (%)	Between group difference (% [95% CI])
CE (TOC)	77.4	78.1	−0.7 (−7.86, 6.39)
cMITT (TOC)	68.8	73.0	−4.2 (−10.76, 2.46)
MITT (EOT)[a]	79.6	80.6	−1.1 (−6.4, 4.3)
MITT (TOC)[a]	67.2	69.1	−1.9 (−8.1, 4.3)
mMITT (EOT)[a]	80.2	87.2	−7.0 (NA)
mMITT (TOC)[a]	67.4	73.3	−6.0 (NA)
mMITT (TOC, CAZ-NS)[a]	75.5	71.2	4.3 (NA)

[a] Clinical cure rates at EOT and TOC visits (ITT and micro-ITT populations).
CAZ-NS, ceftazidime nonsusceptible; CE, clinically evaluable; CI, confidence interval; cMITT, clinically modified intention-to-treat; EOT, end-of-treatment; TOC, test-of-cure; MITT, modified intention-to-treat; mMITT, microbiologically modified intention-to-treat; NA, no answer.

Table 14.20 Per-pathogen clinical cure rates and per-patient microbiological response rates at TOC across treatment groups in the REPROVE study.

	Ceftazidime–avibactam (%)	Meropenem (%)	Between group difference (% [95% CI])
Per-pathogen clinical cure rates			
Patients infected with ceftazidime-nonsusceptible pathogens	80.6	78.0	2.5 (−16.42, 20.74)
Patients infected with ceftazidime-susceptible pathogens	75.0	78.4	−3.4% (−16.18, 9.30)
Per-patient microbiological response rates			
mMITT	55.6	64.1	−8.6 (−18.65, 1.64)
eME	64.0	67.9	−3.9 (−15.49, 7.66)
ME	65.4	70.3	−4.9 (−17.10, 7.28)

CI, confidence interval; eME, extended microbiologically evaluable; ME, microbiologically evaluable; mMITT, microbiologically modified intention-to-treat; TOC, test-of-cure.

of ceftazidime alone and consistent with the known profile of ceftazidime–avibactam with no new safety concerns identified.

Based on the results of this pivotal phase 3 REPROVE study, in February 2018, the US FDA approved the sNDA of the combination which also expanded the approved use of the ceftazidime–avibactam combination (Avycaz®) to include the treatment of HABP/VABP caused by susceptible Gram-negative microorganisms in adult patients.

Ceftazidime and Avibactam Combination Phase 3 Clinical Trials in the Treatment of Pediatric Patients with cUTI and cIAI [159]

The efficacy, safety and tolerability of the ceftazidime and avibactam combination in pediatric patients were also evaluated in two active-controlled studies in children or infants with cIAI (vs. meropenem) and cUTI (vs. cefepime). In the cIAI study, the clinical cure rate in the intent-to-treat (ITT) population at the TOC visit was similar for the ceftazidime and avibactam combination plus metronidazole group (91.8%) and the meropenem group (95.5%). In the cUTI study, the combined clinical and microbiological response rate in the microbiological-ITT population at TOC was slightly better in the combination group (72.2%) than in the cefepime group (60.9%). No new safety concerns were observed in the pediatric population for the combination therapy and the overall safety profile was found to be similar in the pediatric and adult trials. Based on these results, in March 2019, the FDA expanded the approval of the ceftazidime and avibactam combination to include pediatric patients aged ≥3 months for the treatment of cUTI including pyelonephritis, and for cIAI used in combination with metronidazole [159].

14.3.4 Summary of Ceftazidime and Avibactam Combination

The ceftazidime and avibactam combination is a rationally designed combination therapy leveraging the broad-spectrum β-lactamase inhibitor avibactam to increase the susceptibility of resistant pathogens to ceftazidime. It showed good activity against Gram-negative bacteria, especially ESBLs MDR *P. aeruginosa*, and carbapenemase-resistant Enterobacteriaceae which are serious health threats in the USA with few safe and effective treatments currently available. In addition, some recent real-life treatments suggest that the ceftazidime and avibactam combination could be effective in treating other off-label indications such as skin and soft tissue or bone infections, primary or catheter-related bacteremia, and endocarditis. The combination also showed an effect on gut decolonization which could be useful to reduce the risk of a systemic infection and possible relapses of previously cured infections [160].

While this combination showed promising efficacy in treating the above-mentioned Gram-negative bacterial infections, it was not effective in the treatment of infections caused by class B metallo-β-lactamase-producing Enterobacterales, and not as effective in the treatment of infections originated from *A. baumannii*. In addition, a rational use of these new therapeutic options is critically important in managing multidrug resistance since resistance has already been observed with the ceftazidime and avibactam combination. There are several factors attributed to the occurrence of resistance including point mutations in existing β-lactamases. In addition, resistance mechanisms other than β-lactamase modifications can arise such as target site modifications and reduced permeability due to porin alterations. To prevent the rapid development of resistance and maintain its clinical utility, patient populations should be carefully managed, and it should not be used for pathogens that are naturally resistant to this combination [150]. With the possibility of increased resistance toward ceftazidime–avibactam after extended use and its intrinsic ineffectiveness against the above-mentioned pathogens, more effective and powerful antibacterial agents are urgently needed. Novel antibiotic combinations will continue to play a critical role in this regard and may be the only option in the absence of new antibiotics with novel MOA [161].

14.4 Combination Therapy Future Perspectives

As illustrated in this chapter, both the olaparib and bevacizumab combination and ceftazidime and avibactam combination were rationally designed combination therapies based on an understanding of fundamental biological processes. With the increase in cancer diagnoses and a global surge in new cancer cases, and the emergence of difficult to treat MDR, PDR, and XDR bacterial infections, these unmet medical needs require better and more efficacious new drugs, and drug combinations may play an essential role in this regard. In order to achieve better outcomes for drug combinations, a few factors should be taken into consideration: (i) Understanding drug resistance and employing translational biomarkers may hold the key to the development of future successful drug combinations. With the advances in genomic technologies and understanding the underlying biology of human diseases, and recognizing the roles of extensive genetic heterogeneity, clonal selection and adaptive feedback loops in drug resistance will provide valuable information in the design of more effective drug combinations. The availability of predictive and molecular response biomarkers, the understanding of molecular interactions to attack complementary cancer hallmarks or distinct cell populations in heterogeneous tumors, and the implementation of novel combination mechanisms such as "contextual synthetic lethality," will help to increase the translational success of future drug combinations [15]. (ii) In addition to predictive biomarkers, multiple factors may contribute to a better clinical outcome from preclinically predicted combination effects. In this context, factors include drug ratios, the difference in drug PK profiles (drug concentration, exposure time, and amount), drug administration schedule, and analytic methods for evaluating drug interactions, just to name a few. Due to the independent PKs and biodistribution of individual drugs in the clinical setting, the translation of preselected *in vitro* drug ratios to the clinical setting is complex and may result in the loss of therapeutic activity due to insufficient therapeutic target coverage. The difficulty to control drug ratios in the systemic circulation especially in tumor tissues may partly contribute to the less than expected efficacy outcome in clinical trials predicted by preclinical studies, not to mention that preclinical animal models for cancer are notoriously poor predictors of human efficacy. Thus, these factors should be taken into consideration when evaluating the effects of drug combination in preclinical settings based on established cell lines, primary cell cultures, or animal models [11]. (iii) More complex combination strategies employing more than two drug components (three-, four-, or even five-molecule combinations) may be a potential solution to treat human cancer and combat hard-to-treat bacterial infections. Among the currently available treatment options, most of the cancer drugs and almost all new antibiotics will face the threat of resistance. Multiple drug combinations might be a solution to overcome drug resistance and insufficient efficacy originating from alternative pathway-related escape mechanisms. In this regard, and using antibacterial drug combination as an example, a MAGIC (mathematical analysis for general interactions of components) framework was recently applied to evaluate the effectiveness of multiple antibiotic combinations on pathogenic *E. coli*. Working with eight antibiotics, the researchers analyzed every possible four- and five-drug combination (a total of over 18 000 combinations in all). Among the four- and five-drug combinations, there were 1676 and 6443 groupings that performed better than they expected, respectively. While these are promising results, the drug combinations have only been tested in a laboratory setting and it remains to be seen whether these results will translate to viable treatments. On the other hand, this might be worth considering in the absence of new antibiotics. Last but not least, the advent of new technology such as artificial intelligence may help to accelerate the selection and translational process [162].

Since the discovery of the BRCA1 and BRCA2 genes in 1994 and 1995, HR within the DNA repair pathway has become a target for cancer therapy. It took almost 20 years to translate this discovery into viable treatment of human disease. In December 2014, olaparib was granted accelerated approval by the US FDA as a therapy for OC patients with germline BRCA1/2 mutations who have received three or more prior lines of chemotherapy [73]. On the other hand, since Napoleone Ferrara and colleagues at Genentech first isolated and cloned vascular endothelial growth factor in 1989, it took almost 15 years for the first approval of bevacizumab (Avastin®) in combination with i.v. 5-fluorouracil-based chemotherapy for first- or second-line treatment of patients with metastatic colon or rectal carcinoma in 2004. Sixteen years after its first approval, bevacizumab combination therapies were adopted as important therapeutic options and led to the approval of the olapanib and bevacizumab combination in 2020. Similarly for the ceftazidime and avibactam combination, it was approved by the FDA and EMA more than 30 years after the initial approved use of ceftazidime in the early 1980s. Finding cures for cancer, bacterial infections, and other human diseases is time consuming and requires relentless and collaborative efforts from the discovery and development scientists, clinicians, data scientists, and regulatory agencies. Combination therapies will play increasingly important roles in the never-ending quest to provide the best treatment options for life-threatening human diseases.

References

1 Editorial (2017). Rationalizing combination therapies. *Nat. Med.* 23 (10): 1113.

2 Ascierto, P.A. and Marincola, F.M. (2011). Combination therapy: the next opportunity and challenge of medicine. *J. Transl. Med.* 9: 115.

3 Nasrallah, H.A. (2010). Combination therapy is here to stay. *Curr. Psychiatr. Ther.* 9 (5): 11.

4 Toews, M.L. and Bylund, D.B. (2005). Pharmacologic principles for combination therapy. *Proc. Am. Thorac. Soc.* 2: 282–289.

5 Pritcharda, J.R., Brunoa, P.M., Gilberta, L.A. et al. (2012). Defining principles of combination drug mechanisms of action. *Proc. Natl. Acad. Sci. USA*: E170–E179.

6 Tremblay, C.L. and Hirsch, M.S. (2004). The basic principles for combination therapy. In: *Combination Therapy of AIDS. Milestones in Drug Therapy MDT* (eds. E.D.A. De Clercq and A.M.I. Vandamme), 41. Basel: Birkhäuser.

7 Jia, J., Zhu, F., Ma, X. et al. (2009). Mechanisms of drug combinations: interaction and network perspectives. *Nat. Rev. Drug Discov.* 8: 111–128.

8 Lehr, J., Krueger, A.S., Avery, W. et al. (2009). Synergistic drug combinations tend to improve therapeutically relevant selectivity. *Nat. Biotechnol.* 27: 659–666.

9 Huang, X.H. and Chen, D.Y.-K. (2012). A case study of single-pill combination therapy: the ezetimibe/simvastatin combination for treatment of hyperlipidemia. *Chem. Med. Chem.* 7 (11): 1884–1892.

10 Saputra, E.C., Huang, L., Chen, Y., and Tucker-Kellogg, L. (2018). Combination therapy and the evolution of resistance: the theoretical merits of synergism and antagonism in cancer. *Cancer Res.* 78 (9): 2419–2431.

11 Pinto, A.C., Moreira, J.N., and Simões, S. (2011). Combination chemotherapy in cancer: principles, evaluation and drug delivery strategies. http://www.intechopen.com/books/current-cancer-treatment-novel-beyond-conventionalapproaches/combination-chemotherapy-in-cancer-principles-evaluation-and-drug-delivery-strategies (accessed 2 October 2021).

12 Mayer, L.D. and Janoff, A.S. (2007). Optimizing combination chemotherapy by controlling drug ratios. *Mol. Interv.* 7 (4): 216–223.

13 Zoli, W., Ricotti, L., Tesei, A. et al. (2001). in vitro preclinical models for a rational design of chemotherapy combinations in human tumors. *Crit. Rev. Oncol. Hematol.* 37 (1): 69–82.

14 U.S. Department of Health and Human Services Food and Drug Administration Center for Drug Evaluation and Research (CDER) (2013). Guidance for industry: codevelopment of two or more new investigational drugs for use in combination. https://www.fda.gov/media/80100/download (accessed 2 October 2021).

15 Al-Lazikani, A., Banerji, U., and Workman, P. (2012). Combinatorial drug therapy for cancer in the post-genomic era. *Nat. Biotechnol.* 30 (7): 1–13.

16 National Cancer Institute (n.d.) Cancer prevalence and cost of care projections. https://costprojections.cancer.gov (accessed 2 October 2021).

17 Jemal, B.F., Grey, A., Ferlay, N., and Forman, D. (2012). Global cancer transitions according to the Human Development Index (2008–2030): a population-based study. *Lancet Oncol.* 13 https://doi.org/10.1016/S1470-2045(12)70211-52012.

18 Silicon Valley Bank (2020).Impressive performance in an extraordinary year to date. https://www.svb.com/globalassets/library/managedassets/pdfs/healthcare-report-2020-midyear.pdf (accessed 2 October 2021).

19 Tremont, A., Lu, J., and Cole, J.T. (2017). Endocrine therapy for early breast cancer: updated review. *Ochsner J.* 17: 405–411.

20 Student, S., Hejmo, T., Poterała-Hejmo, A. et al. (2020). Anti-androgen hormonal therapy for cancer and other diseases. *Eur. J. Pharmacol.* 866: 172783.

21 Rossari, F., Minutolo, F., and Orciuolo, E. (2018). Past, present, and future of Bcr-Abl inhibitors: from chemical development to clinical efficacy. *J. Hematol. Oncol.* 11: 84.

22 Jebbinka, M., de Langena, A.J., Boelensb, M.C. et al. (2020). The force of HER2 – a druggable target in NSCLC? *Cancer Treat. Rev.* 86: 101996.

23 Scholtens, A., Geukes Foppen, M.H., Blank, C.U. et al. (2015). B Vemurafenib for BRAF V600 mutated advanced melanoma: results of treatment beyond progression. *Eur. J. Cancer* 51: 642–652.

24 Troiani, T., Napolitano, S., Della Corte, C.M. et al. (2016). Therapeutic value of EGFR inhibition in CRC and NSCLC: 15 years of clinical evidence. *ESMO Open* 1: e000088.

25 Dobosz, P. and Dzieciatkowski, T. (2019). The intriguing history of cancer immunotherapy. *Front. Immunol.* 10: 2965.

26 Haslam, A. and Prasad, V. (2019). Estimation of the percentage of US patients with cancer who are eligible for and respond to checkpoint inhibitor immunotherapy drugs. *JAMA Netw. Open* 2: e192535.

27 Marin-Acevedo, J.A., Dholaria, B., Soyano, A.E. et al. (2018). Next generation of immune checkpoint therapy in cancer: new developments and challenges. *J. Hematol. Oncol.* 11: 39.

28 Lawrence, M.S., Stojanov, P., Mermel, C.H. et al. (2014). Discovery and saturation analysis of cancer genes across 21 tumour types. *Nature* 505 (7484): 495.

29 Housman, G., Byler, S., Heerboth, S. et al. (2014). Drug resistance in cancer: an overview. *Cancer* 6: 1769–1792.

30 Daub, H., Specht, K., and Ullrich, A. (2004). Strategies to overcome resistance to targeted protein kinase inhibitors. *Nat. Rev. Drug Discov.* 3: 1001–1010.

31 Leary, M., Heerboth, S., Lapinska, K., and Sarkar, S. (2018). Sensitization of drug resistant cancer cells: a matter of combination therapy. *Cancer* 10: 483.

32 Jeon, M., Kim, S., Park, S. et al. (2018). In silico drug combination discovery for personalized cancer therapy. *Syst. Biol.* 12 (Suppl. 2): 59–67.

33 Drake, C.G. (2012). Combination immunotherapy approaches. *Ann. Oncol.* 23 (Supplement 8): viii41–viii46.

34 Mokhtari, R.B., Homayouni, T.S., Baluch, N. et al. (2017). Combination therapy in combating cancer. *Oncotarget* 8 (23): 38022–38043.

35 Yadav, B., Wennerberg, K., Aittokallio, T., and Tang, J. (2015). Searching for drug synergy in complex dose–response landscapes using an interaction potency model. *Comput. Struct. Biotechnol. J.* 13: 504–513.

36 Berenbaum, M.C. (1989). What is synergy? *Pharmacol. Rev.* 41: 93–141.

37 Loewe, S. (1953). The problem of synergism and antagonism of combined drugs. *Arzneimittelforschung* 3: 285–290.

38 Bliss, C.I. (1939). The toxicity of poisons applied jointly. *Ann. Appl. Biol.* 26: 585–615.

39 Chou, T.-C. and Talalay, P. (1983). Analysis of combined drug effects: a new look at a very old problem. *Trends Pharmacol. Sci.* 4: 450–454.

40 Chou, T.C. and Talalay, P. (1984). Quantitative analysis of dose-effect relationships: the combined effects of multiple drugs or enzyme inhibitors. *Adv. Enzyme Regul.* 22: 27–55.

41 Zimmer, A., Katzir, I., Dekel, E. et al. (2016). Prediction of multidimensional drug dose responses based on measurements of drug pairs. *Proc. Natl. Acad. Sci. USA* 113: 10442–10447.

42 Schindler, M. (2017). Theory of synergistic effects: hill-type response surfaces as "null-interaction" models for mixtures. *Theor. Biol. Med. Model.* 14: 15.

43 Twarog, N.R., Stewart, E., Hammill, C.V., and Shelat, A.A. (2016). BRAID: a unifying paradigm for the analysis of combined drug action. *Sci. Rep.* 6: 25523.

44 Koch, G., Walz, A., Lahu, G., and Schropp, J. (2009). Modeling of tumor growth and anticancer effects of combination therapy. *J. Pharmacokinet. Pharmacodyn.* 36: 179–197.

45 Meyer, C.T., Wooten, D.J., Paudel, B.B. et al. (2019). Quantifying drug combination synergy along potency and efficacy axes. *Cell Syst.* 8: 97–108.

46 Di Veroli, G.Y., Fornari, C., Wang, D. et al. (2016). Combenefit: an interactive platform for the analysis and visualization of drug combinations. *Bioinformatics* 32: 2866–2868.

47 Ianevski, A., Giri, A.K., Gautam, P. et al. (2019). Prediction of drug combination effects with a minimal set of experiments. *Nat. Mach. Intell.* 1: 568–577.

48 Ianevski, A., Timonen, S., Kononov, A. et al. (2020). SynToxProfiler: an interactive analysis of drug combination synergy, toxicity and efficacy. *PLoS Comput. Biol.* 16 (2): e1007604.

49 He, L., Kulesskiy, E., Saarela, J. et al. (2018). Methods for high-throughput drug combination screening and synergy scoring. *Methods Mol. Biol.* 1711: 351–398.

50 Ianevski, A., He, L., Aittokallio, T., and Tang, J. (2017). SynergyFinder: a web application for analyzing drug combination dose–response matrix data. *Bioinformatics* 33: 2413–2415.

51 Mathews Griner, L.A., Guha, R., Shinn, P. et al. (2014). High-throughput combinatorial screening identifies drugs that cooperate with ibrutinib to kill activated B-cell-like diffuse large B-cell lymphoma cells. *Proc. Natl. Acad. Sci. USA* 111: 2349–2354.

52 He, L., Zheng, S., Wennerberg, K. et al. (2020). User instructions and tutorials of the synergyfinder package. http://bioconductor.riken.jp/packages/3.11/bioc/vignettes/synergyfinder/inst/doc/synergyfinder.pdf (accessed 2 October 2021).

53 Kanj, S.S. and Kanafani, Z.A. (2011). Current concepts in antimicrobial therapy against resistant Gram-negative organisms: extended-spectrum β-lactamaseproducing Enterobacteriaceae, carbapenem-resistant Enterobacteriaceae, and multidrug-resistant Pseudomonas aeruginosa. *Mayo Clin. Proc.* 86: 250–259.

54 Boucher, H.W., Talbot, G.H., Bradley, J.S. et al. (2009). Bad bugs, no drugs: no ESKAPE! An update from the Infectious Diseases Society of America. *Clin. Infect. Dis.* 48 (1): 1–12.

55 Centers for Disease Prevention and Control (CDC) (2019). Antibiotic resistance threats in the United States. https://www.cdc.gov/drugresistance/pdf/threats-report/2019-ar-threats-report-508.pdf (accessed 2 October 2021).

56 WHO (2019). Antibacterial agents in clinical development: an analysis of the antibacterial clinical development pipeline. https://apps.who.int/iris/bitstream/handle/10665/330420/9789240000193-eng.pdf (accessed 2 October 2021).

57 Pitout, J.D., Nordmann, P., and Poirel, L. (2015). Carbapenemase-producing Klebsiella pneumoniae, a key pathogen set for global nosocomial dominance. *Antimicrob. Agents Chemother.* 59 (10): 5873–5884.

58 Qureshi, Z.A., Paterson, D.L., Potoski, B.A. et al. (2012). Treatment outcome of bacteremia due to KPC-producing Klebsiella pneumoniae: superiority of combination antimicrobial regimens. *Antimicrob. Agents Chemother.* 56 (4): 2108–2113.

59 Levin, S. and Harris, A. (1975). Principles of combination therapy. *Bull. N. Y. Acad. Med.*: 2010–1038.

60 Baeder, D.Y., Yu, G., Hozé, N. et al. (2016). Antimicrobial combinations: Bliss independence and Loewe additivity derived from mechanistic multi-hit models. *Philos. Trans. R. Soc. B* 371: 20150294. http://dx.doi.org/10.1098/rstb.2015.0294.

61 Wertheimer, A.I. and Morrison, A. (2002). Combination drugs: innovation in pharmacotherapy. *Pharmacol. Ther.* 27 (1): 44–49.

62 Tyers, M. and Wright, G.D. (2019). Drug combinations: a strategy to extend the life of antibiotics in the 21st century. *Nat. Rev. Microbiol. Rev.* 17: 141–155.

63 Moellering, R.C. Jr. (1983). Rationale for use of antimicrobial combinations. *Am. J. Med.* 75: 4–8.

64 Keith, C.T., Borisy, A.A., and Stockwell, B.R. (2005). Multicomponent therapeutics for networked systems. *Nat. Rev. Drug Discov.* 4: 71–78.

65 American Cancer Society (n.d.) Key statistics for ovarian cancer. https://www.cancer.org/cancer/ovarian-cancer/about/key-statistics.html#:~:text=The%20American%20Cancer%20Society%20estimates,will%20die%20from%20ovarian%20cancer (accessed 2 October 2021).

66 Global Cancer Observatory (n.d.) Globocan 2012 statistics. http://globocan.iarc.fr (accessed 2 October 2021).

67 Ferlay, J., Steliarova-Foucher, E., Lortet-Tieulent, J. et al. (2013). Cancer incidence and mortality patterns in Europe: estimates for 40 countries in 2012. *Eur. J. Cancer* 49: 1374–1403.

68 Fung-Kee-Fung, M., Oliver, T., Elit, L. et al. (2007). Optimal chemotherapy treatment for women with recurrent ovarian cancer. *Curr. Oncol.* 14: 195–208.

69 Virag, L. and Szabo, C. (2002). The therapeutic potential of poly(ADP-ribose) polymerase inhibitors. *Pharmacol. Rev.* 54: 375–429.

70 Rouleau, M., Patel, A., Hendzel, M.J. et al. (2010). PARP inhibition: PARP1 and beyond. *Nat. Rev. Cancer* 10: 293–301.

71 Velic, D., Couturier, A.M., Ferreira, M.T. et al. (2015). DNA damage Signalling and repair inhibitors: the long-sought-after Achilles' heel of cancer. *Biomolecules* 5: 3204–3259.

72 Tangutooria, S., Baldwin, P., and Sridhar, S. (2015). PARP inhibitors: a new era of targeted therapy. *Maturitas* 81 (1): 5–9.

73 Walsh, C.S. (2015). Two decades beyond BRCA1/2: homologous recombination, hereditary cancer risk and a target for ovarian cancer therapy. *Gynecol. Oncol.* 137: 343–350.

74 Bryant, H.E., Schultz, N., Thomas, H.D. et al. (2005). Specific killing of BRCA2-deficient tumours with inhibitors of poly(ADP-ribose) polymerase. *Nature* 434: 913–917.

75 Farmer, H., McCabe, N., Lord, C.J. et al. (2005). Targeting the DNA repair defect in BRCA mutant cells as a therapeutic strategy. *Nature* 434: 917–921.

76 Ashworth, A. (2008). A synthetic lethal therapeutic approach: poly(ADP)ribose polymerase inhibitors for the treatment of cancers deficient in DNA double-strand break repair. *J. Clin. Oncol.* 26: 3785–3790.

77 Cancer Genome Atlas Research Network (2011). Integrated genomic analyses of ovarian carcinoma. *Nature* 474: 609–615.

78 Drew, Y. (2015). The development of PARP inhibitors in ovarian cancer: from bench to bedside. *Br. J. Cancer* 113: S3–S9.

79 Dockery, L.E., Gunderson, C.C., and Moore, K.N. (2017). Rucaparib: the past, present, and future of a newly approved PARP inhibitor for ovarian cancer. *Onco. Targets. Ther.* 10: 3029–3037.

80 Donawho, C.K., Luo, Y., Penning, T.D. et al. (2007). ABT-888, an orally active poly(ADP-ribose) polymerase inhibitor that potentiates DNA-damaging agents in preclinical tumor models. *Clin. Cancer Res.* 13 (9): 2728–2737.

81 Jones, P., Altamura, S., Boueres, J. et al. (2009). Discovery of 2-{4-[(3S)-piperidin-3-yl]phenyl}-2H-indazole-7-carboxamide (MK-4827): a novel oral poly(ADP-ribose)polymerase (PARP) inhibitor efficacious in BRCA-1 and -2 mutant tumors. *J. Med. Chem.* 52 (22): 7170–7185.

82 Wang, B., Chu, D., Feng, Y. et al. (2016). Discovery and characterization of (8S,9R)-5-Fluoro-8-(4-fluorophenyl)-9-(1-methyl-1H-1,2,4-triazol-5-yl)-2,7,8,9-tetrahydro-3H-pyrido[4,3,2-de]phthalazin-3-one (BMN 673, Talazoparib), a novel, highly potent, and orally efficacious poly(ADP-ribose) polymerase-1/2 inhibitor, as an anticancer agent. *J. Med. Chem.* 59 (1): 335–357.

83 Menear, K.A., Adcock, C., Boulter, R. et al. (2008). 4-[3-(4-Cyclopropanecarbonylpiperazine1-carbonyl)-4-fluorobenzyl]-2H-phth alazin-1-one: a novel bioavailable inhibitor of poly(ADP-ribose) polymerase-1. *J. Med. Chem.* 51: 6581–6591.

84 Kaufman, B., Shapira-Frommer, R., Schmutzler, R.K. et al. (2014). Olaparib monotherapy in patients with advanced cancer and a germline BRCA1/2 mutation. *J. Clin. Oncol.* 33 (3): 244–250.

85 Fong, P.C., Boss, D.S., Yap, T.A. et al. (2009). Inhibition of poly(ADP-ribose) polymerase in tumors from BRCA mutation carriers. *N. Engl. J. Med.* 361: 123–134.

86 Fong, P.C., Yap, T.A., Boss, D.S. et al. (2010). Poly(ADP)-ribose polymerase inhibition: frequent durable responses in BRCA carrier ovarian cancer correlating with platinum-free interval. *J. Clin. Oncol.* 28: 2512–2519.

87 Balmaña, J., Tung, N.M., Isakoff, S.J. et al. (2014). Phase I trial of olaparib in combination with cisplatin for the treatment of patients with advanced breast, ovarian and other solid tumors. *Ann. Oncol.* 25: 1656–1663.

88 Weil, M.K. and Chen, A. (2011). PARP inhibitor treatment in ovarian and breast cancer. *Curr. Probl. Cancer* 35 (1): 7–50.

89 Ricks, T.K., Chiu, H.-J., Ison, G. et al. (2015). Successes and challenges of PARP inhibitors in cancer therapy. *Front. Oncol.* 5: 222.

90 Center for Drug Evaluation and Research (2014). Pharmacology review(s). https://www.accessdata.fda.gov/drugsatfda_docs/nda/2014/206162Orig1s000PharmR.pdf (accessed 2 October 2021).

91 Oza, A.M., Cibula, D., Benzaquen, A.O. et al. (2015). Olaparib combined with chemotherapy for recurrent platinum-sensitive ovarian cancer: a randomised phase 2 trial. *Lancet Oncol.* 16 (1): 87–97.

92 Penson, R.T., Valencia, R.V., Cibula, D. et al. (2020). Olaparib versus nonplatinum chemotherapy in patients with platinum-sensitive relapsed ovarian cancer and a germline BRCA1/2 mutation (SOLO3): a randomized phase III trial. *J. Clin. Oncol.* 38 (11): 1164–1174.

93 Matulonis, U.A., Harter, P., Gourley, C. et al. (2016). Olaparib maintenance therapy in patients with platinum-sensitive, relapsed serous ovarian cancer and a BRCA mutation: overall survival adjusted for Postprogression poly(adenosine diphosphate ribose) polymerase inhibitor therapy. *Cancer* 122: 1844–1852.

94 Moore, K.N., DiSilvestro, P., Lowe, E.S. et al. (2014). SOLO1 and SOLO2: randomized phase III trials of olaparib in patients (pts) with ovarian cancer and a BRCA1/2 mutation (BRCAm). *Proc. Am. Soc. Clin. Oncol.* 32 (Suppl. 15) Abstract, TPS5616.

95 Poveda, A., Floquet, A., Ledermann, J. et al. (2020). Final overall survival results from SOLO2/ENGOT-ov21: a phase III trial assessing survival after maintenance treatment with the PARP inhibitor olaparib in patients with platinum-sensitive, relapsed ovarian cancer and a BRCA mutation. *J. Clin. Oncol.* 38 (Suppl. 15): 6002. https://doi.org/10.1200/JCO.2020.38.15_suppl.6002.

96 Moore, K., Colombo, N., Scambia, G. et al. (2018). Maintenance Olaparib in atients with newly diagnosed advanced ovarian cancer. *N. Engl. J. Med.* 379: 2495–2505.

97 Bergers, G. and Benjamin, L. (2003). Tumorigenesis and the angiogenic switch. *Nat. Rev. Cancer* 3: 401–410.

98 Lugano, R., Ramachandran, M., and Dimberg, A. (2020). Tumor angiogenesis: causes, consequences, challenges and opportunities. *Cell. Mol. Life Sci.* 77: 1745–1770.

99 Apte, R.S., Chen, D.S., and Ferrara, N. (2019). VEGF in signaling and disease: beyond discovery and development. *Cell* 176 (6): 1248–1264.

100 European Medicines Agency (EMA) (2017). Avastin: Summary of product characteristics (SmPC). https://www.ema.europa.eu/en/documents/product-information/avastin-epar-product-information_en.pdf (accessed 2 October 2021).

101 Presta, L.G., Chen, H., O'Connor, S.J. et al. (1997). Humanization of an anti-vascular endothelial growth factor monoclonal antibody for the therapy of solid tumors and other disorders. *Cancer Res.* 57: 4593–4599.

102 US FDA (2018). Avastin: US prescribing information (USPI). https://www.accessdata.fda.gov/drugsatfda_docs/label/2018/125085s323lbl.pdf (accessed 2 October 2021).

103 Hurwitz, H., Fehrenbacher, L., Novotny, W. et al. (2004). Bevacizumab plus irinotecan, fluorouracil, and leucovarin for metastatic colorectal cancer. *New Engl. J. Med.* 350: 2335–2342.

104 National Comprehensive Cancer Network (NCCN) (2019). NCCN Clinical Practice Guidelines in Oncology (NCCN Guidelines): Colon Cancer, Version 3 (26 September 2019).

105 Garcia, J., Hurwitz, H.I., Sandler, A.B. et al. (2020). Bevacizumab (Avastin®) in cancer treatment: a review of 15 years of clinical experience and future outlook. *Cancer Treat. Rev.* 86: 102017.

106 Pujade-Lauraine, E., Hilpert, F., Weber, B. et al. (2014). Bevacizumab combined with chemotherapy for platinum-resistant recurrent ovarian cancer: the AURELIA open-label randomized phase III trial. *J. Clin. Oncol.* 32 (13): 1302–1308.

107 Coleman, R.L., Brady, M.F., Herzog, T.J. et al. (2017). Bevacizumab and paclitaxel–carboplatin chemotherapy and secondary cytoreduction in recurrent, platinum-sensitive ovarian cancer (NRG Oncology/Gynecologic Oncology Group study GOG-0213): a multicentre, open-label, randomised, phase 3 trial. *Lancet Oncol.* 18 (6): 779–791.

108 Aghajanian, C., Goff, B., Nycum, L.R. et al. (2015). Final overall survival and safety analysis of OCEANS, a phase 3 trial of chemotherapy with or without bevacizumab in patients with platinum-sensitive recurrent ovarian cancer. *Gynecol. Oncol.* 139 (1): 10–16.

109 Tewari, K.S., Burger, R.A., Enserro, D. et al. (2019). Final verall survival of a randomized trial of bevacizumab for primary treatment of ovarian cancer. *J. Clin. Oncol.* 37 (26): 2317–2328.

110 Bindra, R.S., Gibson, S.L., Meng, A. et al. (2005). Hypoxiainduced down-regulation of BRCA1 expression by E2Fs. *Cancer Res.* 65: 11597–11604.

111 Chan, N. and Bristow, R.G. (2010). "Contextual" synthetic lethality and/or loss of heterozygosity: tumor hypoxia and modification of DNA repair. *Clin. Cancer Res.* 16: 4553–4560.

112 Wei, W., Li, Y., Lv, S. et al. (2016). PARP-1 may be involved in angiogenesis in epithelial ovarian cancer. *Oncol. Lett.* 12: 4561–4567.

113 Drean, A., Lord, C.J., and Ashworth, A. (2016). PARP inhibitor combination therapy. *Crit. Rev. Oncol. Hematol.* 108: 73–85.

114 Dean, E., Middleton, M.R., Pwint, T. et al. (2012). Phase I study to assess the safety and tolerability of olaparib in combination with bevacizumab in patients with advanced solid tumors. *Br. J. Cancer* 106 (3): 468–474.

115 Mirza, M.R., Avall-Lundqvist, E., Birrer, M.J. et al. (2019). Combination of niraparib and bevacizumab versus niraparib alone as treatment of recurrent platinum-sensitive ovarian cancer: a randomized controlled chemotherapy-free study: NSGO-AVANOVA2/ENGOT-OV24. *J. Clin. Oncol.* 37 (Suppl): 5505. https://doi.org/10.1200/JCO.2019.37.15_suppl.5505.

116 Liu, J.F., Barry, W.T., Birrer, M. et al. (2014). Combination cediranib and olaparib versus olaparib alone for women with recurrent platinum-sensitive ovarian cancer: a randomised phase 2 study. *Lancet Oncol.* 15: 1207–1214.

117 Liu, J.F., Barry, W.T., Birrer, M. et al. (2019). Overall survival and updated progression-free survival outcomes in a randomized phase II study of combination cediranib and olaparib versus olaparib in relapsed platinum-sensitive ovarian cancer. *Ann. Oncol.* 30: 551–557.

118 Ray-Coquard, I., Pautier, P., Pignata, S. et al. (2019). Olaparib plus bevacizumab as first-line maintenance in ovarian cancer. *N. Engl. J. Med.* 381: 2416–2428.

119 National Institutes of Health/National Cancer Institute (n.d.). Clinical trials using olaparib. https://www.cancer.gov/about-cancer/treatment/clinical-trials/intervention/olaparib (accessed 2 October 2021).

120 National Institutes of Health/National Cancer Institute (n.d.). Clinical trials using bevacizumab. https://www.cancer.gov/about-cancer/treatment/clinical-trials/intervention/bevacizumab (accessed 2 October 2021).

121 Fojo, T. and Bates, S. (2013). Mechanisms of resistance to PARP inhibitors–three and counting. *Cancer Discov.* 3: 20–23.

122 Murai, J., Zhang, Y., Morris, J. et al. (2014). Rationale for poly(ADP-ribose) polymerase (PARP) inhibitors in combination therapy with Camptothecins or Temozolomide based on PARP trapping versus catalytic inhibitions. *J. Pharmacol. Exp. Ther.* 349: 408–416.

123 Miles, D., Cameron, D., Bondarenko, I. et al. (2017). Bevacizumab plus paclitaxel versus placebo plus paclitaxel as first-line therapy for HER2-negative metastatic breast cancer (MERiDiAN): a double-blind placebo-controlled randomised phase III trial with prospective biomarker evaluation. *Eur. J. Cancer* 70: 146–155.

124 Concia, E., Mazzaferri, F., and Cordioli, M. (2016). New antibiotic development: barriers and opportunities. *Ital. J. Med.* 10: 255–271.

125 Bush, K. and Bradford, P.A. (2016). β-Lactams and β-lactamase inhibitors: an overview. *Cold Spring Harb. Perspect. Med.* 6 (8): a025247.

126 Allergan USA, Inc. (2020). Avycaz (ceftazidime and avibactam): highlights of prescribing information. https://media.allergan.com/actavis/actavis/media/allergan-pdf-documents/product-prescribing/Avycaz_Final_PI_CBE-0_10_2019.pdf (accessed 2 October 2021).

127 Ehmann, D.E., Jahic, H., Ross, P.L. et al. (2012). Avibactam is a covalent, reversible, non-b-lactam b-lactamase inhibitor. *Proc. Natl. Acad. Sci. USA* 109: 11663–11668.

128 Mawal, Y., Critchley, I.A., Riccobene, T.A., and Talley, A.K. (2015). Ceftazidime–avibactam for the treatment of complicated urinary tract infections and complicated intra-abdominal infections. *Expert Rev. Clin. Pharmacol.* 8 (6): 691–707. https://doi.org/10.1586/17512433.2015.1090874.

129 Drawz, S.M., Papp-Wallace, K.M., and Bonomo, R.A. (2014). New b-lactamase inhibitors: a therapeutic renaissance in an MDR world. *Antimicrob. Agents Chemother.* 58: 1835–1846.

130 Zasowski, E.J., Rybak, J.M., and Rybak, M.J. (2015). The b-lactams strike back: ceftazidime-avibactam. *Pharmacotherapy* 35 (8): 755–770.

131 Lagacé-Wiens, P., Walkty, A., and Karlowsky, J.A. (2014). Ceftazidime–avibactam: an evidence-based review of its pharmacology and potential use in the treatment of Gram negative bacterial infections. *Core Evid.* 9: 13–25. http://dx.doi.org/10.2147/CE.S40698.

132 Merdjan, H., Girard, A.M., Miossec, C. et al. (2009). Pharmacokinetics (PK) and efficacy of ceftazidime (CAZ)/NXL104 combination in a murine pneumonia model caused by an AmpC-producing *Klebsiella pneumoniae* [abstract A1-006]. 49th Annual Interscience Conference on Antimicrobial Agents and Chemotherapy, San Francisco, CA.

133 Crandon, J.L., Schuck, V.J., Banevicius, M.A. et al. (2012). Comparative in vitro and in vivo efficacies of human simulated doses of ceftazidime and ceftazidime-avibactam against Pseudomonas aeruginosa. *Antimicrob. Agents Chemother.* 56 (12): 6137–6146.

134 Endimiani, A., Hujer, K.M., Hujer, A.M. et al. (2011). Evaluation of ceftazidime and NXL104 in two murine models of infection due to KPC-producing Klebsiella pneumoniae. *Antimicrob. Agents Chemother.* 55: 82–85.

135 Borgonovi, M., Miossec, C., and Lowther, J. (2007). The efficacy of ceftazidime combined with NXL104, a novel β-lactamase inhibitor, in a mouse model of kidney infections induced by β-lactamase producing Enterobacteriaceae [abstract P794]. 17th European Congress of Clinical Microbiology and Infectious Diseases, Munich.

136 Vazquez, J.A., González Patzán, L.D., Stricklin, D. et al. (2012). Efficacy and safety of ceftazidime-avibactam versus imipenem-cilastatin in the treatment of complicated urinary tract infections, including acute pyelonephritis, in hospitalized adults: results of a prospective, investigator-blinded, randomized study. *Curr. Med. Res. Opin.* 28: 1921–1931.

137 Craig, W.A. (1995). Interrelationship between pharmacokinetics and pharmacodynamics in determining dosage regimens for broad-spectrum cephalosporins. *Diagn. Microbiol. Infect. Dis.* 22: 89–96.

138 Craig, W.A. (2003). Basic pharmacodynamics of antibacterials with clinical applications to the use of b-lactams, glycopeptides, and linezolid. *Infect. Dis. Clin. North Am.* 17: 479–501.

139 Craig, W.A. (2007). Pharmacodynamics of antimicrobials: general concepts and applications. In: *Antimicrobial Pharmacodynamics in Theory and Clinical Practice*, 2e (eds. C.H. Nightingale, P.G. Ambrose, G.L. Drusano and T. Murakawa), 129–146. New York: Informa Healthcare.

140 Muller, A.E., Punt, N., and Mouton, J.W. (2013). Optimal exposures of ceftazidime predict the probability of microbiological and clinical outcome in the treatment of nosocomial pneumonia. *J. Antimicrob. Chemother.* 68: 900–906.

141 Berkhout, J., Melchers, M.J.B., Van Mil, A.C. et al. (2016). Pharmacodynamics of ceftazidime and avibactam in neutropenic mice with thigh or lung infection. *Antimicrob. Agents Chemother.* 60 (1): 368–375. https://doi.org/10.1128/AAC.01269-15.

142 Coleman, K., Levasseur, P., Girard, A.M. et al. (2014). Activities of ceftazidime and avibactam against b-lactamase-producing Enterobacteriaceae in a hollow-fiber pharmacodynamic model. *Antimicrob. Agents Chemother.* 58: 3366–3372.

143 US FDA (2014). Intravenous ceftazidime-avibactam (CAZ-AVI) for treatment in adults with cIAI, cUTI, HABP/VABP, or bacteremia where there are limited or no treatment options. https://wayback.archive-it.org/7993/20170405204434/https://www.fda.gov/downloads/AdvisoryCommittees/CommitteesMeetingMaterials/Drugs/Anti-InfectiveDrugsAdvisoryCommittee/UCM426700.pdf (accessed 4 October 2021).

144 MacVane, S.H., Crandon, J.L., Nichols, W.W., and Nicolau, D.P. (2014). in vivo efficacy of humanized exposures of ceftazidime-avibactam in comparison with ceftazidime against contemporary Enterobacteriaceae isolates. *Antimicrob. Agents Chemother.* 58 (11): 6913–6919.

145 Toussaint, K.A. and Gallagher, J.C. (2015). β-lactam/β-lactamase inhibitor combinations: from then to now. *Ann. Pharmacother.* 49 (1): 86–98.

146 Chahine, E.B., Sourial, M., and Ortiz, R. (2015). Ceftazidime/avibactam: a new antibiotic for gram-negative infections. *Consult. Pharm.* 30 (12): 695–705.

147 US FDA (2007). FORTAZ (ceftazidime for injection) prescribing information. http://www.accessdata.fda.gov/drugsatfda_docs/label/2007/050578s053,050634s020lbl.pdf (accessed 2 October 2021).

148 Merdjan, H., Rangaraju, M., and Tarral, A. (2015). Safety and pharmacokinetics of single and multiple ascending doses of avibactam alone and in combination with ceftazidime in healthy male volunteers: results of two randomized, placebo-controlled studies. *Clin. Drug Investig.* 35: 307–317.

149 Lucasti, C., Popescu, I., Ramesh, M.K. et al. (2013). Comparative study of the efficacy and safety of ceftazidime/avibactam plus metronidazole versus meropenem in the treatment of complicated intra-abdominal infections in hospitalized adults: results of a randomized, double-blind, phase II trial. *J. Antimicrob. Chemother.* 68: 1183–1192.

150 Sharma, R., Park, T.E., and Moy, S. (2016). Ceftazidime-avibactam: a novel cephalosporin/β-lactamase inhibitor combination for the treatment of resistant gram-negative organisms. *Clin. Ther.* 38: 431–444.

151 Center for Drug Evaluation and Research (2015). Medical review(s). https://www.accessdata.fda.gov/drugsatfda_docs/nda/2015/206494Orig1s000MedR.pdf (accessed 2 October 2021).

152 Carrothers, T. J., Green, M., Chiu, J., et al. (2014). Population pharmacokinetic modeling of combination treatment of intravenous ceftazidime and avibactam (abstract T-071). Fifth American Conference on Pharmacometrics (ACoP5), Las Vegas, NV.

153 Wagenlehner, F.M., Sobel, J.D., Newell, P. et al. (2016). Ceftazidime-avibactam versus doripenem for the treatment of complicated urinary tract infections, including acute pyelonephritis: RECAPTURE, a phase 3 randomized trial program. *Clin. Infect. Dis.* 63 (6): 754–762.

154 Carmeli, Y., Armstrong, J., Laud, P.J. et al. (2016). Ceftazidime-avibactam or best available therapy in patients with ceftazidime-resistant Enterobacteriaceae and Pseudomonas aeruginosa complicated urinary tract infections or complicated intra-abdominal infections (REPRISE): a randomised, pathogen-directed, phase 3 study. *Lancet Infect. Dis.* 16 (6): 661–673.

155 Mazuski, J.E., Gasink, L.B., Armstrong, J. et al. (2016). Efficacy and safety of ceftazidime-avibactam plus metronidazole versus meropenem in the treatment of complicated intra-abdominal infection: results from a randomized, controlled, double-blind, phase 3 program. *Clin. Infect. Dis.* 62 (11): 1380–1389.

156 Qin, X., Tran, B.G., Kim, M.J. et al. (2017). A randomised, double-blind, phase 3 study comparing the efficacy and safety of ceftazidime/avibactam plus metronidazole versus meropenem for complicated intra-abdominal infections in hospitalised adults in Asia. *Int. J. Antimicrob. Agents* 49 (5): 579–588.

157 Torres, A., Zhong, N., Pachl, J. et al. (2017). Ceftazidime-avibactam versus meropenem in nosocomial pneumonia, including ventilator-associated pneumonia (REPROVE): a randomised, double-blind, phase 3 non-inferiority trial. *Lancet Infect. Dis.* 18 (3): 285–295.

158 Torres, A., Rank, D., Melnick, D. et al. (2019). Randomized trial of ceftazidime-avibactam vs meropenem for treatment of hospital-acquired and ventilator-associated bacterial pneumonia (REPROVE): analyses per US FDA-specified end points. *Open Forum Infect. Dis.* 6 (4): ofz149. https://doi.org/10.1093/ofid/ofz149.

159 Bradley, J.S., Roilides, E., Broadhurst, H. et al. (2019). Safety and efficacy of ceftazidime–avibactam in the treatment of children ≥3 months to <18 years with complicated urinary tract infection: results from a phase 2 randomized, controlled trial. *Pediatr. Infect. Dis. J.* 38 (9): 920–928.

160 Diet, B., Martínez, L.M., Calbo, E., and Garau, J. (2020). Update on the role of ceftazidime avibactam in the management of carbapenemase-producing Enterobacterales. *Future Microbiol.* 15 (7): 473–484. https://doi.org/10.2217/fmb-2020-0012.

161 Wang, Y., Wang, J., Wang, R., and Cai, Y. (2020). Resistance to ceftazidime–avibactam and underlying mechanisms. *J. Glob. Antimicrob. Resist.* 22: 18–27.

162 Tekin, E., White, C., Kang, T.M. et al. (2018). Prevalence and patterns of higher-order drug interactions in *Escherichia coli*. *npj Syst. Biol. Appl.* 4: 31. https://doi.org/10.1038/s41540-018-0069-9.

15

Accelerating Drug Discovery and Development: Translational Medicine in Combating the COVID-19 Pandemic

Xianhai Huang[1], *David Yu-Kai Chen*[2], *and Wayne Haifeng Tang*[3]

[1] *Discovery Chemistry, InventisBio Co., Ltd., 25A Hanover Road #340, NJ 07932, Florham Park, NJ, USA*
[2] *Department of Chemistry, Seoul National University, 1 Gwanak-ro, 151-742, Gwanak Gu, Seoul, South Korea*
[3] *Drug Discovery Group, Schrödinger Inc., 120 West 45 Street, 17th floor, NY 10036, New York, NY, USA*

15.1 Introduction to Translational Medicine

Drug discovery and development is a very complex, challenging, and expensive undertaking. From the beginning of the process when a biological hypothesis relevant to a particular disease is proposed and investigated in a research lab to a marketed drug takes a very long time. The average development time of a marketed drug is over 10 years from lead generation without counting the conceptual stage in academic labs, and the time from phase 1 to phase 3 studies averages 6.5 years [1]. Furthermore, even with the advancements in basic science and technological breakthroughs of recent years, the cost of discovery of a new drug is increasing year on year with a current estimated cost of 2–3 billion dollars [1, 2]. The high research and development (R&D) cost can primarily be associated with several factors occurring after a drug candidate enters development including low clinical success rates due to translational uncertainties, increasingly stringent requirements from regulators, and expensive and prolonged clinical trials, although long discovery and development times due to scientific complexities in the discovery phase can also play a role. In addition, after a drug navigates through the process from discovery to development to market, its regulatory journey may still face difficulties if phase 4 trials (post approval evaluations of the marketed drug) reveal previously unknown adverse events. The high cost and unpredictability of drug R&D places an enormous burden on the economy and society, and solutions to remedy this situation remain unclear and a top priority of the global health agenda.

In an effort to improve this situation, translational medicine was introduced in the early 1990s not only in the context of drug R&D but also in areas such as behavioral interventions and medical devices. By definition, translational medicine is an interdisciplinary branch of the biomedical field that involves various professions, resources, expertise, and techniques within the different specialized groups. The ultimate goal is to improve the diagnosis and treatment of diseases in a shorter time with a reduced cost [3]. In translational research, the collective efforts of doctors, scientific researchers, engineers, philanthropists, healthcare professionals and many others ultimately converge at the patient, and further conveys its impact to the entire community. Translational medicine is a bidirectional platform with continuous feedback functionality intended to close the gap

Contemporary Accounts in Drug Discovery and Development, First Edition. Edited by Xianhai Huang, Robert G. Aslanian, and Wayne H. Tang.

between "what we know" and "what we practice": one direction is termed "bench to bedside" (basic science to clinical science and public health impact) which facilitates the introduction of new diagnostics, medicinal products or new medical knowledge from their conception to marketed products; the other direction is termed "bedside to bench" wherein new knowledge gained in clinical practice informs new scientific hypotheses in the laboratory. Simply put, translational science plays a key bridging role between biomedical discovery and clinical science and patients in the identification of tools to facilitate and accelerate drug product development to improve public health. This is a rather short explanation of translational medicine, but it is often a very lengthy and complex process and usually not a straightforward one. Any negative findings or failed studies during this long process may lead to an abandonment or adjustment of the original theory and a return to the beginning of the process. As with most of the drug discovery process, there will be many more failed translational research projects than the ones that succeed and make it to the market. Successful translational medicine programs will engender meaningful communication between research labs and the clinic so that new research findings and the unmet clinical/medical needs can be shared and mutually understood by research scientists and the clinicians.

Considering its importance, many universities started to offer translational medicine as a new discipline. A 4T model was proposed to illustrate the different stages of translational medicine (Figure 15.1): T1, translation of basic research (T0) to first in human: basic research prompted by unmet medical needs to human phase 1/2 clinical trials; T2, translation to patients: controlled studies leading to effective care in phase 2/3 clinical trials; T3, translation to practice: delivery of care to the right patient in post approval phase 4 clinical trials; T4, translation to the community: true benefit and impact to society [4, 5]. However, instead of a linear model, translational medicine is more accurately and best represented in a bidirectional loop as illustrated in Figure 15.2 [6].

To facilitate the later discussions and to illustrate the aforementioned bidirectional interactions, a few opening remarks regarding translational medicine are summarized here: (i) In order to improve the success rate and reduce the cost of developing a new drug, surrogate endpoints for clinical testing need to be identified in order to pinpoint the patient groups who may have the best clinical outcomes. The use of biological markers as surrogate endpoints to assess the benefits of a given treatment in the early stages of clinical study is one of the key features of translational medicine, which can help to identify the most predictive *in vitro* and *in vivo* biomarkers in order to facilitate the transition of product candidates into the clinic with shortened time, reduced cost, and higher clinical success. (ii) While *in vivo* preclinical animal models are often used to predict human efficacy, human clinical studies are the only way to ascertain the efficacy and effectiveness of a drug candidate. In this context, the track record of animal models to predict human results has been less than satisfactory, and translational medicine may facilitate the translation of efficacy in

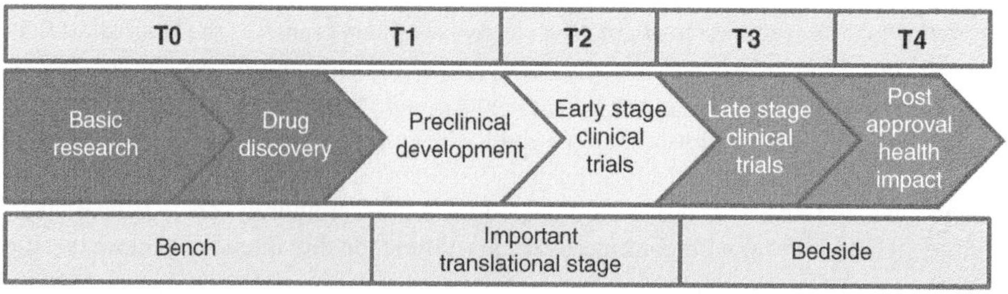

Figure 15.1 Graphic illustration of bench and bedside in connection with translational phase.

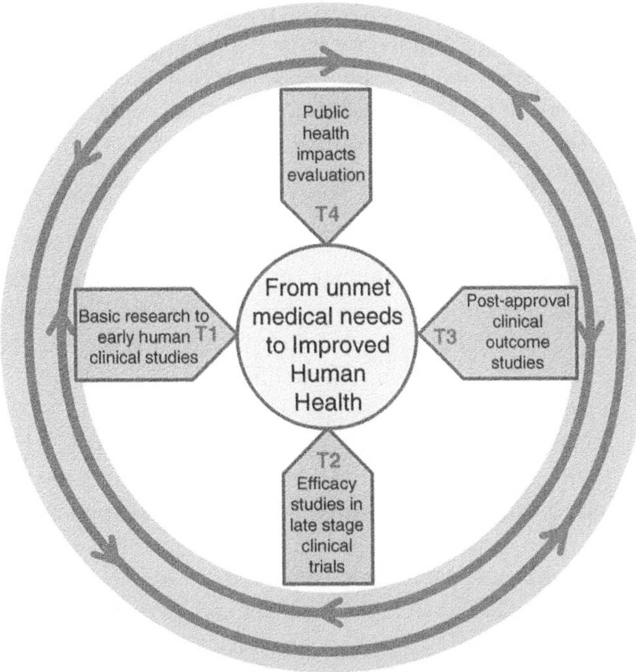

Figure 15.2 Optimal bidirectional interactions between the four phases of translational medicine in a loop manner.

animal models to humans through preclinical studies and early phase clinical trials (Figure 15.1). (iii) The bidirectional nature of translational medicine facilitates the identification of new drug targets and disease relevant biomarkers, in doing so, it accelerates lead compound optimization and provides a better understanding of drug inadequacy and clinical needs. Last but not least, cooperation between basic researchers and clinicians plays a crucial role in the selection of new drug candidates with enhanced efficacy and minimized toxicity, together with the invention of new diagnostic tools and medical devices with better accuracy and efficiency [7].

In this chapter we will use the COVID-19 pandemic battle as a case study of translational medicine to showcase how bench work/basic research is used to facilitate the prediction, diagnosis, treatment, and prevention of COVID-19 infections, how gaps identified in clinical studies help to accelerate the development of new treatment options, and how these can be achieved with speed, accuracy, and efficiency.

15.2 From Bench to Bedside: Translating Basic Research into Desirable Clinical Outcomes for COVID-19 Treatments

On 31 December 2019, the World Health Organization (WHO) was formally notified about a cluster of cases of pneumonia in Wuhan City, home to 11 million people and the cultural and economic hub of central China. By 5 January 2020, 59 cases were known and none had been fatal. Ten days later, WHO was aware of 282 confirmed cases, of which four were in Japan, South Korea, and Thailand. There had been six deaths in Wuhan, 51 people were severely ill and 12 were in a critical condition. The virus responsible was isolated on 7 January 2020 and its genome shared on 12 January 2020. The cause of the severe acute respiratory syndrome that became known as COVID-19 was a novel coronavirus (CoV), SARS-CoV-2.

On 11 February 2020, the WHO announced an official name for the disease as coronavirus disease 2019 (COVID-19), formerly referred to as "2019 novel coronavirus" or "2019-nCoV" [8]. It turned out this virus is highly contagious and is transmitted from person to person. COVID-19 quickly spread to the whole world within a few months. On 11 March 2020, the WHO declared COVID-19 a pandemic. At the time of writing (mid February 2021), there were more than 109 million confirmed COVID-19 cases worldwide and close to 2.42 million people have died with the numbers increasing daily. The COVID-19 pandemic is probably the most devastating human health threat since the 1918–1919 Spanish flu pandemic.

There are many types of human coronaviruses (HCoVs) including alpha, beta, gamma, and delta subtypes. Most CoV strains are narrow in their host range, but zoonotic CoV may infect new host species such as humans. Among the six coronaviruses that affected humans before 2019, four of these are common and cause mild illness (strains of 229E, NL63, OC43, and HKU1) while the other two (severe acute respiratory syndrome coronavirus, SARS-CoV; and Middle East respiratory syndrome coronavirus, MERS-CoV) are associated with more severe symptoms. Like SARS-CoV and MERS-CoV, COVID-19 is a beta type coronavirus caused by SARS-CoV-2. The exact source of this virus is unknown although all three viruses have their origins in bats. So far, COVID-19 has proven more contagious compared with earlier CoV outbreaks and has caused more death and devastating social consequences around the world.

In response to the COVID-19 pandemic, the worldwide scientific community has acted very quickly and worked together collaboratively to take the initiative to share scientific information openly to come up with treatment and preventative options to save lives and stop the spread of the disease. Although the battle is far from over, these efforts are an excellent display of the important role of translational medicine on a grand scale: accelerate the translation of basic science to life saving drugs with unprecedented speed under an urgent pandemic situation. The knowledge collected and experience gained from the efforts to fight COVID-19 will be extremely valuable to future drug discovery and development. We will showcase how translational medicine is applied in a "bidirectional manner" to accelerate the discovery of solutions to the diagnosis, treatment, and prevention of COVID-19 infection. We will also discuss both how basic scientific discovery and knowledge was transformed into diagnostic tools and drug products and how new clinical needs arising from the unpredictable nature of this disease has promoted new drug discovery efforts and diagnostic tool development.

SARS-CoV-2 is a single-stranded RNA-enveloped virus. Its entire genome was characterized through an RNA-based metagenomic next-generation sequencing shortly after the outbreak [9]. The main composition of SARS-CoV-2 includes a single-stranded genomic RNA, four main structural proteins (spike [S], membrane [M], envelope [E], and nucleocapsid [N] proteins) and a few special structural and accessory proteins (Figure 15.3) which are critical to the viral life cycle. The S protein is specifically involved in receptor recognition, viral attachment, and entry into host cells, and is highly conserved among all (HCoVs). This latter feature renders the S protein one of the most important targets for COVID-19 vaccine and therapeutic development, as discussed in later sections. (Many structures of the spike protein of SARS-CoV-2 were reported such as PDB code 6ZGE [10]). The S proteins are glycosylated and cover the surface of viral particle and permit cell entry through its binding to the angiotensin converting enzyme 2 (ACE2) receptor present on the host cell surface. The viral RNA is released and the translation of polyproteins from the RNA genome takes place once the virus enters the host cell. Once viral RNA replication and structural protein synthesis are complete, these components are assembled and packaged in the host cell to be released and cause further infection [11–13].

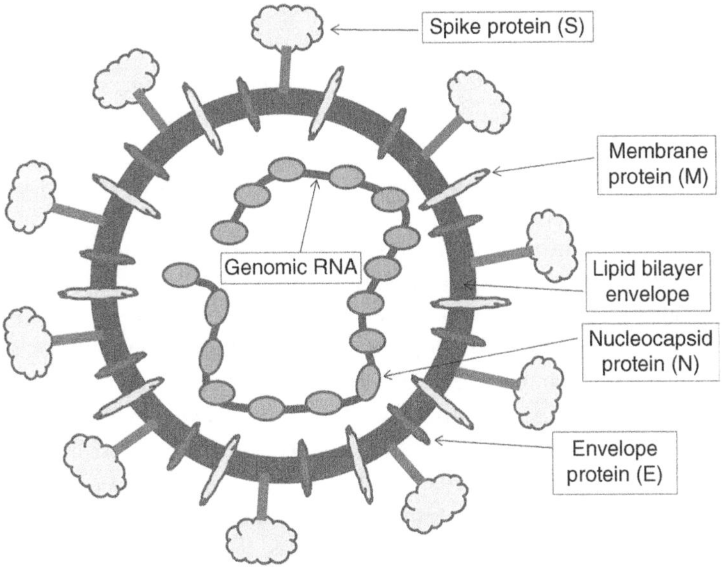

Figure 15.3 Cartoon of the major components of the COVID virus.

15.2.1 The Importance of Diagnostic Biomarkers in Speeding Up Testing to Contain the Spread of the COVID-19 Virus

Nonpharmaceutical control of the pandemic, including good personal hygiene, consistent use of personal protective equipment like masks, early detection, travel restrictions, social distancing to reduce human-to-human contact and isolation, is a broad topic that will not be discussed in depth here. This chapter will focus on pharmaceutical interventions including disease diagnosis and medical treatment to either stabilize or cure infected patients and prevent disease spread to showcase how the principles of translational medicine can help to minimize the impact of this pandemic.

Clearly, the first step in controlling the spread of a virus is to identify those individuals who are infected and likely to spread the disease to uninfected individuals. This can be accomplished using sensitive and selective diagnostic biomarker assays. There are three types of tests available for COVID-19: polymerase chain reaction (PCR), antigen, and antibody (serology) tests. PCR and antigen tests are diagnostic tests that determine whether an individual is presently infected, whereas antibody tests determine whether a person had a prior infection. An antibody test is a limited diagnostic tool since antibodies can take several days to develop after an infection and often stay in the blood for long periods of time after recovery. A list of PCR, antigen, and antibody tests can be found on the Food and Drug Administration (FDA) website [14].

15.2.1.1 The PCR Test

The PCR test is a direct measurement of the viral genetic material and high-sensitivity PCR tests are almost 100% accurate in identifying infected individuals. Under the current SARS-CoV-2 outbreak, identification of the disease etiology and sharing of the viral genetic sequence has led to the development of real-time PCR assays that have contributed immensely in the detection, containment, and tracking of the outbreak. This is an example of translational research that illustrates information sharing between different parties that resulted in significant time reduction in product development. However, PCR tests generally require trained personnel, specific reagents, and costly instruments, not to mention it can take several days for results to become available. Furthermore, logistic delays in sample collection and testing often occurs, as witnessed in the United States. A recent national survey showed that the average waiting time for nasal swab-based quantitative polymerase chain reaction (qPCR) test results was 4.1 days, among these, 31% of tests took more than 4 days and 10% of tests took 10 days or more [15]. These delays, along with a lack of tests, have contributed to the rapid spread of COVID-19 across the United States. The national backlog in processing laboratory-based tests clearly illustrates the need for rapid, point-of-care diagnostic tests with reliable results. An excellent application of translational science in this case is how new products can be quickly developed based on the needs of front-line healthcare workers for novel testing devices to address the issue of the high demand but low throughput of the PCR test. With a clear goal in mind, companies, and academic research labs are producing other PCR replacement tests that are faster, cheaper, and more user-friendly than the standard PCR assay, although they are not being produced on the same scale as antigen tests which will be discussed in the following. Some of these other tests use the gene-editing tool CRISPR to zero in on genetic fragments of the CoV. For example, a paper-based strip test using CRISPR to diagnose COVID was developed by Sherlock Bioscience, a company founded by MIT scientists. It gives fast results similar to a pregnancy test. This product was cleared for emergency use by the US FDA [16]. In a similar way, Indian scientists developed an inexpensive paper-based Feluda test for CoV with 96% sensitivity and 98% specificity. This new test also uses the CRISPR-based gene-editing technology to detect the virus. It has high reliability and can be produced rather quickly on a large scale. This is a simple, affordable, point-of-care test that is not limited by machines and personnel which may help to address the high test demands from a large population such as that in India [17]. More recently, Jennifer A. Doudna, the 2020 Nobel Laureate for her pioneering work in CRISPR gene editing, and her team developed an amplification-free CRISPR-Cas13a-based mobile phone assay for direct detection of SARS-CoV-2 from a nasal swab [18]. This assay accurately detected clinically positive samples in under five minutes and using CRISPR RNAs targeting SARS-CoV-2 RNA improved sensitivity and specificity together with direct quantification of viral load through enzyme kinetics. Reverse transcriptase polymerase chain reaction (RT-PCR) and saliva-based PCR tests are more rapid variants of the PCR test and were also developed and currently used in universities. However, even with the improved PCR tests, providing an adequate supply to meet the short-term demand remains challenging in the midst of the pandemic.

15.2.1.2 The Antigen Test

As discussed above, the PCR test is highly accurate at identifying the genetic material of the virus, but it is very difficult to scale up much further, though the CRISPR-based test may help to modestly improve the situation. However, it is unlikely that there will be a sufficient number of tests available to monitor everybody on a daily basis. Thus, there is a continued need to come up with a higher throughput and affordable COVID-19 test. Diagnostic biomarker-based tests such as the antigen test may help to solve the issue. The antigen test gives results in less than 30 minutes and does not have to be processed in a lab, and it is cheaper to produce. An antigen is a substance that can be identified by the body's immune system to trigger a response by generating proteins (antibodies) that specifically recognize that antigen. As mentioned earlier, the coronavirus has four main structural proteins: nucleocapsid (N), spike (S), membrane (M), and envelope (E). The highly immunogenic S protein is located on the surface of the virus and consists of S1 and S2 subunits. The N protein which is abundantly expressed during infections and also has high immunogenic activity, plays an important role in viral RNA transcription and replication, and packaging of the

Table 15.1 BinaxNOWTM COVID-19 Ag Card performance within seven days of symptom onset.

	Positive	Negative	Total	Agreement
Positive cases	34	1	35	97.1% (95% CI: 85.1–99.9%)
Negative cases	1	66	67	98.5% (95% CI: 92.0–100%)

CI, confidence interval.

encapsulated genome into virions. It also inhibits the cell cycle of the host cells. Therefore, both N and S proteins could be used as diagnostic biomarkers for antigen tests and employed for antibody-based detection of SARS-CoV-2 [19]. These tests could help to readily identify infectious individuals for isolation to avoid further infections, and several rapid and point-of-care antigen tests are being evaluated by WHO for Emergency Use Listing (EUL). Organizations such as The Bill & Melinda Gates Foundation and test producers Abbott and SD Biosensor are working on tests priced at a maximum of $5 for low- and middle-income countries [20]. While many antigen tests are currently available [21], the BinaxNOW™ COVID-19 Ag Card antigen test kit developed by Abbott Labs is worth highlighting in the context of translational research. This test uses nucleocapsid protein antigen (N) as the diagnostic biomarker, and its performance was demonstrated with symptomatic patients (within seven days of onset) who were suspected of COVID-19 infection. It has high sensitivity although it does not differentiate between SARS-CoV and SARS-CoV-2 (Table 15.1) [22], and it is important that patient samples are collected within seven days of symptom onset and tested immediately. This near-person rapid antigen test was engineered for point-of-care settings, and to meet the high demand. Abbott Labs has shipped tens of millions of its tests in September and October 2020.

Although they are fast, scalable, and highly accurate with regards to positive results, antigen tests have a higher chance of false negatives due to variable sensitivity due to the fact that antigen tests need samples with high concentration of virus particles per microliter to produce a result, whereas a typical PCR test can detect a single molecule of RNA in a microliter of solution. Therefore, if a person has low amounts of virus in their body at the time of the test a false-negative result might be obtained. For people who tested positive for SARS-CoV-2 in a standard PCR test, Abbott's antigen assay correctly spotted the virus in 95–100% of cases if the samples were collected within a week of the onset of symptoms. However, if samples were taken more than a week after first signs of symptoms, only 75% accuracy was observed. The sensitivity problem can be improved through repeated testing, and a second or third test will make it likely that anyone missed on the first test will be caught. Confirmation through a PCR test will also help to eliminate the false-negative cases.

15.2.1.3 The Antibody (Serological) Test

Since anti-SARS-CoV-2 antibodies can only be detected in blood by about seven days after the onset of clinical symptoms, antibody (serological) tests cannot be employed to detect acute infections. However, detection of antibodies complements viral testing and plays a very important role in the control and understanding of the COVID-19 pandemic: (i) Validated serologic assays let health care workers know who has acquired immunity from a prior exposure and is crucial for tracking patient contact, identifying the viral reservoir hosts, and epidemiologic studies. These are helpful to evaluate the extent of virus spread in communities to determine control measures, and to evaluate infection fatality rate and burden on the society. (ii) These assays can identify individuals as donors of convalescent plasma to treat infected individuals. (iii) Antibody detection can help to select appropriate individuals for vaccine trials or development of antibody therapy [23, 24].

The next questions are which antibody tests are more relevant for epidemiologic studies and which are better for supporting treatment and vaccine development. Among the four structural proteins of coronavirus, the spike (S) and nucleocapsid (N) proteins are the main immunogens that were used to develop antibody tests as previously discussed. However, the presence of antibodies alone does not suggest the individual is immune from reinfection. It is important to determine whether the antibodies are virus-neutralizing antibodies that block and help to clear viral infection, thus conferring long-term protective immunity. The S1 subunit of the viral entry S protein contains the immunologically crucial receptor binding domain (RBD) and is speculated to be the primary target of neutralizing antibodies [25]. In addition, there is a 90% N protein homology between SARS-CoV-2 and SARSCoV-1, compared with 77% with the S protein, and 66% with the S1 subunit RBD. The low homologies of the S1 subunits could lead to less cross-reactivity among the endemic CoVs, and the S1 subunit could be a specific target antigen for the detection of SARS-CoV-2 antibodies. Among recently developed serology assays for COVID-19, some use only the N protein as the antibody detection antigen. It is not clear whether such immune responses correlate with S-RBD response and COVID-19 immunity since patients with antibodies to the N protein and not the S1-RBD failed to exhibit neutralizing antibodies in some studies. Therefore, antibodies against the N protein indicate a prior exposure to the SARS-CoV-2 or related viruses, but not necessarily an indicator of the presence of neutralizing antibodies. Both N protein and S protein targeting antibody tests can contribute to epidemiologic studies, but the S protein might be better suited for the development of neutralizing antibodies and eventual vaccine development.

There have been many antibody testing kits developed based on the N and S antigens and a list of available tests can be found at the FDA website [26]. ELISA-based immunoglobulin M and immunoglobulin G (IgG) antibody tests can be used for diagnosis of COVID-19 with greater than 95% specificity and seroconversion yielding circulating IgG antibodies against S-RBD may reveal acquired immunity against SARSCoV-2. To illustrate the translational medicine aspects of the development of a serological test for COVID-19 immunity based on the understanding of the fundamental science, one product from PerkinElmer's Euroimmun is selected for discussion. Euroimmun was among the first companies to release an antibody assay based on the S1 subunit of the spike protein. This assay was independently validated by the US Government and gained Emergency Use Authorization (EUA) through collaborations with leading institutions, reference laboratories and academic centers [27]. This is an enzyme-linked immunosorbent assay intended for the qualitative detection of the IgG class of antibodies to SARS-CoV-2 in human serum or plasma and help to identify individuals with recent or prior infection with an adaptive immune response to SARS-CoV-2. The Euroimmun Anti-SARS-CoV-2 ELISA (IgG) displayed a positive agreement of 90% and a negative agreement of 100.0% [28]. More recently on 6 November 2020, GenScript USA Inc. gained US FDA EUA for their cPass SARS-CoV-2 Neutralization Antibody Detection Kit. This represents the first serology test that detects neutralizing antibodies from recent or prior SARS-CoV-2 infection. Before this announcement, the FDA granted EUA to more than 50 antibody (serology) tests, but those early tests only detect the presence of binding antibodies to a pathogen and not necessarily neutralizing antibodies that decrease or eliminate the infection. The cPass™ kit can be used as a biomarker of potential immunity in convalescent patients since it assesses both prior exposure and the presence of neutralizing antibodies, as well as a potentially valuable tool for assessing vaccine performance [29].

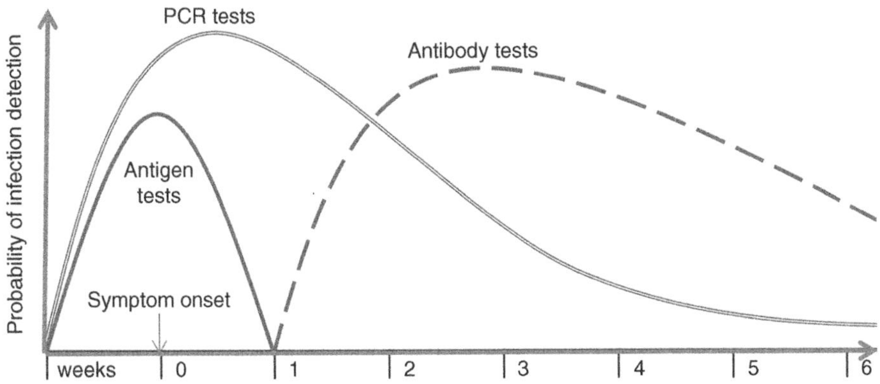

Figure 15.4 The estimated best timeframes for the application of the PCR, antigen, and antibody tests to detect the presence of the SARS-CoV-2 virus depending on the body's response to the infection. Source: Adapted with permission [19]. Copyright© 2020, Springer Nature.

In summary, the PCR, antigen, and antibody tests serve different purposes in testing for COVID-19 and need to be applied at the right time point to achieve the best and conclusive results to help combat the COVID-19 pandemic. Figure 15.4 shows when the best time might be to use these tests to obtain the optimal outcome based on the discussions in this section. For the diagnostic tests, the PCR-based tests can be positive long after a person stops being infectious (up to four weeks and beyond) since they can detect small amounts of viral genetic material; the rapid antigen tests are done within a week of symptom onset and can detect the presence of viral proteins and return positive results when a person is most infectious. For the preventive serological test, antibody tests are not effective at the earliest phase of infection but can detect the body's immune response to the virus when applied one week later after symptom onset with prolonged detection capability up to weeks and even months depending on the body's reaction and antibody types [19, 30–32]. Overall, these tests were developed within a very short time through understanding of basic science, properly applied biomarkers, and stringent validation studies. The bidirectional feature of translational medicine was successfully employed during this process to accelerate the end product development.

15.2.2 The Discovery and Clinical Development of Remdesivir in the Era of the COVID-19 Pandemic

After a person is infected by COVID-19, they are required to stay quarantined typically for two weeks to prevent further spread of the virus. During this period, if symptoms become worse especially with difficulty in breathing, medical treatment becomes necessary. As a new disease, prior to October 2020 there was no clinically proven therapeutics targeting SARS-CoV-2. Although a significant amount of preclinical research was reported in treatments related to the SARS and MERS CoV outbreaks, no existing drug can be adopted from those programs for COVID-19 treatment. Furthermore, considering the development of a new drug can take more than 10 years, repurposed drugs to improve the prognosis of COVID-19 patients is an important alternative. There are a number of such experimental drugs currently under clinical evaluation. Among them, remdesivir has gained the first EUA and eventual approval from the FDA. In this section, we will discuss how basic scientific discovery and a fundamental understanding of the virus's mode of action helped to accelerate the repurposing of remdesivir as a treatment option of the novel COVID-19 virus.

Before considering the identification of potential treatment options, it is important to understand how the virus infects its host and where intervention can be applied to stop the infection and kill the virus. As illustrated in Figure 15.5, the life cycle of the virus is driven by a few sequential steps: (i) Viral entry: the virus enters through direct interaction between the viral S protein and the host cellular receptor ACE2. (ii) Translation and viral protease cleavage: After the virus enters the host cell, the viral genome is released and translated into the viral replicase polyproteins. The viral proteases then cleave these polyproteins into functional proteins. (iii) RNA replication: viral genome replication is processed by the viral replication complex including the RNA-dependent

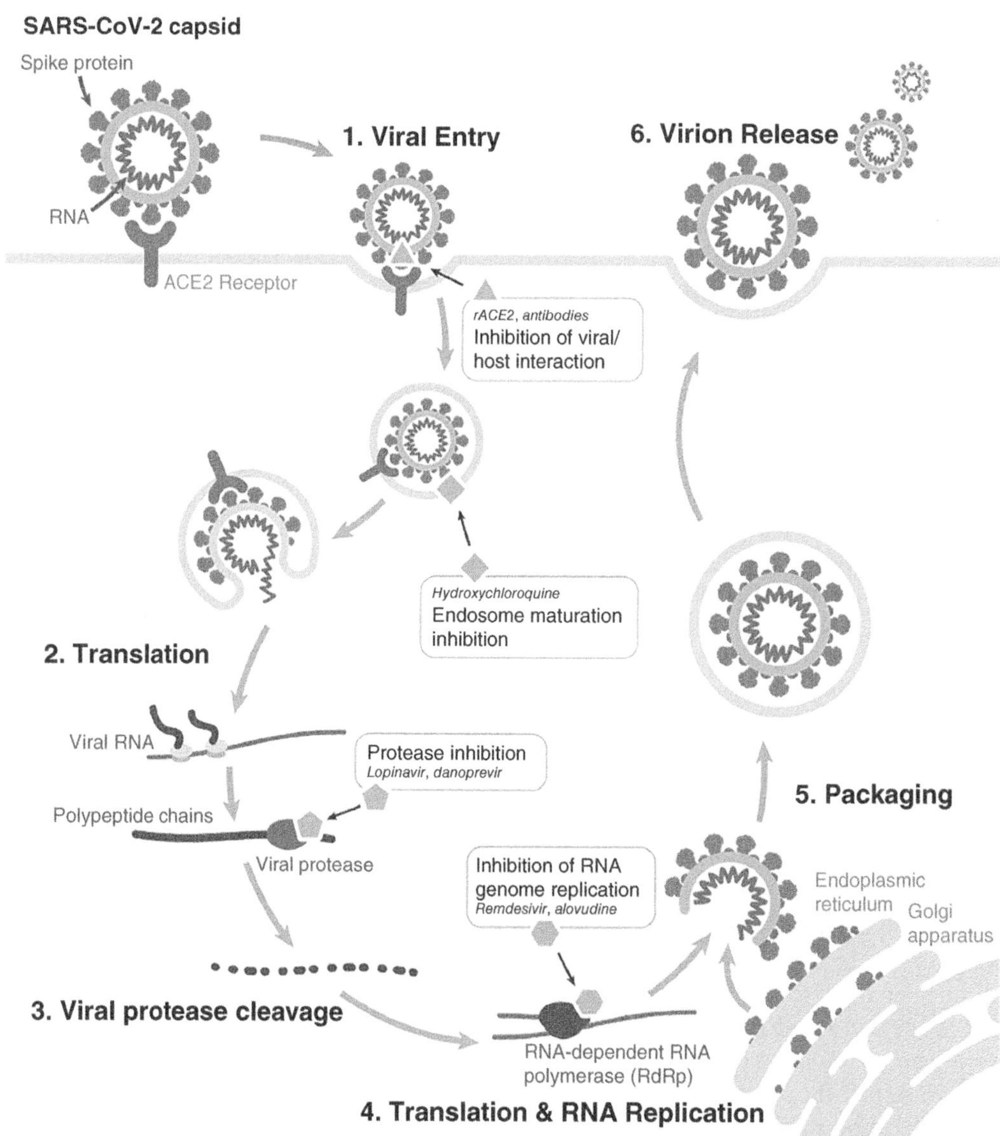

Figure 15.5 Life cycle of SARS-CoV-2 in host cells. Source: Reproduced with permission. Copyright© 2020, American Chemical Society [33], https://pubs.acs.org/doi/10.1021/acscentsci.0c00489, further permissions should be directed to the ACS.

RNA polymerase (RdRp). (iv) Packaging and virion release: viral genomes and translated viral structural proteins are assembled to form viral nucleocapsids and released through exocytosis. During its life cycle, a few interventions can be used to stop the virus infection. Among these potential interception points, remdesivir is targeting the RdRp to inhibit genomic RNA replication [33].

The detailed discovery and development story of remdesivir will only be briefly discussed here to provide some background information. Broad-spectrum therapies capable of inhibiting CoV-related infections may help to address COVID-19 infections since most viruses share a similar life cycle. Antiviral agents including nucleoside analogs targeting the divergent RdRp have proven efficacious against several viruses including hepatitis B and C as well as human immunodeficiency virus [34, 35], and remdesivir is a broad-spectrum nucleotide antiviral drug that was tested in clinical development initially for the treatment of Ebola virus (EBOV) infection.

Remdesivir (GS-5734) is a prodrug of parent drug GS-441524 and is metabolized within cells into an intermediate alanine metabolite which is further processed into the monophosphate derivative and ultimately into the active nucleoside triphosphate (NTP) derivative (Figure 15.6). The NTP can be used to prevent further infection when it is utilized by the viral RdRp for genome replication and misintegrated into viral RNA.

The broad-spectrum activity of remdesivir against EBOV and CoVs is summarized in Table 15.2. Remdesivir was initially evaluated against EBOV infections and effectively inhibited EBOV replication in primary macrophages, liver Huh-7 cells, HMVEC-TERT cells, and HeLa cells, with an EC_{50} value of 86, 60, 53, and 100 nM, respectively [36–38]. The effectiveness of various doses of

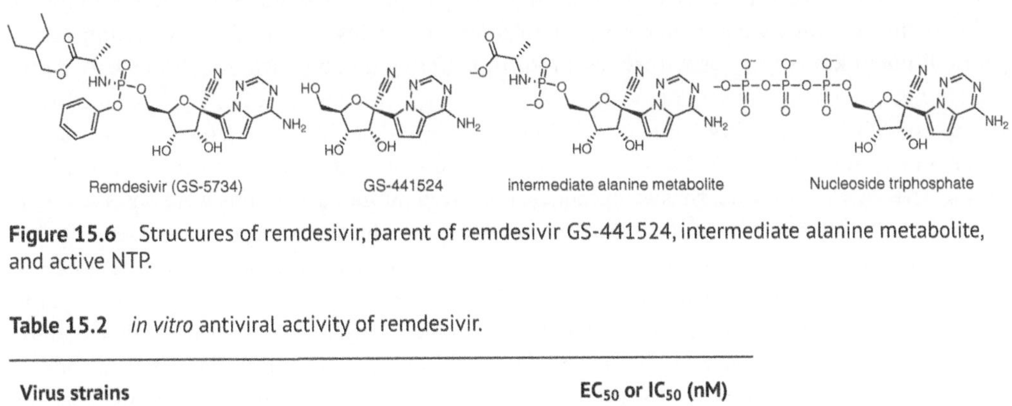

Figure 15.6 Structures of remdesivir, parent of remdesivir GS-441524, intermediate alanine metabolite, and active NTP.

Table 15.2 *in vitro* antiviral activity of remdesivir.

Virus strains	EC_{50} or IC_{50} (nM)
EBOV primary macrophages	86
EBOV Huh-7 cells	60
EBOV HMVEC-TERT cells	53
EBOV HeLa cells	100
MERS-CoV HAE cells	74
MERS-CoV Calu-3 2B4	25
SARS-CoV HAE cells	69
HCoV-229E Huh-7 cells	24
SARS-CoV2 Vero E6 cells	770
SARS-CoV2 human lung cells	1300

remdesivir administered via intramuscular injection was evaluated in a preclinical *in vivo* animal study using Ebola-infected rhesus monkeys [37] and high concentrations of the active form of remdesivir in peripheral blood mononuclear cells were achieved. Protective effects were observed at 10 mg/kg through inhibition of viral replication. In another *in vivo* animal study in Nipah virus-infected African green monkeys (10 mg/kg i.v. dosing), remdesivir also showed protective effects. While those in the control group developed severe respiratory disease, half of the remdesivir-treated monkeys developed mild respiratory symptoms and the other half recovered [39].

Remdesivir was highly efficacious in the control of SARS-CoV and MERS-CoV infections (Table 15.2). It exhibited potent activity against SARS-CoV (EC_{50} = 69 nM) and MERS-CoV (EC_{50} = 74 nM) in cultures of primary human lung epithelial cells. In addition, remdesivir showed potent activity against MERS-CoV in human lung (Calu-3 2B4) cells with an EC_{50} of 25 nM and endemic HCoV-229E with an EC_{50} of 24 nM [40]. In *in vivo* animal model studies, remdesivir was shown to protect rhesus monkeys from infection from MERS-CoV with a decreased presence and severity of lung lesions and reduced lung viral loads. It also effectively reduced lung viral loads in a transgenic mouse model of MERS-CoV. In a similar study in SARS-CoV-infected mice, remdesivir also reduced the viral load in the lung and improved respiratory function [41–43].

The efficacy and safety of remdesivir has been evaluated in clinical trials since 2016 including two EBOV studies. Even though the treatment for EBOV infection did not progress further due to inferior activity compared with the antibody treatment, these clinical trials helped to establish that remdesivir is a safe drug candidate suitable for further development as a potential treatment option for COVID-19 infection [38, 44, 45].

A recent study suggests remdesivir exhibits its effects through a consistent mechanism of action across genetically divergent RNA viruses in humans by targeting a structurally similar region of the EBOV and SARS-CoV-2 polymerases [46]. In addition, remdesivir is likely to be broadly active against different kinds of coronaviruses as an RdRp-targeting therapy including SARS-CoV-2 due to the inherent genetic conservation of the CoV replicase. Indeed, remdesivir showed *in vitro* potency and blocks SARS-CoV-2 infection in VeroE6 cells with an EC_{50} of 770 nM and had an IC_{50} of 1.3 μM in an anti-SARS-CoV-2 screening assay with human lung cells. In a recent *in vivo* animal study in SARS-CoV-2-infected rhesus macaques, improved pulmonary lesions were observed with remdesivir administration. Reduced viral titers in bronchoalveolar lavage were observed after 12 hours of treatment and viral load in the lungs was reduced after seven days of treatment [47].

With these desirable *in vitro* and *in vivo* preclinical results against the SARS-CoV, MERS-CoV, and the new COVID-19 virus infections, and a favorable safety profile in EBOV clinical studies [40], remdesivir was selected as a repurposing drug to treat SARS-CoV-2 infection to address this urgent unmet medical need. However, whether remdesivir can achieve a reasonable level of efficacy remains in question. At least three reasons contribute to this concern: the first reason is that remdesivir is clearly less potent against SARS-CoV-2. The second is that remdesivir was not approved for EBOV treatment due to its inferior activity compared with other agents despite having a very favorable *in vitro* profile. The third is that previous nucleoside prodrugs such as balapiravir failed to translate good preclinical *in vitro* potency and *in vivo* efficacy in animals to humans even though the drug candidates were thoroughly evaluated in the most biologically relevant models of pathogenesis.

In a spirit of urgent need to identify a medical treatment for COVID-19-infected patients, multiple clinical studies of remdesivir were quickly conducted in the United States and worldwide (www.clinicaltrials.gov) [48]. Out of these trials, two phase 3 clinical trials are the most definitive and will be discussed here: one was carried out by the National Institute of Allergy and Infectious Diseases (NIAID), and the other by Gilead Science, the drug maker who discovered remdesivir.

The randomized, controlled phase 3 trial carried out by NIAID was initiated on 21 February 2020 (ClinicalTrials.gov number NCT04280705). This was a double-blind trial of i.v. remdesivir in adults who were hospitalized with COVID-19 and had evidence of lower respiratory tract infection. The primary outcome was the time to recovery as defined by either discharge from the hospital or hospitalization for infection-control purposes only. On 29 April 2020, the NIAID published preliminary results of the effect of remdesivir on COVID-19 illness from this study. The mortality rate for the group of individuals who received remdesivir was 8% as compared with 11.6% in the placebo group. The median time to recovery was 31% quicker for patients who received remdesivir compared with those who received placebo (11 vs. 15 days) for patients who survived the infection [49, 50]. The final results of this trial were reported on 8 October 2020 and further confirmed the results from the preliminary report with the slight adjustment that remdesivir was superior to placebo in shortening the time to recovery in trial patients. The overall outcome is summarized in Table 15.3. Those who received remdesivir had a median recovery time of 10 days compared with 15 days for those who received placebo. For severely ill patients on oxygen, the median time to recovery was 11 days versus 18 days. The Kaplan–Meier estimates of mortality were 6.7% with remdesivir and 11.9% with placebo by day 15, and 11.4% with remdesivir and 15.2% with placebo by day 29. In an analysis that used a proportional-odds model with an eight-category ordinal scale ranging from not hospitalized with/without oxygen to hospitalized with/without oxygen to death, patients receiving remdesivir were more likely to have clinical improvement at day 15 than those on placebo. In this study, remdesivir also helped to prevent patients from progressing to more severe stages of illness. Serious adverse events (SAEs) were reported in 24.6% of patients who received remdesivir and 31.6% of patients who received placebo. Thus, the treatment with remdesivir reduced COVID-19 recovery time by five days compared with patients on placebo. Although results of the overall study suggested the drug may reduce mortality, the benefit may need to be further confirmed in additional clinical studies [50]. The drug appeared to reduce the risk of death by 72% at day 15, and 70% by day 29 in a separate analysis with subgroup patients who received oxygen which suggests that remdesivir may significantly reduce their chances of death compared with other subgroups when it is given to patients on oxygen.

Gilead conducted two randomized phase 3 clinical studies (The SIMPLE trials) to evaluate the safety and efficacy of two dosing durations (5 and 10 days) of remdesivir in adults diagnosed with COVID-19. The first of the two SIMPLE studies (NCT04292899) evaluated the safety and efficacy of both 5-day and 10-day dosing durations of remdesivir in patients with severe COVID-19. The second SIMPLE study (NCT04292730) evaluated the safety and efficacy of 5-day and 10-day dosing

Table 15.3 Overall outcome of phase 3 trial carried out by NIAID (NCT04280705).

	Remdesivir	Placebo
Median time to recovery (days)	10 (95% CI 9–11)	15 (95% CI 13–18)
Rate ratio (95% CI)	1.29 (1.12–1.49 [p < 0.001])	
Kaplan–Meier estimate of mortality by day 15 (%)	6.7 (95% CI 4.8–9.2)	11.9 (95% CI 9.4–15.0)
Hazard ratio day 15 (95% CI)	0.55 (0.36–0.83)	
Kaplan–Meier estimate of mortality by day 29 (%)	11.4 (95% CI 9.0–14.5)	15.2 (95% CI 12.3–18.6)
Hazard ratio day 29 (95% CI)	0.73 (0.52–1.03)	

CI, confidence interval.

regimens of remdesivir in patients with moderate COVID-19 compared with standard-of-care alone.

On 29 April 2020, Gilead disclosed results from the first phase 3 SIMPLE trial in hospitalized patients with severe cases of COVID-19. Patients with pneumonia and reduced oxygen levels that did not require mechanical ventilation at the time of study were included in the trial. Patients who received a 10-day treatment course of remdesivir attained comparable improvement in clinical status compared with those taking a 5-day treatment course. The time to clinical improvement for 50% of patients was 10 days in the 5-day treatment group and 11 days in the 10-day treatment group. More than 50% of patients in both treatment groups (60% for 5-day vs. 52% for 10-day) were discharged from the hospital by day 14. Clinical recovery was achieved at day 14 in 65% of patients in the 5-day treatment group and 54% of patients in the 10-day treatment group (Table 15.4). The use of remdesivir across either treatment group did not cause any unexpected side effects [51, 52].

On 1 June 2020, Gilead announced topline results from the second phase 3 SIMPLE trial in hospitalized patients with moderate COVID-19 pneumonia. The final results were published in August 2020 [53]. This was an open-label study that evaluated 5-day and 10-day courses of remdesivir plus standard-of-care versus standard-of-care alone. Hospitalized patients with confirmed COVID-19 infection and evidence of pneumonia without reduced oxygen levels were randomized (1:1:1) in this study to receive open-label remdesivir for 5 or 10 days or standard-of-care alone. The primary endpoint was the clinical status based on a 7-point ordinal score at day 11, ranging from hospital discharge to increasing levels of oxygen and ventilatory support to death. The secondary objective was to compare the rate of adverse events in each remdesivir treatment group and in standard-of-care. The study results showed that patients in the 5-day remdesivir treatment group were 65% more likely to have clinical improvement at day 11 compared with those in the standard-of-care group. The improvement in clinical status for the 10-day treatment course of remdesivir versus standard-of-care was also favorable, trending toward but not reaching statistical significance. A higher proportion of patients in the 5-day treatment group achieved improvement in clinical status at day 11 versus the standard-of-care group. No new safety signals were identified with remdesivir across either treatment group.

On 1 May 2020, the FDA issued an EUA for emergency use of remdesivir (brand name Veklury®) for the treatment of hospitalized COVID-19 patients based on a review of the topline data from the Gilead-sponsored open-label trial NCT04292899, and from the randomized trial conducted by

Table 15.4 Key efficacy and safety results from the SIMPLE1 study.

Clinical efficacy outcome at day 14		
	5-day remdesivir (%)	**10-day remdesivir (%)**
≥2-point improvement in ordinal scale	65	54
Clinical recovery	65	54
Discharge	60	52
Safety		
Any AE	71	74
≥grade 3 study drug-related AE	4	5
Study drug-related serious AE	2	2

AE, adverse event.

NIAID (NCT04280705). On 28 August 2020, the FDA expanded the EUA enabling the use of remdesivir to treat all hospitalized patients with COVID-19 in addition to the previous authorization for patients hospitalized with severe COVID-19. On 22 October 2020, the FDA approved remdesivir for use in adult and pediatric patients for the treatment of COVID-19 requiring hospitalization [54]. The approval was based on the data from the abovementioned three randomized, controlled clinical trials that included patients hospitalized with mild-to-severe COVID-19. This represents the first treatment for COVID-19 to receive FDA approval.

In summary, remdesivir is a promising drug to treat COVID-19 based on the following facts: it has a broad-spectrum antiviral mechanism of action, demonstrated *in vitro* activity and *in vivo* efficacy in animal models against SARS-CoV-2, and its safety profile has been tested in Ebola patients and in compassionate use in COVID-19 patients. All these results formed a strong translational base to support the development of remdesivir for COVID-19 treatment. However, even though remdesivir demonstrated benefits in the NIAID and Gilead controlled trials, its efficacy is still in question. An authoritative large scale SOLIDARITY trial carried out by the WHO concluded that remdesivir appeared to have little or no effect on hospitalized COVID-19 patients in terms of overall mortality rate, initiation of ventilation, and duration of hospital stay [55]. While these issues are resolved, better treatment options are still needed. For example, the dosing protocol of remdesivir is an i.v. infusion loading dose of 200 mg on day 1 followed by daily i.v. infusion maintenance doses of 100 mg for 5–9 days. This delivery method may not deliver adequate remdesivir and its active metabolites to the target organ (the lung) to kill the SARS-CoV-2 virus based on the observations that plasma exposures of remdesivir and its active metabolite are not correlated with its clinical efficacy. This may have contributed to less than robust efficacy of this drug in COVID-19 patients creating the need to further improve the effectiveness of remdesivir by other means, either through potency improvements with more active analogs, or through formulation to deliver the drug to the target organ with better efficiency, or through combination therapies with other medicinal agents. This exemplifies the bidirectional feature of translational medicine that clinical feedback and patients' needs can promote new drug development. This is further discussed in Section 15.3.

15.2.3 COVID-19 Virus Targeting Antibody Discovery and Development

In addition to small molecule antiviral drugs such as remdesivir, an alternative treatment option for COVID-19 is monoclonal antibodies. Antibodies are proteins produced in the event of an infection and attach to a virus for elimination. Experimental antibody drugs are concentrated versions of specific antibodies identified after a viral infection. As discussed previously, not all antibodies are therapeutically equivalent and only neutralizing antibodies can be employed to counter COVID-19 infections. These neutralizing antibodies not only can prevent the virus from entering the host cell to cause the infection but are also likely to generate immunity against potential reinfection, which is an important factor to determine whether the antibodies have long-term protective immunity. For CoVs and COVID-19 virus infections, neutralizing antibodies primarily target the S protein, in particular the S1 subunit and the RBD contained within the S1 subunit to block the interaction with the host cell's ACE2 receptor and prevent the infection of human cells. These neutralizing antibodies can be identified from the plasma of infected patients or engineered in a lab (monoclonal antibodies).

Many companies are working on the development of antibody drugs against COVID-19. For example, Lilly is testing two different antibodies, LY-CoV555 and LY-CoV016, which bind complementary regions of the SARS-CoV-2 spike protein, individually and in combination. Regeneron is investigating a two-antibody cocktail to enhance the chance its therapy will work even if the virus evades one. Since these are the most advanced in clinical studies, they are selected to illustrate the application of translational medicine in antibody drug discovery.

LY-CoV555 (LY3819253) is a potent and neutralizing IgG1 monoclonal antibody (mAb) directed against the spike protein of SARS-CoV-2 and was identified from a blood sample taken from one of the first US patients who recovered from COVID-19. It is designed to neutralize the virus by blocking viral attachment and entry into human cells. Lilly scientists developed the antibody in less than three months after it was discovered by AbCellera and tested by scientists at the NIAID Vaccine Research Center before it was selected to enter clinical trials. Lilly has completed primary safety assessments of LY-CoV555 in a phase 1 study of hospitalized patients with COVID-19 (NCT04411628) and long-term follow-up is ongoing. As a recombinant fully humanized monoclonal neutralizing antibody, the second Lilly antibody, LY-CoV016 (also known as JS016), specifically binds to the SARS-CoV-2 surface spike protein RBD with high affinity and effectively blocks the binding of the virus to the ACE2 host cell surface receptor. It was jointly developed by Junshi Biosciences and the Institute of Microbiology, Chinese Academy of Science (IMCAS) and later licensed by Lilly. LY-CoV016 is effective against SARS-CoV-2 infection for both prophylactic and therapeutic venues in a SARS-CoV-2 challenge study with rhesus macaques. A phase 1 study (NCT04441931) of LY-CoV016 in healthy US volunteers was conducted to evaluate the safety, tolerability, pharmacokinetics, and immunogenicity. LY-CoV016 was well-tolerated, and no drug-related SAEs were observed.

The BLAZE-1 phase 2 study (NCT04427501) was a randomized clinical study designed to assess the efficacy and safety of LY-CoV555 and LY-CoV016 for the treatment of symptomatic COVID-19 in the outpatient setting. Mild-to-moderate, recently diagnosed COVID-19 patients were enrolled in the monotherapy arms of the trial and three doses of LY-CoV555 (700, 2800, and 7000 mg) were studied versus placebo. Recently diagnosed mild-to-moderate COVID-19 patients were included in the combination arm of the trial and LY-CoV555 2800 mg plus LY-CoV016 2800 mg versus placebo were used. The primary clinical outcome was changed from baseline to day 11 in SARS-CoV-2 viral load. Safety, the percentage of participants who experienced COVID-19-related hospitalizations, emergency room (ER) visits or death from baseline to day 29 are also included as additional endpoints.

The BLAZE-1 trial started in early June 2020 and the interim monotherapy clinical trial data was updated on 16 September 2020 [56]. The primary endpoint was met at the 2800 mg dose level but not at other doses. Near complete viral clearance by day 11 was observed in most patients including those receiving placebo. LY-CoV555 improved viral clearance at an earlier time point (day 3) and reduced the proportion of patients with persistently high viral load at later time points based on additional analyses of viral data. A 72% risk reduction in the pre-specified endpoint of COVID-19-related hospitalizations or ER visits occurred in 1.7% of LY-CoV555 patients as compared with 6% of patients on placebo. Based on an exploratory analyses, a more rapid improvement in symptoms was observed for patients treated with LY-CoV555 versus placebo and no patients progressed to mechanical ventilation or death across all treatment groups including the placebo group. LY-CoV555 was well-tolerated, and no drug-related SAEs were reported. The clinical trial data of LY-CoV555 and LY-CoV016 combination therapy was updated on 7 October 2020 [57]. The primary endpoint of the study was met with the combination therapy which significantly reduced viral load at day 11. The combination treatment significantly reduced the time-weighted average change from baseline from day 1 to 11 and reduced viral levels at earlier time points at day 3 and day 7 during the course of infection. Based on an exploratory analysis, a lower (3.0%) proportion of patients with persistent high viral load at day 7 was observed in combination therapy versus placebo (20.8%). The rate of COVID-related hospitalizations and ER visits was lower for patients treated with combination therapy (0.9%) versus placebo (5.8%) which was similar to that observed in LY-CoV555 monotherapy and represents an 84.5% relative risk reduction. No drug-related SAEs and treatment emergent adverse events were observed in the combination therapy as compared with placebo.

Based on the results of the combination therapy, along with the previously disclosed findings for LY-CoV555 monotherapy, Lilly submitted an initial request for EUA for LY-CoV555 monotherapy in higher-risk patients recently diagnosed with mild-to-moderate COVID-19 as of October 2020. The US FDA granted this EUA request on 9 November 2020 for the treatment of mild to moderate COVID-19 in adults and pediatric patients 12 years and older with a positive COVID-19 test who are at high risk for progressing to severe COVID-19 and/or hospitalization [58]. Additional clinical studies are currently ongoing including a phase 3 study for the prevention of COVID-19 in residents and staff at long-term care facilities (BLAZE-2, NCT04497987) by Lilly and the National Institutes of Health (NIH)-led ACTIV-2 study of LY-CoV555 in ambulatory and hospitalized COVID-19 patients. The ACTIV-3 clinical trial run by NIAID evaluating the efficacy of LY-CoV555 in hospitalized COVID-19 patients stopped recruiting additional patients in this setting based on data suggesting that LY-CoV555 is unlikely to help hospitalized COVID-19 patients recover from this advanced stage of their disease.

Regeneron scientists evaluated thousands of fully human antibodies produced by the company's *VelocImmune* mice (genetically modified to have a human immune system) and antibodies identified from people who have recovered from COVID-19. REGN10933 and REGN10987 were identified and constitute the key ingredients of the antibody cocktail REGN-COV2. These are potent and virus-neutralizing antibodies designed specifically to block infectivity of SARS-CoV-2 by binding noncompetitively to the critical RBD of the viral spike protein. Hence, the ability of mutant viruses to escape treatment can be reduced and spike variants in the human population may be eliminated [59, 60]. REGN-COV2 reduced the amount of virus and associated damage in the lungs of non-human primates in preclinical studies [61]. According to a REGN-COV2 antibody cocktail program update (https://investor.regeneron.com/static-files/a596a85e-e72d-4529-8eb5-d52d87a99070), REGN-COV2 was being evaluated in clinical development programs including: (i) Study 2093 with normal human volunteers for multidose pharmacokinetic/safety evaluation. (ii) Seamless P1/2/3 Study 2066 for hospitalized patients in four cohorts (no oxygen requirement; low flow oxygen; high flow oxygen; and mechanical ventilation). (iii) Seamless P1/2/3 Study 2067 in outpatients for both symptomatic and asymptomatic patients. (iv) Phase 3 Study 2069 for household contact prophylaxis. (v) UK/NHS RECOVERY (Randomized Evaluation of COVID-19 Therapy) phase 3 trial in hospitalized patients. Results from the initial cohort of 275 patients from the seamless P1/2/3 Study 2067 in outpatients were reported in September 2020. All patients had laboratory-confirmed COVID-19 that was being treated in an outpatient setting and were randomized 1:1:1 in this study to receive a one-time infusion of 8 g of REGN-COV2 (high dose), 2.4 g of REGN-COV2 (low dose) or placebo. Patients were prospectively characterized prior to treatment by serology tests to ascertain if they already had antiviral antibodies and were classified as seronegative (no measurable antiviral antibodies, ~41% of patients) or seropositive (measurable antiviral antibodies, ~45% of patients). Approximately 14% of patients did not have clear or known serology status and were categorized as "others". Key virologic endpoints included time-weighted average change from baseline in viral load (log10 copies/ml) from day 1 to day 7 in the seronegative modified full analysis set (mFAS) which was measured by quantitative RT-qPCR in nasopharyngeal (NP) swab samples, and additional analyses on viral reduction.

The study revealed several key findings: (i) Seronegative patients had substantially higher viral levels at baseline and cleared virus more slowly without treatment compared with seropositive patients. How quickly the patients alleviated their COVID-19 clinical symptoms was also predicted by serological status at baseline, with a 7-day median time in seropositive patients on placebo and a 13-day median time in seronegative patients. (ii) REGN-COV2 rapidly reduced viral load through day 7 in seronegative patients and the results are summarized in Table 15.5. (iii) At day 7, patients with increasingly high baseline viral levels had a greater reduction in viral load with REGN-COV2 treatment (Table 15.6). (iv) Greater benefits in the alleviation of symptoms were observed in patients who were seronegative and/or had higher baseline viral levels. Median time to symptom alleviation was 13 days in placebo, 8 days in high dose cohort, and 6 days in low dose cohort among seronegative patients. (v) A small number of medically attended visits with patients in the seronegative group were at higher risk.

In summary, based on the descriptive analysis of the first cohort of 275 patients from the seamless Ph1/2/3 trial Study 2067, REGN-COV2 rapidly reduced viral load through day 7 in seronegative patients and resulted in greater reductions in viral load in patients with higher baseline viral levels. REGN-COV2 also showed positive trends in reducing medical visits. The greatest effect was seen in patients who had not mounted their own immune response prior to treatment based on this interim data. The Independent Data Monitoring Committee (DMC) did not find any unexpected safety findings of REGN-COV2 during the trial.

The REGN-COV2 trial was first posted on www.clinicaltrials.gov on 11 June 2020 and the interim results were reported in late September 2020. On 7 October 72020, Regeneron submitted a request to the US FDA for an EUA for REGN-COV2 investigational antibody combination for COVID-19 and received FDA EUA approval on 21 November 2020 [62]. A preventive trial has started to

Table 15.5 Difference versus placebo of the meantime-weighted-average change from baseline nasopharyngeal viral load through day 7.

	Seronegative group		**Overall population**	
Dose	2.4 g i.v.	8.0 g i.v.	2.4 g i.v.	8.0 g i.v.
Difference vs. placebo by day 7 (log10 copies/ml)	−0.51	−0.60	−0.23	−0.51
95% CI	(−1.05, 0.03)	(−1.14, −0.06)	(−0.58, 0.12)	(−0.87, −0.16)
p-value	0.06	0.03	0.2	0.05

CI, confidence interval.

Table 15.6 Difference versus placebo of the viral load reduction through day 7.

	Baseline viral load >10 × 4 copies/ml		**Baseline viral load >10 × 5 copies/ml**		**Baseline viral load >10 × 6 copies/ml**		**Baseline viral load >10 × 7 copies/ml**	
Dose	2.4 g i.v.	8.0 g i.v.	2.4 g i.v.	8.0 g i.v.	2.4 g i.v.	8.0 g i.v.	2.4 g i.v.	8.0 g i.v.
Log difference vs. placebo at day 7 (log10 copies/ml)	−0.46	−0.60	−0.86	−0.93	−1.65	−1.55	−2.0	−1.79
p-value	0.19	0.1	0.03	0.03	0.001	0.002	0.001	0.002

evaluate REGN-COV2's ability to prevent infection among uninfected people who have had close exposure to COVID-19 from a housemate or family member to test whether the monoclonal drug can prevent infection among people exposed to COVID-19. REGN-COV2 is also included in the RECOVERY trial in the UK which is one of the world's largest randomized clinical trials to evaluate potential COVID-19 treatments. This phase 3 open-label trial in patients hospitalized with COVID-19 will compare the effects of adding REGN-COV2 to the usual standard-of-care versus standard-of-care on its own.

A few important translational medicine principles are evident from the antibody drug development programs: (i) understanding the basic science of viral composition has helped to focus efforts on the right target for antibody development and thus potentially increase the success rate. (ii) With close collaboration and communication between pharmaceutical companies, national health organizations, and regulatory agencies including the FDA, the drug development process can be accelerated. (iii) Breaking the barriers to collaboration between different companies can help to improve the COVID-19 pandemic situation. For example, Eli Lilly and Amgen recently announced a global antibody manufacturing collaboration to significantly increase the supply capacity available for Lilly's potential COVID-19 therapies, and Regeneron and Roche announced a similar collaboration deal to meet the potential demand of their antibody candidates [63, 64].

15.2.4 Accelerated Vaccine Development for COVID-19 Prevention

Vaccines mimic an infection and train the immune system to recognize and combat pathogens through the production of antibodies. Antigens present on all viruses and bacteria can trigger an immune response, and by administering these antigens as a vaccine the immune system can recognize them as hostile invaders, produce antibodies, and memorize them for the future. The ultimate goal for a vaccination program is to protect the entire population, however, once a large enough segment of the population is immunized the virus will no longer have sufficient eligible hosts to establish a foothold, and opportunities for an outbreak become so low that even those who are not immunized benefit. This phenomenon is called "herd immunity" or "community immunity" [65]. An excellent example of this phenomenon are the vaccines used to treat the smallpox and polio viruses. Vaccines may take longer to develop than non-vaccine drugs since they are administered to healthy populations and thus require rigorous safety standards. The discovery and development process for vaccines is performed in a very similar manner to small molecule drugs and involves a series of exploratory and preclinical studies, IND application, phase 1–3 vaccine trials and regulatory approval. Under normal circumstances, this entire process can take up to 10 years, not to mention the Post-Licensure Monitoring of Vaccines that includes phase 4 trials, the Vaccine Adverse Event Reporting System (VEARS), and the Vaccine Safety Datalink (VSD) [66–68]. Many vaccine platforms have been developed such as Live Attenuated Vaccine (LAV) with the whole virus, inactivated virus vaccines, subunit vaccines, viral vector-based vaccines (replicating or non-replicating), DNA vaccines, RNA vaccines, and virus-like particle vaccines [69]. According to the WHO, currently there are more than 180 COVID-19 vaccines under development, [70]. This section will not focus on the detailed vaccine discovery and development process. Instead, three topics will be discussed: the rationale for S protein antigen targeted vaccine development, how to accelerate and shorten the vaccine development timeline under the pandemic situation wherein some developers aggressively aimed to identify and supply a safe and effective vaccine in early 2021 or even by the end of 2020, and how regulatory agencies can properly regulate the process under tight timeline pressure while maintaining the integrity of vaccine development without risking the health of the general population.

The selection and design of protective immunogens against pathogens represents a major challenge in vaccine development especially for newly emerged pathogens. For COVID-19 vaccine development, the critical first step is to identify which antigen can be targeted with neutralizing antibodies to achieve immunity. In this regard, a better understanding of antibody responses against SARS-CoV-2 played a crucial role in the development of effective vaccines, together with the valuable lessons learned from the 2002–2003 SARS outbreak [71]. Coupling the rapid availability of the genome and protein sequence information of SARS-CoV-2 with computational analysis, the characteristics of the virus and the epitopes presented in the pathogen are well understood, which helps to greatly speed up vaccine development [72–74]. As discussed earlier, the S protein is a surface protein of the CoVs and plays a crucial role in mediating virus entry. Based on past experience with SARS-CoV vaccine development, the full-length S protein and its S1 subunit RBD have been frequently used as the vaccine antigens due to their ability to induce antibodies that prevent host cell entry and infection. Studies conducted in a variety of species including transgenic mice expressing human ACE2, ferrets, and non-human primates all showed these vaccines confer protective immunity against viral challenge. In addition, since the S1 subunit of the S protein in the COVID-19 virus displays the lowest homology to the other members of the CoV family, it may provide an opportunity for the development of a more specific vaccine. Based on these scientific rationales, the antigen S protein and the RBD has been targeted extensively in various vaccine development programs [23].

Vaccine development is a lengthy, expensive, and highly complex process. The attrition rate of vaccine development is high, and vaccine approval often requires years of effort involving multiple vaccine candidates [75]. To accelerate vaccine approval, licensure, and distribution, all public, private, and philanthropic sectors need to participate in a strategic and collaborative manner [76]. A vaccine development program must take the necessary measures to ensure the accuracy of data analysis and the efficiency of the manufacturing process in order to reduce failure and minimize financial risks. To shorten this process and to accelerate the delivery of a COVID-19 vaccine under the current pandemic situation, a parallel development paradigm is highly recommended [77]. For example, optimization of vaccine platforms in animal models and phase 1 clinical trials may proceed in parallel, and very early manufacturing scale-up to commercial scale may be carried out in overlapping fashion and before clinical proof of concept (Figure 15.7). This pandemic paradigm certainly helped to accelerate COVID-19 vaccine development and has been exercised by many vaccine developers and manufacturers. However, this process also carries large unpredictable risks if the vaccine fails before it is approved by the regulatory agencies along the way [77]. Different vaccine platforms may also impact the speed of development. For example, DNA- and mRNA-based vaccines can be generated quickly on the basis of viral sequence, thus allowing a rapid pathway to the clinic. This turns out to be the case since the mRNA-based vaccines from both Pfizer/BioNTech and Moderna Therapeutics achieved FDA EUA approval in December 2020 as discussed briefly at the end of this section. There are also other ways to speed up the approval process of a vaccine. For example, if the vaccine was developed based on a platform technology shown to be safe before, previously collected safety and pharmacokinetic data may be used to fast-track its approval. Considering the conflicting goals of speed and safety, rival vaccine manufacturers have jointly pledged not to rush the vaccine approval but to maintain the most rigorous safety standards [78].

To speed up drug development against COVID-19 infection, the FDA created the Coronavirus Treatment Acceleration Program (CTAP) as a special emergency program for possible coronavirus therapies which uses every available method to move new treatments to patients as quickly as possible while at the same time determining their safety and efficacy through supporting clinical

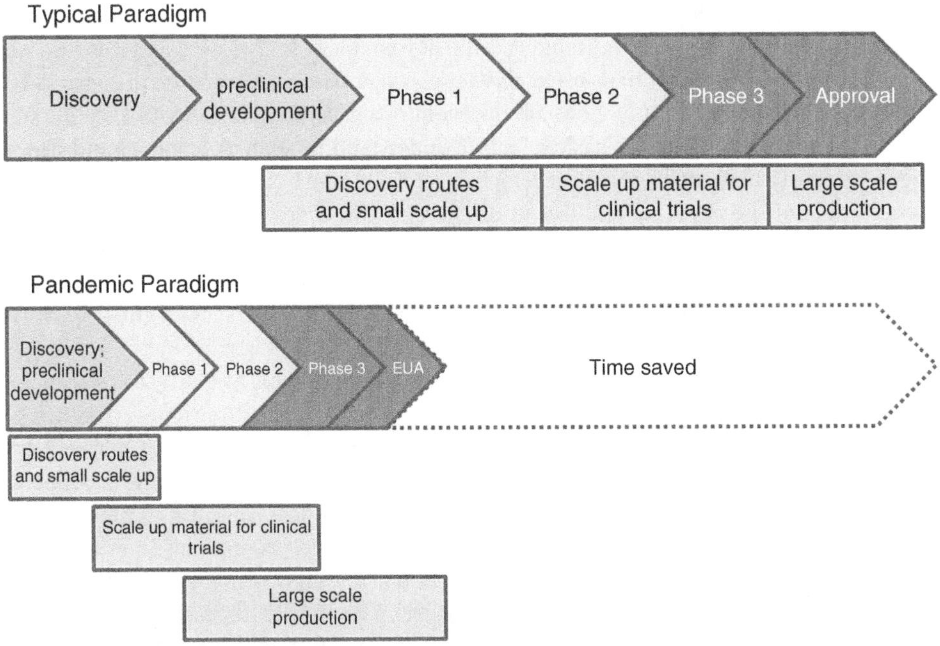

Figure 15.7 Schematic illustration of typical vaccine development stages and development using a pandemic paradigm.

trials [79]. In addition, the US government initiated a public–private partnership called Operation Warp Speed (OWS) to facilitate and accelerate the development, manufacturing, and distribution of COVID-19 vaccines, therapeutics, and diagnostics. OWS's goal was to produce and deliver 300 million doses of safe and effective vaccines with the initial doses available by January 2021. This goal turned to be achievable as the first doses of vaccines were administered in December 2020. OWS served three important roles: (i) It is a partnership among components of the Department of Health and Human Services (HHS), including the Centers for Disease Control and Prevention (CDC), the FDA, the NIH, and the Biomedical Advanced Research and Development Authority (BARDA), and the Department of Defense (DoD). (ii) It engaged with many parties such as private firms and other federal agencies, including the Department of Agriculture, the Department of Energy, and the Department of Veterans Affairs. (iii) It coordinated existing HHS-wide efforts, including the NIH's Accelerating COVID-19 Therapeutic Interventions and Vaccines (ACTIV) partnership, NIH's Rapid Acceleration of Diagnostics (RADx) initiative, and work by BARDA [80]. Through heavy investment and seamless coordination across relevant participants through OWS, two vaccines were delivered by the end of 2020 to patients more rapidly than ever before while adhering to standards for safety and efficacy.

With this ambitious goal and unprecedented speed for vaccine development, how can regulatory agencies ensure that a vaccine will be safe and effective? The FDA issued clear guidance for industry on the steps required for developing and ultimately licensing vaccines to prevent COVID-19 which are the same rigorous safety standards required for all vaccines and set the bar for the primary endpoint of a phase 3 trial as 50% protection for approval of a COVID-19 vaccine [81]. In an emergency situation, such as the COVID-19 pandemic, with new cases and deaths increasing daily, the FDA was authorized to accelerate the approval process and allow the use of unapproved products including vaccines for the diagnosis, treatment, and prevention of disease [82].

In conclusion, the COVID-19 pandemic has forced an almost complete shut-down of the global economy and caused millions of deaths. Rapid vaccine development is critical to end this disease even though the duration of clinical trials ostensibly poses a significant hindrance to this goal [83]. From a translational medicine point of view, development of a vaccine against SARS-CoV-2 coincides with a period of rapid discovery in basic scientific understanding such as genomics and structural biology to support a new era in vaccine development. Combining basic science and technology development with strategic planning and execution, vaccine discovery and development can be dramatically accelerated. On the other hand, other factors, such as political pressure, should not interfere with this process in order to shorten the delivery of vaccine development through any short cuts. Proper regulation and monitoring processes from regulatory agencies should be strictly implemented.

Vaccine development also carries other challenges. Just like small molecule drug development where animal models are not always predictive of human efficacy, vaccine discovery carries translational risks that the assessment of the vaccine-dependent immune enhancement may not be translated from animal models to human. Various viruses can develop antibody dependent enhancement (ADE) as an alternative method to infect a variety of host cells and vaccine associated enhanced respiratory disease (VAERD) can also be induced by a virus-antibody immune complex and T helper type 2 biased responses [84]. In addition, the viral genome is vulnerable to mutations and can undergo antigenic shift and antigenic drift when it spreads from one population to another. The mutations may vary when the environmental conditions and population density changes at different geographical areas which may lead to different subtypes and allow the virus to escape the immune system even after the administration of the vaccine [85, 86]. These challenges have created the need and opportunity for new scientific discoveries and novel product development, an example of bedside to bench translational medicine. It is important to balance the unmet medical needs, efficiency/efficacy, and safety of the vaccines to benefit the patients and eventually the wellbeing of human health.

With all these considerations, faster vaccine development indeed is possible. On 9 November 2020, Pfizer Inc. and BioNTech SE announced their mRNA-based vaccine candidate (BNT162b2) against SARS-CoV-2 had achieved a vaccine efficacy rate above 90% in participants without prior evidence of SARS-CoV-2 infection based on the first interim efficacy analysis conducted by an external, independent DMC from the phase 3 clinical study. They expect to produce up to 50 million vaccine doses in 2020 and up to 1.3 billion doses in 2021 globally [87]. Later that same month, Moderna reported similar efficacy results for their vaccine candidate. Based on the efficacy data as well as an acceptable safety profile, the Pfizer vaccine was granted an EUA by the FDA on 11 December 2020. The first doses of the Pfizer/BioNTech vaccine were administered to patients in late December 2020, thus fulfilling the primary goal of OWS. Furthermore, the Moderna vaccine was granted an EUA on 20 December 2020. The FDA granted an EUA to Johnson & Johnson's single-shot COVID-19 vaccine on 27 February 2021, representing the third approved vaccine in the USA to combat the COVID-19 pandemic.

15.3 From Bedside to Bench: Accelerating Drug Discovery and Development in Treating COVID-19

In the fight against the COVID-19 outbreak, the health science community worldwide has worked together to come up with new treatments for the disease and for the prevention of its spread. Many fundamental scientific discoveries have helped to accelerate the process and identify novel treatment options. On the other hand, much has been learned from the clinical process and patient outcomes which provides feedback to generate better solutions to fighting the disease. The following section will discuss how unmet medical needs have promoted the discovery of more novel medical products.

15.3.1 The Need for an Inhaled Formulation of Remdesivir

As discussed earlier, remdesivir was granted an EUA and then approval from the FDA to treat COVID-19 patients especially at an early stage of the infection. However, the effect was modest and there was no statistically significant benefit of lower mortality. In addition, remdesivir is given by i.v. infusion once daily, which limits accessibility for patients outside a hospital setting and places significant pressure on the healthcare system. The lack of effective delivery methods has hindered the treatment for a broad spectrum of patients at an early stage of infection before the onset of life-threatening symptoms.

The less than ideal efficacy may be partially due to the moderate intrinsic potency of the drug against the target (Table 15.2). This may be improved by the identification of more potent analogs of remdesivir through structure–activity relationship studies in drug discovery labs, although this is a time-consuming process which will not be helpful in the current pandemic situation. A second reason for the limited efficacy could be associated with the i.v. infusion dosing protocol which may not deliver adequate drug to the target organ due to the physicochemical properties of the drug resulting in poor tissue distribution/penetration. One way to overcome this issue is to deliver the drug directly to the infection site via a modification in the formulation of the drug. Furthermore, a more convenient and accessible dosage form may also provide a means to distribute remdesivir to other patients in a non-hospital setting [88].

Since remdesivir is a prodrug and mostly metabolized by the body before it enters systematic circulation, an oral formulation does not appear viable. Furthermore, intramuscular injections also face challenges since the rate of release from the muscles can vary significantly and lead to tissue distribution issues similar to i.v. formulation. This medical need has created the opportunity to identify alternate delivery routes for remdesivir. Based on current scientific understanding, the upper respiratory tract is the most prevalent site of SARS-CoV-2 infection early in the disease, so pulmonary delivery can be considered. Delivering remdesivir directly to the primary site of infection with a nebulized inhaled solution may also potentially lower systemic exposure of the drug to minimize side effects. This represents an important treatment option that can potentially improve patient outcomes and enable more patients access to remdesivir. In this regard, significant research efforts have been undertaken to deliver remdesivir in an inhaled nebulized format [88, 89].

To accelerate the process and validate the effectiveness of inhaled delivery, Gilead Sciences launched a phase 1a/2a clinical trial (NCT04539262) of an inhaled version of remdesivir to treat early stage COVID-19 patients in the outpatient setting. The randomized phase 1a study is evaluating the safety, tolerability, and pharmacokinetics of the new formulation in healthy participants in the US. Data from this trial will help to plan further clinical studies of the inhaled formulation, primarily in patients whose infection has not progressed to hospitalization. A nebulized, inhaled nanoparticle formulation of remdesivir is also in phase 1 clinical study (NCT04480333) [90].

15.3.2 Overcoming Cytokine Storm in COVID-19 Treatment

There are two ways in which viruses make people sick. One is the direct effect of the pathogen itself that causes initial symptoms of infection associated with rapid viral replication. The other is self-inflicted damage from widespread inflammation due to the release of a variety of inflammatory mediators like interferons (IFNs), interleukins (ILs), tumor necrosis factor (TNF-α), and chemokines from an overactive immune response to the pathogen. The second mechanism seems to play a major role in many of the severe COVID-19 cases where the disease is most lethal due to respiratory problems and damage to organs such as the heart and kidneys. Excessive production of proinflammatory cytokines at this stage leads to acute respiratory distress syndrome (ARDS) and systemic hyperinflammation which is usually called "cytokine storm," and widespread tissue damage resulting in multi-organ failure and death [91–93]. Thus, the number of mortalities would drastically decrease if the cytokine storm could be weakened or prevented, and immune modulators may be important treatment options for severe COVID-19 infections and reduce the overall mortality [94].

For an effective treatment, doctors must recognize and intercept the cytokine storm in a timely manner. This prompted an urgent need for biomarkers to detect the beginning of a cytokine storm in order to identify the best treatment options to curb the overactive immune system. Although there is no definitive diagnostic test and the pathogenesis of a cytokine storm remains to be fully elucidated, there are early signs that a cytokine storm may be underway [95–97]. For example, while the WHO does not specify biomarkers for a cytokine storm, interim guidelines from the US CDC (updated on 3 April 32020) suggested that high blood concentrations of the inflammatory indicator C-reactive protein (CRP) and blood ferritin levels may be correlated with more severe illness.

Based on a framework of conserved clinical characteristics using markers that are relevant to COVID-19, and relevant outcomes from related hyperinflammatory syndromes, a six-criterion additive scale was proposed for COVID-19-associated hyperinflammatory syndrome (cHIS) which appears to reflect COVID-19-specific patterns of inflammation (fever, macrophage activation, hematological dysfunction, hepatic injury, coagulopathy, and cytokinaemia) and may help to identify patients for clinical trials of anti-inflammatory therapies; however, further validation in large clinical studies is needed [95].

In addition to the cHIS scale, some specific diagnostic biomarkers can be very important to identify patients. It has been demonstrated that the imbalance of proinflammatory and anti-inflammatory mechanisms, and the interaction of a variety of cells and cytokines, can result in immune regulation disorder and a series of clinical manifestations [91]. IL-6 is a multifunctional cytokine that may promote inflammation in certain clinical conditions. Preliminary studies have shown that IL-6 levels are elevated in patients with severe COVID-19 which may contribute to the severe inflammatory response [98]. Thus, IL-6 could be a diagnostic biomarker for potential cytokine storm and monitoring IL-6 levels in conjunction with the cHIS scale might be an important step in the management of COVID-19 patients [99–101]. IL-6 testing results could be used to identify patients at risk of hyperinflammation such that approved therapies can be implemented to reduce the risk of mortality. To address this urgent need, Beckman Coulter has developed a fully automated Access IL-6 immunoassay to detect IL-6 levels in serum and plasma. Based on serum samples from healthy individuals analyzed with the Access IL-6 assay and a nonparametric reference interval analysis, the IL-6 upper 95% interval of the reference range is <6.4 pg/ml. For PCR-confirmed COVID-19 patients, an Access IL-6 concentration >35 pg/ml indicates they are at increased risk for intubation with mechanical ventilation. IL-6 values should be assessed in conjunction with clinical findings and the results from other laboratory parameters to determine treatment options. This product has received EUA from the US FDA [102]. This is an excellent example of how urgent medical needs promoted the quick development of new products and how biomarkers can help to accelerate the process.

Once a patient has been identified as potentially developing a cytokine storm or is in the midst of one, the corresponding treatment options to control the overactive inflammatory response, such as anti-inflammatory drugs, can be applied [103]. Many anti-inflammatory agents may be considered such as corticosteroids, chloroquine and hydroxychloroquine, cytokine-targeting therapies (IL-6 antagonists, IL-1 blockade, TNF-α blockers, Janus Kinase [JAK] inhibitors, and CC chemokine receptor 5 [CCR5] inhibitors), mesenchymal stem cells, i.v. immunoglobulin therapy, and convalescent plasma therapy [94, 104]. Clinical trials will be necessary to confirm the efficacy and safety of these drugs for treating COVID-19 despite them having been approved for other indications. A number of these drugs are going through clinical trials to decide duration and dose regime of treatment, risks and benefits, and others have already been found to be of no use. Two of these candidates, dexamethasone and IL-6 receptor (IL-6R) targeting mAb tocilizumab, are selected to demonstrate the translational nature of repurposing of these drugs under the urgent situation and how clinical trials can play critical roles to rule out non-effective drugs.

Dexamethasone was first synthesized in 1957 and entered clinical use one year later. It is a corticosteroid used to treat a wide range of inflammatory conditions (Figure 15.8). A controlled, open-label RECOVERY trial was carried out to evaluate the effects of dexamethasone treatment in patients hospitalized with COVID-19 at 176 National Health Service organizations in the UK [105]. The primary outcome was 28-day mortality, and the secondary outcome was hospital discharge within 28 days. Hospitalized patients with clinically suspected or confirmed SARS-CoV-2 infection were eligible for the trial. In total, 22.9% patients in the dexamethasone group and 25.7% in the standard-of-care group died within 28 days after randomization. Among patients receiving invasive mechanical ventilation, the incidence of death was lower in the dexamethasone group than that in the standard-of-care group (29.3 vs. 41.4%). Among those receiving oxygen without invasive mechanical ventilation, the incidence of death was also lower in the dexamethasone group (23.3 vs. 26.2%). This was not the case among those who were receiving no respiratory support at randomization (17.8 vs. 14.0%) (Table 15.7). Thus, the effect of dexamethasone was most pronounced and reduced the risk of death by 30% among the critically ill patients on ventilators and resulted in a reduction of 28-day mortality among patients hospitalized with COVID-19 who were receiving either invasive mechanical ventilation or oxygen alone at randomization. Shortly after these results were released, the UK government immediately

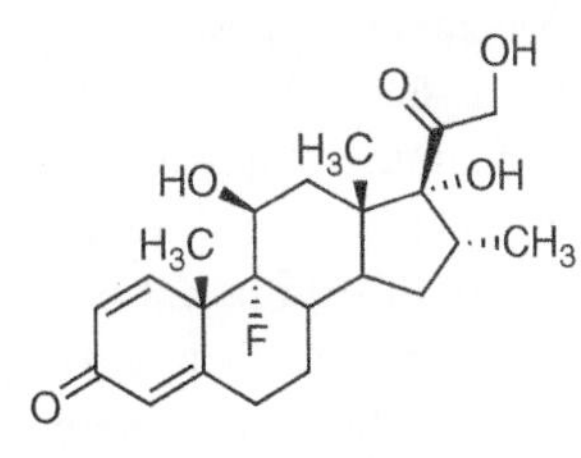

Figure 15.8 Chemical structure of dexamethasone.

Table 15.7 Impact of dexamethasone on mortality of hospitalized patients within 28 days.

	Dexamethasone group (%)	Usual care group (%)	Rate ratio (95% CI)
Mortality in patients receiving invasive mechanical ventilation	29.3	41.4	0.64 (0.51–0.81)
Mortality in patients receiving oxygen without invasive mechanical ventilation	23.3	26.2	0.82 (0.72–0.94)
Mortality in patients receiving no respiratory support at randomization	17.8	14.0	1.19 (0.91–1.55)

CI, confidence interval.

authorized the use of dexamethasone for patients hospitalized with COVID-19 who required oxygen, including those on ventilators [106].

As discussed earlier, IL-6 is one of the key cytokines that might play an important role in the cytokine storm, thus interfering with IL-6 might be potentially beneficial for severe and critical COVID-19-infected patients. As a result, adjunctive therapy with IL-6R antagonists including tocilizumab, sarilumab, and siltuximab has been proposed as a treatment for severe, progressive COVID-19 infection. The IL-6R has two forms: membrane-bound interleukin-6 receptor (mIL-6R) and soluble interleukin-6 receptor (sIL-6R). As a recombinant humanized anti-human IL-6R mAb that is currently used mainly for rheumatoid arthritis, tocilizumab can specifically bind sIL-6R and mIL-6R and inhibit signal transduction [107].

The results of multiple observational and randomized studies with tocilizumab treatment were reported. One of these evaluations was carried out by the Michigan Institute for Clinical & Health Research at the University of Michigan. In this study, patients admitted to the hospital from 9 March 2020 through April 2020 for severe COVID-19 pneumonia with a RT-PCR-positive SARS-CoV-2 test and who required invasive mechanical ventilation were included in the analysis. The primary outcome was survival probability after intubation. Clear clinical benefit was observed with tocilizumab in this study in which 18% of patients on tocilizumab died compared with 36% of those without using tocilizimab by the end of the 28-day period after patients went on a ventilator. Adjusted for health characteristics, this represents a 45% reduction in mortality. For those patients who were still in the hospital at the end of the study period, 82% of the tocilizumab patients and 53% of those who did not receive the drug had come off the ventilator [108].

A different multicenter observational study evaluated patients with laboratory-confirmed COVID-19 who needed support in the intensive care unit (ICU). This study also concluded that tocilizumab exposure among patients with severe SARS-CoV-2 infection requiring ICU support was associated with a reduction in hospital-related mortality [109].

In another study to determine whether tocilizumab improved patient outcomes, a randomized clinical trial investigated COVID-19 patients with moderate or severe pneumonia requiring at least 3 l/min of oxygen but without ventilation or admission to the ICU. The preliminary trial results showed that tocilizumab may reduce the need for mechanical and noninvasive ventilation or death by day 14 but did not reduce mortality by day 28 [110].

Two randomized phase 3 trials of tocilizumab were carried out by Hoffmann-La Roche. The COVACTA trial was the first global phase 3 study to investigate tocilizumab in adult patients hospitalized with severe COVID-19-associated pneumonia. In patients assessed using a 7-category ordinal scale tracking clinical status based on the need for intensive care and/or ventilator use and supplemental oxygen requirements, the difference in clinical status at week four between tocilizumab and placebo was not statistically significant. Thus, the primary endpoint of this trial was not met. The percentage of patients who died by week four was not different between tocilizumab (19.7%) and placebo (19.4%) groups [111]. The phase 3 EMPACTA (Evaluating Minority Patients with Actemra, NCT04372186) clinical trial evaluated the efficacy and safety of tocilizumab in the treatment of COVID-19-associated pneumonia among hospitalized patients who are underrepresented in clinical studies. In this study, 12.2% patients in the tocilizumab arm and 19.3% in the placebo arm progressed to mechanical ventilation or death by day 28, which met its primary endpoint. Compared with the underrepresented patients with COVID-19-associated pneumonia who received placebo plus standard of care, this indicates that patients who received tocilizumab plus standard of care were 44% less likely to progress to mechanical ventilation or death [112]. Considering the different clinical outcomes of these trials involving tocilizumab, further clinical studies are necessary to generate more conclusive results and explain the contradicting trial results between the observational studies and randomized clinical trials. Some of these trials are ongoing including REMDACTA and RECOVERY trials. On 11 February 2021, the RECOVERY website updated the status of tocilizumab trial in a preliminary report based on the primary outcome information available for 92% of patients (https://www.recoverytrial.net/news/tocilizumab-reduces-deaths-in-patients-hospitalised-with-covid-19). It demonstrated that tocilizumab reduced the risk of death for hospitalized patients with severe COVID-19 and shortened the time until patients are successfully discharged and reduced the need for a mechanical ventilator.

In summary, with the increased understanding of COVID-19 disease progression, cytokine storm was identified as a key factor contributing to worsening symptoms and a potential cause of death. Based on the scientific understanding of the causes of the cytokine storm and the body of knowledge of available anti-inflammatory drugs for the treatment of related inflammatory diseases, many such drugs were repurposed for the potential treatment of COVID-19 patients. Dexamethasone and tocilizumab are just two of these examples that have proceeded through vigorous clinical studies to confirm or disprove their effectiveness for treating patients with severe COVID-19 infections. The RECOVERY trial results showed a clear benefit of dexamethasone in patients who require respiratory support but not in patients who do not. The clinical benefit of tocilizumab in treating COVID-19 patients is not currently definitive, but the recent preliminary RECOVERY results mentioned above support the hypothesis that this drug can save lives and reduce the length of hospital stays. These findings suggest that the timing of anti-inflammatory treatments in relation to the stage of disease is important and analysis of subgroup clinical results can be very important to select the best treatment options [113–115].

15.4 Translational Medicine Summary

As described in this chapter, translational medicine played a critical role in the fight against the COVID-19 outbreak. Through multiple collaborations across various scientific communities worldwide, extraordinary speed in the discovery and development of effective diagnostic tools and therapeutics was realized. Furthermore, translational research served a unifying role in this increasingly complex, specialized, and fragmented field of biomedical research. While combating the COVID-19 pandemic, translational research also enabled a better understanding of the underlying biology of human diseases. Collectively, a few valuable lessons in translational medicine are summarized for future directions:

1) Diagnostic, prognostic, and predictive biomarkers play crucial roles in identifying infected individuals, evaluating disease conditions, and providing treatment options. In doing so, a more rational approach in the battle against the pandemic could be implemented. These biomarker strategies impact all drug discovery and development programs and significantly improve the success rate in transforming fundamental scientific discoveries to marketed products [116].
2) The bidirectional nature of translational medicine, despite its complexity, can be effectively executed based on the fundamental principles of translational medicine.
3) Close collaboration and clear channels of communication are essential components of translational medicine. In the fight against the COVID-19 pandemic, the world witnessed an unprecedented scale and level of collaboration and information sharing across the international

community. With that said, the validity of scientific information also proved problematic, particularly at the early stage of the COVID-19 outbreak [117, 118]. For example, the use of hydroxychloroquine and chloroquine for the treatment of COVID-19 infection received extensive media coverage in the beginning of the pandemic [119, 120], claiming "magic bullet" effectiveness across several patient groups [121, 122]. However, the RECOVERY clinical trial carried out in the UK proved that among patients hospitalized with COVID-19, those who received hydroxychloroquine did not have a lower incidence of death at 28 days than those who received usual care [123]. The NIH also halted a clinical trial of hydroxychloroquine since the study showed while the treatment did no harm, it also provided no benefits [124]. In this age of information overload, medical facts need to be rigorously scrutinized and properly communicated in order to provide the best treatment for those in need and minimize unnecessary delays.

4) Translation of fundamental scientific discoveries to commercialized products is a highly complex and data-driven process. At the same time, even the most promising scientific discoveries may not deliver the desired clinical outcome, not to mention that the fundamental biology behind many human diseases remains unclear thus obfuscating efforts to formulate effective treatments and prevention. Despite all the challenges, it is still possible to deliver the end-product in a timely manner, as demonstrated by Pfizer and Moderna in their vaccine development programs. However, in contrast, the phase 3 human trial of Oxford-AstraZeneca's COVID-19 vaccine candidate was placed on hold twice in two months due to safety concerns, while Johnson & Johnson paused all trials of its experimental COVID-19 vaccine in October 2020 after one participant developed an "adverse reaction". Fortunately, both clinical trials resumed shortly after the authorities found no evidence that the vaccine candidates caused the safety events. These examples illustrate that under no circumstances should patients' well-being be compromised. The WHO's chief scientist, Soumya Swaminathan, described this as "a wake-up call" to the global community about the inevitable ups and downs of medical research.
5) Regulatory agencies such as the FDA and EMA will continue to play crucial roles to safeguard the highest standards in all medical products. In this context, despite political pressure, the US FDA maintained its rigorous scientific and regulatory processes to ensure the safety, effectiveness, and quality of COVID-19 vaccines [125, 126]. The US CDC also put safety as the top priority in COVID-19 vaccine development [127].

Since the initial outbreak, enormous progress has been made in the fight against COVID-19 infections, but the fight is far from over (as of January 2021). The understanding of basic science, identification of biomarkers, properly conducted clinical trials, transparent data sharing and feedback, and rigorous regulatory assessment all played important roles amid the COVID-19 pandemic. While many challenges remain, using clinical studies as the axis of translational medicine coupled with the best science, close collaboration, adequate financial support, and properly implemented translational medicine principles, it is possible to accelerate the process to bring scientific discoveries to marketed drugs and we are another giant step closer to deliver the best health care to those in need.

References

1 DiMasia, J.A., Grabowskib, H.G., and Hansen, R.W. (2016). Innovation in the pharmaceutical industry: new estimates of R&D costs. *J. Health Econ.* 47: 20–33.

2 Mestre-Ferrandiz, J., Sussex, J., and Towse, A. (2012). The R&D cost of a new medicine. https://www.ohe.org/publications/rd-cost-new-medicine (accessed 2 October 2021).

3 US NIH(2007). Transforming Translation – Harnessing Discovery for Patient and Public Benefit. Translational Research Working Group of the National Cancer Advisory Board.

4 Khoury, M.J., Gwinn, M., Yoon, P.W. et al. (2007). The continuum of translation research in genomic medicine: how can we accelerate the appropriate integration of human genomic discoveries into health care and disease prevention? *Genet. Med.* 9 (10): 665–674.

5 Westfall, J.M., Mold, J., and Fagnan, L. (2007). Practice-based research–"Blue Highways" on the NIH roadmap. *JAMA* 297: 403–406. https://doi.org/10.1001/jama.297.4.403.

6 Zerhouni, E.A. (2007). Translational research: moving discovery to practice. *Clin. Pharmacol. Ther.* 81: 126–128.

7 Goldblatt, E.M. and Lee, W.-H. (2010). From bench to bedside: the growing use of translational research in cancer medicine. *Am. J. Transl. Res.* 2 (1): 1–18.

8 Chaplin, S. (2020). COVID-19: a brief history and treatments in development. *Prescriber* 31 (5): 23–28.

9 Chen, L., Liu, W., Zhang, Q. et al. (2020). RNA based mNGS approach identifies a novel human coronavirus from two individual pneumonia cases in 2019 Wuhan outbreak. *Emerg. Microbes Infect.* 9: 313–319.

10 Wrobel, A.G., Benton, D.J., Xu, P. et al. (2020). SARS-CoV-2 and bat RaTG13 spike glycoprotein structures inform on virus evolution and furin-cleavage effects. *Nat. Struct. Mol. Biol.* 27: 763–767.

11 Letko, M., Marzi, A., and Munster, V. (2020). Functional assessment of cell entry and receptor usage for SARS-CoV-2 and other lineage B betacoronaviruses. *Nat. Microbiol.* 5: 562–569.

12 Fehr, A.R. and Perlman, S. (2015). Coronaviruses: an overview of their replication and pathogenesis. *Methods Mol. Biol.* 1282: 1–23.

13 Huang, Y., Yang, C., Xu, X.-F. et al. (2020). Structural and functional properties of SARS-CoV-2 spike protein: potential antivirus drug development for COVID-19. *Acta Pharmacol. Sin. 41*: 1141–1149.

14 FDA (2020). in vitro diagnostics EUAs. https://www.fda.gov/medical-devices/coronavirus-disease-2019-covid-19-emergency-use-authorizations-medical-devices/vitro-diagnostics-euas#i%E2%80%A6 (accessed 2 October 2021).

15 Lazer, D., Santillana, M., Perlis, R. H., Ognyanova, K., Baum, M. A., Quintana, A., Druckman, J., Della Volpe, J., Chwe, H., and Simonson, M. (2020). The state of the nation: a 50-state COVID-19 survey. Report #8: failing the test. https://covidstates.net/COVID19%20CONSORTIUM%20REPORT%208%20TEST%20JULY%202020.pdf (accessed 2 October 2021).

16 Sherlock Biosciences (2020). Sherlock Biosciences receives FDA Emergency Use Authorization for CRISPR SARS-CoV-2 rapid diagnostic. https://sherlock.bio/wp-content/uploads/2020/05/Sherlock-EUA-Announcement_FINAL.pdf (accessed 2 October 2021).

17 BBC News (2020). India's new paper Covid-19 test could be a 'game changer'. https://www.bbc.com/news/world-asia-india-54338864?xtor=AL-72-%5Bpartner%5D-%5Bbbc.news.twitter%5D-%5Bheadline%5D-%5Bnews%5D-%5Bbizdev%5D-%5Bisapi%5D&at_custom2=twitter&at_custom1=%5Bpost+type%5D&at_campaign=64&at_custom4=E355E72E-0695-11EB-BC74-69974744363C&at_medium=custom7&at_custom3=%40BBCWorld (accessed 2 October 2021).

18 Fozouni, P., Son, S., de León Derby, M.D. et al. (2020). Direct detection of SARS-CoV-2 using CRISPR-Cas13a and a mobile phone. https://www.medrxiv.org/content/10.1101/2020.09.28.20201947v1.

19 Guglielmi, G. (2020). Fast coronavirus tests: what they can and can't do. *Nature* 585: 496–498.

20 World Health Organization(2020). Global partnership to make available 120 million affordable, quality COVID-19 rapid tests for low- and middle-income countries. https://www.who.int/news/

item/28-09-2020-global-partnership-to-make-available-120-million-affordable-quality-covid-19-rapid-tests-for-low--and-middle-income-countries (accessed 2 October 2021).

21 Johns Hopkins Bloomberg School of Public Health, Center for Health Security (2020). Current molecular and antigen tests with FDA EUA status. https://www.centerforhealthsecurity.org/resources/COVID-19/molecular-based-tests/current-molecular-and-antigen-tests.html (accessed 2 October 2021).

22 FDA (2020). BinaxNOWTM COVID-19 Ag Card. https://www.fda.gov/media/141567/download (accessed 2 October 2021).

23 Okba, N.M.A., Müller, M.A., Li, W. et al. (2020). Severe acute respiratory syndrome coronavirus 2–specific antibody responses in coronavirus disease patients. *Emerg. Infect. Dis.* 26 (7): 1478–1488.

24 Espejo, A.P., Akgun, Y., Al Mana, A.F. et al. (2020). Review of current advances in serologic testing for COVID-19. *Am. J. Clin. Pathol.* 154: 293–304.

25 McAndrews, K.M., Dowlatshahi, D.P., Dai, J. et al. (2020). Heterogeneous antibodies against SARS-CoV-2 spike receptor binding domain and nucleocapsid with implications on COVID-19 immunity. *JCI Insight* 5 (18): e142386. https://doi.org/10.1172/jci.insight.142386.

26 FDA (n.d.). EUA authorized serology test performance. https://www.fda.gov/medical-devices/coronavirus-disease-2019-covid-19-emergency-use-authorizations-medical-devices/eua-authorized-serology-test-performance (accessed 2 October 2021).

27 FDA (2020). Anti-SARS-CoV-2 ELISA (IgG): Instruction for use. https://www.fda.gov/media/137609/download (accessed 2 October 2021).

28 MacDonald, A. (2020). COVID-19 antibody testing: S vs. N protein. https://www.technologynetworks.com/diagnostics/blog/covid-19-antibody-testing-s-vs-n-protein-340327 (accessed 2 October 2021).

29 FDA (2020). Coronavirus (COVID-19) update: FDA authorizes first test that detects neutralizing antibodies from recent or prior SARS-CoV-2 infection. https://www.fda.gov/news-events/press-announcements/coronavirus-covid-19-update-fda-authorizes-first-test-detects-neutralizing-antibodies-recent-or (accessed 2 October 2021).

30 Sethuraman, N., Jeremiah, S.S., and Ryo, A. (2020). Interpreting diagnostic tests for SARS-CoV-2. *J. Am. Med. Assoc.* 323: 2249–2251.

31 Iyer, A.S., Jones, F.K., Nodoushani, A. et al. (2020). Persistence and decay of human antibody responses to the receptor binding domain of SARS-CoV-2 spike protein in COVID-19 patients. *Sci. Immunol.* 5 (52): eabe0367.

32 Isho, B., Abe, K.T., Zuo, M. et al. (2020). Persistence of serum and saliva antibody responses to SARS-CoV-2 spike antigens in COVID-19 patients. *Sci. Immunol.* 5 (52): eabe5511.

33 Eastman, R.T., Roth, J.S., Brimacombe, K.R. et al. (2020). Remdesivir: a review of its discovery and development leading to emergency use authorization for treatment of COVID-19. *ACS Cent. Sci.* 6: 672–683. https://doi.org/10.1021/acscentsci.0c00489.

34 Seley-Radtke, K.L. and Yates, M.K. (2018). The evolution of nucleoside analogue antivirals: a review for chemists and non-chemists. Part 1: early structural modifications to the nucleoside scaffold. *Antiviral Res.* 154: 66–86.

35 Yates, M.K. and Seley-Radtke, K.L. (2019). The evolution of antiviral nucleoside analogues: a review for chemists and non-chemists. Part II: complex modifications to the nucleoside scaffold. *Antiviral Res.* 162: 5–21.

36 Warren, T.K., Jordan, R., Lo, M. et al. (2015). Nucleotide prodrug GS-5734 is a broad-spectrum filovirus inhibitor that provides complete therapeutic protection against the development of Ebola virus disease (EVD) in infected non-human primates. *Open Forum Infect. Dis.* 2 (Suppl l_1): LB-2.

37 Warren, T.K., Jordan, R., Lo, M.K. et al. (2016). Therapeutic efficacy of the small molecule GS- 5734 against Ebola virus in rhesus monkeys. *Nature* 531 (7594): 381–385.

38 Wang, Z. and Yang, L. (2020). GS-5734: FDA potentially approved drug against SARS-Cov-2. *New J. Chem.* https://doi.org/10.1039/D0NJ02656E.

39 Lo, M.K., Feldmann, F., Gary, J.M. et al. (2019). Remdesivir (GS-5734) protects African green monkeys from Nipah virus challenge. *Sci. Transl. Med. 11* (494): eaau9242.

40 Sheahan, T.P., Sims, A.C., Graham, R.L. et al. (2017). Broad-spectrum antiviral GS-5734 inhibits both epidemic and zoonotic coronaviruses. *Sci. Transl. Med. 9* (396): eaal3653. https://doi.org/10.1126/scitranslmed.aal3653.

41 Frediansyah, A., Nainuc, F., Dhamad, K. et al. Remdesivir and its antiviral activity against COVID-19: a systematic review. *Clin. Epidemiol. Glob. Health* https://doi.org/10.1016/j.cegh.2020.07.011.

42 Sheahan, T.P., Sims, A.C., Leist, S.R. et al. (2020). Comparative therapeutic efficacy of Remdesivir and combination lopinavir, ritonavir, and interferon beta against MERS-CoV. *Nat. Commun.* 11 (1): 1–14.

43 de Wit, E., Feldmann, F., Cronin, J. et al. (2020). Prophylactic and therapeutic Remdesivir (GS-5734) treatment in the rhesus macaque model of MERS-CoV infection. *Proc. Natl. Acad. Sci. USA* 117 (12): 6771–6776.

44 Pardo, J., Shukla, A.M., Chamarthi, G., and Gupte, A. (2020). The journey of Remdesivir: from Ebola to COVID-19. *Drugs Context 9*: 2020-4-14.

45 Mulangu, S., Dodd, L.E., Davey, R.T. Jr. et al. (2019). A randomized, controlled trial of Ebola virus disease therapeutics. *N. Engl. J. Med. 381*: 2293–2303. https://doi.org/10.1056/NEJMoa1910993.

46 Lo, M.K., Albariño, C.G., Perry, J.K. et al. (2020). Remdesivir targets a structurally analogous region of the Ebola virus and SARS-CoV-2 polymerases. *Proc. Natl. Acad. Sci. USA* 117 (43): 26946–26954. https://doi.org/10.1073/pnas.2012294117.

47 Williamson, B., Feldmann, F., Schwarz, B. et al. (2020). Clinical benefit of Remdesivir in rhesus macaques infected with SARS-CoV-2. *Nature* 585: 273–276. https://doi.org/10.1038/s41586-020-2423-5.

48 Hendaus, M.A. (2020). Remdesivir in the treatment of coronavirus disease 2019 (COVID-19): a simplified summary. *J. Biomol. Struct. Dyn.* https://doi.org/10.1080/07391102.2020.1767691.

49 NIH (2020). Clinical trial shows Remdesivir accelerates recovery from advanced COVID-19. https://www.nih.gov/news-events/nih-clinical-trial-shows-remdesivir-accelerates-recovery-advanced-covid-19 (accessed 2 October 2021).

50 Beigel, J.H., Tomashek, K.M., Dodd, L.E. et al. (2020). Remdesivir for the treatment of Covid-19 – final report. *N. Engl. J. Med.* https://doi.org/10.1056/NEJMoa2007764.

51 Gilead Sciences (2020). Gilead announces results from phase 3 trial of investigational antiviral Remdesivir in patients with severe COVID-19. https://www.gilead.com/news-and-press/press-room/press-releases/2020/4/gilead-announces-results-from-phase-3-trial-of-investigational-antiviral-remdesivir-in-patients-with-severe-covid-19 (accessed 2 October 2021).

52 Goldman, J.D., Lye, D.C.B., Hui, D.S. et al. (2020). Remdesivir for 5 or 10 days in patients with severe Covid-19. *N. Engl. J. Med.* https://doi.org/10.1056/NEJMoa2015301.

53 Spinner, C.D., Gottlieb, R.L., Criner, G.J. et al. (2020). Effect of Remdesivir vs standard care on clinical status at 11 days in patients with moderate COVID-19 a randomized clinical trial. *JAMA 324* (11): 1048–1057. https://doi.org/10.1001/jama.2020.16349.

54 FDA(2020). FDA approves first treatment for COVID-19. https://www.fda.gov/news-events/press-announcements/fda-approves-first-treatment-covid-19 (accessed 2 October 2021).

55 WHO Solidarity Trial Consortium (2020). Repurposed antiviral drugs for COVID-19 – interim WHO SOLIDARITY trial results. https://www.medrxiv.org/content/10.1101/2020.10.15.20209817v1 (accessed 2 October 2021).

56 Eli Lilly and Company(2020). Lilly announces proof of concept data for neutralizing antibody LY-CoV555 in the COVID-19 outpatient setting. https://investor.lilly.com/news-releases/news-release-details/lilly-announces-proof-concept-data-neutralizing-antibody-ly (accessed 2 October 2021).

57 Eli Lilly and Company (2020). Lilly provides comprehensive update on progress of SARS-CoV-2 neutralizing antibody programs. https://investor.lilly.com/news-releases/news-release-details/lilly-provides-comprehensive-update-progress-sars-cov-2 (accessed 2 October 2021).

58 Eli Lilly and Company (2020). Lilly's neutralizing antibody bamlanivimab (LY-CoV555) receives FDA emergency use authorization for the treatment of recently diagnosed COVID-19. https://investor.lilly.com/news-releases/news-release-details/lillys-neutralizing-antibody-bamlanivimab-ly-cov555-receives-fda (accessed 2 October 2021).

59 Hansen, J., Baum, A., Pascal, K.E. et al. (2020). Studies in humanized mice and convalescent humans yield a SARS-CoV-2 antibody cocktail. *Science 369*: 1010–1014.

60 Baum, A., Fulton, B.O., Wloga, E. et al. (2020). Antibody cocktail to SARS-CoV-2 spike protein prevents rapid mutational escape seen with individual antibodies. *Science* 369: 1014–1018.

61 Baum, A., Ajithdoss, D., Copin, R. et al. (2020). REGN-COV2 antibodies prevent and treat SARS-CoV-2 infection in rhesus macaques and hamsters. *Science* https://doi.org/10.1126/science.abe2402.

62 Regeneron Pharmaceuticals, Inc.(2020). Regeneron's casirivimab and imdevimab antibody cocktail for covid-19 is first combination therapy to receive FDA emergency use authorization. https://investor.regeneron.com/news-releases/news-release-details/regenerons-regen-cov2-first-antibody-cocktail-covid-19-receive (accessed 2 October 2021).

63 Eli Lilly and Company and Amgen (2020). Lilly and Amgen announce manufacturing collaboration for COVID-19 antibody therapies. https://www.amgen.com/newsroom/press-releases/2020/09/lilly-and-amgen-announce-manufacturing-collaboration-for-covid-19-antibody-therapies (accessed 2 October 2021).

64 Regeneron Pharmaceuticals, Inc. and Roche (2020). Regeneron and Roche Collaborate to significantly increase global supply of REGN-COV2 investigational antibody cocktail for COVID-19. https://investor.regeneron.com/index.php/news-releases/news-release-details/regeneron-and-roche-collaborate-significantly-increase-global (accessed 2 October 2021).

65 Jones, D. and Helmreich, S. (2020). A history of herd immunity. *Lancet* 396 (10254): 810–811.

66 FDA. Ensuring the safety of vaccines in the United States (2011). https://www.fda.gov/files/vaccines,%20blood%20&%20biologics/published/Ensuring-the-Safety-of-Vaccines-in-the-United-States.pdf (accessed 2 October 2021).

67 Singh, K. and Mehta, S. (2016). The clinical development process for a novel preventive vaccine: an overview. *J. Postgrad Med. 62* (1): 4–11. https://doi.org/10.4103/0022-3859.173187.

68 Rolling, K.E. and Hayney, M.S. (2016). The vaccine development process. *J. Am. Pharm. Assoc.* 56 (6): 687–689. https://doi.org/10.1016/j.japh.2016.09.009.

69 Wang, N., Shang, J., Jiang, S., and Du, L. (2020). Subunit vaccines against emerging pathogenic human coronaviruses. *Front. Microbiol.* 11: 298. https://doi.org/10.3389/fmicb.2020.00298.

70 WHO(2020). Draft landscape of COVID-19 candidate vaccines. https://www.who.int/publications/m/item/draft-landscape-of-covid-19-candidate-vaccines (accessed 2 October 2021).

71 Subbarao, K. (2020). SARS-CoV-2: a new song recalls an old melody. *Cell Host Microbe* 27 (5): 692. https://doi.org/10.1016/j.chom.2020.04.019.

72 Ong, E., Wong, M.U., Huffman, A., and He, Y. COVID-19 coronavirus vaccine design using reverse vaccinology and machine learning. https://www.biorxiv.org/content/10.1101/2020.03.20.000141v2 (accessed 2 October 2021).

73 Chen, Y., Tong, X., Li, Y. et al. (2020). A comprehensive, longitudinal analysis of humoral responses specific to four recombinant antigens of SARS-CoV-2 in severe and non-severe COVID-19 patients. *PLoS Pathog.* 16 (9): e1008796. https://doi.org/10.1371/journal.ppat.1008796.

74 Chen, H., Tang, L., Yu, X. et al. (2020). Bioinformatics analysis of epitope-based vaccine design against the novel SARS-CoV-2. *Infect. Dis. Poverty* 9: 88. https://doi.org/10.1186/s40249-020-00713-3.

75 The International Federation of Pharmaceutical Manufacturers & Associations (2019). The complex journey of a vaccine: the steps behind developing a new vaccine. https://www.ifpma.org/wp-content/uploads/2019/07/IFPMA-ComplexJourney-2019_FINAL.pdf (accessed 2 October 2021).

76 Corey, L., Mascola, J.R., Fauci, A.S., and Collins, F.S. (2020). A strategic approach to COVID-19 vaccine R&D. *Science* 368 (6494): 948–950. https://doi.org/10.1126/science.abc5312.

77 Lurie, N., Saville, M., Hatchett, R., and Halton, J. (2020). Developing Covid-19 vaccines at pandemic speed. *N. Engl. J. Med.* 382: 1969–1973.

78 *Wall Street Journal* (2020). Covid-19 vaccine developers prepare joint pledge on safety, standards. https://www.wsj.com/articles/covid-19-vaccine-developers-prepare-joint-pledge-on-safety-standards-11599257729 (accessed 2 October 2021).

79 FDA (2020). Coronavirus Treatment Acceleration Program (CTAP). https://www.fda.gov/drugs/coronavirus-covid-19-drugs/coronavirus-treatment-acceleration-program-ctap (accessed 2 October 2021).

80 US Department of Health & Human Services (2020). Explaining Operation Warp Speed https://www.nihb.org/covid-19/wp-content/uploads/2020/08/Fact-sheet-operation-warp-speed.pdf (accessed 2 October 2021).

81 US Department of Health and Human Services, Food and Drug Administration, Center for Biologics Evaluation and Research (2020). Development and licensure of vaccines to prevent COVID-19: Guidance for industry https://www.fda.gov/media/139638/download (accessed 2 October 2021).

82 US Department of Health and Human Services, Food and Drug Administration, Center for Biologics (2020). Emergency Use Authorization for vaccines to prevent COVID-19l: Guidance for industry https://www.fda.gov/media/142749/download (accessed 2 October 2021).

83 Kaur, S.P. and Gupta, V. (2020). COVID-19 vaccine: a comprehensive status report. *Virus Res.* 288: 198114.

84 Haynes, B.F., Corey, L., Fernandes, P. et al. (2020). Prospects for a safe COVID-19 vaccine. *Sci. Transl. Med.* https://doi.org/10.1126/scitranslmed.abe0948.

85 Dorp, L., Acman, M., Richard, D. et al. (2020). Emergence of genomic diversity and recurrent mutations in SARS-CoV-2. *Infect. Genet. Evol.* 83: 104351. https://doi.org/10.1016/j.meegid.2020.104351.

86 Thomson, E.C., Rosen, L.E., Shepherd, J.G. et al. (2020). The circulating SARS-CoV-2 spike variant N439K maintains fitness while evading antibody-mediated immunity. https://www.biorxiv.org/content/10.1101/2020.11.04.355842v1 (accessed 2 October 2021).

87 Pfizer Inc. and BioNTech SE (2020). Pfizer and Biontech announce vaccine candidate against COVID-19 achieved success in first interim analysis from phase 3 study. https://www.pfizer.com/news/press-release/press-release-detail/pfizer-and-biontech-announce-vaccine-candidate-against (accessed 2 October 2021).

88 Sun, D. (2020). Remdesivir for treatment of COVID-19: combination of pulmonary and IV administration may offer Aditional benefit. *AAPS J.* 22: 77. https://doi.org/10.1208/s12248-020-00459-8.

89 Sahakijpijarn, S., Moon, C., Koleng, J.J. et al. (2020). Development of Remdesivir as a dry powder for inhalation by thin film freezing. https://www.biorxiv.org/content/10.1101/2020.07.26.222109v2 (accessed 2 October 2021).

90 ClinicalTrials.gov(2020). Safety, tolerability and pharmacokinetics of inhaled nanoparticle formulation of Remdesivir (GS-5734) and NA-831 (NEUROSIVIR). https://clinicaltrials.gov/ct2/show/NCT04480333 (accessed 2 October 2021).

91 Gao, Y.-M., Xu, G., Wang, B., and Liu, B.-C. (2020). Cytokine storm syndrome in coronavirus disease 2019: a narrative review (review). *J. Intern. Med.* https://doi.org/10.1111/joim.13144.

92 Vabret, N., Britton, G.J., Gruber, C. et al. (2020). The Sinai immunology review project. Immunology of COVID-19: current state of the science. *Immunity 52*: 910–941.

93 Coperchinia, F., Chiovatoa, L., Crocea, L. et al. (2020). The cytokine storm in COVID-19: an overview of the involvement of the chemokine/chemokine-receptor system. *Cytokine Growth Factor Rev.* 53: 25–32.

94 Iannaccone, G., Scacciavillani, R., Giuseppe, M. et al. (2020). Weathering the cytokine storm in COVID-19: therapeutic implications. *Cardiorenal. Med.* 10: 277–287.

95 Webb, B.J., Peltan, I.D., Jensen, P. et al. (2020). Clinical criteria for COVID-19-associated hyperinflammatory syndrome: a cohort study. *Lancet Rheumatol.* https://doi.org/10.1016/S2665-9913(20)30343-X.

96 García, L.F. (2020). Immune response, inflammation, and the clinical spectrum of COVID-19. *Front. Immunol.* 11: 1441. https://doi.org/10.3389/fimmu.2020.01441.

97 Langer-Goulda, A., Smithb, J.B., Gonzalesb, E.G. et al. (2020). Early identification of COVID-19 cytokine storm and treatment with anakinra or tocilizumab. *Int. J. Infect. Dis.* 99: 291–297.

98 Liu, T., Zhang, J., Yang, Y.et al. (2020). The potential role of IL-6 in monitoring severe case of coronavirus disease 2019. https://www.medrxiv.org/content/10.1101/2020.03.01.20029769v2 (accessed 2 October 2021).

99 Ceccon, M., Piovani, D., Brunetta, E. et al. (2020). Early predictors of clinical deterioration in a cohort of 239 patients hospitalized for Covid-19 infection in Lombardy, Italy. *J. Clin. Med. 9* (5): 1548.

100 Mehta, P., McAuley, D.F., Brown, M. et al. (2020). COVID-19: consider cytokine storm syndromes and immunosuppression. *The Lancet 395* (10229): 1033–1034. https://doi.org/10.1016/S0140-6736(20)30628-0.

101 Shibabaw, T. (2020). Inflammatory cytokine: IL-17A signaling pathway in patients present with COVID-19 and current treatment strategy. *J. Inflamm. Res.* 13: 673–680.

102 FDA (2020). Fact sheet for patients: Access IL-6 – Beckman Coulter, Inc.. https://www.fda.gov/media/142599/download (accessed 2 October 2021).

103 Tang, Y., Liu, J., Zhang, D. et al. (2020). Cytokine storm in COVID-19: the current evidence and treatment strategies. *Front. Immunol.* 11: 1708. https://doi.org/10.3389/fimmu.2020.01708.

104 Halacli, B. and Topeli, A. (2020). Treatment of the cytokine storm in COVID-19. *J. Crit. Intensive Care* 11 (Suppl. 1): 36–40.

105 Horby, P., Lim, W.S., Emberson, J.et al. (2020). Dexamethasone in hospitalized patients with Covid-19 — Preliminary report. NEJM.org. doi: 10.1056/NEJMoa2021436.

106 Ledford, H. (2020). Coronavirus breakthrough: dexamethasone is first drug shown to save lives. *Nature* 582: 469. https://doi.org/10.1038/d41586-020-01824-5.

107 Sajna, K.V. and Kamat, S. (2020). Antibodies at work in the time of severe acute respiratory syndrome coronavirus 2. *Cytotherapy*: 1–10.

108 Somers, E.C., Eschenauer, G.A., Troost, J.P. et al. (2020). Tocilizumab for treatment of mechanically ventilated patients with COVID-19. *Clin. Infect. Dis.* 71: 1–10. https://doi.org/10.1093/cid/ciaa954.

109 Biran, N., Ip, A., Ahn, J. et al. (2020). Tocilizumab among patients with COVID-19 in the intensive care unit: a multicentre observational study. *Lancet Rheumatol. 2*: e603–e612. https://doi.org/10.1016/S2665-9913(20)30277-0.

110 Hermine, O., Mariette, X., Tharaux, P.-L. et al. (2020). Effect of tocilizumab vs usual care in adults hospitalized with COVID-19 and moderate or severe pneumonia a randomized clinical trial. *JAMA Intern. Med.* https://doi.org/10.1001/jamainternmed.2020.6820.

111 Roche (2020). Roche provides an update on the phase III COVACTA trial of Actemra/RoActemra in hospitalised patients with severe COVID-19 associated pneumonia. https://www.roche.com/investors/updates/inv-update-2020-07-29.htm (accessed 2 October 2021).

112 Roche (2020). Roche's phase III EMPACTA study showed Actemra/RoActemra reduced the likelihood of needing mechanical ventilation in hospitalised patients with COVID-19 associated pneumonia. https://www.roche.com/media/releases/med-cor-2020-09-18.htm (accessed 2 October 2021).

113 Sinha, P., Mostaghim, A., Bielick, C.G. et al. (2020). Early administration of Interleukin-6 inhibitors for patients with severe Covid-19 disease is associated with decreased intubation, reduced mortality, and increased discharge. *Int. J. Infect. Dis. 99*: 28–33.

114 Xu, X., Han, M., Li, T. et al. (2020). Effective treatment of severe COVID-19 patients with tocilizumab. *Proc. Natl. Acad. Sci. USA* 117: 10970–10975.

115 Scarsi, M., Piantoni, S., Colombo, E. et al. (2020). Association between treatment with colchicine and improved survival in a single-centre cohort of adult hospitalised patients with COVID-19 pneumonia and acuterespiratory distress syndrome. *Ann. Rheum. Dis 79*: 1286–1289. https://doi.org/10.1136/annrheumdis-2020-217712.

116 Luciano, J.S., Andersson, B., Batchelor, C. et al. (2011). The translational medicine ontology and knowledge base: driving personalized medicine by bridging the gap between bench and bedside. *J. Biomed. Semant. 2* (Suppl. 2): S1.

117 Hossen, S., Barek, A., Jahan, N., and Islam, M.S. (2020). A review on current repurposing drugs for the treatment of COVID-19: reality and challenges. *SN Compr. Clin. Med.* 2: 1777–1789.

118 World Health Organization (2020). COVID-19 strategy update. https://www.who.int/docs/default-source/coronaviruse/covid-strategy-update-14april2020.pdf?sfvrsn=29da3ba0_19&download=true (accessed 2 October 2021).

119 Pastick, K.A., Okafor, E.C., Wang, F. et al. (2020). Review: hydroxychloroquine and chloroquine for treatment of SARS-CoV-2 (COVID-19). *Open Forum Infect. Dis.* 7 (4): ofaa130. https://doi.org/10.1093/ofid/ofaa130.

120 Yazdany, J. and Kim, A.H.J. (2020). Use of hydroxychloroquine and chloroquine during the COVID-19 pandemic: what every clinician should know. *Ann. Intern. Med. 172* (11) https://doi.org/10.7326/M20-1334.

121 Arshad, S., Kilgore, P., Chaudhry, Z.S. et al. (2020). Treatment with hydroxychloroquine, azithromycin, and combination in patients hospitalized with COVID-19. *Int. J. Infect. Dis 97*: 396–403.

122 Gautret, F., Lagier, J.C., Parola, P. et al. (2020). Hydroxychloroquine and azithromycin as a treatment of COVID-19: results of an open-label non-randomised clinical trial. *Int. J. Antimicrob. Agents* https://doi.org/10.1016/j.antimicag.2020.105949.

123 The RECOVERY Collaborative Group (2020). Effect of hydroxychloroquine in hospitalized patients with Covid-19. *N. Engl. J. Med.* https://doi.org/10.1056/NEJMoa2022926.

124 NIH (2020). NIH halts clinical trial of hydroxychloroquine. https://www.nih.gov/news-events/news-releases/nih-halts-clinical-trial-hydroxychloroquine (accessed 2 October 2021).

125 US Department of Health and Human Services, Food and Drug Administration, Center for Biologics Evaluation and Research (2020). Emergency Use Authorization for vaccines to prevent COVID-19: Guidance for industry. https://www.fda.gov/media/142749/download (accessed 2 October 2021).

126 US Department of Health and Human Services, Food and Drug Administration, Center for Biologics Evaluation and Research. Development and licensure of vaccines to prevent COVID-19: Guidance for industry. https://www.fda.gov/media/139638/download (accessed 2 October 2021).

127 Centers for Disease Control and Prevention (2019). Key things to know about COVID-19. https://www.cdc.gov/coronavirus/2019-ncov/vaccines/8-things.html (accessed 2 October 2021).

Appendix A

Monoclonal Antibody Drug Discovery and Development Paradigm

Biologic drugs include a wide range of products such as recombinant therapeutic proteins (e.g. antibody-based therapies), cell and gene therapies, blood and blood components, allergenics, and vaccines. A simplified monoclonal antibody (mAb) drug discovery and development paradigm is as follows:

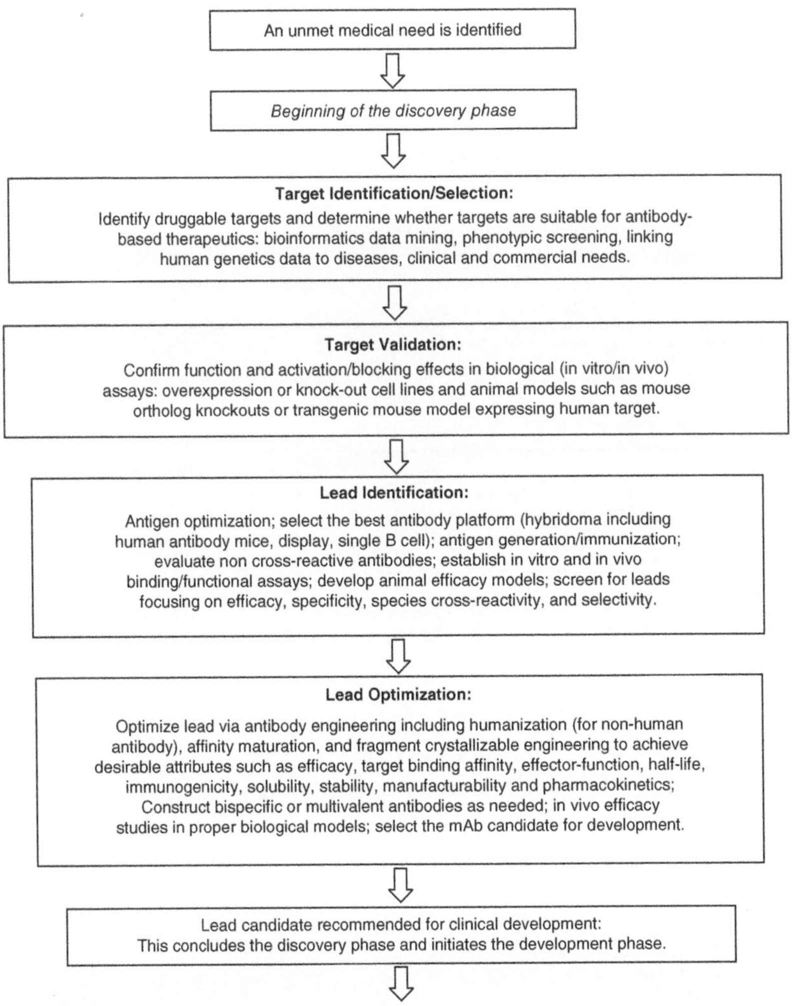

Contemporary Accounts in Drug Discovery and Development, First Edition. Edited by Xianhai Huang, Robert G. Aslanian, and Wayne H. Tang.

Beginning of the development phase

⇩

Development Candidate Preclinical Studies to Investigational New Drug (IND):

Chemistry, Manufacturing and Control development including stable cell line generation, cell culture and purification process development, analytical method development and formulation development; preclinical pharmacokinetics/pharmacodynamics studies; Good Laboratory Practice toxicology studies and early current Good Manufacturing Practice manufacturing; maximize translation of efficacy into the clinic (safety and risk assessment, clinical biomarker plan and clinical plans, commercial assessment and legal evaluation); prepare IND package (scientific data, patient stratification, regulatory, and manufacturing plans) for regulatory review.

⇩

Phase I–III Clinical Trials:

Clinical trial planning and preparation; clinical cohorts plans; phase I assess human PK profile, safety and tolerability in healthy human beings; phase II–III safety, dose, efficacy studies in patients; scale up/process change; clinical trial data collection.

⇩

Biologics License Application (BLA) Filing

⇩

Regulatory Review and Drug Approval

⇩

Post-Marketing Events (phase IV clinical studies if necessary):

Post-approval data collection including reanalysis of premarket data and spontaneous reporting; pharmaco-epidemiological observational studies: pragmatic safety and effectiveness trials.

⇩

Drug Life Cycle Management:

New indications/repurposing; off label use; drug label change.

Appendix B

Glossary

μSv/MBq The "effective dose (ED)" is a biological dose commonly used in radioprotection, as it determines how dangerous an individual's exposure to radiations can be. It takes into consideration not only the nature of the incoming radiation but also the sensitivities of the body parts affected. If the dose is expressed in microsievert (μSv) (millionths of a sievert), the ED is measured in microsievert/megabecquerel (μSv/MBq). MBq is the amount of radioactivity. The same unit as is used for the equivalent dose absorbed locally by an organ, a gland, or any other part of the body.

3k-cutoff membrane filtration The lowest molecular weight solute of 3k in daltons in which 90% of the solute is retained by the membrane.

A negative-sense single-stranded RNA virus or (-)ssRNA virus This is a virus that uses negative sense, single-stranded RNA as its genetic material.

Acute respiratory distress syndrome A life-threatening lung injury that allows fluid to leak into the lungs. It is characterized by rapid onset of widespread inflammation in the lungs and occurs rapidly and leads to lung failure when fluid builds up in the tiny, elastic air sacs (alveoli) in the lungs causing low blood oxygen.

Affinity-based screening A high-throughput screening method to identify biologically active new leads by testing large compound libraries for activity against a target of interest in biochemical *in vitro* tests. Affinity-based screening is only dependent on compound binding to protein targets and does not depend on target function for their readout.

Allosteric site A site, different from the orthosteric site of a receptor, capable of binding with ligands to modulate protein function.

Alzheimer's disease A degenerative brain disease characterized by progressive memory and neuronal loss combined with cognitive impairment with symptoms gradually worsening over a number of years that significantly interferes with social and occupational functioning.

Alzheimer's Disease Assessment Scale A scale specifically designed to evaluate cognitive and behavioral dysfunctions characteristic of Alzheimer's disease which consists of cognitive (ADAS-Cog) and noncognitive (ADAS-Noncog) subscales.

Alzheimer's Disease Cooperative Study Activities of Daily Living Inventory Scale The scale assesses the competence of patients with Alzheimer's disease in basic and instrumental activities of daily living which can be completed in questionnaire format, or as a structured interview.

Amyloid hypothesis A hypothesis suggesting that the accumulation and deposition of oligomeric or fibrillar amyloid β (Aβ) peptide is the primary cause of Alzheimer's disease. An

Contemporary Accounts in Drug Discovery and Development, First Edition. Edited by Xianhai Huang, Robert G. Aslanian, and Wayne H. Tang.

imbalance between Aβ production and Aβ clearance is proposed to result in the rest of the disease process, including formation of neurofibrillary tangles containing tau protein.

ANCOVA model For inferring a treatment effect from the difference between a treated and untreated group on a quantitative outcome measured before and after treatment, a current method is analysis of covariance (ANCOVA) of the outcome with the baseline as covariate.

Angiogenic switch A time-restricted event during tumor progression where the balance between pro- and anti-angiogenic factors tilts toward a pro-angiogenic outcome, resulting in the transition from dormant avascularized hyperplasia to outgrowing vascularized tumor and eventually to malignant tumor progression.

Artificial intelligence The simulation of human intelligence in machines that are programmed to think like humans and mimic their actions. It may also be applied to any machine that exhibits traits associated with a human mind such as learning and problem-solving. It is a wide-ranging branch of computer science concerned with building smart machines capable of performing tasks that typically require human intelligence. Artificial intelligence is an interdisciplinary science where multiple approaches and advancements in machine learning and deep learning are creating a paradigm shift and accelerating the process.

Atypical antipsychotics Also known as second generation antipsychotics, developed after the 1970s. Compared with typical antipsychotics, they are less likely to cause the extrapyramidal side effects.

Base excision repair A DNA repair process that corrects small base lesions that do not significantly distort the DNA helix structure.

Binding potential The ratio of receptor density B_{max} to radioligand equilibrium dissociation constant K_d as defined by Mintun. Binding potential, $BP = B_{max}/K_d$ = receptor density × affinity.

Biomarkers A biomarker is a biological observation that substitutes for and ideally predicts a clinically relevant endpoint or intermediate outcome that is more difficult to observe.

Biotin–streptavidin interaction Streptavidin is a protein that shows considerable affinity for biotin. It has the ability to bind up to four biotin molecules, making this interaction ideal for both purification and detection strategies.

B_{max} This is the total receptor density or concentration in a tissue sample. B_{max} can be estimated with linear regression fitting of saturation binding assay (using a radioligand) data transformed into a Scatchard plot.

BRCA1/2 mutation Mutations in the BRCA1 and BRCA2 genes involved in cell growth, cell division, and the repair of damage to DNA that cause DNA damage in cells to go unrepaired, which increases the chance that a person will develop some types of cancer. Breast and ovarian cancer are the most common diseases linked to BRCA1 and BRCA2 changes, but mutated forms of the BRCA genes may increase the risk for other cancers as well.

Breakpoint A defined antibiotic concentration or zone of inhibition diameter that serves as a gatekeeper for antimicrobial use. The value categorizes organisms as susceptible, susceptible-dose dependent, intermediate, resistant, or nonsusceptible to various antimicrobials. It is evaluated by examining the response of a bacterial isolate to antimicrobial exposure.

Ca^{2+} oscillation assay A fluorescence-based functional assay measuring frequencies of oscillations in cytosolic Ca^{2+} concentration upon receptor activation.

Canonical pocket A generally accepted/recognized pocket, e.g. orthosteric ligand binding pocket.

Capsid assembly This is an essential step in the life cycle of many viruses, during which hundreds to thousands of viral structural proteins (e.g. core protein) assemble around the viral genome to form a protein shell called the capsid.

Carbapenem-resistant Enterobacteriaceae A family of bacteria that has developed resistance to the group of antibiotics called carbapenems.

CC_{50} This is the mean value for cytotoxicity to specific cell lines, e.g. HepDE19 cells.

CellKey assay A label-free, cell-based functional assay measuring complex impedance changes in cells upon activation of cell surface receptors.

CHARM trial The candesartan in heart failure: reduction in mortality and morbidity program (ClinicalTrials.gov registration number NCT00634400) consisted of three trials which enrolled 7599 patients with NYHA class II–IV chronic heart failure and randomized them to candesartan or placebo in addition to standard heart failure therapies. Among a broad spectrum of patients with symptomatic HF, treatment with the angiotensin II type 1 receptor blocker candesartan was associated with a nonsignificant decrease in all-cause mortality compared with placebo, driven primarily by a reduction in cardiovascular mortality.

Child–Pugh A, B, and C The Child–Pugh score (or the Child–Turcotte–Pugh score or Child Criteria) is used to categorize the degree of hepatic impairment in patients with chronic liver disease, mainly cirrhosis.

Clastogenicity A phenomenon of inducing disruption or breakage of chromosomes.

Clinical Dementia Rating Scale This is a five-point scale used to characterize six domains of cognitive and functional performance applicable to Alzheimer's disease and related dementias which includes memory, orientation, judgment and problem solving, community affairs, home and hobbies, and personal care. Based on interview data and clinical judgment, the Clinical Dementia rating (CDR) table will guide the clinician in making appropriate ratings. An overall CDR score may be calculated through the use of an algorithm in addition to ratings for each domain. This score is useful for characterizing and tracking a patient's level of impairment/dementia: 0 = normal; 0.5 = very mild dementia; 1 = mild dementia; 2 = moderate dementia; and 3 = severe dementia.

Clinically evaluable populations The population of patients who meet the definition for the primary analysis population and who follow important components of the protocol as specified.

Clusters of regularly interspaced short palindromic repeats (CRISPR) A specialized region of DNA with two distinct characteristics: the presence of nucleotide repeats and spacers. Repeated sequences of nucleotides are distributed throughout a CRISPR region and spacers are bits of DNA that are interspersed among these repeated sequences.

Common Terminology Criteria for Adverse Events (CTCAE) Also called common toxicity criteria, these are criteria used for the management of chemotherapy administration and dosing, and in clinical trials to provide standardization and consistency in the definition of treatment-related toxicity. In CTCAE, an adverse event (AE) is any unfavorable and unintended sign (including an abnormal laboratory finding), symptom, or disease temporally associated with the use of a medical treatment or procedure that may or may not be considered related to the medical treatment or procedure. The CTCAE displays grades 1 through 5 with unique clinical descriptions of severity for each AE. Grade 1 (mild): asymptomatic or mild symptoms; clinical or diagnostic observations only; intervention not indicated. Grade 2 (moderate): minimal, local or noninvasive intervention indicated; limiting age-appropriate instrumental activities of daily living (ADL). Grade 3 (severe or medically significant but not immediately life-threatening): hospitalization or prolongation of hospitalization indicated; disabling; limiting self-care ADL. Grade 4 (life-threatening consequences): urgent intervention indicated. Grade 5: death related to AE.

Compact modules These are low molecular weight mono-, bis-, or fused-heteroalicyclic scaffolds that can be easily derivatized to provide surrogates for established medicinal chemistry building blocks.

Conformational thermostabilization Membrane proteins are unstable in an hydrophilic environment. Approaches to stabilize the receptors in a particular conformation include for example truncations of the flexible regions, stabilization by antibodies and the use of fusion proteins. Another possibility is conformational stabilization by mutagenesis where the insertion of point mutations leads to increased conformational and thermal stability.

Contextual synthetic lethality A situation where one of the two pathways is lost as a result of the cellular or microenvironmental context and is rendered sensitive to loss of a second pathway. Biological contexts, such as tissue type and disease state, can influence synthetic lethal interactions.

Crystallographic asymmetric unit The smallest portion of a crystal that, when duplicated and moved by crystal symmetry operations, can produce the unit cell of the crystal.

Cytokine storm A life-threatening systemic inflammatory syndrome involving sudden and acute elevated levels of circulating cytokines and immune-cell hyperactivation that can be triggered by various therapies, pathogens, cancers, autoimmune conditions, and monogenic disorders, in which the body starts to attack its own cells.

Cytopathic effect assay An assay to detect the structural changes in host cells that are caused by viral invasion.

DC_{50} The concentration of a compound required to achieve 50% degradation.

D_{max} The maximum degradation achieved by a compound.

DNA damage response A series of processes that take place collectively to repair DNA damage events and enable cells to function normally. If left unrepaired, the level of DNA damage may accumulate to a lethal level and result in cell death.

DNA encoded library screening A new screening platform using DNA-encoded chemical libraries (DELs) commonly employed to discover small-molecules that interfere with the activity of pharmaceutically relevant proteins. DELs are hybrid-type collections of combinatorial organic molecules individually coupled to distinctive DNA fragments. It is an efficient and cost-effective drug discovery tool for the exploration and screening of very large chemical space with up to a billion structures in size.

DNA minor groove The groove occurs where two DNA backbone strands are close together.

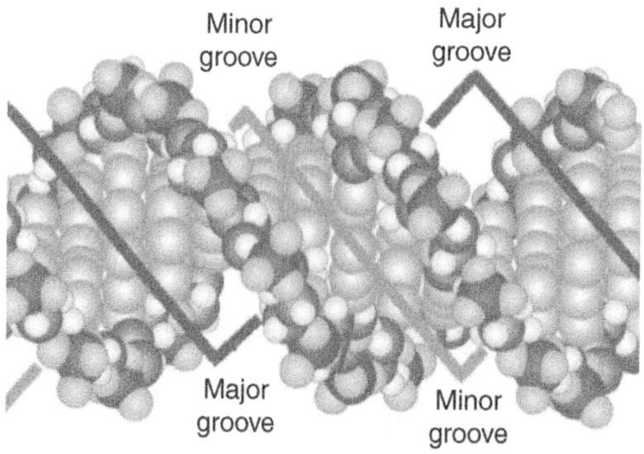

Dopamine hypothesis of schizophrenia A theory explaining the pharmacological mechanism of typical and atypical antipsychotics: blockade of dopamine D2 receptors results in relief of the positive symptoms of psychosis.

Electrostatic potential maps Contour surface within a model or crystal structure of a molecular system depicting regions of relative electron density or charge.

Enzyme-linked immunosorbent assay (ELISA) a commonly used analytical biochemistry assay which is also described as enzyme immunoassay. It is considered the gold standard of immunoassays and is a plate-based assay technique designed for detecting and quantifying peptides, proteins, antibodies, and hormones. In ELISA, an antigen must be immobilized to a solid surface and then complexed with an antibody that is linked to an enzyme.

EQ-5D The EuroQol Group is a network of international multidisciplinary researchers committed to the measurement of health-related quality of life. The group is responsible for the development of EQ-5D, a preference-based measure of health-related quality of life.

ESKAPE An acronym comprising the scientific names of six highly virulent and antibiotic resistant bacterial pathogens capable of "escaping" the biocidal action of antibiotics and mutually representing new paradigms in pathogenesis, transmission and resistance. These pathogens include: *Enterococcus faecium*, *Staphylococcus aureus*, *Klebsiella pneumoniae*, *Acinetobacter baumannii*, *Pseudomonas aeruginosa*, and *Enterobacter* spp.

Explicit-solvent all-atom molecular dynamics analysis Molecular dynamics is a computer simulation method for analyzing the physical movements of atoms and molecules. In this case all atoms of the system are part of the simulation, including solvent water molecules and ions.

Extrapyramidal side effects Antipsychotic-induced movement disorders, which may include akathisia (restlessness), dystonia (involuntary muscle contractions), Parkinsonism (rigid muscles, slow movement, etc.), and tardive dyskinesia (uncontrollable facial movements).

Extrapyramidal symptoms *see* Extrapyramidal side effects.

Familial Alzheimer's disease (FAD) A rare form of Alzheimer's diseases that is entirely passed down through genetics, being inherited from a parent. FAD accounts for 2–3% of all cases of Alzheimer's and usually has a much earlier onset than other types of Alzheimer's, with symptoms developing well before the age of 65 and sometimes in people in their 30s or 40s. People with FAD have a mutation in one to three genes known to aid in the development of Alzheimer's: presenilin 1 (PS1) on chromosome 14, presenilin 2 (PS2) on chromosome 1, and amyloid precursor protein (APP) on chromosome 21.

Five-dimensional framework on research and development productivity In 2011, AstraZeneca embarked on a major revision of its research and development (R&D) strategy with the aim of improving R&D productivity. A cornerstone of the revised strategy was to focus decision-making on five technical determinants: the right target, right tissue, right safety, right patient and right commercial potential (the 5R framework) in the hope that this could be useful to pharmaceutical industry tackling R&D productivity issues.

Five-year survival rate Patients who will be alive five years after their diagnosis.

FLAG-tag a polypeptide protein tag that can be added to a protein, having the sequence motif DYKDDDDK. It is one the most specific tags and an artificial antigen to which specific, high affinity monoclonal antibodies have been developed. A FLAG-tag can be used in many different assays that require recognition by an antibody.

Fluorometric imaging plate reader (FLIPR) assay A technique using fluorescent dyes to detect calcium (Ca^{2+}) flux in cells upon G protein-coupled receptor activation.

Free energy perturbation A method used in computational chemistry for computing free energy differences from molecular dynamics or Metropolis Monte Carlo simulations based on statistical mechanics. These calculations are a useful tool for predicting free energy differences associated with the transfer of molecules from one environment to another employed for predicting ligand–protein binding affinities via molecular simulations. These methods use "bridging" potential energy functions representing alchemical intermediate states that cannot exist as real chemical species. The data collected from these bridging alchemical thermodynamic states allows the efficient computation of transfer free energies (or differences in transfer free energies) with orders of magnitude less simulation time than simulating the transfer process directly.

Frontotemporal dementia A group of disorders caused by progressive nerve cell loss in the brain's frontal lobes or its temporal lobes. It is one of the less common types of dementia.

Generating Antibiotics Incentives Now (GAIN) Act This is an act that passed in 2012 as part of the Food and Drug Administration Safety and Innovation Act (FDASIA) which supports new antibacterial and antifungal drug development by granting priority review status and an expedited review of a drug's application and an additional five years of market exclusivity in addition to the five years of market exclusivity provided by the Food, Drug, and Cosmetic Act.

Generic GPCR residue numbering In these schemes, the first number denotes the helix (1–7) and the second the residue position relative to the most conserved position, which is assigned the number 50. For example, 6.51 denotes a residue in transmembrane helix 6, one position after the most conserved residue (6.50). The reference helix conserved positions differ between the G protein-coupled receptor classes.

Glucagon-like peptide-1 (GLP1) receptor A member of the Family B G protein-coupled receptors.

Glutamate hypothesis of schizophrenia A more recent and alternative theory that relates pathologic mechanisms of schizophrenia to the hypofunction of glutamatergic signaling via N-methyl-D-aspartate receptors.

G protein-coupled receptors Also known as seven-transmembrane domain (7TM) receptors. A type of membrane receptors, coupled with intracellular guanine nucleotide-binding proteins (G proteins), transmitting extracellular signals to intracellular responses.

GTPγS binding assay A radiolabeled functional assay measuring the binding of the non-hydrolyzable GTPγS (an analog of guanosine triphosphate, GTP) to Gα proteins upon G protein-coupled receptor activation.

HBV-infected human liver chimeric uPA-SCID mice model An animal model developed by the HBV infection of human liver chimeric uPA/SCID mice, which can be used to evaluate the potency of antiviral compounds interfering with various stages of the HBV life cycle.

Herd immunity Also known as community immunity or population immunity. It is the indirect protection from an infectious disease that happens when a large portion of a community/population (the herd) becomes immune to the disease, making the spread of disease from person to person unlikely.

High concentration functional screening Also known as high-content analysis or cellomics. It is a method that is used in biological research and drug discovery to identify substances such as small molecules, peptides, or RNAi that alter the phenotype of a cell in a desired manner.

His tag (or polyhistidine tag) It is a string of histidine residues at either the N or C terminus of a recombinant protein. When a protein having a His-tag is contacted with a carrier on which a metal ion such as nickel is immobilized, the histidine residue chelates with metal ion and binds to the carrier.

Homologous recombination and non-homologous end-joining These are mechanistically distinct DNA repair pathways that contribute substantially to double-stranded break (DSB) repair in mammalian cells. Homologous recombination is a type of genetic recombination that requires the presence of homologous DNA sequences to act as templates and direct the repair, which occurs in the S/G2 phases of the cell cycle. Non-homologous end-joining refers to DSB repair in which the broken ends are directly ligated without the direction of homologous templates and can occur in all phases of the cell cycle.

Huntington's disease A rare, inherited disease that causes the progressive breakdown (degeneration) of nerve cells in the brain and results in uncontrolled movements, emotional problems, and loss of thinking ability (cognition).

Hydrodynamic injection This is a method that utilizes the physical force generated by rapidly injecting a large volume of solution into a blood vessel (e.g. tail vein) to facilitate the delivery of genetic material (e.g., plasmid DNA) into hepatocytes.

Hydrophobic hindering mechanism Key hydrophobic residues in Family A G protein-coupled receptors in the core of the helical bundle stabilizing the inactive state. These residues hinder the channel of water seen in active conformation crystal structures that facilitates the formation of the activated state. They consist primarily of L3.43, F6.44 and X6.40 (where X is a bulky hydrophobic residue, I, L, V, or M).

Hyperprolactinemia A medical condition characterized by higher than normal levels of the hormone prolactin in the blood.

IMAC resin Supports such as beaded agarose or magnetic particles can be derivatized with chelating groups to immobilize the desired metal ions which function as ligands for binding biomolecules of interest. His-tags bind best to IMAC resins in near-neutral buffer conditions. Elution and recovery of captured His-tagged protein from an IMAC resin is accomplished by using a high concentration of imidazole or an excess of strong chelators.

Induced-fit A protein conformational change occurring as a consequence of ligand binding. It consists of a conformational relaxation into the bound ground-state conformation.

Intention-to-treat population Patients from the initial treatment assignment and not on the treatment eventually received.

Ionotropic glutamate receptor A type of ligand-gated ion channel activated by glutamate that mediates the majority of excitatory synaptic neurotransmission in the brain.

Kolliphor® ELP A non-ionic oil-in-water emulsifier and/or solubilizer for sensitive active ingredients.

Langendorff experiments Organ bath experiments with isolated rat hearts following the technique of O. Langendorff used to measure heart rate and contractile strength.

Lewy body dementia A type of progressive dementia caused by abnormal microscopic deposits that damage brain cells over time leading to a decline in thinking, reasoning, and independent function.

Ligand dissociation kinetics The rate at which a ligand dissociates from a protein.

Ligand-gated ion channels Membrane proteins with a pore to allow a regulated flow of specific ions across the membrane in response to the binding of a ligand, such as glutamate.

Lipinski's rule of five A guideline for oral drugs proposed by Lipinski: hydrogen bond donors <5, hydrogen bond acceptors <10, molecular weight < 500, LogP <5.

Machine learning A method of data analysis that automates analytical model building through the study of computer algorithms that improve automatically through experience. It is a part of artificial intelligence based on the idea that systems can learn from data, identify patterns and make decisions with minimal human intervention.

Maintenance therapy A type of treatment that is given after initial standard treatment has been completed to try to keep cancer from returning. The goal of maintenance therapy is to extend the length of time before a new recurrence emerges or even to turn a temporary remission into a long-term cure.

Maximum tolerated dose The highest dose of a drug that does not cause unacceptable side effects or overt toxicity.

Metabotropic glutamate receptor A type of membrane receptor activated by glutamate and acting indirectly through a second messenger. They are members of the class 3 G protein-coupled receptors.

Microbiologically evaluable patients Patients who are clinically evaluable and have at least one typical bacterial pathogen identified at study entry from an appropriate microbiological specimen (e.g. blood, sputum, or pleural fluid).

Minimum efficacious dose The lowest dose that provides a biological response and/or a clinically significant response in average efficacy.

MLHF QoL total score The Minnesota Living with Heart Failure Questionnaire (MLHFQ) is a health-related quality of life (QoL) questionnaire for patients with heart failure with scores for physical and emotional dimensions, and a total score.

Modified intent-to-treat population A population used in place of the true intent-to-treat population. It is often defined as all randomized patients who received study treatment.

Molecular mechanics-generalized Born/surface area (MM-GB/SA) scoring A force-field-based scoring method that computes the free energy of binding. It uses molecular mechanics-based scoring functions with the generalized Born model as the implicit solvent (MM-GB/SA) in the rescoring of docking poses which is based on MM-Poisson–Boltzmann (PB)/SA scoring method pioneered by Kollman and co-workers (the combination of molecular mechanics and the PB continuum solvation to compute binding free energies). It is the fastest when compared with the other computational free energy methods, such as free energy perturbation or thermodynamic integration methods.

Negative allosteric modulators Allosteric site binders that decrease functional responses of a receptor.

Neuritic plaques Spherical lesions that contain extracellular aggregates of amyloid-β protein. An array of abnormal dendrites and axons surrounds these plaques.

Neurofibrillary tangles Insoluble twisted fibers found inside the brain's cells which result from the pathological aggregation of the misfolded and abnormally phosphorylated protein tau.

Orthosteric site The site in a receptor where endogenous ligands bind.

Parkinson's disease A neurodegenerative disorder that affects predominately dopamine-producing ("dopaminergic") neurons in a specific area of substantia nigra in the brain which leads to shaking, stiffness, and difficulty with walking, balance, and coordination.

Peptidic mimetics A molecule that mimics active ligands of hormones, cytokines, enzyme substrates, viruses, or other biomolecules. It can be a peptide, a modified peptide or any other molecule.

Pharmacological magnetic resonance imaging A noninvasive technique that allows investigators to study functional activity in the living brain following exposure to drugs.

Pharmacophore analysis The analysis of steric and electronic features of hit compounds that is necessary to ensure the optimal supramolecular interactions with a specific biological target. A validated pharmacophore model serves as a template to search a large database of ligands to derive novel scaffolds bearing chemical features similar to the query pharmacophore. The two major steps in pharmacophore-based virtual screening include handling of the conformational flexibility of database ligands and pharmacophore mapping.

Phencyclidine-evoked locomotion assay A widely used psychosis animal model in which the antipsychotic effects of a drug are assessed by its ability to reverse phencyclidine-induced hyper-locomotion.

Phenotypic drug discovery It uses empirical, target-agnostic lead generation to identify pharmacologically active molecules and novel therapeutics which work through unprecedented drug mechanisms. It implies screening where the molecular mechanism of action is not assumed and does not require knowledge of the molecular target. As such it is comparable with empirical screening, in contrast to the target-based strategies that have been widely used in the pharmaceutical industry in the past three decades. It has been reintroduced as a strategy to provide first-in-class medicines.

Phenotypic screening A screening to identify molecules with particular biological effects in cell-based assays or animal models.

Photoaffinity labeling A method using photoactivatable functional group label proteins using photochemistry. UV irradiation leads to photolysis and the formation of a highly reactive species which covalently cross-links with residues of the target protein in the vicinity of the binding site.

Physiologically based pharmacokinetic analysis This analysis uses models and simulations that combine physiology, population, and drug characteristics to mechanistically describe the parmacokinetic and/or pharmacodynamic behaviors of a drug.

Physiologically based pharmacokinetic modeling These models are mathematical representations of human (and other species) physiology, developed to predict the absorption, distribution, metabolism, and elimination of xenobiotics (drugs). In addition to physiological and anatomical parameters, such as tissue, blood flow, and physiological barriers, they also incorporate the physical and chemical characteristics of the xenobiotics studied. These models are broadly used in drug development to predict the pharmacokinetics of novel molecules in preclinical species and humans.

Plaque-purified virus stocks These are virus stocks prepared from a single plaque.

Platinum-resistant A cancer that recurs less than six months after the end of platinum treatment.

Platinum-sensitive A cancer that recurs more than six months after the end of platinum treatment.

Population-pharmacokinetic (POP-PK)/pharmacodynamic (PD) analyses Population modeling (POP-PK) provides estimates of typical drug levels and drug effects (PK or PK/PD parameters) in a specific population by identifying sources of variability (covariates) in a population and then quantifying the impact of each covariate through a modeling system. Population PK or PK/PD modeling uses the statistical methodology, nonlinear mixed effects modeling, to investigate PK or PK/PD variability, identify significant intrinsic and/or extrinsic factors (covariates) affecting PK or PD and characterize a dose–concentration effect relationship. Thus, it allows forecasting how a drug might interact with other factors before clinical trials begin.

Positive allosteric modulators Allosteric site binders that increase functional responses of a receptor.

Positive and Negative Syndrome Scale A questionnaire widely used in clinical trials for rating severity of positive and negative symptoms of schizophrenic patients.

Probability of pharmacokinetic/pharmacodynamic target attainment In Monte Carlo simulations, the probability that at least a specific value of a pharmacodynamic index (e.g. 30% *f*T>MIC; *f*AUC/MIC of 100) is achieved at a certain (minimum inhibitory) concentration (where MIC is the minimum inhibitory concentration, *f*T>MIC is the cumulative percentage of a 24 hour period that unbound fraction of the drug concentration exceeds the MIC at steady state pharmacokinetic conditions, and *f*AUC is the area under the concentration–time curve for unbound fraction of the drug).

Prodromal Alzheimer's disease This is the very early form (also referred to as mild cognitive impairment) of Alzheimer's when memory is deteriorating but a person remains functionally independent. It can be detected through biomarker tests or a PET scan.

Prodrug A biologically inactive medication or compound that, after administration, is metabolized within the body into a pharmacologically active drug.

Protomer The smallest unique structural unit of an oligomeric protein.

Psychotic-like behaviors These are characterized most commonly by visual and auditory hallucinations. Symptoms of psychosis also include delusions, disordered thinking, disorganized speech, and sometimes physical or social impairment.

Rat rotarod test A performance test used to measure the riding time of a rat on a rotating rod.

Receptor occupancy These assays are designed to quantify and characterize the binding profile of therapeutic drugs to their targets on the cell surface. Receptor occupancy can be calculated from PET signal reduction as the treatment-induced relative change in the concentration of available receptors between treatment and baseline conductions as shown in the equation:

$$\mathrm{RO}(\%) = \frac{\left(\mathrm{Baseline\,BP_{ND}} - \mathrm{Postdrug\,BP_{ND}}\right)}{\mathrm{Baseline\,BP_{ND}}} * 100$$

Reference-tissue model The reference tissue model allows for quantification of receptor kinetics without measuring the arterial input function, thus avoiding arterial cannulation and time-consuming metabolite measurements. The reference tissue model uses the time–activity curve of a reference region with very low specific uptake to estimate binding potential from reversible ligand–receptor PET studies.

Regional K_d K_d is the dissociation constant for a radioligand at equilibrium. Its affinity is the inverse of K_d, and it equals the ratio of rate constants for association K_{on} and dissociation K_{off} of the radioligand to and from the receptors in a tissue sample.

REMDACTA trial A two-armed global phase 3, randomized, double-blind, multicenter clinical trial to evaluate the efficacy and safety of Remdesivir plus Actemra/RoActemra (tocilizumab) compared with Remdesivir plus placebo in hospitalized patients with severe COVID-19 pneumonia receiving standard of care. The primary and secondary endpoints include clinical status, mortality, mechanical ventilation, and intensive care variables.

Ribonuclease A (RNase A) This is a digestive enzyme secreted by the pancreas that specifically hydrolyzes RNA (but not DNA) polymers by endonuclease cleavage of the phosphodiester bonds forming the covalent links between adjacent ribonucleotide residues in these molecules. It belongs to the ribonucleases (RNases) which are a large group of hydrolytic enzymes that degrade RNA molecules and catalyze the breakdown of RNA into smaller components.

Scaffold hopping A method employed by medicinal chemists to discover structurally novel compounds starting from known active compounds by modifying the central core structure of the molecule.

Second shell residues Amino acids that are outside of the ligand binding site and so are not capable of direct interaction with a ligand but which have contact amino acids within the biding site and so have potential to be influenced indirectly by ligand binding.

Serology tests These are also known as antibody tests. The tests are used to detect the presence of antibodies in the blood when the body is responding to a specific infection.

Shake-flask LogD This is the measurement of lipophilicity (distribution coefficient) of a test compound by assessing the ratio of the equilibrium concentrations of a dissolved test compound in a two-phase system consisting of an organic *n*-octanol phase and an aqueous buffer phase. The principal of the method is based on the Nernst partition law.

Shotgun therapy This is a term applied in medicine when multiple therapies that are sometimes not exactly complementary are presented to patients. For example, the choice of a combination of fortified antibiotics used in intensive initial therapy is based on local epidemiological information regarding likely infecting organisms.

Similarity-based virtual screening The principle involved is that molecules with structural similarity are likely to have similar properties.

SiteMap An algorithm for binding site identification and physical–chemical property evaluation.

Steered molecular dynamics A special type of molecular dynamics simulations that applies a directional vector to a molecule or protein in order to study how it responds to external forces.

Surface plasmon resonance A physical process that occurs when plane-polarized light hits a thin metal film under total internal reflection conditions. It is the collective oscillation of conduction band electrons that are in resonance with the oscillating electric field of incident light, which will produce energetic plasmonic electrons through nonradiative excitation. It is an important optical biosensing technology in the areas of biochemistry, biology, and medical sciences. This binding analysis methodology is used to study molecular interactions.

Synthetic lethality The concept that mutation in two genes leads to death, whereas mutation of either alone is compatible with viability.

Target-based screening A screen used to identify molecules with a particular biological effect in a known molecular target (protein, enzyme, receptor, etc.).

Test of cure The re-culturing of a site of initial infection to determine whether the patient is cured.

Therapeutic index A quantitative measurement of the relative safety of a drug. The therapeutic window is the range of drug dosages which can treat disease effectively without having toxic effects and/or acceptable side-effects.

THIOMAB technology This technology introduces Cys residues engineered onto certain positions in the heavy or light chains of antibodies to allow for site-specific conjugation.

Time and metabolism-dependent inhibition of CYP3A4 This is a collective term for a change (most often an increase) in potency of CYP inhibitors during an *in vitro* incubation or dosing period *in vivo*.

Time–activity curve In dynamic PET imaging, this is a curve of the concentration of radioactivity in a region of interest plotted on the Y-axis against the time plotted on the X-axis.

Transient equilibrium tissue ratio For reversible ligand–receptor systems in the brain, an equilibrium analysis might be used to estimate the non-displaceable binding potential (BP_{ND}). Specifically, the reference region time–activity curve (TAC) is subtracted from the TAC of the target region, and the transient equilibrium (TE) occurs at the peak of the resulting bound TAC, as the peak indicates that the rates of association and dissociation are transiently equal at the time. The ratio of the bound concentration and the concentration in the reference region at the time of TE is considered equal to BP_{ND}.

Typical antipsychotics This is also known as first generation antipsychotics. This was first developed in the 1950s and used to treat psychosis and other conditions.

van der Waals interactions These are the molecular interactions that include attraction and repulsions between atoms, molecules, and surfaces, as well as other intermolecular forces. They differ from covalent and ionic bonding in that they are caused by correlations in the fluctuating polarizations of nearby particles (a consequence of quantum dynamics).

Vascular dementia This is dementia caused by problems in the supply of blood to the brain leading to damaged brain cells and worsening cognitive decline due to insufficient oxygen in the brain.

Veber's rule A guideline for oral drugs proposed by Veber: rotatable bond <10 and polar surface area < 140 $Å^2$.

Viral load The number of copies of virus in a given part of the body.

Virtual screening This is the use of computational modeling to evaluate a candidate pool of molecules for affinity. Virtual screening techniques include molecular docking technologies, shape-based screening, and ligand-based methods.

WaterMap A physics-based computational chemistry method that maps the locations and thermodynamic properties of water molecules that solvate protein binding sites and provide rich physical insights into the properties of the pocket.

Watson-Crick base pairing In DNA, adenine (A) forms a base pair with thymine (T) using two hydrogen bonds, and guanine (G) forms a base pair with cytosine (C) using three hydrogen bonds. In RNA, T is replaced by uracil (U).

World Health Organization International Nonproprietary Name (WHO INN) INNs facilitate the identification of pharmaceutical substances or active pharmaceutical ingredients. Each INN is a unique name that is globally recognized and is public property. A nonproprietary name is also known as a generic name. The WHO collaborates closely with INN experts and national nomenclature committees to select a single name of worldwide acceptability for each active substance that is to be marketed as a pharmaceutical.

Appendix C

Abbreviations

Numbers in parentheses after the definitions refer to the chapter numbers where the abbreviations are used.

^{18}F-DOPA	6-[fluoride-8]fluoro-levodopa (9)
5-HIAA	5-hydroxyindoleacetic acid (6)
5-HT	5-hydroxytryptamine (6)
5-HT1A	serotonin (5-hydroxytryptamine) 1A receptor (2)
7-TM	7-transmembrane (6)
β2AR	β2-adrenoceptor (8)
AAV	adeno-associated virus (5)
ABD	agonist binding domain (6)
ABL	Abelson tyrosine-protein kinase (11)
ACC	acetyl-CoA carboxylase (2)
ACE2	angiotensin converting enzyme 2 (15)
Ach	acetylcholine (10)
ACP	acyl-carrier protein (8)
ACTIV	Accelerating COVID-19 Therapeutic Interventions and Vaccines (15)
AD	Alzheimer's disease (4, 10)
ADAMTS	Aggrecan degrading metalloprotease (8)
ADAS	Alzheimer's Disease Assessment Scale (4)
ADC	antibody-drug conjugate (11, 13, 14)
ADCS-ADL	Alzheimer's Disease Cooperative Study Activities of Daily Living inventory scale (4)
ADE	antibody dependent enhancement (15)
ADHF	acute decompensated heart failure (3)
ADME	absorption, distribution, metabolism, and excretion (12)
ADMET	absorption, distribution, metabolism, excretion, and toxicity (5, 14)
AE	adverse event (3, 13, 14)
AICD	APP intracellular domain (4)
ALK	anaplastic lymphoma kinase (11)
ALL	acute lymphoblastic leukemia (13)
ALT	alanine aminotransferease (11)
AM	acute myelogenous (13)

Contemporary Accounts in Drug Discovery and Development, First Edition. Edited by Xianhai Huang, Robert G. Aslanian, and Wayne H. Tang.

AML	acute myeloid leukemia (13)
AMPA	α-amino-3-hydroxy-5-methyl-4-isoxazolepropionic acid (6)
AmpC	ampicillin-type β-lactamases (14)
ANCOVA	analysis of covariance (3)
ANOVA	analysis of variance (11)
ANP	atrial natriuretic peptide (3)
AOP	Fmoc-15-amino-4,7,10,13-tetraoxapentadecanoic acid (8)
Aph-1	anterior pharynx-defective-1 (4)
APL	acute promyelocytic leukemia (13)
APLN	angiopoietins, ephrins, apelin (14)
APP	amyloid-beta precursor protein (4)
AR	androgen receptor (11)
ARB	angiotensin-2 receptor blocker (3)
ARD	androgen receptor degrader (11)
ARDS	acute respiratory distress syndrome (15)
AR-fl	full length androgen receptor (11)
ARG	autoradiography (10)
ARNI	angiotension receptor neprilysin inhibitor (3)
ARNT	aryl hydrocarbon receptor nuclear translocator (11)
ARV-SNEP	ARV-825 loaded self-nanoemulsifying preconcentrate (11)
ASCO	American Society of Clinical Oncology (11)
ASCPT	American Society for Clinical Pharmacology and Therapeutics (10)
ATAD2	ATPase family AAA-domain containing protein 2 (8)
ATD	amino-terminal domain (6)
ATM	ataxia telangiectasia mutated kinase (14)
ATP	adenosine triphosphate (11)
ATR	Rad3-related kinase (14)
AUC	area under the curve (4, 5, 7, 11, 14)
AUC_{last}	area under concentration–time curve from time 0 to time of last measurable plasma concentration (14)
AUC_{ss}	area under the plasma concentration–time curve during any dosing interval at steady state (14)
AUDA	12-[[(tricyclo[3.3.1.13,7]dec-1-ylamino)carbonyl]amino]-dodecanoic acid (8)
BA	bioavailability (4)
BA/AB	basolateral to apical/apical to basolateral (10)
BACE	β-secretase (4)
BALB/c mice	Bagg Albino/c strain mice (5)
B-ALL	B-cell lineage acute lymphoblastic leukemia (13)
BAQ	benzoazepinequinoline (5)
BARDA	Biomedical Advanced Research and Development Authority (15)
BC	biotin carboxylase (2)
BCATm	branched chain aminotransferase mitochondrial (8)
BCL6	B-cell lymphoma 6 protein (11)
Bcl-xL	B-cell lymphoma-extra large (8)
BCL-XL	B-cell lymphoma-extra large (11)
BCMA	B-cell maturation antigen (13)
BCR-ABL1	a kinase protein coded for by the abnormal BCR-ABL1 fusion gene (11)

BCRP breast cancer resistance protein (3, 11)
BER base excision repair (14)
BET bromodomain and extraterminal domain proteins (8, 11)
BETP 4-(3-benzyloxyphenyl)-2-ethylsulfinyl-6-(trifluoromethyl)pyrimidine (9)
BID twice a day (5)
BLK B lymphocyte kinase (11)
B-NHL B-cell lineage non-Hodgkin lymphoma (13)
BOLD blood oxygenation level dependent (6)
BP blood pressure (3)
BPDCN blastic plasmacytoid dendritic cell neoplasm (13)
BP_{ND} non-displaceable binding potential (10)
br. brain (4)
BRCA breast cancer gene (14)
BRCAm BRCA-mutated (14)
BRD bromodomain-containing protein (11)
Btk Bruton's tyrosine kinase (1)
BTK Bruton's tyrosine kinase (8, 11)
cAMP cyclic adenosine monophosphate (6)
CatD cathepsin D (4)
C_{avg} HNSTD concentration average for human standard (13)
CBR clinical benefit rate (13)
CCK-8 cell counting kit-8 (5)
CCR5 CC chemokine receptor 5 (15)
CDC Centers for Disease Control and Prevention (15)
CDK8 cyclin-dependent kinase 8 (11)
CDR-SB clinical dementia rating scale–sum of boxes (4)
CE clinically evaluable (14)
CFU colony-forming unit (14)
cGMP cyclic guanosine monophosphate (3)
CGRP-R calcitonin gene-related peptide receptor (10)
cHIS COVID-19-associated hyperinflammatory syndrome (15)
CHK1 checkpoint kinase 1 (14)
CHK2 checkpoint kinase 2 (14)
CHO Chinese hamster ovary (3)
CI confidence interval (3, 14, 15)
cIAI complicated intra-abdominal infection (14)
cIAP1 cellular inhibitor of apoptosis protein 1 (11)
Cl clearance (7, 14)
CL clearance (5, 8)
CLIPTAC click-formed PROTAC (11)
Cl_R renal clearance (14)
CLSI Clinical & Laboratory Standards Institute (14)
C_{max} maximal drug concentration (4, 5, 14)
CMC Chemistry Manufacturing and Controls (11)
cMITT clinically modified intention-to-treat (14)
Cmpd compound (8)
CNS central nervous system (4, 10)

COPD	chronic obstructive pulmonary disease (8)
CoV	coronavirus (15)
COVID-19	coronavirus disease 2019 (15)
CpAM	core protein assembly modulator (5)
CPE	cytopathic effect or cytopathogenic effect (5)
CPU	central processing unit (2)
CR	complete remission (13)
CRBN	cereblon (11)
CRE	carbapenem-resistant Enterobacteriaceae (14)
CRF	corticotropin-releasing factor (9)
CRi	incomplete count recovery (13)
CRISPR	clustered regularly interspaced short palindromic repeats (11)
CRP	C-reactive protein (15)
CsA	cyclosporine A (12)
CSF	cerebrospinal fluid (4, 6)
$C_{ss,max}$	maximum plasma (peak) concentration in plasma during dosing interval (14)
CT	carboxyltransferase (2)
CT	computerized tomography (10)
CT	threshold concentration (14)
CTAP	Coronavirus Treatment Acceleration Program (15)
CTCAE	Common Terminology Criteria for Adverse Events (14)
CTD	carboxyl-terminal domain (6)
CTEPH	chronic thromboembolic pulmonary hypertension (3)
CTF	C-terminal fragment (4)
CTX-M	cefotaximase (14)
cUTI	complicated urinary tract infection (14)
CV%	coefficient of variation percentage (3)
CV	cardiovascular (3)
CYP	cytochrome p450 (3, 5)
DAR	drug–antibody ratio (13)
DCG-IV	(2*S*,2′*R*,3′*R*)-2-(2′,3′-dicarboxycyclopropyl)glycine (6)
DDB1	DNA binding protein 1 (11)
DDI	drug–drug interaction (3, 5, 11)
DDR	DNA damage response (14)
DEA/NO	diethylamine/nitric oxide complex (3)
DEL	DNA-encoded library (8)
dHepaRG	differentiated HepaRG (5)
DHET	dihydroxyeicosatrienoic acid (8)
DHQ	dihydroquinolizinone (5)
DHT	dihydrotestosterone (11)
DLBCL	diffuse large B-cell lymphoma (13)
DLL3	delta-like protein 3 (13)
DLT	dose limiting toxicity (11)
DMF	dimethylformamide (3, 13)
DMPK	drug metabolism and pharmacokinetics (5, 11)
DNA	deoxyribonucleic acid (11)
DoD	Department of Defense (15)

DOPAC 3,4-dihydroxyphenylacetic acid (6)
DSB double-stranded break (14)
DTS DNA-templated synthesis (8)
DTT dithiothreitol (13)
DUB deubiquitination (11)
DUB deubiquitylating enzyme (8)
EBOV Ebola virus (15)
ec engineered cysteine (13)
ECD extracellular domain (9)
ECG electrocardiogram (6)
ED effective dose (10)
EDHF endothelial-derived hyperpolarizing factor (8)
EET epoxyeicosatriene (8)
EGFR2 epidermal growth factor receptor-2 (13)
ELISA enzyme-linked immunosorbent assay (11)
EMA European Medicines Agency (14)
eME extended microbiologically evaluable (14)
End i.v. end of i.v. therapy (14)
EOC epithelial ovarian cancer (14)
EORTC European Organization for Research and Treatment of Cancer (14)
EOT end-of-treatment (14)
EPSA exposed polar surface area (12)
EPXH epoxide hydrolase (8)
EQ-5D EuroQol-5D (3)
ER emergency room (15)
ER endoplasmic reticulum (4)
ER estrogen receptor (11)
ErbB3 erythroblastosis oncogene B3 (11)
ERK extracellular signal-regulated kinase (11)
ERRα estrogen receptor-related receptor-α (11)
ESBL extended-spectrum β-lactamase (14)
ESKAPE *Enterococcus faecium*, *Staphylococcus aureus*, *Klebsiella pneumoniae*, *Acinetobacter baumanii*, *Pseudomonas aeruginosa*, and *Enterobacter* spp. (14)
ESR1 estrogen receptor 1 (11)
ETV Entecavir (5)
EUA Emergency Use Authorization (15)
EUL Emergency Use Listing (15)
FAD familial Alzheimer's disease (4)
fAD familial forms of Alzheimer's disease (4)
FAK focal adhesion kinase (11)
FBDD fragment-based drug discovery (4)
FBS fetal bovine serum (11)
FD frontotemporal dementia (4)
FDA US Food and Drug Administration (9, 15)
fe fraction of the dose excreted (6)
FEP free energy perturbation (2)
FGF-2 fibroblast growth factor 2 (14)

FKBP	FK506 binding protein (11)
FKBP12^{F36V}	FK506 binding protein with F36V mutation (11)
FL3b	Grade 3b follicular lymphoma (13)
FLT3	FMS-like tyrosine kinase 3 (11)
FR	folate receptor (13)
FTIH	first-time-in-human (8)
GBM	glioblastoma multiforme (13, 14)
gBRCAm	germline BRCA-mutated (14)
GCC	guanylyl cyclase C (13)
GCGR	glucagon receptor (9)
GCN5	general control of amino acid synthesis protein 5 (11)
GCP	Good Clinical Practice (3)
GDP	gross domestic product (15)
GES	Guiana-extended-spectrum β-lactamase (14)
GFP	green fluorescent protein (12)
GI	gastrointestinal (14)
GLP	Good Laboratory Practice (5)
GLP1	glucagon-like peptide-1 (9)
GLP1R	GLP1 receptor (9)
GLS-4	Sunshine Lake Pharm-4 (5)
GlyT-1	glycine transporter-1 (6)
GPCR	G protein-coupled receptor (6, 8, 9, 12)
GPU	graphics processing unit (2)
GRM3	gene encoding $mGlu_3$ receptor (6)
GS	γ-secretase (4)
GSH	glutathione (5)
GSI	γ-secretase inhibitor (4)
GSK	GlaxoSmithKline plc. (8)
GSK-3	glycogen synthase kinase-3 (8)
GSM	γ-secretase modulator (4)
GTN	glycerol trinitrate (3)
GTP	guanosine triphosphate (3)
HAP	heteroaryldihydropyrimidine (5)
HAP	hospital-acquired pneumonia (14)
HAS	highest single agent (14)
HBA	hydrogen bond acceptor (11)
HBD	hydrogen bond donor (11)
HBeAg	hepatitis B e-antigen (5)
HBsAg	HBV surface antigen (5)
HBV	hepatitis B virus (5)
HCoV	human coronavirus (15)
HCS	high concentration functional screening (4)
HCV	hepatitis C (1)
HDCA6	histone deacetylase 6 (11)
HDI	hydrodynamic injection (5)
HECT	homologous to the E6AP C-terminus (11)
HEK cell	human embryonic kidney cell (4)

HEp-2	human epithelial type 2 (5)
HER2	human epidermal growth factor receptor 2 (7, 11)
hERG	human ether-a-go-go-related gene (2)
HES1	hairy and enhancer of split-1 (4)
HF	heart failure (3)
HFpEF	heart failure with preserved ejection fraction (3)
HFrEF	heart failure with reduced ejection fraction (3)
HFSA	Heart Failure Society of America (3)
HH	hyperinsulinaemic hypoglycemia (9)
HHM	hydrophobic hindering mechanism (9)
HHS	Department of Health and Human Services (15)
HIF1α	hypoxia inducible factor 1 alpha (14)
HL	Hodgkin's lymphoma (13)
HLM	human liver microsome (5, 7)
HM	homology modeling (9)
hPepT1	human Peptide transporter 1 (6)
hPepT2	human Peptide transporter 2 (6)
HPLC	high pressure liquid chromatography (8)
hPXR	the human pregnane X receptor (4)
HR	hazard ratio for disease progression or death (14)
HR	heart rate (3)
HRD	homologous recombination deficiency (14)
HSA	human serum albumin (2)
hsEH	human sEH (8)
HTS	high-throughput screening (5, 8, 10)
HVA	homovanillic acid (6)
IAP	inhibitor of apoptosis protein (11)
ICH	International Conference on Harmonization (10)
ICU	intensive care unit (15)
IDE	insulin degrading enzyme (8)
IDMC	Independent Data Monitoring Committee (13)
IDO1	indoleamine 2,3-dioxygenase-1 (8)
IFL	irinotecan, fluorouracil, and leucovorin (14)
IFN	interferon (5, 15)
IFN-α	interferon alpha (5)
IgG	immunoglobulin G (15)
iGluR	ionotropic glutamate receptor (6)
IGN	indolinobenzodiazepine pseudodimer (13)
IHC	immunohistochemical analysis (13)
IL	interleukin (15)
IL-2	interleukin 2 (8)
IL-6R	IL-6 receptor (15)
IMAC	immobilized metal affinity chromatography (8)
IMCAS	Institute of Microbiology, Chinese Academy of Science (15)
IMP	Imipenemase (14)
IND	Investigational New Drug (11, 13)
InhA	enoyl-acyl carrier protein reductase from *Mycobacterium tuberculosis* (8)

IPP	IκBα phosphopeptide (11)
IPSS	International Prognostic Scoring System (13)
IQB	isoquinolidinobenzodiazepine (13)
IR	immediate release (3)
IRAK4	interleukin-1 receptor-associated kinase 4 (11)
ITT	intent-to-treat (3, 14)
IV	dose by intravenous injection (10–13)
IκBα	nuclear factor of kappa light polypeptide gene enhancer in B-cells inhibitor, alpha (11)
JAK	Janus kinase (2, 15)
JM	juxtamembranous (9)
KB-CR	FR-positive cisplatin (13)
KB-PR	FR-positive paclitaxel (13)
KCCQ	Kansas City Cardiomyopathy Questionnaire (3)
KO	knock out (6, 10)
KPC	*Klebsiella pneumoniae* carbapenemase (14)
KRAS	Kirsten rat sarcoma gene (2, 12)
LAV	live attenuated vaccine (15)
LB1	top lobe (6)
LB2	bottom lobe (6)
LBD	Lewy body dementia (4)
L-CCG-1	(2*S*,1′*S*,2′*S*)-2-(carboxycyclopropyl)glycine (6)
LCMS	liquid chromatography–mass spectrometry (8)
LCNEC	large-cell neuroendocrine carcinoma (13)
L-DOPA	L-3,4-dihydroxyphenylalanine (6)
LEPHT	LEft ventricular systolic dysfunction associated with Pulmonary Hypertension riociguat Trial (3)
LFA-1	lymphocyte function-associated antigen-1 (8)
LFU	late follow-up (14)
L-Glu	L-glutamic acid (6)
L-NAME	N_{ω}-nitro-L-arginine methyl ester (3)
LNCaP	lymph node carcinoma of the prostate (11)
LVEF	left ventricular ejection fraction (3)
LYSA	lyophilization solubility assay (5)
M4	muscarinic acetylcholine receptor 4 (10)
mAb	monoclonal antibody (15)
mAChR	muscarinic acetylcholine receptor (10)
MAD	multiple ascending doses (6)
MAGIC	mathematical analysis for general interactions of components (14)
MAP	mitogen-activated protein (8)
MATE	multidrug and toxin extrusion protein (3)
mBC	metastatic breast cancer (14)
MC4R	melanocortin-4 receptor (12)
MCF7	Michigan Cancer Foundation-7 (7)
MCL	mantle cell lymphoma (13)
MCL1	myeloid cell leukemia 1 (8, 11)
mCRC	metastatic colorectal cancer (14)

mCRPC	metastatic castration-resistant prostate cancer (11)
MD	molecular dynamics (9)
MDM2	mouse double minute 2 homolog (11)
MDM2/X	murine double minute 2/X (12)
MDR1	multidrug resistance mutation gene 1 (10, 14)
MDRGN	multidrug-resistant Gram-negative organism (14)
MDS	myelodysplastic syndrome (13)
ME	microbiologically evaluable (14)
MEC	minimal effective concentration (3)
MED	minimum efficacious dose (13)
mEH	microsomal epoxide hydrolase (8)
MERS-CoV	Middle East respiratory syndrome coronavirus (15)
MetAP-2	methionine aminopeptidase-2 (11)
mGluR	metabotropic glutamate receptor (6)
MHPG	3-methoxy-4-hydroxyphenylglycol (6)
MIC	minimum inhibitory concentration (14)
mIL-6R	membrane-bound interleukin-6 receptor (15)
MITT	modified intent-to-treat (14)
MLHF	Minnesota living with heart failure (3)
MLM	mouse liver microsome (5)
MM-GB/SA	molecular mechanics-generalized born/surface area (2)
mMITT	microbiologically modified intent-to-treat (14)
MMP3	matrix metalloproteinase 3 (8)
MMPP	magnesium monoperoxyphthalate (6)
MNT	micronucleus test (5)
MOA	modes of action (14)
MOA	mechanism of action (5)
MOI	multiplicity of infection (5)
MRN	MRE11-RAD50-NSB1 (14)
mRNA	messenger ribonucleic acid (11)
MRT	mean residence time (14)
MSD	Merck, Sharp & Dohme (12)
MSH	melanocyte-stimulating hormone (12)
MTD	maximum tolerated dose (11, 13, 14)
MTG	microbial transglutaminase (13)
mTOR	mammalian target of rapamycin (mechanistic target of rapamycin) (7)
MW	molecular weight (8, 9, 11)
NAM	negative allosteric modulator (6, 9)
NASH	non-alcoholic steatohepatitis (2)
ND	not determined (4)
NDM	New Delhi Metallo-β-lactamase (14)
NEMO	NF-kappa B Essential Modulator (12)
NFT	neurofibrillary tangle (4)
NGS	next generation sequencing (8)
NHEJ	non-homologous end-joining (14)
NHL	non-Hodgkin lymphoma (13)
NIAID	National Institute of Allergy and Infectious Diseases (15)

NIH	National Institutes of Health (15)
NK3	neurokinin-3 (8)
NMDA	*N*-methyl-D-aspartate (6)
NMR	nuclear magnetic resonance (4)
NO	nitric oxide (3)
NOAEL	no observed adverse effect level (5, 10, 11)
NP	nosocomial pneumonia (14)
NPS	neuropsychiatric symptom (10)
nRotB	number of rotatable bonds (11)
NS	nonsusceptible (14)
NS4B	non-structural 4B protein (8)
NSAID	nonsteroidal anti-inflammatory drug (4)
NSCLC	non-small-cell lung cancer (14)
NT	not tested (4)
NTC	no target control (8)
NTP	nucleoside triphosphate (15)
NT-proBNP	N-terminal pro–B-type natriuretic peptide (3)
NYHA	New York Heart Association (3)
OATP1B1	organic anion-transporting polypeptide 1B1 (5)
OATP1B3	organic anion-transporting polypeptide 1B3 (5)
OC	ovarian cancer (14)
OCT	organic cation transporter (3)
ORR	overall response rate (13)
OS	overall survival (13)
OVCAR8	human ovarian cancer cell line 8 (11)
OXA	oxacillinase (14)
OXA-48	oxacillinase-48 (8)
PAC	pulmonary artery compliance (3)
PAD4	protein arginine deiminase 4 (8)
pADPr	poly(ADP)-ribose (14)
PAH	pulmonary artery hypertension (3)
PAMPA	parallel artificial membrane permeability assay (5, 12)
PAM	positive allosteric modulator (6, 8, 9, 10)
PANSS	positive and negative syndrome scale (6)
PAP	poly(A) polymerase (5)
PAP	pulmonary artery pressure (3)
PAPD5 and 7	PAP-associated domain-containing protein 5 and 7 (5)
P_{app}	apparent permeability (7)
PAR2	protease activated receptor 2 (8)
PARP	poly(ADP-ribose) polymerase (8)
PARP1	poly(ADP-ribose) polymerase 1 (11, 14)
PBD	pyrrolobenzodiazepine (13)
PBD-MA	pyrrolo[2,1-*c*][1,4] benzodiazepine monoamide (13)
PBP	penicillin-binding protein (14)
PBPK	physiologically based pharmacokinetic (3, 7)
PBPK	population-based pharmacokinetic (3)
PCAF	P300/CBP-associated factor (11)

PCC preclinical candidate (4, 10, 14)
PCP phencyclidine (6)
PCR polymerase chain reaction (8, 15)
PCSK9 proprotein convertase subtilisin/kexin type 9 (12)
PCWP pulmonary capillary wedge pressure (3)
PD Parkinson's disease (10)
PD pharmacodynamic(s) (3, 4, 6, 11, 14)
PD-1 programmed cell death protein-1 (1)
PDB Protein Data Bank (5)
PDD pyrridinobenzodiazepine (13)
PDE12 phophodiesterase 12 (8)
PDE2A phosphodiesterase 2A (2)
PDGF platelet derived growth factor (14)
PDR pan-drug resistant (14)
PDX patient derived xenograft (11, 13)
PEG polyethylene glycol (8, 11)
Pen-2 presenilin enhancer-2 (4)
PET positron emission tomography (10)
PFS progression-free survival (14)
PFU plaque-forming unit (5)
pGC particulate guanylate cyclase (3)
P-gp P-glycoprotein (3, 10)
PGP P-glycoprotein (4)
pgRNA pregenomic RNA (5)
PH pulmonary hypertension (3)
pH2A.X phospho-histone H2A family member X (13)
PHH primary human hepatocyte (5)
phMRI pharmacological magnetic resonance imaging (6)
PHOTAC photochemically controlled PROTAC (11)
PI3K phosphoinositide 3-kinase (7, 8, 11)
PIP2 phosphatidylinositol-4,5-bisphosphate (7)
PIP3 phosphatidylinositol-3,4,5-trisphosphate (7)
PK pharmacokinetic(s) (3, 4, 6, 8, 11–14)
PKG protein kinase G (3)
pl. plasma (4)
PLC phospholipase C (6)
PLK1 polo-like kinase 1 (11)
PLS physical limitation score (3)
PO dosed orally (12, 13)
p.o. oral (6)
POI protein of interest (11)
POPC 1-palmitoyl-2-oleoyl-sn-glycero-3-phosphocholine (9)
POP-PK population-pharmacokinetic (3)
PPB plasma protein binding (4, 5, 10)
PRLR prolactin receptor (13)
PROTAC proteolysis targeting chimera (1, 11)
PR partial remission (13)

PS	presenilin (4)
PSA	prostate specific antigen (11)
PSA	polar surface area (12)
PSAQ	Patient Scar Assessment Questionnaire (14)
PSEN	presenilin (4)
PSMA	prostate-specific membrane antigen (13)
PTA	PK/PD target attainment (14)
PTEN	phosphatase and tensin homolog (7)
pVEGFA	plasma levels of VEGF-A (14)
PVR	pulmonary vascular resistance (3)
PXR	the pregnane X receptor (4)
q2w	every 14 days (14)
QD	once daily (11)
QLQ	Quality of Life Questionnaire (14)
QM/MM	quantum mechanics and molecular mechanics (4)
QoL	quality of life (3)
qPCR	quantitative polymerase chain reaction (5, 8, 15)
R&D	research and development (15)
R/R	relapsed or refractory (13)
RADx	rapid acceleration of diagnostics (15)
RBD	receptor binding domain (15)
RBR	RING-between-RING (11)
RCC	renal cell carcinoma (13, 14)
RdRp	RNA-dependent RNA polymerase (15)
RECIST	Response Evaluation Criteria in Solid Tumors (11)
RECOVERY	Randomized Evaluation of Covid-19 Therapy (15)
RenTG	renin transgenic (3)
RING	Really Interesting New Gene (11)
RIPK	receptor-interacting serine/threonine-protein kinase (4, 11)
RNA	ribonucleic acid (11)
RNAi	ribonucleic acid interference (11)
RO	receptor occupancy (10)
RSV F	RSV fusion (5)
RSV	respiratory syncytial virus (5)
RT-PCR	reverse transcriptase polymerase chain reaction (15)
SAD	single ascending dose (6)
SAE	serious adverse event (6, 14, 15)
SAR	structure–activity relationship (4–6, 8, 9, 10, 13)
SARD	selective androgen receptor degrader (11)
SARS-CoV	severe acute respiratory syndrome coronavirus (15)
SARS-CoV-2	severe acute respiratory syndrome coronavirus 2 (1)
SBDD	structure-based drug design (7, 9)
sBRCAm	somatic BRCA-mutated (14)
SC	subcutaneous injection (11, 12)
SCF	Skp1-Cullin-F box complex (11)
SCID	severe combined immunodeficient (11, 13)
SCLC	small cell lung cancer (13)

SDPK	single dose pharmacokinetics (5)
sEH	soluble epoxide hydrolase (8)
SERD	selective estrogen receptor degrader (11)
sGC	soluble guanylate cyclase (3)
SGK3	serine/threonine-protein kinase 3 (11)
SGLT2	sodium glucose co-transporter-2 (1, 3)
SHP	Src homology region 2 domain-containing phosphatase (2)
SHV	sulfydryl variant β-lactamase (14)
SI	selectivity index (5)
sIL-6R	soluble interleukin-6 receptor (15)
siRNA	small interfering RNA (11)
SIRT6	Sirtuin 6 (an NAD+-dependent protein deacylase) (8)
SMARCA2	protein coded by SWI/SNF related, matrix associated, actin dependent regulator of chromatin, subfamily A, member 2 (11)
SMDCs	small molecule drug conjugates (13)
SME	*Serratia marcescens* enzyme (14)
SNAP	*S*-nitroso-*N*-acetyl-D,L-penicillamine (3)
sNDA	supplemental new drug application (14)
SNEP	self-nanoemulsifying preconcentrate (11)
SNIPER	specific and nongenetic inhibitors of apoptosis protein dependent protein eraser (11)
SoC	standards of care (13)
SPR	surface plasmon resonance (4)
SSB	single-stranded break (14)
ssDNA	single-stranded DNA (8)
STAT3	signal transducer and activator of transcription 3 (11)
SUV	standard uptake value (10)
SV	stroke volume (3)
SVB	Silicon Valley Bank (14)
$t_{1/2}\alpha$	initial half-life (14)
$t_{1/2}\beta$	terminal half-life (14)
TBK1	TANK binding kinase 1 (11)
TBS	*tert*-butyldimethylsilyl (13)
TCGA	The Cancer Genome Atlas (14)
TEAE	treatment-emergent adverse event (6, 11, 13)
TEM	Temoneira β-lactamase (14)
TEMPO	(2,2,6,6-tetramethylpiperidin-1-yl)oxyl (6)
TE-TR	transient equilibrium tissue ratio (10)
TGI	tumor growth inhibition (14)
TGR	tumor growth regression (14)
TI	therapeutic index (13)
TM	transmembrane (9)
TMD	transmembrane domain (4, 6, 9)
TNBC	triple-negative breast cancer (13)
TNF	tumor necrosis factor (15)
TNKS1	tankyrase 1 (8)
TOC	test-of-cure (14)

TPD	targeted protein degradation (11)
TREM2	triggering receptor expressed on myeloid cells 2 (4)
TrkA	tropomyosin receptor kinase A (11)
TrKC	tropomyosin receptor kinase C (11)
T-RT	test–retest (10)
TYK2	tyrosine kinase 2 (2)
UGT	UDP-glucuronosyltransferase (3)
uPA-SCID	urokinase-type plasminogen activator-severe combined immune deficiency (5)
Upps	undecaprenyl pyrophosphate synthase (8)
UPS	ubiquitin-proteasome system (11)
USPs	ubiquitin-specific proteases (8)
UV–vis	ultraviolet–visible (11)
V_{dss}	volume of distribution at steady state (4)
V_{ss}	volume of distribution at steady state (5, 14)
VAERD	vaccine associated enhanced respiratory disease (15)
VAP	ventilator-associated pneumonia (14)
VCaP	vertebral cancer of the prostate (11)
VD	vascular dementia (4),
VEARS	Vaccine Adverse Event Reporting System (15)
VEGF	vascular endothelial growth factor (8, 14)
VEGFR	VEGF receptor tyrosine kinases (14)
VHL	Von Hippel–Lindau (11)
VIM	Veronaimipenemase (14)
VMA	vanillylmandelic acid (6)
VS	virtual screening (9)
VSD	Vaccine Safety Datalink (15)
WHO INN	World Health Organization International Nonproprietary Name (5)
WHO	World Health Organization (1, 15)
Wip1	wild-type p53-induced phosphatase 1 (8)
XDR	extensively drug resistance (14)
XIAP	X-linked IAP (11)
XIAP-BIR3	X-linked inhibitor of apoptosis protein-baculoviral IAP repeat 3 protein (12)
Y3H	yeast three-hybrid (5)
yR	YoctoReactor (8)
ZIP	zero interaction potency (14)

Index

Contemporary Accounts in Drug Discovery and Development, First Edition. Edited by Xianhai Huang, Robert G. Aslanian, and Wayne H. Tang.

h

i

q

t

Printed and bound by CPI Group (UK) Ltd, Croydon, CR0 4YY